Spectral Approximation of Linear Operators

This is a volume in
COMPUTER SCIENCE AND APPLIED MATHEMATICS
A Series of Monographs and Textbooks

Editor: WERNER RHEINBOLDT

A complete list of titles in this series appears at the end of this volume.

Spectral Approximation of Linear Operators

Françoise Chatelin
Mathématiques Appliquées, IMAG
Université de Grenoble
Grenoble, France

1983

ACADEMIC PRESS
A Subsidiary of Harcourt Brace Jovanovich, Publishers

New York London
Paris San Diego San Francisco São Paulo Sydney Tokyo Toronto

ACADEMIC PRESS, INC.
111 Fifth Avenue, New York, New York 10003

United Kingdom Edition published by
ACADEMIC PRESS, INC. (LONDON) LTD.
24/28 Oval Road, London NW1 7DX

Library of Congress Cataloging in Publication Data

Chatelin, Françoise.
Spectral approximation of linear operators.

(Computer science and applied mathematics)
Includes bibliographical references and index.
1. Linear operators. 2. Approximation theory.
3. Spectral theory (Mathematics) I. Title. II. Series.
QA329.2.C48 515.7'246 82-6744
ISBN 0-12-170620-6 AACR2

PRINTED IN THE UNITED STATES OF AMERICA

83 84 85 86 9 8 7 6 5 4 3 2 1

. . . La corde vibrante sensible oscille, résonne longtemps encore après qu'on l'a pincée. C'est cette oscillation, cette espèce de résonance nécessaire qui tient l'objet présent, tandis que l'entendement s'occupe de la qualité qui lui convient. Mais les cordes vibrantes ont encore une autre propriété, c'est d'en faire frémir d'autres et c'est ainsi qu'une première idée en rappelle une seconde, ces deux-là une troisième, toutes les trois une quatrième, et ainsi de suite, sans qu'on puisse fixer la limite des idées réveillées, enchaînées, du philosophe qui médite ou qui s'écoute dans le silence et l'obscurité. Cet instrument a des sauts étonnants, et une idée réveillée va faire quelquefois frémir une harmonique qui en est à un intervalle incompréhensible. Si le phénomène s'observe entre les cordes sonores, inertes et séparées, comment n'aurait-il pas lieu entre les points vivants et liés, entre les fibres continues et sensibles?

DIDEROT

Entretien entre d'Alembert et Diderot, 1769

Ce fut d'Alembert qui résolut le premier, d'une manière générale, le problème des cordes vibrantes, dont Taylor avait donné auparavant une solution qui n'était que particulière.

POISSON

Mémoires Royales de l'Académie des Sciences de l'Institut de France, tome VIII, p. 358, 1829

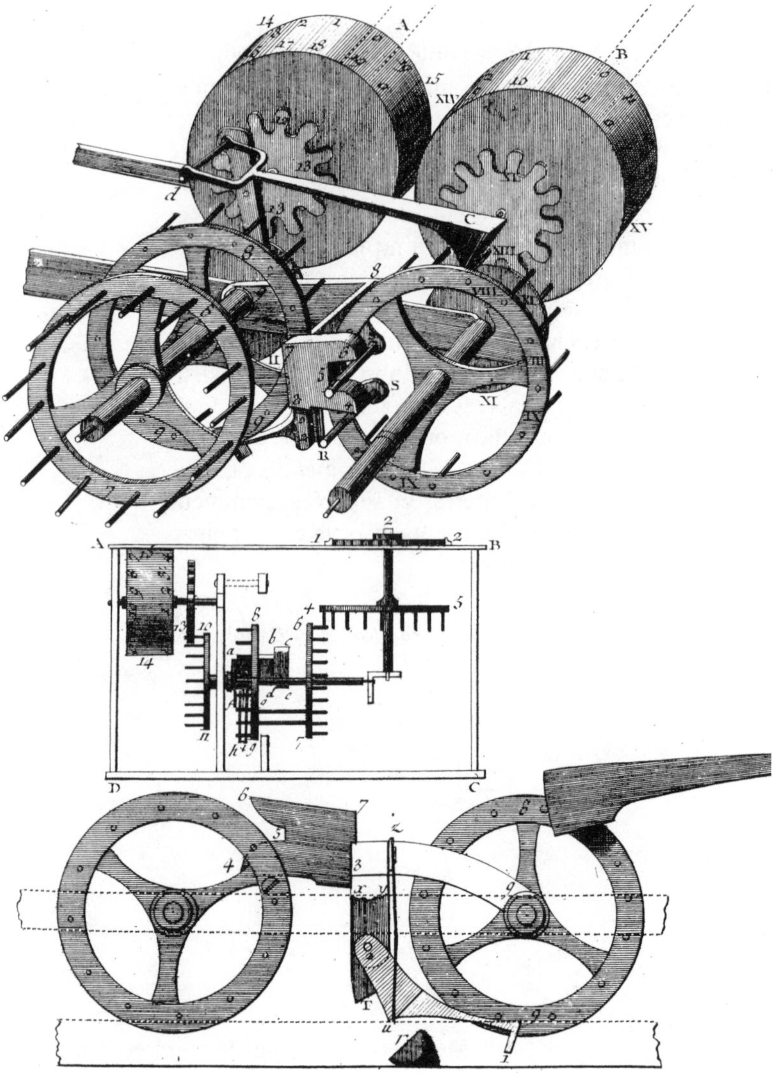

Machine Arithmétique de Pascal

A Jean KUNTZMANN

A Jean LABORDE

Pour plus de précision

Il en revenait toujours là. Un et un—reprenait-il simplement pour se convaincre lui-même—égale deux. Deux et deux: quatre. Mais pourtant le temps de faire cette addition, un intervalle infime s'intercalait, un souffle, et il ressentait le besoin de l'ajouter celui-là aussi, pour plus de précision—un souffle, et derrière cette haleine on distinguait le même chiffre comme inscrit en haut d'une porte ou sur le flanc d'un navire ou comme une étoile derrière un brouillard très fin, à cette heure vacillante, à cette heure si belle, entre le crépuscule et la nuit.

Y. RITSOS
Témoignages

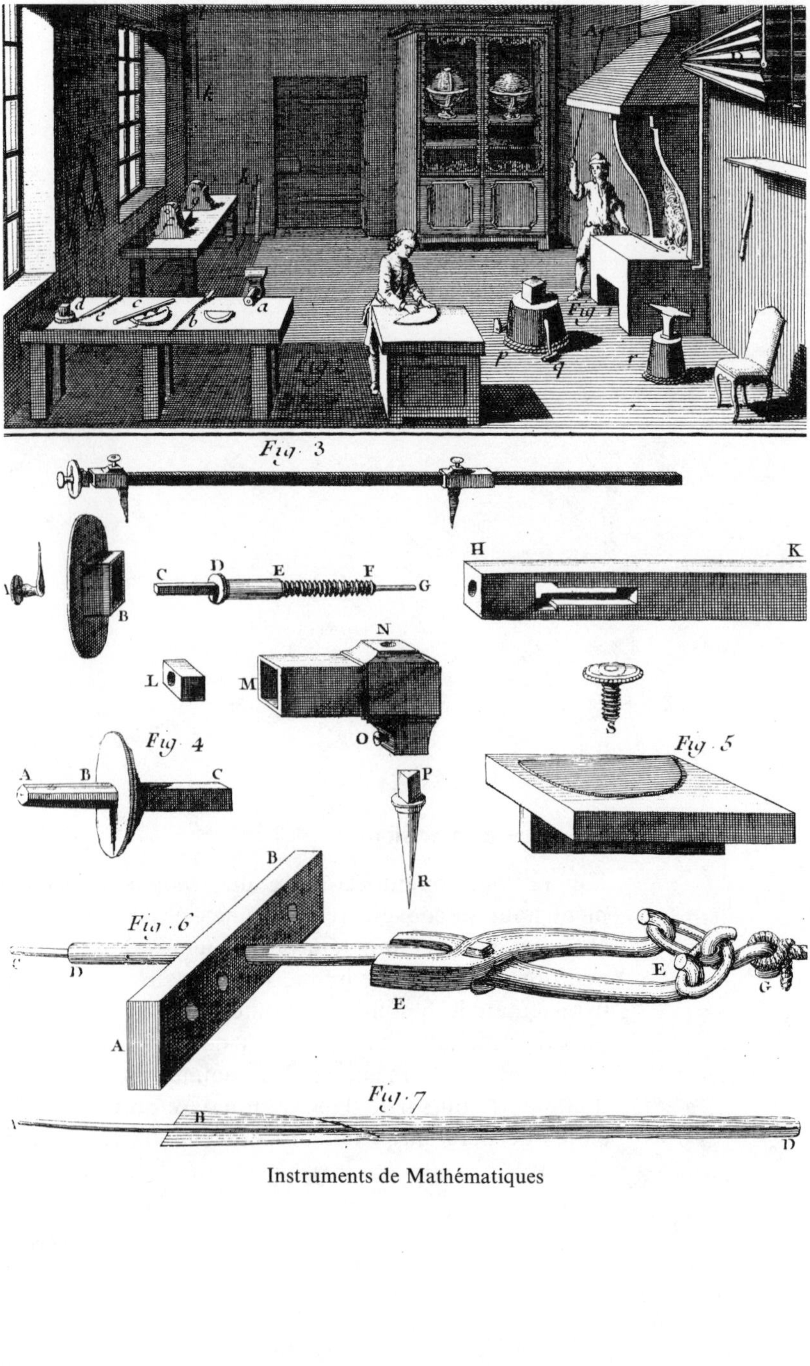

Instruments de Mathématiques

Contents

4. Numerical Approximation Methods for Integral and Differential Operators

5. Spectral Approximation of a Closed Linear Operator

6. Error Bounds and Localization Results for the Eigenelements

7. Some Examples of Applications

Appendix. Discrete Approximation Theory

Foreword

La lumière résultait des vibrations d'un fluide universel extrêmement subtil, agité par des mouvements rapides des particules des corps lumineux, de la même façon que l'air est ébranlé par les vibrations des corps sonores.

FRESNEL

Mémoires Royales de l'Académie des Sciences de l'Institut de France, tome V, p. 340, 1821 et 1822

Many tasks in numerical analysis and in rigorous applied mathematics boil down to a quantification of continuity. Let a problem P be given; more likely than not, in applications, P is too difficult to be solved exactly. But perhaps there is a nearby problem P' that can be solved. By how much does the solution of P deviate from the known solution of P'? This question always arises when P' is a discretized version of P or when a problem originally formulated in an infinite-dimensional space is projected down into a finite-dimensional subspace.

Françoise Chatelin in this work concerns herself with the quantification of continuity in the area of spectral analysis, that is (to use a less fashionable term), in the case where P is an eigenvalue problem. There is no need to emphasize the importance of eigenvalue problems; since the appearance of Courant and Hilbert's classic, "Methods of Mathematical Physics," in 1929, the central position of eigenvalue problems in all of applied mathematics, including engineering sciences and theoretical physics, has been firmly established. Already in the precomputer age, L. Collatz in his unsurpassed book, "Eigenwertprobleme und ihre numerische Behandlung," showed how to deal with eigenvalue problems numerically and demonstrated the effectiveness of numerical methods by dealing explicitly with numerous specific applications, taken mainly from the engineering sciences. Most numerical methods sooner or later require the solution of a finite-dimensional problem. Considerable attention was therefore focused on the numerical solution of matrix eigenvalue problems, and after a period of stormy development J. H. Wilkinson in his authoritative work, "The Algebraic Eigenvalue Problem," summarized the state of the art in a way that even today has lost little of the gloss of perfection.

But in the meantime, mathematics has not stood still. In the area of eigenvalue problems in particular, the concepts, methods, and terminology of functional analysis have made it possible to present the basic facts with a conciseness and clarity that appeared unattainable a few decades ago. To bring this development to the attention of the practicing applied mathematician and numerical analyst, and to make them aware of its implications for the actual computation of the spectral decomposition, is therefore a most timely endeavor. Françoise Chatelin, well trained in the cartesian spirit of French analysis, is equipped as few others are to meet this challenge in an original and well-organized volume. The reader of the present work will enjoy the economy and clarity of her presentation and will be rewarded by a true understanding of the effectiveness of the methods of modern functional analysis in the discussion of qualitative as well as quantitative aspects of eigenvalue problems.

Eidgenössische Technische Hochschule Zürich P. HENRICI

Preface

Some writers, particularly Collatz and Kantorovitch, have shown that a unified view of many results in numerical analysis can be obtained by using the ideas of functional analysis. This may or may not lead to practical advances, but it certainly leads to greater understanding. The aim of such work is far more fundamental than merely to make the subject respectable; it is the characteristically mathematical activity of generalizing from particular cases, and abstracting the common features of whole classes of methods.

L. FOX AND J. WALSH

Numerical Analysis: An Introduction (J. Walsh, ed.), Academic Press, London, 1966

This preface is addressed to those readers who are acquainted with spectral theory in function spaces. The major themes and special features of the book are stated briefly, as is proper in an overture.

The topic covered is the numerical approximation of the eigenvalues of a linear operator T acting in a complex Banach space. T may represent a bounded integral operator or a closed differential operator (with bounded inverse). In practice a great variety of methods are used to solve eigenvalue problems. To compare methods it is necessary to investigate them theoretically in order to answer questions of convergence and of speed of convergence and to provide error estimates. These investigations are made here for each class of equations and for each method. To understand better the different types of convergence it is useful to present the various methods in a unifying framework. Functional analysis provides valuable tools for this study. For example, in the derivation of error bounds for the eigenelements of a non-self-adjoint operator, spectral projection and perturbation theory prove themselves useful. The given T and the linear operator T_n associated with the numerical method are defined in the same Banach space, a common situation in practice. The family $\{T_n\}_{n\in\mathbb{N}}$ is called an approximation of T

if T_n is pointwise convergent to T. In order to get good numerical convergence to the eigenvalues, the preservation of their algebraic multiplicity is essential. The property of strong stability of the approximation $\{T_n\}$ guarantees this preservation.

Error bounds are then established for the eigenelements, with an emphasis on a posteriori error bounds. Since T and T_n act in the same space, T may be considered as the result of the perturbation $T - T_n$ applied to T_n: $T = T_n + (T - T_n)$. Assuming that $\{T_n\}$ is a radial approximation of T permits the application of powerful analytic perturbation theory to get series expansions of the exact eigenelements in terms of the computed eigenelements and in terms of the perturbation $T - T_n$. The same approach can also be used to gain information about the spectrum of T from knowledge of the spectrum of a "smaller" operator PTP, where P is a finite-rank projection on a particular subspace, and then to derive a posteriori error bounds.

The goal of the book is to be comprehensive, running the gamut from practical numerical results to a unifying framework of functional analysis, while simultaneously preserving the energy and excitement of presenting fresh research results. Its contents cover recent results, which either are scattered through the literature or appear here for the first time, as well as providing a nontraditional presentation of standard material. This original view is not given for the sake of novelty, but results from the fact that the study of the approximation of $T\varphi = \lambda\varphi$ requires a more precise knowledge of the way T_n tends to T than is required for the study of approximations to the linear equation $(T - z)x = f$. This is due to the fact the convergence $T_n \to T$ must ensure first that $(T_n - z)^{-1}$ is uniformly bounded in n for z in a neighborhood of λ and secondly that algebraic multiplicities are preserved.

The main lines of the presentation are

(i) a thorough study of the properties of various types of operator convergence,

(ii) a systematic study of the spectral approximation of non-self-adjoint operators,

(iii) a generalization of classical perturbation theory as presented by T. Kato for application to numerical analysis, and

(iv) an emphasis on computable error bounds and iterative refinement techniques.

Chapter 1 deals with the matrix eigenvalue problem, with special attention given to large problems. Chapters 2 and 3 introduce the basic concepts of functional analysis. Chapter 2 presents the classical results, while Chapter 3 contains less widely known or previously unpublished material. Chapter 4 presents the numerical methods used for approximating integral and differential operators. In the analysis of projection methods on finite-dimensional subspaces, the fundamental role projections play is stressed, following the Russian tradition of Kantorovitch–Akilov and Krasnoselskii–Vainikko. Chapter 5 is concerned with characterization of strong stability and the use of radial convergence in studying series expansions for eigenvalues and eigenvectors. Chapter 6 is devoted to the establishment of error bounds, and Chapter 7 presents results on superconvergence and the iterative refinement of eigenelements.

This book is an introduction to current research on numerical spectral approximation. It is also intended to serve as a textbook for advanced courses in numerical analysis. The prerequisites are a good knowledge of numerical and functional analysis. Although the flavor of its contents is mathematical, it is hoped that the book will provide help and guidance to scientists solving differential and integral eigenvalue problems.

Except for some classical results from functional analysis given in Chapter 2 and other minor results which are appropriately referenced, the book is self-contained. It is intended to be a working textbook. The exercises form a vital part of the text. They range from illustrations of minor points to truly difficult problems. Although the results of the exercises are less important than the main text, the exercises fill out the logical development of the text. The solutions for most of them have been provided; they are written in a detailed form and are expected to help the active reader check his understanding.

This text grew from a set of lectures given in the "DEA d'Analyse Numérique" at the University of Grenoble since 1971. It has benefited greatly from fruitful discussions with my colleagues and students. The invaluable help of Dr. Rachid Lebbar, Dr. Jacques Lemordant, and Dr. Youcef Saad is especially acknowledged; Dr. Mario Ahués is heartily thanked for the painstaking job of reading the entire manuscript and completing solutions of the exercises; Dr. Balmohan Limaye and Dr. Rekha Kulkarni from the Indian Institute of Technology at Bombay read part of the manuscript and were kind enough to provide me with their computations. I am indebted to Dr. Ian Duff at AERE, Harwell and Professor Wolfgang Hackbusch at Rhur-Universität Bochum for providing me with computer drawings.

Many friends have helped me at various stages of the work. Peter Henrici gave me the impetus to transform a set of notes into a more polished manuscript. I recall enjoyable discussions on iterative refinement with Lin Qun. Beresford Parlett read several successive versions of the manuscript. The first chapter has obviously been influenced by his intimate knowledge of the matrix eigenvalue problem. He also showed me how audacious I had been to try to master the subtleties of his mother tongue. Chapters 2, 3, and 5 reflect the deep impact that the major book of Tosio Kato (1976) had on me. Above all, I am pleased to acknowledge my debt to Jean Kuntzmann and Jean Laborde to whom this book is dedicated. In 1949, J. Kuntzmann, assisted by my father, J. Laborde, started the first applied mathematics course to be taught by a mathematician to aspiring engineers of the Institut Polytechnique de Grenoble. His influence on the early development of computer science in France, and especially at Grenoble, cannot be overestimated.

The figures on pp. vi and viii are reprinted from the "Dictionnaire raisonné des Sciences, des Arts et des Métiers," edited by Denis Diderot, published by Pellet, Imprimeur Libraire à Genève, rue des Belles-Filles, 1778.

FRANÇOISE CHATELIN

Notation

Symbol	*Definition*
$\mathbb{C}$	Field of complex numbers
$\mathbb{R}$	Field of real numbers
$\mathbb{Q}$	Field of rational numbers
$\mathbb{Z}$	Ring of relative integers
$\mathbb{N}$	Set of natural integers
$\bar{\xi}$	Complex conjugate scalar of ξ in $\mathbb{C}$
δ_{ij}	Kronecker symbol ($\delta_{ij} = 1$ if $i = j$ and $\delta_{ij} = 0$ if $i \neq j$)
$\det A$	Determinant of the square matrix A
A^{-1}	Inverse matrix (resp. operator) of matrix (resp. operator) A
$\dim M$	Algebraic dimension of the (sub)space M
lim	Limit of a function or a sequence
$O(\varepsilon)$	Order of ε: $\lim\limits_{\varepsilon \to 0} \dfrac{O(\varepsilon)}{\varepsilon}$ is finite
$\min\limits_{x \in A} f(x)$	Minimum value (achieved in A) of f over A
$\max\limits_{x \in A} f(x)$	Maximum value (achieved in A) of f over A
$\inf\limits_{x \in A} f(x)$	Greatest lower bound of the set $f(A)$
$\sup\limits_{x \in A} f(x)$	Least upper bound of the set $f(A)$
$\dfrac{df(x)}{dx}$	Ordinary derivative of f at x
$f', f^{(p)}$	First- and pth-order ordinary derivative of f
$\dfrac{\partial f}{\partial x}$	Partial derivative of f with respect to x (classical or generalized sense)
Δ	Laplace operator
$\int_\Gamma f(z)\, dz$	Integral of f over the curve Γ in the complex plane
$T_{\upharpoonright W}$	Restriction of the operator T to the subset W of its domain
$\{x_i\}_1^N$	Set of vectors $x_1, \ldots, x_N$, as well as linear subspace generated by these vectors
$\{x_n\}_{\mathbb{N}}$	Sequence $x_1, x_2, \ldots$
$\{x_n\}_{N_1 \subset \mathbb{N}}$	Subsequence of $\{x_n\}_{\mathbb{N}}$ indexed by N_1
$[a, b]$	Interval of the real numbers set defined by $\{x \in \mathbb{R};\ a \leq x \leq b\}$
card I	Cardinal of set I
$:=$	Is equal by definition
iff	If and only if

CHAPTER 1

The Matrix Eigenvalue Problem

Introduction

This chapter is a survey of both theory and practice for the matrix eigenvalue problem. The theory is presented in a way that will be directly usable in the following chapters, introducing spectral projections, perturbation theory, iterative refinement technique, and a posteriori error bounds; the methods for large sparse matrices are analyzed as projection methods on Krylov subspaces. The aim of this chapter is twofold:

(1) to study the practical methods that one needs to compute the eigenelements of the matrix problem related to the discrete approximation of a differential or integral operator; and

(2) to make the reader familiar with the theoretical tools used throughout the book by presenting them first in the simple context of operators in a finite-dimensional space.

The main reference for matrix eigenvalue problems remains Wilkinson (1965). Parlett (1980a) contains very recent material on the modern treatment of the symmetric eigenvalue problem. For an elementary exposition of general matrix computations, the reader is referred to Stewart (1973a).

1. The Eigenvalue Problem in $\mathbb{C}^N$

1.1. Notation

We first present the general notation and definitions for matrices and vectors that we shall use in this chapter. We denote by $\mathbb{C}^N$ the complex N-dimensional space of column vectors x with components (ξ_i); x^{H} denotes the row vector of components $(\bar{\xi}_i)$.

The main examples of norms on $\mathbb{C}^N$ with which we shall be concerned are the euclidean norm $\|x\|_2 = (\sum_{i=1}^N |\xi_i|^2)^{1/2}$, the sum norm $\|x\|_1 = \sum_{i=1}^N |\xi_i|$, and the max norm $\|x\|_\infty = \max_{1 \le i \le N} |\xi_i|$. Unless otherwise stated, $\|\cdot\|$ will be an arbitrary vector norm on $\mathbb{C}^N$.

The inner product in $\mathbb{C}^N$ is $(x, y) = y^H x$. If $(x, y) = 0$, then x and y are *orthogonal*. Let $\{x_i\}_1^N$ be a basis of $\mathbb{C}^N$, that is, a set of N independent vectors. The basis is *orthonormal* iff

$$(x_i, x_j) = \delta_{ij} \qquad \text{for} \quad i, j = 1, \ldots, N.$$

The expansion of x in the orthonormal basis is then $x = \sum_{i=1}^N (x, x_i)x_i$. When the basis is *not* orthonormal, the coefficients ξ_i of the expansion $x = \sum_{i=1}^N \xi_i x_i$ can be expressed by $\xi_i = (x, y_i)$, $i = 1, \ldots, N$, where $\{y_i\}_1^N$ is a basis of $\mathbb{C}^N$ such that

$$(x_i, y_j) = \delta_{ij} \qquad \text{for} \quad i, j = 1, \ldots, N. \tag{1.1}$$

The proof of the existence and uniqueness of $\{y_i\}_1^N$ is left to the reader. The basis $\{y_i\}_1^N$ defined by (1.1) is called the *adjoint basis* of $\{x_i\}_1^N$. $\{x_i\}$ and $\{y_i\}$ are also said to form a *biorthogonal family* of elements of $\mathbb{C}^N$.

Exercise

1.1 Consider in $\mathbb{C}^2$ the basis

$$x_1 = \begin{pmatrix} 1 \\ 2 \end{pmatrix}, \qquad x_2 = \begin{pmatrix} 1 \\ -2 \end{pmatrix}.$$

Check that the adjoint basis is

$$y_1 = \frac{1}{2}\begin{pmatrix} 1 \\ \frac{1}{2} \end{pmatrix}, \qquad y_2 = \frac{1}{2}\begin{pmatrix} 1 \\ -\frac{1}{2} \end{pmatrix}.$$

Compute the expansion $x = \xi_1 x_1 + \xi_2 x_2$, where

$$x = \begin{pmatrix} 2 \\ 1 \end{pmatrix}.$$

Let $A = (a_{ij})_{i,j=1,\ldots,N}$ be a complex square matrix. $A^H = (\bar{a}_{ji})_{i,j=1,\ldots,N}$ is the conjugate-transposed matrix. A is *normal* iff $AA^H = A^H A$. It is *hermitian* iff $A = A^H$; when A is hermitian, $x^H A x$ is real for $x \in \mathbb{C}^N$. A hermitian matrix A is *positive definite* (resp. *semidefinite*) iff

$$x \neq 0 \Rightarrow x^H A x > 0 \qquad (\text{resp. } x^H A x \geq 0).$$

A matrix Q is *unitary* iff $Q^H Q = I$. Then $Q^{-1} = Q^H$ and $\|x\|_2 = \|Qx\|_2$.

A square matrix A represents a linear operator (or transformation) in the canonical basis of $\mathbb{C}^N$. The null space is $\operatorname{Ker} A = \{x \in \mathbb{C}^N; Ax = 0\}$ and the range space is $\operatorname{Im} A = \{y \in \mathbb{C}^N;\ y = Ax \text{ for } x \in \mathbb{C}^N\}$. Given a vector norm $\|\cdot\|$ on $\mathbb{C}^N$, the operator (or matrix) norm for A is

$$\|A\| := \max_{0 \neq x \in \mathbb{C}^N} \frac{\|Ax\|}{\|x\|} = \max_{\|x\|=1} \|Ax\|.$$

The norm of a rectangular matrix can be defined accordingly. It is easy to verify that

$$\|A\|_1 := \max_{\|x\|_1=1} \|Ax\|_1 = \max_{1 \le j \le N} \sum_{i=1}^{N} |a_{ij}|$$

and

$$\|A\|_\infty = \max_{1 \le i \le N} \sum_{j=1}^{N} |a_{ij}| = \|A^{\mathrm{H}}\|_1.$$

The operator norm satisfies $\|AB\| \le \|A\|\,\|B\|$. The condition number (relative to inversion) of a regular matrix A is $K(A) := \|A\|\,\|A^{-1}\|$.

Exercises

1.2 Show that $K(A) \ge 1$.

1.3 For a unitary matrix Q, show that $\|Qx\|_2 = \|x\|_2$ for $x \in \mathbb{C}^N$, and

$$K_2(Q) = \|Q\|_2\,\|Q^{\mathrm{H}}\|_2 = 1.$$

The unitary matrices have then the smallest condition number relative to inversion.

A matrix P is a *projection* iff it is *idempotent*: $P^2 = P$. It is a projection on the subspace $M := \operatorname{Im} P$ along $W := \operatorname{Ker} P$ (or on M, parallel to W): it defines the direct sum $\mathbb{C}^N = M \oplus W$. M and W are *supplementary* subspaces. Conversely a decomposition $\mathbb{C}^N = M \oplus W$ defines a unique projection on M along W.

We define $M^\perp$, the orthogonal complement of M, by

$$M^\perp = \{y \in \mathbb{C}^N;\ (x, y) = 0 \text{ for all } x \text{ in } M\}.$$

M and $M^\perp$ are particular cases of supplementary spaces. In that case, the corresponding direct sum is denoted $\oplus$: $\mathbb{C}^N = M \oplus M^\perp$ is the orthogonal direct sum decomposition of $\mathbb{C}^N$.

The projection on M along $M^\perp$ is the *orthogonal* projection P_M on M. Then $\|x - P_M x\|_2 = \min_{y \in M} \|x - y\|_2$ for $x \in \mathbb{C}^N$.

For any square matrix A, $\operatorname{Ker}(A^{\mathrm{H}}) = (\operatorname{Im} A)^\perp$ and $\operatorname{Im}(A^{\mathrm{H}}) = (\operatorname{Ker} A)^\perp$.

Exercises

1.4 We define $Q = I - P$. Show that Q is the projection on W along M.

1.5 $\mathbb{C}^N = M \oplus W$. Let P be the projection on M along W. Show that P^{H} is the projection on $M' = W^\perp$ along $W' = M^\perp$.

1.6 Show that P is an orthogonal projection iff $P = P^{\mathrm{H}}$.

1.2. Definitions and Properties

Let A be a square real or complex matrix of order N. We consider the eigenvalue problem

$$\boxed{\text{find} \quad \lambda \in \mathbb{C}, 0 \neq \varphi \in \mathbb{C}^N, \qquad \text{such that} \quad A\varphi = \lambda\varphi}\ .$$

λ is called an *eigenvalue* of A, φ an *eigenvector*. λ and φ are *eigenelements* of A.

The complex number λ is an eigenvalue of A iff $A - \lambda I$ is singular. Therefore any eigenvalue λ of A is a root of the characteristic polynomial $\pi(\lambda) = \det(\lambda I - A)$, the determinant of $\lambda I - A$.

If $A - \lambda I$ is singular, so is $A^{\mathrm{H}} - \bar{\lambda} I$, and $\bar{\lambda}$ is an eigenvalue of A^{H}. Then, if $A\varphi = \lambda\varphi$, there exists $\psi \neq 0$ such that $A^{\mathrm{H}}\psi = \bar{\lambda}\psi$: ψ is an eigenvector of A^{H} associated with $\bar{\lambda}$. Equivalently, $\psi^{\mathrm{H}}A = \lambda\psi^{\mathrm{H}}$: ψ is called a *left* eigenvector of A associated with λ.

According to the Gauss–d'Alembert theorem, the characteristic polynomial $\pi(\lambda)$ of degree N has N complex roots, some of which may be repeated. These roots are the eigenvalues of A. If a root λ is repeated m times, λ is said to be a multiple eigenvalue of *algebraic multiplicity* m. If $m = 1$, the eigenvalue is said to be *simple*.

Let $\lambda_1, \lambda_2, \ldots, \lambda_K$, $1 \leq K \leq N$, be the *distinct* eigenvalues of A. The set $\sigma(A) := \{\lambda_i\}_1^K$ consisting of all the eigenvalues of A is the *spectrum* of A. The *spectral radius* of A is $r_\sigma(A) := \max_{1 \leq i \leq K} |\lambda_i|$.

Remark In numerical linear algebra, $r_\sigma(A)$ is often denoted $\rho(A)$, but this notation introduces a possible confusion with the resolvent set of the matrix A, which is the set of complex numbers z such that $A - zI$ is regular (cf. Chapter 2).

Exercises

1.7 If λ is an eigenvalue of A with eigenvector φ, show that for any integer $k > 0$, λ^k is an eigenvalue of A^k with eigenvector φ.

1.8 Show that if P is idempotent, then $\sigma(P) \subseteq \{0, 1\}$.

1.9 A matrix D is *nilpotent* if $D^k = 0$ for some integer $k > 0$. Show that $\sigma(D) = \{0\}$, and therefore $r_\sigma(D) = 0$.

1.10 Show that if Q is unitary and $\lambda \in \sigma(Q)$, then $|\lambda| = 1$.

1.11 Show that $r_\sigma(A) \leq \|A\|$ for any matrix norm.

1.12 Let λ be an eigenvalue of A with eigenvector φ. Show that $\lambda = \varphi^{\mathrm{H}}A\varphi/\varphi^{\mathrm{H}}\varphi$. λ is the *Rayleigh quotient* of A based on φ.

1.13 Infer from Exercise 1.12 that the eigenvalues of a hermitian matrix are real, and those of a positive definite (resp. semidefinite) matrix are positive (resp. nonnegative).

1.14 Let φ_1 be a right eigenvector of A corresponding to λ_1, and ψ_2 a left eigenvector corresponding to $\lambda_2 \neq \lambda_1$. Prove that $\psi_2^{\mathrm{H}}\varphi_1 = 0$. (*Hint*: compute $\psi_2^{\mathrm{H}}A\varphi_1$.) Deduce that if A is hermitian, the eigenvectors corresponding to distinct eigenvalues are orthogonal.

The eigenvectors associated with λ are solutions of $(A - \lambda I)x = 0$. They span the null space $\text{Ker}(A - \lambda I) = \{x \in \mathbb{C}^N; (A - \lambda I)x = 0\}$, which is called the *eigenspace*; its dimension is the *geometric multiplicity g of* λ, that is, the maximal number of *independent* eigenvectors associated with λ. One can show that $g \leq m$.

Lemma 1.1 *Let $\varphi_1, \ldots, \varphi_K$ be eigenvectors of A corresponding to the distinct eigenvalues $\lambda_1, \ldots, \lambda_K$. Then the vectors $\{\varphi_i\}_1^K$ are independent.*

Proof The proof is by induction on K. For $K = 1$, $\varphi_1 \neq 0$ and $\{\varphi_1\}$ is independent. Assume that the theorem holds for $K - 1$ vectors. Let $X = (\varphi_1, \ldots, \varphi_{K-1})$ be the $N \times (K - 1)$ matrix whose columns are the eigenvectors $\varphi_1, \ldots, \varphi_{K-1}$. We suppose that φ_K is a linear combination of $\{\varphi_i\}_1^{K-1}$; there exists a vector $u \neq 0$, $u \in \mathbb{C}^{K-1}$ such that $\varphi_K = Xu$. Then $\lambda_K \varphi_K = A\varphi_K = AXu = X\Delta u$, where Δ is the diagonal $\Delta = \text{diag}(\lambda_1, \ldots, \lambda_{K-1})$ and $0 = \lambda_K \varphi_K - X\Delta u = X(\lambda_K I - \Delta)u$. Since λ_K is distinct from the other eigenvalues, $\lambda_K I - \Delta$ is regular, and $v = (\lambda_K I - \Delta)u \neq 0$, while $Xv = 0$. This contradicts the independence of the vectors $\{\varphi_i\}_1^{K-1}$. □

We now turn to the theoretical problem of reducing A to simpler forms under similarity transformations.

1.3. Spectral Decomposition

Let $\mu_1, \mu_2, \ldots, \mu_N$ be the N *repeated* eigenvalues of A, numbered according to their *algebraic* multiplicity. For example, λ_1 of multiplicity m_1 is repeated m_1 times: $\mu_1 = \cdots = \mu_{m_1} = \lambda_1$, $\mu_{m_1+1} = \cdots = \lambda_2$.

1.3.1. A is diagonalizable

A is *diagonalizable* iff it is similar to a diagonal matrix. We suppose throughout this section that A is diagonalizable. Let D be the diagonal matrix $D = \text{diag}(\mu_1, \ldots, \mu_N)$.

Theorem 1.2 *A is diagonalizable iff A has N independent eigenvectors $\{\varphi_i\}_1^N$. Then A can be factorized as*

$$A = VDV^{-1}, \tag{1.2}$$

where the ith *column of V (resp.* ith *row of V^{-1}) is the eigenvector φ_i of A (resp. ψ_i of A^{H}) associated with μ_i (resp. $\bar{\mu}_i$).*

Proof If there are N independent eigenvectors $\{\varphi_i\}_1^N$, let V be the regular matrix whose columns are $\varphi_1, \ldots, \varphi_N$. $V^{-1}V = I$ implies that the rows ψ_i^{H} of V^{-1} satisfy $\psi_i^{\text{H}}\varphi_j = \delta_{ij}$ for $i, j = 1, \ldots, N$. Observe that

$A\varphi_i = \mu_i \varphi_i$ for $i = 1, \ldots, N$ may be written $AV = VD$ or equivalently $A = VDV^{-1}$: A is diagonalizable. $AV = VD$ is equivalent to $V^{-1}A = DV^{-1}$; that is, $A^H(V^{-1})^H = (V^{-1})^H D^H$, or $A^H \psi_i = \bar{\mu}_i \psi_i$ for $i = 1, \ldots, N$: the ψ_i are the eigenvectors of A^H normalized by $\psi_i^H \varphi_j = \delta_{ij}$, $i, j = 1, \ldots, N$. Note that $\{\psi_i\}_1^N$ is the adjoint basis of $\{\varphi_i\}_1^N$. The converse is left to the reader. □

Exercises

1.15 Show that any matrix with N distinct eigenvalues is diagonalizable.

1.16 Show that any hermitian matrix is diagonalizable, by means of unitary similarity transformations: $A = QDQ^H$, the eigenvector basis being orthonormal.

1.17 A matrix A is said to be *nondefective* if each eigenvalue λ_i of multiplicity m_i has exactly m_i independent eigenvectors for $i = 1, \ldots, K$. Show that A is nondefective iff it is diagonalizable.

For $i = 1, 2, \ldots, K$, let $M_i = \operatorname{Ker}(A - \lambda_i I)$ be the eigenspace associated with the distinct eigenvalue λ_i, of multiplicity m_i. By hypothesis $\dim M_i = m_i$, and $X = \bigoplus_{i=1}^K M_i$. We consider the decomposition $X = M_i \oplus (\bigoplus_{j \neq i} M_j)$. The *eigenprojection* P_i associated with λ_i is the projection of $\mathbb{C}^N$ on the eigenspace M_i, along $\bigoplus_{j \neq i} M_j$. Clearly

$$\sum_{i=1}^K P_i = I, \qquad P_i P_j = \delta_{ij} P_i,$$

and

$$AP_i = P_i A = P_i A P_i = \lambda_i P_i, \qquad i = 1, \ldots, K. \tag{1.3}$$

Theorem 1.3 *Any diagonalizable matrix A has the spectral decomposition*

$$A = \sum_{i=1}^K \lambda_i P_i. \tag{1.4}$$

Proof The proof of (1.3) and (1.4) is left to the reader. The spectral decomposition (1.4) is unique; the decomposition (1.2) is essentially unique, up to the choice of a basis in each eigenspace M_i. □

Exercises

1.18 If A is hermitian, P_i is an *orthogonal* projection and $\|P_i\|_2 = 1$. Show that $M_i^\perp = \bigoplus_{j \neq i} M_j$.

1.19 λ is simple with associated right and left eigenvectors φ and ψ such that $\|\varphi\|_2 = \psi^H \varphi = 1$. Prove that the eigenprojection P is the matrix (in the canonical basis) of the linear operator $\xi \in \mathbb{C}^N \mapsto P\xi = (\psi^H \xi)\varphi$. (*Hint*: The supplementary subspace of $M = \{\varphi\}$, defined by P, is $\{\psi\}^\perp = \{z \in \mathbb{C}^N; \psi^H z = 0\}$.) Show that $\|P\|_2 = \|\psi\|_2 \geq 1$. Show also that $P = \varphi\psi^H$.

1.20 Let A be diagonalizable. Show that A^H is also diagonalizable and that its spectral decomposition is $A^H = \sum_{i=1}^K \bar{\lambda}_i P_i^H$. Deduce that the multiplicities of λ_i and $\bar{\lambda}_i$ are the same.

1.21 Consider again the right and left eigenvectors bases $\{\varphi_i\}_1^N$ and $\{\psi_i\}_1^N$. Let λ be an eigenvalue of multiplicity m. Let the associated eigenspace be $M = \{\varphi_i, i \in I\}$, where the cardinality of I equals m. And consider the supplementary subspace $W = \{\varphi_i; i \notin I\}$. Show that $W = \{\psi_i; i \in I\}^\perp = \{z \in \mathbb{C}^N;\ \psi_i^H z = 0,\ i \in I\}$. The eigenprojection P associated with λ is then $P = \sum_{i \in I} \varphi_i \psi_i^H$.

Example 1.1

$$A = \begin{pmatrix} 1 & 1 \\ 4 & 1 \end{pmatrix}$$

has the eigenvalues $\lambda_1 = 3$ and $\lambda_2 = -1$ and the associated eigenvectors

$$\varphi_1 = \begin{pmatrix} 1 \\ 2 \end{pmatrix} \quad \text{and} \quad \varphi_2 = \begin{pmatrix} 1 \\ -2 \end{pmatrix}.$$

$$A^H = \begin{pmatrix} 1 & 4 \\ 1 & 1 \end{pmatrix}$$

has the eigenvectors

$$\psi_1 = \frac{1}{2}\begin{pmatrix} 1 \\ \frac{1}{2} \end{pmatrix} \quad \text{and} \quad \psi_2 = \frac{1}{2}\begin{pmatrix} 1 \\ -\frac{1}{2} \end{pmatrix}.$$

The eigenprojection associated with λ_1 is

$$P_1 = \varphi_1 \psi_1^H\colon\ \ P_1\xi = \tfrac{1}{2}(\xi_1 + \tfrac{1}{2}\xi_2)\varphi_1.$$

Similarly for λ_2

$$P_2 = \varphi_2 \psi_2^H\colon\ \ P_2\xi = \tfrac{1}{2}(\xi_1 - \tfrac{1}{2}\xi_2)\varphi_2.$$

Finally, A has the decomposition $A = 3P_1 - P_2$.

The projection P_1 is a projection on $M = \{\varphi_1\}$ along $W = \{\psi_1\}^\perp$ (cf. Fig. 1.1).

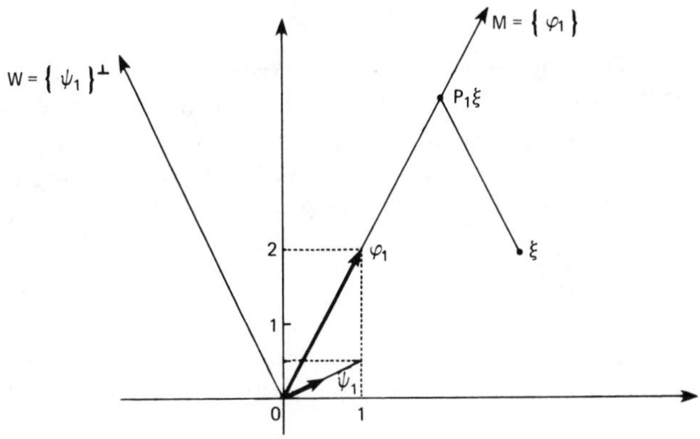

Figure 1.1

The decomposition (1.2) is attractive because the diagonal form is very easy to deal with. In practice, however, the matrix V, although it is nonsingular, may be ill-conditioned with respect to inversion, making $V^{-1}AV$ very difficult to form. In view of that, we may wish to restrict the similarity transformations to a class of well-conditioned matrices, such as unitary matrices, which have a condition number equal to 1.

1.3.2. Unitary similarity transformations

Theorem 1.4 (Schur) *Let A have the eigenvalues $\{\mu_i\}_1^N$ counted with their algebraic multiplicity. Then there exists a unitary matrix Q such that $Q^H AQ$ is upper triangular with diagonal elements $\mu_1, \mu_2, \ldots, \mu_N$, in that order.*

Proof The proof is by induction on N. The theorem is true for $N = 1$. Suppose it is true for matrices of order $N - 1$. We use a deflation technique: knowing the eigenvalue μ_1 and eigenvector φ_1, we produce a matrix of order $N - 1$ that has the same eigenvalues as A, except for μ_1.

We suppose that $\|\varphi_1\|_2 = 1$. There exists an $N \times (N - 1)$ matrix U such that (φ_1, U) is unitary; the columns of U are orthogonal to φ_1, that is, $U^H\varphi_1 = 0$. Since $A\varphi_1 = \mu_1\varphi_1$,

$$A(\varphi_1, U) = (\mu_1\varphi_1, AU)$$

and

$$(\varphi_1, U)^H A(\varphi_1, U) = \begin{pmatrix} \varphi_1^H \\ U^H \end{pmatrix}(\mu_1\varphi_1, AU) = \begin{pmatrix} \mu_1 & \varphi_1^H AU \\ 0 & U^H AU \end{pmatrix}.$$

The eigenvalues of the matrix on the right-hand side are those of A. Hence $U^H AU$ has the eigenvalues $\mu_2, \ldots, \mu_N$. By the induction hypothesis, $U^H AU$ is unitarily similar to an upper-triangular matrix with diagonal elements $\mu_2, \ldots, \mu_N$. The desired result follows. □

Remarks (1) The above proof is an existence theorem. A constructive algorithm, called *QR*, is given in Section 3 for certain classes of matrices.

(2) If A is real, we wish to use real unitary (i.e., orthogonal) transformations. A can then be reduced to a real block upper-triangular matrix, where the diagonal blocks are at most 2×2. The 2×2 diagonal blocks have only conjugate eigenvalues.

Exercises

1.22 Prove that a triangular normal matrix is diagonal.

1.23 Prove that a matrix A is normal iff there exists a unitary matrix Q such that $Q^H AQ$ is diagonal.

1.24 Show that $r_\sigma(A) = \|A\|_2$ for a normal matrix.

1.25 Show that a normal matrix is nilpotent iff $A = 0$.

1.26 For an arbitrary matrix A, show that $\|A\|_2^2 = r_\sigma(A^H A)$.

Using possibly nonunitary transformations, the triangular matrix can be put, for a general matrix, into the simpler *Jordan form.*

1.3.3. The Jordan canonical form

We now suppose that λ_i, of algebraic multiplicity m_i, has fewer than m_i independent eigenvectors, so that g_i, the geometric multiplicity of λ_i, satisfies $g_i < m_i$. To get a basis of $\mathbb{C}^N$, the set of eigenvectors has to be completed. This can be done as follows.

We define $M_i := \operatorname{Ker}(A - \lambda_i I)^{m_i}$ for $i = 1, \ldots, K$.

A subspace M is said to be *invariant* under A if $AM \subseteq M$. For example, the eigenspace $\operatorname{Ker}(A - \lambda_i I)$ is invariant: each eigendirection is invariant.

Theorem 1.5 (Jordan) *For $i = 1, \ldots, K$, the subspaces M_i are invariant under A, $\dim M_i = m_i$, and $X = \bigoplus_{i=1}^K M_i$. There exists a basis in M_i such that A restricted to M_i is represented by the Jordan box of size m_i associated with λ_i:*

$$J_i = \begin{pmatrix} \lambda_i & \varepsilon & & & 0 \\ & \lambda_i & \ddots & & \\ & & \ddots & \ddots & \\ & & & \lambda_i & \varepsilon \\ 0 & & & & \lambda_i \end{pmatrix} \Bigg\updownarrow m_i,$$

where ε is 0 *or* 1.

This theorem states the existence of the Jordan basis. The proof is standard and may be found in any textbook on linear algebra. The Jordan canonical form is block diagonal, with K diagonal boxes J_i, $i = 1, \ldots, K$, associated with the distinct eigenvalues. A box J_i consists of one or several Jordan blocks defined by ε being zero. The Jordan form is unique up to the ordering of the blocks.

M_i is called the *invariant subspace* (or generalized eigenspace) associated with λ_i. The basis of M_i consists of g_i independent eigenvectors associated with λ_i, completed with generalized eigenvectors (or principal vectors) into m_i independent vectors. The principal vectors are invariant globally: they stay in M_i under A, but the direction changes. A principal vector x_k^i is computed from the preceding vector x_{k-1}^i (principal vector or eigenvector) by solving

$$(A - \lambda_i I)x_k^i = x_{k-1}^i.$$

For a complete description of the construction of the Jordan basis, see Wilkinson (1965).

The proof of Theorem 1.5 shows that

$$\{0\} \subset \operatorname{Ker}(A - \lambda_i I) \subset \operatorname{Ker}(A - \lambda_i I)^2 \subset \cdots \subset \operatorname{Ker}(A - \lambda_i I)^{l_i}$$
$$= \cdots = \operatorname{Ker}(A - \lambda_i I)^{m_i} = M_i,$$

the inclusion being proper until M_i is reached. $1 \leq l_i \leq m_i$; l_i is the smallest number l such that $M_i = \mathrm{Ker}(A - \lambda_i I)^l$ [or equivalently $(A - \lambda_i I)^{l_i} M_i = \{0\}$]. It is called the *ascent* (or index or height) of λ_i. If $l_i = 1$, λ_i is said to be *semi-simple*: the invariant subspace reduces to the eigenspace.

Example 1.2 Let λ be an eigenvalue of algebraic multiplicity $m = 7$, geometric multiplicity $g = 3$, and ascent $l = 3$. There are two possible Jordan boxes associated with λ; they consist of three blocks:

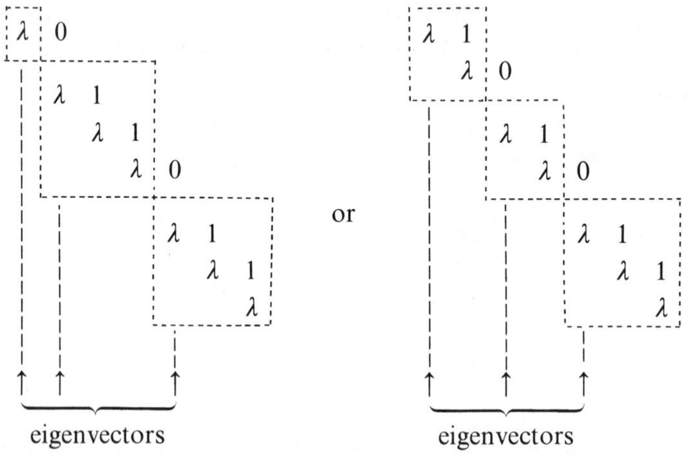

There are three eigenvectors and four principal vectors, and because $l = 3$, there are no more than two successive principal vectors: $(A - \lambda I)^3 x = 0$ for any x in the invariant subspace. Each eigenvector defines a block, the blocks having the size 1, 3, 3 or 2, 2, 3.

Exercises

1.27 Compute the eigenvalues and the Jordan basis of

$$A = \begin{pmatrix} 1 & 0 & 0 \\ -1 & 0 & 1 \\ -1 & -1 & 2 \end{pmatrix}.$$

1.28 In the Jordan box J_i, there are, on the first upper diagonal, exactly $g_i - 1$ zeros (not necessarily consecutive) and no more than $l_i - 1$ consecutive ones.

1.29 Prove that $\lim_{n\to\infty} A^n = 0$ iff $r_\sigma(A) < 1$. (*Hint*: Use the Jordan form.)

1.30 Show that $\lim_{n\to\infty} \|A^n\|^{1/n} = r_\sigma(A)$.

For a general matrix, the notion of eigenprojection is extended to that of *spectral projection*: P_i is the projection on the invariant subspace M_i along

$\bigoplus_{j\neq i} M_j$. Again $\sum_{i=1}^{K} P_i = I$, $P_iP_j = \delta_{ij}P_i$, and

$$AP_i = P_iA = P_iAP_i = \lambda_i P_i + D_i, \qquad i = 1, \ldots, K, \tag{1.5}$$

where D_i is a nilpotent matrix such that $D_i^{l_i} = 0$.

Theorem 1.6 *A general matrix A has the spectral decomposition*

$$A = \sum_{i=1}^{K} (\lambda_i P_i + D_i). \tag{1.6}$$

Exercises

1.31 Prove (1.5) by means of the Jordan canonical form of A. Prove that (1.6) is unique, given a Jordan form for A.

1.32 Prove that $P_iD_j = D_jP_i = \delta_{ij}D_j$, $D_iD_j = 0$ if $i \neq j$.

1.33 Prove that $\bigoplus_{j\neq i} M_j = \mathrm{Im}(A - \lambda_i I)^{l_i}$. Deduce that P_i is the projection associated with the decomposition $\mathbb{C}^N = \mathrm{Ker}(A - \lambda_i I)^{l_i} \oplus \mathrm{Im}(A - \lambda_i I)^{l_i}$.

1.34 Prove that the spectral decomposition of A^{H} is

$$A^{\mathrm{H}} = \sum_{i=1}^{K} (\bar{\lambda}_i P_i^{\mathrm{H}} + D_i^{\mathrm{H}}).$$

Deduce that the multiplicities and ascents of λ_i and $\bar{\lambda}_i$ are the same.

1.35 Let λ be a simple eigenvalue with associated right and left eigenvectors φ and ψ. Prove that the spectral projection is on $\{\varphi\}$, along $\{\psi\}^{\perp}$.

1.36 Let $\{x_i\}_1^N$ be the Jordan basis of A. Using Exercise 1.34, comment on the form of A^{H} in the adjoint basis $\{y_i\}_1^N$. Consider, for example,

$$A = \begin{pmatrix} 1 & 1 \\ 0 & 1 \end{pmatrix}.$$

2. The Stability of the Eigenvalue Problem

It is of prime importance, when solving a problem numerically, to know if small changes in the data produce small (resp. large) changes in the solution: if so, the problem is said to be stable or well-conditioned (resp. unstable or ill-conditioned). There is usually no precise boundary between stable and unstable problems. In any case some appropriate measure of instability will be useful and will enter into most error estimates. This measure of instability is commonly called the *condition number.*

As a basic example, one may think of $K(A) = \|A\|\,\|A^{-1}\|$, the condition number of the regular matrix A relative to inversion. From a mathematical standpoint, the equation $Ax = y$ has a uniquely determined solution $x = A^{-1}y$ if and only if A is regular. But in *numerical* analysis, the picture becomes more complicated. The numerical analyst wants a solution that is not only unique but also, to some extent, insensitive to a perturbation in the data. If A is perturbed by ΔA and y by Δy, this gives rise to a perturbed solution $x + \Delta x$ such that

$$\frac{\|\Delta x\|}{\|x\|} \leq \frac{K(A)}{1 - \|A^{-1}\| \, \|\Delta A\|} \left[\frac{\|\Delta y\|}{\|y\|} + \frac{\|\Delta A\|}{\|A\|} \right] \quad \text{if} \quad \|\Delta A\| < \frac{1}{\|A^{-1}\|}.$$

$K(A)$ is therefore a good measure for the relative sensitivity of the solution x. This explains the central role played in numerical analysis by unitary matrices that have a condition number for the euclidean norm equal to 1.

We now turn to the stability of the eigenvalue problem. The numerical methods that we shall present in Section 3 yield a set of approximate eigenelements of A that are the exact eigenelements of a perturbed matrix $A' = A + H$. Bounds on the difference between the eigenelements of A and A' in terms of H are given in Section 4.

The eigenvalues may be ill-conditioned if they are ill-conditioned roots of the characteristic polynomial. It is known that multiple roots of a polynomial may be ill-conditioned. For example, $\lambda^2 - 2\lambda + (1 - 10^{-8})$ has roots 1 ± 10^{-4}. If we change the constant term to 1, the roots both become 1, so that a change of 10^{-8} in one coefficient caused changes 10^4 times as large in the roots. But well-separated roots may also be ill-conditioned. For the eigenvalues, this may correspond to a nearly degenerate basis of eigenvectors. One may look in Wilkinson (1965, pp. 90–91) at a very illuminating example with the matrix of order 20:

$$A = \begin{pmatrix} 20 & 20 & & & 0 \\ & 19 & \ddots & & \\ & & \ddots & \ddots & \\ & & & 2 & 20 \\ 0 & & & & 1 \end{pmatrix}.$$

We investigate here the effect of a perturbation on a simple eigenvalue λ and its eigenvector φ by means of a perturbation technique. Let A have a *simple* eigenvalue λ with right and left eigenvectors φ and ψ such that $\|\varphi\|_2 = \varphi^{\mathrm{H}}\psi = 1$. The spectral projection is $P = \varphi\psi^{\mathrm{H}}$.

We denote by $\hat{A}_\lambda$ the operator $A - \lambda I$ restricted to $\{\psi\}^\perp$, an invariant subspace: $\hat{A}_\lambda := (A - \lambda I)_{\restriction\{\psi\}^\perp}$. Because λ is a simple eigenvalue of A, $\hat{A}_\lambda$ is bijective, and we denote by S the matrix representing the operator $\hat{A}_\lambda^{-1}(I - P)$. S is the generalized inverse of $A - \lambda I$ relative to the spectral projection P:

$$S(A - \lambda I) = (A - \lambda I)S = I - P.$$

Let the perturbation H be given: $A' := A + H$ with $\varepsilon := \|H\|_2$, which is supposed to be small. λ and φ are approximate eigenelements for A'; the associated *residual vector* for A' is

$$A'\varphi - \lambda\varphi = (A' - A)\varphi = H\varphi.$$

We set

$$\varepsilon' := \|H\varphi\|_2, \qquad \varepsilon' \leq \varepsilon.$$

Theorem 1.7 *With the above definitions, if ε is small enough, there exists a simple eigenvalue λ' of A' with an eigenvector φ' normalized by $\psi^{\mathrm{H}}\varphi' = 1$, such that*

$$\lambda' = \lambda + \psi^{\mathrm{H}}H\varphi + O(\varepsilon^2), \tag{1.7}$$

$$\varphi' = \varphi - SH\varphi + O(\varepsilon^2). \tag{1.8}$$

Proof We define $L = -(1/\varepsilon)H$, $\|L\|_2 = 1$. $A' = A - \varepsilon L$ has eigenelements λ' and φ', which may be expressed as convergent series in ε, for ε small enough. We set $\lambda' := \sum_{i=0}^{\infty} v_i \varepsilon^i$ and $\varphi' := \sum_{i=0}^{\infty} \eta_i \varepsilon^i$. The coefficients v_i and η_i can be iteratively computed by identifying the powers of ε in the formal identity

$$(A - \varepsilon L)\left(\sum_i \eta_i \varepsilon^i\right) = \left(\sum_i v_i \varepsilon^i\right)\left(\sum_i \eta_i \varepsilon^i\right).$$

Equating the constant terms yields $A\eta_0 = v_0\eta_0$; we choose the solutions $v_0 = \lambda$, $\eta_0 = \varphi$.

Equating the coefficients of ε gives the equation

$$(A - \lambda I)\eta_1 = v_1\varphi + L\varphi.$$

Using left multiplication by ψ^{H}, we get

$$\psi^{\mathrm{H}}(A - \lambda I)\eta_1 = 0 = v_1\psi^{\mathrm{H}}\varphi + \psi^{\mathrm{H}}L\varphi.$$

Hence

$$v_1 = -\psi^{\mathrm{H}}L\varphi.$$

The right-hand side of the equation

$$(A - \lambda I)\eta_1 = v_1\varphi + L\varphi$$

lies in $\{\psi\}^{\perp} = (I - P)\mathbb{C}^N$. We define η_1 as the unique solution of this equation in $\{\psi\}^{\perp}$; that is,

$$\eta_1 = S(v_1\varphi + L\varphi) \qquad \text{with} \quad P\eta_1 = 0.$$

Note that $S\varphi = S(I - P)\varphi = 0$ implies $\eta_1 = SL\varphi$.

Equating the coefficients of ε^2 gives the equation

$$(A - \lambda I)\eta_2 = \nu_2\varphi + L\eta_1 + \nu_1\eta_1,$$

from which we get $\nu_2 = -\psi^H L\eta_1$ and $\eta_2 = S[\nu_1\eta_1 + L\eta_1]$. By equating the coefficients of ε^i we get for $i > 2$

$$\nu_i = -\psi^H L\eta_{i-1},$$

$$\eta_i = S\left[\sum_{k=1}^{i-1} \nu_k\eta_{i-k} + L\eta_{i-1}\right], \qquad P\eta_i = 0.$$

For ε small enough, it is easy to prove that the two series converge geometrically:

$$\left|\lambda' - \sum_{k=0}^{i} \nu_k\varepsilon^k\right| = O(\varepsilon^{i+1}),$$

$$\left\|\varphi' - \sum_{k=0}^{i} \eta_k\varepsilon^k\right\|_2 = O(\varepsilon^{i+1}).$$

This result is also a particular case of Lemma 3.30. Since $P\eta_i = 0$ for $i \geq 1$, φ' is such that $P\varphi' = P\varphi = \varphi$, or equivalently $\psi^H\varphi' = 1$.

Letting $i = 1$ in the above equalities yields (1.7) and (1.8) because

$$\nu_0 + \varepsilon\nu_1 = \lambda + \psi^H H\varphi \qquad \text{and} \qquad \eta_0 + \varepsilon\eta_1 = \varphi - SH\varphi. \quad \square$$

Exercise

1.37 Let Π be the orthogonal projection on $\operatorname{Im}(A - \lambda I)$. The *Moore–Penrose inverse* of $A - \lambda I$ is $(A - \lambda I)^\dagger = [(A - \lambda I)\restriction_{(\operatorname{Ker}(A-\lambda I))^\perp}]^{-1}\Pi$. $x^\dagger = (A - \lambda I)^\dagger b$ is the least-squares solution of

$$(A - \lambda I)x = b$$

with minimum euclidean norm (cf. Stewart, 1973a). Show that $S = (A - \lambda I)^\dagger$ iff A is hermitian.

This perturbation technique will be developed in Section 3.5 as an iterative refinement process to compute the eigenelements.

2.1. Condition Number for λ

From (1.7) we get

$$|\psi^H H\varphi| \leq \varepsilon'\|\psi\|_2 \qquad \text{and} \qquad |\lambda' - \lambda| \leq \varepsilon'\|\psi\|_2 + O(\varepsilon^2).$$

So λ is ill-conditioned if $\|\psi\|_2$ is large. $\|\psi\|_2$ is a *condition number for* λ when ψ is normalized by $\psi^H\varphi = \|\varphi\|_2 = 1$; hence $\|\psi\|_2 \geq 1$.

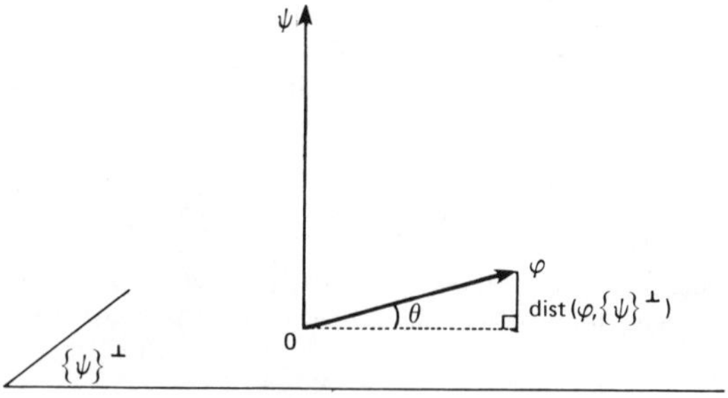

Figure 1.2

The orthogonal complement $\{\psi\}^\perp$ is the subspace of dimension $N - 1$ spanned by the vectors of the Jordan basis for A other than φ (cf. Exercise 1.35). Let θ be the acute angle between φ and the subspace $\{\psi\}^\perp$ (cf. Fig. 1.2):

$$\text{dist}(\varphi, \{\psi\}^\perp) = \min_{u \in \{\psi\}^\perp} \|\varphi - u\|_2 = \sin\theta.$$

Property $\|\psi\|_2 = \|P\|_2 = 1/\text{dist}(\varphi, \{\psi\}^\perp) = 1/\sin\theta.$

Proof The first equality is clear (cf. Exercise 1.19). Set $\hat{\psi} := \psi/\|\psi\|_2$. Then

$$1/\|\psi\|_2 = |\hat{\psi}^{\mathrm{H}}\varphi| = \sin\theta = \text{dist}(\varphi, \{\psi\}^\perp). \quad \square$$

It follows that λ is ill-conditioned if the norm of its spectral projection is large or, equivalently, if θ is small. Then φ is nearly orthogonal to ψ, and the Jordan basis of A is close to degeneracy; that is, φ is nearly a linear combination of the other vectors of the Jordan basis.

Example 1.3 Let

$$A = \begin{pmatrix} 1 & 10^4 \\ 0 & 11 \end{pmatrix}.$$

1 is a simple eigenvalue with the right and left eigenvectors

$$\varphi = \begin{pmatrix} 1 \\ 0 \end{pmatrix} \quad \text{and} \quad \psi = \begin{pmatrix} 1 \\ -10^3 \end{pmatrix}.$$

$$\|\psi\|_2 \approx 10^3, \qquad A + H = \begin{pmatrix} 1 & 10^4 \\ -10^{-5} & 11 \end{pmatrix}$$

has an eigenvalue $\lambda' \approx 1.01$. A perturbation of order 10^{-5} in A generates a perturbation of order 0.01 in λ; that is, the change in λ is 10^3 times greater than the change in A.

Exercises

1.38 Consider $A' = A - \varepsilon L$. Under the hypothesis of Theorem 1.7, we define $\lambda(\varepsilon) = \lambda'$. Prove that

$$\frac{d\lambda}{d\varepsilon}(0) = \lim_{\varepsilon \to 0} \frac{\lambda(\varepsilon) - \lambda}{\varepsilon} = -\psi^{H} L \varphi.$$

1.39 We suppose that the left eigenvector ψ is normalized by $\|\psi\|_2 = 1$. Show that $1/|\psi^{H}\varphi|$ is now a condition number for λ. When no normalization condition is imposed, show that the condition number for λ is

$$\frac{\|\varphi\|_2 \, \|\psi\|_2}{|\psi^{H}\varphi|} = \frac{1}{\sin\theta}.$$

1.40 Prove that $\lambda' = \lambda + \psi^{H} H \varphi - \psi^{H} H S H \varphi + O(\varepsilon^3)$. Give a bound for $|\lambda' - \lambda|$ when $\psi^{H} H \varphi = 0$.

An unsatisfactory feature of the above analysis is that $\|\psi\|_2$ is not invariant by diagonal similarity transformation.

Example 1.4 Consider in $\mathbb{C}^2$ the hermitian matrix

$$A = \begin{pmatrix} 2 & 1 \\ 1 & 2 \end{pmatrix}.$$

The right and left eigenvectors of A associated with $\lambda = 3$ are equal:

$$\psi = \varphi = \frac{1}{\sqrt{2}} \begin{pmatrix} 1 \\ 1 \end{pmatrix} \quad \text{and} \quad \|\varphi\|_2 = 1.$$

The eigenvalue $\lambda = 3$ is well-conditioned. However, if

$$D = \begin{pmatrix} 1 & 0 \\ 0 & \alpha \end{pmatrix},$$

we have

$$D^{-1}AD = \begin{pmatrix} 2 & \alpha \\ 1/\alpha & 2 \end{pmatrix}$$

and now

$$\varphi = \frac{1}{\sqrt{1+\alpha^2}} \begin{pmatrix} \alpha \\ 1 \end{pmatrix}, \quad \psi = \frac{\sqrt{1+\alpha^2}}{2\alpha} \begin{pmatrix} 1 \\ \alpha \end{pmatrix}, \quad \text{and} \quad \|\psi\|_2 = \frac{1+\alpha^2}{2\alpha}.$$

$\|\psi\|_2$ can be made arbitrarily large by choosing a small α. This example is somewhat artificial, because if α is small, $\|D^{-1}AD\|_2 \gg \|A\|_2$. In practice, the relevant values of $\|\psi\|_2$ are those for $D^{-1}AD$, where D has been chosen so that $\|D^{-1}AD\|_2$ is close to its minimum. This is the *balancing* of a matrix (cf. Parlett and Reinsch, 1969).

The high sensitivity of a *simple* eigenvalue to a perturbation when $\|\psi\|_2$ is large may be explained by the following remark. If $\|\psi\|_2$ is large, A is necessarily relatively close to a matrix with a *multiple* eigenvalue (cf. Ruhe, 1970; Golub and Wilkinson, 1976).

2.2. Condition Number for φ

Upon setting $s := \|S\|_2$, we get from (1.8)

$$\|\varphi' - \varphi\|_2 \leq s\varepsilon' + O(\varepsilon^2).$$

φ is ill-conditioned if s is large: s is a *condition number for* φ. Let $d(\lambda)$ be the *distance of isolation* of λ: $d(\lambda) := \min_{\mu \in \sigma(A) - \{\lambda\}} |\mu - \lambda|$.

Property $s \geq 1/d(\lambda)$.

Proof This follows readily from the inequality (Exercise 1.11)

$$r_\sigma(S) = \max_{\mu \in \sigma(A) - \{\lambda\}} \frac{1}{|\lambda - \mu|} \leq \|S\|_2 = s. \quad \square$$

If λ is near another eigenvalue of A, then its eigenvector will be ill-conditioned.

Example 1.5 Let

$$A = \begin{pmatrix} 1 & 0 \\ 0 & 1 - \varepsilon \end{pmatrix}.$$

If $\varepsilon \neq 0$, then 1 is a simple eigenvalue with eigenvector

$$\varphi = \begin{pmatrix} 1 \\ 0 \end{pmatrix}.$$

And $s = 1/|\varepsilon|$. φ will be ill-conditioned if ε is small. Indeed

$$A + H = \begin{pmatrix} 1 & 0 \\ 10^{-5} & 1 - \varepsilon \end{pmatrix}$$

has the eigenvector

$$\varphi' = \begin{pmatrix} 1 \\ 10^{-5}/\varepsilon \end{pmatrix}$$

corresponding to $\lambda' = 1$. If ε is small, φ' differs significantly from φ.

Note that, for nonnormal matrices, s may be large even if λ is well-separated from the rest of the spectrum (cf. Wilkinson, 1965, pp. 90–91).

Exercises

1.41 With the definitions of Theorem 1.7, and for a diagonalizable matrix A, show that

$$\varphi' = \varphi + \sum_k \frac{\psi_k^H H \varphi}{\lambda - \mu_k} \varphi_k + O(\|H\|^2),$$

where the μ_k, φ_k, ψ_k are respectively all the eigenvalues, right and left eigenvectors of A different from λ, φ, ψ, where the ψ_k are normalized by $\psi_k^H \varphi_k = \|\varphi_k\|_2 = 1$.

1.42 Consider the matrix

$$A = \begin{pmatrix} 2 & 0 & 0 \\ 0 & 1 & 1 \\ 0 & 0 & 1 + 10^{-10} \end{pmatrix}.$$

Compute the three condition numbers $\|\psi_1\|_2$, $\|\psi_2\|_2$, and $\|\psi_3\|_2$, where the ψ_i are the left eigenvectors normalized by $\psi_i^H \varphi_i = \|\varphi_i\|_2 = 1$, $i = 1, 2, 3$.

What do you conclude for the condition of the eigenvalues? Solve the same question for the eigenvectors.

1.43 Show that the condition numbers s and $\|\psi\|_2$ of the matrix A are invariant under unitary similarity transformations on A.

Although each individual eigenvector associated with a cluster of eigenvalues is ill-conditioned, the subspace spanned by all these eigenvectors may be insensitive to a perturbation in the matrix, as shown in Example 2.27. If P (resp. P') is the direct sum of the projections associated with all the eigenvalues of A (resp. A') in the cluster, then $\|P' - P\| \to 0$ when $\|A' - A\| \to 0$ (cf. Chapter 5).

The above study does not apply to *multiple* eigenvalues. The perturbation theory adapted to this case requires the use of spectral projections and invariant subspaces. It will be thoroughly studied in the following chapters in the framework of infinite-dimensional spaces. The ascent l of a multiple eigenvalue will play an important role, for both eigenvalues and eigenvectors (cf. also Wilkinson, 1965, Chapter 2; Kato, 1976, Chapter 2).

Example 1.6 Consider in $X = \mathbb{C}^N$, for $\varepsilon > 0$,

$$A(\varepsilon) = \begin{pmatrix} 1 & 1 & & & \\ & 1 & 1 & & \\ & & 1 & \ddots & \\ & & & \ddots & 1 \\ \varepsilon & & & & 1 \end{pmatrix} \underset{\varepsilon \to 0}{\longrightarrow} A = \begin{pmatrix} 1 & 1 & & & \\ & 1 & 1 & & \\ & & 1 & \ddots & \\ & & & \ddots & 1 \\ & & & & 1 \end{pmatrix}.$$

The eigenvalues of $A(\varepsilon)$ are $\lambda_k(\varepsilon) = 1 + \varepsilon^{1/N} e^{2ik\pi/N}$, $|\lambda_k(\varepsilon) - 1| = \varepsilon^{1/N}$ for $k = 0, \ldots, N-1$. Note that the arithmetic mean $\hat{\lambda}$ is such that

$$\hat{\lambda} = \frac{1}{N} \sum_{k=0}^{N-1} \lambda_k(\varepsilon) = 1.$$

A is a Jordan matrix with the eigenvalue 1, of algebraic multiplicity N equal to its ascent: $l = N$.

As for the eigenvector $\varphi_0(\varepsilon)$ associated with $\lambda_0(\varepsilon) = 1 + \varepsilon^{1/N}$, $\varphi_0(\varepsilon) \underset{\varepsilon \to 0}{\longrightarrow} e_1$, the eigenvector of A associated with 1, and we get

$$\|\varphi_0(\varepsilon) - e_1\|_2 = O(\varepsilon^{1/N}).$$

2.3. *A Is Hermitian or Normal*

When A is hermitian or normal, the above study is simplified for two reasons:

(1) $\varphi = \psi$ and $\|\varphi\|_2 = \|\psi\|_2 = 1$; and
(2) $s = 1/d(\lambda)$.

Hence there is only one cause of ill-conditioning for a hermitian (normal) matrix with simple eigenvalues: the existence of *close* eigenvalues. $1/d(\lambda)$ is the condition number of the eigenvectors of a hermitian (normal) matrix, with simple eigenvalues, the latter being well-conditioned with a condition number equal to 1.

3. Some Numerical Methods

We do not intend to describe all the efficient methods to compute the eigenelements of a *full* matrix A of *moderate* size N; we shall describe mainly a class of methods for reducing by unitary similarity transformations a general (resp. hermitian) matrix to an upper-triangular (resp. diagonal) form. The most important among them are the QR algorithm to compute all the eigenvalues and the simultaneous iteration method to compute several of them. This latter method is an extension of the old power method, which is based on the behavior of the (vector) Krylov sequence x, Ax, $A^2x, \ldots$. Given r independent starting vectors $x_1, \ldots, x_r$, which span the subspace $U = \{x_1, \ldots, x_r\}$, we consider the (subspace) Krylov sequence U, AU, $A^2U, \ldots$. We first study the convergence of this sequence.

Let the N repeated eigenvalues of A be ordered by decreasing modulus: $|\mu_1| \geq |\mu_2| \geq \cdots \geq |\mu_N|$. The corresponding Jordan basis is $\{\varphi_i\}_1^N$.

3.1. *Convergence of the Krylov Sequence* A^kU, $k = 0, 1, 2, \ldots$

We suppose that $1 \leq r < N$. Throughout this section, we shall assume that

$$|\mu_1| \geq |\mu_2| \geq \cdots \geq |\mu_r| > |\mu_{r+1}| \geq \cdots \geq |\mu_N|. \tag{1.9}$$

The first r eigenvalues are the r *dominant* eigenvalues. The associated invariant subspace $M = \{\varphi_1, \ldots, \varphi_r\}$ is the *dominant* invariant subspace; $\dim M = r$ and M is unique by assumption (1.9). M is the direct sum of the invariant subspaces associated with the dominant distinct eigenvalues.

Let P be the associated spectral projection. P is a projection on M along $W := (I - P)\mathbb{C}^N$, and $\mathbb{C}^N = M \oplus W$.

Under the assumption (1.9), we wish to give a necessary and sufficient condition on U such that $A^k U$ converges to the dominant invariant subspace M as $k \to \infty$. For that purpose, we define the convergence of a sequence of subspaces $X_k \subset \mathbb{C}^N$ to a subspace $Y \subset \mathbb{C}^N$, and we introduce the following notation:

Given a subspace $Y \subset \mathbb{C}^N$, $\dim Y = r$, $\mathbf{Y}$ is an $N \times r$ matrix whose columns are a basis of Y. The basis of Y is completed by the $N - r$ columns of $\mathbf{Z}$ into a basis of $\mathbb{C}^N$. Given a subspace X_k of $\mathbb{C}^N$, $\dim X_k \leq r$, $\mathbf{X}_k$ is an $N \times r$ matrix whose columns span X_k.

Definition Let $\{X_k\}$ be a sequence of subspaces of $\mathbb{C}^N$, $\dim X_k \leq r$. $X_k \to Y$, $\dim Y = r$, as $k \to \infty$, iff $\mathbf{X}_k = \mathbf{Y}C_k + \mathbf{Z}D_k$, where C_k is an $r \times r$ matrix, regular for k large enough, and D_k is an $(N - r) \times r$ matrix such that $\|D_k C_k^{-1}\| \to 0$ as $k \to \infty$.

The distance between two subspaces need not be defined; instead, the following basic property of matrix norms in a finite-dimensional space is used: a sequence $\{A_k\}$ of complex matrices of fixed finite order converges to the null matrix iff each element sequence $\{a_{ij}^{(k)}\}$ converges to zero.

Exercises

1.44 Show that the above definition is independent of the choice of basis in $\mathbb{C}^N$.
1.45 Show that $\dim X_k = \dim Y = r$ for k large enough.
1.46 Show that if $X_k \to Y$, $X'_k \to Y'$, then $X_k + X'_k \to Y + Y'$.
1.47 Show that, given a sequence $\{X_k\}$, the limit subspace Y is unique.

We now turn to the proof of Theorem 1.8, which is central for the theory of convergence of the numerical methods we have in mind. The general presentation is adapted from Parlett and Poole (1973).

Theorem 1.8 *Under the assumption* (1.9), $A^k U \to M$ *as* $k \to \infty$ *iff* $\{Px_i\}_1^r$ *are independent.*

Proof Let $(\mathbf{M}, \mathbf{W})$ correspond to a basis of $\mathbb{C}^N$, where $\mathbf{M}$ (resp. $\mathbf{W}$) corresponds to a basis of M (resp. W). Let $\mathbf{U}$ be the matrix whose columns are the basis $x_1, \ldots, x_r$ of U. It is clear that $\mathbf{U} = \mathbf{M}C + \mathbf{W}D$, where C is regular iff $\{Px_i\}_1^r$ are independent. Let A_M (resp. A_W) be the matrix of order r (resp. $N - r$), which represents the operator A restricted to M, $A_{\restriction M}: M \to M$

(resp. $A_{\restriction W}: W \to W$) in the chosen basis of M (resp. W). Clearly A_M is regular because only the eigenvalues of A_W may be zero by assumption (1.9). Then, for $k = 1, 2, \ldots, \dim A^k U = r$, and for $X_k := A^k U$:

$$\mathbf{X}_k = A^k \mathbf{U} = \mathbf{M} A_M^k C + \mathbf{W} A_W^k D,$$

$$\|A_W^k D C^{-1} A_M^{-k}\| \leq \|A_W^k\| \, \|A_M^{-k}\| \, \|DC^{-1}\|.$$

Using the fact that $\lim_{k\to\infty} \|A^k\|^{1/k} = r_\sigma(A)$, we deduce that for $\varepsilon > 0$, there exists k_0 such that for $k > k_0$

$$(\|A_W^k\| \, \|A_M^{-k}\|)^{1/k} \leq r_\sigma(A_W) r_\sigma(A_M^{-k}) + \varepsilon = \left|\frac{\mu_{r+1}}{\mu_r}\right| + \varepsilon.$$

Because $|\mu_r| > |\mu_{r+1}|$, we can choose ε such that $|\mu_{r+1}/\mu_r| + \varepsilon < 1$ and this shows that $\|A_W^k D C^{-1} A_M^{-k}\| \to 0$. □

When (1.9) is not satisfied, there is no unique dominant invariant subspace of dimension r. A thorough study of this case is done in Parlett and Poole (1973).

Let $\{e_i\}_1^N$ be the canonical basis of $\mathbb{C}^N$, and E_r be the span of $e_1, \ldots, e_r$. We now show that the starting subspace E_r satisfies the condition of Theorem 1.8 for a particular class of matrices, the unreduced upper-Hessenberg matrices.

Definition The matrix $H = (h_{ij})$ is *upper-Hessenberg* iff $h_{ij} = 0$ for $i > j + 1$. It is *unreduced* iff $h_{i,i-1} \neq 0$ for $i = 2, \ldots, N$.

Any matrix is unitarily similar to an upper-Hessenberg matrix, by the Givens or Householder algorithm (cf. Wilkinson, 1965). If some $h_{i,i-1} = 0$, the problem is reduced to a sequence of subproblems on unreduced Hessenberg matrices.

Exercise

1.48 If H is an unreduced Hessenberg matrix, show that rank $H \geq N - 1$. Conclude that an unreduced Hessenberg matrix with multiple eigenvalues is defective. In particular, a hermitian tridiagonal matrix, with nonzero off-diagonal elements, has simple eigenvalues.

Lemma 1.9 *Under the hypothesis* (1.9), $\{Pe_i\}_1^r$ *are independent, where P is the dominant spectral projection associated with an unreduced upper-Hessenberg matrix H.*

Proof Let $x \in E_r$, but $x \notin E_{r-1}$; then $Hx \in E_{r+1}$, but $Hx \notin E_r$. By repeating the preceding argument, $x, Hx, \ldots, H^{N-r}x$ are independent: any invariant subspace containing x must have a dimension larger than $N - r$. This shows that E_r is disjoint from any invariant subspace of dimension less than or equal to $N - r$, like, for example, the subspace $W = \operatorname{Ker} P$ of dimension $N - r$. Therefore $\mathbb{C}^N = E_r \oplus \operatorname{Ker} P$, and $\dim PE_r = r$. This proves that the r vectors $\{Pe_i\}_1^r$ are independent. □

Corollary 1.10 *Under the hypothesis* (1.9), $H^k E_r \to M$, as $k \to \infty$, *where* H *is an unreduced upper-Hessenberg matrix.*

Proof Clear from Lemma 1.9 and Theorem 1.8. □

The sequence $H^k E_r$ can also converge sometimes when $|\mu_r| = |\mu_{r+1}|$ (cf. Parlett and Poole, 1973).

3.2. The Subspace Iteration Method

The subspace iteration method consists in computing the iterated subspace $A^k U$, where U is an r-dimensional subspace. We distinguish whether $r = 1$ or $r > 1$.

3.2.1. $r = 1$, the power method

We suppose that $|\mu_1| > |\mu_2|$. The right (resp. left) eigenvector associated with the *simple* dominant eigenvalue $\mu_1 = \lambda_1$ is φ_1 (resp. ψ_1). We suppose that $\|\varphi_1\| = \psi_1^H \varphi_1 = 1$. Then $Px = (\psi_1^H x)\varphi_1$. The *power method* is defined by

$$q_0 := \frac{x}{\|x\|}, \qquad q_k := \frac{Aq_{k-1}}{\|Aq_{k-1}\|}, \qquad k = 1, 2, \ldots,$$

where $\|\cdot\|$ is an arbitrary norm on $\mathbb{C}^N$.

If $Px \neq 0$, that is, $\psi_1^H x \neq 0$, the sequence $\{q_k\}$ tends to $\{\varphi_1\}$, by Theorem 1.8, and the scalar $\rho_k := q_k^H A q_k / q_k^H q_k$ tends to μ_1. The convergence is linear, with rate $|\mu_2/\mu_1|$. To improve this rate, one may think of introducing shifts of origin (cf. Wilkinson, 1965, Chapter 9). A similar idea has been used to compute the eigenvector associated with an approximate eigenvalue λ.

Suppose that λ is close to μ_1 but not to any other eigenvalue μ_i, $i = 2, \ldots, N$; then $(A - \lambda I)^{-1}$ has the eigenvalues $(\mu_i - \lambda)^{-1}$, and $1/|\mu_1 - \lambda|$ is much larger than $1/|\mu_i - \lambda|$, $i \geq 2$. This is the basis of the *inverse iteration method* to compute the eigenvector corresponding to a simple eigenvalue for which we know an approximation λ:

$$q_0 := \frac{x}{\|x\|}, \qquad q_k := \frac{z_k}{\|z_k\|},$$

where z_k is solution of

$$(A - \lambda I)z_k = q_{k-1}, \qquad k = 1, 2, \ldots. \tag{1.10}$$

This is the power method on $(A - \lambda I)^{-1}$.

Clearly $q_k \to \varphi_1$, but the nearer λ is to μ_1, the more ill-conditioned is the system (1.10). However, if φ_1 is well-conditioned, most of the error when solving (1.10) will be in the direction of φ_1, and we are interested only in direction $\{\varphi_1\}$. The interested reader may find a nice geometric development of this idea (A being hermitian) in Parlett (1980a, p. 65). The treatment of multiple eigenvalues and ill-conditioned eigenvectors is a more delicate problem (cf. Wilkinson, 1965, Chapter 9).

An important variant of the inverse iteration method is the *Rayleigh quotient iteration*:

$$q_0 := \frac{x}{\|x\|}, \qquad \rho_0 := \lambda,$$

$$q_k := \frac{z_k}{\|z_k\|}, \qquad \rho_k := \frac{z_k^H A z_k}{z_k^H z_k}, \qquad k = 1, 2, \ldots,$$

where z_k is solution of

$$(A - \rho_{k-1} I) z_k = q_{k-1}.$$

If λ is an approximation of a semisimple eigenvalue μ of A and if $\rho_k \to \mu$, the rate of convergence is essentially quadratic:

$$\mu - \rho_{k+1} = O[(\mu - \rho_k)^2] \qquad \text{when} \quad k \to \infty.$$

If A is normal, the convergence is essentially cubic:

$$\mu - \rho_{k+1} = O[(\mu - \rho_k)^3]$$

(cf. Ostrowski, 1957–1959; Parlett, 1974).

3.2.2. $r > 1$, the simultaneous iteration method

Let U be the span of r *orthonormal* vectors $\{x_i\}_1^r$, and let Q_0 be the $N \times r$ matrix $Q_0 := (x_1, \ldots, x_r)$. The algorithm produces a sequence of matrices Q_k whose columns are an orthonormal basis of $X_k = A^k U$ in the following way:

For $k \geq 1$:

(1) compute the $N \times r$ matrix $Y_k = AQ_{k-1}$, and
(2) orthonormalize the column vectors of Y_k to produce

$$Q_k := Y_k R_k,$$

where R_k is an $r \times r$ upper-triangular matrix.

Under the assumption (1.9), the subspace $A^k U$ converges iff the starting vectors are such that $\{Px_i\}_1^r$ are independent.† The convergence rate of the ith column vector is of the order of $|\mu_{i+1}/\mu_i|$ if $|\mu_i| > |\mu_{i+1}|$, which is not an improvement on the power method.

† Further discussion of this convergence appears in the note added in proof on p. 65.

An improvement can be obtained by considering the orthogonal projection of the original eigenvalue problem $A\varphi = \lambda\varphi$ on the subspace X_k. This yields the problem

$$\text{find} \quad \lambda^{(k)} \in \mathbb{C}, \qquad 0 \neq \varphi^{(k)} \in X_k,$$

$$\text{such that} \quad A\varphi^{(k)} - \lambda^{(k)}\varphi^{(k)} \text{ is orthogonal to } X_k.$$

Let $\varphi^{(k)} = Q_k\xi^{(k)}$; $\xi^{(k)}$ is the solution of

$$Q_k^{\mathrm{H}}(AQ_k - \lambda^{(k)}Q_k)\xi^{(k)} = 0;$$

that is,

$$B_k\xi^{(k)} = \lambda^{(k)}\xi^{(k)},$$

where $B_k := Q_k^{\mathrm{H}}AQ_k$ is an $r \times r$ matrix. When $k \to \infty$, then (modulo factors of unit modulus) $B_k \to B := Q^{\mathrm{H}}AQ$, which represents the matrix A_M in the basis Q of M. The ith eigenpair $\mu_i^{(k)}$, $\{\varphi_i^{(k)}\}$ converges to μ_i, $\{\varphi_i\}$; we shall see in Section 5 that the rate of convergence is of the order of $|\mu_{r+1}/\mu_i|$, $i = 1, 2, \ldots, r$ (Theorem 1.32).

3.3. The QR Method

The basic QR algorithm generates a sequence of unitarily similar matrices A_k according to

$$A_1 := A, \qquad A_k := Q_kR_k, \qquad A_{k+1} := R_kQ_k, \qquad \text{for} \quad k = 1, 2, \ldots,$$

where Q_k is unitary and R_k is upper-triangular. $A_{k+1} = Q_k^{\mathrm{H}}A_kQ_k$. We define $\mathbf{Q}_k := Q_1Q_2 \cdots Q_k$, $\mathbf{R}_k := R_k \cdots R_2R_1$; then $A^k = \mathbf{Q}_k\mathbf{R}_k$. There is no restriction in assuming that A is *regular*.

Lemma 1.11 *For $r = 1, 2, \ldots, N$, the r first columns of $\mathbf{Q}_k$ span the subspace $X_k = A^kE_r$.*

Proof It follows from $A^k = \mathbf{Q}_k\mathbf{R}_k$, where $\mathbf{R}_k$ is an upper-triangular regular matrix. □

The QR method is then a set of N subspace iteration methods with the *nested* starting subspaces $E_1 \subset E_2 \subset \cdots \subset E_N = \mathbb{C}^N$.

The convergence of the QR algorithm on a Hessenberg matrix will follow immediately from Corollary 1.10 if the assumption (1.9) is valid for $r = 1, 2, \ldots, N - 1$. We are then led to suppose that the eigenvalues are simple: $\mu_i = \lambda_i$, $i = 1, \ldots, N$.

Theorem 1.12 *Let H be an unreduced regular upper-Hessenberg matrix. If the simple eigenvalues are such that $|\lambda_1| > |\lambda_2| > \cdots > |\lambda_N| > 0$, the QR*

algorithm applied on H converges to an upper-triangular matrix with the eigenvalues $\{\lambda_i\}_1^N$ *on the diagonal, in that order.*

Proof The convergence (modulo factors of unit modulus) follows from Corollary 1.10 applied for $r = 1, 2, \ldots, N-1$. For $r = N$, clearly $E_N = M_N = \mathbb{C}^N$ and $H^k E_N = M_N$ for any k. The triangular limit form comes from the fact that for $r = 1, 2, \ldots, N$, the leading principal $r \times r$ submatrix of H_k tends to the representation of the restriction of H to the dominant invariant subspace of dimension r. □

The complete study of the convergence for Hessenberg matrices in case of multiple or equimodular eigenvalues is done in Parlett (1968). The limit matrix may then be block-triangular.

Remark Theorem 1.12 shows the theoretical role played by the Hessenberg form of a matrix. This form is also of an extreme importance in practice since it remains invariant under the algorithm, which requires many fewer computations when applied to such a matrix than when applied to a full matrix.

The rate of convergence of the basic QR method is again of the order of $\max_{i=1,\ldots,N-1} |\lambda_{i+1}/\lambda_i|$ as for the power method. However, with appropriate shift of origin ($\sigma_k \approx \lambda_N$), the convergence of the lower right corner term $a_{NN}^{(k)}$ to λ_N can become quadratic at least. The algorithm with shift is defined by

$$A_k - \sigma_k I = Q_k R_k, \qquad A_{k+1} := R_k Q_k + \sigma_k I.$$

There is not yet any result on the *global* convergence of certain shifted QR for nonnormal Hessenberg matrices.

Exercises

1.49 Using the relation $\mathbf{Q}_k = A\mathbf{Q}_{k-1}R_k^{-1} = A^k\mathbf{R}_k^{-1}$, for the basic QR algorithm, show that $\{e_N^{\mathrm{H}}\mathbf{Q}_k^{\mathrm{H}}\} = \{e_N^{\mathrm{H}}\mathbf{Q}_{k-1}^{\mathrm{H}}A^{-1}\} = \{e_N^{\mathrm{H}}A^{-k}\}$. Conclude that the Nth column of $\mathbf{Q}_k$ is computed by the power method on $(A^{\mathrm{H}})^{-1}$ starting from the vector e_N.

1.50 We consider the QR method with shift $\sigma_k = a_{NN}^{(k)}$. Using $A_k - \sigma_k I = Q_k R_k$, show that $Q_k e_N$ is proportional to $(A_k^{\mathrm{H}} - \bar{\sigma}_k I)^{-1} e_N$. Conclude to the possible quadratic convergence of $a_{NN}^{(k)}$ to λ_N. (*Hint*: $\sigma_k = a_{NN}^{(k)} = e_N^{\mathrm{H}} A_k e_N$, interpret the computation as one Rayleigh quotient iteration from e_N, an approximate eigenvector of A_k^{H}.)

3.4. A Is Hermitian

We suppose that A is hermitian. For the subspace iteration methods, the rate of convergence of the eigenvalues is the *square* of the rate of convergence of the eigenvectors (cf. Section 5). As for the QR algorithm, it converges to a diagonal matrix. Wilkinson (1968) and Parlett (1980a) have proved the global convergence of the shifted QR on a tridiagonal matrix for certain shift strategies. The convergence is then essentially cubic or better than cubic. If A is *positive definite*, one may use a variant of the QR method based on the Cholesky decomposition:

$$A_k = R_k^H R_k, \qquad R_k R_k^H = A_{k+1}, \qquad k = 0, 1, 2, \ldots.$$

Besides the above algorithms, one has Jacobi's (1846) method where plane rotations are used to annihilate off-diagonal elements in turn. With a suitable strategy, the matrix eventually converges quadratically to a diagonal matrix. The rate of convergence is by no means affected by the existence of close or multiple eigenvalues.

3.5. *Iterative Refinement of Approximate Eigenelements*

We begin to recall the iterative refinement technique to improve an approximate solution of a linear equation.

3.5.1 Iterative refinement of approximate solutions

We wish to compute the exact solution of $Ax = b$, but instead we solve a neighboring equation $A'x' = b$, with $A = A' - H$. x can be recovered as the limit of a sequence of vectors x_k, which are solutions of systems of matrix A'.

Proposition 1.13 *If* $\|H\| < 1/\|A'^{-1}\|$, *then* $x = \lim_k x_k$, *where*

$$x_0 := x' = A'^{-1}b, \qquad x_k := x' + A'^{-1}Hx_{k-1}, \qquad k \geq 1.$$

Proof

$$(A' - H)^{-1} = (1 - A'^{-1}H)^{-1}A'^{-1} = \left(\sum_{k=0}^{\infty} (A'^{-1}H)^k\right)A'^{-1}$$

since $\|A'^{-1}H\| < 1$. Therefore $x = \sum_{k=0}^{\infty}(A'^{-1}H)^k x'$. We set $x_0 = y_0 = x'$, $y_i = A'^{-1}Hy_{i-1}$ for $i \geq 1$, and $x_k = \sum_{i=0}^{k} y_i \to x$ as $k \to \infty$. Note that $x_k = x_{k-1} + y_k$ with $A'y_k = Hy_{k-1} = b - Ax_{k-1}$ for $k \geq 1$. □

This technique may be used when A is close to a matrix that is easily invertible, like a diagonal matrix (cf. Exercise 1.52.). Or it may be used to increase the accuracy of a numerical solution: x', the numerical solution of $Ax = b$, may be regarded as the exact solution of $A'x' = b$. The process is based on the resolution of a sequence of equations

$$A'(x_k - x') = Hx_{k-1}, \qquad k \geq 1.$$

Of course, the matrix A' should not be too ill-conditioned, and the residuals Hx_k should be computed in double precision (cf. Wilkinson and Reinsch, 1971 and Kulisch and Miranker, 1981; cf. also Björck, 1967b, 1968 for the least-squares solution).

Exercises

1.51 Show that $x_k \to x$ at the rate of a geometric progression with quotient $\|H\|$; that is, $\|x - x_k\| \leq C\|H\|^{k+1}$, $k \geq 0$.

1.52 Compute the approximate solution of the system

$$\begin{aligned} x + 0.1y + 0.01z + 0.001t &= 1, \\ 0.1x + y + 0.01z + 0.001t &= 1, \\ 0.1x + 0.01y + z + 0.001t &= 1, \\ 0.1x + 0.01y + 0.001z + t &= 1. \end{aligned}$$

by three successive iterations. Prove that the error in the max norm is less than 2×10^{-4}.

3.5.2. Iterative refinement of approximate eigenelements

We wish to solve $A\varphi = \lambda\varphi$. We consider the neighboring problem $A'\varphi = \lambda'\varphi'$ with $A = A' - H$. We suppose that λ' is *simple*, ψ' is the left eigenvector of A' associated with λ', normalized by $\psi'^{\mathrm{H}}\varphi' = \|\varphi'\|_2 = 1$, P' is the spectral projection $\varphi'\psi'^{\mathrm{H}}$, and S' is the generalized inverse

$$(A' - \lambda' I)^{-1}(I - P') = \hat{A}'^{-1}_{\lambda'}(I - P').$$

Proposition 1.14 *For $\|H\|_2$ small enough, there exists a simple eigenvalue λ of A with an eigenvector φ normalized by $\psi'^{\mathrm{H}}\varphi = 1$ such that the sequence λ_k (resp. φ_k) converges to λ (resp. φ) with*

$$\begin{aligned} &\lambda_0 = \lambda', && \lambda_k = \psi'^{\mathrm{H}} A \varphi_{k-1}, && k \geq 1, \\ &\varphi_0 = \eta_0 = \varphi', && \varphi_k = \varphi' + S'\left[H\varphi_{k-1} + \sum_{i=1}^{k}\sum_{j=1}^{i} \nu_j \eta_{i-j}\right], && k \geq 1, \end{aligned} \tag{1.11}$$

where $\lambda_k = \lambda_{k-1} + \nu_k$ and $\varphi_k = \varphi_{k-1} + \eta_k$, for $k \geq 1$.

Proof Let $H = \varepsilon L$ in the proof of Theorem 1.7 to get the formulas (1.11). A sufficient condition on $\varepsilon = \|H\|_2$ will be established in Section 8, in the more general context of the perturbation of closed operators in a Banach space. □

Formulas (1.11) require the solution of a sequence of equations:

$$\begin{aligned} &(A' - \lambda' I)\varepsilon_k = (I - P')b_k, \qquad \varepsilon_k = \varphi_k - \varphi', \qquad k \geq 1, \\ &P'\varepsilon_k = 0. \end{aligned} \tag{1.12}$$

Exercises

1.53 Prove that the unique solution of the system

$$\begin{aligned}(A' - \lambda' I)x &= (I - P')b,\\ \psi'^{\mathrm{H}} x &= 0\end{aligned} \tag{1.13}$$

is $x = S'b = S'(I - P')b$. Show that the condition of this system depends on $\|S'\|$ and $\|P'\|$.

1.54 Propose a numerical algorithm to solve (1.13). Comment.

1.55 If A' is hermitian, verify that the solution of (1.13) is the least-squares solution of minimum euclidean norm of

$$(A' - \lambda' I)x = b.$$

If the spectral projection $P' = \varphi'\psi'^{\mathrm{H}}$ is ill-conditioned, so is the sequence of systems (1.12) (Exercise 1.53). It may then be advisable to consider another projection Q defined by a vector y different from ψ' such that $y^{\mathrm{H}}\varphi' \neq 0$. We suppose that $y^{\mathrm{H}}\varphi' = 1$; then if $Q = \varphi' y^{\mathrm{H}}$, $\|Q\|_2 = \|y\|_2$. A possible choice is $y = \varphi'$; then $Q = \varphi'\varphi'^{\mathrm{H}}$ is the *orthogonal* projection on $\{\varphi'\}$, $\|Q\|_2 = 1$.

We define $\mathring{A}'_{\lambda'} := [(I - Q)(A' - \lambda' I)]_{\restriction\{y\}^\perp}$. $\mathring{A}'_{\lambda'}$ is regular; hence we denote by Σ' the matrix representing $(I - Q)\mathring{A}'^{-1}_{\lambda'}(I - Q)$.

Now let $\hat{\varphi}$ be the eigenvector of A normalized by $Q\hat{\varphi} = \varphi'$; this is always possible for $\|H\|_2$ small enough. Then λ and $\hat{\varphi}$ are limits of the sequences λ_k, $\hat{\varphi}_k$ defined by

$$\begin{aligned}&\lambda_0 = \lambda', &&\lambda_k = y^{\mathrm{H}} A\hat{\varphi}_{k-1} + y^{\mathrm{H}} A'\hat{\eta}_k, && k \geq 1,\\ &\hat{\varphi}_0 = \hat{\eta}_0 = \varphi', &&\hat{\varphi}_k = \varphi' + \Sigma'\left[H\hat{\varphi}_{k-1} + \sum_{i=1}^{k}\sum_{j=1}^{i} \nu_j \hat{\eta}_{i-j}\right], && k \geq 1,\end{aligned} \tag{1.14}$$

where

$$\lambda_k = \lambda_{k-1} + \nu_k \quad \text{and} \quad \hat{\varphi}_k = \hat{\varphi}_{k-1} + \hat{\eta}_k, \quad \text{for} \quad k \geq 1.$$

Exercises

1.56 Establish formulas (1.14) by the method used in the proof of Theorem 1.7.

1.57 Compare the complexity of formulas (1.11) and (1.14).

1.58 Consider the 4×4 matrix

$$A = \begin{pmatrix} 1 & 0.2 & 0.1 & 0.2\\ 0.1 & 2 & 0.1 & 0.1\\ 0.1 & 0.2 & 3 & 0.5\\ 0.1 & 0.2 & 0.1 & 4\end{pmatrix}.$$

Solve iteratively the eigenvalue problem of A from the one of its diagonal part.

1.59 Study the application of iterative refinement techniques to solve the eigenvalue problem of an almost-triangular matrix.

1.60 Consider

$$A = \begin{pmatrix} 1 & 5\\ \frac{1}{25} & 2\end{pmatrix} \quad \text{and} \quad D = \begin{pmatrix} 1 & 0\\ 0 & 2\end{pmatrix}.$$

Compute iteratively the eigenvalue $\lambda = \frac{3}{2}(1 - 1/\sqrt{5})$ of A from the eigenelements $\lambda' = 1$, $\varphi' = \psi' = e_1$ of D.

One may notice that, in Exercise 1.60,

$$A - D = \begin{pmatrix} 0 & 5 \\ \frac{1}{25} & 0 \end{pmatrix}$$

has no small norm. Nevertheless, the series expansions for the eigenelements are converging. This fact will be explained in Chapter 3, Section 8.

We have assumed for clarity that λ' is simple. Formulas for the iterative refinement of a basis of an approximate invariant subspace are given also in Chapter 3, Section 8.

The iterative refinement of approximate eigenelements that we have presented is of the same type as the iterative refinement of approximate solutions. The rate of convergence depends on $\|H\|$, and crucially on the condition number $\|S'\|$. The pros and cons of the method are then identical for the two problems (Stewart, 1973a). Using the maximal accuracy scalar product defined in Kulisch and Miranker (1981) one can apply refinement techniques to arbitrarily ill-conditioned problems (cf. Kaucher and Rump, 1982; Rump and Böhm, 1982).

The method will be generalized in Chapter 3, Sections 7 and 8, to the perturbation $H = T' - T$ of closed linear operators T and T' in a Banach space. The case in which T' is a numerical approximation of T is treated in Chapters 5 and 7.

3.6. Bibliographical Comments

The *LR* (Rutishauser, 1958), *QR* (Francis, 1961–1962), treppen- and bi-iteration (Bauer, 1957), Rayleigh quotient iteration (Ostrowski, 1957–1959), and the subspace iteration methods are essentially the same. Bauer (1958) showed that some of them could be regarded as Bernoulli processes. They produce the same invariant subspaces; they differ in the basis they construct for the invariant subspace. The complete geometrical study of these methods is in the basic paper of Parlett and Poole (1973); cf. also Buurema (1970) for normal Hessenberg matrices. A comparison between various shift strategies for *QR* on tridiagonal hermitian matrices, together with elegant proofs, is given in Parlett (1980a). For Jacobi's method, the reader is referred to Forsythe and Henrici (1960), Wilkinson (1962), van Kempen (1966) and Ciarlet (1982).

In the simultaneous iteration method, the computation, in the symmetric case, of the eigenelements of the $r \times r$ matrix B_k is in Rutishauser (1970). It is generalized to nonsymmetric matrices in Clint and Jennings (1971) and Stewart (1976a). For the generalized eigenvalue problem $Ax = \lambda Bx$, which is formally equivalent to $B^{-1}Ax = \lambda x$ when B is regular, the QZ method (Moler and Stewart, 1973) is an extension of the QR algorithm that avoids using B^{-1}.

Standard programs to perform these methods are available in ALGOL (Wilkinson and Reinsch, 1971) and FORTRAN (Smith *et al.*, 1976), cf. also the Harwell or NAG routine libraries as well as the IMSL package.

4. Error Analysis

4.1. A Priori Error Bounds

A has an eigenvalue λ with algebraic (resp. geometric) multiplicity m (resp. g) and ascent l, $1 \le l \le m$. Let $A' = A + H$. If $\varepsilon = \|H\|$ is small enough, then A' has m eigenvalues $\{\mu_i'\}_{i \in I}$, card $I = m$, counted with their algebraic multiplicities, in a neighborhood of λ such that

$$\max_{i \in I} |\mu_i' - \lambda| = O(\varepsilon^{1/l}),$$

$$\min_{i \in I} |\mu_i' - \lambda| = O(\varepsilon^{g/m}),$$

$$\left| \frac{1}{m}\left(\sum_{i \in I} \mu_i'\right) - \lambda \right| = O(\varepsilon).$$

And A' has an eigenvector x', $\|x'\| = 1$, such that

$$\operatorname{dist}(x', \operatorname{Ker}(A - \lambda I)) = O(\varepsilon^{1/l}).$$

l is the dimension of the largest Jordan block associated with λ. There are g such blocks. Then $m \le gl$. Moreover, $1/l = g/m$ if the blocks have the same dimension, and $1/l < g/m$ otherwise. In Example 1.2, we had $m = 7$, $g = l = 3$, and $\frac{1}{3} < \frac{3}{7}$.

Let M be the invariant subspace associated with λ, and let M' be the direct sum of the invariant subspaces associated with the distinct approximate eigenvalues of A'. Then

$$\max\left[\sup_{\substack{x \in M \\ \|x\| = 1}} \operatorname{dist}(x, M'), \sup_{\substack{x' \in M' \\ \|x'\| = 1}} \operatorname{dist}(x', M)\right] = O(\varepsilon).$$

The proofs may be found in Wilkinson (1965, Chapter 2), and Kato (1976, Chapter 2). Analogous error bounds, in the general setting of closed linear operators in Banach spaces, are proved in Chapter 6.

The asymptotic behavior of the constants involved in the above equalities will result from an a posteriori analysis.

4.2. A Posteriori Error Analysis

Throughout this section, the norm $\|\cdot\|$ on $\mathbb{C}^N$ is considered to be *monotonic*; that is, for any diagonal matrix $D = \text{diag}(\lambda_1, \ldots, \lambda_N)$, the corresponding operator norm satisfies $\|D\| = \max_{i=1,\ldots,N} |\lambda_i|$. The euclidean norm $\|\cdot\|_2$ and the maximum norm $\|\cdot\|_\infty$ are monotonic.

The above definition is equivalent to the following characterization (Householder, 1964): a norm $\|\cdot\|$ on $\mathbb{C}^N$ is monotonic iff $|\xi_i| \le |\eta_i|$ for $i = 1, \ldots, N$ implies that $\|x\| \le \|y\|$, where $x = (\xi_i)$, $y = (\eta_i)$.

4.2.1. The residual vectors

In a posteriori error analysis, we use the knowledge of given approximate eigenelements to derive error bounds. These error bounds should be easy to compute from the eigenelements at hand.

We suppose that we are given λ and x, $\|x\| = 1$. To test the accuracy of these data as approximate eigenelements of A, it is natural to consider the *residual vector* $r = Ax - \lambda x$. The following result holds:

Proposition 1.15 *Let A be diagonalizable: $A = VDV^{-1}$, with $D = \text{diag}(\mu_1, \ldots, \mu_N)$. Given λ and x such that $r = Ax - \lambda x$ and $\|x\| = 1$, there exists an eigenvalue μ_i of A such that*

$$|\lambda - \mu_i| \le \|V\| \, \|V^{-1}\| \, \|r\|. \tag{1.15}$$

Proof It is trivial if $r = 0$. We then suppose that $D - \lambda I$ is regular. Then $r = Ax - \lambda x = V(D - \lambda I)V^{-1}x$, and

$$1 = \|x\| = \|V(D - \lambda I)^{-1}V^{-1}r\| \le \|V\| \, \|V^{-1}\| \, \frac{\|r\|}{\min_i |\mu_i - \lambda|}.$$

$\|V\| \, \|V^{-1}\| = K(V)$ is the condition number of V (relative to inversion). If A is normal, then $K_2(V) = 1$, and $|\lambda - \mu_i| \le \|r\|_2$. □

Example 1.7 Let

$$A = \begin{pmatrix} 2 & -10^{10} \\ 0 & 2 \end{pmatrix},$$

with $\lambda = 1$ and

$$x = \begin{pmatrix} 1 \\ 10^{-10} \end{pmatrix}, \qquad \|x\|_\infty = 1.$$

Then $\|Ax - \lambda x\|_\infty = 10^{-10}$. But 1 is not an eigenvalue of A, which has only 2 as a double eigenvalue. Check that

$$\begin{pmatrix} 2 & -10^{-10} \\ \varepsilon & 2 \end{pmatrix}$$

has an ill-conditioned eigenvalue problem for $\varepsilon > 0$.

We remark that there is no assumption in Proposition 1.15 that $\|r\|$ is small. On the contrary, we now drop the assumption that A is diagonalizable, and we consider x and y such that $\|x\|_2 = y^{\mathrm{H}}x = 1$, where x is an *approximate* eigenvector of A, while y may or may not be an approximate eigenvector of A^{H}. Note that $\|y\|_2 \geq 1$.

$\zeta := y^{\mathrm{H}}Ax$ is the *generalized Rayleigh quotient* of A, based on x and y. The residual vector associated with ζ and x (resp. $\bar{\zeta}$ and y) is $u := Ax - \zeta x$ (resp. $v := A^{\mathrm{H}}y - \bar{\zeta}y$). $\hat{y} = y/\|y\|_2$ is normalized and $\hat{v} := v/\|y\|_2$ is the corresponding residual.

Let $Q = xy^{\mathrm{H}}$ be the matrix of the projection on $\{x\}$ along $\{y\}^\perp$. If ζ is not an eigenvalue of $[(I - Q)A]_{\restriction\{y\}^\perp}$, then

$$\mathring{A}_\zeta := [(I - Q)(A - \zeta I)]_{\restriction\{y\}^\perp}$$

is regular, and we define

$$\Sigma := (I - Q)(A - \zeta I)^{-1}(I - Q) = (I - Q)\mathring{A}_\zeta^{-1}(I - Q),$$

$$\text{and set } \sigma := \|\Sigma\|_2.$$

We now want to bound the accuracy of ζ as an approximate eigenvalue of A in terms of the norms of the residuals, respectively $\varepsilon := \|u\|_2$, $\varepsilon^* := \|\hat{v}\|_2$. For that purpose, we consider the decomposition of A:

$$\begin{aligned} A &= QAQ + (I - Q)A(I - Q) + (I - Q)AQ + QA(I - Q) \\ &= \underbrace{\zeta Q + (I - Q)A(I - Q)}_{\tilde{A}} + \underbrace{(A - \zeta I)Q + Q(A - \zeta I)}_{\tilde{H}} \end{aligned}$$

A is regarded as the sum of the matrix $\tilde{A}$ and the perturbation $\tilde{H}$.

Exercises

1.61 Check that $[(I - Q)(A - \zeta I)]_{\restriction\{y\}^\perp} = (\tilde{A} - \zeta I)_{\restriction\{y\}^\perp}$.

1.62 Show that ζ is an eigenvalue of $\tilde{A}$, with spectral projection Q if ζ is a simple eigenvalue. Then prove that $(\tilde{A} - \zeta I)\Sigma = \Sigma(\tilde{A} - \zeta I) = I - Q$. Σ is the generalized inverse of $\tilde{A} - \zeta I$, relative to Q.

1.63 Show that the matrix $\tilde{H}$ represents the operator

$$\xi \mapsto \tilde{H}\xi = (y^{\mathrm{H}}\xi)u + (v^{\mathrm{H}}\xi)x \qquad \text{for} \quad \xi \in \mathbb{C}^N.$$

We consider the *formal* series $\sum_{k=0}^{\infty} \nu_k$ and $\sum_{k=0}^{\infty} \eta_k$ defined by

$$\nu_0 = \zeta, \qquad \nu_k = y^{\mathrm{H}}\tilde{H}\eta_{k-1}, \qquad \text{for} \quad k \geq 1,$$

$$\eta_0 = x, \qquad \eta_1 = -\Sigma\tilde{H}x, \qquad \eta_k = \Sigma\left[-\tilde{H}\eta_{k-1} + \sum_{i=1}^{k} \nu_i \eta_{k-i}\right], \qquad \text{for} \quad k \geq 2. \tag{1.16}$$

Lemma 1.16 *If the above series are convergent and such that*

$$\sum_{k=0}^{\infty} \nu_k = \lambda, \qquad \sum_{k=0}^{\infty} \eta_k = \varphi, \qquad \sum_{k=0}^{\infty}\left(\sum_{i=0}^{k} \nu_i \eta_{k-i}\right) = \lambda\varphi,$$

then λ and φ are such that $A\varphi = \lambda\varphi$, with $y^{\mathrm{H}}\varphi = 1$.

Proof For $k \geq 2$,

$$(\tilde{A} - \zeta I)\eta_k + \nu_k x = -(I - Q)\tilde{H}\eta_{k-1} + \sum_{i=1}^{k} \nu_i \eta_{k-i}.$$

Then

$$\tilde{A}\eta_k + (I - Q)\tilde{H}\eta_{k-1} + \nu_k x = \sum_{i=0}^{k} \nu_i \eta_{k-i},$$

and

$$\nu_k x = Q\tilde{H}\eta_{k-1}.$$

Then

$$\tilde{A}\eta_k + \tilde{H}\eta_{k-1} = \sum_{i=0}^{k} \nu_i \eta_{k-i}.$$

By addition for $k \geq 1$, we get

$$\tilde{A}(\varphi - x) + \tilde{H}\varphi = \lambda\varphi - \zeta x,$$

that is, $A\varphi = \lambda\varphi$ with $\varphi \neq 0$ since $y^{\mathrm{H}}\varphi = 1$. □

We set $r := \max(\sigma|v^{\mathrm{H}}\Sigma u|, \|\Sigma u\|_2 \, \|\Sigma^{\mathrm{H}} v\|_2)$. Let g be the function

$$r \mapsto \frac{1 - \sqrt{1 - 4r}}{2r}$$

for $0 \leq r \leq \frac{1}{4}$. $1 \leq g(r) \leq 2$. $g(r) \to 1$ as $r \to 0$, as shown in Fig. 1.3.

For $0 \leq r < \frac{1}{4}$, $g(r)$ has the series expansion

$$g(r) = \sum_{k=0}^{\infty} \gamma_k r^k, \qquad \text{with} \quad \gamma_k = \frac{1}{k+1} C_{2k}^{k} = \sum_{i=1}^{k} \gamma_{i-1}\gamma_{k-i}.$$

We set $g_i(r) := \sum_{k=i}^{\infty} \gamma_k r^k$.

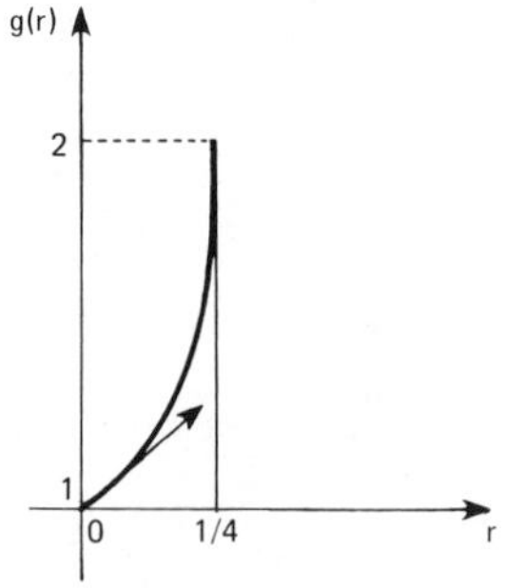

Figure 1.3

Theorem 1.17 (Redont) *If ζ is a simple eigenvalue of $\tilde{A}$, and if $r < \frac{1}{4}$, there exists a simple eigenvalue λ of A and an associated eigenvector φ nomalized by $y^{\mathrm{H}}\varphi = 1$ such that*

$$|\lambda - \zeta| \leq g(r)|v^{\mathrm{H}}\Sigma u|,$$

$$\|\varphi - x\|_2 \leq g(r)\|\Sigma u\|_2.$$

λ is the only eigenvalue of A in the disk $\{z \in \mathbb{C}; |z - \zeta| \leq 1/2\sigma\}$.

Proof We prove that the formal series of Lemma 1.16 converge, by bounding $|v_k|$ and $\|\eta_k\|_2$ for $k \geq 1$. First, it can be proved by induction that, for $k \geq 1$,

(1) $\eta_{2k} = 0$, and
(2) $\eta_{2k+1} = \Sigma M_k \Sigma u$,

where M_k is a matrix such that $\|M_k\|_2 \leq |v^{\mathrm{H}}\Sigma u|\gamma_k r^{k-1}$. It is left to the reader to check the induction ($\eta_1 = -\Sigma\tilde{H}\eta_0$). We deduce easily that, for $k \geq 2$,

$$\nu_{2k-1} = 0, \qquad |\nu_{2k}| = |y^{\mathrm{H}}\tilde{H}\Sigma M_{k-1}\Sigma u| \leq |v^{\mathrm{H}}\Sigma u|\gamma_{k-1}r^{k-1}$$

since $\|\Sigma u\|_2 \; \|\Sigma^{\mathrm{H}}v\|_2 \leq r$ (for $k = 1$, $\nu_1 = 0$ and $|\nu_2| = \gamma_0|v^{\mathrm{H}}\Sigma u|$). Similarly

$$\|\eta_{2k+1}\|_2 \leq \alpha\gamma_k r^k \qquad \text{with} \quad \alpha := \frac{\sigma}{r}\|\Sigma u\|_2\,|v^{\mathrm{H}}\Sigma u| \qquad \text{for} \quad k \geq 1.$$

Then for $r < \frac{1}{4}$, the series $\sum_{k=0}^{\infty}|\nu_k|$ and $\sum_{k=0}^{\infty}\|\eta_k\|_2$ are convergent. Hence, if $\lambda = \sum_{k=0}^{\infty}\nu_k$ and $\varphi = \sum_{k=0}^{\infty}\eta_k$, then $\lambda\varphi = \sum_{k=0}^{\infty}(\sum_{i=0}^{k}\nu_i\eta_{k-i})$, and Lemma 1.16 applies.

By addition on $k \geq 1$, it follows readily from the above bounds that

$$\left|\lambda - \sum_{k=0}^{2i}\nu_k\right| = \left|\lambda - \sum_{k=0}^{2i+1}\nu_k\right| \leq g_i(r)|v^{\mathrm{H}}\Sigma u|, \qquad i \geq 0,$$

$$\left\|\varphi - \sum_{k=0}^{2i-1}\eta_k\right\|_2 = \left\|\varphi - \sum_{k=0}^{2i}\eta_k\right\|_2 \leq g_i(r)\alpha, \qquad i \geq 1.$$

Then for $i = 0$, $|\lambda - \zeta| \le g(r)|v^H\Sigma u|$, and for $i = 1$, $\|\varphi - x - \eta_1\|_2 \le g_1(r)\alpha$; that is,

$$\|\varphi - x\|_2 \le \|\eta_1\|_2 + g_1(r)\alpha \le (1 + g(r) - 1)\|\Sigma u\|_2 = g(r)\|\Sigma u\|_2.$$

We now show that λ is the only point of $\sigma(A)$ in the disk

$$D := \left\{z \in \mathbb{C}; |z - \zeta| \le \frac{1}{2\|\Sigma\|_2}\right\}.$$

First λ belongs to D because

$$g(r)|v^H\Sigma u| < 2|v^H\Sigma u| \le 2r/\sigma < 1/2\sigma$$

is true as soon as $r < \frac{1}{4}$.

We solve the equation $A\xi - z\xi = b$ for $z \in D$. We decompose

$$\xi = (y^H\xi)x + (I - Q)\xi,$$

and set

$$\mathring{A}_z := [(I - Q)(A - zI)]_{\upharpoonright\{y\}^\perp}.$$

We denote by $\Sigma(z)$ the matrix $(I - Q)\mathring{A}_z^{-1}(I - Q)$, and we recall that $\Sigma(\zeta) \equiv \Sigma$. Formally we get

$$\begin{aligned} y^H\xi[(\zeta - z) - v^H\Sigma(z)u] &= y^HQb - v^H\Sigma(z)(I - Q)b, \\ (I - Q)\xi &= \Sigma(z)[(I - Q)b - y^H\xi u]. \end{aligned}$$

Since $\Sigma(\zeta) = \Sigma$ exists, $\Sigma(z)$ exists also for z such that $|z - \zeta| < 1/\|\Sigma\|_2$, in particular for $z = \lambda$; $\Sigma(z) = \sum_{k=0}^{\infty} [(z - \zeta)\Sigma]^k\Sigma$. It remains to show that for $z \in D - \{\lambda\}, |z - \zeta + v^H\Sigma(z)u| > 0$. Since λ is an eigenvalue of $A, (A - \lambda I)^{-1}$ does not exist, therefore $\lambda - \zeta + v^H\Sigma(\lambda)u = 0$, then

$$\begin{aligned} z - \zeta + v^H\Sigma(z)u &= z - \lambda + v^H[\Sigma(z) - \Sigma(\lambda)]u \\ &= (z - \lambda)[1 + v^H\Sigma(z)\Sigma(\lambda)u]. \end{aligned}$$

Now

$$(\Sigma^H(\bar{z})v)^H\Sigma(\lambda)u = \left(\sum_{k=0}^{\infty} [(\bar{z} - \bar{\zeta})\Sigma^H]^k\Sigma^H v\right)^H \sum_{k=0}^{\infty} [(\lambda - \zeta)\Sigma]^k\Sigma u$$

is bounded in modulus by

$$\frac{\|\Sigma u\|_2}{1 - |\lambda - \zeta|\;\|\Sigma\|_2}\,\frac{\|\Sigma^H v\|_2}{1 - |z - \zeta|\;\|\Sigma\|_2} \le 4\|\Sigma u\|_2\,\|\Sigma^H v\|_2 \le 4r < 1.$$

This proves that $(A - zI)^{-1}$ exists for all z in D except for $z = \lambda$. By integrating $(A - zI)^{-1}$ on a Jordan curve around λ (cf. Chapter 2, Section 7) we get the spectral projection P associated with λ: for $\xi \in \mathbb{C}^N$,

$$P\xi = \frac{y^H\xi - v^H\Sigma(\lambda)\xi}{1 + v^H\Sigma^2(\lambda)u}(x - \Sigma(\lambda)u).$$

P is clearly a projection of rank 1, and λ is a simple eigenvalue of A. □

Example 1.8 We consider $\varphi_1 = x + \eta_1$ as defined in the proof of Theorem 1.17. φ_1 is a better approximation of φ than x (which may even not be close to x if $\|\Sigma u\|_2$ is not small but $\|\Sigma^H v\|_2$ is):

$$\|\varphi - \varphi_1\|_2 \leq g_1(r)\alpha = \frac{g_1(r)}{r}\sigma|v^H\Sigma u|\,\|\Sigma u\|_2 \leq \left(\frac{g_1(r)}{r}\right)r\|\Sigma u\|_2,$$

where $g_1(r)/r$ tends to 1 as $r \to 0$. We remark that φ_1 is colinear with $(A - \zeta I)^{-1}x$. Indeed, $\eta_1 = -\Sigma\tilde{H}x$, and

$$(I - Q)(A - \zeta I)\Sigma\tilde{H}x = (I - Q)(A - \zeta I)x$$

proves that $(I - Q)(A - \zeta I)\varphi_1 = 0$. Therefore $(A - \zeta I)\varphi_1 = Q(A - \zeta I)\varphi_1$ is colinear with $x = Qx$. Hence the vector φ_1 corresponds to one step of the inverse iteration method applied on A with ζ as an approximation of λ, and x as an approximation of φ.

We may consider the recursion formulas (1.16) as a numerical method to compute λ and φ from ζ and x by setting

$$\varphi_k = \sum_{i=0}^{k}\eta_i, \qquad \lambda_k = \sum_{i=0}^{k}\nu_i = y^H A\varphi_{k-1}.$$

The difference between this method and the Rayleigh quotient iteration is that, in the present case, we have a sequence of equations involving a *fixed* matrix $A - \zeta I$. Compare to $A - \rho_k I$ for the Rayleigh quotient iteration (Section 3.2). On the other hand, here in the generalized Rayleigh quotient λ_k, only the right vector φ_{k-1} is updated.

Exercises

1.64 Interpret (1.16) as an iterative refinement method. Study the characteristics of this method which follow from the special structure of the perturbation $\tilde{H}$.

1.65 Compare the rates of convergence of (1.16) and of the Rayleigh quotient iteration.

The following corollary is straightforward.

Corollary 1.18 *If ζ is a simple eigenvalue of $\tilde{A}$, and if $r' := \varepsilon\varepsilon^*\|y\|_2\sigma^2 < \frac{1}{4}$, then*

$$|\lambda - \zeta| \leq g(r')|v^H\Sigma u| < 2\sigma\|y\|_2\varepsilon\varepsilon^*,$$

$$\|\varphi - x\|_2 \leq g(r')\|\Sigma u\|_2 < 2\sigma\varepsilon.$$

Proof $r \le r'$, and g is monotone increasing. □

Corollary 1.18 requires only that ε *or* ε^* be small. If ε *and* ε^* are small, the generalized Rayleigh quotient ζ is an approximation of λ of higher order $\varepsilon\varepsilon^*$. In that case, $x \approx \varphi$ and $y \approx \psi$; then $Q = xy^H \approx P = \varphi\psi^H$ and $\Sigma \approx S$. Therefore $\|y\|_2 \approx \|\psi\|_2$ and $\sigma \approx s$. This shows again that the spectral condition numbers are $\|\psi\|_2$ and s. $\|y\|_2$ and σ provide a posteriori estimations of these condition numbers. But the problem of finding an upper bound for σ is not a trivial task in practice.

$-v^H\Sigma u$ is the main term in the error $\lambda - \zeta$ since

$$|\lambda - \zeta + v^H\Sigma u| \le g_1(r)|v^H\Sigma u| = O(r^2).$$

It may be much smaller than ε and ε^*; it may be small even if ε and ε^* are *not* small. This will be exploited later for almost-triangular matrices.

The bounds of Theorem 1.17 are optimal with respect to the data x, y, and $\zeta = y^HAx$. Consider, for example, in $\mathbb{R}^2$,

$$A = \begin{pmatrix} 0 & -a \\ a & 1/b \end{pmatrix}, \qquad \text{with} \quad x = y = e_1, \quad \zeta = 0.$$

Q is the orthogonal projection on $\{e_1\}$,

$$u = \begin{pmatrix} 0 \\ a \end{pmatrix} \qquad \text{and} \qquad v = \begin{pmatrix} 0 \\ -a \end{pmatrix}.$$

Therefore $\varepsilon = \varepsilon^* = |a|$.

$$\Sigma = \begin{pmatrix} 0 & 0 \\ 0 & b \end{pmatrix}, \qquad \|\Sigma\|_2 = \sigma = |b|,$$

$$v^H\Sigma u = -a^2 b, \qquad \text{and} \qquad \|\Sigma u\|_2 = |ab|.$$

The eigenvalues of A are $\lambda = (1/2b)(1 \pm \sqrt{1 - 4a^2b^2})$. Let θ be the acute angle between the eigenvectors and e_1 or e_2. $\tan\theta = |\lambda|/a$. If $r := (ab)^2 < \frac{1}{4}$, the bounds of Theorem 1.17 are sharp: $|\lambda| = g((ab)^2)|b|a^2$ and $\tan\theta = g((ab)^2)|ab|$.

The proof of Theorem 1.17 is due to Redont (1979a) in the more general setting of a closed operator in a Banach space. When T is bounded, the bound of Corollary 1.18 for the eigenvalues is given in Fiedler and Pták (1964) and Pták (1976).

Exercises

1.66 Apply Corollary 1.18 to the matrix

$$A = \begin{pmatrix} a & v^H \\ u & C \end{pmatrix},$$

where $a \in \mathbb{C}$, $u, v \in \mathbb{C}^{N-1}$, and C is $(N-1) \times (N-1)$, assuming that $\|u\|_2 \, \|v\|_2$ is small enough.

1.67 Give the corresponding version of Theorem 1.17 for the case in which A is hermitian.
1.68 Establish a result similar to Theorem 1.17 with an arbitrary norm $\|\cdot\|$ on $\mathbb{C}^N$.

As we have seen in Section 3, the numerical methods to compute all the eigenvalues of a matrix reduce it ultimately, by similarity transformations, to an upper-triangular or diagonal form. Because the process is infinite, we stop the computation when the iterated matrix is "close enough" to the limit matrix. Let Δ be the matrix of simple form for which we know exactly the eigenvalues, and let $A = \Delta + H$ be the iterated matrix, for which the eigenvalues—unknown—are the exact eigenvalues of the original problem. Because we are now interested in the method error, we do not consider the effect of roundoff errors on the computation of A from the original matrix. From a numerical point of view, two cases are interesting: Δ being diagonal or upper-triangular, because their diagonal elements are themselves approximate eigenvalues for A. When A is almost diagonal, all right and left approximate eigenvectors are known as well.

4.2.2. A is almost diagonal

The study is based on the following theorem for an arbitrary matrix A:

Theorem 1.19 *Any eigenvalue of $A = (a_{ij})$ lies in at least one of the N Gershgorin disks $\{z \in \mathbb{C}; |z - a_{ii}| \le \sum_{j \ne i} |a_{ij}|\}$ for $i = 1, 2, \ldots, N$.*

Proof We write $A = D + H$, where $D = (a_{ii})$ is the diagonal of A and H is the off-diagonal part. Let λ be an eigenvalue of A. The theorem is trivial if $\lambda = a_{ii}$ for some i. We then suppose that $\lambda \ne a_{ii}$, $i = 1, \ldots, N$. $\lambda I - D$ is regular and

$$\lambda I - A = \lambda I - D - H = (\lambda I - D)(I - (\lambda I - D)^{-1}H).$$

It is easily seen that $\|(\lambda I - D)^{-1}\| \, \|H\| < 1$ is a sufficient condition for $A - \lambda I$ to be regular. For any eigenvalue λ of A, we get

$$\|(\lambda I - D)^{-1}\|_\infty \, \|H\|_\infty = \max_i \left(|\lambda - a_{ii}|^{-1} \sum_{j \ne i} |a_{ij}| \right) \ge 1. \quad \square$$

Corollary 1.20 *If the N Gershgorin disks are not overlapping, each of them contains one and only one eigenvalue of A, which is then simple.*

Proof We suppose the a_{ii} distinct, and consider $A(\varepsilon) = D + \varepsilon H$, for $0 \le \varepsilon \le 1$. For $\varepsilon = 0$, the N disks are the points a_{ii}. By continuity argument, when ε increases, each disk contains one eigenvalue, as long as the disks remain disjoint. $\square$

The Gershgorin localization applies to a general matrix. It gives error bounds when applied to an almost-diagonal matrix A: the off-diagonal part H is small, and $\sum_{j\neq i}|a_{ij}|$ is small as well. This provides a means to bound $|\lambda - a_{ii}|$ when the disks are not overlapping. In certain cases of overlap, diagonal similarity transformations may be applied to A (cf. Wilkinson, 1965; Varga, 1965).

Exercises

1.69 Let $\{x_i\}_1^N$ be a set of N positive numbers. Show that

$$\forall\lambda \in \sigma(A),\ \exists i: |\lambda - a_{ii}| \le \frac{1}{x_i}\sum_{j\neq i}|a_{ij}|x_j.$$

1.70 For

$$A = \begin{pmatrix} 1 & 10^{-4} & 2\times 10^{-4} \\ -10^{-4} & 2 & 1.1\times 10^{-4} \\ 1.2\times 10^{-4} & -10^{-4} & 3 \end{pmatrix},$$

use diagonal similarity transformations and the Gershgorin circles to localize the eigenvalues around 1, 2, 3 with a precision of the order of 10^{-8}. Interpret this by application of Theorem 1.7 and Exercise 1.40 to $A = D + H$, with $e_i^{\mathrm{H}}He_i = 0$ for $i = 1, 2, 3$.

1.71 Prove the following proposition (Ky Fan), which shows the optimality of the localization by the Gershgorin circles, with diagonal similarity transformations. Let $B = (b_{ij})$ be an irreducible nonnegative matrix such that $b_{ii} = 0$, $b_{ij} \ge 0$, $i, j = 1, \ldots, N$. For any set of N positive numbers $\{\rho_i\}_i^N$ and for any matrix A such that

(i) $|a_{ij}| = b_{ij}$, $i \neq j$,
(ii) any eigenvalue of A lies in at least one of the disks

$$\{z; |z - a_{ii}| \le \rho_i\}, \qquad i = 1, \ldots, N,$$

there exist N positive numbers $\{x_i\}_1^N$ such that

$$\frac{1}{x_i}\sum_{j\neq i} b_{ij}x_j \le \rho_i, \qquad i = 1, \ldots, N.$$

1.72 A *diagonally dominant* matrix A is such that $|a_{ii}| > \sum_{j\neq i}|a_{ij}|$, $i = 1, \ldots, N$. Show that it is regular (Hadamard).

1.73 Prove that a hermitian diagonally dominant matrix with positive diagonal elements is positive definite.

1.74 When A is almost diagonal, compare the localization results given by Corollaries 1.18 and 1.20.

4.2.3. A is almost triangular

If we have only some information on the size of the perturbation H, then for a diagonalizable matrix Δ, of an arbitrary form, the following theorem holds:

Theorem 1.21 (Bauer–Fike) *Let Δ be diagonalizable: $\Delta = VDV^{-1}$ with $D = \operatorname{diag}(\mu_1, \ldots, \mu_N)$. Then for any eigenvalue λ of $A = \Delta + H$, there exists an eigenvalue μ_i of Δ such that*

$$|\lambda - \mu_i| \le \|V^{-1}HV\| \le \|V\|\,\|V^{-1}\|\,\|H\|.$$

Proof It is trivial if $\lambda = \mu_i$ for some i. We then suppose that $\lambda \neq \mu_i$, $i = 1, \ldots, N$. $\lambda I - D$ is regular and

$$\lambda I - \Delta - H = V(\lambda I - D - V^{-1}HV)V^{-1}.$$

For any eigenvalue λ of A,

$$\|(\lambda I - D)^{-1}V^{-1}HV\| = \max_i |\lambda - \mu_i|^{-1} \|V^{-1}HV\| \geq 1. \quad \square$$

Exercise

1.75 Show that Theorem 1.21 is a consequence of Proposition 1.15.

The inequalities of Proposition 1.15 and Theorem 1.21 both show that $K(V) = \|V\| \|V^{-1}\|$ is a global condition number for all the eigenvalues of a diagonalizable matrix Δ. If Δ is either hermitian or normal, then $\|V\|_2 = \|V^{-1}\|_2 = 1$. But in general, to know $K(V)$ requires to know all the eigenvectors of Δ. On the other hand, Corollary 1.18 gives an error bound based on the knowledge of only *one* approximate eigenvector of A.

Often enough in practice A is almost triangular, for example, in the QR method. Let $A := R + H$, where $R = (a_{ij})$, $i \leq j$, is the upper-triangular part of A, and H, the strictly lower triangular part of A, is small. Then e_1 (resp. e_N) is an approximate eigenvector of A (resp. A^H). Theorem 1.17 or 1.19 yield bounds for $|\lambda - a_{11}|$ and $|\lambda - a_{NN}|$, $\lambda \in \sigma(A)$. Bounds for the other eigenvalues of A would require knowledge of the other eigenvectors of R, and this computation might be ill-conditioned, especially if the eigenvalues are close. In this respect, the a posteriori error bounds for a group of eigenvalues given in Lemordant (1977) and Chatelin (1983) for almost-triangular matrices, which *do not* require the computation of approximate eigenvectors, are valuable. These bounds are based on the following theorem, with the notation of Theorem 1.17. $\|\cdot\|$ is now an arbitrary norm on $\mathbb{C}^N$ and a is a scalar such that $a \geq \|\Sigma\|$. We introduce the following condition: there exists $\tilde{\varepsilon}$ such that, for $k \geq 1$,

$$|v^H \Sigma^k u| \leq a^k \|Q\| \tilde{\varepsilon}, \tag{1.17}$$

and we set $\tilde{r} := a^2 \|Q\| \tilde{\varepsilon}$.

Theorem 1.22 (Lemordant) *We suppose that ζ is a simple eigenvalue of $\tilde{A}$ and that* (1.17) *is fulfilled. Then, if $\tilde{r} < \frac{1}{4}$, there exists a simple eigenvalue λ of A such that $|\lambda - \zeta| \leq g(\tilde{r}) |v^H \Sigma u|$, which is the only eigenvalue of A in the disk $\{z \in \mathbb{C}; |z - \zeta| \leq 1/2a\}$. If moreover $y^H P x \neq 0$, where P is the eigenprojection associated with λ, there exists an eigenvector φ normalized by $y^H \varphi = 1$ such that $\|\varphi - x\| \leq g(\tilde{r}) \|\Sigma u\|$.*

Proof This theorem will be proved in Chapter 6 (Theorem 6.27), in the more general context of a closed operator in a Banach space. The proof is based on functional-analysis perturbation techniques. If $\tilde{r} < \frac{1}{8}$, Proposition 6.28 proves that the condition $y^H P x \neq 0$ holds. □

Exercise

1.76 With the euclidean norm on $\mathbb{C}^N$ and $\sigma = \|\Sigma\|_2 = a$, show that the choice $\tilde{\varepsilon} = \varepsilon\varepsilon^*$ yields Corollary 1.18 and that the choice $\tilde{\varepsilon} = r/\sigma^2 \|y\|_2$ yields Theorem 1.17.

We show now that there exists a choice of $\tilde{\varepsilon}$, fulfilling (1.17), adapted to the case where A is *almost triangular*, for which none of the quantities ε, ε^*, and r is small. We choose the norm $\|\cdot\|_1$ on $\mathbb{C}^N$, and $Q = e_i e_i^T$ is the orthogonal projection on $\{e_i\}$. $\zeta = e_i^T A e_i = a_{ii}$. e_i is not an approximate eigenvector of A or A^H, *for* $i \neq 1$ or $i \neq N$.

But A being almost triangular, Σ is also almost triangular with an ith line and ith column being zero. This implies that $v^H \Sigma u$ as well as $v^H \Sigma^k u$ for $k > 1$ are small. More precisely, we have the following lemma.

Lemma 1.23 (Lemordant) *The condition* (1.17) *is satisfied for*

$$x = y = e_i, Q = e_i e_i^T$$

and for the $\|\cdot\|_1$ *norm when A is close enough to a triangular matrix, assuming that no two diagonal elements are equal to* a_{ii}.

Proof Let $A = R - H$, where R is the upper-triangular part of A. We define, for $Q = e_i e_i^T$, $\mathring{A}_{a_{ii}} := [(I - Q)(A - a_{ii} I)]_{\restriction \{e_i\}^\perp}$ and

$$\Sigma := (I - Q)\mathring{A}_{a_{ii}}^{-1}(I - Q).$$

If $\varepsilon := \|H\|_1$ is small enough and no two diagonal elements are equal to a_{ii}, $\mathring{A}_{a_{ii}}$ is regular and Σ is well defined. We define accordingly

$$\mathring{R}_{a_{ii}} := [(I - Q)(R - a_{ii} I)]_{\restriction \{e_i\}^\perp}, \; \mathring{H} := [(I - Q)H]_{\restriction \{e_i\}^\perp}.$$

For $k \geq 1$, $\Sigma^k = (I - Q)\mathring{A}_{a_{ii}}^{-k}(I - Q)$, and in $\{e_i\}^\perp$

$$\mathring{A}_{a_{ii}}^{-k} = [(\mathring{R}_{a_{ii}} - \mathring{H})^{-k} - \mathring{R}_{a_{ii}}^{-k}] + \mathring{R}_{a_{ii}}^{-k},$$

if $\mathring{R}_{a_{ii}}^{-1}$ exists, that is, if no two diagonal elements are equal to a_{ii}. We define $s_1 := \|\mathring{R}_{a_{ii}}^{-1}\|_1$ and $\varepsilon_1 := \|\mathring{H}\|_1$ and we suppose that $\varepsilon_1 s_1 < 1$; note that $\varepsilon_1 \leq \varepsilon$. Using the identity

$$(R - H)^{-k} - R^{-k} = \sum_{j=0}^{k-1} (R - H)^{-j}[(R - H)^{-1} - R^{-1}]R^{-(k-j-1)}$$

$$= \sum_{j=0}^{k-1} (R - H)^{-j-1} H R^{-k+j}$$

and

$$(R - H)^{-1} = R^{-1}(1 - HR^{-1})^{-1} = R^{-1} \sum_{i=0}^{\infty} (HR^{-1})^i,$$

we get

$$\|(\mathring{R}_{a_{ii}} - \mathring{H})^{-k} - \mathring{R}_{a_{ii}}^{-k}\|_1 \leq k\varepsilon_1 s_1 \left(\frac{s_1}{1 - \varepsilon_1 s_1}\right)^k \quad \text{for} \quad k \geq 1.$$

The vectors u and v take the form

$$u = Ae_i - a_{ii}e_i, \qquad v = A^{\mathrm{H}}e_i - \bar{a}_{ii}e_i.$$

Their respective norms are

$$\|u\|_1 = \sum_{j=1}^{i-1} |a_{ji}| + \sum_{i+1}^{N} |a_{ji}|, \qquad \|v\|_\infty = \max_{j \neq i} |a_{ij}|.$$

We define $\varepsilon_u := \sum_{j=i+1}^{N} |a_{ji}|$ and $\varepsilon_v := \max_{j<i} |a_{ij}|$, and also

$$\mathring{u} := [(I - Q)u]_{\restriction\{e_i\}^\perp}, \; \mathring{v} := [(I - Q)v]_{\restriction\{e_i\}^\perp}.$$

Note that $\max(\varepsilon_u, \varepsilon_v) \leq \varepsilon$. Then

$$|\mathring{v}^{\mathrm{H}}[(\mathring{R}_{a_{ii}} - \mathring{H})^{-k} - \mathring{R}_{a_{ii}}^{-k}]\mathring{u}| \leq k\varepsilon_1 s_1 \left(\frac{s_1}{1 - \varepsilon_1 s_1}\right)^k \|u\|_1 \, \|v\|_\infty,$$

$$|\mathring{v}^{\mathrm{H}}\mathring{R}_{a_{ii}}^{-k}\mathring{u}| \leq s_1^k(\|u\|_1\varepsilon_v + \|v\|_\infty \varepsilon_u).$$

Note that $(\frac{3}{2})^k > k$ for $k \geq 1$. We then get

$$|v^{\mathrm{H}}\Sigma^k u| = |\mathring{v}^{\mathrm{H}}\mathring{A}_{a_{ii}}^{-k}\mathring{u}| \leq a^k\tilde{\varepsilon}$$

with

$$a = \frac{3}{2}\frac{s_1}{1 - \varepsilon_1 s_1} > \frac{3}{2}s_1 > s_1 \qquad \text{provided that} \quad \varepsilon_1 s_1 < 1,$$

and

$$\tilde{\varepsilon} = s_1\|u\|_1 \, \|v\|_\infty \varepsilon_1 + \|u\|_1 \varepsilon_v + \|v\|_\infty \varepsilon_u \leq c\varepsilon. \quad \square$$

Exercises

1.77 Check that $a \geq \|\Sigma\|_1$ in Lemma 1.23. Check also that $e_i^{\mathrm{T}}\varphi \neq 0$ for ε small enough, where φ is an eigenvector for A associated with the eigenvalue close to a_{ii}.

1.78 Using the same example as for Theorem 1.17, show that the bounds given in Theorem 1.22 are optimal with respect to the data x and y.

If ε is such that $\tilde{r} = a^2\tilde{\varepsilon} < \frac{1}{4}$ and $\varepsilon_1 s_1 < 1$ and if $e_i^{\mathrm{T}}\varphi \neq 0$ (cf. Exercise 1.77), then the localization results given by Theorem 1.22 apply for almost-triangular matrices that satisfy the assumptions of Lemma 1.23. Note that $e_i^{\mathrm{T}}\varphi \neq 0$ if $\tilde{r} < \frac{1}{8}$, by Proposition 6.28, and that $g(\frac{1}{8}) < 1.2$.

Example 1.9 Let

$$A = \begin{pmatrix} 1 & 1 & 1.6 \\ 0 & 5 & 1 \\ 10^{-3} & 10^{-3} & 9 \end{pmatrix}$$

in $\mathbb{R}^3$. To localize the eigenvalue close to 5, we choose $x = y = e_2$. Then

$$\mathring{u} = \begin{pmatrix} 1 \\ 10^{-3} \end{pmatrix}, \qquad \mathring{v} = \begin{pmatrix} 0 \\ 1 \end{pmatrix}$$

and $\varepsilon_u = 10^{-3}$, $\varepsilon_v = 0$.

$$\mathring{R}_5^{-1} = \begin{pmatrix} -4 & 1.6 \\ 0 & 4 \end{pmatrix}^{-1} = \begin{pmatrix} -0.25 & 0.1 \\ 0 & 0.25 \end{pmatrix}.$$

We deduce that

$$s_1 = \|\mathring{R}_5^{-1}\|_1 = 0.35,\ \varepsilon_1 = \|\mathring{H}\|_1 = 10^{-3},\ \varepsilon_1 s_1 = 35 \times 10^{-5} < 1;$$

then $a = 0.53$ and $\tilde{\varepsilon} = 1.4 \times 10^{-3}$. $\|Q\|_1 = 1$ and $\tilde{r} = a^2\tilde{\varepsilon} = 3.9 \times 10^{-4} < \frac{1}{4}$; then $g(\tilde{r}) = 1.0004$. $w = \Sigma u$ is computed by solving

$$\begin{pmatrix} -4 & 1.6 \\ 10^{-3} & 4 \end{pmatrix} \mathring{w} = \mathring{u}.$$

We conclude that there exists an eigenvalue λ of A such that $|\lambda - 5| \leq g(\tilde{r})|v^{\mathrm{H}}\Sigma u| \leq 3.13 \times 10^{-4}$. λ is the only eigenvalue of A in the disk $\{z \in \mathbb{C};\ |z - 5| \leq 1/1.06 = 0.94\}$. The exact value is $\lambda = 4.9996875$, and $\lambda - 5 = 3.125 \times 10^{-4}$.

Corresponding to λ the exact eigenvector normalized by $e_2^{\mathrm{T}}\varphi = 1$ is $\varphi = (0.24989, 1, -0.00031)^{\mathrm{T}}$. Clearly $\varphi - e_2$ is not small compared to $\tilde{r}$. We correct e_2 by computing $x = e_2 + w$: $w = (-0.24987, 0, 0.00031)^{\mathrm{T}}$ and

$$\varphi - x = (1.953 \times 10^{-5}, 0, -0.002 \times 10^{-5})^{\mathrm{T}}.$$

Therefore $\|\varphi - x\|_1 = 0.20 \times 10^{-4}$, and the bound given in Exercise 6.41, Chapter 6, yields

$$\|\varphi - x\|_1 \leq |g(\tilde{r}) - 1|\ \|w\|_1 \leq 0.25 \times 3.9 \times 10^{-4} = 0.96 \times 10^{-4}.$$

Example 1.10 Let in $\mathbb{R}^5$

$$B(\delta) = \begin{pmatrix} 1 & 0.850 & 0.662 & 0.231 & 0.933 \\ 0 & 3-\delta & 0.726 & 0.216 & 0.214 \\ 0 & 0 & 3 & 0.663 & -0.247 \\ 0 & 0 & 0 & 4 & -0.481 \\ 0 & 0 & 0 & 0 & 5 \end{pmatrix}$$

and

$$H = \begin{pmatrix} 0 & 0 & 0 & 0 & 0 \\ 0.118 & 0 & 0 & 0 & 0 \\ 0.092 & 0.101 & 0 & 0 & 0 \\ 0.032 & 0.030 & 0.092 & 0 & 0 \\ 0.129 & 0.030 & -0.034 & -0.067 & 0 \end{pmatrix}.$$

We are interested in the exact eigenvalue λ of $B(\delta) + \varepsilon H$ which is close to 3. We compare for different values of δ and ε the exact difference $\eta = \lambda - 3$ and its bound $\hat{\eta} = 2|v^{\mathrm{H}}\Sigma u|$ in the notation of Lemma 1.23, $i = 3$. We have $\|H\|_2 = 0.219$. The results are summarized in the following table

ε	δ	η	$\hat{\eta}$
1	1	0.034	0.071
1	0.5	0.085	0.218
1	0.1	0.190	1.726
10^{-3}	0.1	7.4×10^{-4}	1.29×10^{-3}

This example shows the degradation of the sharpness of the bound as the ratio $\delta/\|H\|_2$ decreases. It also shows that $\hat{\eta}$ remains a reliable estimate for η, although the condition $\tilde{r} < \frac{1}{4}$ is not satisfied anymore.

Exercise

1.79 When using the QR algorithm to compute the eigenvalues of A, we stop at a stage k where $A_k = R_k + H_k$. Propose a way to compute H_k (cf. Wilkinson, 1965).

So far we have dealt with simple eigenvalues. The case of multiple or close eigenvalues will be treated in Chapter 6 since it requires more sophisticated techniques.

The practical use of these bounds for almost-triangular matrices requires the estimation of s_1, the norm of the inverse of a known triangular matrix, which requires some computational effort. Strictly speaking, s_1 is required only to know if A is close enough to a triangular matrix. Since $g(\tilde{r}) < 2$, the actual computation of the bounds given in Theorem 1.22 involves only the computation of the vector $w = \Sigma u$, that is, the solution of the equations

$$(I - Q)(A - \zeta I)(I - Q)w = (I - Q)u,$$
$$Qw = 0.$$

4.3. *A Is Hermitian*

When A is hermitian, the theory is simplified because the eigenvalues are real, there exists a basis of orthogonal eigenvectors, and $\|A\|_2 = \max_{\lambda \in \sigma(A)} |\lambda|$.

Let $A = QDQ^H$, with $D = \text{diag}(\mu_1, \ldots, \mu_N)$, ordered by increasing value: $\mu_1 \leq \mu_2 \leq \cdots \leq \mu_N$, $\{\varphi_i\}_1^N$ is a set of orthonormal eigenvectors.

Suppose we are given a vector x, $\|x\|_2 = 1$. The *Rayleigh quotient* of A, based on x, is $x^H A x$. The properties of the Rayleigh quotient play a fundamental role in the theory of hermitian matrices, as we shall see in what follows.

Theorem 1.24 (Courant–Fischer min–max representation) *Let A be a hermitian matrix with eigenvalues $\mu_1 \leq \cdots \leq \mu_N$. Then for $k = 1, 2, \ldots, N$, $\mu_k = \min_{V_k} \max(x^H A x; x \in V_k, \|x\|_2 = 1)$, where V_k is an arbitrary subspace of $\mathbb{C}^N$ of dimension k.*

Proof Let M_k be the span of $\varphi_k, \ldots, \varphi_N$, $\dim M_k = N - k + 1$, so that $\dim V_k = k$ implies $M_k \cap V_k \neq \{0\}$. There exists $x \in M_k \cap V_k$ such that $\|x\|_2 = 1$, and $x = \sum_{i=k}^N \xi_i \varphi_i$.

Then $x^H A x = \sum_k^N \mu_i |\xi_i|^2 \geq \mu_k \sum_k^N |\xi_i|^2 = \mu_k$. Because V_k is arbitrary $\min_{V_k} \max(x^H A x; x \in V_k, \|x\|_2 = 1) \geq \mu_k$.

Let V'_k be the span of $\varphi_1, \ldots, \varphi_k$, $\dim V'_k = k$, and let $x \in V'_k$ with $\|x\|_2 = 1$ and $x = \sum_{i=1}^k \eta_i \varphi_i$. Then $x^H A x = \sum_1^k \mu_i |\eta_i|^2 \leq \mu_k$ and

$$\max(x^H A x; x \in V'_k, \|x\|_2 = 1) \leq \mu_k. \quad \square$$

It should be noted that this result actually goes back to Poincaré (see Weinstein and Stenger, 1972).

Corollary 1.25 $\mu_1 \leq x^H A x \leq \mu_N$ *when* $x^H x = 1$.

Proof This is an easy consequence of Theorem 1.24. $\square$

If x, $\|x\|_2 = 1$, is an approximate eigenvector of A, the Rayleigh quotient $\rho := x^H A x$ may be used for a posteriori error analysis. In the case where $A = A^H$, the choice of $\rho = x^H A x$ (instead of $\zeta = y^H A x$) is self-asserting to take full advantage of the fact that x is an approximate eigenvector for A as well as for A^H.

The two main results that we give (Lemma 1.26 and Theorem 1.28) are valid in the general context of a self-adjoint operator in a Hilbert space. They have been known for a long time; see Lord Rayleigh (1899) and Temple (1928). Lemma 1.26 is often attributed to Krylov and Weinstein. We set $\varepsilon := \|Ax - \rho x\|_2$.

Lemma 1.26 *For any vector x such that $\|x\|_2 = 1$, there exists an eigenvalue λ of A such that $|\lambda - \rho| \leq \varepsilon$.*

Proof It is an easy application of Proposition 1.15, where λ is taken to be $\rho = x^H A x$. Then $\|r\|_2 = \|Ax - \rho x\|_2 = \varepsilon$. $\square$

If we have some information about the distance of λ to the other eigenvalues of A, the above inequality may be improved, as we shall see after proving a preliminary lemma.]a,b[denotes the open interval

$$\{x \in \mathbb{R};\, a < x < b\}.$$

Lemma 1.27 (Kato) *Let a, b be two real numbers such that $a < \rho < b$ and $]a, b[$ contains no eigenvalue of A; then $(b - \rho)(\rho - a) \leq \varepsilon^2$.*

Proof $A = QDQ^{\mathrm{H}}$; set $y := Q^{\mathrm{H}}x$, $\|y\|_2 = \|x\|_2 = 1$. Then $y = \sum_{i=1}^{N} \xi_i \varphi_i$, and hence

$$(Ax - bx)^{\mathrm{H}}(Ax - ax) = (Dy - by)^{\mathrm{H}}(Dy - ay) = \sum_{i=1}^{N} (\mu_i - a)(\mu_i - b)|\xi_i|^2$$

is nonnegative. On the other hand

$$\begin{aligned}(Ax - bx)^{\mathrm{H}}(Ax - ax) &= (Ax - \rho x + (\rho - b)x)^{\mathrm{H}}(Ax - \rho x + (\rho - a)x)^{\mathrm{H}} \\ &= \varepsilon^2 + (\rho - b)(\rho - a) \geq 0. \quad \square\end{aligned}$$

Let $\theta = \theta(x, M)$ be the acute angle between x and the eigenspace M associated with λ.

Theorem 1.28 (Kato–Temple) *We suppose that there exists an interval $]\underline{\lambda}, \bar{\lambda}[$ which contains no eigenvalue of A other than λ, and such that $\underline{\lambda} < \rho < \bar{\lambda}$. Then*

$$\rho - \frac{\varepsilon^2}{\bar{\lambda} - \rho} \leq \lambda \leq \rho + \frac{\varepsilon^2}{\rho - \underline{\lambda}},$$

$$\sin\theta \leq \frac{2}{\bar{\lambda} - \underline{\lambda}}\left[\left(\rho - \frac{\underline{\lambda} + \bar{\lambda}}{2}\right)^2 + \varepsilon^2\right]^{1/2}.$$

Proof It is a straightforward application of Lemma 1.27. (See the proof of Theorem 6.21). $\square$

The larger $\bar{\lambda} - \rho$ and $\rho - \underline{\lambda}$ are compared to ε, the better the Kato–Temple bound is compared to the Krylov–Weinstein inequality. We define

$$\delta(\rho) := \min_{\mu \in \sigma(A) - \{\lambda\}} |\rho - \mu| = \operatorname{dist}(\rho, \sigma(A) - \{\lambda\}).$$

Corollary 1.29 $|\lambda - \rho| \leq \varepsilon^2/\delta(\rho)$ *and* $\sin\theta \leq \varepsilon/\delta(\rho)$.

Proof We choose $\underline{\lambda} = \rho - \delta(\rho)$ and $\bar{\lambda} = \rho + \delta(\rho)$. $\square$

Remarks (1) When ε is small, $\delta(\rho) \approx d(\lambda)$, the isolation distance of λ.
(2) Corollary 1.29 improves on Lemma 1.26 only if $\varepsilon < \delta(\rho)$.
(3) Corollary 1.18 appears as a natural extension, for a simple eigenvalue and a nonhermitian matrix, of Corollary 1.29. With $x = y$, $\|x\|_2 = 1$,

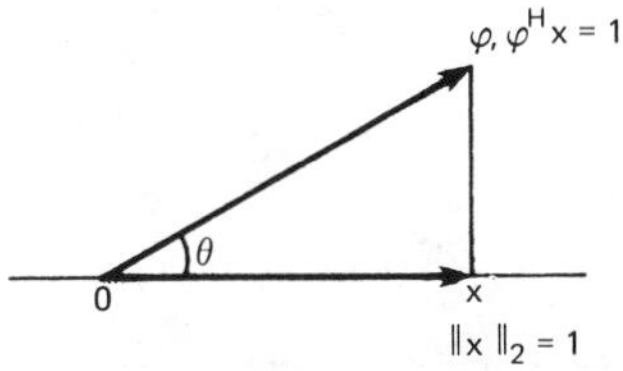

Figure 1.4

$\zeta = \rho$, $\varepsilon = \varepsilon^*$, Corollary 1.29 yields $|\lambda - \rho| \le 2\sigma\varepsilon^2$ and $\|\varphi - x\|_2 = \tan\theta \le 2\sigma\varepsilon$, with $\sigma \approx s = 1/d(\lambda)$; cf. Fig. 1.4.

(4) The Krylov–Weinstein inequality is useful to compute $\underline{\lambda}$ and $\bar{\lambda}$ as defined in Theorem 1.28.

Example 1.11 For

$$A = \begin{pmatrix} 1 & 10^{-5} & 10^{-5} \\ 10^{-5} & 2 & 10^{-5} \\ 10^{-5} & 10^{-5} & 3 \end{pmatrix},$$

we get, with $x = e_i$, $i = 1, 2, 3$, $|\lambda_i - i| \le \sqrt{2} \times 10^{-5}$ by Krylov–Weinstein. Then by application of Theorem 1.28 we get

$$1 - \frac{2 \times 10^{-10}}{1 - \sqrt{2} \times 10^{-5}} \le \lambda_1 \le 1, \qquad |\lambda_2 - 2| \le \frac{2 \times 10^{-10}}{1 - \sqrt{2} \times 10^{-5}},$$

and

$$3 \le \lambda_3 \le 3 + \frac{2 \times 10^{-10}}{1 - \sqrt{2} \times 10^{-5}}.$$

Exercises

1.80 Let x be fixed, $\|x\|_2 = 1$. Show that $\min_{\lambda \in \mathbb{C}} \|Ax - \lambda x\|_2$ is achieved for $\lambda = x^H Ax$.

1.81 Let A be a hermitian matrix with repeated eigenvalues $\mu_1 \le \mu_2 \le \cdots \le \mu_N$, and let A_{N-1} be the leading principal submatrix of order $N - 1$, obtained by deleting the last row and column of A. If $\mu'_1 \le \mu'_2 \le \cdots \le \mu'_{N-1}$ are the repeated eigenvalues of A_{N-1}, show that $\mu_1 \le \mu'_1 \le \mu_2 \le \mu'_2 \le \cdots \le \mu'_{N-1} \le \mu_N$.

This section on a posteriori error analysis may be summed up as follows. The results are either localization results that are valid whatever the norm of the residual, or refined error bounds that are valid only if the residual norm is small enough.

We conclude this section by the following remark. Some methods for hermitian matrices are based on the fact that the eigenvalues of A are the only stationary values of the Rayleigh quotient $\rho(x) := x^H Ax$, so these

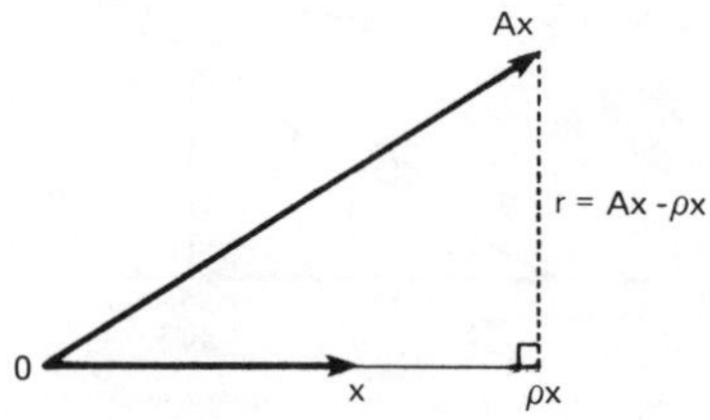

Figure 1.5

methods may then be regarded as methods for optimizing $\rho(x)$. In this respect, the following properties of $\rho(x)$ are interesting:

(i) the gradient of $\rho(x)$ is a multiple of the residual

$$g(x) := \rho'(x) = 2[Ax - \rho(x)x];$$

(ii) the Hessian of $\rho(x)$ is

$$H(x) := \rho''(x) = 2[A - \rho(x)I - g(x)x^{\mathrm{H}} - xg^{\mathrm{H}}(x)].$$

Then $x^{\mathrm{H}}g(x) = 0$, and $H(x)x = -g(x)$; cf. Fig. 1.5. The eigenvalues μ_i are the stationary values of $\rho(x)$; we suppose that

$$\mu_1 \geq \mu_2 \geq \cdots \geq \mu_N.$$

Then $\mu_1 = \max_{x \in \mathbb{C}^N} \rho(x)$, and the rate of convergence of most iterative optimization algorithms to compute the eigenelements μ_1, φ_1 is dependent on the condition number relative to inversion of the Hessian of ρ at φ_1. We suppose μ_1 simple:

$$H(\varphi_1) = 2(A - \mu_1 I).$$

We set $\hat{H} := H(\varphi_1)_{\restriction\{\varphi_1\}^\perp}$:

$$K(\hat{H}) = \|\hat{H}\| \, \|\hat{H}^{-1}\| = \frac{\mu_1 - \mu_N}{\mu_1 - \mu_2}.$$

Similarly for $1 < i \leq N$, $\mu_i = \max(x^{\mathrm{H}}Ax; x \in \{\varphi_1, \ldots, \varphi_{i-1}\}^\perp$ and $x^{\mathrm{H}}x = 1)$, the corresponding condition number is $(\mu_i - \mu_N)/(\mu_i - \mu_{i+1})$, if μ_i is simple. These numbers will play a role in the analysis of the Lanczos method (Section 5.3).

5. Large-Eigenvalue Problems

A rich variety of problems in the natural, engineering, and social sciences require the solution of large-eigenvalue problems, either $A\varphi = \lambda\varphi$, or more often $A\varphi = \lambda B\varphi$, where A and B are sparse. Structured, often symmetric

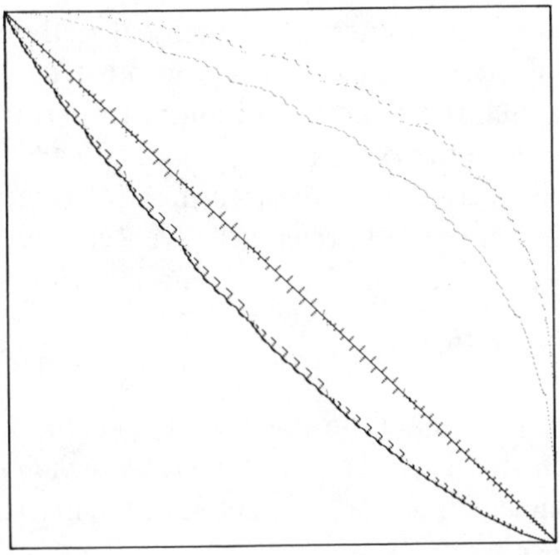

Figure 1.6 Reproduced by courtesy of Dr. I. S. Duff of AERE, Harwell.

matrices commonly arise from finite-difference and finite-element discretizations of continuous problems; and unstructured, unsymmetric matrices arise in social sciences. We cite, for example, a Markov chain model of a queuing system (Cachard (1981)) and dynamic stability of macroeconomic models (d'Almeida, 1980). Patterns of two unsymmetric matrices are given in Figs. 1.6 and 1.7. Figure 1.6 displays the pattern of a Markov matrix of size 1017 (Cachard, 1981), and Fig. 1.7 is the pattern of a basis matrix of size 822 obtained at a certain stage of the application of the simplex method (Duff and Reid, 1979).

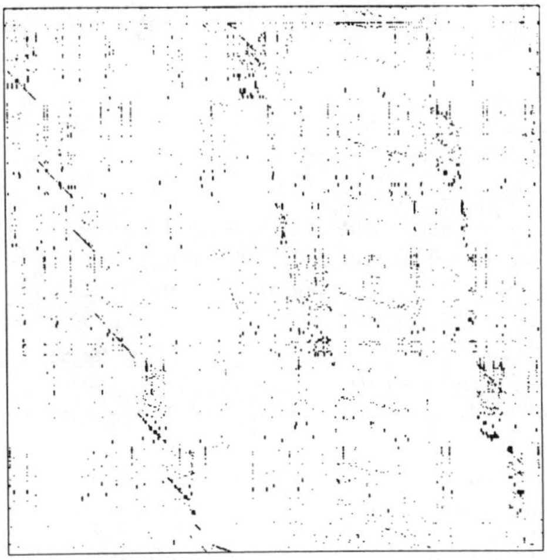

Figure 1.7 Reproduced by courtesy of Dr. I. S. Duff of AERE, Harwell.

We give a review of current iterative numerical methods to compute some of the eigenvalues of $A\varphi = \lambda\varphi$, when A is large, unstructured, and in general unsymmetric. As a rule these methods require to perform only matrix-by-vector multiplications of the type Ax.

An extensive bibliography on the treatment of large sparse matrices is given in Duff (1977). More recent references are Reid (1980) and Duff (1981).

5.1. Principle of the Methods

The idea is to approximate the eigenelements of A, of order N, by those of a matrix of much smaller order ν, obtained by *orthogonal projection* on a ν-dimensional subspace X_n of $X = \mathbb{C}^N$. Let π_n be the matrix of the orthogonal projection of X onto X_n.

The eigenvalue problem

$$\text{find} \quad \lambda \in \mathbb{C}, \quad 0 \neq \varphi \in \mathbb{C}^N, \qquad \text{such that} \quad A\varphi = \lambda\varphi \tag{1.18}$$

is approximated in X_n by

$$\text{find} \quad \lambda_n \in \mathbb{C}, \quad 0 \neq \varphi_n \in X_n, \qquad \text{such that} \quad \pi_n(A\varphi_n - \lambda_n\varphi_n) = 0. \tag{1.19}$$

(1.19) is the *Galerkin* approximation of (1.18) (cf. Chapter 4). The computational scheme consists in generating an orthonormal basis for X_n and solving (1.19) in this basis. Let Q_n be the $N \times \nu$ matrix whose columns are ν orthonormalized vectors of X_n. We set $\varphi_n := Q_n\xi_n$, $\xi_n \in \mathbb{C}^\nu$; ξ_n is the solution of

$$Q_n^H(AQ_n - \lambda_n Q_n)\xi_n = 0.$$

With the $\nu \times \nu$ matrix $B_n := Q_n^H A Q_n$, (1.19) is then equivalent to the eigenvalue problem in $\mathbb{C}^\nu$:

$$\text{find} \quad \lambda_n \in \mathbb{C}, \quad 0 \neq \xi_n \in \mathbb{C}^\nu, \qquad \text{such that} \quad B_n\xi_n = \lambda_n\xi_n. \tag{1.20}$$

Exercises

1.82 Let V_n be an $N \times \nu$ matrix whose columns are an arbitrary basis for X_n (not necessarily orthonormal). Write the eigenvalue problem (1.19) in that basis.

1.83 Check that the matrix B_n represents the operator $\mathscr{A}_n = \pi_n A_{\restriction X_n}: X_n \to X_n$ in the basis Q_n of X_n.

1.84 Prove that $A_n = \pi_n A$, $A_n\pi_n = \pi_n A \pi_n$, and $\mathscr{A}_n$ have the same *nonzero* eigenvalues λ_n and the same associated eigenvectors φ_n in X_n.

1.85 Prove that, if A is hermitian, so are $\mathscr{A}_n$ and B_n. Show that $\lambda_n = \varphi_n^H A_n \varphi_n = \varphi_n^H A \varphi_n$, provided that $\|\varphi_n\|_2 = 1$.

1.86 Propose a similar method to approximate the solution x of the large system $Ax = b$. Write the corresponding system to solve in $\mathbb{C}^\nu$.

In practice the subspaces X_n are generated from Krylov vectors obtained from a given starting vector x, or from a set of r independent vectors $\{x_i\}_1^r$.

Let U be the span of $x_1, x_2, \ldots, x_r$, $1 \le r < N$. There are two main classes of methods derived from the following choices for X_n:

(a) $X_n = A^n U$, for $n = 1, 2, \ldots$, $\dim X_n = r$ is constant throughout the iterations. This choice yields the simultaneous iteration method for $r > 1$ and the power method for $r = 1$.

(b) $X_n = \{U, AU, \ldots, A^{n-1}U\}$, for $n = 1, 2, \ldots, N$, $\dim X_n = nr$ increases over the iterations. This choice corresponds to the generalized block-Lanczos method applied to a general (nonhermitian) matrix. It reduces to the block-Lanczos (Lanczos if $r = 1$) when A is hermitian. In practice, the method is incomplete, which means that n is kept much smaller than N.

Given an eigenvalue λ, with associated eigenvector φ, $\|\varphi\|_2 = 1$, we wish to know if there exists a sequence of eigenelements λ_n, φ_n of $\mathscr{A}_n = \pi_n A_{\restriction X_n}$ (or $A_n = \pi_n A$) that will converge rapidly to λ, φ as n increases.

The approximation method is a *projection* method. It is then reasonable (cf. Chapters 4 and 6) to try to bound the errors $|\lambda - \lambda_n|$ and $\|\varphi - \varphi_n\|_2$ in terms of $\mathrm{dist}(\varphi, X_n) = \|(I - \pi_n)\varphi\|_2 = \sin\theta(\varphi, X_n)$, where $\theta(\varphi, X_n)$ is the acute angle between φ and the subspace X_n. Indeed, the convergence analysis of the methods will be in two steps:

(1) Show that for the choice of X_n under consideration, there exist one or several eigenvectors φ such that $\alpha_n := \|(I - \pi_n)\varphi\|_2$ is small.

(2) Bound $|\lambda - \lambda_n|$ and $\|\varphi - \varphi_n\|_2$ in terms of α_n.

The rate of convergence of the methods will then derive from *estimates* for α_n, which play a key role. We shall assume, in general, that λ is *simple*. If λ is multiple, the analysis is more delicate and the exponent $1/l$ appears, where l is the ascent of λ.

5.2. The Subspace Iteration Method

We suppose that the N repeated eigenvalues of A are such that $|\mu_1| \ge \cdots \ge |\mu_r| > |\mu_{r+1}| \ge \cdots \ge |\mu_N|$. $\{\varphi_i\}_1^N$ is the associated Jordan basis.

Let M be the dominant invariant subspace of dimension r, $1 \le r < N$. P is the associated spectral projection, $W := (I - P)X$, with $X = M \oplus W$.

5.2.1. Convergence

By Theorem 1.8, the method is convergent iff $\{Px_i\}_1^r$ are independent. Then with the notation of Section 3.2, X_n tends to M, and $\mathscr{A}_n$ tends to $A_{\restriction M}$. The eigenelements of $\mathscr{A}_n$ converge to the eigenelements of $A_{\restriction M}$: the eigenvalues are the r dominant eigenvalues of A, with the corresponding eigenvectors.

5.2.2. Estimate of $\|(I - \pi_n)\varphi_i\|_2$

The vectors $\varphi_1, \ldots, \varphi_r$ associated with $\mu_1, \ldots, \mu_r$ are either eigenvectors or generalized eigenvectors. We suppose that φ_i is an eigenvector of A.

Lemma 1.30 *If the r vectors $\{Px_k\}_1^r$ are independent, then for any φ_i, $1 \le i \le r$, which is an eigenvector of A, there exists a unique $u_i \in U$ such that $Pu_i = \varphi_i$ and $\|(I - \pi_n)\varphi_i\|_2 \le \|\varphi_i - u_i\|_2(|\mu_{r+1}/\mu_i| + \varepsilon_n)^n$, where $\varepsilon_n \to 0$ as $n \to \infty$.*

Proof Any $u \in U$ may be written $u = \sum_{k=1}^r t_k x_k$. Then $Pu = \sum_{k=1}^r t_k Px_k$. Given $\varphi_i \in M$, there exists a unique $u_i \in U$ such that $Pu_i = \varphi_i$, because of the independence of the $\{Px_k\}_1^r$. Then u_i may be written $u_i = \varphi_i + v_i$, where $v_i = (I - P)u_i \in W$. By definition

$$\|(I - \pi_n)\varphi_i\|_2 = \min_{x \in X_n} \|\varphi_i - x\|_2 \le \|\varphi_i - \hat{x}_i\|_2,$$

where $\hat{x}_i = (1/\mu_i^n)A^n u_i = \varphi_i + (1/\mu_i^n)A^n v_i$, because φ_i is the eigenvector associated with μ_i. If A is diagonalizable,

$$\left\| \frac{1}{\mu_i^n} A^n v_i \right\|_2 \le \left| \frac{\mu_{r+1}}{\mu_i} \right|^n \|v_i\|_2, \qquad \text{with} \quad v_i \in W.$$

If A is not diagonalizable, we set $B := (1/\mu_i)A_{\restriction W}$. We then have $r_\sigma(B) = |\mu_{r+1}/\mu_i|$. And

$$\|B^n v_i\|_2 \le \|B^n\|_2 \, \|v_i\|_2 = (\|B^n\|_2^{1/n})^n \|v_i\|_2 \le (r_\sigma(B) + \varepsilon_n)^n \|v_i\|_2,$$

where $\varepsilon_n \to 0$ as $n \to \infty$. $\|v_i\|_2 = \|\varphi_i - u_i\|_2 = \|(P - I)u_i\|_2$. □

When $n \to \infty$, $\operatorname{dist}(\varphi_i, X_n) \to 0$ at the rate of $|\mu_{r+1}/\mu_i|$.

The constant $\|\varphi_i - u_i\|_2$ expresses how good the initial subspace U is with respect to φ_i, cf. the case $r = 1$ in Fig. 1.8. If A is hermitian, $\|v_i\|_2 = \tan\theta_i$,

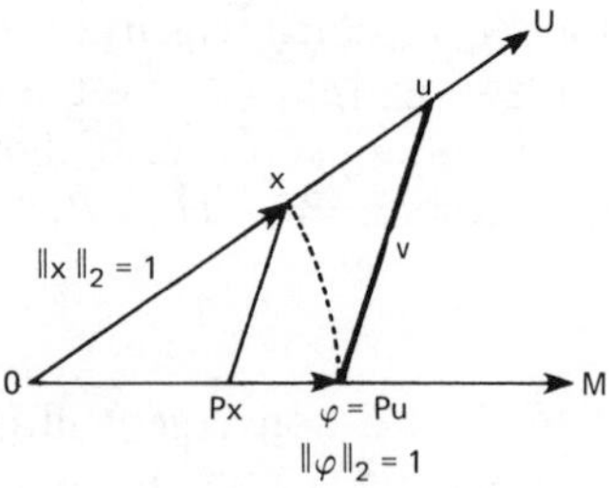

Figure 1.8

where $\theta_i = \theta(\varphi_i, u_i)$ is the acute angle between the eigenvector φ_i and the corresponding vector u_i in U.

5.2.3. Rate of convergence

Let λ (resp. λ_n) be a *simple* eigenvalue of A (resp. $A_n = \pi_n A$) with associated eigenvector φ, $\|\varphi\|_2 = 1$ (resp. φ_n, $\|\varphi_n\|_2 = 1$). λ is chosen among the r dominant eigenvalues of A. c is a generic constant.

Lemma 1.31 *Given λ, φ, then for n large enough, there exist λ_n, φ_n such that $|\lambda - \lambda_n| \leq c\alpha_n$, $\|\varphi - \varphi_n\|_2 \leq c\alpha_n$. If A is hermitian, then $|\lambda - \lambda_n| \leq c\alpha_n^2$.*

Proof We apply Theorem 1.17 to A_n with $x = \varphi$ and $y = \psi$, respectively the right and left eigenvectors of A, normalized by $\psi^H\varphi = \|\varphi\|_2 = 1$. Let $\hat{\varphi}_n$ be the eigenvector of A_n normalized by $\psi^H\hat{\varphi}_n = 1$. The generalized Rayleigh quotient is $\zeta_n = \psi^H A_n \varphi$. The residual vector is $A_n\varphi - \zeta_n\varphi$. We first bound its norm in terms of α_n:

$$A_n\varphi - \zeta_n\varphi = (A_n - \lambda I)\varphi + (\lambda - \zeta_n)\varphi = (A_n - A)\varphi + [\psi^H(A - A_n)\varphi]\varphi$$
$$= \lambda(\pi_n - I)\varphi + \lambda[\psi^H(I - \pi_n)\varphi]\varphi.$$

Therefore

$$\|A_n\varphi - \zeta_n\varphi\|_2 \leq |\lambda|(1 + \|\psi\|_2)\alpha_n,$$

tends to 0 as $n \to \infty$. Similarly $|\lambda - \zeta_n| \leq |\lambda| \, \|\psi\|_2 \alpha_n$ tends to 0 when $n \to \infty$. $Q = \varphi\psi^H$ is the spectral projection for A, associated with λ. If ζ_n is not an eigenvalue of $[(I - Q)A_n]_{\restriction\{\psi\}^\perp}$, we define $\Sigma_n := (I - Q)(A_n - \zeta_n I)^{-1}(I - Q)$. It remains to prove that $\|\Sigma_n\|_2$ is uniformly bounded. Since $A^n U \to M$, the orthogonal projection π_n on $A^n U$ tends to π, the orthogonal projection on M, and $A_n = \pi_n A \to A' := \pi A$. The nonzero eigenvalues of A' are the r dominant eigenvalues of A. Because λ is a simple eigenvalue of A, it is no longer an eigenvalue of $A'(I - Q)$. When $n \to \infty$, $\zeta_n \to \lambda$, and there is no loss in supposing that $\lambda \neq 0$.

Therefore Σ_n is well defined for n large enough, and

$$\Sigma_n \to (I - Q)(A' - \lambda I)^{-1}(I - Q).$$

We may then apply Theorem 1.17 to get

$$|\lambda - \zeta_n| \leq c\alpha_n, \qquad \|\varphi - \hat{\varphi}_n\|_2 \leq c\alpha_n.$$

Then the identity

$$\lambda - \lambda_n = \psi^H A\varphi - \psi^H A_n\hat{\varphi}_n = \psi^H(A - A_n)\varphi + \psi^H A_n(\varphi - \hat{\varphi}_n)$$

implies that $|\lambda - \lambda_n| \leq c\alpha_n$.

Set $\varphi_n := \hat{\varphi}_n / \|\hat{\varphi}_n\|_2$. Because $\|\hat{\varphi}_n\|_2 \to 1$, it is easy to prove that $\|\varphi - \varphi_n\|_2 \leq c\alpha_n$.

If A is hermitian, $|\lambda - \lambda_n| \le c\|(A - A_n)\varphi_n\|_2^2$ since $\lambda_n = \varphi_n^H A \varphi_n$. From $(A - A_n)\varphi_n = \lambda(1 - \pi_n)\varphi + (1 - \pi_n)A(\varphi_n - \varphi)$ follows that $|\lambda - \lambda_n| \le c\alpha_n^2$, λ being possibly multiple. □

Theorem 1.32 (Chatelin–Saad) *If $|\mu_r| > |\mu_{r+1}|$ and if $\{Px_k\}_1^r$ are independent, the simultaneous iteration method on r vectors is convergent. If, moreover, the* i*th dominant eigenvalue is simple, the rate of convergence of the* i*th eigenpair is of the order of $|\mu_{r+1}/\mu_i|$, for $i = 1, \ldots, r$. If A is hermitian, the rate of convergence for the* i*th eigenvalue is squared to $|\mu_{r+1}/\mu_i|^2$.*

Proof It is a straightforward application of Lemmas 1.30 and 1.31. Note that when A is hermitian, the hypothesis that the ith eigenvalue is simple may be dropped. □

Exercise

1.87 Show that $0 \le \mu_1 - \mu_1^{(n)} \le (\mu_1 - \mu_N)|\mu_{r+1}/\mu_1|^{2n} \tan^2(\varphi_1, u_1)$ when A is hermitian, and $\mu_1^{(n)}$ is the dominant eigenvalue of A_n.

5.3. *The Lanczos Method for a Hermitian Matrix*

5.3.1. The algorithm

We suppose that A is hermitian. Given $x \ne 0$, the sequence $\{X_n\}$ is now defined by $X_n = \{x, Ax, \ldots, A^{n-1}x\}$ for $n = 1, 2, \ldots, \nu < N$. The Lanczos method provides a simple way of realizing the Ritz projection of A on X_n. If exact computation is performed, the Lanczos algorithm generates *iteratively* an orthonormal basis $\{v_i\}_1^n$ of X_n, in which $\mathcal{A}_n$ is represented by a *tridiagonal* matrix T_n:

(1) $v_1 := x/\|x\|_2$, $a_1 := v_1^H A v_1$, $b_1 := 0$,
(2) for $j = 1, 2, \ldots, n - 1$, do

$$x_{j+1} := Av_j - a_j v_j - b_j v_{j-1}, \qquad b_{j+1} := \|x_{j+1}\|_2;$$

$$v_{j+1} := x_{j+1}/\|x_{j+1}\|_2, \qquad a_{j+1} := v_{j+1}^H A v_{j+1}.$$

T_n is an $n \times n$ tridiagonal hermitian matrix with diagonal elements a_i, $i = 1, \ldots, n$ and off-diagonal elements b_i, $i = 2, \ldots, n$.

This algorithm was proposed by Lanczos as a means for tridiagonalizing a hermitian matrix, by performing N steps ($X_N = X$). But numerical loss of orthogonality caused it to be abandoned for the Householder method. For large matrices, however, the use of the above incomplete tridiagonalization (ν steps) has revived this old method.

This algorithm is feasible if $x_j \ne 0$ for $j = 2, \ldots, n$. This is achieved if the starting vector x is such that its annihilating polynomial is of degree $\ge n$, that is, if $p(A)x \neq 0$ for any polynomial p of degree less than n. T_n is then an unreduced tridiagonal matrix, and, as a consequence, all its eigenvalues are *simple* (Exercise 1.48). The eigenelements of A_n are often referred to as the *Ritz values* and *vectors* of A.

Let $\{\lambda_i\}_1^K$ be the K *distinct* (possibly multiple) eigenvalues of A. For $1 \leq n_0 \leq n < N$, we are interested in the n_0 largest eigenvalues of A.

The eigenvalues of A (resp. T_n) are ordered by decreasing magnitude:

$$\lambda_1 > \lambda_2 > \cdots > \lambda_K \quad (\text{resp.} \quad \lambda_1^{(n)} > \lambda_2^{(n)} > \cdots > \lambda_n^{(n)}),$$

we set $\lambda_{\min} := \lambda_K$. For $i = 1, \ldots, K$, P_i is the orthogonal eigenprojection associated with λ_i. $\varphi_i^{(n)}$ is the eigenvector of A_n associated with $\lambda_i^{(n)}$, $P_i^{(n)}$ is the corresponding eigenprojection, $i = 1, \ldots, n$.

$\theta(\cdot, \cdot)$ denotes the acute angle between a vector and a subspace (or a vector). The estimate of $\|(I - \pi_n)\varphi_i\|_2 = \sin \theta(\varphi_i, X_n)$ will be given below through bounds on $\tan \theta(\varphi_i, X_n)$.

5.3.2. Estimate of $\tan \theta(\varphi_i, X_n)$

Let $\mathbb{P}_{n-1}$ be the set of polynomials of degree less than or equal to $n - 1$. We suppose that $P_i x \neq 0$, and set

$$\varphi_i := \frac{P_i x}{\|P_i x\|_2},$$

$\hat{x}_i := (I - P_i)x/\|(I - P_i)x\|_2$ if $(I - P_i)x \neq 0$, $\hat{x}_i := 0$ otherwise.

Lemma 1.33

$$\tan \theta(\varphi_i, X_n) = \left[\min_{\substack{p \in \mathbb{P}_{n-1} \\ p(\lambda_i) = 1}} \|p(A)\hat{x}_i\|_2\right] \tan \theta(\varphi_i, x).$$

Proof Because $X_n = \{x, Ax, \ldots, A^{n-1}x\}$, any $u \in X_n$ may be written $u = q(A)x$, $q \in \mathbb{P}_{n-1}$. Now from $\sum_{j=1}^N P_j = I$ follows that $x = P_i x + \sum_{j \neq i} P_j x$, $u = q(\lambda_i)P_i x + \sum_{j \neq i} q(\lambda_j)P_j x$.

$$\tan^2 \theta(P_i x, u) = \left(\textstyle\sum_{j \neq i} q^2(\lambda_j)\|P_j x\|_2^2\right)/q^2(\lambda_i)\|P_i x\|_2^2$$

If $(I - P_i)x \neq 0$, then

$$\sum_{j \neq i} q^2(\lambda_j)\|P_j x\|_2^2 = \|q(A)\hat{x}_i\|_2^2 \, \|(I - P_i)x\|_2^2, \quad \|\hat{x}_i\|_2 = 1.$$

If $x = P_i x$, we set $\hat{x}_i = 0$: $\theta(P_i x, x) = 0$. We define $p(t) = q(t)/q(\lambda_i)$: $p \in \mathbb{P}_{n-1}$ and $p(\lambda_i) = 1$:

$$\tan \theta(P_i x, X_n) = \min_{u \in X_n} \tan \theta(P_i x, u) = \min_{\substack{p \in \mathbb{P}_{n-1} \\ p(\lambda_i) = 1}} \|p(A)\hat{x}_i\|_2 \frac{\|(I - P_i)x\|_2}{\|P_i x\|_2}$$

and

$$\|(I - P_i)x\|_2/\|P_i x\|^2 = \tan\theta(P_i x, x). \quad \square$$

We now define for $i \le n_0$,

$$K_1 := 1, \qquad K_i := \prod_{j=1}^{i-1} \frac{\lambda_j - \lambda_{\min}}{\lambda_j - \lambda_i}, \quad i > 1,$$

and

$$\gamma_i := 1 + 2\frac{\lambda_i - \lambda_{i+1}}{\lambda_{i+1} - \lambda_{\min}}, \qquad t_{in} := \min_{\substack{p \in \mathbb{P}_{n-1} \\ p(\lambda_i) = 1}} \|p(A)\hat{x}_i\|_2.$$

$C_m(t) = \frac{1}{2}[(t + \sqrt{t^2 - 1})^m + (t - \sqrt{t^2 - 1})^m]$ is the mth-degree Chebyshev polynomial of the first kind in t.

Theorem 1.34 (Saad) *If $P_i x \neq 0$, then with $\varphi_i = P_i x/\|P_i x\|_2$:*

$$\tan\theta(\varphi_i, X_n) \le \frac{K_i}{C_{n-i}(\gamma_i)} \tan\theta(\varphi_i, x) \qquad \text{for} \quad i \le n_0.$$

Proof We wish to bound t_{in}. We define $\beta_j = \|P_j \hat{x}_i\|_2$, $\beta_i = 0$, hence $\sum_{j \neq i} \beta_j^2 = 1$.

(a) Case $i = 1$.

$$t_{1n} = \min_{\substack{p \in \mathbb{P}_{n-1} \\ p(\lambda_1) = 1}} \left[\sum_{j \neq 1} \beta_j^2 p^2(\lambda_j)\right]^{1/2}.$$

Now

$$\left(\sum_{j \neq 1} \beta_j^2 p^2(\lambda_j)\right)^{1/2} \le \left(\sum_{j \neq 1} p^2(\lambda_j)\right)^{1/2} \le \max_{t \in [\lambda_{\min}, \lambda_2]} |p(t)|.$$

$$\min_{\substack{p \in \mathbb{P}_{n-1} \\ p(\lambda_1) = 1}} \max_{t \in [\lambda_{\min}, \lambda_2]} |p(t)| = \frac{1}{C_{n-1}(\gamma_1)}$$

with $\gamma_1 = 1 + 2(\lambda_1 - \lambda_2)/(\lambda_2 - \lambda_{\min})$ (cf. Cheney, 1966).

(b) Case $i > 1$.

$$\sum_{j \neq i} \beta_j^2 p^2(\lambda_j) \le \max_{j \neq i} |p(\lambda_j)|^2.$$

$$\min_{\substack{p \in \mathbb{P}_{n-1} \\ p(\lambda_i) = 1}} \max_{j \neq i} |p(\lambda_j)| \le \min_{\substack{p \in \mathbb{P}_{n-1} \\ p(\lambda_1) = \cdots = p(\lambda_{i-1}) = 0 \\ p(\lambda_i) = 1}} \max_{j \neq i} |p(\lambda_j)|.$$

Now such a p may be decomposed:

$$p(t) = \left(\prod_{l=1}^{i-1} \frac{t - \lambda_l}{\lambda_i - \lambda_l}\right) \frac{q(t)}{q(\lambda_i)}$$

where $q \in \mathbb{P}_{n-i}$. Then

$$\max_{j>i} |p(\lambda_j)| = \max_{j>i} \left| \left(\prod_{l=1}^{i-1} \frac{\lambda_j - \lambda_l}{\lambda_i - \lambda_l} \right) \frac{q(\lambda_j)}{q(\lambda_i)} \right| \leq \left(\prod_{l=1}^{i-1} \frac{\lambda_l - \lambda_{\min}}{\lambda_l - \lambda_i} \right) \max_{j>i} \frac{|q(\lambda_j)|}{|q(\lambda_i)|}$$

Therefore

$$t_{in} \leq K_i \min_{\substack{q \in \mathbb{P}_{n-i} \\ q(\lambda_i)=1}} \max_{t \in [\lambda_{\min}, \lambda_{i+1}]} |q(t)| = \frac{K_i}{C_{n-i}(\gamma_i)}. \quad \square$$

We define $\hat{t}_{in} := K_i / C_{n-i}(\gamma_i)$, $i \leq n_0$. Theorem 1.34 shows that $\theta(\varphi_i, X_n)$ decreases at least as rapidly as $\hat{t}_{in}$. $\gamma_i > 1$ depends on the gap $\lambda_i - \lambda_{i+1}$. We define $\tau_i := \gamma_i + \sqrt{\gamma_i^2 - 1} > 1$. For n large enough, $C_{n-i}(\gamma_i) \approx \frac{1}{2}\tau_i^{n-i}$ and the rate of decay of $\theta(\varphi_i, X_n)$ is $1/\tau_i$; the larger $\gamma_i = 1 + 2(\lambda_i - \lambda_{i+1})/(\lambda_{i+1} - \lambda_{\min})$, the better.

This theorem also indicates that for any eigenvalue λ_i, $i \leq n_0$, there exists at least one vector in X_n that is close to the eigenvector $\varphi_i = P_i x / \|P_i x\|_2$. We show now that there is *only one*. This means that a multiple eigenvalue λ_i can be approximated by at most *one* simple eigenvalue $\lambda_i^{(n)}$. Let E be the invariant subspace spanned by the K vectors $\{P_i x\}_1^K$ that we suppose nonzero, $\dim E = K$. Let A' be the matrix of order K representing $A_{\restriction E}$ in an orthonormal basis of E.

Proposition 1.35 *The Lanczos process amounts to approximating the eigenelements of A', whose eigenvalues are simple.*

Proof Let $x = \sum_{i=1}^K P_i x$. A is hermitian; then $A = \sum_{i=1}^K \lambda_i P_i$ and we get $A^k x = \sum_{i=1}^K \lambda_i^k P_i x$, for $k = 1, \ldots, n-1$. Therefore $X_n \subseteq E$ for all n. Thus the Lanczos method applied to A or to A' yields the same matrices $\mathscr{A}_n$ and T_n. $A' P_i x = A P_i x = \lambda_i P_i x$: λ_i is an eigenvalue of A' corresponding to the eigenvector $P_i x$, $P_i x \neq 0$. A', which is of order K, has then K distinct eigenvalues, which have to be simple. $\square$

5.3.3. Rate of "convergence"

Because n takes a finite number of values, we cannot, rigorously speaking, talk about the convergence of the method. But $\lambda_i - \lambda_i^{(n)}$ and $\|\varphi_i - \varphi_i^{(n)}\|_2$ will be bounded in Proposition 1.37 by means of $\beta_{in} := \tan(\varphi_i, x)\hat{t}_{in}$, the bound of $\tan(\varphi_i, X_n)$. This gives the accuracy of the Lanczos method as function of n and x.

We set

$$K_{1n} := 1, \qquad K_{in} := \prod_{j=1}^{i-1} \frac{\lambda_j^{(n)} - \lambda_{\min}}{\lambda_j^{(n)} - \lambda_i} \qquad \text{for} \quad 1 < i \leq n_0$$

(defined if $\lambda_{i-1}^{(n)} > \lambda_i$), $d_{in} := \min_{j \neq i} |\lambda_i - \lambda_j^{(n)}|$.

Lemma 1.36 *If $P_i x \neq 0$, then for $i \leq n_0$,*

$$0 \leq \lambda_i - \lambda_i^{(n)} \leq (\lambda_i - \lambda_{\min})\left(\frac{K_{in}}{K_i}\right)^2 \beta_{in}^2,$$

$$\|(I - \pi_n)\varphi_i\|_2 \leq \|(I - P_i^{(n)})\varphi_i\|_2 \leq \sqrt{1 + \|A\|_2^2/d_{in}^2}\,\beta_{in}.$$

Proof It is based on the min–max characterization of the eigenvalues (cf. Saad, 1980a). □

Proposition 1.37 *If $P_i x \neq 0$, then $0 \leq \lambda_i - \lambda_i^{(n)} \leq k\beta_{in}^2$ and*

$$\|(I - P_i^{(n)})\varphi_i\|_2 \leq k\beta_{in},$$

where k is a generic constant.

Proof The constants in Lemma 1.36 can be bounded independently of n if β_{in} is small enough. For the eigenvalues, this may be done by induction on i: for $i = 1$, $K_{1n} = K_1 = 1$, then $\lambda_1^{(n)} \to \lambda_1$. Now assuming that $\lambda_j^{(n)} \to \lambda_j$ for $j = 1, \ldots, i - 1$ implies that $\lambda_{i-1}^{(n)} > \lambda_i$, $K_{in} \to K_i$ and $\lambda_i^{(n)} \to \lambda_i$. As for the eigenvectors, if $\lambda_j^{(n)} \to \lambda_j$ for $j = 1, \ldots, i + 1$, $d_{in} \to d_i := \min_{j \neq i} |\lambda_i - \lambda_j|$. □

The above bounds may be weak in the case in which λ_i is close to λ_{i+1}. The bounds may be sharpened by taking advantage of the structure of the spectrum, that is, by choosing more appropriately the polynomials used in the proofs of Theorem 1.34 (cf. Saad, 1980a).

A few of the largest eigenvalues are then approximated with a good accuracy when n is kept much smaller than N. Alternatively, if we order the eigenvalues by increasing magnitude, we see that the smallest eigenvalues are also well approximated. The Lanczos method is consequently well suited to compute a few extreme eigenvalues of the spectrum. Several variants of the method have been designed to

(1) avoid full reorthogonalization (Kahan and Parlett, 1976; Parlett and Scott, 1979);

(2) compute interior eigenvalues, by performing $n > N$ iterations (Cullum and Willoughby, 1978a, 1979a,b; Parlett and Reid, 1981).

The computation of the eigenvectors normally requires us to store in memory all the vectors $\{v_i\}_1^n$. A variant is given in Saad (1979a), which requires only the storage of five vectors instead of n.

Another way to cope with a cluster of dominant eigenvalues or a multiple eigenvalue is to use the block-Lanczos method.

5.4. The Block-Lanczos Method

Given a set of r orthonormal vectors $\{x_k\}_1^r$, let U be the span $U = \{x_1, \dots, x_r\}$ and $X_n = \{U, AU, \dots, A^{n-1}U\}$. The block-Lanczos algorithm realizes a projection on X_n in the following way. Let Q_0 be the $N \times r$ matrix $Q_0 := (x_1, \dots, x_r)$. The algorithm produces a sequence of orthonormal $N \times r$ matrices $Q_j, j = 1, \dots, n-1$ such that the columns of $Q_0, Q_1, \dots, Q_{n-1}$ are an orthonormal basis of X_n, in which $\mathscr{A}_n$ is represented by a *block-triangular* matrix $\overset{\square}{T}_n$, the blocks being $r \times r$:

(1) $\overset{\square}{A}_1 := Q_0^H A Q_0$; $\overset{\square}{B}_1 := 0$,
(2) For $j = 1, 2, \dots, n-1$ do $D_j := AQ_{j-1} - Q_{j-1}\overset{\square}{A}_j - Q_{j-2}\overset{\square}{B}{}_j^H$;

perform the orthonormalization of D_j, $D_j := Q_j R_j$, where R_j is an $r \times r$ regular triangular matrix; and set $\overset{\square}{B}_{j+1} := R_j$, $\overset{\square}{A}_{j+1} := Q_j^H A Q_j$.

$\overset{\square}{T}_n$ is an $nr \times nr$ band matrix, the bandwidth being $r + 1$.

If $\dim X_n = nr$, the algorithm is feasible (R_j regular for $j = 1, \dots, n-1$). This condition is satisfied if the starting vectors $\{x_k\}_1^r$ are such that $\sum_{k=1}^r p_k(A)x_k \neq 0$ for all $p_k \in \mathbb{P}_{n-1}$, $k = 1, \dots, r$. A result similar to Proposition 1.35 can be proved: the block-triangular matrix has eigenvalues of multiplicities not larger than r, and we can assume, without loss of generality, that the eigenvalues of A are of multiplicities *not larger than* r.

Let $\{\mu_i\}_1^N$ be the *repeated* eigenvalues of A. The associated eigenvectors (resp. eigenprojections) are $\{\varphi_i\}_1^N$ (resp. $\{P_i\}_1^N$). Note the difference with the notation of Section 5.3. We suppose that

$$\mu_1 \geq \cdots \geq \mu_{i-1} > \mu_i \geq \mu_{i+1} \geq \cdots \geq \mu_{i+r-1} > \mu_{i+r} \geq \cdots \geq \mu_N.$$

Let I be the set of indices $\{i, i+1, \dots, i+r-1\}$, $P := \sum_{j \in I} P_j$.

The rates of convergence can be studied again by means of an estimate of $\tan \theta(\varphi_l, X_n)$, $l \in I$. We define for $l \in I$

$$\hat{\gamma}_l := 1 + 2\,\frac{\mu_l - \mu_{i+r}}{\mu_{i+r} - \mu_N}.$$

K_i as introduced in Theorem 1.34 is well defined since $\mu_{i-1} > \mu_i$.

Lemma 1.38 *If the r vectors $\{Px_k\}_1^r$ are independent, then given φ_l, $l \in I$, there exists a unique $u_l \in U$ such that $Pu_l = \varphi_l$.*

Proof $u \in U$ may be written $u = \sum_{k=1}^r t_k x_k$, $Pu = \sum_{k=1}^r t_k P x_k$. The existence and uniqueness of u_l follow from the independence of $\{Px_k\}_1^r$. We set $v_l := (I - P)u_l$: $u_l = \varphi_l + v_l$ and $\|\varphi_l - u_l\|_2 = \tan \theta(\varphi_l, u_l)$. □

Theorem 1.39 (Saad) *We suppose that $\{Px_k\}_1^r$ are independent and $\mu_{i-1} > \mu_l > \mu_{i+r}$ for $l \in I = \{i, i+1, \dots, i+r-1\}$. Then*

$$\tan\theta(\varphi_l, X_n) \le \frac{K_i}{C_{n-i}(\hat{\gamma}_l)} \tan\theta(\varphi_l, u_l) \qquad \textit{for} \quad l \in I.$$

Proof Given φ_l, we write $u_l = \varphi_l + \sum_{j \notin I} P_j u_l$. We consider $u \in X_n$ of the form $u = q(A)u_l$, with $q \in \mathbb{P}_{n-1}$. Then $u = q(\mu_l)\varphi_l + \sum_{j \notin I} q(\mu_j)P_j u_l$.

(a) Case $i = 1$.

$$\frac{\|(I - P_1)u\|_2^2}{\|P_1 u\|_2^2} = \sum_{j \ge 1+r} \frac{q^2(\mu_j)\|P_j u_1\|_2^2}{q^2(\mu_1)}.$$

The minimum of the right-hand side over $q \in \mathbb{P}_{n-1}$ is achieved for $p \in \mathbb{P}_{n-1}$. We set

$$\bar{u} = p(A)u_1 \in X_n, \qquad \alpha_1 := \frac{2}{\mu_{1+r} - \mu_N}, \qquad \beta_1 := \frac{\mu_{1+r} + \mu_N}{\mu_{1+r} - \mu_N}.$$

Then for $j \ge 1 + r$, $\alpha_1\mu_j - \beta_1 = 1 - 2(\mu_{1+r} - \mu_j)/(\mu_{1+r} - \mu_N) = \theta_j$, $|\theta_j| \le 1$, and $|C_{n-1}(\theta_j)| \le 1$ because $|C_{n-1}(t)| \le 1$ on $[-1, 1]$.

$$\tan^2\theta(\varphi_1, X_n) \le \frac{\|(I - P_1)\bar{u}\|_2^2}{\|P_1\bar{u}\|_2^2} = \sum_{j \ge 1+r} \frac{p^2(\mu_j)\|P_j u_1\|_2^2}{p^2(\mu_1)}$$

$$\le \sum_{j \ge 1+r} \frac{C_{n-1}^2(\alpha_1\mu_j - \beta_1)\|P_j u_1\|_2^2}{C_{n-1}^2(\alpha_1\mu_1 - \beta_1)} \le \sum_{j \ge 1+r} \frac{\|P_j u_1\|_2^2}{C_{n-1}^2(\hat{\gamma}_1)},$$

and

$$\sum_{j \ge 1+r} \|P_j u_1\|_2^2 = \|(I - P)u_1\|_2^2 = \|\varphi_1 - u_1\|_2^2.$$

(b) Case $i > 1$.

We now set

$$\alpha_i := \frac{2}{\mu_{i+r} - \mu_N}, \qquad \beta_i := \frac{\mu_{i+r} + \mu_N}{\mu_{i+r} - \mu_N},$$

then $\alpha_i\mu_l - \beta_i = \hat{\gamma}_l$. We define

$$p_i(t) := \left[\prod_{j=1}^{i-1}(t - \mu_j)\right] C_{n-i}(\alpha_i t - \beta_i),$$

$\bar{u} = p_i(A)u_l \in X_n$, with $p_i(\mu_j) = 0$ for $j = 1, \dots, i-1$.

$$\tan^2\theta(\varphi_l, X_n) \le \frac{\|(I - P_l)\bar{u}\|_2^2}{\|P_l\bar{u}\|_2^2} \le \sum_{j \ge i+r} \frac{p_i^2(\mu_j)\|P_j u_l\|_2^2}{p_i^2(\mu_l)}$$

$$\le \left(\prod_{j=1}^{i-1} \frac{\mu_j - \mu_N}{\mu_j - \mu_i}\right)^2 \left(\sum_{j \ge i+r} \|P_j u_l\|_2^2\right) \Big/ C_{n-i}^2(\hat{\gamma}_l)$$

$$\le \frac{K_i^2}{C_{n-i}^2(\hat{\gamma}_l)} \|\varphi_l - u_l\|_2^2. \quad \square$$

The above bound reduces to that of Theorem 1.34 if $r = 1$. $\theta(\varphi_l, X_n)$ decreases at least as rapidly as $1/C_{n-i}(\hat{\gamma}_l)$. This quantity depends on the gap $\mu_l - \mu_{i+r}$. The extension from the Lanczos to the block-Lanczos method is, in many respects, similar to the extension from the power method to the simultaneous iteration method.

The condition $\{Px_k\}_1^r$ independent is given in Saad (1980a) in the equivalent form $\{\Pi\varphi_l\}_{l\in I}$ independent, where Π is the orthogonal projection onto U.

From the above estimate of $\tan\theta(\varphi_l, X_n)$, one may derive the rates of convergence of $|\mu_l - \mu_l^{(n)}|$ and $\|\varphi_l - \varphi_l^{(n)}\|_2$ in a manner similar to that of Proposition 1.37 (cf. Saad, 1980a). Different approaches may be found in Cullum and Donath (1974) and Underwood (1975).

5.5. *The Arnoldi Method for a Nonhermitian Matrix*

We now describe a method that generalizes the Lanczos method to nonhermitian matrices. The simplest generalization of the Lanczos method is by means of the Arnoldi algorithm (cf. Wilkinson, 1965), as advocated by Saad (1980b).

5.5.1. The algorithm

Consider again $X_n = \{x, Ax, \ldots, A^{n-1}x\}$, the Arnoldi algorithm computes *iteratively* an orthonormal basis $\{v_i\}_1^n$ of X in which $\mathscr{A}_n$ is represented by an *upper-Hessenberg* matrix $H_n = (h_{ij})$, $h_{ij} = 0$ for $i > j + 1$:

(1) $v_1 := x/\|x\|_2, \quad h_{11} := v_1^{\mathrm{H}} A v_1, \quad h_{21} := \|x\|_2,$

(2) for $j = 1, \ldots, n-1$, do:

$$x_{j+1} := Av_j - \sum_{i=1}^{j} h_{ij} v_i, \quad h_{j+1,j} := \|x_{j+1}\|_2,$$

$$v_{j+1} := x_{j+1}/\|x_{j+1}\|_2, \quad h_{i,j+1} := v_i^{\mathrm{H}} A v_{j+1} \quad \text{for} \quad i \le j+1.$$

It is feasible if $x_j \neq 0$ for $j = 1, \ldots, n-1$, that is, $p(A)x \neq 0$ for all $p \in \mathbb{P}_{n-1}$. If this condition is satisfied, H_n is an unreduced Hessenberg matrix: $h_{j+1,j} \neq 0$ for $j = 1, \ldots, n-1$. Therefore, if A is diagonalizable, so are $\mathscr{A}_n$ and H_n; and H_n has n simple eigenvalues (Exercise 1.48).

5.5.2. Estimate of $\|(I - \pi_n)\varphi_i\|_2$

Let $\{\lambda_i\}_1^K$ be the K *distinct* eigenvalues of A, $\{P_i\}_1^K$ are the associated spectral projections. If A is *diagonalizable*, $A = \sum_{i=1}^K \lambda_i P_i$, and Proposition 1.35 holds for the Arnoldi process. We may then suppose, without loss of generality, that the N eigenvalues of A are *simple*; we make this assumption in the rest of this section.

The eigenvalues of A (resp. H_n) are ordered by decreasing magnitude:

$$\lambda_1 > \lambda_2 > \cdots > \lambda_N \quad (\text{resp.} \quad \lambda_1^{(n)} > \lambda_2^{(n)} > \cdots > \lambda_n^{(n)}).$$

$\varphi_i^{(n)}$ is the eigenvector of $A_n = \pi_n A$ associated with $\lambda_i^{(n)}$, $P_i^{(n)}$ is the spectral projection. We decompose $x = \sum_1^N P_i x$. If $P_i x \neq 0$, we define

$$C_i := \left(\sum_{j \neq i} \|P_j x\|_2\right) \Big/ \|P_i x\|_2 \geq \frac{\|(I - P_i)x\|^2}{\|P_i x\|_2}.$$

$$\varepsilon_i^{(n)} := \min_{\substack{p \in \mathbb{P}_{n-1} \\ p(\lambda_i) = 1}} \max_{j \neq i} |p(\lambda_j)|$$

is the degree of approximation of the null function on the set $\{\lambda_j\}_1^N$, $j \neq i$, by polynomials p of degree less than or equal to $n - 1$, satisfying $p(\lambda_i) = 1$ (cf. Lorentz, 1966).

Theorem 1.40 *If A is diagonalizable and $P_i x \neq 0$, then*

$$\|(I - \pi_n)\varphi_i\|_2 \leq C_i \varepsilon_i^{(n)}, \qquad \textit{with } \varphi_i = P_i x / \|P_i x\|_2.$$

Proof Any $u \in X_n$ may be written $u = q(A)x$, $q \in \mathbb{P}_{n-1}$.

$$x = P_i x + \sum_{j \neq i} P_j x, \; u = q(\lambda_i) P_i x + \sum_{j \neq i} q(\lambda_j) P_j x.$$

For $p \in \mathbb{P}_{n-1}$ such that $p(\lambda_i) = 1$,

$$\frac{1}{\|P_i x\|_2} u - \varphi_i = \sum_{j \neq i} p(\lambda_j) \frac{P_j x}{\|P_i x\|_2}.$$

$$\min_{u \in X_n} \|u - \varphi_i\|_2 \leq \left\| \sum_{j \neq i} \frac{P_j x}{\|P_i x\|_2} p(\lambda_j) \right\|_2$$

$$\leq \max_{j \neq i} |p(\lambda_j)| \underbrace{\left(\sum_{j \neq i} \|P_j x\|_2\right) \Big/ \|P_i x\|_2}_{C_i}. \qquad \square$$

C_i expresses how good x is, with respect to φ_i; cf. Fig. 1.9. $\varepsilon_i^{(n)}$ decreases as n increases, but the analysis of the rate of decay is a difficult problem of

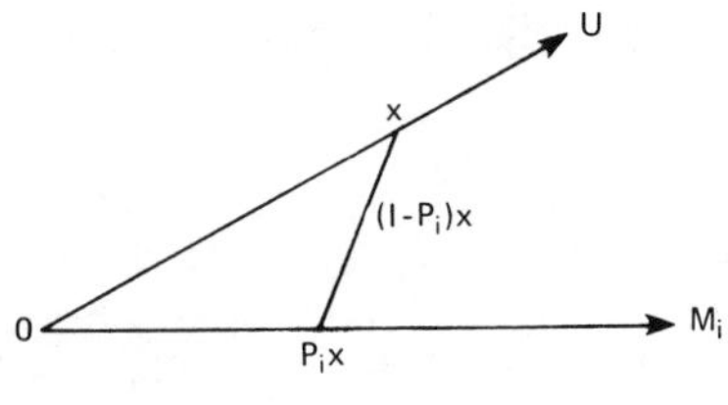

Figure 1.9

approximation theory in the complex variable. Except for some particular shapes of spectra, such as purely real or almost-real spectra, it is not easy to establish bounds on $\varepsilon_i^{(n)}$ that are both sharp and simple. For example, with the notation of Theorem 1.34 we have the following proposition.

Proposition 1.41 *We suppose that A is diagonalizable with real eigenvalues:* $\lambda_1 > \lambda_2 > \cdots > \lambda_N$. *Then* $\varepsilon_i^{(n)} \leq K_i/C_{n-i}(\gamma_i)$.

Proof The proof is analogous to that of Theorem 1.34. A similar result holds if the spectrum lies on a straight line of the complex plane. □

Exercise

1.88 If the dominant eigenvalue λ_1 is such that $|\lambda_1| > |\lambda_2| \geq |\lambda_i|$, $i \geq 3$, show that $\varepsilon_1^{(n)} \leq |\lambda_2/\lambda_1|^{n-1}$. (*Hint*: consider the disk $(0, |\lambda_2|)$ and the polynomial $p(z) = (z/\lambda_1)^{n-1}$.)

More generally, one can use elliptic domains and obtain sharper bounds than that of Exercise 1.88. If the spectrum is almost real, the result becomes close to the bound in Proposition 1.41 (see Saad, 1980b).

5.5.3. "Convergence"

We wish to show that $|\lambda_i - \lambda_i^{(n)}|$ and $\|(I - P_i^{(n)})\varphi\|_2$ can be bounded by $\varepsilon_i^{(n)}$ for n large enough. Since A is not hermitian, we cannot use the min–max characterization of the eigenvalues, as was done in Lemma 1.36. Let λ be a *simple* eigenvalue of A; with the notation of Lemma 1.31, we consider again $\Sigma_n = (I - Q)(A_n - \zeta_n I)^{-1}(I - Q)$. c is a generic constant.

Lemma 1.42 *Given the eigenelements* λ, φ *for* A, λ *being simple, we assume that* Σ_n *is uniformly bounded. Then for* n *large enough, there exist* λ_n, φ_n *eigenelements of* A_n *such that* $|\lambda - \lambda_n| < c\alpha_n$, $\|\varphi - \varphi_n\|_2 < c\alpha_n$.

Proof The proof of Lemma 1.31 applies when α_n is small enough. □

The number $\|\Sigma_n\|_2$ plays the role of a condition number for the approximate problem. The errors $|\lambda_i - \lambda_i^{(n)}|$ and $\|\varphi_i - \varphi_i^{(n)}\|_2$ are of the order of $\varepsilon_i^{(n)}$, provided that the approximate problem is not too ill-conditioned.

Exercises

1.89 Propose a method based on Arnoldi's algorithm to approximate the solution of $Ax = b$ when A is large and unsymmetric. Study its accuracy under the assumption that $\|(\pi_n A_{\restriction X_n})^{-1}\|_2$ is uniformly bounded.

1.90 Let x_0 be an initial guess for $x^* = A^{-1}b$. Set $r_0 := b - Ax_0$ and $x^* := x_0 + z^*$. Let $K_n := \{r_0, Ar_0, \ldots, A^{n-1}r_0\}$. z^* is approximated by the solution $z^{(n)} \in K_n$ of $\pi_n A z^{(n)} = r_0$, where π_n is the orthogonal projection on K_n. Prove that $\operatorname{dist}(z^*, K_n) = \min_{p \in \mathbb{P}_n,\, p(0)=1} \|p(A)z^*\|_2$.

1.91 Prove that the solution $x^{(n)} \in K_n$ of $\pi_n(Ax^{(n)} - b) = 0$ satisfies $x^{(n)} = x_0 + z^{(n)}$. Deduce that $x^* - x^{(n)} = z^* - z^{(n)}$.

1.92 Suppose that A is diagonalizable. Prove that

$$\operatorname{dist}(z^*, K_n) \le k \underbrace{\min_{\substack{p \in \mathbb{P}_n \\ p(0)=1}} \max_{j=1,\dots,N} |p(\lambda_j)|}_{\varepsilon^{(n)}}.$$

1.93 If the eigenvalues of A are real and positive, let $\lambda_{\min}$ (resp. $\lambda_{\max}$) be the smallest (resp. largest). Prove that

$$\varepsilon^{(n)} \le \frac{1}{C_n(\gamma)} \quad \text{with} \quad \gamma = \frac{\lambda_{\max} + \lambda_{\min}}{\lambda_{\max} - \lambda_{\min}}.$$

5.5.4. A practical algorithm

The incomplete Arnoldi algorithm has been suggested by Saad (1979b, 1980b) to deal with large nonhermitian matrices. The method is stable when one uses reorthogonalization, but it requires a lot of storage. In the second paper, Saad has devised several variants with incomplete orthogonalization, which partially overcome this drawback. They are based on the observation that the upper right corner of H_n tends to vanish when n increases.

In the case of close or multiple eigenvalues, a block-Arnoldi method ($r > 1$) can be considered as well.

5.6. "*Oblique*" *Projection Methods*

In the preceding paragraphs we have considered the approximation of (1.18) by an *orthogonal* projection method. When A is not hermitian, non-orthogonal (or oblique) projections can be considered as well. The problem can be presented in an abstract setting as follows. Given two sequences of subspaces X_n and Y_n with $\dim X_n = \dim Y_n$, (1.18) is approximated by the problem

$$\begin{aligned} &\text{find} \quad \lambda_n \in \mathbb{C}, \qquad 0 \neq \varphi_n, \\ &\text{such that} \quad A\varphi_n - \lambda_n \varphi_n \quad \text{is orthogonal to} \quad Y_n. \end{aligned} \tag{1.21}$$

(1.21) is the *Petrov* approximation of (1.18) (cf. Chapter 4).

Exercises

1.94 Let π'_n be the orthogonal projection on Y_n. Write (1.21) in the form of (1.19), where the orthogonal projection on X_n is replaced by $\varpi_n = (\pi'_{n\restriction X_n})^{-1}\pi'_n$, assuming that

$$\sup_n \|(\pi'_{n\restriction X_n})^{-1}\|_2 < \infty.$$

1.95 Let P_n (resp. Q_n) be an $N \times \nu$ matrix, the columns of which form an orthonormal basis for X_n (resp. Y_n). Write (1.21) in this basis.

Example 1.12 The best example is provided by the incomplete bi-orthogonalization method (Lanczos, 1952). In this method, given x and y such that $x^{\mathrm{H}}y \neq 0$, the subspaces are $X_n = \{x, Ax, \ldots, A^{n-1}x\}$ and $Y_n = \{y, A^{\mathrm{H}}y, \ldots, (A^{\mathrm{H}})^{n-1}y\}$. The algorithm computes iteratively biorthogonal bases for X_n and Y_n in which the matrix associated with $\pi_n' A \pi_n$ in (1.21) has a tridiagonal form. This method has been neglected for a long time because of its instability as a means of tridiagonalizing a nonhermitian matrix and computing its eigenvalues. However, this fact has been reconsidered (Parlett and Taylor, 1981), and the method is useful to solve large unsymmetric systems, in conjunction with a preconditioning technique. Various other algorithms of this type are presented in Saad (1982a).

5.7. *Bibliographical Comments*

The incomplete Lanczos method has been used for hermitian matrices by physicists and engineers from 1958 onwards (see, e.g., Sebe and Nachamkin, 1969; Godunov and Propkopov, 1970). The first mathematical analysis was done in Paige (1971). The bounds for the Lanczos and the block-Lanczos methods are due to Saad (1980a), generalizing Kaniel's (1966) bounds. A thorough study of the Lanczos method can be found in Parlett (1980a). Often enough in practice, one has a generalized eigenvalue problem $Ax = \lambda Bx$ to solve, where A and B are hermitian, B being possibly positive definite. To apply the Lanczos method, we need to factorize $B = R^{\mathrm{H}}R$, by Cholesky, and work on $C = (R^{\mathrm{H}})^{-1}AR^{-1}$. To avoid this, one may use iterative methods such as coordinate relaxation (Schwarz, 1974), successive overrelaxation (Ruhe, 1974), and conjugate gradient optimization on $\rho(x)$ (Ruhe, 1977). For the use of Lanczos on the general eigenvalue problem, the reader is referred to Parlett (1980a, 1981) and Scott (1981a).

For nonsymmetric problems, the incomplete Arnoldi method is applied in Saad (1980b) to the computation of the eigenelements, and in Saad (1981) to the solution of large linear systems.

Note added in proof: We have seen on p. 23 that under the assumption (1.9), the subspace A^kU converges iff the starting vectors are such $\{Px_i\}_1^r$ are independent. Let us assume further that $|\mu_1| > |\mu_2| > \cdots > |\mu_r| > |\mu_{r+1}| \geq \cdots \geq |\mu_N|$, and for $s = 1, \ldots, r$, let U_s (resp. M_s) be the subspace spanned by $\{x_i\}_1^s$ (resp. $\{\varphi_i\}_1^s$). Clearly, $A^kU_s \to M_s$ for $s = 1, \ldots, r$, and the simultaneous iteration method on r vectors is a set of r subspace iteration methods with the *nested* starting subspaces $\{U_s\}_1^r$. Let $\{q_i\}_1^r$ be the set of vectors in $M = M_r$ orthonormalized from $\{\varphi_i\}_1^r$. We denote by $q_1^k, \ldots, q_r^k$ the columns of Q_k. From $\{q_1^k\} \to \{q_1\}$ as $k \to \infty$, we deduce that there exists a scalar $z_1^{(k)}$ such that $|z_1^{(k)}| = 1$ and $z_1^{(k)}q_1^k \to q_1$. It follows by induction that there exists an $r \times r$ unitary diagonal matrix Z_k such that $Q_kZ_k \to Q$, where $Q := (q_1, \ldots, q_r)$.

CHAPTER 2

Elements of Functional Analysis: Basic Concepts

Introduction

This chapter is a survey of classical results in functional analysis, which can be found in any textbook on the subject. When proofs are not provided, we give precise references to a selection of books, namely, Kato (1976), Yosida (1965), Taylor (1958), and Dunford and Schwartz (1958, 1963).

In Part A we present the properties of bounded and closed linear operators in a complex Banach space. In Part B, we introduce the spectral theory of a closed linear operator T, with emphasis on the behavior of T in a neighborhood of an isolated eigenvalue with finite algebraic multiplicity.

A. BOUNDED AND CLOSED OPERATORS

1. Banach and Hilbert Spaces

The setting for the theory to follow is an arbitrary, real or complex, Banach space X (i.e., a complete normed vector space). When the norm is generated by an inner product (i.e., $\|x\| := (x, x)^{1/2}$), we get a Hilbert space. In general, the letters X, Y denote Banach spaces; the letters H, V denote Hilbert spaces.

1.1. Examples of Banach Spaces

We recall the definitions of a few functional spaces that we shall use.

1.1.1. Sequence spaces

For $1 \leq p < \infty$, l^p is the set of real or complex sequences $x = \{x_i\}_{i \in \mathbb{N}}$ such that $\|x\|_p := (\sum_{i=1}^{\infty} |x_i|^p)^{1/p}$ is finite.

With the norm $\|\cdot\|_p$, l^p is a Banach space. l^p is a proper subspace of l^q if $1 \leq p < q \leq \infty$ because $\|x\|_q \leq \|x\|_p$ if $x \in l^p$. l^∞ is the space of all bounded sequences, with the norm $\|x\|_\infty := \sup_i |x_i|$. c_0 is the subspace of l^∞ of sequences converging to 0. c is the subspace of convergent sequences $c_0 \subset c \subset l^\infty$.

1.1.2. Function spaces

Let Ω be a bounded open set in $\mathbb{R}^n$.

$C(\Omega)$ is the space of continuous real or complex functions $t \mapsto x(t)$ defined in Ω with the norm

$$\|x\|_\infty := \sup_{t \in \Omega} |x(t)|.$$

$C(0, 1)$ is the space of continuous functions on $[0, 1]$. $C^1(0, 1)$ is the space of functions x with continuous derivative x' on $[0, 1]$ with the norm

$$\|x\| := \|x\|_\infty + \|x'\|_\infty.$$

For $1 \leq p < \infty$, the *Lebesgue* space $L^p(\Omega)$ is the space of classes of complex or real functions x such that $|x(t)|^p$ is Lebesgue-integrable for $t \in \Omega$, with the norm

$$\|x\|_p := \left(\int_\Omega |x(t)|^p \, dt \right)^{1/p}, \qquad dt = dt_1 \cdots dt_n.$$

$L^\infty(\Omega)$ is the space of classes of essentially bounded complex or real functions with the norm

$$\|x\|_\infty := \operatorname{ess\,sup}_{t \in \Omega} |x(t)| = \inf(\alpha; |x(t)| \leq \alpha \text{ almost everywhere in } \Omega).$$

Clearly $C(\Omega) \subset L^\infty(\Omega)$.

For $1 \leq p, k < \infty$, the *Sobolev* space $W^{k,p}(\Omega)$ is the space of classes of real or complex functions x such that x and the generalized partial derivatives $D^s x$, of order $|s| := \sum_{j=1}^n s_j \leq k$, belong to $L^p(\Omega)$, with the norm

$$\|x\|_{k,p} := \left(\sum_{|s| \leq k} \int_\Omega |D^s x(t)|^p \, dt \right)^{1/p},$$

with

$$D^s := \frac{\partial^s}{\partial t_1^{s_1} \partial t_2^{s_2} \cdots \partial t_n^{s_n}}.$$

$W^{0,p}(\Omega) = L^p(\Omega)$. For Sobolev inclusion theorems, one may look, for example, in Ciarlet (1978).

1.2. Examples of Hilbert Spaces

Let $p = 2$ in the above definitions. The corresponding norms are generated by the following inner products:

l^2: $(x, y)_{l^2} := \sum_{i=1}^{\infty} x_i \bar{y}_i, \quad x, y \in l^2$;
$L^2(\Omega)$: $(x, y)_{L^2} := \int_\Omega x(t)\overline{y(t)}\, dt, \quad x, y \in L^2$;
$H^k(\Omega) = W^{k,2}(\Omega)$: $(x, y)_{H^k} := \sum_{|s| \le k} \int_\Omega D^s x(t) \overline{D^s y(t)}\, dt, \quad x, y \in H^k$.

2. Adjoint Space

Let X be a Banach space on $\mathbb{C}$.

2.1. Definitions

A complex-valued function $f[x]$ defined on X is called a *semilinear* form (or functional) iff

$$f[\alpha x + \beta y] = \bar{\alpha} f[x] + \bar{\beta} f[y], \qquad x, y \in X, \quad \alpha, \beta \in \mathbb{C}.$$

f is also called antilinear or conjugate linear.

X^*, the *adjoint* space of X, is the space of all bounded semilinear forms on X. To treat X and X^* at the same level, $f[x]$ will be denoted by $\langle f, x \rangle$: this is the *scalar product* of $f \in X^*$ and $x \in X$, and is, by definition, linear in f and semilinear in x. The norms on X and X^* will be denoted by the same symbol $\|\cdot\|$.

With the norm

$$\|f\| := \sup_{0 \neq x \in X} \frac{|\langle f, x \rangle|}{\|x\|} = \sup_{\|x\| = 1} |\langle f, x \rangle|,$$

X^* is a Banach space. From the definition of $\|f\|$ follows the *Schwarz inequality*:

$$|\langle f, x \rangle| \le \|f\| \, \|x\|, \qquad f \in X^*, \quad x \in X.$$

The adjoint space X^{**} of X^* is again a Banach space. X is *reflexive* iff it can be identified with X^{**} via a norm-preserving one-to-one map between X and X^{**}. In general, X is identified with a subspace of X^{**} so that $\langle x, f\rangle = \overline{\langle f, x\rangle}$ for $x \in X, f \in X^*$ (Theorem 2.1 below is needed).

Example 2.1 $X = \mathbb{C}^n$ is the space of vectors $x = (\xi_i)_{i=1,\ldots,n}$, and X^* is the space of vectors $f = (\alpha_i)_{i=1,\ldots,n}$ with the scalar product

$$\langle f, x\rangle := \sum_{i=1}^{n} \alpha_i \bar{\xi}_i = x^{\mathrm{H}} f.$$

This is the inner product (f, x) (see Section 2.4), and $X = X^*$ is a Hilbert space.

Example 2.2 For $1 \leq p < \infty$ and $p^{-1} + q^{-1} = 1$, $(l^p)^*$ is identified with l^q, with the scalar product $\langle f, x\rangle := \sum_{i=1}^{\infty} f_i \bar{x}_i$, where $x = \{x_i\}_{\mathbb{N}} \in l^p$, $f = \{f_i\}_{\mathbb{N}} \in l^q$. $(c_0)^* = c^* = l^1$.

Example 2.3 For $1 \leq p < \infty$ and $p^{-1} + q^{-1} = 1$, $(L^p(\Omega))^*$ is identified with $L^q(\Omega)$, with the scalar product $\langle f, x\rangle := \int_\Omega f(t)\overline{x(t)}\, dt$.

Example 2.4 $(C(a, b))^* = BV(a, b)$, the space of all functions of bounded variation properly normalized (cf. Yosida, 1965, p. 119), with the scalar product $\langle f, x\rangle = \int_a^b \bar{x}(t)\, df(t)$.

2.2. Properties

Of basic importance in Banach-space theory is the *Hahn–Banach extension theorem*:

Theorem 2.1 *Any bounded semilinear form defined in X can be extended to a bounded semilinear form defined on X without increasing the norm.*

For a proof see Yosida (1965, p. 106).

We shall need the following *Riesz lemma* on "nearly orthogonal" elements.

Lemma 2.2 *Let M be a closed subspace of X, $M \neq X$. For any ε, $0 < \varepsilon < 1$, there exists an $x_\varepsilon \in X$ such that $\|x_\varepsilon\| = 1$ and*

$$\operatorname{dist}(x_\varepsilon, M) := \inf_{m \in M} \|x_\varepsilon - m\| \geq 1 - \varepsilon.$$

If $\dim M < \infty$, $\operatorname{dist}(x_\varepsilon, M) = 1$ *can be achieved.*

Proof Let $y \in X - M$. Since M is closed,

$$\operatorname{dist}(y, M) = \inf_{m \in M} \|y - m\| > d > 0.$$

There exists an $m_\varepsilon \in M$ such that $\|y - m_\varepsilon\| \le d/(1 - \varepsilon)$. The vector $x_\varepsilon := (y - m_\varepsilon)/\|y - m_\varepsilon\|$ satisfies $\|x_\varepsilon\| = 1$, and for $m \in M$;

$$\|x_\varepsilon - m\| = \frac{1}{\|y - m_\varepsilon\|}(y - m_\varepsilon - \|y - m_\varepsilon\|m) \ge \frac{d}{\|y - m_\varepsilon\|} \ge 1 - \varepsilon.$$

If dim $M < \infty$, let Y be the span of M and y. We can apply the above result to the subspace M of Y. With $\varepsilon = 1/n$, there exists $x_n \in Y$ such that $\|x_n\| = 1$ and $\operatorname{dist}(x_n, M) \ge 1 - 1/n$. Since dim $Y < \infty$, there is a subsequence of $\{x_n\}$ converging to x such that $\|x\| = 1$ and $\operatorname{dist}(x, M) = 1$. □

Example 2.5 Let X be the space of all continuous functions x on $[0, 1]$ such that $x(0) = 0$ with the max norm. Let M be the subset of all $x \in X$ such that $\int_0^1 x(t)\,dt = 0$. Let us see that there is no point on the surface of the unit sphere in X at unit distance from M. We suppose that $x_1 \in X$, $\|x_1\| = 1$, and $\|x_1 - x\| \ge 1$ for $x \in M$. For $y \in X - M$, let

$$\alpha := \int_0^1 x_1(t)\,dt \Big/ \int_0^1 y(t)\,dt.$$

$x_1 - \alpha y \in M$ and

$$1 \le \|x_1 - (x_1 - \alpha y)\| = |\alpha|\,\|y\|,$$

$$\left|\int_0^1 y(t)\,dt\right| \le \left|\int_0^1 x_1(t)\,dt\right| \|y\|$$

for each $y \in X - M$. $|\int_0^1 y(t)\,dt|$ may be arbitrarily close to 1 with $\|y\| = 1$. Thus

$$1 \le \left|\int_0^1 x_1(t)\,dt\right|.$$

But $x_1(0) = 0$, $\|x_1\| = 1$, and the continuity of x_1 imply the contradiction

$$\left|\int_0^1 x_1(t)\,dt\right| < 1.$$

•

Corollary 2.3 *Let M be a subspace of X with* dim $M = m$. *Then M has a basis $\{x_i\}_1^m$ such that $\|x_i\| = 1$ and $\|x_i - \sum_{j \ne i} \alpha_j x_j\| \ge 1$ for $i = 1, \ldots, m$ and any choice of the scalars α_j.*

Proof This is a straightforward application of Lemma 2.2. □

Theorem 2.4 *Let M be a closed subspace of X, and let $x_0 \in X$ not belong to M. There exists $f \in X^*$ such that*

$$\langle f, x_0\rangle = 1, \qquad \langle f, x\rangle = 0 \quad \text{for} \quad x \in M, \qquad \|f\| = 1/\operatorname{dist}(x_0, M).$$

Proof Let M' be the span of M and x_0. Each $x \in M'$ has the form $x = \xi x_0 + y$, $y \in M$. ξ is determined by x. The function $f[x] = \bar{\xi}$ defined on M' is semilinear and bounded: $|\xi| \le \|x\|/d$, where $d := \operatorname{dist}(x_0, M)$. The bound is trivial when $\xi = 0$, and it follows from

$$\|\xi^{-1}x\| = \|x_0 + \xi^{-1}y\| \ge d \qquad \text{when} \quad \xi \ne 0.$$

Hence $\|f\| \le d^{-1}$. By applying Lemma 2.2 to M' for any ε, $0 < \varepsilon < 1$, there is $x \in M'$ such that $\|x\| = 1$ and $\operatorname{dist}(x, M) \ge 1 - \varepsilon$. For this x,

$$1 - \varepsilon \le \operatorname{dist}(x, M) = \operatorname{dist}(\xi x_0 + y, M) = |\xi| \operatorname{dist}(x_0, M) = |\xi|\, d;$$

that is

$$|f[x]| \ge (1 - \varepsilon)\|x\|/d.$$

Then $\|f\| = d^{-1}$. This f can be extended to X preserving the norm. Clearly $f[x_0] = 1$ and $f[y] = 0$ for $y \in M$. □

2.3. *Annihilators*

Let M be a subspace of X. The *annihilator* $M^\perp := \{f \in X^*; \langle f, x\rangle = 0$ for any $x \in M\}$ is a closed subspace of X^*.

Let M and N be two *closed* subspaces of X. They are *supplementary* subspaces iff $X = M \oplus N$. Then X^* has the decomposition $X^* = N^\perp \oplus M^\perp$.

Exercises

2.1 Let $M \subset X$. Prove that $M^\perp$ is closed.

2.2 Let M and N be closed supplementary subspaces of X. Prove that $X^* = N^\perp \oplus M^\perp$.

2.3 Each finite-dimensional subspace M of X has a supplementary subspace N. (*Hint*: $\dim M = 1$; $0 \ne x \in M$. Let $f \in X^*$: $\langle f, x\rangle = 1$ and $N = \{y \in X; \langle f, y\rangle = 0\}$.)

Lemma 2.5 *Let M be a finite-dimensional subspace of X such that $X = M \oplus N$, where N is a closed subspace of X. Then $N^\perp$ is the adjoint space M^*.*

Proof Let $\{x_i\}_1^m$ be a basis of M. For $j = 1, \dots, m$, let

$$N_j := \{x_1, \dots, x_{j-1}, x_{j+1}, \dots, x_m\} \oplus N.$$

N_j is closed (left to the reader) and $x_j \notin N_j$; therefore there exists, by Theorem 2.4, an $x_j^* \in X^*$ such that

$$\langle x_j^*, x_j\rangle = 1, \qquad \langle x_j^*, x_i\rangle = 0, \quad i \ne j, \qquad \|x_j^*\| = 1/\operatorname{dist}(x_j, N_j),$$

and

$$\langle x_j^*, y\rangle = 0, \qquad y \in N,$$

i.e., $x_j^* \in N^\perp$ and $\langle x_j^*, x_i \rangle = \delta_{ij}$, $i, j = 1, \ldots, m$. This proves that the $\{x_j^*\}_1^m$ are independent. We now prove that they are a basis of $N^\perp$. For any $f \in N^\perp$, let

$$\alpha_j := \langle f, x_j \rangle, \qquad j = 1, \ldots, m,$$

and

$$g = f - \sum_{j=1}^{m} \alpha_j x_j^*.$$

Then

$$\langle g, x_i \rangle = \langle f, x_i \rangle - \sum_{j=1}^{m} \langle f, x_j \rangle \delta_{ij} = 0, \qquad i = 1, \ldots, m,$$

and

$$\langle g, y \rangle = 0 \qquad \text{for} \quad y \in N.$$

Hence $\langle g, x \rangle = 0$ for $x \in X$; that is, $f = \sum_{j=1}^m \alpha_j x_j^*$ and $\dim N^\perp = m$. $N^\perp$ can then be identified with M^*, the adjoint space of M considered as a Banach space, via the map $x^* \in N^\perp \mapsto x^*_{\upharpoonright M} \in M^*$, because $\dim M^* = m$ and $\{x^*_{j\upharpoonright M}\}_1^m$ are independent. This justifies the notation $N^\perp = M^*$. □

The basis $\{x_j^*\}_1^m$ of M^* defined by $\langle x_j^*, x_i \rangle = \delta_{ij}$, $i, j = 1, \ldots, m$, is called the *adjoint basis* of the basis $\{x_i\}_1^m$ of M. In these bases, $x \in M$ has the representation $x = \sum_{i=1}^m \xi_i x_i$ with $\xi_i = \langle x, x_i^* \rangle$, $i = 1, \ldots, m$.

Example 2.6 Let $X = C(a, b)$. $[a, b]$ is divided into $m - 1$ intervals at the points t_i, $i = 1, \ldots, m$, $t_1 = a$, $t_m = b$. Let M be the subspace of X consisting of piecewise-linear functions on $[a, b]$, each function being linear on every subinterval $[t_i, t_{i+1}]$, $i = 1, \ldots, m - 1$. For $i = 1, \ldots, m$, let e_i be the piecewise-linear function taking the value 1 at t_i and 0 at all other points t_j, $j \neq i$. $\{e_i\}_1^m$ is a basis of M. The adjoint basis consists of $\{e_i^*\}_1^m$, where e_i^* is the evaluation functional at t_i: $\langle x, e_i^* \rangle = x(t_i)$. Then, according to the basis $\{e_i\}_1^m$, $x \in M$ has the representation $x = \sum_{i=1}^m x(t_i) e_i$.

Exercise

2.4 If $\dim M < \infty$, $(M^\perp)^\perp = M$.

In general $(M^\perp)^\perp \cap X = \overline{M}$, the closure of M.

2.4. *Inner Product in a Hilbert Space*

Let H be a Hilbert space on $\mathbb{C}$. The *inner product* (x, y), $x, y \in H$, is a complex-valued sesquilinear form (linear in x and semilinear in y) which is hermitian $\overline{(x, y)} = (y, x)$. Moreover, $(x, x) > 0$ for $x \neq 0$. The norm is defined by $\|x\| := (x, x)^{1/2}$.

Also of basic importance in Hilbert space theory is the *Riesz representation theorem*:

Theorem 2.6 *Given a bounded semilinear form f defined on a Hilbert space H, there exists a uniquely determined vector $x_f \in H$ such that*

$$(f, y) = (x_f, y) \qquad \text{for all} \quad y \in H$$

and

$$\|f\| = \|x_f\|.$$

Conversely, any vector $x \in H$ defines a bounded semilinear form f_x on H by

$$(f_x, y) = (x, y) \qquad \text{for all} \quad y \in H$$

and

$$\|f_x\| = \|x\|.$$

For a proof see Yosida (1965, p. 90).

Theorem 2.6 establishes the existence of a norm-preserving one-to-one correspondence $f \leftrightarrow x_f$ between H^* and H. H^* can then be identified with H (in particular, $H^{**} = H^* = H$), and H is reflexive. The inner product (x, y) may be regarded as a special case of the scalar product defined between $x \in H^*$ and $y \in H$.

We say that x and y are *orthogonal* if $(x, y) = 0$.

For every closed subspace M of H, the subspace $M^\perp = \{y \in H, (x, y) = 0$ for any x in $M\}$ is such that $H = M \oplus M^\perp$. $M^\perp$ is called *the orthogonal complement* of M, and $(M^\perp)^\perp = M$ (cf. Kato, 1976, p. 252). To distinguish between any direct sum $H = M \oplus N$ and the orthogonal direct sum $H = M \oplus M^\perp$, we use the notation ⦶ for the latter, that is, $H = M$ ⦶ $M^\perp$.

If $\dim M = m < \infty$, let $\{x_i\}_1^m$ be a basis of M (not necessarily orthonormal). The adjoint basis $\{x_i^*\}_1^m$ in $M^* \equiv M$ is defined by $(x_j^*, x_i) = \delta_{ij}$, $i, j = 1, \dots, m$. $\{x_i\}_1^m$ and $\{x_j^*\}_1^m$ form a biorthogonal family. $x_i^* = x_i$, $i = 1, \dots, m$, iff $\{x_i\}_1^m$ is an orthonormal basis.

We did not define the notion of a basis in a Banach space. In a Hilbert space, the role of an orthonormal basis is played by a *complete* orthonormal family, that is, a family $\{x_i\}$, where i runs over a certain set, such that

$$(x_i, x_j) = \delta_{ij} \qquad \text{and} \qquad x = \sum_i \xi_i x_i = \sum_i (x, x_i) x_i.$$

If H is separable, H contains a dense countable subset, and any orthonormal family is at most countable. Therefore a complete orthonormal family can be constructed by the Schmidt orthogonalization process applied to any sequence dense in H.

Example 2.7 $L^2(0, 1)$ is a separable Hilbert space. The set $\{e_k\}_{\mathbb{N}}$ is an orthonormal basis with $e_1(t) = 1$, $e_k(t) = \sqrt{2}\cos(k - 1)\pi t$ for $k = 2, 3, \dots$ and $0 \le t \le 1$.

3. Compact Sets in a Banach Space

Our analysis of numerical approximation methods will depend heavily on compactness properties. Some simplification occur in our context because the space is complete.

3.1. Definitions

Let A be a subset of the Banach space X. A is *compact* iff each open cover of A has a finite subcover; A is *relatively compact* iff $\bar{A}$ is compact. A is *sequentially compact* iff each sequence in A has a convergent subsequence with limit in X. A is *precompact* (or *totally bounded*) if for all $\varepsilon > 0$, A has a finite cover with sets of diameter less than ε.

The following three properties are equivalent:

(1) A is relatively compact;
(2) A is sequentially compact;
(3) A is precompact.

Hence A is compact iff each sequence in A has a convergent subsequence with limit in A.

3.2. Examples

We shall make heavy use of the following two theorems:

(1) In $\mathbb{R}^n$, we have the *Borel–Lebesgue theorem*:

$A \subset \mathbb{R}^n$ is relatively compact iff A is bounded.

The property is characteristic of finite-dimensional spaces: a Banach space X has a finite dimension iff its unit sphere is compact.

(2) In $C(\Omega)$, where Ω is a bounded open set in $\mathbb{R}^n$, we have the *Ascoli–Arzela theorem*:

$A \subset C(\Omega)$ is relatively compact iff any bounded sequence $\{x_i\}_{\mathbb{N}} \subset A$ is

(i) *equicontinuous*:

$$\lim_{\delta \to 0} \left(\sup_{\substack{i \geq 1 \\ \mathrm{dist}(t,t') \leq \delta}} |x_i(t) - x_i(t')| \right) = 0;$$

(ii) *equibounded*:

$$\sup_{i \geq 1} \sup_{t \in \bar{\Omega}} |x_i(t)| < \infty.$$

For a proof see Yosida (1965, p. 85).

We cite two examples of application of the Ascoli–Arzela theorem:

(1) The unit sphere of $C^1(a, b)$ is a relatively compact set of $C(a, b)$.

(2) Let Ω be a bounded open set of $\mathbb{R}^n$ with a "smooth" boundary. The unit sphere of $H^1(\Omega)$ is a relatively compact set of $L^2(\Omega)$ (*Rellich* theorem). The unit sphere of $H^s(\Omega)$ is a relatively compact set of $H^{s'}(\Omega)$, for $s' < s$ (proof in Aubin, 1972, p. 203).

The interested reader can find characterizations of compact sets in the spaces l^p and L^p for $1 \leq p < \infty$, for example, in Natanson (1964, Vol. 2, p. 202) or in Yosida (1965, p. 275).

4. Bounded Linear Operators

4.1. Definitions

Let T be a linear operator from X to Y, where X and Y are complex Banach spaces. We set

Dom $T := \{x \in X; Tx \in Y\}$, the *domain* (of definition) of T,

Im $T := \{y \in Y;$ there exists x in Dom T such that $y = Tx\}$, the *image* of Dom T under T, also called the *range* of T,

Ker $T := \{x \in X; Tx = 0\}$, the *null space* (or *kernel*) of T.

T is *bounded* iff $\|T\| := \sup_{0 \neq x \in \text{Dom}\, T} \|Tx\|/\|x\| < \infty$.

Exercise

2.5 Prove that

$$\begin{aligned}\|T\| &= \sup(\|Tx\|; x \in \text{Dom}\, T, \|x\| = 1)\\ &= \inf(\alpha > 0; \|Tx\| \leq \alpha\|x\| \text{ for all } x \text{ in Dom}\, T).\end{aligned}$$

$\mathscr{L}(X, Y)$ is the space of bounded linear operators on X into Y (i.e., Dom $T = X$). With the norm $\|T\| := \sup_{0 \neq x \in X} \|Tx\|/\|x\|$, $\mathscr{L}(X, Y)$ is a Banach space. When $X = Y$, $\mathscr{L}(X, Y)$ is denoted $\mathscr{L}(X)$.

Let $T \in \mathscr{L}(X)$. T is invertible iff Ker $T = \{0\}$. Its *inverse* is denoted T^{-1}.

Example 2.8 Let $X = Y = C(a, b)$. The integral operator K is defined by $(Kx)(s) = \int_a^b k(s, t)\,dt$, where the kernel k is continuous on $[a, b] \times [a, b]$. T is a bounded linear operator in $\mathscr{L}(X)$; its norm is

$$\|K\| = \sup_{0 \neq x \in X} \left(\max_{s \in [a, b]} \left| \int_a^b k(s, t)x(t)\,dt \right| \Big/ \max_{t \in [a, b]} |x(t)| \right)$$

$$\leq \underbrace{\left(\max_{s, t \in [a, b]} |k(s, t)| \right)}_{M} (b - a) < \infty.$$

The integral equation $Kx = y$,

$$\int_a^b k(s, t)x(t)\,dt = y(s), \qquad a \leq s \leq b,$$

is a *Fredholm* equation of the *first kind*. The integral equation $x - Kx = y$ is a *Fredholm* equation of the *second kind.*

Example 2.9 The identity operator on X is denoted $1 \in \mathscr{L}(X)$ if there is no ambiguity. For example, the Fredholm equation of the second kind may be written $Tx = y$ where $T = 1 - K$.

Example 2.10 Let $X = L^1(0, 2\pi)$. Let $\mathscr{A}$ be the class of functions f analytic in the unit disk $\{z; |z| < 1\}$. For $x \in X$, we define

$$x(\theta) \overset{T}{\mapsto} f(z) = \frac{1}{2\pi} \int_0^{2\pi} \frac{x(\theta)}{1 - ze^{-i\theta}}\,d\theta.$$

T is a linear bounded operator on X into $\mathscr{A}$. Characterization of Im T is difficult. A simple *sufficient* condition on $f \in \mathscr{A}$ to be represented as the above integral with some $x \in X$ is the following: $f \in \mathscr{A}$ is continuous for $|z| \leq 1$. Using the Cauchy formula

$$f(z) = \frac{1}{2i\pi} \int_\Gamma \frac{f(t)}{t - z}\,dt,$$

where Γ is the circle $\{t \in \mathbb{C}; |t| = 1\}$, we get by setting $t := e^{i\theta}$, $0 \leq \theta \leq 2\pi$,

$$f(z) = \frac{1}{2\pi} \int_0^{2\pi} \frac{f(e^{i\theta})}{1 - ze^{-i\theta}}\,d\theta,$$

which gives $Tx = f$ with $x(\theta) = f(e^{i\theta})$.

Theorem 2.7 *Let $T \in \mathscr{L}(X)$. The limit*

$$r_\sigma(T) := \lim_{k \to \infty} \|T^k\|^{1/k} = \inf_k \|T^k\|^{1/k}$$

exists. It is called the spectral radius *of T.*

Proof Set $a_k := \log \|T^k\|$. We have to show that $a_k/k \to b := \inf_k a_k/k$. $\|T^{m+k}\| \le \|T^m\| \, \|T^k\|$ gives $a_{m+k} \le a_m + a_k$. For a fixed positive integer m, set $k := mq + r$, where q, r are integers such that $0 \le r < m$. Then $a_k \le qa_m + a_r$ and

$$a_k/k \le (q/k)a_m + (1/k)a_r.$$

If $k \to \infty$ for m fixed, $q/k \to 1/m$. Hence $\limsup_k(a_k/k) \le a_m/m$, where m is arbitrary. Then $\limsup_k(a_k/k) \le b$. On the other hand, $a_k/k \ge b$; therefore $\liminf_k(a_k/k) \ge b$. □

Theorem 2.7 shows that $r_\sigma(T) \le \|T^k\|^{1/k}$ for any positive integer k. In particular, $r_\sigma(T) \le \|T\|$. Conversely, one can construct on X a norm $\|\cdot\|_*$ that is equivalent to $\|\cdot\|$ and such that $\|T\|_*$ is arbitrarily close to the spectral radius $r_\sigma(T)$. This norm depends on T.

Let $\varepsilon > 0$ be given. Determine k such that

$$\|T^k\|^{1/k} \le r_\sigma(T) + \varepsilon.$$

For any $x \in X$, we set $r := r_\sigma(T)$ and

$$\|x\|_* := (r + \varepsilon)^{k-1}\|x\| + (r + \varepsilon)^{k-2}\|Tx\| + \cdots + \|T^{k-1}x\|.$$

Then

$$(r + \varepsilon)^{k-1}\|x\| \le \|x\|_* \le [(r + \varepsilon)^{k-1} + (r + \varepsilon)^{k-2}\|T\| + \cdots + \|T^{k-1}\|]\|x\|;$$

that is,

$$m(\varepsilon, T)\|x\| \le \|x\|_* \le M(\varepsilon, T)\|x\|.$$

Clearly

$$\|Tx\|_* = (r + \varepsilon)^{k-1}\|Tx\| + \cdots + \|T^k x\| \le (r + \varepsilon)\|x\|_*;$$

that is,

$$r \le \|T\|_* \le r + \varepsilon.$$

An operator T such that $r_\sigma(T) = 0$ is said to be *quasi-nilpotent*.

Example 2.11 We consider in $X = C(a, b)$ the *Volterra* integral operator defined by

$$(Kx)(s) = \int_a^s k(s, t)x(t)\,dt, \qquad a \le s \le b,$$

where the kernel k is continuous, $a \le t, s \le b$. To see that K is quasi-nilpotent, one may prove by induction that

$$\|K^k\| \le \frac{M^k(b - a)^k}{(k - 1)!}, \qquad k \ge 1,$$

where $M = \sup_{s, t \in [a, b]} |k(s, t)|$.

Exercises

2.6 Show that the definition of the Volterra operator of Example 2.11 can be written in the form given in Example 2.8, where the kernel k is supposed to be continuous on the triangle $a \leq t \leq s \leq b$ and to vanish for $a \leq s < t \leq b$.

2.7 Let $T, U \in \mathscr{L}(X)$ be such that $TU = UT$. Prove that $r_\sigma(UT) = r_\sigma(TU) \leq r_\sigma(T) r_\sigma(U)$. If $UT \neq TU$ with $U \in \mathscr{L}(X, Y)$ and $T \in \mathscr{L}(Y, X)$, show that $r_\sigma(UT) = r_\sigma(TU)$.

2.8 Show by linearity that T is continuous everywhere in Dom T if it is continuous at $x = 0$. T is continuous iff it is bounded.

2.9 Show that the nullspace of $T \in \mathscr{L}(X)$ is a closed subspace of X.

2.10 Let $T: X \to Y$ be a linear operator defined on X, a finite-dimensional space. Show that T is continuous.

2.11 In $X = C(a, b)$, we consider the operator T defined by $x \in X \mapsto y = Tx$, which is the unique solution of the differential equation

$$y'' = x, \qquad y(a) = y(b) = 0,$$

where y'' is the second derivative of y. Is T linear? Show that

$$y(s) = \int_a^s du \int_a^u x(t)\,dt - \frac{s-a}{b-a}\int_a^b du \int_a^u x(t)\,dt$$

or

$$y(s) = \int_a^b k(s, t)x(t)\,dt$$

with

$$k(s, t) = \begin{cases} (s-b)(t-a)/(b-a) & \text{if} \quad a \leq t \leq s, \\ (s-a)(t-b)/(b-a) & \text{if} \quad s \leq t \leq b. \end{cases}$$

Conclude that $T \in \mathscr{L}(X)$.

2.12 Let $T \in \mathscr{L}(X, Y)$, where X and Y have the same finite dimension. T^{-1} exists iff Im $T = Y$. Find an example where Im $T = Y$ but T^{-1} does not exist when X and Y are infinite dimensional.

The next theorem gives important information about the existence and nature of the inverse $(1 - T)^{-1}$ under certain conditions.

Theorem 2.8 *Let $T \in \mathscr{L}(X)$ be such that $\|T\| < 1$. Then* $\operatorname{Im}(1 - T) = X$, $(1 - T)^{-1}$ *exists and is bounded on X, and*

$$(1 - T)^{-1} = 1 + T + T^2 + \cdots + T^k + \cdots, \tag{2.1}$$

where the series converges in $\mathscr{L}(X)$, and $\|(1 - T)^{-1}\| \leq 1/(1 - \|T\|)$.

Proof Since $\|T\| < 1$, the series $\sum_{k=0}^{\infty} \|T\|^k$ converges. Since $\|T^k\| \leq \|T\|^k$, the series $\sum_{k=0}^{\infty} T^k$ converges in $\mathscr{L}(X)$. Denote by R its limit. $RT = TR = \sum_{k=0}^{\infty} T^{k+1}$; therefore $(1 - T)R = R(1 - T) = 1$, which proves (2.1). □

Identity (2.1) is known as the *Neumann series expansion* of $(1 - T)^{-1}$.

The solution x of the equation $x - Tx = y$ may be represented by the series

$$x = y + Ty + T^2y + \cdots$$

for any y in X. It can be computed by the *Picard* iteration method, also known as the method of *successive approximations*:

$$x = \lim_{k\to\infty} x_k \qquad \text{with} \quad x_0 = y, \quad x_k = Tx_{k-1} + y, \quad \text{for} \quad k \geq 1.$$

x_k converges to x at the rate of a geometric progression with quotient $\|T\|$:

$$\|x_k - x\| \leq \frac{\|T\|^{k+1}}{1 - \|T\|}\|y\| \qquad \text{for} \quad k \geq 0.$$

Exercises

2.13 Show that the integral equation

$$x(t) + 0.1\int_0^1 e^{ts}x(s)\,ds = 1, \qquad 0 \leq t \leq 1,$$

has a unique continuous solution. Give an algorithm to compute an approximate solution accurate to 0.1 in the max norm on $C(0, 1)$.

2.14 Consider the Volterra operator of Example 2.11. Using $\|K^k\| \leq M^k(b - a)^k/(k - 1)!$, show that $\sum_{k=0}^{\infty} K^k$ converges in $\mathscr{L}(X)$. Deduce that the Volterra equation of second kind,

$$x(s) - \int_a^s k(s, t)x(t)\,dt = y(s),$$

has a unique solution x in $C(a, b)$ for each y in $C(a, b)$.

As another example of a bounded operator in a Hilbert space H, we now prove a variant of the Riesz representation theorem known as the *Lax–Milgram lemma.* It is a useful tool in the theory of linear partial differential equations of elliptic type. $\mathscr{Re}\, z$ denotes the real part of the complex number z.

Theorem 2.9 *Let $a(x, y)$ be a complex-valued form defined on the product Hilbert space $H \times H$ that satisfies the following conditions:*

sesquilinearity: $a(\cdot, \cdot)$ is linear in x and semilinear in y;
boundedness:

$$|a(x, y)| \leq \beta\|x\|\,\|y\| \qquad \text{for all} \quad x, y \text{ in } H; \tag{2.2}$$

coercivity:

$$\mathscr{Re}\, a(x, x) \geq \alpha\|x\|^2 \qquad \text{for all} \quad x \text{ in } H, \quad \alpha > 0. \tag{2.3}$$

Then there exists a uniquely determined bounded operator $S: H \to H$ with a bounded inverse S^{-1} such that $(x, y) = a(Sx, y)$ whenever x and $y \in H$ and $\|S\| \leq 1/\alpha$, $\|S^{-1}\| \leq \beta$.

Proof Let D be the set of all $x \in H$ for which there exists x^* such that $(x, y) = a(x^*, y)$ for all $y \in H$. D is not empty, because $0 \in D$. x^* is uniquely determined by x, because if z is such that $a(z, y) = 0$ for all y, then $z = 0$ by $0 = a(z, z) \geq \alpha\|z\|^2$. $x \mapsto x^*$ defines a linear operator S with domain $\operatorname{Dom} S = D$, $Sx = x^*$. If

$$\alpha\|Sx\|^2 \leq \mathcal{R}e\, a(Sx, Sx) \leq |(x, Sx)| \leq \|x\|\,\|Sx\|,$$

then

$$\|Sx\| \leq 1/\alpha\|x\| \qquad \text{for} \quad x \in D.$$

D is a closed subspace of H. Let $x_n \in D$ with $x = \lim_n x_n$. $\{Sx_n\}$ is a Cauchy sequence and has a limit $t = \lim_n Sx_n$. Then $a(Sx_n, y) = (x_n, y) \to (x, y)$, and also $a(Sx_n, y) \to a(t, y)$. Therefore $(x, y) = a(t, y)$, which proves that $x \in D$ and $Sx = t$.

Let us suppose that $D \neq X$. There exists $z_0 \in H$ such that $z_0 \neq 0$ and $z_0 \in D^\perp$. Consider the semilinear form $f[u] = a(z_0, u)$ defined on H. $|f[u]| = |a(z_0, u)| \leq \beta\|z_0\|\,\|u\|$. f is bounded on H. By Theorem 2.5, there exists $z'_0 \in H$ such that $a(z_0, u) = f[u] = (z'_0, u)$ for all $u \in H$. Then $z'_0 \in D$ and $Sz'_0 = z_0$. But

$$\alpha\|z_0\|^2 \leq \mathcal{R}e\, a(z_0, z_0) = \mathcal{R}e(z'_0, z_0) = 0,$$

and $z_0 = 0$, which contradicts $z_0 \neq 0$. $Sx = 0$ implies $(x, y) = a(Sx, y) = 0$ for all $y \in H$; that is, $y = 0$. Therefore S^{-1} exists. As above, for every $x \in H$, there exists x' such that $(x', y) = a(x, z)$ for all $z \in H$. Hence $x = Sx'$ and S^{-1} is defined on H. Now

$$|(S^{-1}x, z)| = |a(x, z)| \leq \beta\|x\|\,\|z\|$$

implies $\|S^{-1}\| \leq \beta$. □

4.2. Adjoint Operator

For all y in Y^*, $x \mapsto \langle y, Tx\rangle$ is a semilinear form bounded on X since

$$|\langle y, Tx\rangle| \leq \|y\|\,\|Tx\| \leq \|T\|\,\|x\|\,\|y\|.$$

There exists $z \in X^*$ such that $\langle y, Tx\rangle = \langle z, x\rangle$, and z depends on y. Therefore the linear map $y \mapsto z = T^*y$ defines the *adjoint operator* T^* from Y^* into X^*:

$$\langle y, Tx\rangle = \langle T^*y, x\rangle \qquad \text{for} \quad x \in X, \quad y \in Y^*.$$

The following properties have been proved by Banach:

$\|T^*\| = \|T\|$.
$\operatorname{Ker}(T^*) = (\operatorname{Im} T)^\perp$.
If Im T is closed, $\operatorname{Im}(T^*) = (\operatorname{Ker} T)^\perp$ (closed range theorem).
$(T^*)^*$ is an extension of T if X is identified with a subspace of X^{**}.

The proofs are in Kato (1976, pp. 24, 154, 155, and 234).

Example 2.12 Let $X = \mathbb{C}^m$ with the inner product $(x, y) = y^{\mathrm{H}}x$. We consider the bases $\{x_i\}_1^m$ and $\{y_i\}_1^m$, which form a biorthogonal family $(x_i, y_j) = \delta_{ij}$ for $i, j = 1, \ldots, m$. Let T be a linear operator on X, represented in the given bases by the matrix A of elements

$$a_{ij} = (Tx_j, y_i) = y_i^{\mathrm{H}} Tx_j, \qquad i, j = 1, \ldots, m.$$

The adjoint T^* is represented in the same bases by the conjugate transposed matrix A^{H}. Indeed

$$(T^*y_j, x_i) = (y_j, Tx_i) = (Tx_i)^{\mathrm{H}} y_j = \overline{y_j^{\mathrm{H}} Tx_i} = \bar{a}_{ji}.$$

Example 2.13 Let $X = L^p(a, b)$, with $1 \leq p < \infty$. T is the integral operator defined by

$$(Tx)(s) = \int_a^b k(s, t)x(t)\, dt,$$

where $k(s, t)$ is such that

$$\int_a^b |k(s, t)|\, dt < c, \quad s \in [a, b], \qquad \int_a^b |k(s, t)|\, ds < c', \quad t \in [a, b].$$

The adjoint T^* is bounded on $X^* = L^q(a, b)$, with $p^{-1} + q^{-1} = 1$. Actually T^* is an integral operator with kernel k^* defined by $k^*(s, t) = \overline{k(t, s)}$ for $s, t \in [a, b]$. This follows from

$$\langle T^*y, x\rangle = \langle y, Tx\rangle = \int_a^b y(s)\, ds \int_a^b \overline{k(s, t)x(t)}\, dt = \int_a^b \overline{x(t)}\, dt \int_a^b \overline{k(s, t)}y(s)\, ds.$$

Therefore

$$(T^*y)(t) = \int_a^b k^*(t, s)y(s)\, ds = \int_a^b \overline{k(s, t)}\, y(s)\, ds.$$

Example 2.14 (The adjoint relative to a coercive continuous sesquilinear form) Let the sesquilinear form $a(x, y)$ satisfy the conditions (2.2) and (2.3) of Theorem 2.9, and let $T \in \mathscr{L}(H)$. We define $T^\times$, the adjoint of T relative to a, by

$$a(Tx, y) = a(x, T^\times y) \qquad \text{for} \quad x, y \in H.$$

Now let S be the isomorphism defined by Lax–Milgram. Its adjoint S^* is such that $a^*(S^*x, y) = (x, y)$ for all x, y in H, where we have set $a^*(x, y) := \overline{a(y, x)}$. We show that $T^\times = (S^{-1}TS)^* = S^*T^*(S^{-1})^*$. On one hand

$$a(Tx, y) = (S^{-1}Tx, y) = (x, (S^{-1}T)^*y),$$

and on the other

$$a(x, T^\times y) = (S^{-1}x, T^\times y) = (x, (S^{-1})^*T^\times y).$$

Therefore $(S^{-1})^*T^\times = (S^{-1}T)^*$ and $T^\times = (S^{-1}TS)^*$. It is easily seen that $S^\times = S^*$. Note that $T^\times$ is the adjoint of T when the adjoint space H^* is identified with H by means of the isomorphism $I: H^* \to H$ defined by $a(x, Iy) = \overline{y[x]}$ for $x \in H$, $y \in H^*$: $T^\times = IT^*I^{-1}$. $T^\times$ is called the a adjoint of T in H.

Exercises

2.15 Let $T \in \mathscr{L}(X, Y)$, where dim X and dim Y are finite. Show that Im $T = (\mathrm{Ker}(T^*))^\perp$, Ker $T = (\mathrm{Im}(T^*))^\perp$, and $\mathrm{Im}(T^*) = (\mathrm{Ker}\ T)^\perp$. If dim X = dim Y, Im $T = Y$ iff T^* has an inverse.

2.16 If $T \in \mathscr{L}(H)$, $\|T^*T\| = \|T\|^2$.

In a Hilbert space H, $T \in \mathscr{L}(H)$ is *normal* iff $T^*T = TT^*$; it is *self-adjoint* iff $T^* = T$. When T is self-adjoint, (Tx, x) is real for $x \in H$. If $(Tx, x) \geq 0$ (resp. > 0) for $x \in H$ (resp. $0 \neq x \in H$), T is said to be *nonnegative* (resp. *positive*) *definite*.

Proposition 2.10 *If $T \in \mathscr{L}(H)$ is self-adjoint or normal, then $r_\sigma(T) = \|T\|$.*

Proof We begin by showing that $\|T^n\| = \|T\|^n$, for $n = 2^m, m = 1, 2, \ldots$. For that, we first suppose that T is self-adjoint. Using $\|T^*T\| = \|T\|^2$ (Exercise 2.16), we get $\|T^2\| = \|T\|^2$, and similarly $\|T^n\| = \|T\|^n$ for $n = 2^m$, $m = 1, 2, \ldots$. T being now normal, we have $\|T^n\|^2 = \|(T^n)^*T^n\|$, and $(T^n)^*T^n = (T^*T)^n$, where T^*T is self-adjoint. This proves $\|T^n\| = \|T\|^n$ for $n = 2^m$. Now

$$r_\sigma(T) = \lim_{n \in \mathbb{N}} \|T^n\|^{1/n} = \lim_{\substack{n = 2^m \\ m \in \mathbb{N}}} \|T^n\|^{1/n} = \|T\|. \quad \square$$

4.3. Compact Operators

Definitions $T \in \mathscr{L}(X, Y)$ is *compact* iff the image of any bounded set in X is a relatively compact set in Y. Equivalently $T \in \mathscr{L}(X, Y)$ is *compact* iff the image $\{Tx_n\}$ of any bounded sequence $\{x_n\}_\mathbb{N}$ of X has a convergent subsequence in Y.

Example 2.15 Let $T \in \mathscr{L}(X, Y)$ be such that $\dim(\operatorname{Im} T) < \infty$. T is compact by the Borel–Lebesgue theorem.

Example 2.16 The integral operator with continuous kernel k defined in Example 2.8 is compact by the Ascoli–Arzela theorem: Let B be the unit sphere of $C(a, b)$. For $x \in B$, $\|Tx\| \leq M(b - a)\|x\|$, so that TB is equibounded. By the Schwarz inequality,

$$|(Tx)(s_1) - (Tx)(s_2)|^2 \leq \left(\int_a^b |k(s_1, t) - k(s_2, t)|^2 \, ds\right)\left(\int_a^b |x(t)|^2 \, dt\right),$$

and TB is equicontinuous; that is,

$$\lim_{\delta \to 0} \sup_{|s_1 - s_2| \leq \delta} |(Tx)(s_1) - (Tx)(s_2)| = 0 \quad \text{uniformly in} \quad x \in B.$$

Therefore TB is relatively compact in $C(a, b)$.

Example 2.17 The injection of $C^1(a, b)$ in $C(a, b)$, the injection of $H^s(\Omega)$ in $H^{s'}(\Omega)$, $s' < s$, when Ω has a "smooth" boundary, are compact operators. This follows from the examples given in Section 3.2.

The following properties hold for compact operators:

(1) The set of all compact operators of $\mathscr{L}(X, Y)$ is a closed subspace of $\mathscr{L}(X, Y)$.

(2) The product of a compact operator by a bounded operator is a compact operator, regardless of the order in which the product is made.

(3) The adjoint of a compact operator is compact.

Proofs are in Kato (1976, pp. 158–159).

Exercise

2.17 Show that if $X = l^2$ and (a_{ij}) is an infinite matrix such that $\alpha := \sum_{i,j=1}^{\infty} |a_{ij}|^2 < \infty$, then the equations $y_i = \sum_{j=1}^{\infty} a_{ij} x_j$, $i = 1, 2, \ldots$, for $\sum_1^{\infty} |y_i|^2 < \infty$ define a compact linear operator T of X into itself with $y = Tx$ and $\|T\| \leq \alpha^{1/2}$. (*Hint*: Consider a diagonal subsequence.)

4.4. Bounded Operators of Finite Rank

$T \in \mathscr{L}(X, Y)$ is of *finite rank* iff rank $T := \dim(\operatorname{Im} T) = m < \infty$.

Suppose that $T \in \mathscr{L}(X)$ with $\dim(\operatorname{Im} T) = m$. $M = \operatorname{Im} T$ is invariant under T; that is, $TM \subset M$. Let $\{x_i\}_1^m$ be a basis of M, and $\{x_j^*\}_1^m$ be the adjoint basis in M^*. Let T_M denote the restriction of T to the subspace M:

$$x \in M \overset{T_M}{\mapsto} Tx \in M.$$

T_M is represented in the bases $\{x_i\}_1^m$ and $\{x_i^*\}_1^m$ by the matrix

$$t_{ij} = \langle Tx_j, x_i^* \rangle, \qquad i, j = 1, \ldots, m.$$

The *trace* of T is defined by the trace of T_M, that is, the trace of the matrix (t_{ij}):

$$\operatorname{tr} T := \sum_{i=1}^{m} \langle Tx_i, x_i^* \rangle.$$

Exercises

2.18 Show that if $T \in \mathscr{L}(X, Y)$ is of finite rank, then it is compact.

2.19 Let $\{T_n\}_{\mathbb{N}}$ be a sequence of finite-rank operators in $\mathscr{L}(X, Y)$ such that $\|T - T_n\| \to 0$. Show that T is compact but not necessarily of finite rank. (*Hint*: Consider a diagonal subsequence.)

2.20 Let $X = Y$, and let $A \in \mathscr{L}(X)$. Show that if $T \in \mathscr{L}(X)$ is of finite rank, then so are TA and AT. Deduce that $\operatorname{tr} TA = \operatorname{tr} AT$.

2.21 Show that $\operatorname{tr} T$ does not depend on the choice of the basis $\{x_i\}_1^m$ in M.

4.5. *Projections*

A bounded operator $P \in \mathscr{L}(X)$ is a *projection* iff $P^2 = P$. Associated with P, we have the decomposition

$$X = M \oplus N \qquad \text{with} \quad M := PX = \operatorname{Ker}(1 - P)$$

and

$$N := \operatorname{Ker} P = (1 - P)X.$$

P projects X on M along N, and $1 - P$ is a projection on N along M:

$$(1 - P)^2 = 1 - P - P + P^2 = 1 - P.$$

M and N are closed subspaces of X, being the null spaces of $1 - P$ and P, respectively. Conversely, the decomposition $X = M \oplus N$ into two closed subspaces defines a projection P on M along N. [The boundedness of P is a corollary of the closed graph theorem; see p. 93 and Kato (1976, p. 167).] Moreover, given a closed subspace M of X, there exists a projection from X onto M iff there is a closed subspace N such that $X = M \oplus N$.

Exercises

2.22 Show that a projection P is compact iff $\dim(\operatorname{Im} P) < \infty$.

2.23 Show that the adjoint P^* of a projection P on M along N is a projection in X^* such that $M^* := P^*X^* = \operatorname{Ker}(1 - P^*) = (\operatorname{Im}(1 - P))^\perp = N^\perp$ and $N^* := (1 - P^*)X^* = \operatorname{Ker}(P^*) = (\operatorname{Im} P)^\perp = M^\perp$.

2.24 Show that $(z1 - P)^{-1} = (1/z)1 + [1/z(z - 1)]P$ if $z(z - 1) \neq 0$.

Let M be a closed subspace of a Hilbert space H. The decomposition $H = M \oplus M^{\perp}$ defines the *orthogonal projection* P_M on M along $M^{\perp}$.

Proposition 2.11 *An orthogonal projection is self-adjoint and nonnegative definite.*

Proof For

$$x, y \in H, (P_M x, y) = (P_M x, P_M y + P_{M^{\perp}} y) = (P_M x, P_M y)$$
$$= (P_M x + P_{M^{\perp}} x, P_M y) = (x, P_M y).$$

And

$$(P_M x, x) = (P_M x, P_M x + P_{M^{\perp}} x) = (P_M x, P_M x)$$
$$= \|P_M x\|^2 \geq 0 \qquad \text{for} \quad x \in H. \quad \square$$

Exercises

2.25 Show that $\|(1 - P_M)x\| = \operatorname{dist}(x, M)$ where $\operatorname{dist}(x, M) := \min_{y \in M} \|x - y\|$ for any $x \in H$.

2.26 "Oblique" (i.e., nonorthogonal) projections may also be considered in a Hilbert space. Show that for such projections $\|P\| \geq 1$ if $P \neq 0$ and that $\|P\| = 1$ iff P is an orthogonal projection.

Example 2.18 (The projection relative to a coercive continuous sesquilinear form) Let $a(x, y)\colon H \times H \to \mathbb{C}$ be a sesquilinear form satisfying (2.2) and (2.3). Let V be a closed subspace of H. We set $W := \{x \in H; a(x, y) = 0$ for all $y \in V\}$. The coercivity of $a(\cdot, \cdot)$ on V implies that $V \cap W = \{0\} = V^{\perp} \cap W^{\perp}$. Therefore $H = V \oplus W$. The projection π^a on V along W is the projection on V *relative to* $a(\cdot, \cdot)$; it is also called the a projection (or elliptic projection) from H onto V. It is characterized by

$$a((1 - \pi^a)x, y) = 0 \qquad \text{for all} \quad y \text{ in } V \quad \text{and} \quad x \text{ in } H.$$

Let π_V be the orthogonal projection on V associated with $H = V \oplus V^{\perp}$. For any $x \in H$,

$$a((1 - \pi^a)x, (1 - \pi^a)x) = a((1 - \pi^a)x, (1 - \pi_V)x).$$

Therefore

$$\alpha\|(1 - \pi^a)x\|^2 \leq \mathcal{Re}\, a((1 - \pi^a)x, (1 - \pi^a)x) \leq |a((1 - \pi^a)x, (1 - \pi_V)x)|$$
$$\leq \beta\|(1 - \pi^a)x\| \, \|(1 - \pi_V)x\|.$$

This proves that

$$\|(1 - \pi^a)x\| \leq (\beta/\alpha) \operatorname{dist}(x, V) \qquad \text{for all} \quad x \text{ in } H.$$

Similarly, using $a(\pi^a x, \pi^a x) = a(x, \pi^a x)$ for x in H and using (2.2) and (2.3) on H, it is easily proved that $\|\pi^a\| \le \beta/\alpha$, a constant that does not depend on V. The adjoint $\pi^{a\times}$ relative to the form $a(\cdot, \cdot)$ is such that

$$a(x, (1 - \pi^{a\times})y) = 0 \qquad \text{for all} \quad x \text{ in } V, \quad y \text{ in } H.$$

If a is hermitian, $a(x, y) = \overline{a(y, x)}$ for all x, y in H, then due to its coercivity $a(x, x) \ge \alpha\|x\|^2$ on H, $a(\cdot, \cdot)$ defines an inner product $(\cdot, \cdot)_a$ on H. The associated norm is $\|x\|_a = \sqrt{a(x, x)}$ for x in H. It is clear that $\pi^a = \pi^{a\times}$: the a projection is the projection on V, orthogonal with respect to the inner product $(\cdot, \cdot)_a$:

$$\|(1 - \pi^a)x\|_a = \min_{y \in V} \|x - y\|_a \qquad \text{for all} \quad x \text{ in } H.$$

5. Pairs of Projections and the Gap between Subspaces

Let P, Q be a pair of projections on closed subspaces of a Banach space X or a Hilbert space H on $\mathbb{C}$. $M := PX$, $N := QX$. If P^* and Q^* are the adjoint projections, then

$$M^* := \operatorname{Im}(P^*) = [(1 - P)X]^\perp \qquad \text{and} \qquad N^* := \operatorname{Im}(Q^*) = [(1 - Q)X]^\perp.$$

5.1. Projections in a Hilbert Space

M and N are closed subspaces of the Hilbert space H.

Let P and Q be the *orthogonal* projections on M and N. The *gap* (or *aperture*) between M and N is

$$\Theta(M, N) := \max\left(\sup_{\substack{x \in M \\ \|x\| = 1}} \|(1 - Q)x\|, \ \sup_{\substack{x \in N \\ \|x\| = 1}} \|(1 - P)x\| \right).$$

We also have

$$\begin{aligned}\Theta(M, N) &= \max\left(\sup_{\substack{x \in M \\ \|x\| = 1}} \operatorname{dist}(x, N), \ \sup_{\substack{x \in N \\ \|x\| = 1}} \operatorname{dist}(x, M) \right) \\ &= \max(\|(P - Q)P\|, \|(P - Q)Q\|) = \|P - Q\|.\end{aligned}$$

We may interpret geometrically the meaning of $\Theta(M, N)$ in a finite-dimensional space:

$$\Theta(M, N) = \sin\theta,$$

where θ is the acute angle between the hyperplanes M and N, cf. Fig. 2.1.

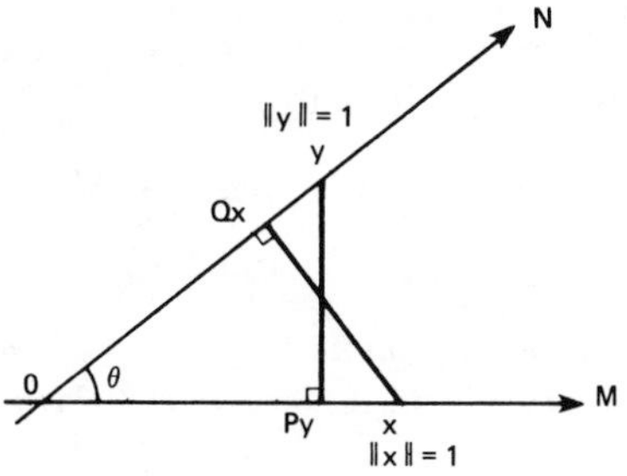

Figure 2.1

If $\|P - Q\| < 1$, then $\dim M = \dim N$ (Sz.-Nagy, 1946/47). It is also true for oblique projections. If P' and Q' are oblique projections such that $\operatorname{Im} P' = M$ and $\operatorname{Im} Q' = N$, then $\|P - Q\| \leq \|P' - Q'\|$ (Kato, 1976, p. 58).

5.2. *Projections in a Banach Space*

M and N are now closed subspaces of the Banach space X. Let

$$\delta(M, N) := \sup_{\substack{x \in M \\ \|x\| = 1}} \operatorname{dist}(x, N) \leq \|(P - Q)P\|.$$

$\delta(M, N)$ is the smallest number δ such that $\operatorname{dist}(x, N) \leq \delta\|x\|$, $x \in M$, $\delta \geq 0$. $\operatorname{dist}(x, N) \leq \|x\|$; hence $\delta \leq 1$. Note that $\delta(M, N) \neq \delta(N, M)$.

The *gap* between M and N is

$$\Theta(M, N) := \max(\delta(M, N), \delta(N, M)).$$

Lemma 2.12 *If* $\dim N < \infty$ *and* $\dim M > \dim N$, *then there exists an* $x \in M$ *such that* $\operatorname{dist}(x, N) = \|x\| > 0$; *hence* $\delta(M, N) = 1$.

Proof This lemma is nontrivial in a Banach space (cf. Kato, 1976, p. 199); in a Hilbert space, there is only the orthogonal projection of x on N to be considered. □

Proposition 2.13 $\Theta(M, N)$ *has the following properties*:

(i) $\Theta(M, N) \leq \max(\|(P - Q)P\|, \|(P - Q)Q\|)$.
(ii) $\delta(M, N) = \delta(M^*, N^*)$.

We suppose now that $\dim N < \infty$. *We then have the additional properties*:

(iii) $\delta(M, N) < 1$ *implies* $\dim M \leq \dim N$.
(iv) $\Theta(M, N) < 1$ *implies* $\dim M = \dim N$.
(v) $\dim M = \dim N$ *implies* $\delta(N, M) \leq \delta(M, N)/[1 - \delta(M, N)]$.

Proof (i) and (ii) are easy. (iii) is a consequence of Lemma 2.12. (iv) then follows. The proof of (v) is in Kato (1958, p. 265). □

Proposition 2.14 $\|P - Q\| < 1$ *implies* $\dim M = \dim N$.

For a proof see Kato (1976, pp. 33 and 156).

Corollary 2.15 *Let $P(t)$ be a family of projections depending continuously on a parameter t varying in a connected region of $\mathbb{R}$ or $\mathbb{C}$. Then* $\dim(P(t)X)$ *is constant.*

Proof $t \mapsto \dim(P(t)X)$ is a continuous function of t with values in $\mathbb{N}$. Hence it is constant. □

We suppose that X has two decompositions: $X = M \oplus N = M' \oplus N$, with $M \neq M'$.

Lemma 2.16 *Let P (resp. P') be the projection on M (resp. M') along N. If*

$$a := \|1 - P'\|\delta(M, M') < 1,$$

then

$$\|P - P'\| \leq \|P'\|a/(1 - a).$$

Proof Let

$$d := \sup_{\substack{x \in M \\ \|x\| = 1}} \|x - P'x\|.$$

We choose $\delta > \delta(M, M')$ and $x \in M$, $\|x\| = 1$. There exists $y \in M'$ such that $\|x - y\| < \delta$,

$$\|x - P'x\| = \|(1 - P')(x - y)\| < \|1 - P'\|\delta.$$

Thus

$$d \leq \|1 - P'\|\delta \qquad \text{for all} \quad \delta > \delta(M, M')$$

implies

$$d \leq \|1 - P'\|\delta(M, M') = a.$$

By hypothesis, $d \leq a < 1$. For all z in M, $\|z - P'z\| \leq d\|z\|$ and $P' = P'P$. Then for any t in X,

$$\|Pt\| \leq \|Pt - P'Pt\| + \|P'Pt\| \leq d\|Pt\| + \|P't\|,$$

and we get

$$\|Pt\| \leq \frac{1}{1 - d}\|P't\| \leq \frac{1}{1 - a}\|P't\|.$$

Therefore

$$\|Pt - P't\| = \|Pt - P'Pt\| \leq d\|Pt\| \leq \frac{a}{1-a}\|P't\| \leq \frac{a}{1-a}\|P'\|\,\|t\|. \qquad \square$$

6. Closed Linear Operators

It is essential for the application to differential equations to consider unbounded linear operators. An important class of them has a property that, in some respects, compensates for the lack of continuity. This property on the linear operator T from X into Y is expressed in terms of the *graph* of T, denoted $G(T)$, which is defined in $X \times Y$ by $G(T) := \{(x, Tx); x \in \operatorname{Dom} T\}$.

6.1. Definition

T is *closed* iff its graph $G(T)$ is a closed set in $X \times Y$, with the topology of the product space. We consider on $X \times Y$, the product norm $\|(x, y)\|_{X\times Y} = (\|x\|_X^2 + \|y\|_Y^2)^{1/2}$. Other choices are possible, but the euclidean norm allows the identification $(X \times Y)^* \equiv X^* \times Y^*$.

$\mathscr{C}(X, Y)$ is the set of closed operators from X into Y. If $X = Y$, $\mathscr{C}(X, X)$ is denoted $\mathscr{C}(X)$. $T \in \mathscr{C}(X, Y)$ iff for any sequence $x_n \in D$ such that $x_n \to x$ and $Tx_n \to y$, one has $x \in D$ and $Tx = y$.

Example 2.19 Let $X = C(0, 1)$. Let $D = C^1(0, 1)$ be the set of $x \in X$ such that the derivative x' is in $C(0, 1)$. Let T be the linear operator with domain D defined by $x \mapsto Tx = x'$ for $x \in D$. If $x(t) = t^n$, $x'(t) = nt^{n-1}$, so that $\|x\| = 1$ and $\|x'\| = n$. T is not bounded, but it is closed. Let $x_n \in D$ such that $x_n \to x$ and $Tx_n \to y$. That is, $x_n' \to y$. The pointwise convergence $x_n \to x$ and the uniform convergence $x_n' \to y$ prove that $x \in D$ and $x' = y$.

The following two properties are easy to prove:

(1) A bounded operator T with domain $D \subset X$ is closed iff D is closed. In particular $\mathscr{L}(X, Y) \subset \mathscr{C}(X, Y)$.

(2) If T^{-1} exists, T is closed iff T^{-1} is closed.

These two properties show that an unbounded operator whose inverse is in $\mathscr{L}(X)$ is itself closed. This is the case of many differential operators appearing in mathematical physics.

Example 2.20 Let $X = C(a, b)$, and let D be the set of $x \in X$ such that the second derivative x'' is in $C(a, b)$ and $x(a) = x(b) = 0$. Consider the differential operator T_1 with domain D defined by $x \in D \mapsto x''$. T_1 is the inverse of the integral operator T defined in Exercise 2.11: $T_1 T x = x$ for $x \in X$ and $T T_1 x = x$ for $x \in D$.

Example 2.21 Let $X = C(a, b)$, and let D be the set of $x \in X$ such that $x', x'' \in X$ and $x(a) = 1$, $x'(a) = 1$. We consider the *initial-value* problem

$$x''(s) + a_1(s)x'(s) + a_2(s)x(s) = y(s), \qquad a \le s \le b,$$
$$x'(a) = x(a) = 1,$$

and the associated differential operator

$$T : x \in D \mapsto x'' + a_1 x' + a_2 x.$$

We write $x''(s) = z(s)$, assuming that x is a solution. Then

$$\int_a^u z(t)\, dt = x'(u) - 1,$$

$$\int_a^s du \int_a^u z(t)\, dt = x(s) - 1 - (s - a).$$

The last integral takes the form $\int_a^s z(t)\,dt \int_t^s du = \int_a^s (s - t)z(t)\,dt$. Putting back x, x', x'' in terms of z in the equation gives

$$z(s) + a_1(s)\left(1 + \int_a^s z(t)\, dt\right) + a_2(s)\left(1 + s - a + \int_a^s (s - t)z(t)\, dt\right) = y(s).$$

We define $k(s, t) = -a_1(s) - a_2(s)(s - t)$, $w(s) = y(s) - a_2(s) - a_1(s) - a_2(s)(s - a)$. It yields $z(s) - \int_a^s k(s, t)z(t)\,dt = w(s)$. Then $z(s) = x''(s)$ is the solution of a Volterra integral equation of the second kind. The inverse T^{-1} is then easily seen to be bounded on X, $T^{-1} \in \mathscr{L}(X)$, and therefore $T \in \mathscr{C}(X)$.

As another fundamental example, we consider now the *boundary-value* problem for the linear ordinary differential equation of order p, $p \ge 2$:

$$[\mathfrak{L}(x)](t) := a_0(t)x^{(p)}(t) + a_1(t)x^{(p-1)}(t) + \cdots + a_p(t)x(t) = f(t),$$
$$a \le t \le b,$$

where the functions $a_0, a_1, \ldots, a_p, f$ are continuous on $[a, b]$, $a_0(t) \ne 0$ on $[a, b]$, and where the function x is subjected to the p homogeneous linear boundary conditions

$$\mathfrak{l}_k(x) := \sum_{i=0}^{p-1} [a_{ik} x^{(i)}(a) + b_{ik} x^{(i)}(b)] = 0, \qquad k = 1, \ldots, p,$$

and a_{ik} and b_{ik} are real or complex constants. The linear forms $\{\mathfrak{l}_k\}_1^p$ in $x(a), \ldots, x^{(p-1)}(a), x(b), \ldots, x^{(p-1)}(b)$ are supposed to be independent.

Let D be the set of all functions x in $L^\infty(a, b)$ such that $x \in C^{p-1}(a, b)$,

$x^{(p)} \in L^\infty(a, b)$ and $\mathfrak{l}_k(x) = 0$ for $k = 1, \ldots, p$. To the set

$$(\mathfrak{L}, \mathfrak{l}_k) = \begin{cases} \mathfrak{L}(x) \\ \mathfrak{l}_k(x) = 0, \qquad k = 1, \ldots, p, \end{cases}$$

is associated the differential operator T defined in $X = L^\infty(a, b)$, with domain D, by $x \in D \overset{T}{\mapsto} Tx = \mathfrak{L}(x)$.

Definition A *Green function* of the set $(\mathfrak{L}, \mathfrak{l}_k)$ is a function $g(t, s)$ defined for $a \leq s, t \leq b$ and such that

(i) $g(t, s)$ is continuous and has continuous partial derivatives in t up to order $p - 2$, for $a \leq s \leq b$;

(ii) $(\partial^{p-1}/\partial t^{p-1})g(t, s)$ has a discontinuity at $t = s$ such that

$$\frac{\partial^{p-1}}{\partial t^{p-1}} g(t, s)\bigg|_{t=s^+} - \frac{\partial^{p-1}}{\partial t^{p-1}} g(t, s)\bigg|_{t=s^-} = \frac{1}{a_0(s)};$$

(iii) in each interval $[a, s[$ and $]s, b]$ the function $t \mapsto g_s(t) = g(t, s)$ is a solution of the equation $\mathfrak{L}(g_s) = 0$.

(iv) $g(t, s)$ satisfies the boundary conditions $\mathfrak{l}_k(g_s) = 0$, $k = 1, \ldots, p$, for any fixed s, $a \leq s \leq b$.

We define

$$(P_0) = \begin{cases} \mathfrak{L}(x) = 0 \\ \mathfrak{l}_k(x) = 0, \qquad k = 1, \ldots, p, \end{cases}$$

to be the homogeneous problem associated with $(\mathfrak{L}, \mathfrak{l}_k)$.

Proposition 2.17 *If (P_0) has only the solution $x = 0$, there exists a unique Green function $g(t, s)$.*

Proof Let $x_1, \ldots, x_p$ be the p linearly independent solutions of $\mathfrak{L}(x) = 0$. Because of (iii), $g(t, s)$ should satisfy for a fixed s, $a \leq s \leq b$,

$$g_s(t) = \sum_{i=1}^{p} \alpha_i x_i(t), \qquad a \leq t < s,$$

$$g_s(t) = \sum_{i=1}^{p} \beta_i x_i(t), \qquad s < t \leq b,$$

where α_i and β_i are functions of s. Because of the continuity at $t = s$ of $g_s(t)$ and its $p - 2$ first derivatives and because of (ii), we get

$$\sum_{i=1}^{p} (\beta_i(s) - \alpha_i(s))x_i^{(j)}(s) = 0, \qquad j = 0, \ldots, p - 2,$$

$$\sum_{i=1}^{p} (\beta_i(s) - \alpha_i(s))x_i^{(p-1)}(s) = \frac{1}{a_0(s)}.$$

We set $\gamma_i(s) := \beta_i(s) - \alpha_i(s)$, $i = 1, \ldots, p$. It yields the $p \times p$ system

$$\sum_{i=1}^{p} \gamma_i(s) x_i^{(j)}(s) = 0, \qquad j = 0, \ldots, p-2,$$

$$\sum_{i=1}^{p} \gamma_i(s) x_i^{(p-1)}(s) = \frac{1}{a_0(s)}.$$

Its determinant is the value at $t = s$ of the wronskian $W(x_1, \ldots, x_p)$, which is nonzero. Therefore the p functions $\{\gamma_i\}_1^p$ are uniquely determined by the above linear system, for each s, $a \le s \le b$. We now determine $\beta_i(s)$ and $\alpha_i(s) = \beta_i(s) - \gamma_i(s)$ from the boundary conditions (iv). It is easily seen that

$$\sum_{i=1}^{p} \beta_i \mathfrak{l}_k(x_i) = \sum_{i=1}^{p} \gamma_i \left(\sum_{l=0}^{p-1} a_{lk} x_i^{(l)}(a) \right), \qquad k = 1, \ldots, p.$$

The determinant of this system is $\det(\mathfrak{l}_k(x_i))$, $i, k = 1, \ldots, p$, which is nonzero because of the independence of the forms $\{\mathfrak{l}_k\}_1^p$. The function $g(t, s)$ is then uniquely determined. □

Exercises

2.27 Consider the set $x''(t)$ and $x(0) = x(1) = 0$ on $[0, 1]$. Show that the Green function is

$$g(t, s) = \begin{cases} (s-1)t, & t \le s \\ (t-1)s, & t > s. \end{cases}$$

2.28 For $x''(t)$ and $x(0) = x'(1)$ on $[0, 1]$, $g(t, s)$ is not uniquely defined. Show, for example, that if $\beta_2(s) = 0$,

$$g(t, s) = \begin{cases} -t, & t \le s \\ -s, & t > s. \end{cases}$$

2.29 Show that for $x''(t)$, $x(0) = x'(0)$, $x(1) = -x'(1)$ on $[0, 1]$,

$$g(t, s) = \begin{cases} \frac{1}{3}(t+1)(s-2), & t \le s \\ \frac{1}{3}(s+1)(t-2), & t > s. \end{cases}$$

The importance of the Green function may be fully appreciated with the resolution of the boundary-value problem

$$(P) = [\mathfrak{L}(x) = f, \qquad \mathfrak{l}_k(x) = 0, \qquad k = 1, \ldots, p].$$

Theorem 2.18 *If $g(t, s)$ is a Green function associated with the boundary-value problem (P), its solution x is given by*

$$x(t) = \int_a^b g(t, s) f(s)\, ds, \qquad a \le t \le b.$$

The proof is left to the reader.

Theorem 2.18 establishes that, under the conditions of Proposition 2.17, the differential operator T has an inverse $T^{-1} = A$. A is a Fredholm integral operator whose kernel is the Green function. A is bounded in $C(a, b)$; therefore T is closed.

In Example 2.21, the Green function of an initial-value problem has been given. In the case of partial differential equations with homogeneous boundary conditions, the Green function can again be introduced as the kernel of an equivalent integral equation [cf. Courant and Hilbert (1953, Vol. 1, pp. 363–394) or Garabedian (1967)].

As we have already seen, a bounded operator with domain X is closed. The *closed graph* theorem, which follows, is a converse.

Theorem of the Closed Graph *A closed operator from X to Y with domain X is bounded.*

For a proof see Kato (1976, p. 166).

Exercise

2.30 Let $T \in \mathscr{C}(X, Y)$ with Im $T = Y$. If T^{-1} exists, then $T^{-1} \in \mathscr{L}(Y, X)$.

6.2. *The Graph Norm*

Let $T \in \mathscr{C}(X, Y)$ with domain D. When D is equipped with the graph norm $\|x\|_D := \|x\|_X + \|Tx\|_Y$, for $x \in D$, it becomes a Banach space $\hat{D}$, due to the closedness of T. Hence T is a bounded operator from $\hat{D}$ into X with $\|T\|_{\hat{D}\to X} < 1$. Depending on the point of view, a differential operator may then be regarded as an unbounded operator with domain D or as a bounded operator defined on $\hat{D}$.

Exercise

2.31 Consider the differential operator defined by

$$(Tx)(t) = x''(t) + tx'(t) + t^2x(t), \qquad 0 \leq t \leq 1,$$

with domain $D = C^2(0, 1)$ and $X = C(0, 1)$. Check that T is bounded from $\hat{D}$ into X.

6.3. *Adjoint of a Densely Defined Operator*

Let T be a linear operator from X into Y, and let S be an operator from Y^* into X^*. T and S are said to be *adjoint* to each other if

$$\langle y, Tx\rangle = \langle Sy, x\rangle \quad \text{for} \quad x \in \text{Dom } T \quad \text{and} \quad y \in \text{Dom } S.$$

For a T, there can be, in general, many operators from Y^* into X^* that are adjoint to T. If $\overline{\text{Dom } T} = X$, there is a unique maximal operator T^* adjoint to T, which can be constructed as follows: $\text{Dom}(T^*) = \{y \in Y^*; \exists z \in X^*$ such that $\langle y, Tx \rangle = \langle z, x \rangle$ for $x \in \text{Dom } T\}$. For a given y, $z \in X^*$ is unique because $\langle z, x \rangle = \langle z', x \rangle$ for $x \in \text{Dom } T$ implies $z = z'$, since $\text{Dom } T$ is dense in X. An operator T^* from Y^* into X^* can be defined by setting $T^*y = z$. The operator T^* is called the *adjoint of* T, which we define *only* if $\overline{\text{Dom } T} = X$.

The following properties can be proved (cf. Kato, 1976, pp. 168–169 and 234):

T^* is closed even if T is not.

$\text{Ker } T^* = (\text{Im } T)^\perp$.

Let $T \in \mathscr{C}(X, Y)$. If T^{-1} exists and $T^{-1} \in \mathscr{L}(Y, X)$, then $(T^*)^{-1}$ exists and belongs to $\mathscr{L}(X^*, Y^*)$, with $(T^*)^{-1} = (T^{-1})^*$. The converse is true.

$\text{Im } T$ is closed in $Y \Leftrightarrow \text{Im } T = (\text{Ker } T^*)^\perp \Leftrightarrow \text{Im } T^* = (\text{Ker } T)^\perp$ (closed-range theorem of Banach).

Exercises

2.32 Let $X := L^2(0, 1)$, $D := \{x \in X;\ x$ is absolutely continuous on $[0, 1]$, $dx/dt \in X$, $x(0) = 0\}$. The operator defined by $x \overset{T}{\mapsto} dx/dt$ for $x \in D$ is closed in X. Prove that the adjoint T^* is defined by:

$$\text{Dom}(T^*) = \left\{x \in X,\ x \text{ is absolutely continuous}, \frac{dx}{dt} \in X,\ x(1) = 0\right\} \quad \text{and} \quad T^*x = -\frac{dx}{dt}.$$

2.33 Let $X = L^2(0, 1)$, $D := \{x \in X; x'$ is absolutely continuous, $x'' \in X$, $x(0) = x(1) = 0\}$. Show that the operator $T: x \in D \mapsto -x'' + 2x' - x$ has for adjoint $T^*: y \in D \mapsto -y'' - 2y' - y$.

2.34 Let $\Omega =]0,1[\times]0,1[$. Show that the following boundary-value problems in Ω correspond to operators that are adjoint to each other in appropriate spaces:

$$(P) = \begin{cases} -\Delta u = f & \text{for } (x, y) \in \Omega, \\ \left.\begin{array}{l} u(1, y) = u(0, y) \\ u(x, 1) = u(x, 0) = 0 \\ \dfrac{\partial u}{\partial x}(0, y) = 0 \end{array}\right\} & \text{for } 0 \le x, y \le 1; \end{cases}$$

$$(P^*) = \begin{cases} -\Delta v = g & \text{for } (x, y) \in \Omega, \\ \left.\begin{array}{l} v(1, y) = 0 \\ v(x, 0) = v(x, 1) = 0 \\ \dfrac{\partial v}{\partial x}(0, y) = \dfrac{\partial v}{\partial x}(1, y) \end{array}\right\} & \text{for } 0 \le x, y \le 1. \end{cases}$$

B. INTRODUCTION TO SPECTRAL THEORY

Throughout this part, X (resp. H) is a complex Banach (resp. Hilbert) space. T is either in $\mathscr{C}(X)$ or $\mathscr{L}(X)$.

The eigenvalue problem considered in Chapter 1 in the finite-dimensional case requires essential modification when we deal with operators T in a Banach space. The notion of characteristic polynomial has no meaning anymore. Nevertheless, part of the results of the finite-dimensional case can be generalized, as we shall see. The central role is played by the operator $(T - z1)^{-1}$ whenever it exists in $\mathscr{L}(X)$ for $z \in \mathbb{C}$, and by the behavior of the function $z \mapsto (T - z1)^{-1}$ with values in the Banach space $\mathscr{L}(X)$.

7. Resolvent and Spectrum

If $T \in \mathscr{C}(X)$, then for any $z \in \mathbb{C}$, $T - z \in \mathscr{C}(X)$, where z stands for $z1$. We define

the *resolvent set* $\rho(T) := \{z \in \mathbb{C}; (T - z)^{-1} \in \mathscr{L}(X)\}$;

the *resolvent operator* $R(T, z) := (T - z)^{-1}$ for $z \in \rho(T)$; $R(T, z)$ has domain X and range Dom T for any $z \in \rho(T)$;

the *spectrum* of T: $\sigma(T)$ is the complementary set in $\mathbb{C}$ of $\rho(T)$.

For any z in $\rho(T)$ and any f in X, we remark that the equation $(T - z)x = f$ has a unique solution $x = R(T, z)f$; this fact has given its name to $R(T, z)$, in short the resolvent. When there is no ambiguity, $R(T, z)$ is simply denoted $R(z)$. Spectral theory is, roughly speaking, the study of the relations between the operators T and $R(z)$, the sets $\rho(T)$ and $\sigma(T)$, and other related operators and subspaces to be defined later.

7.1. *The Resolvent Operator*

$R(z)$ satisfies the two following identities, often referred to as the *first* and *second resolvent equations* (Hilbert).

$$(1)\quad R(z_1) - R(z_2) = (z_1 - z_2)R(z_1)R(z_2) = (z_1 - z_2)R(z_2)R(z_1) \qquad (2.4)$$

for z_1, z_2 in $\rho(T)$,

$$(2)\quad \begin{aligned} R(T_1, z) - R(T_2, z) &= R(T_1, z)(T_2 - T_1)R(T_2, z) \\ &= R(T_2, z)(T_2 - T_1)R(T_1, z) \end{aligned} \qquad (2.5)$$

for z in $\rho(T_1) \cap \rho(T_2)$, assuming that Dom T_1 = Dom T_2.

To prove (2.4) we write $z_1 - z_2 = T - z_2 - (T - z_1)$, and to prove (2.5) we use $T_2 - T_1 = T_2 - z - (T_1 - z)$.

Proposition 2.19 *$\rho(T)$ is an open set of $\mathbb{C}$ and $\sigma(T)$ is closed in $\mathbb{C}$; $z \mapsto R(z)$ is holomorphic for $z \in \rho(T)$ and $\|R(z)\| \geq 1/\text{dist}(z, \sigma(T))$.*

Proof Let $z_0 \in \rho(T)$. By Theorem 2.8, $|z - z_0| < \|R(z_0)\|^{-1}$ is a sufficient condition for the existence of

$$L(z) := [1 - (z - z_0)R(z_0)]^{-1} = \sum_{k=0}^{\infty} (z - z_0)^k R(z_0)^k.$$

Now

$$T - z = T - z_0 + z_0 - z = (T - z_0)[1 - (z - z_0)R(z_0)]$$

implies that $L(z)R(z_0) = R(z)$. Hence $z \in \rho(T)$ and

$$R(z) = \sum_{k=0}^{\infty} (z - z_0)^k R(z_0)^{k+1}.$$

$R(z)$ has a converging Taylor expansion in the neighborhood of z_0. We say that the function $z \mapsto R(z)$ with values in $\mathscr{L}(X)$ is holomorphic in z for z in $\rho(T)$ (cf. Section 7.3). $\rho(T)$ is an open set of $\mathbb{C}$, and $\sigma(T)$ is closed. $\rho(T)$ contains the set $\{z; |z_0 - z| < \|R(z_0)\|^{-1}\}$; therefore

$$\text{dist}(z_0, \sigma(T)) \geq \|R(z_0)\|^{-1}. \quad \square$$

Proposition 2.20 *$z \mapsto r_\sigma(R(z))$ is upper semicontinuous in $\rho(T)$.*

Proof The function $z \mapsto \|R^n(z)\|^{1/n}$ is continuous in $\rho(T)$ for $n \in \mathbb{N}$. By Theorem 2.7, for any $\varepsilon > 0$ and any z in $\rho(T)$, there exists $\nu \in \mathbb{N}$ such that $\|R^\nu(z)\|^{1/\nu} \leq r_\sigma[R(z)] + \varepsilon/2$ and there exists $\alpha > 0$ such that

$$|z' - z| < \alpha \Rightarrow \|R^\nu(z')\|^{1/\nu} \leq \|R^\nu(z)\|^{1/\nu} + \varepsilon/2 \leq r_\sigma[R(z)] + \varepsilon.$$

Since $r_\sigma[R(z')] = \inf_{n \geq 1} \|R^n(z')\|^{1/n}$, we get

$$|z' - z| < \alpha \Rightarrow r_\sigma[R(z')] \leq r_\sigma[R(z)] + \varepsilon. \quad \square$$

7.2. Subdivisions of the Spectrum

The spectrum is divided into three mutually exclusive parts, $P\sigma(T)$, $C\sigma(T)$, and $R\sigma(T)$, with the following definitions:

(1) $P\sigma(T) := \{z \in \sigma(T); T - z$ has no inverse$\}$ is the *point spectrum* of T.

(2) $C\sigma(T) := \{z \in \sigma(T); T - z$ has an (unbounded) inverse with domain dense in $X\}$ is the *continuous spectrum* of T.

(3) $R\sigma(T) := \{z \in \sigma(T);\ T - z$ has an inverse (bounded or not) whose domain is not dense in $X\}$ is the *residual spectrum* of T.

A necessary and sufficient condition for λ to belong to $P\sigma(T)$ is the the equation $Tx = \lambda x$ has a nonzero solution $x \in \operatorname{Dom} T$. In this case λ is called an *eigenvalue* and x is a corresponding *eigenvector*. The null space $\operatorname{Ker}(T - \lambda)$ is the *eigenspace* of T corresponding to the eigenvalue λ. Its dimension g is the *geometric multiplicity* of λ.

Example 2.22 If X is finite dimensional, any linear operator T is represented by a matrix A. $P\sigma(T)$ is the set of eigenvalues of A, whereas $C\sigma(T)$ and $R\sigma(T)$ are empty.

Example 2.23 In $X = C(a, b)$, let $D_1 := \{x \in X;\ x' \in X\}$; then the operator $T_1 : x \in D_1 \mapsto x'$ has for spectrum $\sigma(T_1) = P\sigma(T_1) = \mathbb{C}$. The equation $(T - z)x = 0$ has the solution $t \mapsto x(t) = e^{zt}$, which belongs to D_1 for all z in $\mathbb{C}$. Let $D_2 := \{x \in D_1; x(a) = 0\}$; then the operator $T_2 : x \in D_2 \mapsto x'$ has an empty spectrum. The resolvent is in $\mathscr{L}(X)$ for every z in $\mathbb{C}$ and is given by

$$R_2(z)x(t) = e^{zt} \int_a^t e^{-zs} x(s)\, ds.$$

Let $D_3 := \{x \in D_1; x(b) = kx(a),\ k = \text{const} \neq 0\}$; then the operator T_3: $x \in D_3 \mapsto x'$ has for spectrum

$$\sigma(T_3) = P\sigma(T_3) = \left\{\lambda_n = \frac{1}{b - a}(\log k + 2in\pi),\ n \in \mathbb{Z}\right\}.$$

For $z \neq \lambda_n$ the resolvent is in $\mathscr{L}(X)$ and is given by

$$R_3(z)x(t) = \frac{e^{zt}}{k - e^{(b-a)z}} \left[k \int_a^t e^{-zs} x(s)\, ds + e^{(b-a)z} \int_t^b e^{-zs} x(s)\, ds \right].$$

The spectrum of the differential operator $x \mapsto x'$ depends heavily on the boundary conditions.

Example 2.24 In $X = L^2(0, 1)$, the multiplication operator $T: x(t) \mapsto (\cos \pi t)x(t)$ has for spectrum $\sigma(T) = C\sigma(T) = [-1, 1]$. We consider the equation $(T - z)x(t) = (\cos \pi t - z)x(t) = 0$. For z complex or z real such that $|z| > 1$, $x(t) = 0$ and $(T - z)^{-1}$ is bounded. For z real, $|z| \leq 1$, $x(t) = 0$ if $t \neq (1/\pi) \arccos z$. $(T - z)^{-1}$ exists and by solving $(\cos \pi t - z)x = y$, one sees that $\operatorname{Dom}[(T - z)^{-1}]$ is the set of $y \in X$ that vanish in a neighborhood of $t = (1/\pi) \arccos z$; the neighborhood may vary with y. Hence $\operatorname{Dom}[(T - z)^{-1}]$ is dense in X, and $C\sigma(T) = [-1, 1]$.

Example 2.25 In $X = l^2$, the left-shift operator T,

$$(x_1, x_2, \ldots, x_k, x_{k+1}, \ldots) \mapsto (x_2, x_3, \ldots, x_{k+1}, x_{k+2}, \ldots),$$

has for spectrum $\sigma(T) = P\sigma(T) \cup C\sigma(T) = \{z;\ |z| \leq 1\}$ with $P\sigma(T) = \{z; |z| < 1\}$ and $C\sigma(T) = \{z; |z| = 1\}$. $(T - z)x = y$ means $x_{k+1} - zx_k = y_k$, $k = 1, 2, \ldots$. By induction,

$$x_{k+1} = z^k x_1 + z^{k-1} y_1 + \cdots + y_k$$

and

$$x_1 = z^{-k} x_{k+1} - (z^{-1} y_1 + \cdots + z^{-k} y_k).$$

$z^{-k} x_{k+1} \to 0$ as $k \to \infty$ if $|z| \geq 1$. This allows to find x in terms of y and shows that $(T - z)^{-1}$ is bounded if $|z| > 1$. Therefore $\{z; |z| > 1\} \subset \rho(T)$. For $|\lambda| < 1$, λ is an eigenvalue, with associated eigenvector $(1, \lambda, \lambda^2, \ldots, \lambda^k, \ldots)$. If $|\lambda| = 1$, $(T - \lambda)^{-1}$ exists because $(1, \lambda, \lambda^2, \ldots)$ is not an element of l^2. Because $\sigma(T)$ is closed, we conclude that $\sigma(T) = \{z; |z| \leq 1\}$. When $|z| = 1$, $\text{Dom}(T - z)^{-1} = \text{Im}(T - z)$ is dense, for it contains y if the number of nonzero components of y is finite.

Example 2.26 In $X = l^2$, the right-shift operator T',

$$(x_1, x_2, \ldots, x_k, x_{k+1}, \ldots) \mapsto (0, x_1, \ldots, x_{k-1}, x_k, \ldots),$$

has the same spectrum $\sigma(T') = \{z;\ |z| \leq 1\}$. (Note that T' is the adjoint of T defined in Example 2.25.) But this time $R\sigma(T') = \{z; |z| < 1\}$ and $C\sigma(T') = \{z; |z| = 1\}$. The equation $T'x - zx = y$ yields $-zx_1 = y_1, x_{k-1} - zx_k = y_k$, $k \geq 2$. When $z = 0$, T'^{-1} exists and is bounded on a domain that is not dense: $T'x = y$ implies $y_1 = 0$; therefore $0 \in R\sigma(T')$. When $z \neq 0$, $x = (T' - z)^{-1} y$ exists and is given by

$$x_k = z^{-k} y_1 + z^{-k+1} y_2 + \cdots + z^{-1} y_k.$$

The resolvent is bounded when $|z| > 1$, and $\{z; |z| > 1\} \subset \rho(T')$. We leave it to the reader to check that for $|z| < 1$, $\text{Dom}(T' - z)^{-1}$ is not dense in X, and for $|z| = 1$, $\text{Dom}(T' - z)^{-1}$ is dense (cf. Taylor, 1958, p. 266).

In the canonical basis $\{e_i\}_{\mathbb{N}}$ of l^2, T and T' are represented respectively by the infinite matrices

$$A = \begin{pmatrix} 0 & 1 & 0 & 0 & \cdots \\ 0 & 0 & 1 & 0 & \cdots \\ 0 & 0 & 0 & 1 & \cdots \\ 0 & 0 & 0 & 0 & \cdots \\ \vdots & \vdots & \vdots & \vdots & \ddots \end{pmatrix}$$

and

$$A' = \begin{pmatrix} 0 & 0 & 0 & 0 & \cdots \\ 1 & 0 & 0 & 0 & \cdots \\ 0 & 1 & 0 & 0 & \cdots \\ 0 & 0 & 1 & 0 & \cdots \\ \vdots & \vdots & \vdots & \vdots & \ddots \end{pmatrix}.$$

Note that the upper left corner square matrices of order n, $n = 2, 3, \ldots$, are *nilpotent*.

Exercises

2.35 Show that the characterization of $\sigma(T)$ in Example 2.25 is still valid in l^p, $1 \le p < \infty$. In l^∞, show that $P\sigma(T) = \{z; |z| \le 1\} = \sigma(T)$.

2.36 Use Exercise 2.24 to analyze the spectrum of a projection P, $P \neq 1$ or 0.

7.3. *Functions of a Complex Variable with Values in a Banach Space*

The key fact, when T is closed, is that $R(z)$ depends analytically on z as z varies in $\rho(T)$. We shall then use the theory of analytic functions of a complex variable with value in a complex Banach space (namely, X or $\mathscr{L}(X)$). Most results of classical theory can be taken over, except that norms replace absolute values. We list below the most important properties that we shall use. For a more detailed treatment, the reader is referred to Dunford and Schwartz (1958, Part 1, pp. 224–232), or to Dieudonné (1960, Chapter IX).

Let X denote a complex Banach space. Let G be an open set of $\mathbb{C}$. A function f defined in G, with values in X is said to be *analytic* (or *holomorphic*) in G if f is continuous and df/dz exists at each point of G.

Let Γ be a closed Jordan curve (i.e., a rectifiable simple curve), positively oriented and enclosing λ. We suppose that f is analytic in a neighborhood of λ containing Γ. Then the Cauchy integral formula is valid; namely,

$$f(\lambda) = \frac{1}{2i\pi} \int_\Gamma \frac{f(z)}{z - \lambda}\, dz.$$

By differentiation,

$$f^{(k)}(\lambda) = \frac{k!}{2i\pi} \int_\Gamma \frac{f(z)}{(z - \lambda)^{k+1}}\, dz.$$

For such a function, the *Taylor expansion*

$$f(z) = \sum_{k=0}^{\infty} \frac{f^{(k)}(z_0)}{k!} (z - z_0)^k$$

about z_0 enclosed by Γ is valid and the series converges absolutely and uniformly for z inside any circle centered at z_0 which lies inside Γ.

Conversely, any power series $f(z) = \sum_{k=0}^{\infty} a_k(z - z_0)^k$ defines an analytic function in the open set $\{z; |z - z_0| < r\}$, where

$$r = \left(\limsup_{k\to\infty} \|a_k\|^{1/k}\right)^{-1}.$$

The series converges absolutely and uniformly in z on any set $\{z; |z - z_0| \leq \alpha\}$ with $\alpha < r$. Furthermore, the series is uniquely defined by f; that is,

$$a_k = \frac{f^{(k)}(z_0)}{k!}, \qquad k = 0, 1, 2, \ldots.$$

We suppose now that f is analytic in the annulus $\{z; \alpha < |z - z_0| < \beta\}$. It has a unique *Laurent expansion*

$$f(z) = \sum_{k=-\infty}^{+\infty} a_k(z - z_0)^k,$$

which converges absolutely and uniformly for z in $\{z; \alpha + \varepsilon < |z - z_0| < \beta - \varepsilon\}$ with $\varepsilon > 0$. The coefficients a_k are given by the formula

$$a_k = \frac{1}{2i\pi} \int_\Gamma \frac{f(z)}{(z - z_0)^{k+1}}\, dz, \qquad k \in \mathbb{Z},$$

where Γ is any closed positively oriented Jordan curve in the annulus $\alpha < |z - z_0| < \beta$ that separates the two boundary circles.

If f is analytic in $\{z; 0 < |z - z_0| < \beta\}$, but not analytic in $\{z; |z - z_0| < \beta\}$, z_0 is an *isolated singularity* of f. The Laurent expansion that converges in $\{z; 0 < |z - z_0| < \beta\}$ is the Laurent expansion of f about z_0. If an infinite number of coefficients a_k with $k < 0$ are nonzero, z_0 is an *essential singularity* of f. If only a finite number of a_k with $k < 0$ are nonzero, z_0 is a *pole* of f. The largest integer l such that $a_{-l} \neq 0$ is the *order* of the pole.

The maximum modulus principle and Liouville's theorem are valid for vector-valued analytic functions. Let f be analytic in the open disk $\{z; |z - z_0| < \beta\}$. Let $r < \beta$. The coefficients a_k of the expansion

$$f(z) = \sum_{k=0}^{\infty} a_k z^k$$

can be bounded by the *Cauchy inequalities*

$$\|a_k\| \leq M r^{-k}, \qquad k \geq 0,$$

if

$$\|f(z)\| \leq M \quad \text{for} \quad |z| = r.$$

7.4. *The Spectrum of a Bounded Operator*

The main feature of the spectrum of $T \in \mathscr{L}(X)$ is that it is bounded and nonempty. We recall that $\lim_{k\to\infty} \|T^k\|^{1/k}$ exists and is equal to the spectral radius $r_\sigma(T)$ (Theorem 2.7.).

Theorem 2.21 *Let T be bounded. If $|z| > r_\sigma(T)$, the resolvent $R(z)$ exists and is given by*

$$R(z) = -\sum_{k=0}^{\infty} z^{-k-1}T^k, \tag{2.6}$$

which converges in $\mathscr{L}(X)$.

Proof $\|T^k\|^{1/k} \to r_\sigma(T)$. If $|z| \geq r_\sigma(T) + \varepsilon$, with $\varepsilon > 0$, then

$$|z|^{-1}\|T^k\|^{1/k} \leq (r_\sigma(T) + \varepsilon)^{-1}(r_\sigma(T) + \varepsilon/2)$$

and

$$\|(z^{-1}T)^k\| \leq \left(\frac{r_\sigma(T) + \varepsilon/2}{r_\sigma(T) + \varepsilon}\right)^k$$

for n large enough. The series $z^{-1}\sum_{k=0}^{\infty}(z^{-1}T)^k$ is then convergent in the norm of operators when $|z| > r_\sigma(T)$. Multiplication by $T - z$ on the left and right of the series gives 1, so that the series is actually the resolvent $R(z)$. □

Identity (2.6) is a Taylor expansion about $z = \infty$: $R(z) = -z^{-1}\sum_{k=0}^{\infty}(z^{-1}T)^k$. The standard formula for the radius of convergence of a power series tells us that the radius of convergence of (2.6) is

$$\limsup_k \|T^k\|^{1/k} = r_\sigma(T).$$

Then (2.6) diverges for $|z| < r_\sigma(T)$.

Corollary 2.22 *For a bounded operator T, $\rho(T)$ and $\sigma(T)$ are nonempty and $\sigma(T)$ is compact.*

Proof Clearly $\rho(T) \supset \{z; |z| > r_\sigma(T)\}$ and $\sigma(T) \subset \{z; |z| \leq r_\sigma(T)\}$. This last set reduces to the single point $z = 0$ if T is quasi-nilpotent ($r_\sigma(T) = 0$). $\rho(T)$ is nonempty and $\sigma(T)$ is bounded since $r_\sigma(T) \leq \|T\| < \infty$; therefore $\sigma(T)$ is a compact subset of $\mathbb{C}$. It follows from Theorem 2.8 that $\|R(z)\| \leq 1/(|z| - \|T\|)$ if $|z| > \|T\|$; therefore $\|R(z)\| \to 0$ when $|z| \to \infty$. If $\sigma(T)$ were empty, it would follow that $R(z)$ would be analytic and bounded on the whole plane. Then it would be a constant by Liouville's theorem, the constant being the zero operator. This gives the contradiction $1 = (T - z)R(z) = 0$. □

Corollary 2.23 *For $T \in \mathscr{L}(X)$, $r_\sigma(T) = \max_{\lambda\in\sigma(T)}|\lambda|$.*

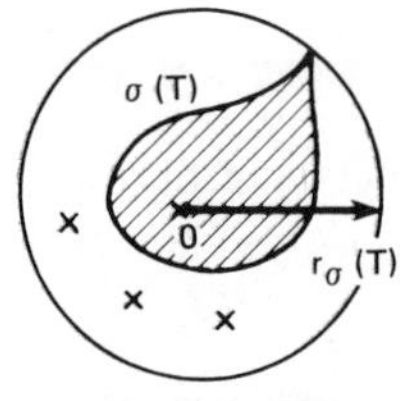

Figure 2.2

Proof We show that there is at least one point of $\sigma(T)$ on the circle $\{z; |z| = r_\sigma(T)\}$, cf. Fig. 2.2. Because the convergence domain of (2.6) is $|z| > r_\sigma(T)$, there is at least one singularity of $R(z)$ on the convergence circle $|z| = r_\sigma(T)$ provided that $r_\sigma(T) > 0$. If $r_\sigma(T) = 0$, $z = 0$ belongs to $\sigma(T)$, because otherwise $\sigma(T)$ would be empty. □

Exercise

2.37 $T \in \mathscr{L}(X)$ is quasi-nilpotent iff $\sigma(T) = \{0\}$.

For T in $\mathscr{L}(X)$, one may define the *approximate point spectrum* $\sigma_a(T)$ as follows: $\sigma_a(T) := \{\lambda \in \mathbb{C};$ there exists a sequence $\{x_n\}$ in X with $\|x_n\| = 1$ and $\lim_n \|(T - \lambda I)x_n\| = 0\}$. We prove the following property.

Proposition 2.24 *Let $T \in \mathscr{L}(X)$. $\sigma_a(T)$ is a nonempty subset of $\sigma(T)$ and the boundary of $\sigma(T)$ is contained in $\sigma_a(T)$. Both $P\sigma(T)$ and $C\sigma(T)$ are contained in $\sigma_a(T)$.*

Proof Suppose that $\lambda \in \rho(T)$. $T - \lambda$ has a bounded inverse and

$$\|x\| = \|(T - \lambda)^{-1}(T - \lambda)x\| \le \|(T - \lambda)^{-1}\| \, \|(T - \lambda)x\| \qquad \text{for} \quad x \in X.$$

This means $\|(T - \lambda)x\| \ge \alpha\|x\|$, where $\alpha := 1/\|(T - \lambda)^{-1}\|$. This shows that $\rho(T) \subset \mathbb{C} - \sigma_a(T)$; hence $\sigma_a(T) \subseteq \sigma(T)$. Let λ' be such that $|\lambda - \lambda'| < \alpha/2$. Then for $x \in X$, $\|x\| = 1$,

$$\|(T - \lambda')x\| \ge \|(T - \lambda)x\| - |\lambda - \lambda'| \ge \alpha/2,$$

so that $\lambda' \in \mathbb{C} - \sigma_a(T)$. This proves that $\sigma_a(T)$ is closed. Let λ be a point on the boundary of $\sigma(T)$. For a given $\varepsilon > 0$, there exists $z \in \rho(T)$ such that $|\lambda - z| < \varepsilon/2$. Then, by Proposition 2.19, $\|(T - z)^{-1}\| \ge 2/\varepsilon$. There exists $x \in X$, $\|x\| = 1$, such that $\|(T - z)^{-1}x\| \ge 1/\varepsilon$. Set

$$x_0 := \|(T - z)^{-1}x\|^{-1}(T - z)^{-1}x.$$

Then $\|x_0\| = 1$ and $\|(T - \lambda)x_0 - (T - z)x_0\| < \varepsilon/2$. Therefore

$$\|(T - \lambda)x_0\| \le \|(T - \lambda)x_0 - (T - z)x_0\| + \|(T - z)x_0\|$$

$$\le \frac{\varepsilon}{2} + \frac{\|x\|}{\|(T - z)^{-1}x\|} < \frac{3\varepsilon}{2}.$$

This proves that $\lambda \in \sigma_a(T)$, and $\sigma_a(T)$ is nonempty since it contains the boundary of $\sigma(T)$, a nonempty compact subset of $\mathbb{C}$. Clearly, $P\sigma(T) \subset \sigma_a(T)$. Now if $z \in C\sigma(T)$, $(T - z)^{-1}$ exists but is not bounded; hence $\|(T - z)x\|$ is not bounded from below for all x such that $\|x\| = 1$; that is, $z \in \sigma_a(T)$. □

We conclude from Corollary 2.23 and Proposition 2.24 that any λ in $\sigma(T)$ such that $|\lambda| = r_\sigma(T)$ belongs to $\sigma_a(T)$.

We are now able to weaken the assumption $\|T\| < 1$ required in Theorem 2.8 for the convergence of the Neumann expansion (2.1) of $(1 - T)^{-1}$.

Proposition 2.25 *Given $T \in \mathscr{L}(X)$ such that $r_\sigma(T) < 1$. Then $(1 - T)^{-1}$ exists and its Neumann expansion* (2.1) *converges in $\mathscr{L}(X)$.*

Proof We apply Theorem 2.21 with $z = 1$. If $r_\sigma(T) < 1$, $R(1)$ exists and $-R(1) = (1 - T)^{-1} = \sum_{k=0}^{\infty} T^k$, which converges in $\mathscr{L}(X)$. □

A necessary and sufficient condition for the convergence of (2.1) is then that for some positive integer k, $\|T^k\| < 1$. If $r_\sigma(T) < 1$, the solution of the equation $x - Tx = y$ may be computed by the method of successive approximations:

$$x_0 = y, \qquad x_k = Tx_{k-1} + y, \qquad \text{for} \quad k \geq 1,$$

which converges to $x = (1 - T)^{-1}y$ at the rate of a geometric progression whose quotient is arbitrarily close to $r_\sigma(T)$ (cf. Exercise 2.38).

Exercises

2.38 For any positive ε such that $r_\sigma(T) + \varepsilon < 1$, show that $\|x_k - x\| \leq C(\varepsilon, T)(r_\sigma(T) + \varepsilon)^{k+1}$. (*Hint*: Use $\|T\|_* \leq r_\sigma(T) + \varepsilon$.)

2.39 Given $T \in \mathscr{L}(X)$ and $t \in \mathbb{C}$, show that $(1 - tT)^{-1} = \sum_{k=0}^{\infty} t^k T^k$ is convergent in $\mathscr{L}(X)$ if t is such that $|t| < 1/r_\sigma(T)$.

2.40 We suppose that $T \in \mathscr{L}(X)$ is of finite rank and quasinilpotent. We set $M := \operatorname{Im} T$, and define $T_M : M \xrightarrow{T} M$. Show that T_M is nilpotent, that is, there exists an integer k such that $(T_M)^k = 0$.

7.5. *The Spectrum of a Closed Operator*

So far we have not considered the point at infinity for the partition of the complex plane into $\sigma(T)$ and $\rho(T)$. For a closed operator, it will be useful to consider it. If $\sigma(T)$ is bounded, then either T is bounded and $R(z)$ is holomorphic at $z = \infty$ (converse of Corollary 2.22) or $R(z)$ has an essential singularity at $z = \infty$ (Kato, 1976, p. 176). It is then natural to include $z = \infty$ in $\rho(T)$ if $T \in \mathscr{L}(X)$ or in $\sigma(T)$ if $T \in \mathscr{C}(X)$. We shall then speak of the *extended spectrum* $\sigma_e(T) = \sigma(T) \cup \{\infty\}$ and of the *extended resolvent* $\rho_e(T) = \rho(T) \cup \{\infty\}$. We are now able to relate the spectra of T and T^{-1} (if T^{-1} exists) through the following theorem.

Theorem 2.26 *Let T be a closed invertible operator in X. The extended spectra $\sigma_e(T)$ and $\sigma_e(T^{-1})$ are mapped onto each other by the mapping $z \mapsto 1/z$ defined on the extended complex plane.*

Proof Let $0 \neq z \in \rho(T)$, so that $R(z)$ exists. Set $S(z) := TR(z) = 1 + zR(z) \in \mathscr{L}(X)$. For all $x \in X$, $S(z)x = TR(z)x$ and $T^{-1}S(z)x = R(z)x = z^{-1}(S(z) - 1)x$. Then

$$zT^{-1}S(z)x - S(z)x = z(T^{-1} - 1/z)S(z)x = -x.$$

This shows that $\operatorname{Im}(T^{-1} - 1/z) = X$. This operator is invertible since $(T^{-1} - 1/z)y = 0$ implies $y = zT^{-1}y$, or equivalently $(T - z)y = 0$, that is, $y = 0$. Therefore $(T^{-1} - 1/z)^{-1} = -zS(z) \in \mathscr{L}(X)$ and $1/z \in \rho(T^{-1})$. If $z = 0$ belongs to $\rho(T)$, $T^{-1} \in \mathscr{L}(X)$ so that $0^{-1} = \infty$ belongs to $\rho_e(T^{-1})$ by definition. If $z = \infty$ belongs to $\rho_e(T)$, $T \in \mathscr{L}(X)$ and $0 = \infty^{-1}$ belongs to $\rho(T^{-1})$. This proves that $\rho_e(T)$ is mapped by $z \mapsto 1/z$ onto $\rho_e(T^{-1})$. The same holds for the extended spectra by complementarity. □

Similarly it can be shown that, for any polynomial p, $p[\sigma(T)] = \sigma[p(T)]$ (Dunford and Schwartz, 1958 Part 1, p. 604). We conclude easily that $r_\sigma^2(T) = r_\sigma(T^2)$ for T in $\mathscr{L}(X)$.

Exercises

2.41 For $z_0 \in \rho(T)$, the spectrum of $R(z_0) = (T - z_0)^{-1} \in \mathscr{L}(X)$ is the bounded set obtained from the extended spectrum of $T \in \mathscr{C}(X)$ by the mapping $z \mapsto 1/(z - z_0)$ and $R(z_0) - 1/(z - z_0) = [1/(z - z_0)](z - T)R(z_0)$. Deduce that

$$\left(R(z_0) - \frac{1}{z - z_0}\right)^{-1} = z_0 - z - (z - z_0)^2 R(z) \qquad \text{for} \quad z \in \rho(T), z \neq z_0$$

and

$$r_\sigma[R(z_0)] = \frac{1}{\operatorname{dist}(z_0, \sigma(T))}.$$

2.42 Give the spectra of the integral operators T_2^{-1} and T_3^{-1}, where T_2 and T_3 are as defined in Example 2.23.

7.6. Separation of the Spectrum

Let $T \in \mathscr{C}(X)$ with domain $\operatorname{Dom} T$. A subspace M of X is *invariant* under T if $T[(\operatorname{Dom} T) \cap M] \subset M$. The operator $(\operatorname{Dom} T) \cap M \xrightarrow{T} M$ is then the restriction T_M of T to the invariant subspace M. Let (M, N) be a pair of closed subspaces invariant under T such that $X = M \oplus N$. Let P be the projection on M along N. The operator T is said to be *completely reduced* by the pair of invariant supplementary subspaces (M, N) if $P(\operatorname{Dom} T) \subset \operatorname{Dom} T$.

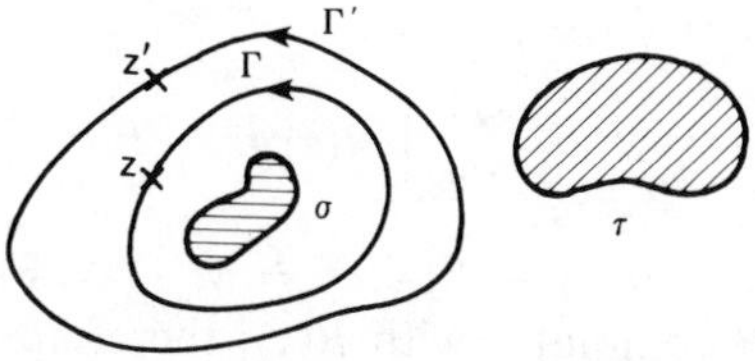

Figure 2.3

This definition can be extended readily to a finite number of invariant subspaces. For example, if $X = \operatorname{Dom} T = \mathbb{C}^N$, a diagonalizable matrix is completely reduced by the set of its N eigendirections, the projections being the eigenprojections.

When T is completely reduced by the pair (M, N), the study of the spectrum of T will reduce to the study of the spectra of T_M and T_N, respectively, as we shall see. This happens when, for example, the spectrum $\sigma(T)$ contains a bounded part σ separated from the rest τ in such a way that a closed Jordan curve—or more generally a finite number of such curves—can be drawn in $\rho(T)$ around σ, having τ in its exterior. In most applications, σ will consist of a finite number of points.

Theorem 2.27 *Suppose that $\sigma(T) = \sigma \cup \tau$, as described above. Then there exists a decomposition of X in $X = M \oplus N$ such that T is completely reduced by the pair (M, N), and the spectra of T_M and T_N are σ and τ, respectively, with $T_M \in \mathcal{L}(M)$.*

Proof Set

$$P := \frac{-1}{2i\pi} \int_\Gamma R(z)\, dz \in \mathcal{L}(X),$$

where Γ is the positively oriented Jordan curve around σ. We first show that P is a projection. Let Γ' be the boundary curve of a neighborhood of σ, containing Γ and leaving τ in its exterior, cf. Fig. 2.3.

$$\begin{aligned} P^2 &= \frac{1}{(2i\pi)^2} \int_\Gamma \int_{\Gamma'} R(z)R(z')\, dz'\, dz \\ &= \frac{1}{(2i\pi)^2} \int_\Gamma \int_{\Gamma'} \frac{R(z') - R(z)}{z' - z}\, dz'\, dz, \end{aligned}$$

using (2.4). We exchange the order on integration and note that for $z \in \Gamma$, $z' \in \Gamma'$,

$$\int_{\Gamma'} \frac{dz'}{z' - z} = 2i\pi \qquad \text{and} \qquad \int_\Gamma \frac{dz}{z' - z} = 0.$$

We get

$$P^2 = \frac{-1}{2i\pi}\int_\Gamma R(z)\,dz = P.$$

Now we set $M := PX$, $N := (1 - P)X$. $X = M \oplus N$. Furthermore, $PR(z) = R(z)P$ for $z \in \rho(T)$. P commutes with $R(z)$; therefore it commutes with T. For $x \in \operatorname{Dom} T$, $Px \in \operatorname{Dom} T$, and $PTx = TPx$; hence M is invariant under T, as well as N. T is completely reduced by (M, N) and the restrictions T_M and T_N are defined. The restrictions of $R(z)$ to M and N are denoted $R_M(z)$ and $R_N(z)$. Clearly $R_M(z) = (T_M - z)^{-1}$. This shows that both $\rho(T_M)$ and $\rho(T_N)$ contain $\rho(T)$. We prove that $\rho(T_M)$ also contains τ. For any z in $\rho(T)$ not on Γ, we have

$$\begin{aligned} R(z)P &= \frac{-1}{2i\pi}\int_\Gamma R(z)R(z')\,dz' \\ &= \frac{-1}{2i\pi}\int_\Gamma (R(z) - R(z'))\frac{dz'}{z - z'}. \end{aligned}$$

If z is outside Γ, this gives

$$R(z)P = \frac{1}{2i\pi}\int_\Gamma R(z')\frac{dz'}{z - z'}.$$

The right member is holomorphic outside Γ; it follows that $R(z)P$ and $R_M(z)$ have an analytic continuation outside Γ. $\rho(T_M)$ contains the exterior of Γ (in particular, τ). Hence $\sigma(T_M) \subset \sigma$. If z is inside Γ,

$$R(z)P = R(z) + \frac{1}{2i\pi}\int_\Gamma R(z')\frac{dz'}{z - z'}.$$

This shows that $R(z)(1 - P)$ and $R_N(z)$ have an analytic continuation inside Γ, and $\sigma(T_N) \subset \tau$.

If $z \in \sigma(T)$, then $z \notin \rho(T_M) \cap \rho(T_N)$; otherwise z would belong to $\rho(T)$ since $R_M(z)P + R_N(z)(1 - P)$ would be the inverse of $T - z$. Therefore $\sigma(T) \subset \sigma(T_M) \cup \sigma(T_N)$. This establishes that $\sigma(T_M) = \sigma$ and $\sigma(T_N) = \tau$. Finally, because $TR(z) = 1 + zR(z)$,

$$TP = \frac{-1}{2i\pi}\int_\Gamma zR(z)\,dz \in \mathscr{L}(X) \qquad \text{and} \qquad T_M \in \mathscr{L}(M). \quad \square$$

7.7. *Isolated Eigenvalues*

We suppose in this section that the spectrum of $T \in \mathscr{C}(X)$ has an *isolated point* λ. Obviously $\sigma(T)$ is separated into $\sigma \cup \tau$ with $\sigma = \{\lambda\}$. Any closed curve enclosing λ but no other point of $\sigma(T)$ may be chosen as Γ. The projection $P = (-1/2i\pi)\int_\Gamma R(z)\,dz$ does not depend on Γ but on λ only, since λ is the only singularity of $R(z)$ inside Γ. P (resp. $M = PX$) is called the *spectral projection* (resp. *invariant subspace*) associated with λ. T_M has the single point λ for spectrum. Hence $T_M - \lambda$ is quasi-nilpotent since

$$r_\sigma(T_M - \lambda) = \max_{z \in \sigma(T_M)} |z - \lambda| = 0$$

and the series

$$R_M(z) = -\sum_{k=0}^{\infty} (z - \lambda)^{-k-1}(T_M - \lambda)^k$$

converges except for $z = \lambda$ (Theorem 2.21). For $x \in M$, $R(z)Px = R_M(z)x$; therefore

$$R(z)P = -\frac{P}{z - \lambda} - \sum_{k=1}^{\infty} \frac{D^k}{(z - \lambda)^{k+1}},$$

where

$$D := (T - \lambda)P = \frac{-1}{2i\pi}\int_\Gamma (z - \lambda)R(z)\,dz$$

is quasi-nilpotent just as $T_M - \lambda$ (i.e., $r_\sigma(D) = 0$), and $D = DP = PD$.

For $x \in N$, $R(z)(1 - P)x = R_N(z)x$; therefore $R(z)(1 - P)$ is holomorphic at $z = \lambda$ and has the development given in the proof of Proposition 2.19, where z_0 is replaced by λ:

$$R(z)(1 - P) = \sum_{k=0}^{\infty} (z - \lambda)^k S^{k+1}$$

with

$$S := R_N(\lambda)(1 - P) = \lim_{z \to \lambda} [R(z)(1 - P)].$$

We define $S(z) := R(z)(1 - P)$ if $z \neq \lambda$ and S if $z = \lambda$; $S(z)$ is the *reduced resolvent* of T with respect to λ. Note that

$$S(\lambda) := S = \frac{1}{2i\pi}\int_\Gamma \frac{R(z)}{z - \lambda}\,dz \in \mathscr{L}(X)$$

and

$$(T - \lambda)S = 1 - P, \quad SP = PS = 0.$$

λ is an isolated singularity of $R(z)$, which has the *Laurent-series* expansion about λ

$$R(z) = -\frac{P}{z - \lambda} - \sum_{k=1}^{\infty} \frac{D^k}{(z - \lambda)^{k+1}} + \sum_{k=0}^{\infty} (z - \lambda)^k S^{k+1},$$

since $R(z) = R(z)P + R(z)(1 - P)$.

Exercise

2.43 All the above-defined operators that are functions of T commute, namely, T, $R(z)$ for $z \in \rho(T)$, P, S, D. For example, for $x \in \operatorname{Dom} T$ and $z \in \rho(T)$, $TR(z)x = R(z)Tx$. Establish the remaining identities.

A particular case of great importance is the case where $m := \dim M$ is *finite*. Then λ constitutes the spectrum of the finite-rank operator T_M, which is isomorphic to an $m \times m$ matrix: λ is an eigenvalue of T_M of algebraic multiplicity m. It is then an *eigenvalue* of T, and $m = \dim M$ is again called the *algebraic multiplicity* of the eigenvalue λ of T. All the relevant definitions are taken over from T_M to T.

$T_M - \lambda$ is actually *nilpotent*, just as D (Exercise 2.40). Moreover,

$$M \supset (T_M - \lambda)M \supset \cdots \supset (T_M - \lambda)^l M = \cdots = (T_M - \lambda)^m M = \{0\},$$

with all inclusions being proper until $\{0\}$ is reached; therefore $D^l = 0$. l is the *ascent* of the eigenvalue, and $R(z)$ has the Laurent expansion

$$R(z) = -\frac{P}{z - \lambda} - \sum_{k=1}^{l-1} \frac{D^k}{(z - \lambda)^{k+1}} + \sum_{k=0}^{\infty} (z - \lambda)^k S^{k+1}. \tag{2.7}$$

λ is a pole of $R(z)$ of order l.

TP has the spectral representation $TP = \lambda P + D$, $D^l = 0$. In the case of K isolated eigenvalues $\{\lambda_i\}_1^K$ of finite multiplicities, we get the spectral representation

$$P = \sum_{i=1}^{K} P_i, \qquad TP = \sum_{i=1}^{K} (\lambda_i P_i + D_i), \qquad PX = \bigoplus_{i=1}^{K} P_i X.$$

(Compare Theorem 1.6 in Chapter 1.)

This is a restricted representation of T, but it gives a fairly complete description of T if one is interested in a limited fraction of the plane containing only a finite number of eigenvalues with finite algebraic multiplicities.

In full generality the eigenvalues are not isolated points of the spectrum (cf. Example 2.23), nor are they of finite multiplicity (cf. Exercise 2.49). Nevertheless, *isolated* eigenvalues of *finite* multiplicities occur commonly enough in mathematical physics. This is the case, for example, for the eigenvalues of a compact operator, or of an operator with compact resolvent, as we shall see in Section 9.

We define $Q\sigma(T)$ as the part of $P\sigma(T)$ consisting of isolated eigenvalues of finite multiplicities. In this book we shall be concerned exclusively with the numerical computation of points in $Q\sigma(T)$.

The eigenspace $E := \operatorname{Ker}(T - \lambda)$ is a subspace of M, the invariant subspace associated with the eigenvalue λ. It is different from M if λ is not semisimple, that is, if $l > 1$. A basis of M consists of the g independent eigenvectors associated with λ, completed by $m - g$ generalized eigenvectors (cf. Chapter 1, Theorem 1.5). Because E is finite dimensional, it has at least one supplementary subspace F such that $X = E \oplus F$ (cf. Exercise 2.3). The projection Q from X onto E along F is called an *eigenprojection*. Note that Q is not uniquely defined. When λ is semisimple, $E = M := PX$ and P and Q are identical with the choice $F = (1 - P)X$.

Unlike P, Q cannot be easily expressed as a function of T. As a consequence, the dependence of the eigenvectors on T will be more difficult to study.

Example 2.27 In $X = \mathbb{R}^2$, let

$$A(\varepsilon) := \begin{pmatrix} 1 & 1 \\ \varepsilon & 1 \end{pmatrix},$$

where ε is a real positive parameter.

$$A(\varepsilon) \to A = \begin{pmatrix} 1 & 1 \\ 0 & 1 \end{pmatrix}$$

when $\varepsilon \to 0$. $A(\varepsilon)$ has the eigenvalues $\lambda_{1\varepsilon} = 1 + \sqrt{\varepsilon}$ and $\lambda_{2\varepsilon} = 1 - \sqrt{\varepsilon}$. $\lambda_{1\varepsilon} \neq \lambda_{2\varepsilon}$ if $\varepsilon > 0$. The associated eigenvectors are (cf. Fig. 2.4)

$$\varphi_{1\varepsilon} = \begin{pmatrix} 1 \\ \sqrt{\varepsilon} \end{pmatrix} \quad \text{and} \quad \varphi_{2\varepsilon} = \begin{pmatrix} 1 \\ -\sqrt{\varepsilon} \end{pmatrix}.$$

$$R(\varepsilon, z) = (A(\varepsilon) - zI)^{-1} = \frac{1}{(1 - z)^2 - \varepsilon} \begin{pmatrix} 1 - z & -1 \\ -\varepsilon & 1 - z \end{pmatrix}.$$

For any $\varepsilon > 0$, let Γ_1 (resp. Γ_2) be a closed Jordan curve enclosing $\lambda_{1\varepsilon}$ (resp. $\lambda_{2\varepsilon}$) and leaving $\lambda_{2\varepsilon}$ (resp. $\lambda_{1\varepsilon}$) in its exterior; cf. Fig. 2.5.

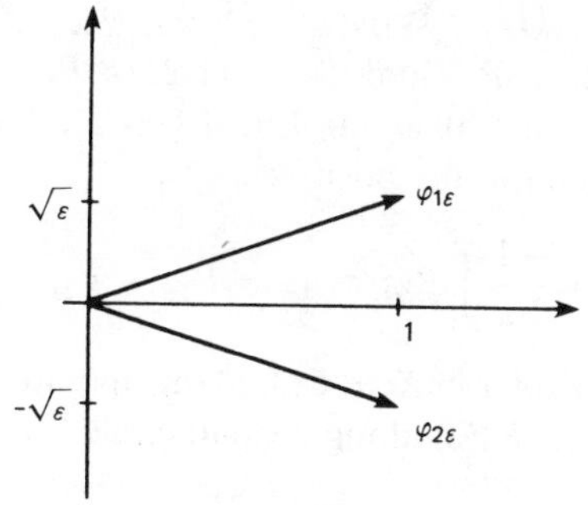

Figure 2.4

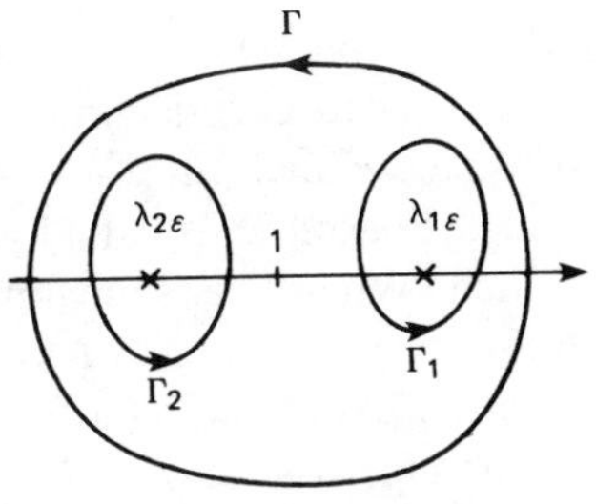

Figure 2.5

The spectral projection $P_{1\varepsilon}$ associated with $\lambda_{1\varepsilon}$ is

$$P_{1\varepsilon} = \frac{-1}{2i\pi}\int_{\Gamma_1} R(\varepsilon, z)\,dz = \frac{1}{2}\begin{pmatrix} 1 & 1/\sqrt{\varepsilon} \\ \sqrt{\varepsilon} & 1 \end{pmatrix}.$$

For

$$x = \begin{pmatrix} x_1 \\ x_2 \end{pmatrix} \in \mathbb{R}^2,$$

we have

$$P_{1\varepsilon}x = \tfrac{1}{2}(x_1 + (1/\sqrt{\varepsilon})x_2)\varphi_{1\varepsilon}.$$

The adjoint basis of $\{\varphi_{1\varepsilon}, \varphi_{2\varepsilon}\}$ is $\{\psi_{1\varepsilon}, \psi_{2\varepsilon}\}$ with

$$\psi_{1\varepsilon} = \frac{1}{2}\begin{pmatrix} 1 \\ 1/\sqrt{\varepsilon} \end{pmatrix} \quad \text{and} \quad \psi_{2\varepsilon} = \frac{1}{2}\begin{pmatrix} 1 \\ -1/\sqrt{\varepsilon} \end{pmatrix}.$$

It is then easy to check that $P_{1\varepsilon}x = (x, \psi_{1\varepsilon})\varphi_{1\varepsilon}$. Similarly

$$P_{2\varepsilon} = \frac{1}{2}\begin{pmatrix} 1 & -1/\sqrt{\varepsilon} \\ -\sqrt{\varepsilon} & 1 \end{pmatrix};$$

that is, $P_{2\varepsilon}x = \frac{1}{2}(x_1 - (1/\sqrt{\varepsilon})x_2)\varphi_{2\varepsilon} = (x, \psi_{2\varepsilon})\varphi_{2\varepsilon}$. We have also $D_{1\varepsilon} = D_{2\varepsilon} = 0$, $S_{1\varepsilon} = -(1/2\sqrt{\varepsilon})P_{2\varepsilon}$ and $S_{2\varepsilon} = (1/2\sqrt{\varepsilon})P_{1\varepsilon}$. Note that $P_{1\varepsilon}$ and $P_{2\varepsilon}$ have no limit when $\varepsilon \to 0$, $\varepsilon > 0$. But if Γ is a closed Jordan curve enclosing $\lambda_{1\varepsilon}$ and $\lambda_{2\varepsilon}$, then for any positive ε

$$P_\varepsilon = \frac{-1}{2i\pi}\int_\Gamma R(\varepsilon, z)\,dz = P_{1\varepsilon} + P_{2\varepsilon} = I.$$

Example 2.28 Consider in $X = C(a, b)$ the operator T_3 of Example 2.23. Integrating the resolvent $R_3(z)$ along a small circle Γ around $z = \lambda_n$; we get

$$P_n x(t) = \frac{-1}{2i\pi}\int_\Gamma R_3(z)x(t)\,dz = \frac{e^{\lambda_n t}}{b-a}\int_a^b e^{-\lambda_n s}x(s)\,ds, \qquad n = 1, 2, \ldots.$$

P_n is an integral operator of rank one with kernel

$$p_n(s, t) = \frac{1}{b - a} e^{\lambda_n(t - s)}.$$

Each λ_n is an isolated eigenvalue of T_3 with algebraic multiplicity 1.

Example 2.29 Let X and Y be two complex Banach spaces. We consider the generalized eigenvalue problem defined by $A \in \mathscr{C}(X, Y)$ and $B \in \mathscr{L}(X, Y)$:

$$\begin{aligned} &\text{find} \quad \lambda \in \mathbb{C}, \quad 0 \neq \varphi \in \operatorname{Dom} A \\ &\text{such that} \quad A\varphi = \lambda B\varphi. \end{aligned}$$

We suppose that $A^{-1} \in \mathscr{L}(Y, X)$. There are two equivalent formulations of the problem:

in X: $\varphi = \lambda A^{-1} B\varphi$,
in Y: $\psi = A\varphi$, $\psi = \lambda BA^{-1}\psi$.

The corresponding spectral projections are

in X:

$$P = \frac{-1}{2i\pi} \int_\Gamma (A - zB)^{-1} B \, dz \in \mathscr{L}(X),$$

in Y:

$$Q = \frac{-1}{2i\pi} \int_\Gamma B(A - zB)^{-1} \, dz \in \mathscr{L}(Y).$$

Note that $QY = APX$.

Exercises

2.44 For $f \in X$, $A \in \mathscr{C}(X, Y)$, $B \in \mathscr{L}(X, Y)$, consider the equation $(A - zB)x = Bf$. We set $y = Bx$, $g = Bf$, check that P (resp. Q) is associated with the resolvent defined by $x = (A - zB)^{-1}Bf$ in X (resp. $y = B(A - zB)^{-1}g$ in Y).

2.45 Consider the differential eigenvalue problem

$$x''(t) + tx'(t) + t^2x(t) = \lambda[x'(t) + tx(t)], \qquad 0 \leq t \leq 1,$$
$$x(0) = x(1) = 0.$$

Show that this problem can be put into the above abstract setting.

2.46 In $X = C(0, 2\pi)$, set $D := \{x; x' \in X \text{ and } x(0) = x(2\pi)\}$, and consider $T: x(t) \mapsto ix'(t)$ defined on D. Show that $\sigma(T) = P\sigma(T) = \{0, \pm 1, \pm 2, \ldots\}$. Compute the corresponding spectral projections. Conclude that the eigenvalues are simple.

2.47 In $X = L^2(0, 1)$, set $D = \{x \in X; x' \text{ is absolutely continuous}; x'' \in X \text{ and } x(0) = x(1), x'(0) = x'(1)\}$, and consider $T: x \in D \mapsto -x'' + 2x' - x$. Give the spectra of T and T^*.

2.48 Let T be a closed operator in X, and let λ be an isolated eigenvalue of finite multiplicitly with ascent l. Show that

$$X = \mathrm{Ker}[(T - \lambda)^k] \oplus \mathrm{Im}[(T - \lambda)^k] \qquad \text{for} \quad k \geq l.$$

2.49 In $X = C(-1, 1)$, T is defined by $(Tx)(s) = \int_{-1}^{1} k(s, t)x(t)\, dt$ with kernel $k(s, t) = a(t)b(s)$; a and b are continuous on $[-1, 1]$; moreover, a has no zero in $[-1, 1]$. Check that 0 is an eigenvalue of infinite multiplicity. Give the other eigenvalues and eigenfunctions.

In the case of an isolated eigenvalue, Exercise 2.41 can be completed as follows:

Proposition 2.28 *Let $T \in \mathscr{C}(X)$ and $z_0 \in \rho(T)$. $\lambda \in Q\sigma(T)$ iff $v = (\lambda - z_0)^{-1} \in Q\sigma(R(z_0))$. The associated spectral projections for T and $R(z_0)$ are identical, as well as the null spaces of $T - \lambda$ and $R(z_0) - v$.*

Proof λ is an isolated point of $\sigma(T)$ iff v is an isolated point of the spectrum of $R(z_0)$ (cf. Exercise 2.41). Using the identity

$$\left(R(z_0) - \frac{1}{z - z_0}\right)^{-1} = z_0 - z - (z - z_0)^2 R(z) \qquad \text{for} \quad z \in \rho(T)$$

and setting $t := 1/(z - z_0)$, we write the spectral projection of $R(z_0)$ associated with v,

$$P(v, R(z_0)) = \frac{-1}{2i\pi} \int_{\Gamma(v)} (R(z_0) - t)^{-1}\, dt,$$

where $\Gamma(v)$ is a Jordan curve around v,

$$\frac{-1}{2i\pi} \int_{\Gamma(\lambda)} \left(\frac{1}{z - z_0} + R(z)\right) dz = P(\lambda, T)$$

if the Jordan curve $\Gamma(\lambda)$ around λ does not enclose z_0. This is always possible because $z_0 \neq \lambda$.

Also, $T - \lambda$ and $R(z_0) - v$ have the same null space, which follows from the equivalence

$$Tx = \lambda x \Leftrightarrow (T - z_0)x = (\lambda - z_0)x \Leftrightarrow R(z_0)x = vx. \quad \square$$

This proposition applies in particular to $T \in \mathscr{C}(X)$ and $T^{-1} \in \mathscr{L}(X)$ if $z_0 = 0 \in \rho(T)$. It will be used later to relate the spectral properties of differential operators to those of their corresponding integral inverses.

7.8. Resolvent and Spectrum of the Adjoint

We suppose that T is densely defined in X so that T^* is uniquely defined.

Proposition 2.29 *$\rho(T^*)$ and $\sigma(T^*)$ are the conjugate sets of $\rho(T)$ and $\sigma(T)$, respectively, in $\mathbb{C}$.*

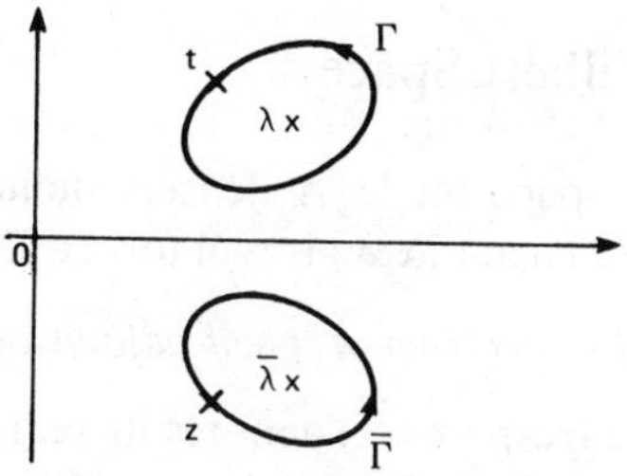

Figure 2.6

Proof The identity

$$\langle y, (T - z)x \rangle = \langle (T^* - \bar{z})y, x \rangle \qquad \text{for} \quad x \in \text{Dom } T, \quad y \in \text{Dom}(T^*)$$

implies that $R^*(z, T) = R(\bar{z}, T^*)$ for $z \in \rho(T)$. The result follows. □

Corollary 2.30 *Let $\lambda \in Q\sigma(T)$ have spectral projection P. Then $\bar{\lambda} \in Q\sigma(T^*)$ has spectral projection P^*. λ and $\bar{\lambda}$ have the same algebraic and geometric multiplicities as well as the same ascent.*

Proof Let $\lambda \in Q\sigma(T)$ be isolated by Γ. Let $\bar{\Gamma}$ be the conjugate curve (around $\bar{\lambda}$) positively oriented. The spectral projection of T^* associated with $\bar{\lambda}$ is

$$P(\bar{\lambda}, T^*) = \frac{-1}{2i\pi} \int_{\bar{\Gamma}} R(z, T^*)\, dz = \frac{-1}{2i\pi} \int_{\bar{\Gamma}} R^*(\bar{z}, T)\, dz.$$

We set $t = \bar{z}$ and note that t runs negatively on Γ; cf. Fig. 2.6. This proves that

$$\frac{-1}{2i\pi} \int_{\bar{\Gamma}} R^*(\bar{z}, T)\, dz = \frac{1}{2i\pi} \int_{\Gamma} R^*(t, T)\, dt = P(\lambda, T)^*,$$

the adjoint of the spectral projection of T associated with λ. The rest follows easily if one notes that the problem is reduced to a finite-dimensional one for the parts T_M and $T^*_{M^*}$, each of which may be considered the adjoint of the other. □

Let $\{x_i\}_1^m$ be a basis of the invariant subspace $M = PX$, and let $\{x_j^*\}_1^m$ be the adjoint basis in $M^* = P^*X^*$. Then

$$Px = \sum_{i=1}^{m} \langle x, x_i^* \rangle x_i, \qquad P^*y = \sum_{i=1}^{m} \langle y, x_i \rangle x_i^*.$$

If λ is semisimple, the $\{x_i\}$ are the eigenvectors of T and the $\{x_j^*\}$ are the eigenvectors of T^* such that

$$\langle x_i, x_j^* \rangle = \delta_{ij}, \qquad i, j = 1, \ldots, m.$$

8. Operators in a Hilbert Space

Let H be a Hilbert space on $\mathbb{C}$. A densely defined closed operator T is *self-adjoint* iff $T^* = T$. Then (Tx, x) is real for $x \in \operatorname{Dom} T$.

Proposition 2.31 *The spectrum of a self-adjoint operator T is real.*

Proof If $z \in \mathbb{C}$, $\mathscr{Re}\, z$ (resp. $\mathscr{Im}\, z$) denotes its real (resp. imaginary) part. $T - \mathscr{Re}\, z$ is self-adjoint, as is T. By developing $\|Tx - (\mathscr{Re}\, z + i\,\mathscr{Im}\, z)x\|^2$, we get

$$\|(T - z)x\|^2 = \|(T - \mathscr{Re}\, z)x\|^2 + (\mathscr{Im}\, z)^2 \|x\|^2 \qquad \text{for} \quad x \in \operatorname{Dom} T.$$

Then

$$\|(T - z)x\| \geq |\mathscr{Im}\, z|\, \|x\| \qquad \text{for} \quad x \in \operatorname{Dom} T.$$

And $T - z$ has a bounded inverse for nonreal z, $\mathscr{Im}(T - z)$ is closed and equal to H since $\operatorname{Ker}(T^* - z) = (\mathscr{Im}(T - z))^\perp = \{0\}$. Hence $R(z) \in \mathscr{L}(H)$ for $\mathscr{Im}\, z \neq 0$. □

Example 2.30 In $L^2(a, b)$, the integral operator defined by

$$T\colon x(s) \mapsto \int_a^b k(s, t)x(t)\, dt,$$

where $k(s, t)$ is continuous on $[a, b] \times [a, b]$, is self-adjoint iff $k(s, t) = \overline{k(t, s)}$ for almost every s, t in $[a, b]$ (cf. Example 2.13).

Example 2.31 Let $H = L^2(-\infty, +\infty)$, and let $D = \{x(t); x(t)$ and $tx(t) \in X\}$. The operator T defined by $x(t) \in D \mapsto tx(t)$ is self-adjoint. Let $y \in \operatorname{Dom}(T^*)$, and set $T^*y = y^*$. Then for all $x \in D = \operatorname{Dom} T$,

$$\int_{-\infty}^{+\infty} t\, x(t)\overline{y(t)}\, dt = \int_{-\infty}^{+\infty} x(t)\overline{y^*(t)}\, dt.$$

Let $x(t)$ be the characteristic function of the interval $[\alpha, t_0]$. We have

$$\int_\alpha^{t_0} t\,\overline{y(t)}\, dt = \int_\alpha^{t_0} y^*(t)\, dt.$$

Then by differentiation $t_0\, y(t_0) = \overline{y^*(t_0)}$ for almost all t_0. Therefore $y \in D$ and $(T^*y)(t) = ty(t)$. Conversely, $y \in D$ implies that $y \in \operatorname{Dom}(T^*)$ and $(T^*y)(t) = ty(t)$.

Example 2.32 Let $H = L^2(0, 2\pi)$ and let $D := \{x \in H;\ x$ is absolutely continuous on $[0, 2\pi]$, $x(0) = x(2\pi)$ and $x' \in H\}$. The operator T defined by $x \in D \mapsto -ix'$ is symmetric; for, if $x, y \in D$,

$$i\frac{d}{dt}[x(t)\overline{y(t)}] = ix(t)\overline{y'(t)} + ix'(t)\overline{y(t)}$$

and

$$(x, Ty) - (Tx, y) = ix(t)\overline{y(t)}\Big|_{t=0}^{t=2\pi} = 0.$$

Example 2.33 (The second-order ordinary differential equation) Let $H = L^2(a, b)$. We consider the formal differential operator T,

$$(Tx)(t) = -\frac{d}{dt}[p(t)x'(t)] + q(t)x(t),$$

where p and q are real-valued functions. There must be conditions on p, q, and Dom T ensuring that Tx is a well-defined element of H when $x \in \text{Dom } T$. If $Tx = u$ and $Ty = v$, we easily check that

$$u(t)\overline{y(t)} - x(t)\overline{v(t)} = \frac{d}{dt}[p(t)(x(t)\overline{y'(t)} - x'(t)\overline{y(t)})].$$

Therefore

$$(Tx, y) - (x, Ty) = p(t)(x(t)\overline{y'(t)} - x'(t)\overline{y(t)})\Big|_{t=a}^{t=b}.$$

The operator T is symmetric provided that the definition of Dom T ensures that

$$p(b)(x(b)\overline{y'(b)} - x'(b)\overline{y(b)}) = p(a)(x(a)\overline{y'(a)} - x'(a)\overline{y(a)}).$$

For example, the operator associated with the problem

$$\begin{aligned} -(px') + qx &= f, \\ \alpha x(a) + \beta x'(a) &= 0, \\ \gamma x(b) + \delta x'(b) &= 0 \end{aligned}$$

(where p is of class C^1, q is of class C on $[a, b]$, $p \neq 0$, the functions p and q and the coefficients α, β, γ, δ are real), is self-adjoint.

A self-adjoint operator T is *positive definite* iff

$$(Tx, x) > 0 \qquad \text{for all} \quad x \neq 0 \text{ in } H.$$

Example 2.34 Let T be a self-adjoint positive definite operator in H with domain D. For $x, y \in D$, $(x, y)_T := (Tx, y)$ defines an inner product on D. The corresponding norm is

$$\|x\|_T = \sqrt{(Tx, x)} \qquad \text{for} \quad x \in D.$$

Exercise

2.50 Prove that the spectrum of a self-adjoint positive definite operator lies in the set of real positive numbers.

Proposition 2.32 *If T is self-adjoint, then* $\|R(z)\| = [\text{dist}(z, \sigma(T))]^{-1}$.

Proof It follows from the proof of Proposition 2.29 that $R^*(z) = R(\bar{z})$; therefore $R^*(z)R(z) = R(z)R^*(z)$ and $R(z)$ is a *normal* operator in $\mathscr{L}(X)$ for $z \in \rho(T)$. Hence, by Proposition 2.10,

$$\|R(z)\| = r_\sigma[(T - z)^{-1}] = \max_{\lambda \in \sigma(T)} |\lambda - z|^{-1}$$

$$= \left(\min_{\lambda \in \sigma(T)} |\lambda - z|\right)^{-1} = (\text{dist}(z, \sigma(T)))^{-1}. \quad \square$$

Note that $R(z)$ is self-adjoint for z real in $\rho(T)$.

Proposition 2.33 *If λ is an isolated eigenvalue of a self-adjoint operator T, the associated spectral projection is orthogonal: $P^* = P$. A multiple eigenvalue is semisimple.*

Proof Let Γ be a closed Jordan curve isolating Γ. We may assume that Γ is symmetric with respect to the real axis. Then the proof of Corollary 2.30 shows immediately that $P = P^*$. Then $D = (T - \lambda)P = D^*$; D is a self-adjoint quasi-nilpotent operator; therefore $\|D\| = r_\sigma(D) = 0$. D is null, which implies that $l = 1$. λ is semisimple, and PH is the eigenspace associated with λ. $\square$

A self-adjoint operator T with domain D has a spectral representation that generalizes the spectral decomposition (1.4) $A = \sum_{i=1}^{K} \lambda_i P_i$ for a hermitian matrix (cf. Chapter 1, Section 1).

It can be shown that there exists a uniquely determined family of orthogonal projections $E(\lambda)$ for $\lambda \in \mathbb{R}$ such that

(i) $E(\lambda)E(\mu) = E(\mu)E(\lambda) = E(\lambda)$ if $\lambda \leq \mu$;
(ii) $\lim_{\mu \to \lambda^+} E(\mu) = E(\lambda)$;
(iii) $E(\lambda)T = TE(\lambda)$;
(iv) $\lim_{\lambda \to -\infty} E(\lambda) = 0$, $\lim_{\lambda \to +\infty} E(\lambda) = 1$;
(v) for any polynomial p,

$$(p(T)x, y) = \int_{-\infty}^{+\infty} p(\lambda)\, d(E(\lambda)x, y) \qquad \text{for} \quad x \in D, \quad y \in H,$$

where the integral is a Stieljes integral over $\mathbb{R}$ (cf. Kato, 1976, p. 356; Yosida, 1965, p. 313). In the particular case $p(\lambda) = \lambda$, we have

$$(Tx, y) = \int_{-\infty}^{+\infty} \lambda\, d(E(\lambda)x, y) \qquad \text{for} \quad x \in D, \quad y \in H.$$

We write symbolically

$$T = \int_{-\infty}^{+\infty} \lambda \, dE(\lambda) \tag{2.8}$$

and call this the *spectral representation* of the self-adjoint operator T.

The following properties can be proved:

$d(E(\lambda)x, x) = d\|E(\lambda)x\|^2 \geq 0$.

If $\mu \in P\sigma(T)$, then $E(\mu) \neq E(\mu^-)$ and $E(\mu) - E(\mu^-) = P$ is the eigenprojection associated with μ.

Conversely, if $E(\lambda^-) = E(\lambda)$, then λ is *not* an eigenvalue of T.

$R\sigma(T)$ is empty.

9. Spectrum of Compact Operators and Operators with Compact Resolvent

The spectrum of compact operators (resp. operators with compact resolvent) has a simple structure.

Theorem 2.34 *Let $T \in \mathscr{L}(X)$ be compact. $\sigma(T)$ is a countable set with no accumulation point other than zero. Each nonzero $\lambda \in \sigma(T)$ is an isolated eigenvalue of T with finite algebraic multiplicity.*

Proof This is a classical result, for which the reader is referred to Kato (1976, pp. 185–187; cf. also Taylor, 1958, pp. 274–285; Dunford and Schwartz, 1958, Part 1, pp. 577–580; Zaanen, 1960, pp. 311–455). □

Any nonzero eigenvalue of T then has properties quite similar to an eigenvalue of a finite-dimensional operator. $\bar{\lambda}$ is an eigenvalue of T^* with the same algebraic and geometric multiplicities as λ. The equation $(T - z)x = y$ is solvable iff $y \in (\mathrm{Ker}(T^* - \bar{\lambda}))^\perp$, and $(T^* - \bar{z})f = g$ is solvable iff $g \in (\mathrm{Ker}(T - \lambda))^\perp$. This is known as the *Riesz–Schauder theorem*, and generalizes the *Fredholm alternative*.

Let $\{\lambda_k\}_{\mathbb{N}}$ be the nonzero distinct eigenvalues of a compact operator T, with associated spectral projections P_k. For each λ_k,

$$TP_k = \lambda_k P_k + D_k,$$

where the D_k are nilpotent, and $P_k P_l = P_k \delta_{kl}, k, l \in \mathbb{N}$. If we set $Q_k = \sum_{p=1}^{k} P_p$, then

$$TQ_k = \sum_{p=1}^{k} (\lambda_p P_p + D_p).$$

This suggests an infinite expansion for T, but this is not correct without a supplementary assumption. Note, for example, that a compact quasi-nilpotent operator has no nonzero eigenvalues. Nevertheless, the expansion

$$T = \sum_{k=1}^{\infty} (\lambda_k P_k + D_k)$$

is valid if T is a compact normal or self-adjoint operator in a Hilbert space. Let T be a compact self-adjoint operator in a Hilbert space H. Let $\{\lambda_k\}_{\mathbb{N}}$ be the real nonzero distinct eigenvalues ordered by decreasing order of magnitude. Let $\{P_k\}$ be the associated orthogonal eigenprojections.

Proposition 2.35 *A compact self-adjoint T has the spectral representation in $\mathscr{L}(H)$*

$$T = \sum_{k=1}^{\infty} \lambda_k P_k, \qquad P_k = P_k^*, \qquad \dim P_k < \infty, \qquad P_k P_l = P_k \delta_{kl}.$$

Proof With $Q_k = \sum_{p=1}^{k} P_p$,

$$TQ_k = \sum_{p=1}^{k} \lambda_p P_p.$$

$T(1 - Q_k)$ is self-adjoint and has the eigenvalues $\lambda_{k+1}, \lambda_{k+2}, \ldots,$ and possibly 0. Then

$$\|T(1 - Q_k)\| = r_\sigma[T(1 - Q_k)] = |\lambda_{k+1}| \to 0 \qquad \text{as} \quad k \to \infty.$$

This gives

$$\left\| T - \sum_{p=1}^{k} \lambda_p P_p \right\| \to 0.$$

We now show that the eigenvectors plus a basis of Ker T form a complete family of H. We first show that $Q = \sum_{k=1}^{\infty} P_k$ exists. For $x \in H$,

$$\sum_{l=1}^{k} \|P_l x\|^2 = \left\| \sum_{l=1}^{k} P_l x \right\|^2 \leq \|x\|^2$$

since the P_l are orthogonal projections and $P_k P_l = \delta_{kl} P_k$. Therefore $\sum_{l=1}^{k} \|P_l x\|^2$ is convergent, and

$$\left\| \sum_{l=k}^{k+n} P_l x \right\|^2 = \sum_{l=k}^{k+n} \|P_l x\|^2 \to 0 \qquad \text{as} \quad n \to \infty$$

establishes that $Qx = \sum_{k=1}^{\infty} P_k x$ is convergent in H for every x in H. QH and $(1 - Q)H$ are invariant under T since T commutes with all the Q_k. It is easily seen that the restriction of T to $(1 - Q)H$ is the null operator and that $(1 - Q)H = \text{Ker } T$. This shows that

$$H = \text{Ker } T \oplus \left[\bigoplus_{k=1}^{\infty} P_k H \right]. \quad \square$$

Let $\{\mu_k\}_{\mathbb{N}}$ be the repeated eigenvalues of T. The associated eigenvectors $\{x_k\}_{\mathbb{N}}$ form an orthonormal basis of QH and

$$T = \sum_{k=1}^{\infty} \mu_k(\cdot, x_k)x_k.$$

Exercise

2.51 For $\lambda = \mu_j$ (supposed to be simple), show that the range of $T - \lambda$ consists of all vectors y in $\{x_j\}^{\perp}$. For such y, the general solution of $(T - \lambda)x = y$ is

$$x = -\frac{1}{\lambda} y + \frac{1}{\lambda} \sum_{\mu_k \neq \lambda} \mu_k \frac{(y, x_k)}{\mu_k - \lambda} x_k + \alpha x_j, \qquad \alpha \in \mathbb{C}.$$

We now turn to operators with compact resolvent.

Proposition 2.36 *Let $T \in \mathscr{C}(X)$ be such that $R(z)$ exists and is compact for some z. Then the spectrum of T is a countable set of isolated eigenvalues with finite algebraic multiplicities, and $R(z)$ is compact for any $z \in \rho(T)$.*

Proof It follows easily from Proposition 2.28 and from the first resolvent equation

$$R(z) = R(z_0)[1 + (z - z_0)R(z)],$$

which shows that if $R(z_0)$ is compact, so is $R(z)$ for all $z \in \rho(T)$. $\square$

Most differential operators that appear in classical boundary-value problems have compact resolvents.

Exercise

2.52 Let $X = C(0, 1)$ and let $D = \{x \in X; x', x'' \in X, x(0) = x(1) = 0\}$. Show that the operator T defined by $x \in D \mapsto -x''$ has a compact resolvent.

An operator with compact resolvent is sometimes called *anticompact*. From now on, we shall use this name for the sake of brevity. We recall that $\hat{D}$ is the domain D equipped with the graph norm.

Proposition 2.37 *A closed operator T with domain D is anticompact iff the injection of $\hat{D}$ into X is compact.*

Proof We suppose that $R(z_0)$ is compact for some $z_0 \in \rho(T)$. Let $\{x_n\} \in D$ such that $\|x_n\|_D = \|Tx_n\| + \|x_n\| \leq 1$. Then

$$\|(T - z_0)x_n\| \leq \|Tx_n\| + |z_0| \, \|x_n\| \leq 1 + |z_0|.$$

$\{(T - z_0)x_n\}$ is a bounded sequence, and $x_n = R(z_0)(T - z_0)x_n$ converges for $n \in N_1 \subset \mathbb{N}$. This shows the compactness of the imbedding i of $\hat{D}$ in X. Conversely, let $z_0 \in \rho(T)$, $R(z_0) = i \circ R(z_0)$, where $i: \hat{D} \to X$ is compact, and $R(z_0): X \to \hat{D}$ is bounded. This shows that $R(z_0): X \to X$ is compact. □

Examples will be given in Chapter 4.

CHAPTER 3

Elements of Functional Analysis: Convergence and Perturbation Theory

Introduction

In this chapter, we present results in functional analysis, many of which have not yet appeared in a textbook, and some of which have even not been published before.

In Part A we present a systematic study of the various notions of convergence of a sequence of bounded or closed operators. These notions will be used later in the analysis of numerical methods.

In Part B the classical analytic perturbation theory of Rellich, Kato, and others, which is valid when the parameter is small enough, is extended to the unit disk. The iterative computation of the series expansions of the eigenelements is presented in the case of simple, multiple, or close eigenvalues.

A. CONVERGENCE OF A SEQUENCE OF OPERATORS

Let $\mathscr{C}(X)$ be the space of closed operators, and let $T \in \mathscr{C}(X)$ have domain D. Practical solutions of $(T - z)x = f$ or $T\varphi = \lambda\varphi$ often involve a family of operator approximations T_n in $\mathscr{C}(X)$, Dom $T_n = D, n = 1, 2, \ldots$. Knowledge of the mathematical properties of the convergence $T_n \to T$ as $n \to \infty$ greatly eases the study of the numerical approximation. In this part, we define various types of convergence that cover most practical methods.

Often enough in discretization methods T_n is of *finite rank*. It is useful to consider the following class of operators. We are given $\{X_n\}_{\mathbb{N}}$, a sequence of *finite-dimensional subspaces* of X such that there exists a sequence of projections π_n on X_n, which satisfy $\pi_n x \xrightarrow[n\to\infty]{} x$ for all $x \in X$. Given $D \subseteq X$, set $D_n := \pi_n D$, and let $\mathcal{T}_n: X_n \to X_n$ be a linear operator with domain D_n. The operator $T_n := \mathcal{T}_n \pi_n$ is said to be of *class* $\mathfrak{D}$. Such an operator is a natural extension of $\mathcal{T}_n$ to X; clearly T_n is bounded with domain D. In particular, when $D = X$ and $D_n = X_n$, let $\mathcal{T}_n$ be a linear operator on X_n. Then $T_n = \mathcal{T}_n \pi_n$ is bounded with domain X: $T_n \in \mathcal{L}(X)$.

Example 3.1 Let $X = C(a, b)$. $[a, b]$ is divided into $n - 1$ intervals at the points t_i, $i = 1, \ldots, n$, $t_1 = a$, $t_n = b$. Let X_n be the subspace of X consisting of piecewise linear functions on $[a, b]$. With the basis $\{e_i^{(n)}\}_1^n$ defined in Example 2.6 (Chapter 2), $\pi_n x = \sum_{i=1}^n x(t_i) e_i^{(n)}$. Let $\mathcal{T}_n: X_n \to X_n$ be defined by the $n \times n$ matrix $A_n = (a_{ij}^{(n)})$ in the chosen basis; then

$$T_n x = \mathcal{T}_n \pi_n x = \sum_{i=1}^{n} \left(\sum_{j=1}^{n} a_{ij}^{(n)} x(t_j) \right) e_i^{(n)}.$$

1. Convergence of a Sequence of Operators in $\mathcal{L}(X)$

B denotes the unit sphere of X, $B := \{x \in X;\ \|x\| \le 1\}$. The sequence $\{x_n\}_{\mathbb{N}}$, $x_n \in X$, is *relatively compact* in X iff the set of its values $\bigcup_{n=1}^\infty x_n$ is relatively compact in X, that is, iff each subsequence $\{x_n\}_{N_1 \subset \mathbb{N}}$ has a converging subsequence $\{x_n\}_{N_2 \subset N_1}$. For example, let T be compact. The sequence $\{Tx_n\}_{\mathbb{N}}$ is relatively compact if $\{x_n\}_{\mathbb{N}}$ is an arbitrary bounded sequence in X.

Let $\{T_n\}_{\mathbb{N}}$ be a sequence of operators in $\mathcal{L}(X)$ converging to $T \in \mathcal{L}(X)$ according to one of the following definitions:

(a) *pointwise* convergence $T_n \xrightarrow{\text{p}} T$ iff for all x in X, $T_n x \to Tx$ as $n \to \infty$.
(b) *uniform* (or norm) convergence iff $\|T - T_n\| \to 0$ as $n \to \infty$.
(c) *collectively compact* convergence $T_n \xrightarrow{\text{cc}} T$ iff

(i) $T_n \xrightarrow{\text{p}} T$, and
(ii) the following condition is satisfied:

$$\text{the set } K := \bigcup_{n=1}^{\infty} (T - T_n)B \text{ is relatively compact in } X. \tag{3.1}$$

(d) *compact* convergence $T_n \xrightarrow{\text{c}} T$ iff

(i) $T_n \xrightarrow{\text{p}} T$, and
(ii) the following condition is satisfied:

for any sequence $\{x_n\}_{\mathbb{N}}$ in B, the sequence $\{(T - T_n)x_n\}_{\mathbb{N}}$ is relatively compact in X. (3.2)

(e) *discrete-compact* convergence $T_n \xrightarrow{\text{d-c}} T$ for $T_n = \mathscr{T}_n \pi_n$ of class $\mathfrak{D}$ iff

(i) $T_n \xrightarrow{\text{p}} T$, and
(ii) the following condition is satisfied:

for any sequence $\{x_n\}_{\mathbb{N}}$ such that $x_n \in X_n$, $\|x_n\| \le c$, the sequence $\{(T - T_n)x_n\}_{\mathbb{N}} = \{(T - \mathscr{T}_n)x_n\}_{\mathbb{N}}$ is relatively compact in X. (3.3)

In the following four examples, X is taken to be l^2; $\{e_i\}_1^\infty$ is the canonical basis, $X_n = \{e_1, e_2, \ldots, e_n\}$, $n = 1, 2, \ldots$.

Example 3.2 $Tx = x$, $T_n x = \sum_{i=1}^n (x, e_i)e_i$, $T_n x \to x$: $T_n \xrightarrow{\text{p}} T$.

Example 3.3 $Tx = 0$, $T_n x = (1/n)x$, $\|T_n - T\| = 1/n \to 0$, and $T_n \xrightarrow{\text{c}} T$, but $T_n \stackrel{\text{cc}}{\not\to} T$.

Example 3.4 $Tx = 0$, $T_n x = (x, e_n)e_1$, $\bigcup_{n=1}^\infty T_n B$, which is a bounded set of $\{e_1\}$, is relatively compact: $T_n \xrightarrow{\text{cc}} \mathscr{T}$, but $\|T - T_n\| = 1$.

Example 3.5 Let $x = \sum_{i=1}^\infty x_i e_i$, with $x_i = (x, e_i)$. $x \mapsto Tx = \sum_{i=1}^\infty x_{i+1} e_i$. For $x \in X_n$, let $x \mapsto \mathscr{T}_n x = \sum_{i=1}^{n-1} x_{i+1} e_i$. Let π_n be the orthogonal projection on X_n, $T_n = \mathscr{T}_n \pi_n$ with $x \mapsto T_n x = \sum_{i=1}^{n-1} x_{i+1} e_i = T\pi_n$. For $x_n \in X_n$, $(T - T_n)x_n = 0$: $T_n \xrightarrow{\text{d-c}} T$. Note that T is not compact: its spectrum has been given in Chapter 2, Example 2.25, and the eigenvalues are not isolated.

The convergence notions defined above are not strictly ordered for arbitrary bounded operators T, T_n. The mutual relationships are given now. The results are summarized in Table 3.1, p. 129.

2. Properties of the Convergences in $\mathscr{L}(X)$

2.1. About Pointwise Convergence

We begin with the *Banach–Steinhaus theorem.*

Theorem 3.1 *Let X and Y be complex Banach spaces. The necessary and sufficient conditions for the sequence $\{T_n\}$ of bounded operators of $\mathscr{L}(X, Y)$ to be pointwise convergent on X to a linear operator are*

(i) $\sup_n \|T_n\| < M$, *and*
(ii) *$T_n x$ is convergent for all x in some set F dense in X.*

For a proof see Yosida (1965, p. 69) or Banach and Steinhaus (1927).

For example, let $\{\pi_n\}$ be a sequence of projections from X onto X_n such that $\pi_n \xrightarrow{\text{p}} 1$; then there exists M such that $\|\pi_n\| \le M$. The π_n are uniformly bounded in n. Conversely, if $\sup_n \|\pi_n\| < \infty$, if for any n_0 there exists $N(n_0)$

such that $X_{n_0} \subset X_n$ for $n > N(n_0)$, and if $\bigcup_{n \geq 1} X_n$ is dense in X, then $\pi_n x \to x$ for all x in X.

Theorem 3.2 *Suppose that $T_n x \to Tx$, $x \in X$. Then for any relatively compact set U, $\sup_{x \in U} \|(T_n - T)x\| \to 0$.*

Proof Let U be a relatively compact set. For any $\varepsilon > 0$, U has a finite cover with sets of diameter less than ε. For each $x \in U$, there exists x_ε such that $\|x - x_\varepsilon\| < \varepsilon$; there are a finite number of such x_ε. Then

$$\begin{aligned}\|(T_n - T)x\| &\leq \|(T_n - T)(x - x_\varepsilon)\| + \|(T_n - T)x_\varepsilon\| \\ &\leq (\|T_n\| + \|T\|)\varepsilon + \max_{x_\varepsilon} \|(T_n - T)x_\varepsilon\|.\end{aligned}$$

The desired result follows. □

Exercises

In the following two exercises, T_n is of class $\mathfrak{D}$.

3.1. Show that the conditions

(i) $$\|(T_n - \pi_n T)_{\restriction X_n}\| = \sup_{\substack{x \in X_n \\ \|x\| = 1}} \|(\mathscr{T}_n - \pi_n T)x\| \to 0,$$

(ii) $$\|(1 - \pi_n)T\| \to 0,$$

given in Kantorovitch and Akilov (1964, p. 543) imply that $T_n \overset{\text{d-c}}{\to} T$.

3.2 Show that the condition

$$\|(T - T_n)_{\restriction X_n}\| = \sup_{\substack{x \in X_n \\ \|x\| = 1}} \|(T - \mathscr{T}_n)x\| \to 0$$

given in Descloux *et al.* (1978a) imply that $T_n \overset{\text{d-c}}{\to} T$. Compare these conditions with those given in Exercise 3.1.

3.3 Show that if $T_n \overset{\text{cc}}{\to} T$, then $\|(T_n - T)^2\| \to 0$ as $n \to \infty$.

3.4 Show that if $T_n \overset{\text{p}}{\to} T$ in $\mathscr{L}(X)$, then $(T - T_n)x_n \to y$ for a relatively compact sequence $\{x_n\}$ implies that $y = 0$.

2.2. *About Collectively Compact Convergence*

2.2.1. Relation with compact convergence

It is clear that $T_n \overset{\text{cc}}{\to} T$ implies $T_n \overset{\text{c}}{\to} T$, but the reciprocal is not true in general, as shown by Example 3.3. From a practical point of view, cases when the collectively compact convergence lets itself be characterized by the compact convergence are interesting, because (3.2) is easier to prove than (3.1).

We first note that $T_n \xrightarrow{cc} T$ implies that each $T - T_n$ is compact, since for $n \in \mathbb{N}$, $(T - T_n)B \subset \bigcup_{n=1}^{\infty}(T - T_n)B$, which is relatively compact. An arbitrary sequence in $K := \bigcup_{n=1}^{\infty}(T - T_n)B$ has the form $y_n = (T - T_{\sigma(n)})x_n$, $x_n \in B$, where σ is a mapping from $\mathbb{N}$ into itself. K is relatively compact iff any sequence $\{y_n\}_{\mathbb{N}}$ in K has a convergent subsequence in X. We look for conditions on T, T_n under which $T_n \xrightarrow{c} T$ would imply that $T_n \xrightarrow{cc} T$.

Proposition 3.3 *The following are equivalent:*

(i) *$T - T_n$ is compact for any integer n and $T_n \xrightarrow{c} T$; and*
(ii) *$T_n \xrightarrow{cc} T$.*

Proof From the above discussion it remains to prove that (i) implies (ii). Let σ be an arbitrary mapping: $\mathbb{N} \xrightarrow{\sigma} \mathbb{N}$. We first suppose that $\sigma(\mathbb{N})$ is not bounded. There exists an infinite subset $N_1 \subset \mathbb{N}$ such that $\sigma_1 = \sigma_{\restriction N_1}$ is injective. σ_1 is bijective on $\sigma(N_1)$. We define $x'_n = 0$ if $n \in \mathbb{N} - \sigma(N_1)$ and $x'_n = x_{\sigma_1^{-1}(n)}$ if $n \in \sigma(N_1)$, $y'_n = (T_n - T)x'_n$. For $n \in N_1$,

$$y_n = (T - T_{\sigma(n)})x_n = (T - T_{\sigma(n)})x_{\sigma_1^{-1}(\sigma(n))} = (T - T_{\sigma(n)})x'_{\sigma(n)} = y'_{\sigma(n)}.$$

Hence

$$\bigcup_{n \in N_1} y_n = \bigcup_{\sigma(n) \in \sigma(N_1)} y'_{\sigma(n)} \subset \bigcup_{n \in \mathbb{N}} y'_n,$$

which is relatively compact by (3.2). And the sequence y_n has a converging subsequence. If $\sigma(\mathbb{N})$ is bounded, there exists an infinite subset $N_1 \subset \mathbb{N}$ and $p \in \mathbb{N}$ such that $\sigma(N_1) = \{p\}$. For $n \in \mathbb{N}$, $y_n = (T - T_p)x_n$, where $T - T_p$ is compact. □

Proposition 3.4 *Suppose that $V_n \xrightarrow{cc} V$, V compact, and for $U_n \in \mathscr{L}(X)$, $U_n \xrightarrow{p} U$, then $V_n U_n \xrightarrow{cc} VU$ and $U_n V_n \xrightarrow{cc} UV$.*

Proof V and $V - V_n$ are compact; hence V_n, VU, and UV are compact. $V_n U_n x \to VUx$ and $U_n V_n x \to UVx$ for $x \in X$. Let $x_n \in B$; it is then clear that for any infinite subset N_1 of $\mathbb{N}$, each sequence $\{V_n U_n x_n\}_{N_1}$ and $\{U_n V_n x_n\}_{N_1}$ has a converging subsequence. □

Very often in practice T_n is of finite rank (but T is not).

Proposition 3.5 *If T_n is compact, $T_n \xrightarrow{cc} T$ implies T compact.*

Proof $T_n \xrightarrow{cc} T$ implies $T_n - T$ compact for $n \in \mathbb{N}$, which in turn implies that $T = T_n - (T_n - T)$ is compact. □

T_n is often compact in practice since it is usually of finite rank; hence the notion of collectively compact convergence will be useful only if T_n is the approximation of an operator T itself compact. In that case, (3.1) can be proved by showing (3.2) only; that is, $T_n \xrightarrow{cc} T \Leftrightarrow T_n \xrightarrow{c} T$.

2.2.2. Relation with uniform convergence

In general, $\|T_n - T\| \to 0$ does not imply $T_n \xrightarrow{cc} T$, as shown by Example 3.3. Moreover, $T_n \xrightarrow{cc} T$ does not imply $T_n^* \xrightarrow{p} T^*$, as seen in Example 3.4, where $T_n^* e_1 = e_n$, which does not converge in l^2.

Proposition 3.6 *If $T_n - T$ is compact, $n \in \mathbb{N}$, then $\|T_n - T\| \to 0$ implies $T_n \xrightarrow{cc} T$.*

Proof $\|T - T_n\| \to 0$ implies (3.2). Apply Proposition 3.3. □

Proposition 3.7 *If $T_n \xrightarrow{cc} T$ and $T_n^* x \to T^* x$, $x \in X^*$, then $\|T - T_n\| \to 0$.*

Proof Suppose that $\|T - T_n\| \not\to 0$. There exist $\{x_n\}_{N_1 \subset \mathbb{N}}$ with $x_n \in B$, $\delta > 0$, and $y \in X$ such that

$$\|(T - T_n)x_n\| \geq \delta \qquad \text{and} \qquad \lim_n (T - T_n)x_n = y, \qquad \|y\| \geq \delta.$$

Let $y^* \in X^*$ be such that $\|y\| = \langle y^*, y\rangle$. Then

$$\|y\| = \lim_n \langle y^*, (T - T_n)x_n\rangle = \lim_n \langle (T^* - T_n^*)y^*, x_n\rangle = 0,$$

which contradicts $\|y\| \geq \delta$. □

Proposition 3.8 *Let $S \in \mathscr{L}(X)$ be compact and suppose that $S_n \xrightarrow{cc} S$. If $T_n \xrightarrow{p} T$, then $\|(T - T_n)S_n\| \to 0$.*

Proof $\bigcup_{n=1}^{\infty} S_n B$ is relatively compact since S is compact. The result follows by Theorem 3.2. □

2.2.3. Relation with discrete-compact convergence

We suppose that T_n is of class $\mathfrak{D}$. It is clear that $T_n \xrightarrow{cc} T$ implies $T_n \xrightarrow{d\text{-}c} T$ since $X_n \subset X$. In Example 3.5, $T_n \xrightarrow{d\text{-}c} T$, but $T_n \not\xrightarrow{cc} T$. Since T_n is of finite rank and T is not compact, the collectively compact convergence is impossible by Proposition 3.5. The discrete-compact convergence will then play a role when approximating a *noncompact* bounded operator T.

Proposition 3.9 *If T is compact, T_n of class $\mathfrak{D}$, then $T_n \xrightarrow{d\text{-}c} T$ implies $T_n \xrightarrow{cc} T$.*

Proof $T - T_n$ is compact; then (3.2) implies (3.1). We then wish to show that (3.3) implies (3.2). Recall that $T_n \pi_n = T_n$. Let $x_n \in B$,

$$(T - T_n)x_n = (T - T_n)\pi_n x_n + (T - T_n)(1 - \pi_n)x_n.$$

Since $\{\pi_n x_n\}_{\mathbb{N}}$ is a bounded sequence in X_n, $\{(T - T_n)\pi_n x_n\}$ is relatively compact by (3.3) and $\bigcup_{n=1}^{\infty} (T - T_n)\pi_n B$ is relatively compact by Proposition 3.3. Since T is compact, and $\{(1 - \pi_n)x_n\}$ is a bounded sequence in X,

$\bigcup_{n=1}^{\infty} T(1-\pi_n)B$ is relatively compact. With $T_n(1-\pi_n)x_n = 0$, this establishes that $\bigcup_{n=1}^{\infty} (T - T_n)B$ is relatively compact. $\square$

It follows that if T_n is of class $\mathfrak{D}$, $T_n \xrightarrow{\text{d-c}} T$ implies $T_n \xrightarrow{\text{c}} T$ when T is compact. On the other hand $T_n \xrightarrow{\text{c}} T$ always implies $T_n \xrightarrow{\text{d-c}} T$.

2.3. Convergence of a Sequence of Compact Projections

To deal with spectral projections later, we shall be interested in the behavior of a sequence of compact projections. Let $\{P_n\}_{\mathbb{N}}$ be a sequence of projections in $\mathscr{L}(X)$, pointwise convergent to the projection P. We set $M_n := P_n X$, $M := PX$, and we suppose that $m := \dim M < \infty$. P is a *compact* projection.

Proposition 3.10 *Let P, P_n, $n \in \mathbb{N}$, be projections such that $P_n \xrightarrow{\text{p}} p$, and* $\dim PX < \infty$. *Then* $\|(P - P_n)P\| \to 0$ *and, for n large enough,*

$$\dim P_n X \geq \dim PX.$$

Proof $\|(P - P_n)P\| \to 0$ follows from Theorem 3.2. *Let* $\{x_i\}_1^m$ be a basis of M, and let $\{x_i^*\}_1^m$ be the adjoint basis in M^*: $\langle x_i, x_j^* \rangle = \delta_{ij}$, $i, j = 1, \dots, m$. Clearly $\langle P_n x_i, x_j^* \rangle \to \delta_{ij}$. Hence, for n large enough, the m vectors $x_{in} = P_n x_i$ are independent in M_n. $\dim P_n X \geq \dim PX$. $\square$

We give an example where $\dim P_n X > \dim PX$, $n \in \mathbb{N}$. Let $X := l^2$ with the canonical basis $\{e_i\}_{\mathbb{N}}$. For $x \in X$, $P_n x = \langle x, e_1 \rangle e_1 + \langle x, e_n \rangle e_n$; $P_n x \to Px = \langle x, e_1 \rangle e_1$; and $\dim P_n X = 2$, $\dim PX = 1$.

The preservation of the dimensions of M_n and M, which is required in the case where P and P_n are spectral projections (cf. Chapter 5), calls for another hypothesis in addition to the pointwise convergence of P_n toward P. We suppose that $\dim P_n X = m$ for $n = 1, 2, \dots$. For n large enough, the m vectors $\{P_n x_i\}_1^m$, which are independent, form a basis of M_n. We study the adjoint basis in $M_n^* := P_n^* X^*$, defined by

$$\langle P_n x_i, x_{jn}^* \rangle = \delta_{ij}, \qquad i, j = 1, \dots, m.$$

Lemma 3.11 *The adjoint basis $\{x_{in}^*\}_1^m$ is such that* $\lim_n \|x_{in}^* - P_n^* x_i^*\| = 0$ *and* $\|x_{in}^*\| \leq c$.

Proof For n large enough, the $\{x_{in}^*\}_1^m$ can be built by Schmidt orthogonalization from the $\{P_n^* x_i^*\}_1^m$, as we see now.

For $i = 1$, $x_{1n}^* = \alpha_{11}^{(n)} P_n^* x_1^*$, where

$$1 = \langle x_{1n}^*, P_n x_1 \rangle = \alpha_{11}^{(n)} \langle x_1^*, P_n x_1 \rangle.$$

Since $P_n x_1 \to x_1$, we conclude that $\alpha_{11}^{(n)} \to 1$ as $n \to \infty$.

For $i = 2$, $x_{2n}^* = \alpha_{21}^{(n)} P_n^* x_1^* + \alpha_{22}^{(n)} P_n^* x_2^*$, where

$$0 = \langle x_{2n}^*, P_n x_1 \rangle = \alpha_{21}^{(n)} \langle x_1^*, P_n x_1 \rangle + \alpha_{22}^{(n)} \langle x_2^*, P_n x_1 \rangle,$$
$$1 = \langle x_{2n}^*, P_n x_2 \rangle = \alpha_{21}^{(n)} \langle x_1^*, P_n x_2 \rangle + \alpha_{22}^{(n)} \langle x_2^*, P_n x_2 \rangle.$$

Solving this 2×2 linear system, we get $\alpha_{22}^{(n)} \to 1$ and $\alpha_{21}^{(n)} \to 0$ as $n \to \infty$. By induction on i, $i = 1, 2, \ldots, m$, it is easily shown that

$$x_{in}^* = \sum_{j=1}^{i} \alpha_{ij}^{(n)} P_n^* x_j^*$$

with $\alpha_{ij}^{(n)} \to \delta_{ij}$ for $j \le i$, $i = 1, \ldots, m$. Therefore

$$| \|x_{in}^*\| - \|P_n^* x_i^*\| | \le \|x_{in}^* - P_n^* x_i^*\| \to 0.$$

Since $\|P_n\| = \|P_n^*\| \le c$, we get $\|x_{in}^*\| \le c'$. □

Lemma 3.12 *Let M, M_n be closed subspaces of X such that* $\dim M < \infty$ *and, for any bounded sequence $\{x_n\}_{\mathbb{N}}$, $x_n \in M_n$, there exists a subsequence $\{x_n\}_{N_1 \subset \mathbb{N}}$ converging to $x \in M$. Then* $\dim M_n \le \dim M$ *for n large enough.*

Proof Suppose that $\dim M_n \ge m$ for $n \in N_1$ an infinite subset of $\mathbb{N}$, where m is a given positive integer. By Corollary 2.3 there exists $x_{n,i} \in M_n$ such that $\|x_{n,i}\| = 1$, $\|x_{n,i} - \sum_{j=1}^{i-1} \alpha_j x_{n,j}\| \ge 1$ for $n \in N_1$, $i = 1, \ldots, m$, and for any choice of the α_j. There exists a subsequence $x_{n,i} \to x_i \in M$ for $n \in N_2 \subset N_1$, $i = 1, \ldots, m$, such that $\|x_i\| = 1$ and $\|x_i - \sum_{j=1}^{i-1} \alpha_j x_j\| \ge 1$. The m vectors $\{x_i\}$ are then independent, and $\dim M \ge m$. Therefore, if $\dim M < m$, then $\dim M_n < m$ for n large enough. Equivalently, $\dim M_n \le \dim M$. □

We are now able to prove the following fundamental characterization:

Proposition 3.13 (Anselone) *For projections P and P_n such that P is compact, the following are equivalent*:

(i) $P_n \xrightarrow{\text{p}} P$ *and* $\dim P_n X = \dim PX < \infty$ *for n large enough*; *and*
(ii) $P_n \xrightarrow{\text{cc}} P$.

Proof We first show that (i) implies (ii). Since $P_n - P$ is compact for $n \in \mathbb{N}$, $P_n \xrightarrow{\text{cc}} P$ iff for any sequence $\{x_n\}_{\mathbb{N}}$ in B, $\{P_n x_n\}_{\mathbb{N}}$ is a relatively compact sequence in B. Let N_1 be an arbitrary infinite set of $\mathbb{N}$. We wish to show that $\{P_n x_n\}_{N_1}$ has a converging subsequence. In the adjoint bases $\{P_n x_i\}_1^m$ and $\{x_{in}^*\}_1^m$, $P_n x_n = \sum_{i=1}^m \beta_{in} P_n x_i$, with $\beta_{in} = \langle x_n, x_{in}^* \rangle$. By Lemma 3.11, $\sup_n \|x_{in}^*\| < \infty$ and $|\beta_{in}| \le c$, $i = 1, \ldots, m$, then $y(n) = \sum_{i=1}^m \beta_{in} x_i$ is a bounded sequence in M. There is a converging subsequence $y(n) \to y$, and $P_n y(n) \to y$ for $n \in N_2 \subset N_1$.

Conversely, if $P_n \xrightarrow{\text{cc}} P$, it follows that $\dim P_n X \ge \dim PX$ for n large enough, by Proposition 3.10, and $\dim P_nX \leq \dim PX$ for n large enough, by Lemma 3.12 with $M = PX$, $M_n = P_nX$. □

Proposition 3.14 *Let P and P_n be projections such that* $\dim PX < \infty$. *Then $P_n \overset{cc}{\to} P$ implies $\|(P - P_n)P_n\| \to 0$ and $\Theta(PX, P_nX) \to 0$.*

Proof This is a consequence of Propositions 3.8, 3.10, and 2.13. □

Proposition 3.15 *If P and P_n are orthogonal projections with $\dim PX < \infty$ in a Hilbert space, $\|P - P_n\| \to 0$ iff $P_n \overset{cc}{\to} P$.*

Proof It is clear from Propositions 3.6 and 3.7. □

3. Summary

The five notions of convergence $T_n \to T$, namely, p (pointwise), c (compact), d-c (discrete-compact), cc (collectively compact), and $\|\cdot\|$ (uniform) can be ordered as shown in Table 3.1., depending on the hypotheses on T and T_n. When the discrete-compact convergence is considered, T_n is assumed to be of class $\mathfrak{D}$. Examples of the convergences defined in Section 1 will be given in Chapter 4 for practical methods.

The collectively compact convergence, introduced by Sobolev (1956), was used extensively by Anselone (1971) for integral operators. The discrete-compact convergence was introduced in Vainikko (1969a) for application to finite difference methods. In the context of discrete approximation theory (cf. Appendix), where X_n, the space in which $\mathcal{T}_n$ is defined, is *not a subspace* of X, the discrete-compact convergence was first defined by Stummel (1970), who proved Proposition 3.13 in a general abstract setting (Stummel, 1971, pp. 253–254). In its present form, Proposition 3.13 was proved by Anselone (1971).

Table 3.1

Convergences in $\mathcal{L}(X)$

Hypotheses on T and T_n	Order on the convergences in $\mathcal{L}(X)$
$T, T_n \in \mathcal{L}(X)$	$\|\cdot\|$ ⇒ c ⇐ cc; c ⇒ d-c ⇒ p
$T - T_n$ compact	$\|\cdot\|$ ⇒ cc ⇔ c ⇒ d-c ⇒ p
T, T_n compact	$\|\cdot\|$ ⇒ cc ⇔ c ⇔ d-c ⇒ p
T, T_n self-adjoint	$\|\cdot\|$ ⇔ cc ⇒ c ⇒ d-c ⇒ p
$\dim T_nX = \dim TX < \infty$	$\|\cdot\|$ ⇒ cc ⇔ c ⇔ d-c ⇔ p

The compact convergence is defined in Goldberg (1974). It has been generalized in Anselone and Ansorge (1979) to nonlinear operators with applications to integral equations. The reader interested in the topological properties of the above convergences is refered, for the four first notions to Anselone and Ansorge (1979) and for the fifth to Stummel (1970).

4. Convergence of a Sequence of Operators in $\mathscr{C}(X)$

Let T_n, $n = 1, 2, \ldots$, be a sequence of operators in $\mathscr{C}(X)$ converging to $T \in \mathscr{C}(X)$ according to one of the following definitions. We suppose that $\text{Dom } T_n = \text{Dom } T = D$ for $n = 1, 2, \ldots$.

(a) *pointwise* convergence $T_n \xrightarrow{\text{p}} T$ iff, for all x in D, $T_n x \to Tx$ as $n \to \infty$.
(b) *stable* convergence $T_n \xrightarrow{\text{s}} T$ iff

(i) $T_n \xrightarrow{\text{p}} T$, and
(ii) the following stability condition is satisfied:

$$\exists M > 0, \exists N\colon \text{for } n > N,\ T_n^{-1} \in \mathscr{L}(X) \text{ and } \|T_n^{-1}\| \leq M. \tag{3.4}$$

(c) *regular* convergence $T_n \xrightarrow{\text{r}} T$ iff

(i) $T_n \xrightarrow{\text{p}} T$, and
(ii) the following regularity condition is satisfied:

any bounded sequence $\{x_n\}_{\mathbb{N}}$ in D such that $T_n x_n \to y$, $n \in N_1 \subset \mathbb{N}$, is itself such that $x_n \to x$, $n \in N_2 \subset N_1$, and $Tx = y$. (3.5)

Note that the above definitions apply also when T and T_n are bounded; in that case $\text{Dom } T_n = \text{Dom } T = D \equiv X$. The condition $Tx = y$ in (3.5) is a consequence of what precedes.

Example 3.6 Consider in $X = L^2(0, 1)$ the operator T of Example 2.24 (Chapter 2). T is self-adjoint with spectrum $\sigma(T) = [-1, 1]$. Let π_n be the orthogonal projection on $X_n = \{e_1, \ldots, e_n\}$, where $e_i(t) = \sqrt{2} \sin i\pi t$. $\{e_i\}_1^\infty$ is an orthonormal basis of $L^2(0, 1)$. We consider $T_n = \pi_n T \pi_n$; clearly $T_n \xrightarrow{\text{p}} T$ as $n \to \infty$. The restriction $T_{n \restriction X_n}$ is represented in the chosen basis of X_n by the $n \times n$ matrix A_n defined by the elements $a_{ij} = (\cos \pi t e_j(t), e_i(t)) = 0$ except for $a_{i,i+1} = a_{i+1,i} = \frac{1}{2}$, $i = 1, \ldots, n-1$. Except possibly for zero, A_n and T_n have the same spectrum. And

$$\sigma(A_n) = \left\{\cos \pi \frac{i}{n+1}, i = 1, \ldots, n\right\} \subset \sigma(T).$$

Then

$$\|(T_n - 2)^{-1}\| = \frac{1}{\text{dist}(2, \sigma(A_n))} \leq 1.$$

Therefore $T_n - 2 \xrightarrow{s} T - 2$ as $n \to \infty$. Show that $T'_n = \pi_n T$ also satisfies $T'_n - 2 \xrightarrow{s} T - 2$.

Example 3.7 In the preceding example, $T_n - 2 \xrightarrow{r} T - 2$. Indeed, let $\{x_n\}$ be a bounded sequence in X such that $(T_n - 2)x_n \to y$. 2 is in $\rho(T)$. We set $x := (T - 2)^{-1}y$. Then

$$x_n - x = (T_n - 2)^{-1}[(T_n - 2)x_n - y + (T - 2)x - (T_n - 2)x]$$

shows that $x_n \to x$ with $(T - 2)x = y$.

Example 3.8 We suppose that T and T_n are bounded and $\|T_n - T\| \to 0$. For any fixed z in $\rho(T)$, we write

$$T_n - z = T - z - (T - T_n) = [1 - (T - T_n)R(z)](T - z).$$

For any $\varepsilon > 0$, there exists N such that, for $n > N$, $\|T - T_n\| < \varepsilon$. Then for $\varepsilon < 1/\|R(z)\|$, $\|(T - T_n)R(z)\| < 1$ and the Neumann-series expansion for $[1 - (T - T_n)R(z)]^{-1}$ converges in $\mathscr{L}(X)$. By Theorem 2.7,

$$\|[1 - (T - T_n)R(z)]^{-1}\| \leq [1 - \|(T - T_n)R(z)\|]^{-1} \leq \frac{1}{1 - \varepsilon\|R(z)\|}.$$

It follows that the series

$$R_n(z) := (T_n - z)^{-1} = R(z) \sum_{k=0}^{\infty} [(T - T_n)R(z)]^k \tag{3.6}$$

is also convergent in $\mathscr{L}(X)$. We conclude easily that $\|R_n(z)\|$ is uniformly bounded in n, $T_n - z \xrightarrow{s} T - z$, and $\|R_n(z) - R(z)\| \to 0$ for any fixed z in $\rho(T)$.

Example 3.9 We now suppose that T and T_n are bounded and $T_n \xrightarrow{cc} T$. For any fixed z in $\rho(T)$,

$$r_\sigma[(T - T_n)R(z)] \leq \|[(T - T_n)R(z)]^2\|^{1/2} \to 0$$

as $n \to \infty$. The series (3.6) converges in $\mathscr{L}(X)$, and

$$R_n(z) - R(z) = R(z)[1 + (T - T_n)R(z)](T - T_n)R(z) \sum_{k=0}^{\infty} [(T - T_n)R(z)]^{2k}.$$

This shows that $T_n - z \xrightarrow{s} T - z$; $R_n(z) \xrightarrow{cc} R(z)$ follows from the second resolvent equation (2.5) applied to T and T_n.

Other types of convergence can be considered. When T and T_n are closed, $T_n \to T$ in $\mathscr{C}(X)$ may be defined by means of the convergence in $\mathscr{L}(X)$ of the resolvents $R_n(z) \to R(z)$ for some z in $\rho(T)$, assuming that $z \in \rho(T_n)$ for n large enough. For example, $\|R_n(z) - R(z)\| \to 0$ or $R_n(z) \xrightarrow{\text{cc}} R(z)$ for some $z \in \rho(T)$. The *gap* convergence is defined by $\Theta(T_n, T) \to 0$, where

$$\Theta(T, U) := \Theta(G(T), G(U))$$

is the gap between the closed subspaces of $X \times X$ consisting of the graphs $G(T)$ and $G(U)$ of the operators T and U.

Exercises

3.5 Assuming that T and $T_n \in \mathscr{C}(X)$ are invertible, show that $\Theta(T, T_n) \to 0$ iff

$$\|T_n^{-1} - T^{-1}\| \to 0.$$

3.6 Let T and T_n be bounded. Show that for $\|T_n - T\| \to 0$ (resp. $T_n \xrightarrow{\text{cc}} T$), it is necessary that, for each $z \in \rho(T)$, $z \in \rho(T_n)$ for large enough n and $\|R_n(z) - R(z)\| \to 0$ (resp. $R_n(z) \xrightarrow{\text{cc}} R(z)$), while it is sufficient that this be true for some $z \in \rho(T)$.

3.7 Show that if $T_n \xrightarrow{\text{s}} T$, then $T^{-1} \in \mathscr{L}(X)$.

3.8 If $T_n \xrightarrow{\text{s}} T$, show that $T_n - z \xrightarrow{\text{s}} T - z$ for all z such that $|z| < 1/M$, where M is defined by $\|T_n^{-1}\| \le M$ for n large enough.

5. Properties of the Convergences in $\mathscr{C}(X)$

5.1. About Stable Convergence

For any given z in $\rho(T)$, we consider the well-posed equation in X

$$(T - z)x = f, \qquad x \in D. \tag{3.7}$$

It is approximated in X by

$$(T_n - z)x_n = f, \qquad x_n \in D. \tag{3.8}$$

Clearly $x = R(z)f$ and $x_n = R_n(z)f$.

Lemma 3.16 *The following are equivalent for all $z \in \rho(T)$:*

(i) $T_n - z \xrightarrow{\text{s}} T - z$, *and*
(ii) $R_n(z) \xrightarrow{\text{p}} R(z)$ *and* $T_n \xrightarrow{\text{p}} T$.

Proof It is clear from the identity

$$R_n(z) - R(z) = R_n(z)(T - T_n)R(z). \quad \square$$

The importance of the stable convergence follows obviously from the above characterization, since $x_n = R_n(z)f$ converges to $x = R(z)f$ for any right-hand side f in X. It is said classically that T_n is a *stable approximation* of T at a given z in $\rho(T)$ iff $T_n - z \xrightarrow{s} T - z$. On the other hand, $T_n - z \xrightarrow{s} T - z$ shows that $z \in \rho(T)$ (cf. Exercise 3.7). To find a stable approximation of T at z is then a constructive way, first, to show that Eq. (3.7) has a unique solution x, and second, to approximate x by x_n.

5.2. *About Regular Convergence*

We prove the following characterization:

Proposition 3.17 *The following are equivalent for all z in $\rho(T)$:*

(i) $T_n - z \xrightarrow{r} T - z$, *and*
(ii) $T_n - z \xrightarrow{s} T - z$.

Proof Let z be fixed in $\rho(T)$. We first look at the direct implication. The stability of $T_n - z$ can be written

$$\|(T_n - z)x_n\| \geq M^{-1}(z)\|x_n\| \qquad \text{for} \quad x_n \in D, \quad n > N(z).$$

We suppose that it does not hold: there exists a sequence $\{x_n\}_{N_1 \subset \mathbb{N}}$, $x_n \in D$, $\|x_n\| = 1$ such that $\|(T_n - z)x_n\| \to 0$. By regularity, $x_n \to x$, $x \in D$, $\|x\| = 1$ for $n \in N_2 \subset N_1$, and $(T - z)x = 0$, which contradicts $z \in \rho(T)$.

Conversely, if $(T_n - z)x_n \to y$ for $x_n \in D$, set $x := (T - z)^{-1}y \in D$; then

$$x_n - x = (T_n - z)^{-1}[(T_n - z)x_n - y + (T - z)x - (T_n - z)x]$$

and $x_n \to x$. This shows that $T_n - z \xrightarrow{r} T - z$. □

For Eq. (3.8) to be an approximation of (3.7) in X such that $x_n \to x$, it is necessary and sufficient that $T_n - z \xrightarrow{r} T - z$. This condition is easier to check than the stability condition (ii). Surprisingly enough, it has been much less used than (ii) to prove the convergence of practical approximation methods.

Exercises

3.9 Show that if $z_n \to z$ and $T_n - z \xrightarrow{r} T - z$, then $T_n - z_n \xrightarrow{r} T - z$. If $z \neq 0$ and $T_n \xrightarrow{r} T$, show that $z_n T_n \xrightarrow{r} zT$.

3.10 We assume that $T_n \xrightarrow{r} T$ and $U_n \xrightarrow{r} U$ in $\mathscr{L}(X)$. Deduce that $T_n U_n \xrightarrow{r} TU$.

3.11 Set $t = 1/(z - z_0)$, for $z \neq z_0 \in \rho(T)$, $v = 1/(\lambda - z_0)$ for $\lambda \in Q\sigma(T)$. Prove the following four implications, for $T, T_n \in \mathscr{C}(X)$, and $z \in \rho(T)$.

(i) $$\left.\begin{matrix} T_n - z \xrightarrow{s} T - z \\ R_n(z_0) \xrightarrow{p} R(z_0) \end{matrix}\right\} \Rightarrow R_n(z_0) - t \xrightarrow{s} R(z_0) - t,$$

(ii) $$\left.\begin{matrix} R_n(z_0) - t \xrightarrow{s} R(z_0) - t \\ T_n \xrightarrow{p} T \end{matrix}\right\} \Rightarrow T_n - z \xrightarrow{s} T - z,$$

(iii) $\left.\begin{array}{l} T_n - z \xrightarrow{r} T - z \\ R_n(z_0) \xrightarrow{p} R(z_0) \end{array}\right\} \Rightarrow R_n(z_0) - t \xrightarrow{r} R(z_0) - t,$

(iv) $\left.\begin{array}{r} R_n(z_0) - t \xrightarrow{r} R(z_0) - t \\ T_n \xrightarrow{p} T \end{array}\right\} \Rightarrow T_n - z \xrightarrow{r} T - z.$

Study (iii) and (iv) if $z = \lambda$ and $t = \nu$.

We are interested in conditions on T and T_n which are sufficient to guarantee that $T_n - z \xrightarrow{r} T - z$ for *any* z in $\rho(T)$.

Proposition 3.18 *$T_n - z \xrightarrow{r} T - z$ for any z in $\rho(T)$ if one of the three following properties holds:*

(i) $\|T_n - T\| \to 0$;
(ii) $T_n \xrightarrow{cc} T$;
(iii) $T_n \xrightarrow{c} T$.

Proof It was proved in Section 2 that (i) or (ii) implies (iii). We then suppose that $T_n \xrightarrow{c} T$ and show that $T_n - z \xrightarrow{r} T - z$ for any z in $\rho(T)$. Let $x_n \in B$ be such that $(T_n - z)x_n \to y$. Then $(T - T_n)x_n \to u$ for $n \in N_1 \subset \mathbb{N}$ and $(T - z)x_n \to y + u$, $n \in N_1$. The convergence of $\{x_n\}_{N_1}$ follows since $z \in \rho(T)$, $x_n \to x$ and $u = 0$ by Exercise 3.4. Therefore $(T - z)x = y$. □

Let T be closed with domain D. D becomes the Banach space $\hat{D}$ with the graph norm, and T is bounded in $\mathscr{L}(\hat{D}, X)$. Let i be the injection of $\hat{D}$ into X. If i is compact, recall that T is anticompact in $\mathscr{C}(X)$.

Proposition 3.19 *If the injection $i\colon \hat{D} \to X$ is compact, then in $\mathscr{L}(\hat{D}, X)$, $T_n \xrightarrow{r} T$ implies $T_n - z \xrightarrow{r} T - z$ for all $z \in \mathbb{C}$.*

Proof Let $\{x_n\}$ be a bounded sequence in $\hat{D}$ such that $\{(T_n - z)x_n\}$ is relatively compact in X for $z \in \mathbb{C}$. The compactness of i implies that $\{T_n x_n\}$ is also relatively compact in X; thus $\{x_n\}$ is relatively compact in $\hat{D}$ because $T_n \xrightarrow{r} T$ in $\mathscr{L}(\hat{D}, X)$. This proves that $T_n - z \xrightarrow{r} T - z$ as well. □

5.3. T_n is of Class $\mathfrak{D}$

We consider a family of approximations T_n of class $\mathfrak{D}$, a very common situation in practice. We define the notion of discrete-stable (resp. discrete-regular) convergence. It generalizes the stable (resp. regular) convergence in the same way that the discrete-compact convergence generalizes the compact convergence. We recall that $T_n = \mathscr{T}_n \pi_n$, and $D_n = \pi_n D$:

(a) *discrete-stable* convergence $T_n \xrightarrow{d\text{-}s} T$ iff

(i) $T_n \xrightarrow{p} T$, and

(ii) the following stability condition is satisfied:

$$\exists M > 0, \exists N\colon \text{for } n > N\colon \|T_n x_n\| \geq M^{-1}\|x_n\| \text{ for all } x_n \text{ in } D_n. \quad (3.9)$$

(b) *discrete-regular* convergence $T_n \xrightarrow{\text{d-r}}$ iff

(i) $T_n \xrightarrow{\text{p}} T$, and

(ii) the following regularity condition is satisfied:

any sequence $\{x_n\}_{\mathbb{N}}$ where $x_n \in D_n$, $\|x_n\| \leq c$, and which is such that $T_n x_n \to y$, $n \in N_1 \subset \mathbb{N}$, is itself such that $x_n \to x \in D$, $n \in N_2 \subset N_1$, and $Tx = y$. (3.10)

These notions are useful in studying the discrete approximation in X_n of Eq. (3.7) by

$$(\mathscr{T}_n - z)x_n = \pi_n f, \qquad x_n \in D_n. \quad (3.11)$$

Let 1_n be the identity on X_n; we write $z1_n = z$, and set $\mathscr{R}_n(z) := (\mathscr{T}_n - z)^{-1}$, the inverse in X_n. $x_n = \mathscr{R}_n(z)\pi_n f$ converges to $x = R(z)f$ iff $T_n - z\pi_n \xrightarrow{\text{d-s}} T - z$, as can be seen from the identity

$$\pi_n R(z) - \mathscr{R}_n(z)\pi_n = \mathscr{R}_n(z)(\pi_n T - \mathscr{T}_n \pi_n)R(z).$$

It is easy to prove the following characterization, which is the discrete analog of Proposition 3.17.

Proposition 3.20 *The following are equivalent for z in $\rho(T)$:*

(i) $T_n - z\pi_n \xrightarrow{\text{d-r}} T - z$, *and*

(ii) $T_n - z\pi_n \xrightarrow{\text{d-s}} T - z$.

Proof It is an easy adaptation of the proof of Proposition 3.17, where $\{x_n\}$ now belongs to D_n. □

Again the regularity condition (i) is easier to deal with than the stability condition (ii).

Example 3.10 Let $1_n\colon X_n \to X_n$ be the identity on X_n, and let π_n be a projection on X_n such that $\pi_n \xrightarrow{\text{p}} 1$. $\pi_n = 1_n \pi_n$ is an operator of class $\mathfrak{D}$. $\pi_n \xrightarrow{\text{d-r}} 1$ as well as $\pi_n \xrightarrow{\text{d-s}} 1$.

Example 3.11 The notion of regular convergence originates in the work of Polskii (1962) and his followers on the solvability–approximability of the equation $Lx = f$, where L is a bounded operator on a Hilbert space H, such that $0 \in \rho(L)$, by *projection* on a finite-dimensional subspace E_n of H. Let π_n be the orthogonal projection on E_n and assume that $\pi_n \xrightarrow{\text{p}} 1$. The approximate equation in E_n is taken to be

$$\pi_n L x_n = \pi_n f, \qquad x_n \in E_n.$$

This equation corresponds to the discrete approximation $\pi_n L_{\upharpoonright E_n}$ and its extension $L_n := \pi_n L \pi_n$ of class $\mathfrak{D}$.

By Proposition 3.20, $x_n \to x$ iff $L_n \xrightarrow{\text{d-r}} L$. The characterization given by Polskii is $x_n \to x$ iff $\sup_n \Theta(LE_n, E_n) < 1$. It is then said that L is a *regular* operator (Vainikko and Umanskii, 1968). Vainikko (1965a) showed that $L \in \mathscr{L}(H)$ such that $0 \in \rho(L)$ is regular iff it is representable as $\alpha L = 1 + S + K$, where α is a nonzero constant, $\|S\| < 1$, and K is compact.

We shall verify in Chapter 4 that if $L = 1 + K$, K compact, $1 \in \rho(K)$, then $\sup_n \Theta(LE_n, E_n) < 1$ is satisfied (cf. also Krasnoselskii *et al.*, 1972).

Example 3.12 The operator $T_n = \pi_n T \pi_n$ defined in Example 3.6 is of class $\mathfrak{D}$. Clearly $T_n - 2\pi_n \xrightarrow{\text{d-s}} T - 2$ as well as $T_n - 2 \xrightarrow{\text{s}} T - 2$. This is a particular case of the following lemma.

Lemma 3.21 *If T_n is of class $\mathfrak{D}$, then, for any nonzero z in $\rho(T)$, the following are equivalent:*

(i) $T_n - z \xrightarrow{\text{s}} T - z$, *and*
(ii) $T_n - z\pi_n \xrightarrow{\text{d-s}} T - z$.

Proof Consider the equation $(T_n - z)x = y$ for $z \in \rho(T)$, $y \in X$. It yields $\pi_n x = (\mathscr{T}_n - z)^{-1}\pi_n y$, and $(1 - \pi_n)x = -(1/z)(1 - \pi_n)y$. Therefore $x = R_n(z)y$ is defined for $0 \neq z \in \rho(T)$ by $R_n(z) = (\mathscr{T}_n - z)^{-1}\pi_n - (1/z)(1 - \pi_n)$. For $z \neq 0$, $R_n(z)$ is uniformly bounded if $(\mathscr{T}_n - z)^{-1}$ is so. For $z = 0$, note that $T_n \xrightarrow{\text{s}} T$ is not defined when T_n is of class $\mathfrak{D}$. □

Exercises

3.12 Show that for T_n of class $\mathfrak{D}$, $T_n - z\pi_n \xrightarrow{\text{d-r}} T - z$ is equivalent to $T_n - z \xrightarrow{\text{r}} T - z$ for any nonzero z in $\rho(T)$.

3.13 Show that for T_n of class $\mathfrak{D}$, $T_n \xrightarrow{\text{d-s}} T$ implies that $T_n - z\pi_n \xrightarrow{\text{d-r}} T - z$ for any z in $\rho(T)$.

3.14 Let the injection $i: \hat{D} \to X$ be compact. Show that for $T, T_n \in \mathscr{L}(\hat{D}, X)$ and T_n of class $\mathfrak{D}$, $T_n \xrightarrow{\text{d-r}} T$ implies $T_n - z\pi_n \xrightarrow{\text{d-r}} T - z$ for all z in $\mathbb{C}$.

3.15 If the stability condition (3.9) is satisfied, show that the problem of solving $T_n x_n = y_n$, $\|y_n\| \leq C$, is stable uniformly in n. This means that the perturbed equation

$$(T_n + \Delta T_n)(x_n + \Delta x_n) = y_n + \Delta y_n$$

such that $\|\Delta T_n\| \leq \varepsilon$, $\|\Delta y_n\| \leq \alpha$, is solvable if $\varepsilon < 1/M$, and there exists β such that $\|\Delta x_n\| \leq \beta$.

6. Summary

The various connections between the convergences are summarized in Tables 3.2 and 3.3, z does not appear in the conditions lying in the boxes. Some of the characterizations could be given in a different form (Grigorieff, 1973; Vainikko, 1978a), but the chosen form is the most directly usable for

Table 3.2

Stability in $\rho(T)^a$

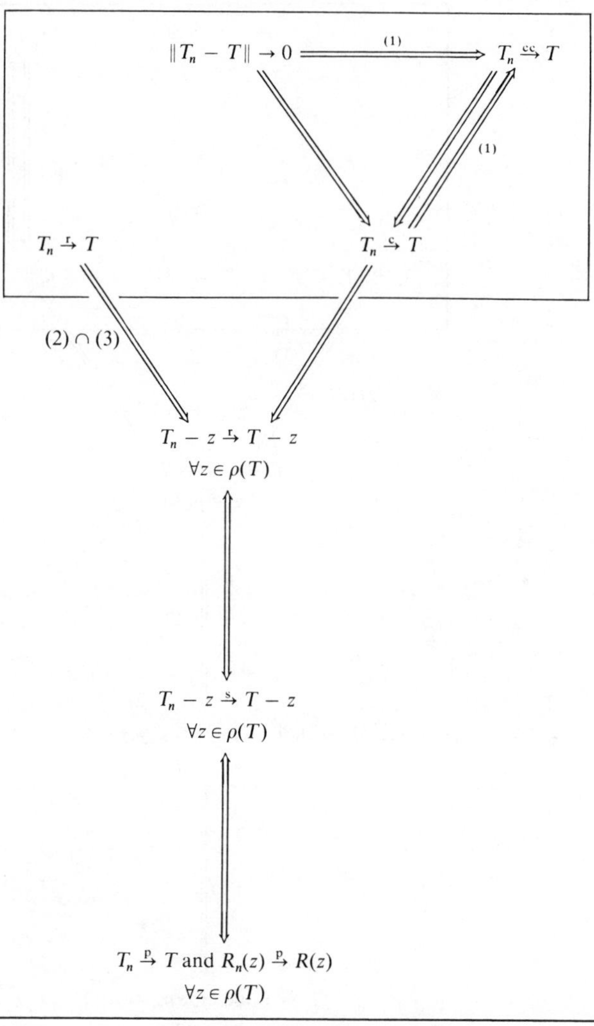

[a] (1) valid if T and T_n are compact; (2) valid if T is anticompact; (3) valid in $\mathscr{L}(\hat{D}, X)$.

the study of spectral convergence to be done in Chapter 5, where the spectral projection P is defined by integration of the resolvent $R(z)$ for z on $\Gamma \subset \rho(T)$.

Lemma 3.16 has been known for a long time, and the stable convergence has been used more or less explicitly for solving integral or differential equations. In the context of discrete approximation theory, the discrete-regular convergence was introduced in Karma's (1971) thesis and in Grigorieff

Table 3.3

Discrete stability in $\rho(T)$[a]

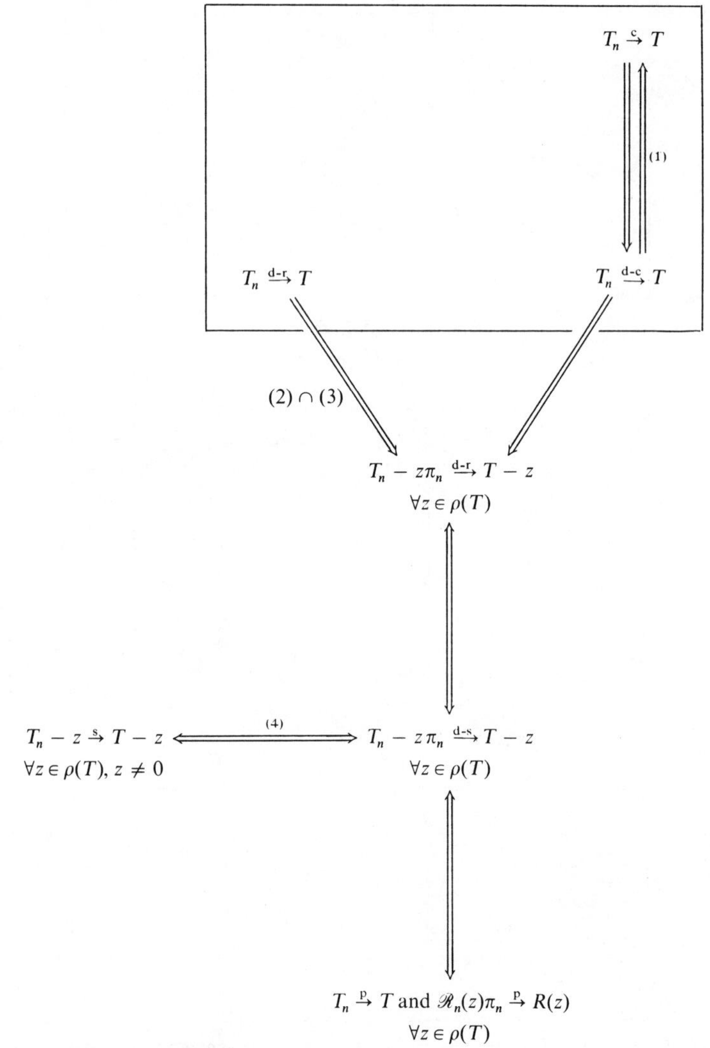

[a] (1) valid if T and T_n are compact; (2) valid if T is anticompact; (3) valid in $\mathscr{L}(D, X)$; (4) valid if z nonzero.

(1973), generalizing the works of Polskii (1962), Browder (1967), and Petryshyn (1968) for a special class of projection methods. Criteria for discrete-regular convergence are given in Grigorieff (1973). A thorough study of the discrete-stable and discrete-regular convergences for operators in $\mathscr{L}(\hat{D}, X)$ is in Vainikko (1978a) together with applications to nonlinear operators. The notion of regular convergence, given by Anselone and Ansorge (1979) for nonlinear continuous operators, reduces to the above definition for linear operators. The concept of discrete-regular convergence has been applied successfully to finite difference methods for differential equations (cf. Grigorieff, 1970a,b, 1972, 1973, 1975a,b; Grigorieff and Jeggle, 1973; Vainikko, 1975, 1976a,b, 1977a,b, 1978a,b; Vainikko and Karma, 1974a,b). The property given in Exercise 3.14 has been used in essence in the content of partial differential equations by Grigorieff, Vainikko, and their co-workers to prove the discrete stability of finite difference schemes. The characterization of the discrete stability by the discrete regularity appeared in Grigorieff (1973).

The gap convergence $\Theta(T_n, T) \to 0$ is applied in Descloux *et al.* (1981) to the approximation of a closed operator with a noncompact resolvent by the finite-element method.

B. ANALYTIC PERTURBATION THEORY

Let T be a closed operator with domain D. We suppose that $T = T' - H$. T is the result of a perturbation H applied to T', where H and T' are closed operators with domain D. Our problem is to solve

$$(T - z)x = f \qquad \text{or} \qquad T\varphi = \lambda\varphi.$$

But we know only how to solve equations in which T is replaced by T'. This is a very common situation. We shall be concerned mainly with the two fundamental examples:

(a) T' is a *numerical* approximation T_n of T; therefore $T = T_n - (T_n - T)$;

(b) T' is the part $\tilde{T}$ of T defined by $\tilde{T} := QTQ + (1 - Q)T(1 - Q)$, where Q is a given projection, and $T = \tilde{T} + \tilde{H}$, with $\tilde{H} := QT(1 - Q) + (1 - Q)TQ$.

To relate T and T', we consider the family of operators $T(t) = T' - tH$ depending on the complex parameter t. Clearly $T(1) = T$ and $T(0) = T'$. We consider the problems

$$(T(t) - z)x(t) = f,$$

$$T(t)\varphi(t) = \lambda(t)\varphi(t).$$

We wish to know if $x(t)$, $\lambda(t)$, and $\varphi(t)$ are analytic functions of t in a disk centered at zero, which includes $t = 1$. When the answer is positive, the solutions x, λ, and φ will be iteratively computed from the solutions x', λ', and φ' of the respective equations

$$(T' - z)x' = f, \tag{3.12}$$

$$T'\varphi' = \lambda'\varphi'. \tag{3.13}$$

Let Γ be a closed positively oriented Jordan curve around λ', an isolated eigenvalue of T' of multiplicity m and ascent l. We define formally

$$R(t, z) = (T(t) - z)^{-1} \qquad \text{for} \quad z \in \rho(T');$$

$$P(t) = \frac{-1}{2i\pi} \int_\Gamma R(t, z)\, dz \qquad \text{for} \quad \Gamma \subset \rho(T');$$

$$\hat{\lambda}(t) = (1/m) \operatorname{tr} T(t)P(t).$$

We study the analyticity in t of these functions.

7. Analyticity of $R(t, z)$, $P(t)$, and $\hat{\lambda}(t)$

7.1. *Analyticity of* $R(t, z)$

Let z be given in $\rho(T')$. $R'(z) := (T' - z)^{-1}$, $TR'(z)$, and $T'R'(z)$ are closed operators with domain X. They are therefore bounded by the closed-graph theorem, as is $HR'(z)$. We define

$$r'_z := r_\sigma[HR'(z)],$$

$$\delta'_z := \{t \in \mathbb{C}; |t| < 1/r'_z\}.$$

Lemma 3.22 *$T(t)$ is closed with domain D for all t in δ'_z.*

Proof From the identity

$$T(t) - z = T' - z - tH = [1 - tHR'(z)](T' - z)$$

follows formally the Neumann-series expansion for $R(t, z)$:

$$R(t, z) = R'(z) \sum_{k=0}^{\infty} t^k [HR'(z)]^k. \tag{3.14}$$

The series in (3.14) converges in $\mathscr{L}(X)$ iff $r_\sigma[tHR'(z)] < 1$, by Proposition 2.25, that is, iff $|t| < 1/r'_z$. $T(t) - z$, which has a bounded inverse with range D, is itself closed with domain D. □

Proposition 3.23 *$R(t, z)$ is analytic for all t in δ'_z, and its Neumann-series expansion* (3.14) *is convergent in* $\mathscr{L}(X)$.

Proof Clear from the proof of Lemma 3.22. □

Corollary 3.24 *$x(t)$ is analytic for all t in δ'_z.*

Proof $x(t) = R(t, z)f = \sum_{k=0}^{\infty} t^k y_k$, with $y_k := R'(z)[HR'(z)]^k f$, $k \geq 0$. □

7.2. *Analyticity of* $P(t)$ *and* $\hat{\lambda}(t)$

Let Γ be a closed Jordan curve around λ', an isolated eigenvalue of T' of finite algebraic multiplicity m. We recall that the spectral radius $r_\sigma(HR'(z))$ is upper semicontinuous in $\rho(T')$ (Proposition 2.20). It then achieves its maximum on the compact set $\Gamma \subset \rho(T')$. We define

$$r'_\Gamma := \max_{z \in \Gamma} r_\sigma[HR'(z)],$$

$$\delta'_\Gamma := \{t \in \mathbb{C};\, |t| < 1/r'_\Gamma\}.$$

Then $r'_z \leq r'_\Gamma$ and $\delta'_z \supset \delta'_\Gamma$ for all $z \in \Gamma$.

Exercises

3.16 Prove that δ'_Γ is the largest open disk in $\mathbb{C}$, centered at 0, such that, for any t in δ'_Γ, $T(t)$ is closed with domain D and Γ lies in $\rho(T(t))$.

3.17 Show the following characterization in $\mathscr{L}(\hat{D}, X)$ (Lemordant): the set

$$\left\{H \in \mathscr{L}(\hat{D}, X);\, \max_{z \in \Gamma} r_\sigma[HR'(z)] < 1\right\}$$

is the largest balanced set W in $\mathscr{L}(\hat{D}, X)$ such that, for all H in W, $T' - H \in \mathscr{C}(X)$ and $\Gamma \subset \rho(T' - H)$.

Proposition 3.25 *$P(t)$ and $\hat{\lambda}(t)$ are analytic functions of t in δ'_Γ.*

Proof $P(t)$ is well defined in δ'_Γ, and we prove its analyticity around t_0 fixed in δ'_Γ. If $|t - t_0| < (\max_{z \in \Gamma}\|HR(t_0, z)\|)^{-1}$, the series

$$R(t, z) = R(t_0, z) \sum_{k=0}^{\infty} (t - t_0)^k [HR(t_0, z)]^k$$

converges uniformly for z on Γ. By integration on Γ, this shows that $P(t)$ is analytic around any point t_0 in δ'_Γ.

Since $P(t)$ is a continuous function of t in δ'_Γ, the dimension of its range $P(t)X$ is constant by Corollary 2.15. This proves that $\dim P(t)X = m$. $T(t)$ has m isolated eigenvalues inside Γ, counted with their algebraic multiplicities, $\hat{\lambda}(t) := (1/m)\,\mathrm{tr}\, T(t)P(t)$ is their arithmetic mean.

We now show that $\hat{\lambda}(t)$ is analytic in a neighborhood of t_0 in δ'_Γ. Let $\{x_i\}$ (resp. $\{x_i^*\}$) be a basis of $M_0 := P(t_0)X$ (resp. $M_0^* := P^*(t_0)X^*$) such that $\langle x_i^*, x_j\rangle = \delta_{ij}$:

$$P(t_0) = \sum_{i=1}^{m} \langle\cdot, x_i^*\rangle x_i.$$

For $|t - t_0|$ small enough, the vectors $\{P(t)x_i\}_1^m$ form a basis of $P(t)X$, by Proposition 3.10. The adjoint basis is defined by $\{\sum_{k=1}^m b_{jk}(t)x_k^*\}_1^m$, where the functions b_{jk} are solutions of the set of equations

$$\sum_{k=1}^{m} b_{jk}(t)\langle x_k^*, P(t)x_i\rangle = \delta_{ji} \qquad \text{for} \quad i, j = 1, \ldots, m.$$

The $m \times m$ matrix $A(t)$ defined by the coefficients $\langle x_i^*, P(t)x_j\rangle$ is invertible around t_0, since $A(t_0) = I$. The coefficients of $A^{-1}(t)$ are the $b_{jk}(t)$, which are therefore analytic around t_0. Now

$$m\hat{\lambda}(t) = \operatorname{tr} T(t)P(t) = \sum_{i=1}^{m} \left\langle (T' - tH)P(t)x_i, \sum_{k=1}^{m} b_{ik}(t)x_k^* \right\rangle$$

$$= \sum_{i=1}^{m} \sum_{k=1}^{m} \overline{b_{ik}(t)}\langle (T' - tH)P(t)x_i, x_k^*\rangle.$$

This shows that $\hat{\lambda}(t)$ is analytic around any point t_0 in δ'_Γ. □

Proposition 3.26 *The following expansions hold for t in δ'_Γ:*

$$P(t) = P' - \sum_{k=2}^{\infty} t^{k-1} \sum_{*} S'^{(p_1)}HS'^{(p_2)} \cdots HS'^{(p_k)}, \tag{3.15}$$

$$\hat{\lambda}(t) = \lambda' + \frac{1}{m}\sum_{k=1}^{\infty} t^k \frac{1}{k}\sum_{*} \operatorname{tr}[HS'^{(p_1)} \cdots HS'^{(p_k)}], \tag{3.16}$$

with

$$* = \left\{p_i \geq 1 - l, i = 1, \ldots, k; \sum_{i=1}^{k} p_i = k - 1\right\},$$

$$S'^{(0)} := -P', \qquad S'^{(-p)} := -D'^p, \qquad S'^{(p)} := S'^p, \qquad \textit{for} \quad p > 0,$$

$$D' := (T' - \lambda')P', \qquad \textit{and} \qquad S' := \lim_{z \to \lambda'} R'(z)(1 - P').$$

Proof The above expansions are the well-known perturbation series for a family of closed operators $T(t) = T' - tH$. They are given, for example, in Kato (1976, pp. 74–80, 379, and 380) for *t small enough.* □

7.3. *Analyticity of $\phi(t)$ when $m = 1$*

When $m = 1$, if $P'\varphi(t) \neq 0$, the eigenvector $\phi(t)$ of $T(t)$ normalized by $P'\phi(t) = \varphi'$ is well defined. And we assume now on that the domain D is dense, so that the adjoint T'^* is uniquely defined.

Proposition 3.27 *When $m = 1$, the eigenvector $\phi(t)$ normalized by $P'\phi(t) = \varphi'$ is analytic for $|t|$ small enough.*

Proof We recall that for $m \geq 1$, $\hat{\lambda}(t) = \sum_{k=0}^{\infty} t^k v_k$ for all t in δ'_Γ, by (3.16), where

$$v_0 := \lambda', \qquad v_k := \frac{1}{mk} \sum_{*} \operatorname{tr}[HS'^{(p_1)} \cdots HS'^{(p_k)}] \qquad \text{for} \quad k \geq 1.$$

But the situation is more complicated for the eigenvectors, even if $m = 1$. Let φ' be an eigenvector of T' associated with λ', $\|\varphi'\| = 1$. $P(t)\varphi'$ is analytic for t in δ'_Γ, but is an eigenvector of $T(t)$ only if it is *nonzero*. For $|t|$ small enough, $P(t)\varphi'$ is nonzero because it is close to $P'\varphi' = \varphi' \neq 0$. $\phi(t)$ and $P(t)\varphi'$ are colinear. We set $\phi(t) := \gamma(t)P(t)\varphi'$, where $1/\gamma(t) = \langle P(t)\varphi', \varphi'^* \rangle$. γ is analytic in t as long as $\langle P(t)\varphi', \varphi'^* \rangle$ is nonzero. So is $\phi(t)$. For $|t|$ small enough, $1/\gamma(t)$ is nonzero since $1/\gamma(0) = \langle P'\varphi', \varphi'^* \rangle = 1$. Therefore, for $|t|$ small enough, $\phi(t) = \sum_{k=0}^{\infty} t^k \eta_k$, where the coefficients η_k result from the product of the series expansions of $\gamma(t)$ and $P(t)\varphi'$. □

Under the condition

$$\langle P(t)\varphi', \varphi'^* \rangle \text{ is nonzero for } t \text{ in } \delta'_\Gamma, \tag{3.17}$$

the eigenvector $\phi(t)$ normalized by $P'\phi(t) = \varphi'$ is analytic for t in δ'_Γ. Note that $P'P(t)\varphi' = \langle P(t)\varphi', \varphi'^* \rangle \varphi$.

Exercise

3.18 Show that $\langle P(t)\varphi', \varphi'^* \rangle$ is nonzero for $|t| \leq 1$ if

$$\max_{z \in \Gamma} \|HR'(z)\| < \left(1 + \frac{\text{meas}\,\Gamma}{2\pi} \max_{z \in \Gamma} \|R'(z)\| \, \|\varphi'^*\|\right)^{-1},$$

where meas Γ denotes the Lebesgue measure of Γ.

8. Iterative Computation of the Coefficients of the Series Expansions

We present an *iterative* computation of the coefficients y_k, v_k, η_k of the series expansion for $x(t)$, $\hat{\lambda}(t)$, and $\phi(t)$. For the eigenelements, we assume that $m = 1$, so that $\hat{\lambda}(t) = \lambda(t)$, the only eigenvalue of $T(t)$ inside Γ.

The formulas to get v_k and η_k will be considerably simpler than the one obtained from (3.15) and (3.16).

8.1. Computation of the Coefficients of $x(t)$

Lemma 3.28 *The series $x(t) = \sum_{k=0}^{\infty} t^k y_k$ converges for t in δ'_z, where the coefficients y_k are given by*

$$y_0 = x', \qquad y_k = R'(z)Hy_{k-1}, \qquad \textit{for} \quad k \geq 1.$$

Proof It follows readily from (3.14). □

Let x be the solution of

$$(T - z)x = f. \tag{3.7}$$

If $t = 1$ belongs to δ'_z, we may regard T' as an approximation of T such that x can be computed by iterative improvement from x', by computations that require only the inversion of $T' - z$. For that purpose, we define

$$x_k := \sum_{i=0}^{k} y_i = x_{k-1} + y_k \qquad \text{for} \quad k \geq 1, \quad x_0 = y_0 = x'.$$

Then

$$\begin{aligned} Hy_{k-1} &= [T' - z - (T - z)]y_{k-1} = f - (T - z)\left(\sum_{i=0}^{k-1} y_i\right) \\ &= f - (T - z)x_{k-1}; \end{aligned}$$

that is, y_k is solution of

$$(T' - z)y_k = f - (T - z)x_{k-1} \qquad \text{for} \quad k \geq 1, \tag{3.18}$$

where the right-hand side is the residual for (3.7) computed at x_{k-1}.

Proposition 3.29 *If $r'_z < 1$, $x = \lim_k x_k$, where the x_k are solutions of*

$$x_0 = x', \qquad (T' - z)(x_k - x') = (T' - T)x_{k-1} \qquad \textit{for} \quad k \geq 1. \tag{3.19}$$

Proof If $r'_z < 1$, $t = 1$ belongs to δ'_z, and (3.19) follows readily from Lemma 3.28. □

x is the limit of solutions of systems (3.19), where only the right-hand side varies. It requires the evaluation Tx_{k-1}.

Example 3.13 In $X = \mathbb{C}^N$, an example is provided by the method of iterative refinement applied to the solution x of $Ax = b$, under the condition $\|H\| \, \|A'^{-1}\| < 1$, which is a stronger assumption than $r_\sigma(HA'^{-1}) < 1$.

Example 3.14 The iterative refinement method is a special case of the *iterative defect correction* method presented in Stetter (1978): $R'(z)$ is an approximate inverse of the *linear* operator $T - z$,

$$HR'(z) = 1 - (T - z)R'(z) = K(z).$$

Since $K(z) \in \mathscr{L}(X)$, it is a contraction iff its spectral radius is less than 1, which is the condition $r'_z < 1$. We shall go back to this question when T' is a numerical approximation of T (Chapters 5 and 7).

Example 3.15 Consider, in $\mathbb{R}^2$, the 2×2 matrices

$$A = \begin{pmatrix} 1 & 0 \\ 0 & 2 \end{pmatrix} \quad \text{and} \quad A_n = \begin{pmatrix} 1 & \sqrt{n} \\ 1/n & 2 \end{pmatrix}$$

for $n \in \mathbb{N}$. Then

$$A_n - A = \begin{pmatrix} 0 & \sqrt{n} \\ 1/n & 0 \end{pmatrix}, \qquad A^{-1} = \begin{pmatrix} 1 & 0 \\ 0 & 1/2 \end{pmatrix},$$

$$(A_n - A)A^{-1} = \begin{pmatrix} 0 & 1/2\sqrt{n} \\ 1/n & 0 \end{pmatrix},$$

$$[(A_n - A)A^{-1}]^2 = \begin{pmatrix} 1/2\sqrt{n} & 0 \\ 0 & 1/2\sqrt{n} \end{pmatrix},$$

and

$$r_\sigma[(A_n - A)A^{-1}] = \left(\frac{1}{2\sqrt{n}}\right)^{1/2} = \frac{1}{\sqrt{2}} n^{-1/4} \xrightarrow[n \to \infty]{} 0.$$

Note that $\|A_n - A\|_2 \to \infty$ as $n \to \infty$.
The system

$$x + \sqrt{n}y = 1, \qquad (1/n)x + 2y = 0$$

has the solution

$$x = \left(1 - \frac{1}{2\sqrt{n}}\right)^{-1}, \qquad y = \frac{-1}{2n}\left(1 - \frac{1}{2\sqrt{n}}\right)^{-1}.$$

The solution has a converging expansion in powers of $n^{-1/2}$ if $1/2\sqrt{n} < 1$, which is the condition $r_\sigma[(A_n - A)A^{-1}] < 1$.

Exercises

3.19 Let

$$B = \begin{pmatrix} 1 & 1 \\ 1 & 2 \end{pmatrix}.$$

Compute iteratively the solution of the system

$$x + y = 1, \qquad x + 2y = 0$$

from the solution of

$$x = 1, \qquad 2y = 0.$$

Verify that the rate of convergence is geometric with quotient $1/\sqrt{2}$.

3.20 Show that the rate of convergence $x_k \to x$ in Proposition 3.29 is arbitrarily close to $r'_z < 1$. (*Hint*: Proceed as for Exercise 2.38.)

8.2. Computation of the Coefficients of $\lambda(t)$ and $\phi(t)$ when $m = 1$

We begin with an analog of Lemma 3.28.

Lemma 3.30 *When $m = 1$, under the assumption* (3.17), *the series $\lambda(t) = \sum_{k=0}^{\infty} t^k \nu_k$ and $\phi(t) = \sum_{k=0}^{\infty} t^k \eta_k$ converge for $t \in \delta'_\Gamma$. The coefficients ν_k and η_k are given by*

$$\begin{aligned} \nu_0 &= \lambda', \qquad & \nu_k &= \langle -H\eta_{k-1}, \varphi'^* \rangle \qquad & &\text{for} \quad k \geq 1, \\ \eta_0 &= \varphi', \qquad & \eta_k &= S'\left[H\eta_{k-1} + \sum_{i=1}^{k} \nu_i \eta_{k-i}\right], \qquad & &\text{for} \quad k \geq 1. \end{aligned} \tag{3.20}$$

Proof We use the method of undetermined coefficients. We identify formally the coefficients of t^k in

$$T(t)\phi(t) = (T' - tH)\phi(t) = \lambda(t)\phi(t),$$

where $\phi(t)$ is normalized by $P'\phi(t) = \varphi'$.

For $k = 0$, $T'\eta_0 = \nu_0\eta_0$ is satisfied with the choice $\nu_0 = \lambda'$ and $\eta_0 = \varphi'$. Since $P'\phi(t) = \varphi' = \phi(0)$ it follows that $P'\eta_k = 0$ for $k > 0$.

For $k = 1$, we get

$$(T' - \lambda')\eta_1 - H\varphi' = \nu_1\varphi'.$$

By left multiplication by φ'^* such that $T'^*\varphi'^* = \bar{\lambda}'\varphi'^*$ and $\langle \varphi'^*, \varphi' \rangle = 1$, we get

$$\bar{\nu}_1 = \langle \varphi'^*, -H\varphi' \rangle \Leftrightarrow \nu_1 = \langle -H\varphi', \varphi'^* \rangle.$$

ν_1 has been chosen such that the right-hand side of the equation

$$(T' - \lambda')\eta_1 = H\varphi' + \nu_1\varphi'$$

lies in $(1 - P')X$. There is a unique solution η_1 such that

$$\eta_1 = S'[H\varphi' + \nu_1\varphi'] = S'H\varphi', \qquad P'\eta_1 = 0.$$

For $k = 2$, we get

$$(T' - \lambda')\eta_2 - H\eta_1 - \nu_1\eta_1 = \nu_2\varphi'.$$

Since $P'\eta_1 = 0$, we get by multiplication by φ'^*

$$\nu_2 = \langle -H\eta_1, \varphi'^* \rangle.$$

Now

$$(T' - \lambda')\eta_2 = H\eta_1 + \nu_1\eta_1 + \nu_2\varphi'$$

has the unique solution η_2 such that

$$\eta_2 = S'[H\eta_1 + \nu_1\eta_1], \qquad P'\eta_2 = 0.$$

By similarly equating the coefficients of t^k, $k > 2$, we prove (3.20). □

If $t = 1$ belongs to δ'_Γ, λ and ϕ, solutions of $T\phi = \lambda\phi$ where ϕ is normalized by $P'\phi = \varphi'$, can be iteratively computed from λ' and φ'. We define $\lambda_k := \sum_{i=0}^{k} \nu_i = \lambda_{k-1} + \nu_k$ for $k \geq 1$, $\lambda_0 := \nu_0 = \lambda'$. Similarly

$$\phi_k := \sum_{i=0}^{k} \eta_i = \varphi_{k-1} + \eta_k \qquad \text{for} \quad k \geq 1, \quad \varphi_0 := \eta_0 = \varphi'.$$

Proposition 3.31 *If $m = 1$ and $r'_\Gamma < 1$, then under the assumption* (3.17), $\lambda = \lim_k \lambda_k$, $\phi = \lim_k \varphi_k$, *where the λ_k, φ_k are solutions of*

$$\lambda_0 = \lambda', \qquad \varphi_0 = \eta_0 = \varphi', \qquad \lambda_k = \langle T\varphi_{k-1}, \varphi'^* \rangle,$$

$$(T' - \lambda)(\varphi_k - \varphi') = (1 - P')\left[(T' - T)\phi_{k-1} + \sum_{i=1}^{k} \sum_{j=1}^{i} \nu_j \eta_{i-j}\right], \qquad (3.21)$$

for $k \geq 1$.

Proof If $r'_\Gamma < 1$, $t = 1$ belongs to δ'_Γ, and (3.21) follows readily from Lemma 3.30. □

Proposition 3.32 *Under the assumptions of Proposition* 3.31, *the rate of convergence of the sequences λ_k, φ_k is at least the rate of a geometric progression with quotient arbitrarily close to r'_Γ.*

Proof $\lambda(t)$ and $\phi(t)$ are analytic in δ'_Γ. We may then apply the Cauchy inequalities on the circle $\{t; |t| = 1/q\}$ with $q > r'_\Gamma$ (cf. Chapter 2, Section 7.3). We get $|\nu_i| \leq Cq^i$ and $\|\eta_i\| \leq C'q^i$ for $i \geq 0$ with $C = \sup_{|t|=1/q} |\lambda(t)|$ and $C' = \sup_{|t|=1/q} \|\phi(t)\|$. The result for $\lambda - \lambda_k = \sum_{k+1}^{\infty} \nu_i$ and $\phi - \varphi_k = \sum_{k+1}^{\infty} \eta_i$ follows at once. □

Example 3.16 We have seen in Chapter 1, two examples of the iterative refinement method applied to matrices.

(a) In Proposition 1.14, it is applied to $A = A' - H$. Condition (3.17) is shown to be fulfilled if $\|H\|$ is small enough by Exercise 3.18 (cf. also Theorem 1.7, where $A' = A - \varepsilon L$).

(b) In Theorem 1.17, it is applied to the decomposition $A = \tilde{A} + \tilde{H}$ corresponding to the projection $Q = xy^{\mathrm{H}}$, $\zeta = y^{\mathrm{H}}Ax$. The very special

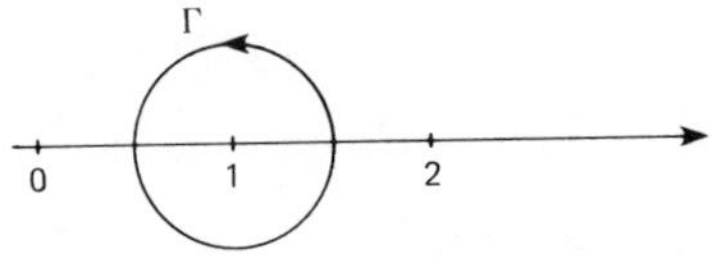

Figure 3.1

structure of $\tilde{H}$ implies the particularities of the series, namely, $\eta_{2k} = 0$ and $\nu_{2k-1} = 0$ for $k \geq 1$. In the proof of Theorem 1.17, the convergences $\lambda_k \to \lambda$ and $\varphi_k \to \varphi$ (where φ is normalized by $Q\varphi = x$) are proved directly, without considering $r'_\Gamma = \max_{z \in \Gamma} r_\sigma[\tilde{H}(\tilde{A} - zI)^{-1}]$, Γ being a Jordan curve around ζ.

Example 3.17 In the framework of quantum mechanics the series expansion of $\lambda(t)$ and $\phi(t)$ given by (3.20) are known as the *Rayleigh–Schrödinger series* (Reed and Simon, 1978). When T' is a numerical approximation T_n of a compact T, other types of iterations will be given in Chapter 5, in Sections 6.5 and 6.6, in connection with the iterative defect correction method on a *nonlinear* operator.

Example 3.18 Consider in $\mathbb{C}^2$

$$B(t) = \begin{pmatrix} 1 & 5t \\ t/25 & 2 \end{pmatrix}.$$

Then

$$B(0) = A = \begin{pmatrix} 1 & 0 \\ 0 & 2 \end{pmatrix}, \qquad R(z) = (A - zI)^{-1} = \begin{pmatrix} 1/(1-z) & 0 \\ 0 & 1/(2-z) \end{pmatrix},$$

$$(B(t) - A)R(z) = \begin{pmatrix} 0 & 5t/(2-z) \\ t/25(1-z) & 0 \end{pmatrix}.$$

A has two eigenvalues, 1 and 2. Let Γ be the circle $|z - 1| = \frac{1}{2}$ around $\lambda = 1$; cf. Fig. 3.1. Then

$$\max_{z \in \Gamma} r_\sigma[(B(t) - A)R(z)] = \sqrt{\frac{4}{5}t^2} = (2/\sqrt{5})|t|.$$

$B(t)$ has the eigenvalues

$$\lambda(t) = \frac{3}{2} \pm \frac{1}{2}\sqrt{1 + \frac{4}{5}t^2}.$$

They have a converging series expansion in powers of t^2 if $t^2 < \frac{5}{4}$, which is the condition

$$\max_{z \in \Gamma} r_\sigma[(B(t) - A)R(z)] < 1.$$

Note that for $t = 1$,

$$B(1) = \begin{pmatrix} 1 & 5 \\ 1/25 & 2 \end{pmatrix},$$

and

$$B(1) - A = \begin{pmatrix} 0 & 5 \\ 1/25 & 0 \end{pmatrix}$$

has no small norm.

Exercise

3.21 Show that the stability of the computation of φ_k depends on $\|S'\|$ and $\|P'\|$.

To compute λ and ϕ in Proposition 3.31, we have assumed for clarity that λ' is simple. The case of a multiple eigenvalue or of a group of close eigenvalues is treated in Lemordant (1980). We present now the method of Lemordant, rewritten to fit our framework.

8.3. The Case $m > 1$

We now suppose that the spectral projection P' is of rank m, considering the two possibilities: Γ encloses a single eigenvalue λ' with algebraic multiplicity m, or Γ encloses several eigenvalues μ_i' with total algebraic multiplicity m. With the notation of Section 7.2, $P(t)$ and $\hat{\lambda}(t)$ are analytic functions of t in δ_Γ'. We define the invariant subspace $M(t) := P(t)X$. We shall show that there exists a basis $\{u_i(t)\}_1^m$ of $M(t)$ such that $u_i(t) = \sum_{k=0}^{\infty} t^k u_{ik}$, $i = 1, \ldots, m$, where the vectors $\{u_{ik}\}$ can be iteratively constructed from a basis of $M' := P'X$.

The basis of $M(t)$ is defined by means of a projection $Q(t)$ onto $M(t)$, along a *fixed* subspace W. For example, let $x_1, \ldots, x_m$ be m independent vectors in M'. We consider m vectors $\{y_i\}_1^m$ in X^* such that

$$\langle y_i, x_j \rangle = \delta_{ij}, \qquad i, j = 1, \ldots, m.$$

$Q' := \sum_{i=1}^m \langle \cdot, y_i \rangle x_i$ is the projection on M' along $W := \{y_1, \ldots, y_m\}^\perp \cap X$. Similarly, we wish to define $Q(t)$ as the projection on $M(t)$ along W. We introduce the condition

$$r_\sigma[(P(t) - P')Q'] < 1 \qquad \text{for} \quad t \text{ in } \delta_\Gamma' \tag{3.22}$$

Lemma 3.33 *Under the condition* (3.22), $P(t)(Q'P(t)_{\restriction M'})^{-1}Q'$ *is a function of t analytic in δ_Γ'. It defines the projection on $M(t)$ along W.*

Proof We set $S(t) := [Q'(P(t) - P')Q']_{\restriction M'}$. Since Q' is a projection, $r_\sigma[Q'(P(t) - P')Q'] = r_\sigma[(P(t) - P')Q'] < 1$; therefore $(1 + S(t))^{-1}$ exists

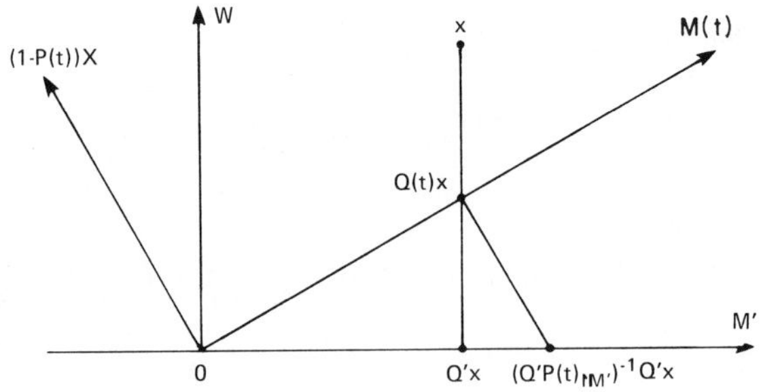

Figure 3.2

in $\mathscr{L}(M')$ and is analytic for t in δ'_Γ. $1 + S(t) = Q'P(t)_{\restriction M'}$, and we set $Q(t) := P(t)(Q'P(t)_{\restriction M'})^{-1}Q'$. Clearly $Q^2(t) = Q(t)$ (cf. Fig. 3.2), $Q(t)X = M(t)$, and $Q(t)(1 - Q') = 0$. □

Lemma 3.34 *Under the condition* (3.22) *the expansion* $Q(t) = \sum_{k=0}^{\infty} t^k Q_k$ *is convergent in* $\mathscr{L}(X)$. *The* Q_k *are solution of the recursion*

$$
\begin{aligned}
Q_0 = Q', \qquad (1 - Q')T'Q_k - Q_k T'Q' &= B_k, \\
Q'Q_k = Q_k(1 - Q') = 0, \qquad k &\geq 1
\end{aligned}
\tag{3.23}
$$

with $B_1 := (1 - Q')HQ'$,

$$
B_k := (1 - Q')HQ_{k-1} - \sum_{i=1}^{k-1} Q_i H Q_{k-i-1} + \sum_{i=1}^{k-1} Q_i T' Q_{k-i}, \qquad k \geq 2.
$$

Proof $Q(t)$ is analytic and the coefficient Q_k can be obtained by the method of undetermined coefficients applied to the equation

$$
(1 - Q(t))(T' - tH)Q(t) = 0.
$$

This yields $Q_0 = Q'$, and Eq. (3.23) with $Q'Q_k = 0$, $Q_k Q' = Q_k$, $k \geq 1$ by using $Q'Q(t) = Q$ and $Q(t)(1 - Q') = 0$. Note that we also have

$$
Q'B_k = B_k(1 - Q') = 0. \quad \square
$$

We now concentrate on the resolution of Eq. (3.23) for a fixed $k \geq 1$. For that purpose, we consider the general equation in the unknown $\mathbf{X}$, a bounded operator,

$$
(1 - Q')T'\mathbf{X} - \mathbf{X}T'Q' = B, \qquad Q'\mathbf{X} = \mathbf{X}(1 - Q') = 0, \tag{3.24}
$$

where $Q'B = B(1 - Q') = 0$.

Lemma 3.35 *If Eq.* (3.24) *has a solution* $\mathbf{X}$, *then it is unique and it can be written* $\mathbf{X} = \sum_{i=1}^{m} \langle \cdot, y_i \rangle v_i$, *with* $v_i \in (1 - Q')D$.

Proof We suppose that (3.24) has two solutions, and we consider $\mathbf{Y} = \mathbf{X}_1 - \mathbf{X}_2$. $\mathbf{Y}$ is solution of

$$(1 - Q')T'\mathbf{Y} - \mathbf{Y}T'Q' = 0, \qquad Q'\mathbf{Y} = \mathbf{Y}(1 - Q') = 0. \tag{3.25}$$

The conditions $Q'\mathbf{Y} = \mathbf{Y}(1 - Q') = 0$ imply that $\mathbf{Y}$ can be written $\mathbf{Y} = \sum_{i=1}^{m} \langle \cdot, y_i \rangle w_i$ with $w_i = \mathbf{Y}x_i \in (1 - Q')D$. From (3.25), we derive $(1 - Q')T'w_j - \sum_{i=1}^{m} \langle T'Q'x_j, y_i \rangle w_i = 0$, $j = 1, \ldots, m$. Let V be the span of $\{w_1, \ldots, w_m\}$: it is invariant under $(1 - Q')T'$. Let Π be a projection on V. $\Pi\mathbf{Y} = \mathbf{Y}$, and (3.25) implies

$$\Pi(1 - Q')T'\Pi\mathbf{Y} - \mathbf{Y}Q'T'Q' = 0. \tag{3.26}$$

Π and Q' being projections of rank at most m, (3.26) defines a matrix equation having 0 for only solution, according to Gantmacher (1959), if the spectra of the matrices corresponding to $\Pi(1 - Q')T'\Pi$ and $Q'T'Q'$ are disjoint. This is indeed true, since the first spectrum lies outside Γ and the second inside (Exercise 3.24). □

Exercises

3.22 When $m = 1$ and $Q' = P'$, check that (3.23) reduces to (3.20).

3.23 Show that $r_\sigma[Q'(P(t) - P')Q'] < 1$ implies $Q'P(t)x_i \neq 0$ for $i = 1, \ldots, m$. Compare with the condition in (3.17) when $m = 1$.

$\mathbf{X} = \sum_{i=1}^{m} \langle \cdot, y_i \rangle v_i$ is determined if the m vectors $\{v_i\}_1^m$ are known. Note that $v_i = \mathbf{X}x_i$; solving Eq. (3.24) in the unknown $\mathbf{X} \in \mathscr{L}(X)$ is equivalent to solving the system of m equations and m unknowns $v_i \in X$: $(1 - Q')T'v_i - \sum_{j=1}^{m} \langle T'x_i, y_j \rangle v_j = Bx_i$, $i = 1, \ldots, m$. We remark that if the x_i, $i = 1, \ldots, m$, are eigenvectors for T', then $\langle T'x_i, y_j \rangle = \mu_i' \delta_{ij}$, and the system reduces to the set of m equations $(1 - Q')(T' - \mu_i')v_i = Bx_i$, $i = 1, \ldots, m$.

We now give an algorithm for solving (3.24) when the eigenvalues of T' inside Γ form a *group* in a way that will be made more precise. We denote by $\mu_1', \ldots, \mu_m'$ the m repeated eigenvalues of T' inside Γ, and $\hat{\lambda}' = (1/m) \sum_{i=1}^{m} \mu_i'$ is their arithmetic mean. We define

$$a := \max_i |\mu_i' - \hat{\lambda}'|,$$

$$b := \operatorname{dist}(\hat{\lambda}', \sigma(T') - \{\mu_i'\}_1^m),$$

and

$$U' := (T' - \hat{\lambda}')Q'.$$

If $b > 0$, $\hat{\lambda}'$ is not in the spectrum of $(1 - Q')T'_{\restriction W}$ (Exercise 3.25) and we set $\Sigma' := (1 - Q')[\mathring{T}'_{\hat{\lambda}'}]^{-1}(1 - Q')$, where $\mathring{T}'_{\hat{\lambda}'} := (1 - Q')(T' - \hat{\lambda}')_{\restriction W}$. Then $\Sigma' \in \mathscr{L}(X)$.

Exercises

3.24 Show that $\{0\} \cup \sigma(T') = \sigma(Q'T'Q') \cup \sigma((1 - Q')T'(1 - Q'))$ (*Hint*: $(1 - Q')T'Q' = 0$). Prove that the two spectra are disjoint, except for 0.

3.25 Study the spectra of $U'_{\restriction M}$, and of $\mathring{T}'_{\hat{\lambda}'}$. Show that the first lies in the disk

$$\{z; |z - \hat{\lambda}'| \leq a\}$$

and the second lies inside

$$\{z; |z - \hat{\lambda}'| \geq b\}.$$

Conclude that $\hat{\lambda}'$ is not in the spectrum of $(1 - Q')T'_{\restriction W}$ if $b > 0$.

Lemma 3.36 *If $b > 0$, the solution $\mathbf{X}$ of Eq.* (3.24) *is the fixed point of* $\mathbf{X} = \Sigma'(\mathbf{X}U' + B)$ *such that* $Q'\mathbf{X} = \mathbf{X}(1 - Q') = 0$.

Proof From (3.24) we get $(1 - Q')(T' - \hat{\lambda}')\mathbf{X} - \mathbf{X}(T' - \hat{\lambda})Q' = B$. Equivalently $(1 - Q')(T' - \hat{\lambda}')(1 - Q')\mathbf{X} - (1 - Q')\mathbf{X}U' = (1 - Q')B$. Therefore $\mathbf{X} = \Sigma'(\mathbf{X}U' + B)$. □

We remark that if there is a single eigenvalue λ' inside Γ with ascent equal to 1, U' reduces to $(T' - \lambda')Q' = 0$. In that case $\mathbf{X}$ is simply given by $\mathbf{X} = \Sigma'B$, and $b > 0$ is automatically satisfied since λ' is then an isolated eigenvalue. In the most general case, $\mathbf{X}$ can be *computed* by a fixed-point iteration under the stronger assumption that $0 \leq a < b$.

Proposition 3.37 *If $a < b$, the solution $\mathbf{X}$ of Eq.* (3.24) *is the limit of the fixed-point iteration*

$$\mathbf{X}_0 = \Sigma'B,$$

$$\mathbf{X}_{i+1} = \Sigma'(\mathbf{X}_i U' + B), \qquad Q'\mathbf{X}_i = \mathbf{X}_i(1 - Q') = 0, \qquad i = 0, 1, 2, \ldots.$$

The rate of convergence of $\mathbf{X}_i$ to $\mathbf{X}$ is arbitrarily close to a/b.

Proof The fixed-point iteration converges if there exists a constant $\alpha < 1$ such that

$$\|\Sigma'\mathbf{X}^1 U' - \Sigma'\mathbf{X}^2 U'\| \leq \alpha \|\mathbf{X}^1 - \mathbf{X}^2\|.$$

This condition is satisfied if $\|\Sigma'\| \, \|U'\| < 1$. We know that $r_\sigma(U') \leq a$ and $r_\sigma(\Sigma') \leq 1/b$. Therefore, for any $\varepsilon > 0$, there exists an equivalent norm $\|\cdot\|_*$ such that $\|\Sigma'\|_* \|U'\|_* < (a + \varepsilon)/(b - \varepsilon)$. If $a < b$, there exists $\varepsilon > 0$ such that $(a + \varepsilon)/(b - \varepsilon) < 1$, and the rate of convergence, bounded by $(a + \varepsilon)/(b - \varepsilon)$ can be made arbitrarily close to a/b. □

It is left to the reader to derive from Lemma 3.36 and Proposition 3.37 a method to compute the m vectors $\{v_i\}_1^m$ that define $\mathbf{X}$.

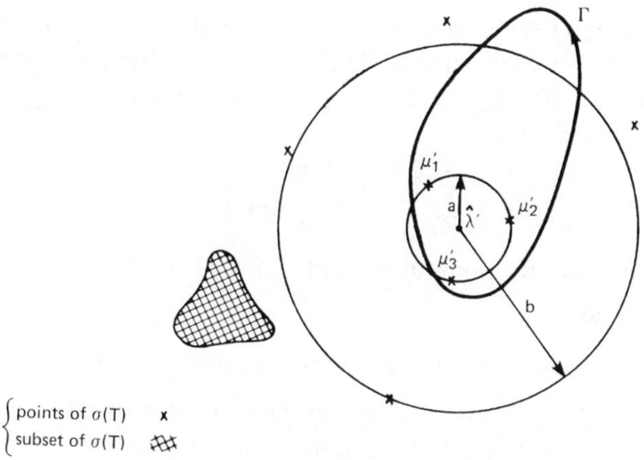

Figure 3.3

We then say that the eigenvalues of T' inside Γ form a *group* when $a < b$ (Fig. 3.3). We may remark that the choice of $\hat{\lambda}'$ in Lemma 3.36 is somewhat arbitrary. We may choose $\mu \in \rho((1 - Q')T'_{\restriction W})$ such that a/b is as small as possible (Fig. 3.4).

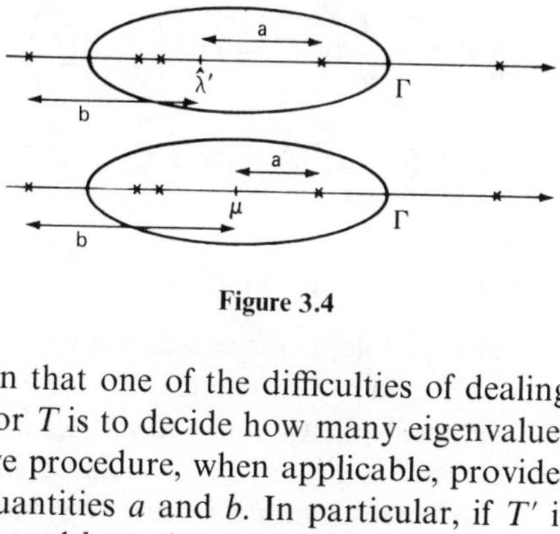

Figure 3.4

It is well known that one of the difficulties of dealing with multiple or close eigenvalues for T is to decide how many eigenvalues for T' should be grouped. The above procedure, when applicable, provides a strategy based on the available quantities a and b. In particular, if T' is a numerical approximation of T, a and b can be computed a posteriori.

Theorem 3.38 *Under the condition (3.22) $Q(t) = \sum_{k=0}^{\infty} t^k Q_k$ converges for $t \in \delta'_\Gamma$, where Q_k is the solution of (3.23). If $a < b$, then for a fixed $k \geq 1$, Q_k is the limit of the fixed-point iteration defined in Proposition 3.37, the rate of convergence being a/b. $Q(t)$ defines for $M(t)$ a basis*

$$u_i(t) = \sum_{k=0}^{\infty} t^k u_{ik}, \qquad Q'u_i(t) = x_i, \qquad i = 1, \ldots, m.$$

Proof Since $Q'P(t)x_i \neq 0$ for $i = 1, \ldots, m$, the vectors $u_i(t)$ normalized by $Q'u_i(t) = x_i$ are well defined. The rest is clear from Lemma 3.34 and Proposition 3.37. For $k \geq 1$, each Q_k can be written $Q_k = \sum_{i=1}^{m} \langle \cdot, y_i \rangle u_{ik}$ with $u_{ik} \in (1 - Q')D$. Then

$$Q(t) = \sum_{k=0}^{\infty} t^k Q_k = \sum_{i=1}^{m} \langle \cdot, y_i \rangle \left(\sum_{k=0}^{\infty} t^k u_{ik} \right) = \sum_{i=1}^{m} \langle \cdot, y_i \rangle u_i(t). \quad \square$$

We now turn to the computation of $\hat{\lambda}(t) = (1/m) \operatorname{tr} T(t)P(t)$.

Lemma 3.39

$$\hat{\lambda}(t) = (1/m) \operatorname{tr} T(t)P(t) = (1/m) \operatorname{tr} T(t)Q(t).$$

Proof Clear since the trace of the finite-rank operator $T(t)P(t)$ is the trace of $T(t)_{\restriction M(t)}$, and $P(t)$, $Q(t)$ have the same range $M(t)$. $\square$

Theorem 3.40 $\hat{\lambda}(t) = \sum_{k=0}^{\infty} t^k \hat{\nu}_k$ *converges for* $t \in \delta'_\Gamma$, *where the coefficient* $\hat{\nu}_k$ *are given by*

$$\hat{\nu}_0 = \hat{\lambda}', \qquad \hat{\nu}_k = \frac{1}{m} \sum_{i=1}^{m} \langle (T'Q_k - HQ_{k-1})x_i, y_i \rangle, \qquad k \geq 1.$$

Proof By Proposition 3.26, $\hat{\lambda}(t)$ is analytic in δ'_Γ and therefore has a unique expansion,

$$\hat{\lambda}(t) = \sum_{k=0}^{\infty} t^k \hat{\nu}_k = \frac{1}{m} \operatorname{tr}(T' - tH) \left(\sum_{k=0}^{\infty} t^k Q_k \right)$$

$$= \frac{1}{m} \operatorname{tr} T'Q' + \sum_{k=1}^{\infty} t^k \frac{1}{m} (\operatorname{tr} T'Q_k - \operatorname{tr} HQ_{k-1}).$$

The desired result follows. Note that we use the expansion $\sum_{k=0}^{\infty} t^k Q_k$, which does not necessarily converge for all t in δ'_Γ. $\square$

We consider now the problem $T\varphi = \lambda\varphi$, with $T = T' - H$. If $t = 1$ belongs to δ'_Γ, the curve Γ lies in $\rho(T)$ and encloses one or several eigenvalues of T with total algebraic multiplicity equal to m. Let $\hat{\lambda}$ be their arithmetic mean, and let P be the spectral projection associated with the eigenvalues inside Γ. $M = PX$ is the corresponding invariant subspace. Let Q be the projection on M along W. $\hat{\lambda}$ and a basis $\{u_i\}_1^m$ of M can be iteratively computed according to the following theorem.

Theorem 3.41 *If* $r'_\Gamma < 1$, *then under the condition* (3.22) $\hat{\lambda} = \sum_{k=0}^{\infty} \hat{\nu}_k$ *and* $u_i = \sum_{k=0}^{\infty} u_{ik}$, $i = 1, \ldots, m$, *where* $\hat{\lambda}_k$ *and* u_{ik} *are defined in Theorems* 3.40 *and* 3.38, *respectively.*

Proof Straightforward, because $t = 1$ belongs to δ'_Γ. Note that $\hat{\nu}_0 = \hat{\lambda}'$, $u_{i0} = x_i$ and $Q'u_i = x_i$. We write $Q = \sum_{i=1}^{m} \langle \cdot, y_i \rangle u_i$, then $\hat{\lambda} = (1/m) \operatorname{tr} TQ$. $\square$

8.4. Iterative Refinement on the Eigenelements of an Almost-Triangular Matrix

We now deal with an application on matrices in $\mathbb{C}^N$. Let $A = T + H$, where T is the upper-triangular part of A and H is strictly lower-triangular. We set $\varepsilon := \|H\|_2$, a_{11}(resp. a_{NN}) is an approximate eigenvalue of A with e_1 (resp. e_N) as an approximate right (resp. left) eigenvector. An algorithm is devised in Exercise 1.59 to compute the exact eigenelement of A (resp. A^{H}) from a_{11}, e_1 (resp. $\bar{a}_{NN}$, e_N). To refine the approximate eigenvalue a_{ii}, $1 < i < N$, we now propose an algorithm which does not require the knowledge of an approximate (left or right) eigenvector for a_{ii}. We distinguish whether a_{ii} is well separated from the other diagonal elements or there exist m diagonal elements close to a_{ii}. Let $\sigma > 0$ be a given threshold.

8.4.1. $\min_{j \neq i} |a_{ii} - a_{jj}| \geq \sigma$

We interchange the ith row and column with the first. We get, after permutation,

$$A' = i \rightarrow \left[\begin{array}{c|ccccc} a_{ii} & \varepsilon & - & \overset{i \downarrow}{\varepsilon} & \times & \times \\ \hline 0 & a_{22} & \times & \varepsilon & \times & \times \\ \vdots & & \ddots & | & & \\ 0 & & & a_{11} & & \\ \vdots & & & & \ddots & \times \\ 0 & 0 & & & & a_{NN} \end{array}\right]$$

$$+ \left[\begin{array}{c|ccccc} 0 & 0 & & \overset{i \downarrow}{0} & & 0 \\ \hline \times & 0 & & & & \\ | & & \ddots & & 0 & \\ \times & \times & - & \times & 0 & \\ & & & & \ddots & \\ \varepsilon & \varepsilon & & & \varepsilon & 0 \end{array}\right] \leftarrow i$$

$$= T' + H' = \left[\begin{array}{c|c} a_{ii} & T'_2 \\ \hline 0 & T'_4 \end{array}\right] + \left[\begin{array}{c|c} 0 & 0 \\ \hline H'_3 & H'_4 \end{array}\right],$$

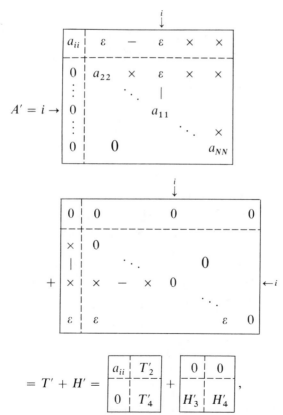

where ε denotes an element of order ε and × denotes an element of order 1.

a_{ii} is a simple eigenvalue of T' with eigenvector e_1. P' is the associated eigenprojection, $Q' = e_1 e_1^T$ is the orthogonal projection on $\{e_1\}$. Let Γ be a curve isolating a_{ii}. Set $R'(z) := (T' - zI)^{-1}$. $A(t) := A' + tH'$, and $P(t)$ its spectral projection.

Lemma 3.42 *The conditions*

$$\max_{z \in \Gamma} r_\sigma[H'R'(z)] < 1$$

and

$$\sup_{|t| \leq 1} r_\sigma[Q'(P(t) - P')Q'] < 1$$

are fulfilled for ε small enough.

Proof We first prove that $\max_{z \in \Gamma} \|[H'R'(z)]^3\| < 1$ for ε small enough. For any z on Γ, $\|[H'R'(z)]^3\| \leq c \max_{1 \leq j \leq N} \|[H'R'(z)]^3 e_j\|$. Owing to the structure of $R'(z)$ and H', it is easy to check that it suffices to prove that $\|[H'R'(z)]^3 e_1\|$ is small:

$$\begin{aligned}
R'(z)e_1 &= (\times\ 0\ \cdots\ 0)^{\mathrm{T}} & x_1 &= H'R'(z)e_1 = (0\ \times\ \cdots\ \times\ \overset{\overset{i}{\downarrow}}{\times}\ \varepsilon \cdots \varepsilon)^{\mathrm{T}},\\
R'(z)x_1 &= (\varepsilon\ \times\ \cdots\ \times\ \varepsilon \cdots \varepsilon)^{\mathrm{T}}, & x_2 &= H'R'(z)x_1 = (0\ \varepsilon\ \cdots\ \varepsilon\ \times\ \varepsilon \cdots \varepsilon)^{\mathrm{T}},\\
R'(z)x_2 &= (\varepsilon\ \varepsilon\ \cdots\ \times\ \varepsilon \cdots \varepsilon)^{\mathrm{T}}, & x_3 &= H'R'(z)x_2 = (0\ \varepsilon\ \cdots\ \varepsilon\ \varepsilon\ \varepsilon \cdots \varepsilon)^{\mathrm{T}},
\end{aligned}$$

that is, $\|[H'R'(z)]^3 e_1\| < c\varepsilon$. Now

$$Q'(P(t) - P')Q' = \frac{-1}{2i\pi} \int_\Gamma Q'[\underbrace{R(t, z) - R'(z)}_{A(t,z)}]Q'\, dz$$

with

$$A(t, z) = tR'(z)H'R'(z) + R'(z)[tH'R'(z)]^2 + R'(z) \sum_{k=1}^{\infty} ([tH'R'(z)]^3)^k.$$

$$\begin{aligned}
\sup_{|t| \leq 1} r_\sigma[Q'(P(t) - P')Q'] &= \sup_{|t| \leq 1} |e_1^{\mathrm{T}}(P(t) - P')e_1| \\
&\leq c \sup_{|t| \leq 1} \max_{z \in \Gamma} |e_1^{\mathrm{T}} A(t, z) e_1| < 1
\end{aligned}$$

for ε small enough. □

Exercise

3.26 Set $L = H' - H$. Prove that L is a nilpotent matrix such that $L^3 = 0$ (resp. $L^2 = 0$) for $i > 2$ (resp. $i = 2$). Prove that L is of rank 2 (resp. 1) for $i > 2$ (resp. $i = 2$).

$A\varphi = \lambda\varphi$, where λ is the eigenvalue of A close to a_{ii}. We know that the eigenvector φ is such that $e_i^{\mathrm{T}}\varphi \neq 0$ (Exercise 1.77). The interchange of the first and ith component in φ defines φ' the eigenvector of A' such that $A'\varphi' = \lambda\varphi'$, and $e_1^{\mathrm{T}}\varphi' \neq 0$. This is also a consequence of Lemma 3.42. We denote by ϕ' the eigenvector of A' normalized by $e_1^{\mathrm{T}}\phi' = 1$.

We define

$$\lambda_0 = a_{ii}, \qquad \lambda_k = a_{ii} + \sum_{j=1}^{k} v_j,$$

and

$$\varphi_0 = e_1, \qquad \varphi_k = e_1 + \sum_{j=1}^{k} \eta_j, \qquad e_1^T\eta_j = 0.$$

Set

$$\eta_k = \begin{pmatrix} 0 \\ \mathring{\eta}_k \end{pmatrix},$$

where $\mathring{\eta}_k \in \mathbb{C}^{N-1}$. v_k and $\mathring{\eta}_k$ are computed by the iteration $v_k = T'_2\mathring{\eta}_k$ for $k \geq 1$,

$$(T'_4 - a_{ii}I)\mathring{\eta}_k = \begin{cases} -H'_3 & \text{for} \quad k = 1 \\ -H'_4\mathring{\eta}_{k-1} + \sum_{j=1}^{k-1} v_j\mathring{\eta}_{k-j} & \text{for} \quad k \geq 2. \end{cases} \tag{3.27}$$

Then

$$\varphi_k = e_1 + \sum_{j=1}^{k} \eta_j, \qquad \text{and} \qquad \lambda_k = e_1^T T'\varphi_k, \qquad k \geq 1.$$

The system (3.27) has a fixed *triangular* matrix with diagonal elements $a_{jj} - a_{ii}$, for $j \neq i$, $|a_{jj} - a_{ii}| \geq \sigma$.

Application of Theorem 3.41 shows that $\lambda_k \to \lambda$ and $\varphi_k \to \phi'$ as $k \to \infty$ (Exercise 3.27).

Exercise

3.27 Prove that (3.23) in the above particular setting reduces to (3.27). Prove that the recursion defined above is also the one given in (1.14).

Example 3.19 We consider the 4×4 matrix $A = T + H$ with

$$T = \begin{pmatrix} 1 & 2 & 3 & 4 \\ 0 & 5 & 6 & 7 \\ 0 & 0 & 8 & 9 \\ 0 & 0 & 0 & 10 \end{pmatrix}, \qquad H = 10^{-3}\begin{pmatrix} 0 & 0 & 0 & 0 \\ 0.1 & 0 & 0 & 0 \\ 3 \times 10^{-5} & 0.12 & 0 & 0 \\ 0.046 & 0.09 & 0.09 & 0 \end{pmatrix}.$$

We compute the exact eigenvalue close to 8. The iteration converges in 10 steps to the exact value with 15 decimal digits. For the first few iterates, the convergence is irregular, as shown by the sequence of rounded v_j:

$$-2 \times 10^{-4}, \quad -10^{-3}, \quad -10^{-4}, \quad 2 \times 10^{-7}, \quad 2 \times 10^{-8},$$
$$7 \times 10^{-11}, \quad -7 \times 10^{-11}, \quad -7 \times 10^{-12}, \quad -2 \times 10^{-13}.$$

This is a result of the norm of $H'R'(z)$ raised to the *third* power being small.

If 8 is changed to 5.1, close to 5, the rate of convergence drastically slows down: 15 steps give the exact answer to 6 decimal digits only. The sequence of rounded v_j is now:

$$7 \times 10^{-3},\quad -10^{-2},\quad -10^{-3},\quad -10^{-3},\quad -5 \times 10^{-5},\quad -2 \times 10^{-4},$$
$$5 \times 10^{-7},\quad -6 \times 10^{-5},\quad 3 \times 10^{-6},\quad -2 \times 10^{-5},\quad 2 \times 10^{-6},$$
$$-6 \times 10^{-6},\quad 9 \times 10^{-7},\quad -2 \times 10^{-6},\quad 4 \times 10^{-7}.$$

In that case one should use an algorithm which takes care of close diagonal elements.

8.4.2. m diagonal elements are such that $|a_{ii} - a_{jj}| \leq \sigma$

We suppose for simplicity that $m = 2$ and a_{ii}, $a_{i+1,i+1}$ are the two close diagonal elements. We interchange the ith and $(i + 1)$th rows and columns with the first and second. We get, after permutation,

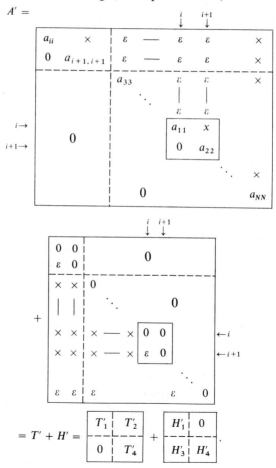

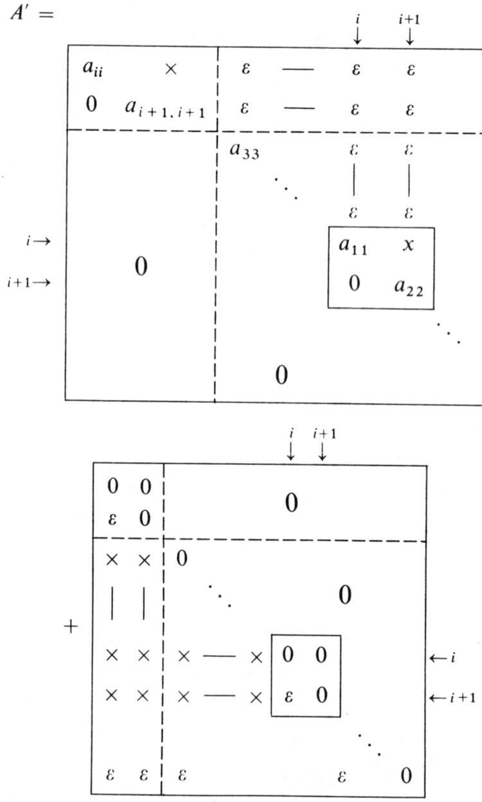

$$= T' + H' = \begin{array}{|c|c|} \hline T'_1 & T'_2 \\ \hline 0 & T'_4 \\ \hline \end{array} + \begin{array}{|c|c|} \hline H'_1 & 0 \\ \hline H'_3 & H'_4 \\ \hline \end{array}.$$

For $m \geq 2$, $e_1, \ldots, e_m$ span the invariant subspace for T' associated with the m first diagonal elements. $Q' = \sum_{i=1}^{m} e_i e_i^T$ is the orthogonal projection on $\{e_1, \ldots, e_m\}$.

We leave it to the reader to check that Lemma 3.42 is still valid, and Theorem 3.41 applies. Formula (3.23) involves matrices Q_k (resp. B_k) such that $Q'Q_k = Q_k(I - Q') = 0$ (resp. $Q'B_k = B_k(I - Q') = 0$). We denote by $\mathring{Q}_k$ (resp. $\mathring{B}_k$) the $(N - m) \times m$ matrix corresponding to the nonzero block in Q_k (resp. B_k). We set $\mathring{Q} = \sum_{k=1}^{\infty} \mathring{Q}_k$. Then the $N \times m$ matrix $\hat{Q}$ defined by

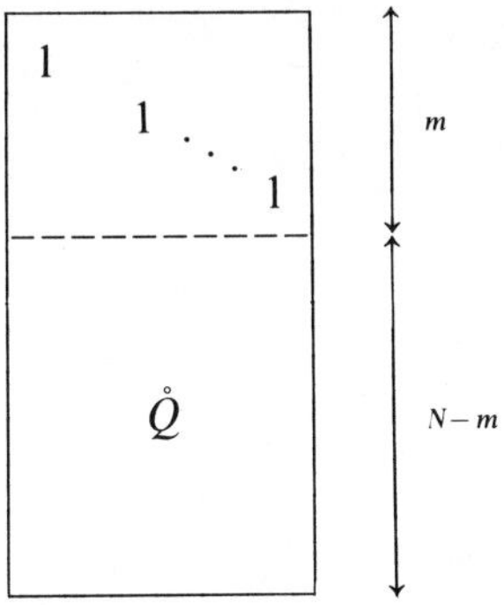

is such that its columns form a basis for the invariant subspace for T'. For $k \geq 1$, $\mathring{Q}_k$ is solution of

$$T'_4 \mathring{Q}_k - \mathring{Q}_k T'_1 = \mathring{B}_k, \tag{3.28}$$

with

$$\mathring{B}_1 = -H'_3, \qquad \mathring{B}_k = -H'_4 \mathring{Q}_{k-1} + \mathring{Q}_{k-1} H'_1 + \sum_{i=1}^{k-1} \mathring{Q}_i T'_2 \mathring{Q}_{k-i},$$

and

$$\hat{\lambda}_k = \sum_{j=0}^{k} \hat{v}_j \frac{1}{m} \operatorname{tr} T'_1 + \frac{1}{m} \operatorname{tr} T'_2 \left(\sum_{j=1}^{k} \mathring{Q}_j \right).$$

The resolution of (3.28) is straightforward since T'_4 and T'_1 are already triangular. There exists a unique solution $\mathring{Q}_k$ iff $\sigma(T'_1) \cap \sigma(T'_4) = \varnothing$, which results from the hypothesis.

For $m = 2$, let $\mathring{Q}_k = (y_1^k, y_2^k)$ and $\mathring{B}_k = (b_1^k, b_2^k)$. Then (3.28) yields the two *triangular* systems in $N - 2$ unknowns

$$(T'_4 - a_{ii} I) y_1^k = b_1^k, \qquad (T'_4 - a_{i+1,i+1} I) y_2^k = b_2^k + a_{i,i+1} y_1^k.$$

And $\operatorname{dist}(a_{ll}, \sigma(T'_4)) \geq \sigma$ for $l = i$ or $i + 1$.

Exercises

3.28 Let $A = T + H$. Prove directly that for $\|H\|$ small enough, there exists a basis $\{u, v\}$ of the invariant subspace for A associated with the eigenvalues close to a_{ii}, $a_{i+1,i+1}$, such that $e_i^T u \neq 0$ and $e_{i+1}^T v \neq 0$.

3.29 Consider the general equation $AX + XB = C$, where the matrix A is $n \times n$, B is $m \times m$, X and C are $n \times m$. Give a sufficient condition on the eigenvalues of A and B such that $X = \lim_i X_i$ with

$$X_0 = (A - \alpha I)^{-1}C, \qquad X_{i+1} = (A - \alpha I)^{-1}[-X_i(B - \alpha I) + C], \qquad i \geq 0,$$

where $\alpha \notin \sigma(A)$. Compare with the algorithm given in Bartels and Stewart (1972).

Example 3.20 For the matrix

$$A = \begin{pmatrix} -6 & 2 & 3 & 3 & 1 \\ 10^{-6} & 3.9998 & 1 & 2 & 2 \\ 10^{-6} & 10^{-4} & 4.0002 & 1 & 1 \\ 10^{-6} & 10^{-4} & 10^{-4} & 4 & 1 \\ 10^{-6} & 10^{-6} & 10^{-6} & 10^{-6} & 14 \end{pmatrix}$$

the computation gives

$$\hat{Q} = \begin{pmatrix} 0.2 & 0.28 & 0.23 \\ 1 & 0 & 0 \\ 0 & 1 & 0 \\ 0 & 0 & 1 \\ -1.2 \times 10^{-7} & -1.4 \times 10^{-7} & -1.6 \times 10^{-7} \end{pmatrix}$$

and

$$\hat{\lambda} = \tfrac{1}{3} \operatorname{tr} A\hat{Q} = 4 + 5.7 \times 10^{-8}.$$

The three column vectors of $\hat{Q}$ are orthonormalized; this yields a 5×3 matrix Q such that the 3×3 matrix $Q^{\mathrm{T}}AQ$ has for eigenvalues the three eigenvalues of A close to 4. The problem of computing the eigenvalues of A, three of which are close, has been reduced to its truly difficult part.

We wish to warn the reader that the algorithms we have just described are not necessarily the best in practice for matrices. We have presented them as a simple illustration of the most general algorithm given in Theorem 3.41, which is to be applied with T' chosen as a numerical approximation T_n of T (Chapters 5 and 7).

In numerical linear algebra, the *finite dimension* of the setting allows much more freedom in the choice of the decomposition $A = A' + H$. For example, several algorithms to refine a, e_1 have been based on the splitting

$$A = \begin{pmatrix} a & v^H \\ u & C \end{pmatrix}$$

with $a \in \mathbb{C}$, $u, v \in \mathbb{C}^{N-1}$, and on a fixed-point formulation of the eigenvalue problem $A\phi = \lambda\phi$ with

$$\phi = \begin{pmatrix} 1 \\ d \end{pmatrix}, \qquad d \in \mathbb{C}^{N-1},$$

that is, $\lambda = a - v^H(C - \lambda I)^{-1}u$, $d = -(C - \lambda I)^{-1}u$ provided that $\|u\|_2 \|v\|_2$ is small enough. This condition can clearly be weakened to cover the case in which A is the result of a permutation on an almost-triangular matrix, as in Section 8.4.1. We get the iteration

$$d_0 = 0, \qquad d_k = -(C - \lambda_{k-1} I)^{-1} u, \qquad \lambda_k = a - v^H d_k, \qquad k \geq 1.$$

Except for its $(i-1)$th row, C is an almost-triangular matrix.

For m close diagonal elements, we consider the splitting

$$A = \begin{pmatrix} A_1 & V \\ U & C \end{pmatrix},$$

where A_1 and C are square blocks of order m and $N - m$, respectively. To compute an invariant basis for A from the basis $\{e_1, \ldots, e_m\}$ for

$$\begin{pmatrix} A_1 & O \\ O & C \end{pmatrix}$$

requires the computation of an $m \times (N - m)$ matrix D and an $m \times m$ matrix M such that

$$A \begin{pmatrix} I \\ D \end{pmatrix} = \begin{pmatrix} I \\ D \end{pmatrix} M;$$

that is,

$$A_1 + VD = M, \qquad CD - DM = -U.$$

A Newton's iteration to solve this problem is presented in Dongarra *et al.* (1981).

8.5. Bibliographical Comments

Our framework in Sections 7 and 8 has been *analytic* perturbation theory. The convergence of the Neumann-series expansion for $R(t, z)$ plays a fundamental role in allowing the series expansion of the resolvent, spectral projections, and trace to be convergent. This is of great practical importance because this is the basis for the Rayleigh–Schrödinger expansions of the eigenelements often used in quantum mechanics. In the context of differential equations, perturbations such that the Neumann-series expansion of the resolvent is convergent are called *regular* perturbations (Reed and Simon, 1978).

Rellich (1936–1942), Sz.-Nagy (1951), Kato (1976), and others have developed a theory of analytic perturbation of a closed operator when the parameter t is *small enough*. In our case, however, relating T and T', we need that $t = 1$ belongs to the analyticity domain. One may think that the introduction of the parameter t in this context is somewhat artificial. Indeed, Lemordant (1980) has presented a theory of the analytic dependence on a closed perturbation H. Most of our presentation is adapted from Redont (1979a) and Lemordant (1980).

As we have already seen in Chapter 1 and in this chapter for matrices, the series expansions of the solution and of the eigenelements can be both a theoretical and a practical tool. On one hand they are a theoretical tool when used in connection with a decomposition $T = \tilde{T} + \tilde{H}$. In Chapter 6 we shall use such a decomposition to get some information on the spectrum of a closed operator T from the knowledge of the spectrum of a "smaller" operator QTQ, where Q is a finite-rank projection on a given subspace. In practice, this subspace may be chosen to be an approximate invariant subspace for T, but this is not necessary to get a posteriori error bounds. A similar decomposition of an approximation T_n by means of the spectral projection of T yields convergence rates for the eigenelements. On the other hand, they are a practical tool when used in connection with the decomposition $T = T_n - (T_n - T)$. The case in which T' is a numerical approximation T_n of T will be developed in Chapters 5 and 7. A sufficient condition on $T_n - T$ will be given to fulfill (3.17). That condition will be easily satisfied for the practical cases of interest ($\|T_n - T\| \to 0$ or $T_n \overset{cc}{\to} T$ when T is an integral operator; cf. Chapter 7). A summary of the use of analytic perturbation theory throughout the text is given in Table 3.4.

Table 3.4

Use of Analytic Perturbation Theory

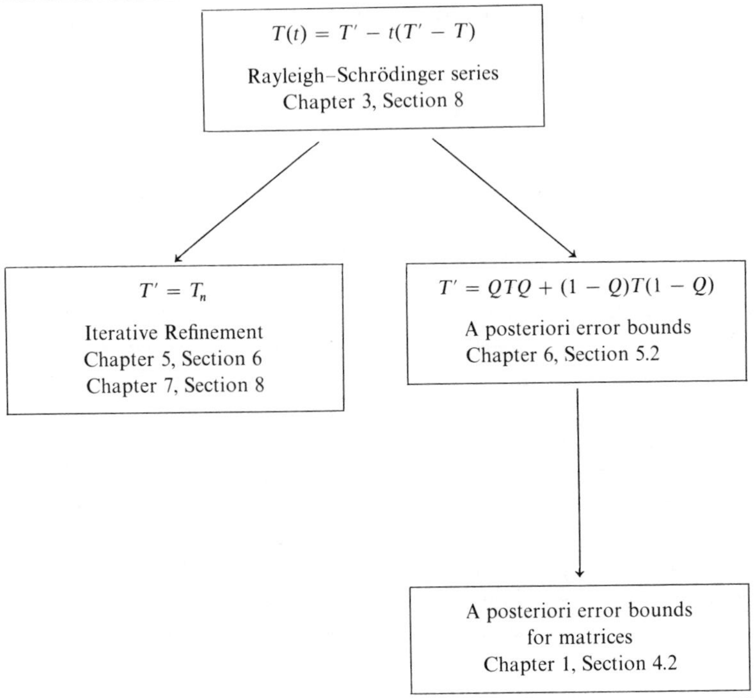

CHAPTER 4

Numerical Approximation Methods for Integral and Differential Operators

> Dans sa conférence sur "L'avenir des mathématiques" au Congrès de Rome en 1908, Poincaré a remarqué que l'on devait pouvoir appliquer la théorie des équations intégrales linéaires à la théorie des équations différentielles ordinaires non linéaires. Un premier pas pour réaliser l'idée de Poincaré a été fait par Fredholm dans une note dans les *Comptes rendus* 23 août 1920.
>
> *TORSTEN CARLEMAN*
>
> *Acta Matem., 18 mars 1932*

Introduction

We present some families of numerical methods to approximate the equation

$$(T - z)x = f, \qquad (*)$$

for a given z in $\rho(T)$ and the associated eigenvalue problem $T\varphi = \lambda\varphi$, $\varphi \neq 0$. We have chosen to study the equation under the above form $(*)$ rather than under the form $Tx = f$, for the obvious link between $(*)$ and the spectral projection

$$P = \frac{-1}{2i\pi}\int_\Gamma (T - z)^{-1}\, dz.$$

These numerical methods are put into an abstract setting where an approximation operator T_n is defined in the same Banach space X as T. The solution x of (∗) is approximated by the solution x_n of $(T_n - z)x_n = f$.

In most cases, the convergence $x_n \to x$ has been proved directly in the literature. Here we present a study of the convergence $T_n \to T$ from which the convergence $x_n \to x$ is straightforward. And, what is more interesting, the convergence of the eigenelements (which is always more difficult to establish than $x_n \to x$) will also derive from the convergence $T_n \to T$. This will be done in Chapter 5.

An important class of methods is provided by the projection method on a finite-dimensional subspace X_n of X; it includes such methods as the Galerkin, least-squares, collocation, and finite-element methods. The error $\|x - x_n\|$ can be expressed in terms of $\text{dist}(x, X_n)$, the distance of the solution x to X_n. The estimation of $\|x - x_n\|$ is then reduced to the problem, in approximation theory, of the evaluation of $\text{dist}(x, X_n)$. Examples given in Chapters 6 and 7 will show the essential role played by the projection.

A. FREDHOLM INTEGRAL OPERATORS

1. The Problem

Let X be a complex Banach space, and let T be bounded on X. We are concerned with the numerical solution of the problems

$$Tx - zx = f, \qquad \text{for a given } f \text{ in } X,\ z \text{ in } \rho(T), \tag{4.1}$$

$$T\varphi = \lambda\varphi, \qquad 0 \neq \varphi \in X \tag{4.2}$$

The main example of operator T will be the integral operator

$$x(t) \mapsto (Tx)(t) = \int_\Omega k(t, s)x(s)\, ds, \qquad t \in \Omega, \tag{4.3}$$

where Ω is a bounded domain in $\mathbb{R}^N$, $N \geq 1$. X is often in practice $C(\Omega)$ or $L^2(\Omega)$. (4.1) is a Fredholm equation of the second kind for $z \neq 0$, and (4.2) is the associated eigenvalue problem.

Defining, for each t in Ω, the function k_t as

$$k_t(s) = k(t, s), \qquad s \in \Omega,$$

we can rewrite (4.3) as

$$(Tx)(t) = \int_{\Omega} k_t(s)x(s)\,ds.$$

We assume that x and k_t are Lebesgue-measurable functions so that (4.3) is well defined.

We shall consider T as an operator defined on $L^r(\Omega)$, $1 \le r \le \infty$. Many integral operators are compact as well as bounded. We give a characterization of the kernel k such that T is compact from $L^r(\Omega)$ into $C(\Omega)$.

Theorem 4.1 (Graham and Sloan) *Let* $1 \le p \le \infty$, $p^{-1} + q^{-1} = 1$. *The integral operator* T *defined by* (4.3) *is compact from* $L^r(\Omega)$ *into* $C(\Omega)$ *for all* r *such that* $q \le r \le \infty$ *iff* k *satisfies*

(*i*) $\sup_{t\in\Omega} \|k_t\|_p < \infty$;
(*ii*) $\lim_{t'\to t} \|k_{t'} - k_t\|_p = 0$ *for all* t', $t \in \Omega$.

For a proof see Graham and Sloan (1979).

It follows from the inclusions $C \subset L^\infty \subset L^r \subset L^1$ for $1 \le r \le \infty$ that k also induces a compact operator from L^r to L^r, from L^r to C, and from L^∞ to C. The last is often the most important case for applications.

We now give an example of a noncompact operator.

Example 4.1 The operator T defined by

$$(Tx)(s) = \int_0^s \frac{x(t)\,dt}{\sqrt{s^2 - t^2}} \qquad \text{for} \quad 0 < s \le 1,$$

with $(Tx)(0) = (\pi/2)x(0)$, is bounded but not compact. Indeed the scalar 1 can be shown to belong to the point spectrum of T, with *infinite* multiplicity.

Exercise

4.1 Let $X = C(a, b)$ and suppose that

(i) $\sup_{a\le t\le b} \int_a^b |k(t, s)|\,ds < \infty$,

(ii) $\lim_{\delta\to 0} \int_a^b |k(t + \delta, s) - k(t, s)|\,ds = 0, \qquad a \le t \le b.$

Prove directly that T is compact in X by the Ascoli–Arzela theorem. Check the result by Theorem 4.1.

Modern practical methods consist mainly in approximating T by projection on a finite-dimensional subspace or by numerical quadrature. We begin by a review of these techniques.

2. Projections and Numerical Quadratures

2.1. Projections in Numerical Analysis

Let X_n be a finite-dimensional subspace of the complex Banach space X. For computational reasons, X_n is often a set of polynomials, either piecewise polynomials or polynomials defined by a single expression over Ω. We consider a sequence of projections π_n from X onto X_n. We set

$$\delta_n(x) := \text{dist}(x, X_n) = \inf_{y \in X_n} \|x - y\|.$$

For any x in X,

$$\delta_n(x) \leq \|(1 - \pi_n)x\| \leq \|1 - \pi_n\|\delta_n(x).$$

The pointwise convergence $\pi_n x \to x$ depends therefore on $\delta_n(x)$ and $\|\pi_n\|$. $\|\pi_n\|$ may not be bounded as $n \to \infty$.

As a basic example we consider in $X = C(a, b)$ the interpolatory projection L_n, which associates with $x \in C(a, b)$ its $(n - 1)$th-degree *Lagrange* polynomial interpolant $L_n x$ at the points $\{t_{in}\}_1^n, a \leq t_{1n} < t_{2n} < \cdots < t_{nn} \leq b$. X_n is $\mathbb{P}_{n-1}$, the set of polynomials of degree less than n on $[a, b]$. Then for any choice of $\{t_{in}\}_1^n$,

$$\|L_n\|_\infty \geq (2/\pi^2) \log(n - 1) + b(n),$$

where b is a bounded function of n (Laurent, 1972, p. 301). Therefore there exists at least one function f in $C(a, b)$ such that $L_n f \not\to f$. Some choices of $\{t_{in}\}_1^n$, however, are better than others. If the $\{t_{in}\}_1^n$ are the *Chebyshev points* (that is, the zeros of the nth-degree Chebyshev polynomial of the first kind on $[a, b]$), then $\|L_n\|_\infty \leq (4/\pi) \log n + 8$. For evenly spaced nodes, $e^{n/2} \leq \sup_n \|L_n\|_\infty \leq 2^n$ (Laurent, 1972, p. 302).

Note that if a projection is not bounded in a certain space, it may be bounded in a different space. For example, if the $\{t_{in}\}_1^n$ are the Chebyshev points on $[-1, 1]$, the L_n are uniformly bounded as operators from $C(-1, 1)$ into $L^2_\rho(-1, 1)$, the space of functions x that are square integrable with weight function $\rho(t) := (1 - t^2)^{-1/2}$, for $t \in [-1, 1]$ (cf. Section 7).

In our example where $X = C(a, b)$ and $X_n = \mathbb{P}_{n-1}$, the *best approximation* $p^* \in \mathbb{P}_{n-1}$ that achieves $\|x - p^*\|_\infty = \text{dist}_\infty(x, X_n)$ can be bounded in terms of the *modulus of continuity* $\omega(\delta, x)$ of the continuous function x for $\delta > 0$

$$\omega(\delta, x) := \sup_{|t - t'| \leq \delta} (|x(t) - x(t')|, \qquad t, t' \in [a, b]).$$

We have for $x \in C(a, b)$, $\text{dist}_\infty(x, X_n) = 0(\omega(1/n, x))$, and more generally for $x \in C^p(a, b)$ and $n \geq p$, $\text{dist}_\infty(x, X_n) \leq (C/n^p)\|x^{(p)}\|_\infty$. (See the Jackson theorems in Cheney, 1966, pp. 142–145.)

Exercises

4.2 Let $0 \neq x \in X$; show that the gap $\delta(\{x\}, X_n)$ is equal to $(1/\|x\|)\,\delta_n(x)$.

4.3 Let π_n be the orthogonal projection on E_n, a finite-dimensional subspace of a Hilbert space H. Show that $\pi_n x \to x$ iff $\mathrm{dist}(x, E_n) \to 0$.

Let Δ be a strict partition of $[a, b]$: $a = t_0 < t_1 < \cdots < t_n = b$. We set $\Delta_i := [t_{i-1}, t_i[$, $i = 1, \ldots, n$, and $h := \max_{1 \le i \le n}(t_i - t_{i-1})$. $\mathbb{P}_{r,\Delta}$ is the set of piecewise polynomials of degree less than $r + 1$ on each Δ_i, $i = 1, \ldots, n$, the value at b being defined by continuity.

Exercises

4.4 In $H = L^2(a, b)$, let π_n be the projection $x \mapsto \pi_n x$ that associates with $x_{\restriction \Delta_i}$ its least-squares approximation in the set of polynomials of degree less than $r + 1$ on Δ_i, $i = 1, \ldots, n$, $(\pi_n x)(b)$ being defined by continuity. Show that $\pi_n: L^2(a, b) \to \mathbb{P}_{r,\Delta}$ is orthogonal. Prove that $\pi_n x \to x$ in $L^2(a, b)$ if $h \to 0$, r being fixed.

4.5 Show that in $L^2(a, b)$, if $x \in C^\alpha(a, b)$, then $\mathrm{dist}_2(x, \mathbb{P}_{r,\Delta}) \le h^\beta \|f^{(\beta)}\|_\infty$ with

$$\beta := \min(r + 1, \alpha).$$

The projection π_n defined in Exercise 4.4 is such that $\|\pi_n\|_2 = 1$ in $L^2(a, b)$. The same projection can be considered as an operator from $L^\infty(a, b)$ onto $\mathbb{P}_{r,\Delta} \subset L^\infty(a, b)$, and $\sup_n \|\pi_n\|$ may be bounded under some assumption on the partition Δ (de Boor, 1976; Graham, 1982).

With $\bar{\Delta}_i = [t_{i-1}, t_i]$, we define $C_\Delta := \prod_{i=1}^n C(\bar{\Delta}_i)$. $f \in C_\Delta$ consists of n components $f_i \in C(\bar{\Delta}_i)$, f is a piecewise continuous function having (possibly) different left and right values at the partition points t_i, $i = 2, \ldots, n - 1$. With the norm $\|\cdot\|_\Delta$ defined by $\|f\|_\Delta = \max_i \|f_i\|_\infty$, C_Δ is a Banach space and $C_\Delta \subset L^\infty(a, b)$ since $\|f\|_\Delta = \|f\|_\infty$ for $f \in C_\Delta$. For $f \in \mathbb{P}_{r,\Delta}$, if the value at t_i^- is defined by continuity, then $\mathbb{P}_{r,\Delta} \subset C_\Delta$.

Exercise

4.6 For $x \in C_\Delta$, we define its piecewise Lagrange interpolant $L_n x$ of degree less than $r + 1$ at the points $\{\tau_i^j\}_{j=1}^{r+1}$, on each $\bar{\Delta}_i$, $i = 1, \ldots, n$: $L_n x \in \mathbb{P}_{r,\Delta}$. Show that $x \in C_\Delta \mapsto L_n x$ defines a bounded operator in $L^\infty(a, b)$ with domain C_Δ such that $L_n^2 = L_n$. Prove that $L_n x \to x$ on $C(a, b)$ as $h \to 0$. Deduce that $\sup_n(\|L_n x\|_\infty, x \in C(a, b), \|x\|_\infty = 1) < \infty$.

Another computationally important space is the spline space $S_{r,\Delta}$, that is, the subspace of $C^{r-2}(a, b)$ consisting of functions that belong to $\mathbb{P}_{r-1,\Delta}$: $S_{r,\Delta}$ defines the splines of *order* r on Δ. When $h \to 0$, we have

$$\mathrm{dist}_\infty(x, S_{r,\Delta}) = O(\omega(h, x)) \qquad \text{if} \quad x \in C(a, b)$$

and

$$\mathrm{dist}_\infty(x, S_{r,\Delta}) = O(h^r \omega(h, x^{(p)})) \qquad \text{if} \quad x \in C^p(a, b), \qquad 0 \le p \le r - 1.$$

Exercise

4.7 Prove that $\dim S_{r,\Delta} = n + r - 1$.

2.2. Numerical Quadratures

We consider the approximate evaluation of the linear functional

$$I(x) = \int_a^b x(t)\,dt,$$

with $x \in C(a, b)$, by means of the quadrature formula

$$I_n(x) = \sum_{i=1}^{n} w_{in} x(t_{in}), \qquad a \le t_{in} \le b$$

where the weights w_{in} are real or complex.

We want the formula I_n to be convergent for any continuous function x: $I_n(x) \to I(x)$ for all $x \in C(a, b)$.

Lemma 4.2 *A necessary and sufficient condition for $I_n(x) \to I(x)$ to hold for any x in $C(a, b)$ is that*

(*i*) $\lim_{n\to\infty} I_n(p) = I(p)$ *for any polynomial* p,
(*ii*) $\sup_n \sum_{i=1}^n |w_{in}| < \infty$.

Proof This is a straightforward consequence of the Banach–Steinhaus theorem since $\|I_n\| = \sum_{i=1}^n |w_{in}|$. □

If the weights w_{in} are nonnegative for all i and n, (ii) is a consequence of (i). Indeed, for the polynomial p defined by $p(t) = 1$ for $t \in [a, b]$,

$$b - a = \int_a^b dt = \lim_n \sum_{i=1}^{n} w_{in}.$$

In condition (i), the set of all polynomials can be replaced by any other set dense in $C(a, b)$, like the set of all piecewise-linear continuous functions.

We describe now a fundamental method to obtain quadrature formulas. Given $\{t_{in}\}_1^n$ in $[a, b]$ and n independent functions $\{e_i^{(n)}\}_1^n$ such that

$$e_i^{(n)}(t_{jn}) = \delta_{ij}, \qquad i, j = 1, \ldots, n,$$

we consider the interpolatory projection

$$x \mapsto \pi_n x = \sum_{i=1}^{n} x(t_{in}) e_i^{(n)},$$

defined for $x \in C(a, b)$, and we define

$$I_n(x) := \int_a^b (\pi_n x)(t)\,dt = \sum_{i=1}^{n} w_{in} x(t_{in})$$

with

$$w_{in} = \int_a^b e_i^{(n)}(t)\,dt, \qquad i = 1, \ldots, n.$$

Let $X_n = \{e_1^{(n)}, \ldots, e_n^{(n)}\}$. $I_n(x) = I(x)$ for $x \in X_n$; that is, I_n is exact on X_n.

Example 4.2 The trapezium rule with $n + 1$ equally spaced points on $[0, 1]$ is

$$I_n(f) = \frac{1}{n}\left[\frac{1}{2} f(0) + \frac{1}{2} f(n) + \sum_{j=1}^{n-1} f\left(\frac{j}{n}\right)\right].$$

X_n is the subset of $C(0, 1)$ consisting of piecewise-linear functions.

Let us consider $X_n = \mathbb{P}_{n-1}$. Then condition (i) of Lemma 4.2 is satisfied by construction. A necessary and sufficient condition for the convergence of these formulas is then condition (ii). In particular, the convergence is satisfied if $w_{in} \geq 0$. Such convergence does not occur for all choices of $\{t_{in}\}_1^n$.

Example 4.3 If the $\{t_{in}\}_1^n$ are the Chebyshev points on $[a, b]$, the weights are positive and the quadrature formula is convergent (Laurent, 1972, p. 317).

Example 4.4 If the $\{t_{in}\}_1^n$ are the *Gauss* points (that is, the zeros of the Legendre polynomial of degree n), the n-point Gauss quadrature is exact for polynomials up to degree $2n - 1$ and the weights are positive.

In certain cases, the direct application of a simple quadrature formula is unsatisfactory; this happens, for example, if x oscillates rapidly. The technique of *product integration* can be used to overcome these difficulties. We write $x(t) = r(t)\sigma(t)$, where $r(t)$ is well behaved and $\sigma(t)$ contains the part causing the difficulty. $I(x)$ is approximated by

$$I_n(x) = \int_a^b \sigma(t)(\pi_n r)(t)\,dt = \sum_{i=1}^n \left[\int_a^b \sigma(t) e_i^{(n)}(t)\,dt\right] r(t_{in}) = \sum_{i=1}^n w_{in} r(t_{in}).$$

$\sigma(t)$ should be simple enough that the weights w_{in} can be evaluated in closed form (Atkinson, 1976a; Sloan, 1980a).

Exercise

4.8 Consider $I(r) = \int_0^1 [r(t)/\sqrt{t}]\,dt$. Compute the weights corresponding to the trapezium rule with $n + 1$ equally spaced points applied to $r(t)$. If $r \in C^2(0, 1)$, prove that

$$|(I - I_n)(r)| \leq \frac{2}{n^2} \max_{t \in [0, 1]} |r''(t)|.$$

For more on numerical integration, the reader is referred to Davis and Rabinowitz (1974) and Stroud (1971), for example.

3. Projection Methods

Let X be a complex Banach space, X_n a finite-dimensional subspace of X, and π_n a bounded projection from X onto X_n: $\pi_n X = X_n$. (4.1) and (4.2) are approximated *in* X_n respectively by the finite-dimensional problems

$$\pi_n T x_n - z x_n = \pi_n f, \qquad x_n \in X_n, \tag{4.4}$$

and

$$\pi_n T \varphi_n = \lambda_n \varphi_n, \qquad 0 \neq \varphi_n \in X_n. \tag{4.5}$$

This is the general *projection* method on the subspace X_n. x_n (resp. λ_n, φ_n) is computed such that the residual $(T - z)x_n - f$ (resp. $(T - \lambda_n)\varphi_n$) has a zero projection in X_n. Note that $\pi_n x_n = x_n$ (resp. $\pi_n \varphi_n = \varphi_n$). This method is often called the *Galerkin* method.

A Remark on the Terminology The name Galerkin is often given to the projection method in a Banach space in the context of integral equations. On the contrary, in the context of differential equations, the name Galerkin is often used in a more restrictive way, namely, when the projection is orthogonal in a Hilbert space. We are then led to introduce two definitions: the Galerkin method (either in a Banach space or with an oblique projection in a Hilbert space) and the orthogonal-Galerkin, abbreviated $\perp$-Galerkin (in a Hilbert space when the projection is orthogonal).

3.1. The Associated Discrete Problems

The problem in X_n is defined by $\mathscr{T}_n := \pi_n T_{\restriction X_n}$:

$$(\mathscr{T}_n - z)x_n = \pi_n f, \qquad \mathscr{T}_n \varphi_n = \lambda_n \varphi_n.$$

The dimension of X_n depends on n. For simplicity we set $\dim X_n =: n$. Given a basis $\{e_i^{(n)}\}_1^n$ of X_n and the adjoint basis $\{e_j^{(n)*}\}_1^n$ of $X_n^* = \pi_n^* X^*$, the projection π_n is such that $\pi_n x = \sum_{i=1}^m \langle x, e_i^{(n)*} \rangle e_i^{(n)}$. We define the matrix associated with $\mathscr{T}_n$:

$$\tilde{A}_n := (a_{ij}^{(n)}) = \langle T e_j^{(n)}, e_i^{(n)*} \rangle, \qquad i, j = 1, \ldots, n,$$

and the vectors

$$\xi_n := (\xi_{in}), \qquad u_n := (u_{in}), \qquad \eta_n := (\eta_{in}), \qquad i = 1, \ldots, n,$$

where

$$\xi_{in} = \langle x_n, e_i^{(n)*} \rangle, \qquad u_{in} = \langle \varphi_n, e_i^{(n)*} \rangle, \qquad \eta_{in} = \langle f, e_i^{(n)*} \rangle.$$

ξ_n and u_n are solutions in $\mathbb{C}^n$ of the respective matrix problems

$$(\tilde{A}_n - zI)\xi_n = \eta_n \qquad \text{and} \qquad \tilde{A}_n u_n = \lambda_n u_n.$$

The solutions x_n and φ_n in X_n are then deduced from their components in the basis $\{e_i^{(n)}\}_1^n$.

From a computational point of view, it may be more interesting to consider bases in X_n and X_n^* that are *not* adjoint, that is, such that

$$\langle e_i^{(n)*}, e_j^{(n)} \rangle \neq \delta_{ij}.$$

For example, if X_n is a subset of cubic spline functions, then the B-splines (de Boor, 1968; Prenter, 1975) are an excellent basis of this kind.

Let us then consider a basis $\{x_i^{(n)}\}_1^n$ of X_n and a basis $\{y_i^{(n)}\}_1^n$ of X_n^* with corresponding Gram matrix

$$\tilde{B}_n = (b_{ij}^{(n)}) = (\langle x_j^{(n)}, y_i^{(n)} \rangle), \qquad i, j = 1, \ldots, n.$$

π_n is now defined by $\pi_n x = \sum_{i=1}^m \langle x, \tilde{B}_n^{-1\,\mathrm{H}} y_i^{(n)} \rangle x_i^{(n)}$. $\tilde{A}_n$, ξ_n, u_n and η_n are defined respectively by $\langle Tx_j^{(n)}, y_i^{(n)} \rangle$, $\langle x_n, y_i^{(n)} \rangle$, $\langle \varphi_n, y_i^{(n)} \rangle$, and $\langle f, y_i^{(n)} \rangle$. The matrix problems to be solved are

$$(\tilde{A}_n - z\tilde{B}_n)\xi_n = \eta_n \qquad \text{and} \qquad \tilde{A}_n u_n = \lambda_n \tilde{B}_n u_n.$$

3.2. *Examples*

If there is no ambiguity, the index n will be dropped now on for t_i and e_i, although it remains in force.

Example 4.5 Let $X = C(a, b)$. $[a, b]$ is divided into $n - 1$ intervals at the points t_i, $i = 1, \ldots, n$, $t_1 = a$, $t_n = b$. Let X_n be the subset of X consisting of piecewise-linear functions on $[a, b]$, with the basis $\{e_i\}_1^n$ defined in Example 2.6 of Chapter 2. π_n is the piecewise-linear interpolatory projection at the points $\{t_i\}_1^n$, defined by $\pi_n x = \sum_{i=1}^n x(t_i)e_i$. Clearly $\pi_n \xrightarrow{\mathrm{p}} 1$ if

$$\max_{2 \leq i \leq n} |t_i - t_{i-1}| \to 0.$$

The associated matrix $\tilde{A}_n$ is defined by

$$a_{ij}^{(n)} = \int_{t_{j-1}}^{t_{j+1}} k(t_i, s)e_j(s)\, ds, \qquad i, j = 1, \ldots, n$$

(with $t_0 = a$, $t_{n+1} = b$). This is an example of *collocation method* at the collocation points $\{t_i\}_1^n$, with the basis functions $\{e_i\}_1^n$ (cf. Section 3.5).

Example 4.6 Let $X = L^\infty(a, b)$. $[a, b]$ is divided into n intervals at the points t_i, $a = t_0 < t_1 < \cdots < t_n = b$. Set $\Delta_i := [t_{i-1}, t_i[$, $i = 1, \ldots, n$. X_n is the set of piecewise-constant functions on $[a, b]$, and the basis $\{e_i\}_1^n$ is the set of n characteristic functions of Δ_i, $e_i(t) = 1$ on Δ_i, 0 elsewhere. e_i^* is the functional on X defined by

$$\langle e_i^*, x\rangle = \frac{1}{t_i - t_{i-1}} \int_{\Delta_i} \overline{x(t)}\, dt.$$

π_n is the averaging projection

$$(\pi_n x)(t) = \frac{1}{t_i - t_{i-1}} \int_{\Delta_i} x(t)\, dt, \qquad t \in \Delta_i.$$

If $\max_{1 \le i \le n} |t_{i-1} - t_i| \to 0$, $\pi_n x \to x$ for $x \in C(a, b)$, but there exist functions f in $L^\infty(a, b)$ such that $\pi_n f \not\to f$. The matrix $\tilde{A}_n$ is defined by

$$a_{ij}^{(n)} = \frac{1}{t_i - t_{i-1}} \int_{\Delta_i} \int_{\Delta_j} k(s, t)\, dt\, ds \qquad \text{for} \quad i, j = 1, \ldots, n.$$

This is an example of the method of *separating regions* (Mikhlin and Smolitskii, 1967, p. 254).

Example 4.7 Let $H = L^2(a, b)$. Let $\{e_i\}_{\mathbb{N}}$ be an *orthonormal* basis of H. Then $e_i = e_i^*$ for $i \in \mathbb{N}$. E_n is spanned by $e_1, \ldots, e_n$. π_n is the *orthogonal* projection on E_n. We set

$$a_{ij} := \int_a^b \overline{e_i(t)} \int_a^b k(t, s) e_j(s)\, ds\, dt,$$

for $i, j = 1, 2, \ldots$. The elements of $\tilde{A}_n$ do not depend on n. $\tilde{A}_n$ is the upper left square matrix of size n of the infinite matrix with coefficients a_{ij}. This is the original *Galerkin* method. The method replaces T by an operator T_n defined on E_n by the kernel

$$k_n(t, s) = \sum_{i=1}^n \left[\int_a^b k(t, s) e_i(s)\, ds \right] \overline{e_i(s)} = \sum_{i=1}^n u_i(t) \overline{e_i(s)}.$$

For $x \in E_n$, let

$$\xi_j := \int_a^b \overline{e_j(s)}\, x(s)\, ds, \qquad j = 1, \ldots, n.$$

Then $x(s) = \sum_{j=1}^n \xi_j e_j(s)$ and

$$(T_n x)(t) = \int_a^b k_n(t, s) x(s)\, ds = \sum_{j=1}^n \xi_j u_j(t).$$

T_n is represented in $\{e_i\}_1^n$ by the matrix $\tilde{A}_n$. The kernel k_n is a finite sum of functions that are the product of a function of t times a function of s. Such a kernel is said to be *degenerate*.

Example 4.8 Another example of the $\perp$-Galerkin method in $H = L^2(a, b)$ is given by the method of moments (Vorobyev, 1965), where π_n is the orthogonal projection on $\{u, Tu, \ldots, T^{n-1}u\}$ for a given u in H.

Example 4.9 When T is self-adjoint positive definite in a Hilbert space H, the $\perp$-Galerkin method takes the particular form of the *Rayleigh–Ritz* method. Let E_n be a finite-dimensional subspace of H. π_n is the orthogonal projection on E_n.

The Rayleigh–Ritz method consists in approximating the solution x of $Tx = f$ by the solution $x_n \in E_n$ of $\pi_n[Tx_n - f] = 0$. $(x, y)_T := (Tx, y)$ defines a scalar product on H and the corresponding norm $\|x\|_T^2 := (Tx, x)$. Then

$$Tx_n - f \in E_n \Leftrightarrow x_n - x \perp_T E_n,$$

where $\perp_T$ denotes the orthogonality relative to the scalar product $(\cdot, \cdot)_T$. x_n is the least-squares fit in E_n of x, relative to the norm $\|\cdot\|_T$.

$$\|x_n - x\|_T = \min_{y \in E_n} \|x_n - y\|_T.$$

Exercise

4.9 Consider the infinite system of equations $\sum_{j=1}^{\infty} a_{ij}x_j - x_i = b_i$, $i = 1, 2, \ldots$, with the assumption that

$$\sum_{i,j=1}^{\infty} |a_{ij}|^2 < \infty, \qquad \sum_{i=1}^{\infty} |b_i|^2 < \infty.$$

Set the problem in l^2 with the canonical basis $\{e_i\}_{\mathbb{N}}$. Characterize the approximate solution for the$\perp$-Galerkin method defined by $E_n = \{e_1, \ldots, e_n\}$. Give a sufficient condition on (a_{ij}) under which the original system has a unique solution.

3.3. The Petrov Method

The projection method can be generalized by considering *two* subspaces X_n and Y_n of X with the same finite dimension. X_n is the *right* subspace, and Y_n is the *left* subspace. Let π_n be a projection on Y_n. (4.1) and (4.2) are approximated respectively by

$$\pi_n[(T - z)x_n - f] = 0, \qquad x_n \in X_n, \tag{4.6}$$

and

$$\pi_n[(T - \lambda_n)\varphi_n] = 0, \qquad \varphi_n \in X_n. \tag{4.7}$$

Note that now $\pi_n x_n \neq x_n$ (resp. $\pi_n \varphi_n \neq \varphi_n$). We define $\tilde{\pi}_n := \pi_{n \restriction X_n}: X_n \to Y_n$ and introduce the condition

$$\text{for } n \text{ large enough, } \tilde{\pi}_n^{-1} \text{ exists and is uniformly bounded.} \tag{4.8}$$

Exercises

4.10 Write the discrete problems associated with (4.6) and (4.7) in given bases of X_n and Y_n.

4.11 Consider $\pi'_n := \tilde{\pi}_n^{-1}\pi_n$. Prove that π'_n is a projection on X_n. Deduce that, under the condition (4.8), the Petrov method is equivalent to the Galerkin method on X_n defined with the projection π'_n.

4.12 Let E_n and F_n be subspaces of a Hilbert space H such that $\dim E_n = \dim F_n < \infty$. π_n is the orthogonal projection on F_n, cf. Fig. 4.1. We define

$$\theta_n := \Theta(E_n, F_n),$$

$$\tau_n := \inf_{\substack{x \in E_n \\ \|x\| = 1}} \|\pi_n x\| = (\|\tilde{\pi}_n^{-1}\|)^{-1}.$$

Prove that $\theta_n^2 + \tau_n^2 = 1$. Deduce that, for n large enough,

$$\|\tilde{\pi}_n^{-1}\| < \infty \Leftrightarrow \Theta(E_n, F_n) < 1$$

(Krasnoselskii *et al.*, 1972).

Remark When condition (4.8) is satisfied, the Petrov method is not theoretically different from the Galerkin method (Exercise 4.11). The Petrov method is often called *generalized* Galerkin method. We also distinguish the Petrov method (in a Banach space or in a Hilbert space with an oblique projection π_n) and the $\perp$-Petrov method in a Hilbert space with an orthogonal projection π_n. Note that the $\perp$-Petrov method is equivalent to an (oblique) Galerkin method when $\sup_n \Theta(E_n, F_n) < 1$. In the Russian literature, the Petrov, $\perp$-Petrov, and $\perp$-Galerkin methods are respectively called projection, Petrov–Galerkin, and Bubnov–Galerkin methods (Krasnoselskii *et al.*, 1972).

Example 4.10 Let T be a bounded operator in a complex Hilbert space H. For $y \in H$, we consider the equation $(T - z)x = y$ for $z \in \rho(T)$. Let

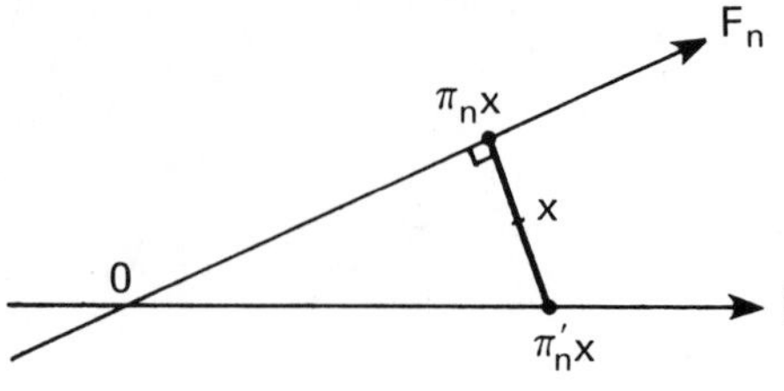

Figure 4.1

two dense sequences $\{e_i\}_{\mathbb{N}}$ and $\{f_i\}_{\mathbb{N}}$ be given in H. The approximate solution x_n is sought in the form $x_n = \sum_{i=1}^{n} \xi_{in} e_i$. The coefficients ξ_{in} are determined by the condition that the residual $(T - z)x_n - y$ be orthogonal to $f_1, \ldots, f_n$, that is,

$$((T - z)x_n - y, f_i) = 0, \qquad i = 1, \ldots, n.$$

This gives

$$\sum_{j=1}^{n} ((T - z)e_j, f_i)\xi_{jn} = (y, f_i), \qquad i = 1, \ldots, n.$$

$E_n := \{e_1, \ldots, e_n\}$ and $F_n := \{f_1, \ldots, f_n\}$ are subspaces of H. Let π_n be the *orthogonal* projection from H onto F_n. The above $n \times n$ system is equivalent to the equation

$$\pi_n[(T - z)x_n - y] = 0, \qquad x_n \in E_n.$$

This is the original *Petrov* method. The corresponding approximate eigenvalue problem is

$$((T - \lambda_n)\varphi_n, f_i) = 0, \qquad i = 1, \ldots, n,$$

$$\varphi_n = \sum_{j=1}^{n} u_{jn} e_j.$$

This gives

$$\sum_{j=1}^{n} (Te_j, f_i)u_{jn} = \lambda_n \sum_{j=1}^{n} (e_j, f_i)u_{jn}, \qquad i = 1, \ldots, n.$$

Example 4.11 We choose in the preceding example $F_n = TE_n$. The equation $Tx = f$ with $T^{-1} \in \mathscr{L}(H)$ is approximated by

$$\pi_n[Tx_n - f] = 0, \qquad x_n \in E_n,$$

where π_n is the orthogonal projection on $F_n = TE_n$. This defines the *least-squares method.* It is easy to check that $x_n \in E_n$ is the solution of the minimization problem

$$\|Tx_n - f\| = \min_{y \in E_n} \|Ty - f\|.$$

Exercise

4.13 Let T be self-adjoint positive definite. Interpret the least-squares method as an $\perp_T$-Galerkin method defined by a projection that is orthogonal with respect to the scalar product $(\cdot, \cdot)_T$.

In $\mathbb{C}^N$, examples of the $\perp$-Petrov method have been given in Chapter 1, Section 5, with the oblique projection methods (see Saad, 1982a). Other examples of the Petrov method will be considered in Part B.

3.4. Projection Method on the Kantorovitch Regularized Equation

For those equations (4.1) for which $\pi_n f \nrightarrow f$ (see Section 2.1), Kantorovitch (1948) suggested the following modification, when nevertheless $\pi_n Tf \to Tf$. Set $y := Tx$. Then y satisfies $y - zx = f$ by (4.1). By multiplication by T we get the "regularized" equation

$$(T - z)y = Tf, \qquad x = (1/z)(y - f)$$

to which we apply the projection method (if $z \neq 0$). It yields

$$\begin{aligned} (\pi_n T - z)y_n &= \pi_n Tf, \qquad && y_n \in X_n, \\ x'_n &= (1/z)(y_n - f) && \text{if} \quad z \neq 0. \end{aligned} \tag{4.9}$$

x'_n does not belong to X_n. Note that (4.9) and (4.4) yield the same matrix $\tilde{A}_n$. This method requires the evaluation of Tf.

Note that the above regularization has nothing to do with Tichonov's regularization of ill-posed problems.

Exercise

4.14 Let $\tilde{x}$ be an approximate solution of (4.1) and set $r := f - (T - z)\tilde{x}$, $v := x - \tilde{x}$. Then consider the equation $(T - z)v = r$. Write the regularized equation, its approximation by projection. Deduce another approximation for x.

3.5. Collocation Method

Given a set of n independent functions $\{l_i\}_1^n$, and given n collocation points $\{\tau_i\}_1^n$ on $[a, b]$ such that $\det(l_j(\tau_i)) \neq 0$, the solutions x_n and φ_n are sought in the form

$$x_n = \sum_{i=1}^{n} \xi_{in} l_i, \qquad \varphi_n = \sum_{i=1}^{n} u_{in} l_i.$$

The index n is again dropped for τ_i and l_i, $i = 1, \ldots, n$.

The basic idea of the collocation method is to determine the coefficients ξ_{in} and u_{in} such that the respective residuals $(T - z)x_n - f$ and $T\varphi_n - \lambda_n \varphi_n$ vanish at the n points $\{\tau_i\}_1^n$. It yields two $n \times n$ systems:

$$\sum_{j=1}^{n} \left[\int_a^b k(\tau_i, s) l_j(s)\, ds - z l_j(\tau_i) \right] \xi_{jn} = f(\tau_i), \qquad i = 1, \ldots, n; \tag{4.10}$$

$$\sum_{j=1}^{n} \left(\int_a^b k(\tau_i, s) l_j(s)\, ds \right) u_{jn} = \lambda_n \sum_{j=1}^{n} l_j(\tau_i) u_{jn}, \qquad i = 1, \ldots, n. \tag{4.11}$$

We define the matrices $\tilde{A}_n$ and $\tilde{B}_n$ by their elements:

$$a_{ij}^{(n)} = \int_a^b k(\tau_i, s) l_j(s)\, ds, \qquad b_{ij}^{(n)} = l_j(\tau_i), \qquad i, j = 1, \ldots, n.$$

(4.10) and (4.11) may then be written

$$(\tilde{A}_n - z\tilde{B}_n)\xi_n = (f(\tau_i)), \qquad \tilde{A}_n u_n = \lambda_n \tilde{B}_n u_n.$$

To show that this corresponds to a projection method, begin by letting $X_n := \{l_1, \ldots, l_n\}$. Define the projection π_n onto X_n such that $\pi_n x$ is the element in X_n that interpolates x at the points $\{\tau_i\}_1^n$. Let

$$(\pi_n x)(t) = \sum_{j=1}^{n} c_{jn} l_j(t).$$

The c_{jn} are uniquely determined by

$$\sum_{j=1}^{n} c_{jn} l_j(\tau_i) = x(\tau_i), \qquad i = 1, \ldots, n$$

since $\det(l_j(\tau_i)) \neq 0$. It is left to the reader to check that (4.10) and (4.11) can be written in the form (4.4) and (4.5) with the projection π_n. Collocation methods are therefore projection methods defined with *interpolatory* projections. There are two main classes of functions from which the basis set $\{l_i\}_1^n$ is chosen in practice: first, piecewise polynomials, and second, polynomials defined by a single expression over $[a, b]$. In the first case, the pointwise convergence $\pi_n \xrightarrow{\text{p}} 1$ usually takes place, from which the convergence of the method may be established. In the second case, convergence depends on the choice of the collocation points $\{\tau_i\}_1^n$ (cf. Section 2). For the Chebyshev points and the Lagrange interpolation, $\|L_n\|_\infty$ is proportional to $\log n$, which is the slowest by which it can increase. In addition, the error formula for interpolation at the $\{\tau_i\}_1^n$, for a smooth enough x,

$$x(t) - (L_n x)(t) = \frac{1}{n!} \prod_{i=1}^{n} (t - \tau_i) x^{(n)}(s), \qquad a \leq s \leq b,$$

suggests also the Chebyshev zeros, since $\max_{t\in[a,b]} |\prod_{i=1}^{n} (t - \tau_i)|$ is minimized by that choice (see Cheney, 1966, p. 61).

One should note that, unlike the $\perp$-Galerkin method in $L^2(a, b)$, the collocation method does not require the computation of integrals for the scalar products in $\tilde{A}_n$ and $\tilde{B}_n$ (compare (4.10) and (4.11) with the general formulas given in Section 3.1).

Exercises

4.15 Prove that the $\perp$-Galerkin method on (4.1) in $L^2(a, b)$ where the scalar products are computed with approximate quadrature formulas amounts to a collocation method whenever the collocation points are defined to be the nodes of the quadrature rule chosen for Galerkin's method.

4.16 Interpret the collocation method as a Petrov method with $X_n = \{l_1, \ldots, l_n\}$, π_n being an interpolatory projection at $\{\tau_i\}_1^n$ defined by $Y_n = \{l'_1, \ldots, l'_n\}$ such that $\det(l'_i(\tau_j)) \neq 0$, the l'_i being otherwise arbitrary.

The Bland's collocation method for the generalized airfoil equation may be interpreted as a Petrov method on $T: X \to Y$, defined by X_n, Y_n subspaces of X, Y, respectively (see Bland, 1970; Fromme and Golberg, 1978).

3.6. Bibliographical Comments

Projection methods have been used for a long time. Their abstract treatment goes back at least to Kantorovitch (1948) and to Kantorovitch and Akilov (1964). A stability analysis of the discrete systems with matrices $\tilde{A}_n - zI$ or $\tilde{A}_n - z\tilde{B}_n$ is given in Atkinson (1976a), where the reader can also find a study of degenerate kernel methods (cf. Example 4.13). Collocation methods were introduced by Kantorovitch (1948) for ordinary differential equations (cf. Section 8). Polynomial collocation at Chebyshev points was studied in Karpilovskaia (1963) and Vainikko (1965). More recently, collocation has been revived because of the use of piecewise polynomials. The Petrov method is thoroughly studied in Krasnoselskii *et al.* (1972). Comparative testing of automatic and nonautomatic integral-equation-solving routines is done in Riddell and Delves (1980).

4. Approximate Quadrature Methods

In approximate quadrature methods, the integral in (4.3) is approximated by some quadrature rule. We treat here the case where Ω is a finite interval $[a, b]$, $X = C(a, b)$. We suppose that the kernel of the integral operator defined by (4.3) is continuous on $[a, b] \times [a, b]$. The index n is again dropped for t_i, e_i, w_i, and we suppose that we are given an interpolatory formula

$$I_n(x) = \sum_{i=1}^{n} w_i x(t_i) = \sum_{i=1}^{n} \left(\int_a^b e_i(t)\, dt \right) x(t_i),$$

which is convergent, for all continuous x, to $I(x) = \int_a^b x(t)\, dt$.

4.1. The Fredholm Method

In (4.1) and (4.2) the integral is numerically approximated using the quadrature formula I_n. This gives

$$\sum_{j=1}^{n} w_j k(s, t_j) x_n(t_j) - z x_n(s) = f(s), \qquad a \le s \le b, \tag{4.12}$$

$$\sum_{j=1}^{n} w_j k(s, t_j) \varphi_n(t_j) = \lambda_n \varphi_n(s), \qquad a \le s \le b. \tag{4.13}$$

The *Fredholm* method consists in discretizing in s with the same nodes to get the $n \times n$ systems

$$\sum_{j=1}^{n} w_j k(t_i, t_j) x_n(t_j) - z x_n(t_i) = f(t_i), \qquad i = 1, \ldots, n, \tag{4.14}$$

$$\sum_{j=1}^{n} w_j k(t_i, t_j) \varphi_n(t_j) = \lambda_n \varphi_n(t_i), \qquad i = 1, \ldots, n. \tag{4.15}$$

Upon setting

$$\tilde{A}_n := (w_j k(t_i, t_j)), \qquad \xi_n := (x_n(t_i)), \qquad u_n := (\varphi_n(t_i)),$$

(4.14) and (4.15) may be rewritten

$$(\tilde{A}_n - zI)\xi_n = (f(t_i)) \qquad \text{and} \qquad \tilde{A}_n u_n = \lambda_n u_n.$$

Upon computing the discrete solutions ξ_n and u_n in $\mathbb{C}^n$, the solutions of (4.14) and (4.15) in $X_n := \{e_1, \ldots, e_n\}$ are taken to be

$$x_n = \sum_{i=1}^{n} \xi_{in} e_i \qquad \text{and} \qquad \varphi_n = \sum_{i=1}^{n} u_{in} e_i.$$

x_n and φ_n are the functions computed from the values at the points $\{t_i\}_1^n$ by means of the interpolation associated with the quadrature formula I_n.

Exercise

4.17 Compare the matrix of the Fredholm method to the matrix of the projection method defined by π_n such that $\pi_n x = \sum_{i=1}^{n} x(t_i) e_i$.

4.2. The Nyström Method

Knowing $x_n(t_i)$ and $(\lambda_n, \varphi_n(t_i))$ from Eqs. (4.14) and (4.15), we can solve Eqs. (4.12) and (4.13) for $s \neq t_i$, $i = 1, \ldots, n$, by the evaluation of

$$x_n(s) = \frac{1}{z}\sum_{j=1}^{n} w_j k(s, t_j)x_n(t_j) - \frac{1}{z}f(s), \tag{4.16}$$

$$\varphi_n(s) = \frac{1}{\lambda_n}\sum_{j=1}^{n} w_j k(s, t_j)\varphi_n(t_j), \tag{4.17}$$

provided that z and λ_n are nonzero. This observation was first made by Nyström (1930). Equations (4.16) and (4.17) are called the "natural" interpolation formulas, since they define $x_n(t)$ and $\varphi_n(t)$ for all t, given $x_n(t_i)$ and $\varphi_n(t_i)$ for $i = 1, \ldots, n$.

Exercise

4.18 Show that (4.16) is an interpolation means to get $x_n(t)$ from $x_n(t_i)$ only if $x_n(t_i)$ is solution of (4.14).

The natural interpolation contains some information on the operator T, and as a consequence the Nyström solution will often be significantly better than the Fredholm solution (cf. Chapter 6). The Nyström and Fredholm solutions may be viewed as two different interpolated functions between the values at $\{t_i\}_1^n$.

4.3. Product Integration

If the kernel k is smooth, but the right-hand side f is not, the regularization technique of Section 3.4 indicates the use of the Nyström method on the regularized equation

$$(T - z)y = Tf, \qquad x = (1/z)(y - f),$$

where Tf is at least as smooth as the kernel k.

When the kernel k is discontinuous or badly behaved, the quadrature rule should be of *product integration* form. Let us write $k(s, t) = r(s, t)\sigma(s, t)$, where r is a regular function of t for each s in $[a, b]$, whereas $\sigma(s, t)$ is singular or badly behaved.

With the notation of Section 2.2, the integral in (4.3) is approximated by

$$\int_a^b \sigma(s, t)\left[\sum_{i=1}^{n} x(t_i)r(s, t_i)e_i(t)\right] dt = \sum_{i=1}^{n} w_i(s)r(s, t_i)x(t_{in}),$$

where

$$w_i(s) = \int_a^b \sigma(s, t)e_i(t)\,dt.$$

The most important practical point is to have a function $\sigma(s, t)$ for which the weights $w_i(s)$ can be computed for some quadrature formulas. This is done in Atkinson (1976a) for

$$\sigma(s, t) = 1/|s - t|^\alpha, \qquad \alpha < 1 \qquad \text{and} \quad \sigma(s, t) = \log|s - t|.$$

The product integration solution $x_n(s)$ (resp. $\varphi_n(s)$) is then computed by Eqs. (4.14) and (4.16) (resp. (4.15) and (4.17)), where the weights w_j are replaced by $w_j(t_i)$ and $w_j(s)$.

The full discretization method was proposed by Fredholm (1903). Since its introduction in 1930, the Nyström variant has been widely used. The product integration was devised by Young (1954) and studied by Atkinson (1972, 1976a). The use of Kantorovitch's regularized equation as well as iterative variants of the Nyström method together with numerical experiments can be found also in Atkinson (1976a).

5. The Iterated Solution and Eigenvector

Both z and λ are assumed to be *nonzero*. This is certainly not a restriction when T is compact, since z, which belongs to $\rho(T)$, cannot be zero, and any isolated eigenvalue of T is nonzero as well. If $\lambda_n \to \lambda$ as $n \to \infty$, then λ_n is nonzero for n large enough. If z or λ is zero, see Exercises 4.20 and 4.21.

Let x_n (resp. λ_n, φ_n) be an approximate solution of (4.1) (resp. (4.2)). We define the *iterated solution* $\tilde{x}_n$ (resp. *iterated eigenvector* $\tilde{\varphi}_n$) from x_n (resp. λ_n, φ_n) by

$$\tilde{x}_n = (1/z)(Tx_n - f), \tag{4.18}$$

$$\tilde{\varphi}_n = (1/\lambda_n)T\varphi_n. \tag{4.19}$$

$\tilde{x}_n$ can be regarded as the result of one fixed-point iteration applied to the equation $x = (1/z)Tx - (1/z)f$, from the approximate solution x_n.

Exercises

4.19 Consider in $\mathbb{C} \times X$ the nonlinear operator F:

$$(z, x) \overset{F}{\mapsto} \left(z, \frac{1}{z}Tx\right).$$

Prove that (λ, φ) is a fixed point of F, and interpret (4.19).

4.20 Consider the equation $Tx = f$, $0 \in \rho(T)$. Let $0 \neq z_0 \in \rho(T)$ be given. If x_n is an approximate solution, prove that an iterated solution can be defined by

$$\tilde{x}_n = (1/z_0)[f - (T - z_0)x_n]$$

4.21 $\lambda = 0$ is an eigenvalue of T: $T\varphi = 0$, $0 \neq \varphi$. Let φ_n be an approximate eigenvector: $T_n\varphi_n = \lambda_n\varphi_n$, $0 \neq \varphi_n$, such that $\lambda_n \to 0$ as $n \to \infty$. Prove that for any given $\alpha \neq 0$, an iterated eigenvector can be defined by

$$\tilde{\varphi}_n = \frac{1}{\lambda_n + \alpha}(T + \alpha)\varphi_n$$

The computation of $\tilde{x}_n$ and $\tilde{\varphi}_n$ requires the evaluation of Tu, where u is a known function, either x_n or φ_n. When this cannot be done in closed form, an approximation T_M of higher order $M \gg n$ may be used to approximate Tu by $T_M u$ (cf. Example 4.15). Example 4.12 deals with the case where x_n and φ_n result of the projection method. Of course, the additional computation of $\tilde{x}_n$ (resp. $\tilde{\varphi}_n$) after x_n (resp. λ_n, φ_n) is worthwhile only if $\tilde{x}_n$ (resp. $\tilde{\varphi}_n$) is a better approximation of x (resp. φ) than x_n (resp. φ_n). This question will be considered in Chapters 6 and 7.

Example 4.12 Let x_n (resp. φ_n) be the solution of (4.4) (resp. 4.5)). T is approximated by the projection method. The use of the iterated Galerkin solution $\tilde{x}_n$ (resp. $\tilde{\varphi}_n$) has been proposed by Sloan (1976a,b) as a variant of the Galerkin method. $\tilde{x}_n$ (resp. $\tilde{\varphi}_n$) is an approximate solution with a nonzero component outside X_n. Note that $\pi_n\tilde{x}_n = x_n$ (resp. $\pi_n\tilde{\varphi}_n = \varphi_n$). Note also that the evaluation of $\tilde{x}_n$ (resp. $\tilde{\varphi}_n$) is fairly inexpensive. Indeed if $x_n = \sum_{i=1}^n \xi_{in} e_i$, then $Tx_n = \sum_{i=1}^n \xi_{in} Te_i$, where the $\{Te_i\}_1^n$ have already been computed to get $\tilde{A}_n$. Therefore only some additional storage is required to store these vectors for preassigned values of the variable.

Example 4.13 Let $\{e_i\}_1^\infty$ be an orthonormal basis of $H = L^2(a, b)$, and set $E_n := \{e_1, \ldots, e_n\}$. Given the integral operator T, let T_n be defined by the degenerate kernel $k_n(s, t) = \sum_{i=1}^n (Te_i)(s)e_i(t)$. The method of *degenerate kernels* consists in solving

$$(T_n - z)\tilde{x}_n = f \qquad \text{for} \quad \tilde{x}_n \in X.$$

Hence

$$\tilde{x}_n(s) = -\frac{1}{z} f(s) + \frac{1}{z}\sum_{j=1}^n \xi_{jn}(Te_j)(s)$$

$$= \frac{1}{z}(Tx_n - f)(s), \qquad a \leq s \leq b,$$

where $x_n = \sum_j \xi_{jn} e_j$ and the ξ_{jn} are solutions of

$$\sum_{j=1}^n (Te_j, e_i)\xi_{jn} - z\xi_{in} = (f, e_i), \qquad i = 1, \ldots, n.$$

Considering the $\perp$-Galerkin method defined by E_n, we see that $\tilde{x}_n$ is the iterated Galerkin solution.

Example 4.14 Let x_n (resp. φ_n) be the collocation solution (resp. eigenvector) defined by (4.10) (resp. (4.11)). The iterated solution (resp. eigenvector) satisfies

$$\tilde{x}_n(s) = \frac{1}{z}\sum_{j=1}^{n}\left[\int_a^b k(s,t)l_j(t)\,dt\right]\xi_{jn} - \frac{1}{z}f(s), \qquad a \le s \le b,$$

$$\text{(resp.}\quad \tilde{\varphi}_n(s) = \frac{1}{\lambda_n}\sum_{j=1}^{n}\left[\int_a^b k(s,t)l_j(t)\,dt\right]u_{jn}, \qquad a \le s \le b).$$

Then

$$\tilde{x}_n(\tau_i) = \sum_{j=1}^{n}\xi_{jn}l_j(\tau_i) = x_n(\tau_i) \qquad (\text{resp.}\quad \tilde{\varphi}_n(\tau_i) = \varphi_n(\tau_i)).$$

The iterated solutions agree at the collocation points. x_n and $\tilde{x}_n$ (resp. φ_n and $\tilde{\varphi}_n$) may be regarded as two different functions obtained by interpolating between the values of x_n (resp. φ_n) at the points $\{\tau_i\}_1^n$.

We remark that $\tilde{x}_n(s)$ is identical with the product integration solution if $r(s,t) = 1$ for all s, t, and $l_i = e_i$, $i = 1, \ldots, n$. The iterated collocation is then a special case of product integration in which $k = \sigma$ and the set $\{l_i\}_1^n$ satisfies $l_j(t_i) = \delta_{ij}$. It is always possible to work with such a set by means of a change of basis in $X_n = \{l_1, \ldots, l_n\}$, because of the assumption that $\det(l_j(t_i)) \neq 0$.

Example 4.15 When x_n is the Nyström solution defined by (4.16), Brakhage (1960) has considered $\tilde{x}_M = (1/z)(T_M x_n - f)$, where T_M is an approximation of T of order $M \gg n$. $\tilde{x}_M$ is therefore an approximation of the iterated Nyström solution.

Exercises

4.22 Write the iterated solution associated with the approximate solution defined in Exercise 4.9.

4.23 Prove that $\tilde{x}_n$ and $\tilde{\varphi}_n$ defined by (4.18) and (4.19) are solutions of $(T\pi_n - z)\tilde{x}_n = f$ and $T\pi_n\tilde{\varphi}_n = \lambda_n\tilde{\varphi}_n$.

4.24 Let x_n and φ_n be defined by (4.6) and (4.7). Prove that $\tilde{x}_n$ and $\tilde{\varphi}_n$ satisfy, respectively, $\pi_n x_n = \pi_n \tilde{x}_n \neq x_n$ and $\pi_n\varphi_n = \pi_n\tilde{\varphi}_n \neq \varphi_n$.

6. Abstract Setting for the Approximation Operators

To distinguish between the various solutions defined in the two previous sections we introduce the notations of Table 4.1, where G, S, P, F, and N stand, respectively, for Galerkin, Sloan, projection, Fredholm and Nyström. To establish the mutual relationships between these solutions and eigenvectors, we begin by the preliminary study to follow.

Table 4.1

	Projection methods					Approximate quadrature methods			
Equation number	(4.4)	(4.5)	(4.18)[a]	(4.19)[a]	(4.9)	(4.14)	(4.15)	(4.16)	(4.17)
Notations	$x_n = x_n^{\mathrm{G}}$	$\varphi_n = \varphi_n^{\mathrm{G}}$	$\tilde{x}_n = x_n^{\mathrm{S}}$	$\tilde{\varphi}_n = \varphi_n^{\mathrm{S}}$	$x'_n = x_n^{\mathrm{P}}$	$x_n = x_n^{\mathrm{F}}$	$\varphi_n = \varphi_n^{\mathrm{F}}$	$x_n = x_n^{\mathrm{N}}$	$\varphi_n = \varphi_n^{\mathrm{N}}$

[a] When $T_n = \pi_n T$.

6.1. Decomposition of T according to a Decomposition of X

Let $T \in \mathscr{L}(X)$, and let X have the decomposition $X = Y \oplus Z$, where Y and Z are two supplementary closed subspaces of X. Let π be the projection from X onto Y, along Z. T is decomposed into four parts by the projection π. $T = T_1 + T_2 + T_3 + T_4$ with $T_1 := \pi T \pi$, $T_2 := \pi T(1 - \pi)$, $T_3 := (1 - \pi)T\pi$, and $T_4 := (1 - \pi)T(1 - \pi)$. T has the block-representation

$$T = \left(\begin{array}{c|c} T_1 & T_2 \\ \hline T_3 & T_4 \end{array}\right).$$

We consider the following three operators in $\mathscr{L}(X)$:

$$\begin{aligned} T_\alpha &:= \pi T \pi, &\quad \text{i.e.,}\quad T_\alpha &= T_1, \\ T_\beta &:= \pi T, &\quad \text{i.e.,}\quad T_\beta &= T_1 + T_2, \\ T_\gamma &:= T\pi, &\quad \text{i.e.,}\quad T_\gamma &= T_1 + T_3. \end{aligned}$$

Remark When π is an orthogonal projection in a Hilbert space, $T_\alpha = \pi T \pi$ is called a *compression* of T, and, conversely, T is called a *dilation* of T_α (see Halmos, 1950; Sz.-Nagy and Foias, 1970).

We set $\mathscr{T} := \pi T_{\restriction Y}: Y \to Y$ and define, in $\mathscr{L}(Y)$, $\mathscr{R}(z) := (\mathscr{T} - z)^{-1}$ for $z \in \rho(\mathscr{T})$. Let $\mu \neq 0$ be an eigenvalue of finite multiplicity of $\mathscr{T}$, isolated by the Jordan curve Γ, the spectral projection associated with μ is

$$\mathscr{P} = \frac{-1}{2i\pi} \int_\Gamma \mathscr{R}(z)\, dz \in \mathscr{L}(Y);$$

$\mathscr{D} = (\mathscr{T} - \mu)\mathscr{P}$ is the corresponding nilpotent, and l is the ascent of μ.

Theorem 4.3 *For $\kappa \in \{\alpha, \beta, \gamma\}$, any nonzero z in $\rho(\mathscr{T})$ belongs to $\rho(T_\kappa)$ and any nonzero eigenvalue μ of $\mathscr{T}$ of finite multiplicity isolated by Γ is an* *eigenvalue of T_κ with the same multiplicity. The following formulas hold if Γ does not enclose* 0:

$$R_\alpha(z) = \mathscr{R}(z)\pi - (1/z)(1 - \pi), \qquad P_\alpha = \mathscr{P}\pi, \tag{4.20}$$

$$R_\beta(z) = (1/z)\mathscr{R}(z)T_\beta - 1/z, \qquad P_\beta = \mathscr{E}\mathscr{P}T_\beta, \tag{4.21}$$

$$R_\gamma(z) = (1/z)T_\gamma\mathscr{R}(z)\pi - 1/z, \qquad P_\gamma = T_\gamma\mathscr{E}\mathscr{P}\pi, \tag{4.22}$$

with

$$\mathscr{E} = \sum_{k=0}^{l-1} \frac{1}{\mu}\left(\frac{-\mathscr{D}}{\mu}\right)^k.$$

Proof We first solve formally $(T - z)x = f$ to get $x = R(z)f$ in the three cases.

Case 1
$$(T_\alpha - z)x = f \Leftrightarrow \begin{cases} (\mathscr{T} - z)\pi x = \pi f \\ -z(1 - \pi)x = (1 - \pi)f \end{cases}$$

$$\Leftrightarrow \begin{cases} \pi x = \mathscr{R}(z)\pi f \\ (1 - \pi)x = -(1/z)(1 - \pi)f. \end{cases}$$

Hence any $z \neq 0$ in $\rho(\mathscr{T})$ is in $\rho(T_\alpha)$.

Case 2
$$(T_\beta - z)x = f \Leftrightarrow \begin{cases} (\mathscr{T} - z)\pi x + \pi T(1 - \pi)x = \pi f \\ -z(1 - \pi)x = (1 - \pi)f \end{cases}$$

$$\Leftrightarrow x = \mathscr{R}(z)\left[\pi f + \frac{1}{z}\pi T(1 - \pi)f\right] - \frac{1}{z}(1 - \pi)f.$$

$\mathscr{R}(z)\pi T\pi f = \mathscr{R}(z)\mathscr{T}\pi f = \pi f + z\mathscr{R}(z)\pi f$. Therefore

$$x = \frac{1}{z}\mathscr{R}(z)T_\beta f - \frac{1}{z}f = R_\beta(z)f,$$

and $0 \neq z$ in $\rho(\mathscr{T})$ belongs to $\rho(T_\beta)$.

Case 3
$$(T_\gamma - z)x = f \Leftrightarrow \begin{cases} (\mathscr{T} - z)\pi x = \pi f \\ (1 - \pi)T\pi x - z(1 - \pi)x = (1 - \pi)f \end{cases}$$

$$\Leftrightarrow x = \mathscr{R}(z)\pi f - \frac{1}{z}(1 - \pi)f + \frac{1}{z}(1 - \pi)T\mathscr{R}(z)\pi f$$

and

$$\pi T\mathscr{R}(z)\pi f = \mathscr{T}\mathscr{R}(z)\pi f = \pi f + z\mathscr{R}(z)\pi f.$$

Hence

$$x = \frac{1}{z} T_\gamma \mathscr{R}(z)\pi f - \frac{1}{z} f = R_\gamma(z) f \qquad \text{and} \qquad 0 \neq z \in \rho(T_\gamma).$$

Since Γ does not enclose 0 and T_κ is bounded, it is clear that in each case μ is the only point of $\sigma(T_\kappa)$ inside Γ. The associated spectral projection is obtained by integration of $R_\kappa(z)$ on Γ. We infer that $\dim P_\kappa X = \dim \mathscr{P} Y$. Hence μ is an eigenvalue of T_κ. $P_\alpha = \mathscr{P}\pi$ follows easily from

$$\frac{-1}{2i\pi} \int_\Gamma \mathscr{R}(z)\pi \, dz = \mathscr{P}\pi \qquad \text{and} \qquad \int_\Gamma \frac{dz}{z} = 0$$

since Γ does not enclose 0.

$$\frac{1}{z}\mathscr{R}(z) = -\frac{\mathscr{P}}{z(z-\mu)} - \sum_{k=1}^{l-1} \frac{\mathscr{D}^k}{z(z-\mu)^{k+1}} + \sum_{k=0}^{\infty} \frac{1}{z}(z-\mu)^k \mathscr{S}^{k+1}.$$

By integrating on Γ around μ, we get

$$P_\beta = \left[\frac{\mathscr{P}}{\mu} + \sum_{k=1}^{l-1} \frac{1}{\mu}\left(\frac{-\mathscr{D}}{\mu}\right)^k\right] T\pi = \mathscr{E}\mathscr{P}T_\beta.$$

Similarly $P_\gamma = T_\gamma \mathscr{E}\mathscr{P}\pi$. □

For $\kappa \in \{\alpha, \beta, \gamma\}$, the reduced resolvent is $S_\kappa = \lim_{z \to \lambda} R_\kappa(z)(1 - P_\kappa)$ (resp. $\mathscr{S} = \lim_{z\to\lambda} \mathscr{R}(z)(1 - \mathscr{P})$).

Corollary 4.4 *The following formulas hold if* $\mu \neq 0$

$$S_\alpha = \mathscr{S}\pi - \frac{1}{\mu}(1 - \pi), \tag{4.23}$$

$$S_\beta = \frac{1}{\mu}\mathscr{S}T_\beta + \frac{1}{\mu}\mathscr{E}\mathscr{P}T_\beta - \frac{1}{\mu}, \tag{4.24}$$

$$S_\gamma = \frac{1}{\mu} T_\gamma \mathscr{S}\pi + \frac{1}{\mu} T_\gamma \mathscr{E}\mathscr{P}\pi - \frac{1}{\mu}. \tag{4.25}$$

Proof We write the identities:

$$R_\alpha(z)(1 - P_\alpha) = \mathscr{R}(z)(1 - \mathscr{P})\pi - \frac{1}{z}(1 - \pi)$$

and

$$R_\beta(z)(1 - P_\beta) = \frac{1}{z}\mathscr{R}(z)T_\beta - \frac{1}{z} - \frac{1}{z}\mathscr{R}(z)\mathscr{T}\mathscr{E}\mathscr{P}T_\beta + \frac{1}{z}\mathscr{E}\mathscr{P}T_\beta.$$

$\mathcal{T}\mathcal{E} = \mathcal{E}\mathcal{T}$; therefore

$$\mathcal{R}(z)\mathcal{T}\mathcal{E}\mathcal{P} = \mathcal{R}(z)\mathcal{E}(\mu\mathcal{P} + \mathcal{D}) = \mathcal{R}(z)\left[\sum_{k=0}^{l-1}\left(\frac{-\mathcal{D}}{\mu}\right)^k\right]\left(1 + \frac{\mathcal{D}}{\mu}\right)\mathcal{P} = \mathcal{R}(z)\mathcal{P}$$

since $\mathcal{D}^l = 0$, and

$$R_\beta(z)(1 - P_\beta) = \frac{1}{z}\mathcal{R}(z)(1 - \mathcal{P})T_\beta - \frac{1}{z} + \frac{1}{z}\mathcal{E}\mathcal{P}T_\beta.$$

Similarly

$$R_\gamma(z)(1 - P_\gamma) = \frac{1}{z}T_\gamma\mathcal{R}(z)\pi - \frac{1}{z} - \frac{1}{z}T_\gamma\mathcal{R}(z)\mathcal{T}\mathcal{E}\mathcal{P}\pi + \frac{1}{z}T_\gamma\mathcal{E}\mathcal{P}\pi$$

$$= \frac{1}{z}T_\gamma\mathcal{R}(z)(1 - \mathcal{P})\pi + \frac{1}{z}T_\gamma\mathcal{E}\mathcal{P}\pi - \frac{1}{z}. \quad \square$$

Exercise

4.25 Suppose that Γ encloses K distinct nonzero eigenvalues $\{\mu_i\}_1^K$ of $\mathcal{T}$ with finite multiplicities. Let $\mathcal{P}_i$ (resp. $\mathcal{D}_i$) be the spectral projection (resp. nilpotent) associated with μ_i of ascent l_i. Prove that

$$P_\alpha = \left(\sum_{i=1}^K \mathcal{P}_i\right)\pi, \quad P_\beta = \left(\sum_{i=1}^K \mathcal{E}_i\mathcal{P}_i\right)T_\beta, \qquad P_\gamma = T_\gamma\left(\sum_{i=1}^K \mathcal{E}_i\mathcal{P}_i\right)\pi$$

with

$$\mathcal{E}_i = \sum_{k=0}^{l_i-1}\frac{1}{\mu_i}\left(\frac{-\mathcal{D}_i}{\mu_i}\right)^k.$$

6.2. *The Projection Method*

6.2.1. The Galerkin method

With $\mathcal{T}_n := \pi_n T_{\restriction X_n}: X_n \to X_n$, Eqs. (4.4) and (4.5) can be written in X_n with notation of Table 4.1:

$$(\mathcal{T}_n - z)x_n^{\mathrm{G}} = \pi_n f, \qquad \mathcal{T}_n\varphi_n^{\mathrm{G}} = \lambda_n\varphi_n^{\mathrm{G}}.$$

If we set $\mathcal{R}_n(z) := (\mathcal{T}_n - z)^{-1}: X_n \to X_n$, then $x_n^{\mathrm{G}} = \mathcal{R}_n(z)\pi_n f$. But the operator $\pi_n T: X \to X$ also enters the picture very naturally. We set $T_n^{\mathrm{P}} := \pi_n T$ and $R_n^{\mathrm{P}}(z) := (T_n^{\mathrm{P}} - z)^{-1}$. By Theorem 4.3, applied with $X_n = \pi_n X$, $X = \pi_n X \oplus (1 - \pi_n)X$, we get $R_n^{\mathrm{P}}(z) = (1/z)\mathcal{R}_n(z)T_n^{\mathrm{P}} - (1/z)$, for $z \neq 0$. Clearly $R_n^{\mathrm{P}}(z)\pi_n = \mathcal{R}_n(z)\pi_n$, and $x_n^{\mathrm{G}} = R_n^{\mathrm{P}}(z)\pi_n f$.

Exercises

4.26 Prove that φ_n^G is an eigenvector of T_n^P associated with λ_n.

4.27 If we set $T_n^G := \pi_n T \pi_n$, then $T_n^G = \mathscr{T}_n \pi_n$ is an extension of $\mathscr{T}_n$ of class $\mathfrak{D}$. Prove that

$$R_n^G(z) = \mathscr{R}_n(z)\pi_n - \frac{1}{z}(1 - \pi_n) \qquad \text{and} \qquad x_n^G = R_n^G(z)\pi_n f.$$

Show that φ_n^G is also an eigenvector of T_n^G associated with λ_n. Be careful that $x_n^G \neq R_n^G(z)f = x_n^G - (1/z)(1 - \pi_n)f$. This is a consequence of the condition $x_n^G \in X_n$.

4.28 Consider Eq. (4.9). Prove that $x_n' = x_n^P = R_n^P(z)f$.

4.29 Let $\mathscr{M}_n$ (resp. M_n^G, M_n^P) be the invariant subspace for $\mathscr{T}_n$ (resp. T_n^G, T_n^P) associated with the eigenvalue $\lambda_n \neq 0$, isolated by the Jordan curve Γ not enclosing 0. Show that

$$\mathscr{M}_n = M_n^G = M_n^P.$$

4.30 Let φ be an eigenvector of T associated with λ. Prove that $P_n^P\varphi = (\lambda/\lambda_n)P_n^G\varphi$, where P_n^P and P_n^G are the spectral projections of T_n^P and T_n^G, respectively, associated with a semisimple nonzero eigenvalue λ_n.

6.2.2. The Sloan variant

We set $T_n^S := T\pi_n$ and $R_n^S(z) := (T_n^S - z)^{-1}$. For $z \neq 0$,

$$R_n^S(z) = \frac{1}{z} T\mathscr{R}_n(z)\pi_n - \frac{1}{z} \qquad \text{and} \qquad x_n^S = \frac{1}{z} Tx_n^G - \frac{1}{z} f = R_n^S(z)f.$$

Exercises

4.31 Prove that $\varphi_n^S = (1/\lambda_n)T\varphi_n^G$ is an eigenvector of T_n^S associated with the eigenvalue $\lambda_n \neq 0$. Prove that $P_n^S\varphi = (1/\lambda_n)TP_n^G\varphi$ if λ_n is semisimple.

4.32 Let M_n^S be the invariant subspace of T_n^S associated with λ_n. Prove that $M_n^S = T\pi_n\mathscr{M}_n$.

6.3. Approximate Quadrature Methods

The operator associated with the Nyström method is denoted T_n^N and is defined by

$$(T_n^N x)(s) = \sum_{j=1}^{n} w_j k(s, t_j)x(t_j).$$

If $R_n^N(z) := (T_n^N - z)^{-1}$, the solution x_n^N of (4.16) is then $x_n^N = R_n^N(z)f$. φ_n^N is an eigenvector of T_n^N associated with λ_n.

To define the Fredholm operator T_n^F associated with the Fredholm method, we suppose that I_n is an interpolatory quadrature formula defined by the points $\{t_i\}_1^n$ and the functions $\{e_i\}_1^n$ such that $e_i(t_j) = \delta_{ij}$. Then T_n^F is defined by

$$T_n^F x = \sum_{i=1}^{n} \left[\sum_{j=1}^{n} w_j k(t_i, t_j)x(t_j) \right] e_i.$$

Exercises

4.33 Let $\mathscr{T}'_n: X_n \to X_n$ be the operator defined in the basis $\{e_i\}_1^n$ of X_n and the adjoint basis $\{e_i^*\}_1^n$ of X_n^* by the matrix

$$\tilde{A}_n = (w_j k(t_i, t_j)), \qquad i, j = 1, \ldots, n$$

with $\langle x, e_i^* \rangle = x(t_i)$. Show that $T_n^{\mathrm{F}} = \mathscr{T}'_n \pi_n = \pi_n T_n^{\mathrm{N}}$.

4.34 Define $\mathscr{R}'_n(z) = (\mathscr{T}'_n - z)^{-1}$; $X_n \to X_n$ and $R_n^{\mathrm{F}}(z) = (T_n^{\mathrm{F}} - z)^{-1}$. $\mathscr{P}'_n$, $\mathscr{D}'_n$, and P_n^{F} are defined accordingly. Prove that $R_n^{\mathrm{F}}(z) = \mathscr{R}'_n(z)\pi_n - (1/z)(1 - \pi_n)$ if $z \neq 0$ and that $P_n^{\mathrm{F}} = \mathscr{P}'_n \pi_n$. Deduce that $x_n^{\mathrm{F}} = \mathscr{R}'_n(z)\pi_n f = R_n^{\mathrm{F}}(z)\pi_n f$ and that φ_n^{F} is an eigenvector of T_n^{F} associated with λ_n.

4.35 Prove that

$$R_n^{\mathrm{N}}(z) = \frac{1}{z} T_n^{\mathrm{N}} \mathscr{R}'_n(z)\pi_n - \frac{1}{z} \quad \text{and} \quad P_n^{\mathrm{N}} = T_n^{\mathrm{N}} \mathscr{E}'_n \mathscr{P}'_n \pi_n \quad \text{with} \quad \mathscr{E}'_n = \sum_{k=0}^{l_n - 1} \frac{1}{\lambda_n}\left(-\frac{\mathscr{D}'_n}{\lambda_n}\right)^k$$

Define M_n^{N} (resp. $\mathscr{M}'_n$), the invariant subspace of T_n^{N} (resp. $\mathscr{T}'_n$) associated with λ_n, express M_n^{N} in terms of $\mathscr{M}'_n$ and T_n^{N}.

Equations (4.16) and (4.17) can be rewritten

$$x_n^{\mathrm{N}} = \frac{1}{z}(T_n^{\mathrm{N}} x_n^{\mathrm{F}} - f), \qquad \varphi_n^{\mathrm{N}} = \frac{1}{\lambda_n} T_n^{\mathrm{N}} \varphi_n^{\mathrm{F}}.$$

These formulas are very similar to the corresponding ones for the projection method (4.18) and (4.19), which can be written

$$x_n^{\mathrm{S}} = \frac{1}{z}(T_n^{\mathrm{S}} x_n^{\mathrm{G}} - f), \qquad \varphi_n^{\mathrm{S}} = \frac{1}{\lambda_n} T_n^{\mathrm{S}} \varphi_n^{\mathrm{G}},$$

with the choice $x_n = x_n^{\mathrm{G}}$ and $\varphi_n = \varphi_n^{\mathrm{G}}$.

7. Convergence of the Numerical Approximation Methods

We are not ready yet to study the convergence of the eigenelements. This is postponed until Chapter 5. As for the convergence of the solution x_n of $(T_n - z)x_n = f$ to the solution x of $(T - z)x = f$, for z in $\rho(T)$, we know by Lemma 3.16 that this convergence is equivalent to $T_n - z \xrightarrow{\mathrm{s}} T - z$, for z in $\rho(T)$.

We assume in this section that T is *compact* in the complex Banach space X. Then under suitable assumptions on X_n, we shall see that either $\|T_n - T\| \to 0$ or $T_n \xrightarrow{\mathrm{cc}} T$ for the approximation methods defined in Section 6. Upon setting $R(z) := (T - z)^{-1}$ and $R_n(z) := (T_n - z)^{-1}$, we recall that

(i) $\|T_n - T\| \to 0 \Rightarrow \|R_n(z) - R(z)\| \to 0$ for any z in $\rho(T)$, and
(ii) $T_n \xrightarrow{\mathrm{cc}} T \Rightarrow R_n(z) \xrightarrow{\mathrm{cc}} R(z)$ for any z in $\rho(T)$.

Exercise

4.36 If $\pi_n \xrightarrow{p} 1$ and $R_n(z) \xrightarrow{p} R(z)$, show that $R_n(z)\pi_n \xrightarrow{p} R(z)$.

7.1. Convergence of Projection Methods When $\pi_n \xrightarrow{p} 1$

When $\pi_n \xrightarrow{p} 1$, the convergence of the methods is easy to establish because of the compactness of T.

Theorem 4.5 *If T is compact and $\pi_n \xrightarrow{p} 1$ in X, then $\|T_n^P - T\| \to 0$ and $T_n^S \xrightarrow{cc} T$. If, moreover, $\pi_n^* \xrightarrow{p} 1$ in X^*, then $\|T_n^S - T\| \to 0$.*

Proof $T_n^P - T = (1 - \pi_n)T$. The uniform convergence follows by Theorem 3.2. Now, to show that $T\pi_n \xrightarrow{cc} T$, we apply Proposition 3.4 with $V_n := T$ and $U_n := \pi_n$. The remark

$$\|T_n^S - T\| = \|T(\pi_n - 1)\| = \|(\pi_n^* - 1)T^*\|$$

ends the proof. □

An obvious example of a sequence of projections π_n such that $\pi_n \xrightarrow{p} 1$ and $\pi_n^* \xrightarrow{p} 1$ is provided by a sequence of orthogonal projection pointwise converging in a Hilbert space. Nonorthogonal projections in a Hilbert space such that π_n and π_n^* converge to 1 are not uncommon in practice. Such an example will be given in the context of the finite-element method (Section 9, Part B). This will be a sequence of projections associated with a nonhermitian coercive continuous sesquilinear form.

For any z in $\rho(T)$ and for any f in X, the convergence of the solutions x_n^G, x_n^P, and x_n^S is deduced from the stable convergence of the approximations in $\rho(T)$, and the following error bounds hold. c is a generic constant.

Proposition 4.6 *If T is compact and $\pi_n \xrightarrow{p} 1$, then*

$$c_1\|(1 - \pi_n)x\| \le \|x - x_n^G\| \le c_2\|(1 - \pi_n)x\|,$$
$$\|x - x_n^P\| \le c\|(1 - \pi_n)Tx\|,$$
$$\|x - x_n^S\| \le c\|T(1 - \pi_n)x\|.$$

Proof Upon developing $(\pi_n T - z)(x - x_n^G)$, we get

$$z(1 - \pi_n)x = (z - \pi_n T)(x - x_n^G) \qquad \text{and} \qquad x - x_n^G = zR_n^P(z)(\pi_n - 1)x.$$

The other two inequalities are derived by application of the identity

$$x - x_n = (R(z) - R_n(z))f = R_n(z)(T_n - T)R(z)f = R_n(z)(T_n - T)x$$

to the operators T_n^P and T_n^S, respectively. □

The Galerkin solution x_n^G achieves the optimal order $\delta_n(x)$ relative to X_n, and this order cannot be improved. Since

$$\|T(1 - \pi_n)x\| \leq \|T(1 - \pi_n)\| \, \|(1 - \pi_n)x\|,$$

x_n^S improves on x_n^G if $\|T(1 - \pi_n)\| \to 0$, and sometimes even if $\|T(1 - \pi_n)\| \nrightarrow 0$ (cf. Chapters 6 and 7). On the other hand, Tx is often smoother than x, so that x_n^P may improve on x_n^G.

Exercises

4.37 Let π_n be orthogonal in a Hilbert space. Show that the constants c_1, c_2 in Theorem 4.6 can be chosen such that $c_1 = 1$ and $c_2 \leq 1 + \varepsilon_n$, where $\varepsilon_n \to 0$ as $n \to \infty$.

4.38 Let $\pi_n: L^2(a, b) \to \mathbb{P}_{r,\Delta}$ be the projection defined in Exercise 4.4. Prove that if $h \to 0$, then $\|(1 - \pi_n)T\|_2 \to 0$, $x_n^G \to x$ if $x \in C(a, b)$, and $\|x - x_n^G\|_2 = O(h^{r+1})$ if $x \in C^p(a, b)$, $p \geq r + 1$.

4.39 $X = L^\infty(a, b)$. T is compact from X into $C(a, b)$. L_n is the projection defined in Exercise 4.6. Prove that if $h \to 0$, then $\|(1 - L_n)T\|_\infty \to 0$ and $(L_n T - z)^{-1}$ is uniformly bounded in $L^\infty(a, b)$ for z in $\rho(T)$. Deduce that $x_n^G \to x$ if $x \in C(a, b)$ and $\|x - x_n^G\|_\infty = O(h^{r+1})$ if

$$x \in C^p(a, b), \; p \geq r + 1.$$

4.40 For the projection method defined in Exercise 4.39, prove that in $C(a, b)$, $TL_n \xrightarrow{cc} T$ and $x_n^S \to x$.

4.41 Give a sufficient condition on k and f for x, the solution of $(T - z)x = f$, to be in $C^p(a, b)$.

7.2. *Convergence of the Petrov Method*

We consider the Petrov method defined by the subspaces X_n and Y_n of the complex Banach space X. Let π_n (resp. ϖ_n) be a projection onto Y_n (resp. X_n).

Lemma 4.7 *We suppose that T is compact, $\varpi_n \xrightarrow{p} 1$ in X, π_n is uniformly bounded and (4.8) is fulfilled. Then the Petrov method is convergent and the rate of convergence is given by*

$$\|x - x_n\| \leq c\|(1 - \varpi_n)x\| \leq c \operatorname{dist}(x, X_n).$$

Proof $\pi'_n := \tilde{\pi}_n^{-1}\pi_n$ is a projection on X_n. Then

$$\pi_n[(T - z)x_n - f] = 0, \qquad x_n \in X_n,$$

is equivalent to the Galerkin method defined by π'_n:

$$\pi'_n[(T - z)x_n - f] = 0, \quad x_n \in X_n.$$

It remains to show that $\pi'_n \xrightarrow{p} 1$. For any x in X, the identity

$$1 - \pi'_n = (1 - \tilde{\pi}_n^{-1}\pi_n)(1 - \varpi_n)$$

proves that $\|(1 - \pi_n')x\| \leq c\|(1 - \varpi_n)x\|$. The result follows from Theorem 4.6. □

In the case of an $\perp$-Petrov method in a Hilbert space H defined by the subspaces E_n and F_n, condition (4.8) is equivalent to the condition

$$\sup_n \Theta(E_n, F_n) < 1 \tag{4.26}$$

(cf. Exercise 4.12). We note that for the $\perp$-Galerkin method, condition (4.26) is trivially satisfied since $E_n = F_n$ implies $\Theta(E_n, F_n) = 0$.

Lemma 4.7 expresses the convergence of the Petrov method in terms of the convergence $\varpi_n \xrightarrow{p} 1$ of some projection ϖ_n onto X_n. But there are cases where the properties of the projection π_n on Y_n are better known than those of ϖ_n. It is then useful to be able to study the convergence of the Petrov method in terms of the convergence $\pi_n \xrightarrow{p} 1$. We prove now that this is the case when the relationship

$$X_n = (K - \alpha)Y_n$$

holds for some given compact operator K and $\alpha \in \rho(K)$.

Proposition 4.8 *If T and K are compact and $\pi_n \xrightarrow{p} 1$, then the Petrov method on $(T - z)x = f$, $z \in \rho(T)$, defined by the subspaces X_n, Y_n with $X_n = (K - \alpha)Y_n$, $\alpha \in \rho(K)$, is convergent, and*

$$\|x - x_n\| \leq c\|(1 - \pi_n)(K - \alpha)^{-1}x\| \leq c\,\mathrm{dist}(y, Y_n)$$

with $x = (K - \alpha)y$.

Proof To prove that $\pi_n' \xrightarrow{p} 1$, we apply the Galerkin method defined by Y_n to the equation $(K - \alpha)u = g$, for an arbitrary g in X. It yields the equation

$$\pi_n(K - \alpha)u_n = \pi_n g, \qquad u_n \in Y_n, \tag{4.27}$$

which has, for n large enough, a unique solution u_n converging to $u = (K - \alpha)^{-1}g$, as $n \to \infty$, since K is compact and $\pi_n \xrightarrow{p} 1$. Setting

$$v_n := (K - \alpha)u_n,$$

Eq. (4.27) can be rewritten

$$\pi_n v_n = \pi_n g.$$

For any g in X, the equation has a unique solution $v_n \in X_n$. This proves that $\tilde{\pi}_n = \pi_{n\restriction X_n}$ is invertible on X_n: $v_n = \tilde{\pi}_n^{-1}\pi_n g = \pi_n' g$.

Now $v_n \to g$ can be rewritten

$$\pi_n' g = \tilde{\pi}_n^{-1}\pi_n g \to g$$

for any g in X. By the Banach–Steinhaus theorem, $\|\tilde{\pi}_n^{-1}\pi_n\| \le c$. In particular, for $g_n \in Y_n$,

$$\|\tilde{\pi}_n^{-1}g_n\| = \|\tilde{\pi}_n^{-1}\pi_n g_n\| \le c\|g_n\|.$$

Therefore $\|\tilde{\pi}_n^{-1}\| \le c$. The convergence of the Petrov method follows by Lemma 4.7. The bound $\|x - x_n\| \le c\|(1 - \pi_n')x\| \le c \operatorname{dist}(x, X_n)$ is clear. The desired bound follows from the identity

$$(1 - \pi_n')g = (K - \alpha)(u - u_n) = \alpha(K - \alpha)(\pi_n K - \alpha)^{-1}(\pi_n - 1)(K - \alpha)^{-1}g.$$

Then

$$\|(1 - \pi_n)(K - \alpha)^{-1}x\| \le c \operatorname{dist}((K - \alpha)^{-1}x, (K - \alpha)^{-1}X_n) \le c \operatorname{dist}(y, Y_n).$$

□

An important application of Proposition 4.8 is the analysis of the convergence of the least-squares method on $E_n \subset H$ applied to the equation

$$(T - z)x = f, \qquad z \in \rho(T), \quad T \text{ compact}.$$

It is equivalent to an $\perp$-Petrov method defined by the subspaces E_n, F_n such that $F_n = (T - z)E_n$. Then if we define $S := (1/z)T(T - z)^{-1}$, S is compact and $S - 1/z = (T - z)^{-1}$ is invertible. Therefore $(S - 1/z)F_n = E_n$, and the Petrov method is convergent when $\pi_n \xrightarrow{p} 1$.

Remark We have just proved that $\pi_n \xrightarrow{p} 1$ and T compact imply $\sup_n \Theta((T - z)E_n, E_n) < 1$; $T - z$ is then regular (cf. Chapter 3, Example 3.11).

Exercise

4.42 How can the condition K compact in Proposition 4.8 be weakened?

Another application of Proposition 4.8 will be given in Section 9.

7.3. *An Example Where* $\pi_n \stackrel{p}{\not\to} 1$

In $C(a, b)$, as already pointed out, piecewise polynomial interpolatory projections are pointwise convergent, but this is not the case for polynomial interpolatory projection on the whole interval $[a, b]$. Therefore the choice of the collocation points will be of importance to prove the convergence of such collocation methods.

We specify here $\{\tau_i\}_1^n$ to be the *Chebyshev* points. When $a = -1, b = 1$, the Chebyshev points are given by

$$\tau_i = \cos[(2i - 1)/2n]\pi, \qquad i = 1, \ldots, n,$$

and we denote by L_n the projection mapping every continuous function on $[-1, 1]$ onto its $(n-1)$th-degree interpolant polynomial at the Chebyshev points $\{\tau_i\}_1^n$. The Banach space X may be chosen in various ways. The following result pertains to the space $L_\rho^2(-1, 1)$, with weight function $\rho(t) = 1/\sqrt{1-t^2}$ for $t \in [-1, 1]$.

Proposition 4.9 (Vainikko) *If T is a compact operator from $L_\rho^2(-1, 1)$ into $C(-1, 1)$, then $\|L_n T - T\|_{2,\rho} \to 0$. The error formula is*

$$\|x - x_n\|_{2,\rho} \le c \operatorname{dist}_\infty(x, \mathscr{P}_{n-1}) \qquad \textit{for} \quad x \in C(-1, 1).$$

Proof Clearly $C(-1, 1) \subset L_\rho^2(-1, 1)$ and $\|x\|_{2,\rho} \le c\|x\|_\infty$ for $x \in C(-1, 1)$. Let i be the operator embedding $C(-1, 1)$ in $L_\rho^2(-1, 1)$. By the Erdös–Turan theorem (see Cheney, 1966, p. 137), for each $x \in C(-1, 1)$, $L_n x$ converges to x in quadratic mean with weight $\rho(t)$, that is,

$$\int_{-1}^{1} |(L_n x - x)(t)|^2 \rho(t)\, dt \to 0.$$

In other words, $L_n \to i$ in $L_\rho^2(-1, 1)$. Now, T is a compact operator mapping $L_\rho^2(-1, 1)$ into $C(-1, 1)$, and $\|(L_n - i)T\|_{2,\rho} \to 0$.

Let $z_n \in \mathbb{P}_{n-1}$ and $x \in C(-1, 1)$. Then

$$\begin{aligned} \|x - L_n x\|_{2,\rho} &= \|(x - z_n) - L_n(x - z_n)\|_{2,\rho} \\ &\le \|x - z_n\|_{2,\rho} + \|L_n(x - z_n)\|_{2,\rho} \\ &\le \left(\left[\int_{-1}^{1} \rho(t)\, dt\right]^{1/2} + \|L_n\|_{C \to L_\rho^2}\right) \|x - z_n\|_\infty . \end{aligned}$$

$\|L_n\|_{C \to L_\rho^2}$ is uniformly bounded. Since z_n is arbitrary in $\mathbb{P}_{n-1}$, we get the claimed bound by Proposition 4.6. □

This result establishes only a convergence in *quadratic mean* with weight $\rho(t)$ for the solution x_n of the collocation method based on Chebyshev points. We are now able to prove the complementary result that the solution $\tilde{x}_n$ of the *iterated* collocation method converges in $L^\infty(-1, 1)$.

Proposition 4.10 (Sloan and Burn) *Let the kernel k satisfy the conditions*

$$\text{(i)} \quad \sup_{t \in [-1, 1]} \|k_t\|_p < \infty \tag{4.28}$$

$$\text{(ii)} \quad \lim_{t' \to t} \|k_{t'} - k_t\|_p = 0, \qquad -1 \le t, t' \le 1, \tag{4.29}$$

for some $p > 1$. Then the integral operator T is compact on $C(-1, 1)$ and $TL_n \overset{cc}{\to} T$ in $C(-1, 1)$. The error formula is $\|x - \tilde{x}_n\|_\infty \le c \operatorname{dist}_\infty(x, \mathbb{P}_{n-1})$ for $x \in C(-1, 1)$.

Proof It is based on the following result (Erdös and Feldheim, 1936): for every x in $X = C(-1, 1)$ and every real q, $0 < q < \infty$,

$$\int_{-1}^{1} |(L_n x - x)(t)|^q \rho(t)\, dt \to 0 \qquad \text{as} \quad n \to \infty.$$

We take p such that $p^{-1} + q^{-1} = 1$, $q < \infty$ for $p > 1$. We first show that $TL_n x \to Tx$ for $x \in X$:

$$\begin{aligned}
|TL_n x(s) - Tx(s)| &= \left| \int_{-1}^{1} k(s, t)[L_n x(t) - x(t)]\, dt \right| \\
&\leq \int_{-1}^{1} \left| k(s, t) \frac{1}{\rho(t)} \right| |(L_n x - x)(t)| \rho(t)\, dt \\
&\leq \left[\int_{-1}^{1} \left| k(s, t) \frac{1}{\rho(t)} \right|^p \rho(t)\, dt \right]^{1/p} \\
&\quad \times \left[\int_{-1}^{1} |(L_n x - x)(t)|^q \rho(t)\, dt \right]^{1/q}
\end{aligned}$$

by the Hölder inequality with the weight function $\rho(t)$. For $p > 1$,

$$\max_{t \in [-1, 1]} \rho(t)^{1-p} = 1.$$

By (4.28) and the Erdös–Feldheim theorem, we conclude that $TL_n \xrightarrow{\text{p}} T$ in X. T is compact in X by Theorem 4.1. To see that $TL_n \xrightarrow{\text{cc}} T$, we need only to prove that $K = \bigcup_{n=1}^{\infty} TL_n B$ is relatively compact, where B is the unit sphere of X. K is equibounded since $\sup_n \|TL_n\| < \infty$ follows from the pointwise convergence. To prove the equicontinuity of K, let $\delta > 0$ be given and let $s, s' \in [-1, 1]$ with $|s' - s| \leq \delta$. Let $x \in B$. Then

$$\begin{aligned}
&|TL_n x(s) - TL_n x(s')| \\
&\quad = \left| \int_{-1}^{1} [k(s, t) - k(s', t)](L_n x)(t)\, dt \right| \\
&\quad \leq \left[\int_{-1}^{1} |k(s, t) - k(s', t)|^p \rho(t)^{1-p}\, dt \right]^{1/p} \|L_n x\|_{q,\rho} \\
&\quad \leq \sup \left\{ \left[\int_{-1}^{1} |k(\sigma, t) - k(\sigma', t)|^p\, dt \right]^{1/p} ; \sigma\sigma' \in [-1, 1], |\sigma - \sigma'| \leq \delta \right\} \\
&\qquad \times \sup\{\|L_n x\|_{q,\rho} ; n \geq 1, x \in B\}, \qquad (4.30)
\end{aligned}$$

where $\|\cdot\|_{q,\rho}$ denotes the weighted L^q norm with the weight ρ. By (4.29),

$s \mapsto k_s$ is a continuous function from $[-1, 1]$ into $L^p[-1, 1]$. It is then uniformly continuous:

$$\lim_{\delta \to 0} \sup_{|\sigma - \sigma'| \leq \delta} \|k_\sigma - k_{\sigma'}\|_p = 0.$$

This shows that the first factor on the right-hand side of (4.30) tends to zero. To prove that the second factor is finite, we write $\lim_n \|L_n x - x\|_{q,p} = 0$, from which follows $\sup_n (\|L_n x\|_{q,p}, x \in B) < \infty$ by the principle of uniform boundedness. $\bar{K}$ is therefore compact in X by the Ascoli–Arzela theorem. Let z_n be any polynomial in $\mathbb{P}_{n-1}$. Then

$$\|T(L_n - 1)x\|_\infty = \|T(L_n - 1)(x - z_n)\|_\infty \leq (\|TL_n\|_\infty + \|T\|_\infty)\|x - z_n\|_\infty .$$

Because z_n is arbitrary in $\mathbb{P}_{n-1}$, the claimed bound follows by Proposition 4.6. □

When are conditions (4.28) and (4.29) satisfied? First, it can easily be shown that they are satisfied for every continuous kernel, for any $p > 1$. Many of the weakly singular kernels that arise in practice also satisfy (4.28) and (4.29) for some $p > 1$. (See Graham and Sloan, 1979.)

Exercise

4.43 What is the behavior of the collocation solution at the Chebyshev points? Could it be deduced from Proposition 4.9?

7.4. *Convergence of Approximate Quadrature Methods*

Theorem 4.11 (Anselone) *If T is an integral operator with continuous kernel on $[a, b] \times [a, b]$, then $T_n^N \overset{cc}{\to} T$ in $X = C(a, b)$.*

Proof For s in $[a, b]$, $T_n^N x(s) \to Tx(s)$ as $n \to \infty$, by assumption on the numerical integration method. Also

$$|T_n^N x(s)| \leq \sum_{j=1}^{n} |w_j| \, |k(s, t_j)| \, |x(t_j)|$$

$$\leq \max_{a \leq s, t \leq b} |k(s, t)| \sum_{j=1}^{n} |w_j| \, \|x\|_\infty .$$

Then the sequence $\{T_n^N x\}$ is uniformly bounded for $x \in B$, the unit ball of $C(a, b)$. It is equicontinuous since, for $a \leq s, s' \leq b$,

$$|T_n^N x(s) - T_n^N x(s')| \leq c\|x\|_\infty \max_{a \leq t \leq b} |k(s, t) - k(s', t)|,$$

which tends to zero as $|s - s'| \to 0$, independently of n. $\{T_n^N x\}$ is then a uniformly bounded and equicontinuous sequence of functions converging to Tx for each s in $[a, b]$; it must converge uniformly on $[a, b]$, that is, $T_n^N \xrightarrow{p} T$. This also proves that $\bigcup_{n=1}^{\infty} T_n^N B$ is relatively compact and $T_n^N \xrightarrow{cc} T$ since T is compact. □

Corollary 4.12 *If $\pi_n \xrightarrow{p} 1$, then $T_n^F \xrightarrow{cc} T$.*

Proof It follows from $T_n^F = \pi_n T_n^N$ and application of Proposition 3.4 with $V_n := T_n^N$ and $U_n := \pi_n$. □

Exercise

4.44 Show that if the kernel k is continuous, $\|T_n^N - T\|_\infty \geq \|T\|_\infty$ (Anselone).

In general $\|T_n^N - T\|_\infty$ does not tend to zero. Nevertheless, if the kernel is smooth enough, a uniform convergence can be proved in an appropriate Banach space, when using a repeated formula for approximate quadrature.

Definition A *repeated* quadrature formula on $[0, 1]$ is the linear functional

$$J_n(f) = \sum_{j=1}^{n} \frac{1}{n}\left[\sum_{i=1}^{p} A_i f\left(\frac{j + t_i}{n}\right)\right] \quad \text{for} \quad 0 \leq t_1 \leq \cdots \leq t_n \leq 1,$$

where the integer p and the complex numbers $\{A_i\}_1^p$ are given.

The rectangle, trapezium, Simpson and Gauss formulas are repeated formulas.

For a given integer $m \geq 1$, we consider the set of functions x defined on $[0, 1]$, with continuous derivative up to order m. With the norm $|||x||| := \max_{i=0,\ldots,m} \sup_{0 \leq t \leq 1} |x^{(i)}(t)|$, the set is a Banach space denoted C_*^m.

Proposition 4.13 (Baker) *Let the integral operator T have a kernel k such that the partial derivatives $\partial^{i+j}k(s, t)/\partial s^i \, \partial t^j$ are continuous on $[0, 1] \times [0, 1]$ for $0 \leq i, j \leq m$. Then T is a compact operator from C_*^m into C_*^m. If the Nyström method is applied with a repeated quadrature formula such that $\sum_{i=1}^p A_i = 1$, then $|||T_n^N - T||| \to 0$.*

For a proof see Baker (1971) and Baker and Hodgson (1971).

For T_n^N to be a uniform approximation of T in C_*^1, it is sufficient that the functions k, $\partial k/\partial s$, $\partial k/\partial t$, and $\partial^2 k/\partial s \, \partial t$ be continuous on $[0, 1] \times [0, 1]$. This is not the case if the kernel is the Green function of a second-order differential equation.

If I_n is an interpolatory quadrature formula, we study the relationship between the approximate quadrature methods T_n^F, T_n^N and the corresponding projection methods T_n^P, T_n^S.

Theorem 4.14 *If the functions* $\{e_i\}_1^n$ *satisfy the conditions*

$$\text{(i)}\quad \max_{1\le i\le n}\ \sup_{a\le t\le b}\{|t-t_i|;\, e_i(t)\neq 0\}\to 0 \quad \text{as} \quad n\to\infty; \tag{4.31}$$

$$\text{(ii)}\quad \sup_n \sum_{i=1}^n \int_a^b |e_i(t)|\,dt<\infty, \tag{4.32}$$

and if the kernel is continuous, then $\|T_n^{\mathrm{S}}-T_n^{\mathrm{N}}\|_\infty\to 0$. *If, moreover,* $\pi_n \xrightarrow{\mathrm{p}} 1$, *then* $\|T_n^{\mathrm{G}}-T_n^{\mathrm{F}}\|_\infty\to 0$.

Proof

$$(T_n^{\mathrm{N}}-T_n^{\mathrm{S}})x=\sum_{j=1}^n\left[w_j k(\cdot,t_j)-\int_a^b k(\cdot,t)e_j(t)\,dt\right]x(t_j).$$

We set $D_j=\{t\in[a,b];\, e_j(t)\neq 0\}$. Then

$$\|T_n^{\mathrm{N}}-T_n^{\mathrm{S}}\|_\infty\le \sup_{a\le s\le b}\ \max_{1\le j\le n}\ \sup_{t\in D_j}|k(s,t_j)-k(s,t)|\sum_{j=1}^n\int_a^b|e_j(t)|\,dt.$$

The result follows from (4.31) and (4.32). Now

$$T_n^{\mathrm{G}}-T_n^{\mathrm{F}}=T_n^{\mathrm{P}}\pi_n-T_n^{\mathrm{F}}=\pi_n(T_n^{\mathrm{S}}-T_n^{\mathrm{N}})\qquad\text{and}\qquad \|\pi_n\|_\infty\le M$$

since $\pi_n\xrightarrow{\mathrm{p}}1$. □

Most quadrature formulas satisfy (4.31) and (4.32). If the functions e_i are positive, then (4.32) is satisfied:

$$w_i=\int_a^b e_i(t)\,dt \qquad\text{and}\qquad \sup_n\sum_{i=1}^n w_i<\infty \quad \text{(Lemma 4.2)}.$$

Exercises

4.45 Under the assumptions of Theorem 4.14 and if $\pi_n\xrightarrow{\mathrm{p}}1$, show that $T_n^{\mathrm{F}}=\mathscr{T}'_n\pi_n$ satisfy the two Kantorovitch conditions given in Exercise 3.1, Chapter 3.

4.46 The product integration applied to $k(s,t)=r(s,t)\sigma(s,t)$ defines the operator

$$(T_nx)(s)=\sum_{i=1}^n r(s,t_i)x(t_i)w_i(s)=\sum_{i=1}^n r(s,t_i)x(t_i)\int_a^b e_i(t)\sigma(s,t)\,dt.$$

Under which conditions does $T_n\xrightarrow{\mathrm{cc}}T$ hold?

4.47 X is a complex Banach space with unit ball B and T is compact. Let T^m, T_n, $T_n^m\in\mathscr{L}(X)$. We assume that $K:=\bigcup_{n=1}^\infty T_nB$ and $K^m:=\bigcup_{n=1}^\infty T_n^mB$, $m\in\mathbb{N}$, are relatively compact and that $\|T_n^m-T_n\|\to 0$ as $m\to\infty$, uniformly in n. Prove that $\bigcup_{m,n=1}^\infty T_n^mB$ is relatively compact.

Most of the time, projections methods, as we have seen in Section 3, yield an approximation T^m, which contains integrals that cannot be computed in closed form. We suppose that $\|T^m-T\|\to 0$ as $m\to\infty$. Now let n index the numerical integration of T and T^m, this defines the operators T_n and T_n^m.

If the quadrature rule is convergent, $T_n \overset{cc}{\to} T$ and $T_n^m \overset{cc}{\to} T^m$ for each m. It results in a basic change in approximation mode: $T_n^m \overset{cc}{\to} T$ (Anselone and Lee, 1976).

The convergence of projection methods has been widely studied (Kantorovitch and Akilov, 1964; Krasnoselskii *et al.*, 1972). Proposition 4.9 is due to Vainikko (1965), and Proposition 4.10 to Sloan and Burn (1979). The convergence of the Nyström method has received a definitive treatment in Anselone (1971), by the introduction of the collectively compact convergence concept. Vainikko (1969b) first proved directly the convergence of the Fredholm method, as a perturbation of a Galerkin method. Convergence of the product integration method is studied in Atkinson (1976a). A very comprehensive treatment of methods for integral equations is given in Baker (1977). The influence of approximate quadrature in the projection method is studied in Anselone and Lee (1976), Prenter (1975), and Chandler (1979), for example.

B. DIFFERENTIAL BOUNDARY-VALUE PROBLEMS

8. Boundary-Value Problems in Differential Equations

We shall consider boundary-value problems where an ordinary differential equation, an elliptic partial differential equation, or possibly a system of such equations arises.

8.1. Ordinary Differential Equation

Consider the pth-order differential equation ($p \geq 2$)

$$\mathfrak{L}u := u^{(p)} - \sum_{i=0}^{p-1} a_i(t)u^{(i)} = f(t), \qquad a \leq t \leq b, \tag{4.33}$$

with the p homogeneous boundary conditions

$$\mathfrak{l}_k(u) := \sum_{i=0}^{p-1} [\alpha_{ik} u^{(i)}(a) + \beta_{ik} u^{(i)}(b)] = 0, \qquad k = 1, \ldots, p. \tag{4.34}$$

The coefficients $\{a_i\}_0^{p-1}$ are continuous real or complex functions, the constant coefficients α_{ik}, β_{ik} are real or complex. To the differential problem

$$(P) = [\mathfrak{L}u = f, \mathrm{l}^k(u) = 0, k = 1, \ldots, p],$$

is associated the differential operator T defined as follows. Let $X = L^\infty(a, b)$. Set $D := \{u \in C^{p-1}(a, b);\ u^{(p)} \in L^\infty(a, b) \text{ and } \mathrm{l}_k(u) = 0 \text{ for } k = 1, \ldots, p\}$. Then for $u \in D$, $Tu = \mathfrak{L}u$. T is a closed operator with domain D.

An integral formulation of the problem (P) can be obtained in the following way. We suppose that the problem

$$(Q_0) = [u^p = 0, \mathrm{l}_k(u) = 0, k = 1, \ldots, p],$$

has $u = 0$ for unique solution. Then by Theorem 2.17 there exists a unique Green function $g(t, s)$ such that the solution u of

$$(Q) = [u^p = x, \mathrm{l}_k(u) = 0],$$

is given by

$$u(t) = \int_a^b g(t, s)x(s)\, ds.$$

This defines the integral operator G by $x \mapsto u = Gx$.

Proposition 4.15 *If under the boundary conditions* (4.34) *the equation* $u^{(p)} = 0$ *has the solution* $u = 0$, *then the equation* $Tu = f$ *is equivalent to* $x = Kx + f$, *and* $u(t) = \int_a^b g(t, s)x(s)\, ds$, *where* $g(t, s)$ *is the Green function of the equation* $u^{(p)} = x$ *with conditions* (4.34), *and* K *is the integral operator with kernel*

$$k(t, s) = \sum_{i=0}^{p-1} a_i(t) \frac{\partial^i g(t, s)}{\partial t^i}.$$

Proof

$$u^{(p)} = x \qquad \text{and} \qquad u^{(i)} = \int_a^b \frac{\partial^i g(t, s)}{\partial t^i} x(s)\, ds.$$

Hence x is solution of the Fredholm equation of the second kind:

$$x = Kx + f,$$

with the integral operator

$$Kx = \int_a^b k(\cdot, s)x(s)\, ds$$

where

$$k(t, s) = \sum_{i=0}^{p-1} a_i(t) \frac{\partial^i g(t, s)}{\partial t^i}.$$

Note that k has a first-kind discontinuity on $s = t$. K is compact as an operator from $L^r(a, b)$ into $C(a, b)$, $2 \leq r \leq \infty$, by Theorem 4.1. Indeed

(i) $\sup_t \|k_t\|_2 < \infty$,
(ii) $\|k_{t'} - k_t\|_2 \to 0$ as $t' \to t$. □

We assume that $T^{-1} \in \mathscr{L}(X)$, that is $(1 - K)^{-1} \in \mathscr{L}(X)$.

Example 4.16 The eigenvalue problem $T\psi = \lambda\psi$ is similarly equivalent to the generalized eigenvalue problem $\psi = G\varphi$ and $(1 - K)\varphi = \lambda G\varphi$, if we assume that $0 \in \rho(T)$. Let

$$P = \frac{-1}{2i\pi} \int_\Gamma (T - z)^{-1}\, dz$$

be the spectral projection of T associated with λ.

$$Q = \frac{-1}{2i\pi} \int_\Gamma (1 - K - zG)^{-1} G\, dz$$

is the spectral projection of the generalized eigenvalue problem (cf. Chapter 2, Example 2.28), $(T - z)^{-1} = G(1 - K - zG)^{-1}$ and $P = GQG^{-1}$. Q is also the spectral projection of the (simple) eigenvalue problem

$$\varphi = \lambda(1 - K)^{-1} G\varphi,$$

associated with the eigenvalue $1/\lambda$. Indeed, if we set $U := (1 - K)^{-1}G$, we have

$$\begin{aligned}
\int_{\Gamma(\lambda)} (1 - K - zG)^{-1} G\, dz &= \int_{\Gamma(\lambda)} (1 - zU)^{-1} U\, dz \\
&= \int_{\Gamma(\lambda)} \left[-\frac{dz}{z} + \left(\frac{1}{z} - U\right)^{-1} \frac{dz}{z^2} \right] \\
&= \int_{\Gamma(1/\lambda)} (U - t)^{-1}\, dt,
\end{aligned}$$

with the change of variable $t = 1/z$, since $\Gamma(\lambda)$ does not enclose 0. Note that $GUG^{-1} = T^{-1}$.

Example 4.17 Let $-z \neq 0$ be given in $\rho(T)$. The equation $Tu = f$ can be written $(T + z)u - zu = f$, where $T + z$ has a bounded inverse.

Let $h(t, s)$ be the Green function of $(T + z)u = f$, h has continuous derivatives up to the order $p - 2$. If A_z is the integral operator with kernel $h(t, s)$, the equation $Tu = f$ is equivalent to the Fredholm equation $y = zA_z y + f$ and $u = A_z y$.

Exercises

4.48 If $0 \in \rho(T)$ and $A = T^{-1}$, $T\psi = \lambda\psi$ is equivalent to $\theta = \lambda A\theta$ and $\psi = A\theta$. The spectral projections of T and A associated with λ and $1/\lambda$ are identical.

4.49 Give the integral formulation of the problem (P)

$$u'' - u' = f, \qquad u(0) - u'(0) = 0, \qquad u(1) + u'(1) = 0.$$

4.50 Consider the problem given by

$$\mathfrak{L}\psi = \lambda\mathfrak{N}\psi \qquad \text{and} \qquad (4.34),$$

where $\mathfrak{L}$ is defined by (4.33) and $\mathfrak{N}u := \sum_{i<p} b_i(t)u^{(i)}$. Assuming that $0 \in \rho(T)$, give the integral formulation of the problem. If the functions b_i are continuous, how smooth is the resulting kernel?

8.2. Elliptic Partial Differential Equations

Let Ω be a bounded domain in $\mathbb{R}^N$ with boundary $\partial\Omega$, which will be assumed to be of class C^∞. $H^1(\Omega)$ is the Sobolev space $W^{1,2}(\Omega)$ with the norm $\|\cdot\|_1$, $H_0^1(\Omega)$ is the subspace of $H^1(\Omega)$ consisting of those functions "vanishing on $\partial\Omega$." $H_0^1(\Omega)$ is the closure in $H^1(\Omega)$ of $\mathscr{D}(\Omega)$, the set of indefinitely differentiable functions with compact support in Ω.

We begin to look at the following *Neumann* problem.

Example 4.18 The classical formulation of the homogeneous Neumann problem is the following. For a given f in $L^2(\Omega)$, find u in $H^1(\Omega)$ such that

$$\begin{aligned} -\Delta u + zu &= f \qquad \text{in } \Omega, \quad z \in \mathbb{C}, \\ \frac{\partial u}{\partial v} &= 0 \quad \text{on } \partial\Omega, \end{aligned} \tag{4.35}$$

where

$$\Delta u = \sum_{i=1}^{N} \frac{\partial^2 u}{\partial t_i^2} \qquad \text{and} \qquad \frac{\partial u}{\partial v} = \sum_{i=1}^{N} v_i \frac{\partial u}{\partial t_i},$$

$v = (v_i)_1^N$ being the unit outer normal to the boundary $\partial\Omega$.

Suppose that (4.35) is satisfied, and let $v \in H^1(\Omega)$. Multiplying by $\bar{v}$, integrating over Ω, and integrating by parts yields

$$-\int_\Omega \Delta u\, \bar{v}\, dt = \sum_{i=1}^{N} \int_\Omega \frac{\partial u}{\partial t_i} \overline{\frac{\partial v}{\partial t_i}}\, dt - \int_{\partial\Omega} \frac{\partial u}{\partial v} \bar{v}\, d\gamma,$$

where $dt = dt_1 \cdots dt_N$ and $d\gamma$ is the elementary area on $\partial\Omega$. Because of the boundary condition, we get the equation

$$\int_\Omega \sum_{i=1}^N \frac{\partial u}{\partial t_i} \overline{\frac{\partial v}{\partial t_i}} \, dt + z \int_\Omega u\bar{v} \, dt = \int_\Omega f\bar{v} \, dt \qquad \text{for all} \quad v \in H^1(\Omega). \quad (4.36)$$

Upon setting

$$a(u, v) := \int_\Omega \left(\sum_{i=1}^N \partial_i u \, \overline{\partial_i v} + zu\bar{v} \right) dt,$$

where $\partial u/\partial t_i$ is abbreviated in $\partial_i u$, (4.36) may be written

$$a(u, v) = (f, v) \qquad \text{for all} \quad v \in H^1(\Omega). \quad (4.37)$$

This is the *weak* formulation of (4.35). Conversely, if (4.37) is satisfied, then (4.35) is so since $\int_{\partial\Omega} (\partial u/\partial v)\bar{v} \, d\gamma = 0$ for all $v \in H^1(\Omega)$ implies $\partial u/\partial v = 0$ on $\partial\Omega$. The classical and weak forms of the Neumann problem are therefore equivalent.

We now define the abstract setting for the weak form of elliptic partial differential boundary-value problems. Let V and H be two complex Hilbert spaces, with inner products denoted, respectively, by $(\cdot, \cdot)_V$ and $(\cdot, \cdot)_H$. The corresponding norms are $\|u\|_V = [(u, u)_V]^{1/2}$ and $\|u\|_H = [(u, u)_H]^{1/2}$. We suppose that $\|u\|_H \leq c\|u\|_V$ for $u \in V$, so that $V \subset H$ and the embedding of V into H is continuous. We also suppose that V is dense in H.

Let $(u, v) \mapsto a(u, v)$ be a sesquilinear form defined on $V \times V$; $a(u, v)$ is supposed to be continuous,

$$|a(u, v)| \leq \beta\|u\|_V\|v\|_V \qquad \text{for} \quad u, v \in V, \quad (4.38)$$

and *coercive*,

$$\mathcal{Re} \; a(u, u) \geq \alpha\|u\|_V^2 \qquad \text{for} \quad u \in V, \quad (4.39)$$

for some $\alpha > 0$. Given $a(u, v)$ and $f \in H$, we consider the problems

$$\text{find} \quad u \in V \qquad \text{such that} \quad a(u, v) = (f, v)_H \qquad \text{for all} \quad v \in V, \quad (4.40)$$

$$\text{find } \lambda \in \mathbb{C}, 0 \neq \psi \in V \quad \text{such that} \quad a(\psi, v) = \lambda(\psi, v)_H \quad \text{for all } v \in V. \quad (4.41)$$

Given V, H, and $a(u, v)$ we may define a nonbounded operator T in H. We define D as the set of u in V such that the antilinear form $v \mapsto a(u, v)$ is continuous on V for the norm $\|\cdot\|_H$. Let T be the operator defined by $u \in D \mapsto Tu$ such that

$$a(u, v) = (Tu, v)_H \qquad \text{for all} \quad v \in V.$$

When a is *coercive* on V, T is said to be *elliptic* in D.

Proposition 4.16 *T is a linear operator from H into H with domain D.*

Proof Let $u \in D$. Since V in dense in H, the bounded antilinear form $v \mapsto a(u, v)$ can be extended in a bounded form on H. H is a Hilbert space; then by the Riesz theorem there exists a unique $Tu \in H$ such that

$$a(u, v) = (Tu, v)_H.$$

The mapping is easily seen to be linear. □

Example 4.19 $H = L^2(\Omega)$, $V = H_0^1(\Omega)$, and

$$a(u, v) = \int_\Omega \left(\sum_{i=1}^n \partial_i u \, \overline{\partial_i v} + u\bar{v} \right) dt.$$

$v \mapsto a(u, v)$ is continuous on $H_0^1(\Omega)$ for the norm of $L^2(\Omega)$. Therefore

$$D = H_0^1(\Omega)$$

and the operator T is defined by

$$u \in H_0^1(\Omega) \mapsto Tu = -\Delta u + u \in L^2(\Omega).$$

This is the differential operator associated with the homogeneous *Dirichlet* problem

$$-\Delta u + u = f \quad \text{in} \quad \Omega, \qquad f \in L^2(\Omega),$$
$$u = 0 \quad \text{on} \quad \partial\Omega.$$

By means of the unbounded operator T, Eqs. (4.40) and (4.41) can be written in H, respectively,

$$Tu = f, \qquad T\psi = \lambda\psi.$$

Another formulation of these problems in V is given by the Lax–Milgram theorem. There exists an isomorphism $S \in \mathscr{L}(V)$ such that

$$a(Su, v) = (u, v)_V \qquad \text{for} \quad u, v \in V.$$

The form $v \mapsto (f, v)_H$ is continuous on H, hence on V; therefore

$$(f, v)_H = (Jf, v)_V,$$

which defines $J \in \mathscr{L}(H, V)$.

(4.40) and (4.41) are then equivalent to the equations in V

$$S^{-1}u = Jf, \qquad S^{-1}\psi = \lambda J\psi.$$

We set $A := SJ$, $A \in \mathscr{L}(H, V)$, and $u = Af$ is the solution of (4.40). The eigenvector ψ defined by (4.41) is the eigenvector of A associated with the eigenvalue $1/\lambda$.

Exercises

4.51 Prove that D is dense in H, and that T is closed.

4.52 Consider the norm on D: $|||u|||_D := (\|u\|_H^2 + \|Tu\|_H^2)^{1/2}$ for $u \in D$. Prove that D equipped with this norm is a Hilbert space $\hat{\hat{D}}$, and $T \in \mathscr{L}(\hat{\hat{D}}, H)$.

4.53 Let T' be the operator defined by

$$a^*(u, v) := \overline{a(v, u)} = (f, v)_H \Leftrightarrow a(v, u) = (v, f)_H \Leftrightarrow T'u = f.$$

Show that T' is the adjoint of T in H.

4.54 Prove that $T^{-1} = A \in \mathscr{L}(H)$ with range $D \subset V$.

4.55 Let a, b be given continuous sesquilinear forms on $V \times V$, where a is supposed to be coercive on V. Analyze the eigenvalue problem

$$\text{find} \quad \lambda \in \mathbb{C}, \quad \psi \in V, \quad \psi \neq 0 \qquad \text{such that} \quad a(\psi, v) = \lambda b(\psi, v) \qquad \text{for all} \quad v \in V,$$

by means of the Lax–Milgram theorem.

4.56 Let $A = T^{-1}$, Im $A \subset V$. We define $B \in \mathscr{L}(V)$ by $Bu = Au$ for $u \in V$. Prove that $(A')_{\upharpoonright V} = B^\times$, the a-adjoint of B in V (cf. Example 2.14 of Chapter 2).

4.57 If a is hermitian, prove that $A = A'$ in H and $B = B^\times$ in V. If, moreover, a is coercive prove that A is positive definite in H.

4.58 Let Ω be the square $]0, 1[\times]0, 1[$. Consider the differential equations

$$\mathfrak{M}u := \sum_{|\alpha| \le p} (-1)^{|\alpha|}[D^\alpha p_\alpha(t) D^\alpha u],$$

$$\mathfrak{N}u := \sum_{|\alpha| \le q} (-1)^{|\alpha|}[D^\alpha q_\alpha(t) D^\alpha u],$$

where $0 \le q < p$, p_α, q_α are real-valued functions in $C^{|\alpha|}(\Omega)$, $\alpha = (\alpha_1, \alpha_2) \in \mathbb{N}^2$, $|\alpha| = \alpha_1 + \alpha_2$. The boundary conditions are p independent functionals

$$\mathfrak{l}_i(u) := \sum_{|\alpha| \le 2p-1} b_{i\alpha} D^\alpha x = 0 \qquad \text{for} \quad t \in \partial\Omega, \quad b_{i\alpha} \in \mathbb{C}, \quad i = 1, \dots, p.$$

Let $\mathscr{D}$ be the space of functions in $C^{2p}(\Omega)$ satisfying the boundary conditions. If $(\cdot, \cdot)_2$ is the inner product in $L^2(\Omega)$, we assume that $(\mathfrak{M}x, y)_2$ and $(\mathfrak{N}x, y)_2$ are hermitian, positive-definite sesquilinear forms defined for $x, y \in \mathscr{D}$. Prove that $(x, y)_H = (\mathfrak{M}x, y)_2$ and $(x, y)_V = (\mathfrak{N}x, y)_2$ for $x, y \in \mathscr{D}$ define two inner products in $\mathscr{D}$. If H and V are the Hilbert space completions of $\mathscr{D}$ with respect to $\|\cdot\|_H$ and $\|\cdot\|_V$, prove that V is dense in H and that the eigenvalue problem

$$\mathfrak{M}u = \lambda \mathfrak{N}u \quad \text{in} \quad \Omega,$$
$$\mathfrak{l}_i(u) = 0 \qquad \text{on} \quad \partial\Omega, \qquad i = 1, \dots, p,$$

may be written $(u, v)_H = \lambda(u, v)_V$ for all $v \in V$.

Example 4.20 The partial differential form

$$\mathfrak{L}u := -\sum_{i,j=1}^{N} \frac{\partial}{\partial t_i}\left[a_{ij}(t)\frac{\partial u}{\partial t_j}\right] + a_0(t)u,$$

where a_{ij}, a_0 are functions in $C^\infty(\Omega)$, is associated with the sesquilinear form

$$a(u, v) := \int_\Omega \left[\sum_{i,j=1}^{N} a_{ij}\, \partial_i u\, \partial_j \bar{v} + a_0 u \bar{v}\right] dt.$$

If there exist $\alpha > 0$ and $\alpha_0 > 0$ such that

$$\mathcal{Re} \sum_{i,j=1}^{N} a_{ij}(t)\xi_i\bar{\xi}_j \geq \alpha \sum_{i=1}^{N} |\xi_i|^2, \qquad \xi_i \in \mathbb{C}, \quad \text{almost everywhere in } \Omega,$$

$$\mathcal{Re}\, a_0(t) \geq \alpha_0 \qquad \text{almost everywhere in } \Omega,$$

then

$$\mathcal{Re}\, a(u, u) \geq \min(\alpha, \alpha_0)\|u\|_1^2 \qquad \text{for} \quad u \in H^1(\Omega),$$

by the Gärding's inequality (Yosida, 1965, p. 175). Under these assumptions, $a(u, v)$ is coercive on $H^1(\Omega)$. The boundedness of the coefficients in Ω implies that $a(u, v)$ is continuous on $H^1(\Omega) \times H^1(\Omega)$.

Proposition 4.17 *If the embedding i: $V \to H$ is compact and if, for some $z_0 \in \mathbb{C}$ and $\alpha > 0$,*

$$\mathcal{Re}\,(a(u, u) - z_0\|u\|_H^2) \geq \alpha\|u\|_V^2 \qquad \text{for} \quad u \in V, \tag{4.42}$$

then the operator T has a compact resolvent in H.

Proof If $T: D \to H$ is the operator associated with $a(u, v)$, then $T - z_0$: $D \to H$ is associated with $a(u, v) - z_0(u, v)_H$. Then the equation $(T - z_0)u = f$ has a unique solution $u = (T - z_0)^{-1}f = R(z_0)f$, where $R(z_0) \in \mathcal{L}(H, V)$ by the Lax–Milgram theorem. Therefore $R(z_0) = i \circ R(z_0)$ is compact in H and $R(z_0)_{\restriction V} = R(z_0) \circ i$ is compact in V. □

Example 4.21 The embeddings $H_0^1(\Omega) \to L^2(\Omega)$ and $H^1(\Omega) \to L^2(\Omega)$ are compact. Then the differential operators associated with the homogeneous Dirichlet and Neumann problems have a compact resolvent.

The interested reader is referred to Coddington and Levinson (1955) for more on the ordinary differential equations, and to Lions and Magenes (1968), Aubin (1972), and Ciarlet (1978) for more on elliptic partial differential equations.

We now turn to the presentation of some numerical methods used to solve the types of differential boundary problems that we have introduced.

9. Projection Methods for an Ordinary Differential Equation

We consider again the pth-order differential equation

$$\mathfrak{L}u := u^{(p)} - \sum_{i=0}^{p-1} a_i(t)u^{(i)} = f(t), \qquad a \leq t \leq b, \tag{4.33}$$

with the p homogeneous boundary conditions

$$\mathfrak{l}_k(u) := \sum_{i=0}^{p-1} [\alpha_{ik} u^{(i)}(a) + \beta_{ik} u^{(i)}(b)] = 0, \qquad k = 1, \dots, p. \quad (4.34)$$

Let T be the differential operator in $X = L^\infty(a, b)$ or $L^2(a, b)$ defined for $u \in D$ by $Tu = \mathfrak{L}u$, where

$$D = \{u \in C^{p-1}(a, b);\, u^{(p)} \in X \text{ and } \mathfrak{l}_k(u) = 0,\, k = 1, \dots, p\}.$$

We consider the equation

$$Tu = f \qquad (\text{resp. } T\psi = \lambda\psi).$$

The principle of the projection method is to compute an approximate solution u_n (resp. eigenvector ψ_n) in a subspace X_n of dimension n (say) by imposing n linear conditions on the residual $Tu_n - f$ (resp. $T\psi_n - \lambda_n\psi_n$). For example, let Y_n be a subspace of X of the same dimension n, and let π_n be a projection from X onto Y_n. The projection method requires that

$$\pi_n[Tu_n - f] = 0, \qquad u_n \in X_n,$$

$$(\text{resp. } \pi_n[T\psi_n - \lambda_n\psi_n] = 0, \qquad \psi_n \in X_n).$$

In the methods that we describe now, X_n consists of polynomials or piecewise polynomials.

9.1. The Method of Moments

The problem is set in $H = L^2(a, b)$. The approximate solution of (4.33) and (4.34) is sought in the form

$$u_n(t) = \sum_{i=1}^{n} \xi_{in} e_i(t),$$

where $\{e_i\}_1^n$ are polynomials of degree $p + i - 1$ satisfying (4.34), for $i = 1, \dots, n$. We write $\{e_1, \dots, e_n\} = \mathbb{P}_{p+n-1} \cap (4.34)$.

Besides $\mathfrak{L}u$, another differential form is given

$$\mathfrak{M}u := u^{(p)} - \sum_{i=0}^{p-1} b_i(t) u^{(i)}, \qquad a \le t \le b.$$

We denote by D^p the operator $u \mapsto D^p u = u^{(p)}$. The coefficients ξ_{in} are determined by the conditions

$$\int_a^b (\mathfrak{L}u_n - f)\overline{\mathfrak{M}e_i}\, dt = 0, \qquad i = 1, \dots, n. \qquad (4.43)$$

They are solution of the $n \times n$ linear system

$$\sum_{j=1}^{n} \left(\int_a^b \mathfrak{L}e_j \overline{\mathfrak{M}e_i}\, dt \right) \xi_{jn} = \int_a^b f\overline{\mathfrak{M}e_i}\, dt, \qquad i = 1, \dots, n.$$

Two choices for $\mathfrak{M}$ are usually made in actual computations:

(i) $\mathfrak{M} = \mathfrak{L}$: this yields the least-squares method.
(ii) $\mathfrak{M} = D^p$: one can assume that $\mathfrak{M}e_i = t^{i-1}$, $i = 1, 2, \ldots, n$.

This latter method is sometimes called the "essential" least-squares method.

We define $E_n = \{e_1, \ldots, e_n\}$ and $F_n = \mathfrak{M}E_n = \{\mathfrak{M}e_1, \ldots, \mathfrak{M}e_n\}$ as subspaces of $L^2(a, b)$. Let π_n be the *orthogonal* projection on F_n; (4.43) can be rewritten

$$\pi_n[Tu_n - f] = 0, \qquad u_n \in E_n.$$

This means that the residual $Tu_n - f$ is required to be orthogonal to F_n.

Exercises

4.59 Study the approximation of $T\psi = \lambda\psi$ by the method of moments with the choice (ii).
4.60 Prove that for the least-squares method $F_n = \mathfrak{L}E_n = (1 - K)\mathbb{P}_{n-1}$, whereas for the essential least-squares method $F_n = D^pE_n = \mathbb{P}_{n-1}$.

In the sequel, the method of moments will refer to the particular choice (ii); that is $\mathfrak{M} = D^p$.

9.2. The Collocation Method

The problem is now set in $L^\infty(a, b)$. Let $e_i(t) = \sum_{j=0}^{p-1} c_{ij}t^j + t^{p+i-1}$, $i = 1, 2, \ldots$, be a sequence of polynomials of degree $p, p + 1, p + 2, \ldots$, which satisfy (4.34). The approximate solution u_n is again sought in the form

$$u_n(t) = \sum_{i=1}^{n} \xi_{in} e_i(t).$$

The collocation method determines the ξ_{in} by the condition that the residual $\mathfrak{L}u_n - f$ must vanish at n prescribed collocation points $\{\tau_i\}_1^n$,

$$a \le \tau_1 < \tau_2 < \cdots < \tau_n \le b.$$

This yields

$$u_n^{(p)}(\tau_i) - \sum_{k=0}^{p-1} a_k(\tau_i)u_n^{(k)}(\tau_i) = f(\tau_i), \qquad i = 1, \ldots, n. \tag{4.44}$$

The ξ_{in} are solution of the $n \times n$ linear system

$$\sum_{j=1}^{n}\left[e_j^{(p)}(\tau_i) - \sum_{k=0}^{p-1} a_k(\tau_i)e_j^{(k)}(\tau_i)\right]\xi_{jn} = f(\tau_i), \qquad i = 1, \ldots, n.$$

There is a unique solution if the collocation points $\{\tau_i\}_1^n$ are such that $\det(\mathfrak{L}e_j(\tau_i)) \neq 0$.

We now define $X_n = \{e_1, \ldots, e_n\}$ and $Y_n = \mathbb{P}_{n-1}$ as subspaces of $L^\infty(a, b)$. Let L_n be the Lagrange interpolation mapping every continuous function onto its $(n-1)$th-degree interpolant polynomial at the points $\{\tau_i\}_1^n$. (4.44) can be rewritten

$$L_n(Tu_n - f) = 0, \qquad u_n \in X_n.$$

Exercises

4.61 Study the approximation of $T\psi = \lambda\psi$ by collocation.

4.62 Compare the discrete problems arising in solving $Tu = f$ by collocation and by the method of moments.

4.63 Check that the discrete problem (4.44) depends only on the collocation points and on X_n. The choice $Y_n = \mathbb{P}_{n-1}$ is then arbitrary. What other choice could be done?

9.3. *The Use of Piecewise Polynomials*

The same methods can be defined when the functions $e_i(t)$ are chosen to be *piecewise* polynomials. Let $\Delta = \{t_i\}_0^n$ be a strict partition of $[a, b]$, $a = t_0 < t_1 < \cdots < t_n = b$. Set $\Delta_i := [t_{i-1}, t_i[$, $i = 1, \ldots, n$. Let $\mathbb{P}_{p+r,\Delta}^{p-1}$ be the set of functions in $C^{p-1}(a, b)$ that are polynomials of degree less than $p + r + 1$ on each Δ_i, $i = 1, \ldots, n$. We define C_n to be the set of functions in $\mathbb{P}_{p+r,\Delta}^{p-1}$ that satisfy (4.34): $C_n = \mathbb{P}_{p+r,\Delta}^{p-1} \cap (4.34)$. Let $\{e_i\}_1^{n(r+1)}$ be a basis for C_n; $u_n = \sum_{i=1}^{n(r+1)} \xi_{in} e_i$, where the ξ_{in} are determined so that the residual $\mathfrak{L}u_n - f$

(1) is orthogonal to the subspace $F_n = \mathfrak{M}C_n = \{\mathfrak{M}e_i\}_1^{n(r+1)}$ in the method of moments,

(2) vanishes at the $r + 1$ collocation points $\{\tau_i^j\}_{j=1}^{r+1}$ chosen on each Δ_i, $i = 1, \ldots, n$, in the collocation method.

Exercises

4.64 We choose $\mathfrak{M} = D^p$ for the method of moments. Show that $D^pC_n = \mathbb{P}_{r,\Delta}$. Rewrite the condition $\mathfrak{L}u_n - f \perp F_n$ by means of the projection π_n defined in Exercise 4.4.

4.65 Prove that the condition $(\mathfrak{L}u_n - f)(\tau_i^j) = 0$ can be rewritten by means of the projection L_n defined in Exercise 4.6.

9.4. *Interpretation on the Integral Formulation: A Galerkin Method*

The system of (4.33) and (4.34) is equivalent by Proposition 4.15 to

$$u = Gx, \qquad x = Kx + f. \tag{4.45}$$

We show that the projection method on a differential equation can be interpreted as a Galerkin method on its integral formulation. We consider the method of moments (with $\mathfrak{M} = D^p$) and collocation. The subspaces X_n and Y_n are then defined to be

$$X_n = \mathbb{P}_{p+n-1} \cap (4.34) \qquad \text{and} \qquad Y_n = \mathbb{P}_{n-1}$$

in case of polynomials, or

$$X_n = \mathbb{P}^{p-1}_{p+r,\Delta} \cap (4.34) \qquad \text{and} \qquad Y_n = \mathbb{P}_{r,\Delta}$$

in case of piecewise polynomials, and π_n is a projection on Y_n.

Proposition 4.18 *With X_n, Y_n, π_n defined above, the projection method on T is equivalent to the Galerkin method on $1 - K$ defined by Y_n.*

Proof It follows readily from $D^p X_n = Y_n$. The equation

$$\pi_n(Tu_n - f) = 0, \qquad u_n \in X_n$$

is therefore equivalent to

$$u_n = Gx_n, \qquad x_n = \pi_n K x_n + \pi_n f, \qquad x_n \in Y_n. \quad \square \tag{4.46}$$

Proposition 4.18 may appear as a justification of the choice of Y_n made for the collocation method.

Corollary 4.19 *If $\pi_n \xrightarrow{\text{p}} 1$, then $u_n \to u$ as $n \to \infty$.*

Proof $u - u_n = G(x - x_n)$, where G is bounded. $x_n \to x$ since $\pi_n \xrightarrow{\text{p}} 1$ and K is compact. $\square$

We define the (Galerkin) iterated solution $\tilde{x}_n = Kx_n + f$ for (4.46).

Proposition 4.20 *The following identity holds:*

$$u - u_n = T^{-1}(1 - \pi_n)\tilde{x}_n. \tag{4.47}$$

Proof

$$\begin{aligned} x - x_n &= (1 - K)^{-1} f - (1 - \pi_n K)^{-1} \pi_n f \\ &= (1 - K)^{-1}(1 - \pi_n)K(1 - \pi_n K)^{-1}\pi_n f + (1 - K)^{-1}(1 - \pi_n) f \\ &= (1 - K)^{-1}(1 - \pi_n)(Kx_n + f) = (1 - K)^{-1}(1 - \pi_n)\tilde{x}_n. \end{aligned}$$

Then

$$u - u_n = G(1 - K)^{-1}(1 - \pi_n)\tilde{x}_n = T^{-1}(1 - \pi_n)\tilde{x}_n. \quad \square$$

The identity (4.47) will be used in Chapter 7, Section 6, for the analysis of the superconvergence properties of $u - u_n$, with piecewise polynomials.

Example 4.22 The equation $Tu = f$ is approximated by collocation at Chebyshev points $\{\tau_i\}_1^n$ in $C(-1, 1)$. With the change of variable $u^{(p)} = x$, we recover the collocation method on $1 - K$ defined in Section 7.3. By Proposition 4.9

$$\|u^{(p)} - u_n^{(p)}\|_{2,\rho} \leq c \operatorname{dist}_\infty (u^{(p)}, \mathbb{P}_{n-1}),$$

and by integration

$$\|u^{(i)} - u_n^{(i)}\|_\infty \leq c \operatorname{dist}_\infty(u^{(p)}, \mathbb{P}_{n-1}), \qquad 0 \leq i \leq p - 1.$$

This proves that $u_n \to u$ in the uniform norm. Note that Proposition 4.10 shows that $\tilde{x}_n \to x$ in $C(-1, 1)$.

Example 4.23 Consider the equation $(T - z)u = f$ and its approximate solution u_n by projection. We may [at least formally (cf. Chapter 7)] consider the iterated solution $\tilde{u}_n$ defined by $T\tilde{u}_n = zu_n + f$. $\tilde{u}_n$ is related to the integral formulation as follows:

$$(T - z)u = f \Leftrightarrow \begin{cases} u = Gx \\ (1 - K - zG)x = f \end{cases} \Leftrightarrow \begin{cases} u = Gx \\ x = (1 - K)^{-1}[zGx + f] \end{cases}$$

Then $\tilde{u}_n = Gx'_n$ where $x'_n = (1 - K)^{-1}[zGx_n + f]$. We note that x'_n is an iterated solution different from the one we already considered when $z = 0$ in (4.47), which corresponds to $\tilde{x}_n = (K + zG)x_n + f$ for $z \neq 0$.

Exercises

4.66 Interpret the least-squares method on $Tu = f$ as an $\perp$-Petrov method on $1 - K$ with right subspace Y_n and left subspace $\mathfrak{L}X_n$.

4.67 Prove that the projection method on $T\psi = \lambda\psi$ is equivalent to

$$\psi_n = G\varphi_n, \qquad (1 - \pi_n K)\varphi_n = \lambda_n \pi_n G\varphi_n, \qquad \varphi_n \in Y_n.$$

4.68 Prove that, when $\pi_n \xrightarrow{p} 1$, the difference of the resolvents satisfies for n large enough and $z \in \rho(T)$,

$$\|(1 - \pi_n K - z\pi_n G)^{-1}\pi_n G - (1 - K - zG)^{-1}G\| \to 0 \qquad \text{as} \quad n \to \infty.$$

4.69 Prove the identities

$$\begin{aligned}(1 - P)\psi_n &= (T - \lambda_n)^{-1}(1 - P)(1 - K - \lambda_n G)\varphi_n \\ &= S(\lambda_n)(\varphi_n - \tilde{\varphi}_n) \qquad (\text{with } \tilde{\varphi}_n = K\varphi_n + \lambda_n G\varphi_n) \\ &= S(\lambda_n)(\pi_n - 1)\tilde{\varphi}_n.\end{aligned}$$

The projection method on $Tu = f$ has a very simple interpretation on (4.45) as a Galerkin method, whereas for the eigenvalue problem, this is not as easy because the integral formulation yields a *generalized* eigenvalue problem (cf. Example 4.16 and Exercise 4.67). We now give another possible interpretation by means of the Petrov method rather than Galerkin's.

9.5. Another Interpretation: A Petrov Method

Let $0 \neq -z \in \rho(T)$ be given and set $A_z := (T + z)^{-1}$. We consider the equivalence defined in Example 4.17:

$$Tu = f \Leftrightarrow \begin{cases} u = A_z y \\ y = zA_z y + f \end{cases}$$

With X_n, Y_n, π_n defined as in Section 9.4, we set $X'_n := TX_n = (1 - K)Y_n$.

Proposition 4.21 *The projection method on T is equivalent to the Petrov method on $1 - zA_z$ defined by the right subspace X'_n and the left subspace Y_n.*

Proof $\pi_n[Tu_n - f] = 0$, $u_n \in X_n$ is equivalent to

$$u_n = A_z y_n, \qquad \pi_n[(1 - zA_z)y_n] = \pi_n f, \qquad y_n \in X'_n. \quad \square \tag{4.48}$$

The convergence given by Corollary 4.19 can be proved in this setting in the following way:

$$u - u_n = A_z(y - y_n),$$

and $y_n \to y$ follows from the next lemma.

Lemma 4.22 *If $\pi_n \xrightarrow{\text{p}} 1$, the above-defined Petrov method is convergent.*

Proof We apply Proposition 4.8 with X'_n, Y_n such that $X'_n = (1 - K)Y_n$ where K is compact. $\square$

We define now the (Petrov) iterated solution $\tilde{y}_n = zA_z y_n + f$ for (4.48).

Proposition 4.23 *The following identity holds*

$$u - u_n = (1/z)(y - \tilde{y}_n). \tag{4.49}$$

Proof

$$u - u_n = A_z(y - y_n) = \frac{1}{z}(y - f) - \frac{1}{z}(\tilde{y}_n - f) = \frac{1}{z}(y - \tilde{y}_n). \quad \square$$

Exercises

4.70 If $0 \in \rho(T)$ and $A = T^{-1}$, the projection method on $T\psi = \lambda\psi$ is equivalent to

$$\psi_n = A\theta_n, \qquad \pi_n[\theta_n - \lambda_n A\theta_n] = 0, \qquad \theta_n \in X'_n.$$

4.71 Prove the identity, if $\lambda_n \neq 0$,

$$(1 - P)\psi_n = \frac{1}{\lambda_n}(1 - P)\tilde{\theta}_n \qquad \text{with} \quad \tilde{\theta}_n = \lambda_n A\theta_n.$$

9.6. Bibliographical Comments

The projection method on an unbounded operator T is studied in Krasnoselskii *et al.* (1972), with application to the method of moments and collocation. See Karpilovskaia (1963) and Vainikko (1966) for collocation with polynomials, and Russel and Shampine (1972) for collocation with piecewise polynomials.

10. The Projection Method for Partial Differential Equations

The notations are those of Section 8.2. We recall that H and V are complex Hilbert spaces, $V \subset H$ and V is dense in H, the injection of V into H is continuous. We consider the equation $Tu = f$ and its weak formulation

$$a(u, v) = (f, v)_H \qquad \text{for all} \quad v \in V. \tag{4.40}$$

Let a family of finite-dimensional subspaces $\{V_n\}_{\mathbb{N}}$ of V be given. The projection method on V_n consists in approximating u by $u_n \in V_n$, the solution of the equation in V_n:

$$a(u_n, v_n) = (f, v_n)_H \qquad \text{for all} \quad v_n \in V_n. \tag{4.50}$$

If we suppose that $a(\cdot, \cdot)$ is coercive on V, the projection $\pi_n^a\colon V \to V_n$ is well defined by

$$a((1 - \pi_n^a)u, v) = 0 \qquad \text{for all} \quad v \in V_n, \quad \text{all} \quad u \in V$$

(cf. Example 2.18).

Exercises

4.72 Prove that (4.50) has a unique solution u_n.
4.73 Write the matrix problem associated with a choice of basis in V_n.

Theorem 4.24 *If $a(\cdot, \cdot)$ is coercive on V, the solution u_n of* (4.50) *is the projection $\pi_n^a u$ of the exact solution of* (4.40).

Proof (4.40) defines a bounded operator $A\colon H \to V$ such that $a(Af, v) = (f, v)_H$ for all f in H, v in V, and $u = Af$. Then

$$a((1 - \pi_n^a)Af, v_n) = 0 \qquad \text{for all} \quad f \text{ in } H, \quad v_n \text{ in } V_n$$

proves that the bounded operator $A_n\colon H \to V$ defined by

$$a(A_n f, v_n) = (f, v_n)_H$$

for all v_n in V_n, is such that $A_n = \pi_n^a A$. Therefore

$$u_n = A_n f = \pi_n^a A f = \pi_n^a u. \quad \square$$

In the method of projection applied on the weak formulation of

$$Tu = f \Leftrightarrow u = Af,$$

the operator A with range in V is approximated by $\pi_n^a A$. This is a Galerkin method defined (as in Part A) on the bounded operator A by means of the projection π_n^a. When the subspace V_n consists of continuous piecewise polynomials the method is better known under the name of conforming *finite-element method* (f.e.m.)

Example 4.24 The finite-element method, in its simplest form, is a specific process of constructing the subspaces V_n of V. The reader may find practical examples in Ciarlet (1978), for example.

A and A_n are defined by

$$a(Af, v) = (f, v)_H, \quad \forall v \in V,$$
$$a(A_n f, v_n) = (f, v_n)_H, \quad \forall v_n \in V_n.$$

Similarly, the H-adjoints A' and $A'_n: H \to V$ are defined (Exercise 4.53) by the relations

$$a(v, A'f) = (v, f)_H, \quad \forall v \in V,$$
$$a(v_n, A'_n f) = (v_n, f)_H, \quad \forall v_n \in V_n.$$

We recall that $(A')_{\restriction V} = (A_{\restriction V})^\times$, the a-adjoint of $A_{\restriction V}$ in V, and

$$(A'_n)_{\restriction V} = (A_{n \restriction V})^\times = (\pi_n^a A_{\restriction V})^\times$$

(Exercise 4.56).

Exercise

4.74 Study the approximation of (4.41) by the problem

$$\text{find} \quad \lambda_n \in \mathbb{C}, \quad \psi_n \neq 0: a(\psi_n, v_n) = \lambda_n(\psi_n, v_n)_H, \qquad \forall v_n \in V_n.$$

Remark that $\psi_n \neq \pi_n^a \psi$ in general.

10.1. Convergence of the Method

We set the problem in V. Since $u - u_n = (1 - \pi_n^a)u$, the problem is to bound $\|(1 - \pi_n^a)u\|_V$ in terms of $\delta_n(u) = \text{dist}_V(u, V_n)$.

Lemma 4.25 *If $a(\cdot, \cdot)$ is coercive on V, then, for all $u \in V$,*

$$\max(\|(1 - \pi_n^a)u\|_V, \|(1 - \pi_n^{a\times})u\|_V) \leq (\beta/\alpha)\delta_n(u)$$

where β and α are the constants in (4.38) *and* (4.39), *respectively.*

Proof This has been proved in Example 2.18, Chapter 2. □

The resulting bound $\|u - u_n\|_V \le (\beta/\alpha)\delta_n(u)$ is attributed to Céa (see Ciarlet, 1978, where no use of the projection π_n^a is made). The problem of estimating $\|u - u_n\|_V$ is reduced, by Lemma 4.25, to a problem in approximation theory, that is, to evaluate the distance in the norm of V of the solution u to the subspace V_n of V. We shall go back to this question in Chapter 6, where we deal with error bounds. Because $\|u\|_H \le c\|u\|_V$ for any u in V, we may deduce from Lemma 4.25 the bound $\|u - u_n\|_H \le c\,\delta_n(u)$, where c is a generic constant. Under mild assumptions, the bound may be improved, thanks to the following lemma.

Lemma 4.26 (Aubin–Nitsche) *If $V \subset H$ and $\overline{V} = H$, then*

$$\|u - u_n\|_H \le c\|u - u_n\|_V\, \delta_V(A'H, V_n).$$

Proof $A' : H \to V$, as already seen, satisfies

$$a(v, A'g) = (v, g)_H, \qquad \forall v \in V,\ g \in H.$$

Now $u - u_n = (1 - \pi_n^a)u$ and

$$((1 - \pi_n^a)u, g)_H = a((1 - \pi_n^a)u, A'g) = a((1 - \pi_n^a)u, (1 - \pi_n^{a\times})A'g).$$

Then

$$\|(1 - \pi_n^a)u\|_H = \sup_{0 \ne g \in H} \frac{|((1 - \pi_n^a)u, g)_H|}{\|g\|_H}$$
$$\le c\|(1 - \pi_n^a)u\|_V \sup_{0 \ne g \in H} \frac{\|(1 - \pi_n^{a\times})A'g\|_V}{\|g\|_H}.$$

Now

$$\|(1 - \pi_n^{a\times})A'g\|_V \le c\, \mathrm{dist}_V(A'g, V_n),$$

$$\delta_V(A'H, V_n) = \sup_{g \in H} (\mathrm{dist}_V(A'g, V_n);\ \|A'g\|_V = 1),$$

and

$$\|A'g\|_V < c\|g\|_H$$

since $A' \in \mathscr{L}(H, V)$. Therefore

$$\sup_{0 \ne g \in H} \frac{\|(1 - \pi_n^{a\times})A'g\|_V}{\|g\|_H} \le c \sup_{0 \ne g \in H} \left(\frac{1}{\|A'g\|_V} \mathrm{dist}_V(A'g, V_n) \right)$$
$$\le c\,\delta_V(A'H, V_n).$$

This yields an improvement if $A'H$ is well approximated by V_n. For example, if A is compact, so is A' and $\delta_V(A'H, V_n) \to 0$. □

Similar techniques to bound scalar products will be used later to get improved results, in particular for the error $\lambda - \hat{\lambda}_n$ in Chapter 6.

Because Eqs. (4.40) and (4.41) are considered in V, we define B to be the restriction of A to V: $B = A_{\restriction V} \in \mathscr{L}(V)$. Similarly $(\pi_n^a A)_{\restriction V} = \pi_n^a B$.

Theorem 4.27 *We suppose that*

(i) $a(\cdot, \cdot)$ *is coercive on* V,
(ii) *for any* u *in* V, $\operatorname{dist}_V(u, V_n) \to 0$,
(iii) B *is compact in* V.

Then $\|(1 - \pi_n^a)B\|_V \to 0$.

The proof is clear. When the embedding of V into H is not compact, A and hence B are not compact either. We define

$$\mathscr{B}_n := \pi_n^a B_{\restriction V_n} = \pi_n^a A_{\restriction V_n}: V_n \to V_n,$$
$$B_n := \mathscr{B}_n \pi_n^a.$$

B_n is an approximation of B of class $\mathfrak{D}$.

Theorem 4.28 (Descloux, Nassif, and Rappaz) *Under the hypotheses* (i) *and* (ii) *of Theorem* 4.27 *the following are equivalent*:

(a) $B_n \xrightarrow{\text{d-c}} B$,
(b) $\sup_{v \in V_n}(\|(B - B_n)v\|_V;\ \|v\|_V = 1) \to 0$.

Proof We know that $B_n \xrightarrow{\text{p}} B$ and (b) $\Rightarrow$ (a) by Exercise 3.2. We prove the converse. Suppose that (a) $\not\Rightarrow$ (b). There exists a bounded sequence $\{u_n\}_{\mathbb{N}}$ in V such that $\|(B - B_n)u_n\|_V \geq \delta > 0$ for $n \in N_1 \subset \mathbb{N}$ and

$$\lim_{\substack{n \to \infty \\ n \in N_1}} (B - B_n)u_n = v$$

such that $\|v\|_V \geq \delta$. We define v_n to be the V-orthogonal projection of v on V_n: $v_n \to v$ as $n \to \infty$. Then

$$\begin{aligned} \alpha\|v\|_V^2 \leq \mathscr{Re}\, a(v, v) &\leq \lim_{n \in N_1} |a((B - B_n)u_n, v)| \\ &= \lim_{n \in N_1} |a((B - B_n)u_n, v - v_n)| \\ &\leq \beta \lim_{n \in N_1} (\|(B - B_n)u_n\|_V \|v - v_n\|_V) = 0. \end{aligned}$$

This contradicts $\|v\|_V \geq \delta > 0$. $\square$

By Proposition 3.20 and Exercise 3.13 $B_n \xrightarrow{\text{d-c}} B$ implies $B_n - z\pi_n^a \xrightarrow{\text{d-s}} B - z$ in V for any z in $\rho(B)$.

Exercises

4.75 Prove that $\rho(A) = \rho(B)$ and $Q\sigma(A) = Q\sigma(B)$ (Descloux–Nassif–Rappaz).

4.76 For z in $\rho(B_n)$ for all n, prove that

$$\sup_{u \in V_n} (\|(B_n - z)^{-1}u\|_V;\ \|u\|_V = 1) < c, \quad \forall n,$$

$$\Leftrightarrow \sup_{u \in V_n} (\|(A_n - z)^{-1}u\|_H;\ \|u\|_H = 1) < c, \quad \forall n,$$

where c is a generic constant (Descloux–Nassif–Rappaz–Tartar).

4.77 Prove that $\|(1 - \pi_n^a)A\|_H \to 0$ under the hypotheses of Theorem 4.27.

10.2. The Rayleigh–Ritz Method When $a(\cdot\,,\cdot)$ Is Hermitian

We suppose in this paragraph that $a(\cdot,\cdot)$ is hermitian:

$$a(u, v) = \overline{a(v, u)} \qquad \text{for} \quad u, v \in V.$$

When the sesquilinear form $a(u, v)$ is coercive on V, it defines an inner product on V and its associated norm:

$$(u, v)_a := a(u, v), \qquad \|u\|_a := \sqrt{a(u, u)} \qquad \text{for} \quad u, v \in V.$$

The projection π_n^a is orthogonal with respect to this inner product (Exercise 2.18), and the solution u_n of (4.50) is the *best approximation* of u in the norm $\|\cdot\|_a$:

$$\|u - u_n\|_a = \min_{v \in V_n} \|u - v\|_a.$$

The Galerkin method on $u = Af$ reduces to the computation of the Rayleigh–Ritz approximate $u_n = \pi_n^a u$.

Exercises

4.78 Let T be as defined in Proposition 4.16. Show that T is self-adjoint positive definite in H and $(u, v)_a = (Tu, v)_H$ for $u \in D$, $v \in V$. Same question with A.

4.79 Let S be as defined in Theorem 2.9. Prove that $S = S^* = S^\times$ is self-adjoint positive definite in V and $(u, v)_a = (S^{-1}u, v)$.

4.80 B is a-self-adjoint; deduce that $\pi_n^a B \pi_n^a$ is a-self-adjoint in V. Check that the matrix of the discrete problem (4.50) is hermitian positive definite.

When a is hermitian, the solution u of (4.40) is the solution of the minimization problem

$$\text{find} \quad u \in V \qquad \text{such that} \quad J(u) = \min_{v \in V} J(v), \tag{4.51}$$

where $J: V \to \mathbb{C}$ is the functional defined by

$$J: v \in V \mapsto J(v) := \tfrac{1}{2}a(v, v) - (f, v)_H.$$

u_n is the solution of (4.51) in V_n: $J(u_n) = \min_{v \in V_n} J(v)$. In elasticity, u represents the displacement of a mechanical system and J represents the energy. The norm $\|\cdot\|_a$ is often called the "*energy*" *norm*; it is equivalent to $\|\cdot\|_V$ on V. (4.40) is called the *variational* formulation of (4.51). This name comes from the point of view of differential calculus: let $J'(u)$ be the Fréchet derivative of J at $u \in V$. (4.51) expresses that

$$J'(u)v = a(u, v) - (f, v)_H = 0 \qquad \text{for all} \quad v \text{ in } V,$$

that is, the first variation of J (i.e., the first-order term in its Taylor expansion) vanishes for all v when $J(u)$ is a minimum of J. This condition is also sufficient when J is convex, as is the case here. When a is not hermitian, there is no more minimization interpretation associated with (4.40).

We now deal with an important property of the Rayleigh–Ritz method applied to the eigenvalue problem (4.41) when $a(\cdot,\cdot)$ is hermitian and coercive on V. The approximation is

$$\begin{aligned} &\text{find} \quad \lambda_n \in \mathbb{C},\, 0 \neq \psi_n \in V_n \\ &\text{such that} \quad a(\psi_n, v_n) = \lambda_n(\psi_n, v_n)_H \quad \text{for any} \quad v_n \text{ in } V_n. \end{aligned}$$

We suppose that the lower parts of the spectra contain a finite number of isolated eigenvalues with finite multiplicity: they are numered μ_i (resp. $\mu_i^{(n)}$), $i = 1, \ldots, k$, by increasing value according to their algebraic multiplicity.

Proposition 4.29 *If a is hermitian and coercive on V, then the repeated eigenvalues in the lower part of the spectra are such that*

$$\mu_i^{(n)} \geq \mu_i, \qquad i = 1, \ldots, k.$$

Proof One form of the min–max representation of the eigenvalues is

$$\mu_i = \max_{\substack{F \subset V \\ \operatorname{codim} F = i-1}} \min_{u \in F} \frac{a(u, u)}{(u, u)_H}, \qquad i = 1, \ldots, k,$$

where $\operatorname{codim} F = \dim F^{\perp}$ (Kato, 1976, p. 60). Let F_0 be a subspace of V such that

$$\mu_i = \min_{u \in F_0} \frac{a(u, u)}{(u, u)_H}$$

with codim $F_0 = i - 1$. Then

$$\mu_i \leq \min_{u \in F_0 \cap V_n} \frac{a(u, u)}{(u, u)_H}$$

and $F_0 \cap V_n$ has codimension not larger than $i - 1$ relatively to V_n because codim $F_0 = i - 1$. Thus

$$\mu_i \leq \min_{u \in F_0 \cap V_n} \frac{a(u, u)}{(u, u)_H} \leq \mu_i^{(n)}, \qquad i = 1, \ldots, k. \quad \square$$

10.3. Other Examples of the Finite-Element Method

We present the abstract setting for several examples, which differ from the basic examples (4.40) and (4.41) treated at the beginning of this section.

Example 4.25 (The conforming f.e.m. on a generalized weakly posed elliptic eigenvalue problem) Let V be a complex Hilbert space. Let a, b: $V \times V \to \mathbb{C}$ be continuous sesquilinear forms, a is supposed to be coercive on V. By Lax–Milgram in V,

$$a(u, v) = (S_a^{-1} u, v)_V = (u, S_a^{-1*} v)_V,$$
$$b(u, v) = (S_b^{-1} u, v)_V = (u, S_b^{-1*} v)_V,$$

where S_a^{-1}, S_b^{-1}, and S_a are bounded on V.

The generalized eigenvalue problem

$$\text{find} \quad \lambda \in \mathbb{C}, 0 \neq u \in V; a(u, v) = \lambda b(u, v) \qquad \text{for all} \quad v \in V$$

is equivalent to

$$Bu = (1/\lambda)u \qquad \text{with} \quad B = S_a S_b^{-1} \in \mathscr{L}(V).$$

$a(S_a S_b^{-1} u, v) = (S_b^{-1} u, v)_V = b(u, v) = a(u, S_a^* S_b^{-1*} v)$ proves that the a-adjoint of B is $B^\times = S_a^* S_b^{-1*}$.

Let $\{V_n\}_{\mathbb{N}}$ be a family of subspaces of V. We restrict the original weakly posed problem to $V_n \times V_n$:

$$\text{find} \quad \lambda_n \in \mathbb{C}, 0 \neq u_n \in V_n; a(u_n, v_n) = \lambda_n b(u_n, v_n) \qquad \text{for all} \quad v_n \in V_n.$$

This is equivalent to $B_n u_n = (1/\lambda_n) u_n$ with $B_n = \pi_n^a B$, where π_n^a is the elliptic projection on V_n.

Remark The problem (4.41) is a particular case of the above problem with $b(u, v) = (u, v)_H$. The above analysis confirms the fact that V is the natural space to set the problem; H is secondary.

Example 4.26 (Mixed and hybrid f.e.m. on a nonelliptic equation) When a is hermitian but noncoercive, u is a stationary point of the functional $J(v)$, the form $\mathcal{R}e\, a(v, v)$ can take either sign and v may include two different types of unknowns—both displacements and stresses—in the "mixed" and "hybrid" methods. The problem can be put into the following abstract setting.

Let H_1 and H_2 be two complex Hilbert spaces with respective norms $\|\cdot\|_1$ and $\|\cdot\|_2$. Let be given two continuous sesquilinear forms a and b from $H_1 \times H_2$ into $\mathbb{C}$.

The coercivity condition on a is now replaced by the weaker condition

$$\begin{gathered}\inf_{\substack{u \in H_1 \\ \|u\|_1 = 1}} \sup_{\substack{v \in H_2 \\ \|v\|_2 = 1}} |a(u, v)| \geq \alpha > 0, \\ \sup_{\substack{u \in H_1 \\ \|u\|_1 = 1}} |a(u, v)| > 0 \qquad \text{for all} \quad v \neq 0 \text{ in } H_2.\end{gathered} \tag{4.52}$$

The Riesz representation theorem implies that bounded operators A and B from H_1 into H_2 exist such that $a(u, v) = (Au, v)_2$ and $b(u, v) = (Bu, v)_2$ for all $v \in H_2$. The adjoint operators A^* and B^* are from H_2 into H_1 and satisfy $a(u, v) = (u, A^*v)_1$ and $b(u, v) = (u, B^*v)_1$ for all $u \in H_1$. Hypotheses (4.52) imply that A is an isomorphism from H_2 onto H_1.

Let $T = A^{-1}B$ and $T^\times = A^{*-1}B^*$. T is a bounded operator on H_1 that satisfies $a(Tu, v) = b(u, v)$ for all $v \in H_2$. Similarly $T^\times$ is a bounded operator on H_2 that satisfies $a(u, T^\times v) = b(u, v)$ for all u in H_1. We note that for all u in H_1 and v in H_2, $a(Tu, v) = a(u, T^\times v)$: $T^\times$ is the a-adjoint of T in H_1.

To construct an approximation for T, we introduce two families of finite-dimensional subspaces $H_{1n} \subset H_1$ and $H_{2n} \subset H_2$. We suppose that the following assumptions hold:

$$\inf_{\substack{u_n \in H_{1n} \\ \|u_n\|_1 = 1}} \sup_{\substack{v_n \in H_{2n} \\ \|v_n\|_2 = 1}} |a(u_n, v_n)| = \alpha \geq 0,$$

$$\sup_{\substack{u_n \in H_{1n} \\ \|u_n\|_1 = 1}} |a(u_n, v_n)| > 0 \qquad \text{for all} \quad v_n \neq 0 \text{ in } H_{2n}, \text{ and all } n,$$

and

$$\delta_{1n}(u) = \text{dist}_1(u, H_{1n}) \to 0 \qquad \text{for every} \quad u \in H_1.$$

Let $W_{1n} := \{u \in H_1;\ a(u,\ v) = 0\ \forall v \in H_{2n}\}$. $H_1 = H_{1n} \oplus W_{1n}$ and the projection π_{1n}^a from H_1 onto H_{1n} along W_{1n} is well defined and satisfies

$$\begin{aligned} a((1 - \pi_{1n}^a)u, v_n) &= 0 && \text{for all} \quad v_n \text{ in } H_{2n}, \\ \|(1 - \pi_{1n}^a)u\|_1 &\leq c\, \delta_{1n}(u) && \text{for} \quad u \in H_1. \end{aligned}$$

Similarly a projection π_{2n}^a from H_2 onto H_{2n} can be defined such that

$$a(u_n, (1 - \pi_{2n}^a)v) = 0 \qquad \text{for all} \quad u_n \text{ in } H_{1n}, \text{ and } \pi_{1n}^{a\,\times} = \pi_{2n}^a.$$

Let f be given in H_1. The solution u of the equation $a(u, v) = b(f, v)$ for all v in H_2 is approximated in H_{1n} by

$$a(u_n, v_n) = b(f, v_n) \qquad \text{for all} \quad v_n \text{ in } H_{2n}.$$

Clearly $u = Tf$ and $u_n = T_n f = \pi_{1n}^a Tf$. Note that $(\pi_{1n}^a T)^\times = T^\times \pi_{2n}^a$. By hypothesis $\pi_{1n}^a \xrightarrow{\text{p}} 1$ in H_1 (Exercise 4.81) and if T is compact in H_1, then $\|(1 - \pi_{1n}^a)T\|_1 \to 0$.

Exercise

4.81 Prove that π_{1n}^a is well defined and satisfies $\|(1 - \pi_{1n}^a)u\| \le c\delta_{1n}(u)$ for $u \in H_1$. What can be proved for π_{2n}^a?

Example 4.27 (A nonconforming f.e.m.) We present a situation where the finite-element space V_n is *not* a subspace of the working space V. Let $G \subset H$ be two complex Hilbert spaces with norms $\|\cdot\|_G$ and $\|\cdot\|_H$, respectively. The injection of G into H is continuous but not necessarily compact. Let V be a closed subspace of G, and let a and b be bounded sesquilinear forms on $G \times G$ and $H \times H$, respectively.

We consider the equation for $f \in H$.

$$a(u, v) = b(f, v) \qquad \text{for all} \quad v \text{ in } V. \tag{4.53}$$

To construct its approximation, we are given a family $\{V_n\}$ of finite-dimensional subspaces of G: The equation

$$a(u_n, v_n) = b(f, v_n) \qquad \text{for all} \quad v_n \text{ in } V_n \tag{4.54}$$

is a nonconforming approximation when $V_n \not\subset V$. Concerning the form a, we further assume that

$$\mathcal{Re}\, a(u, u) \ge \alpha\|u\|_G^2, \qquad u \in V, \qquad \alpha > 0,$$
$$\mathcal{Re}\, a(u_n, v_n) > 0 \qquad \text{for all} \quad u_n \ne 0 \text{ in } V_n \text{ and all } n.$$

(4.53) defines $A: H \to V$, which satisfies $a(Af, v) = b(f, v)$ for all v in V and $u = Af$. Similarly (4.54) defines $A_n: H \to V_n$, $u_n = A_n f$.

The form a being coercive on V_n, the projection π_n^a from G onto V_n is defined, but it cannot be proved to be uniformly bounded because of the lack of *uniform* coercivity on V_n. The convergence $A_n \to A$, if it takes place, has to be proved by other means than the use of the projection π_n^a. Note that if a is coercive on G, there is no problem in working with π_n^a.

10.4. References and Comments

The use of the uniform convergence $\|T_n - T\| \to 0$ to study the conforming f.e.m. when the injection of V into H is compact first appears in Bramble and Osborn (1973). The case of a bounded but noncompact injection is treated in Rappaz (1977), Descloux *et al.* (1977, 1978a,b), and Mills (1979a,b) by means of a discrete-compact convergence.

When the sesquilinear form $a(\cdot, \cdot)$ is not coercive, the condition (4.52) is due to Babuška (1973) and Brezzi (1974). The abstract setting introduced in Example 4.26 is used in Kolata (1978, 1979) and in Mercier *et al.* (1981) with an assumption of compactness on T. It is extended to reflexive Banach spaces in Mills (1979c).

Discrete compactness properties are proved for the analysis of mixed f.e.m. (Kikuchi, 1980a,b) as well as nonconforming and hybrid f.e.m. (Stummel, 1980).

In the presentation we have given, an emphasis has been put on the elliptic projection and on the operator convergence. This setting will be used in subsequent chapters to study the spectral convergence and to establish optimal orders of convergence on the eigenelements. Usually the analysis of the convergence of the solution is done by estimating directly $\|u - u_n\|_V$ (see, among many others, Strang and Fix, 1973; Ciarlet, 1978).

From a computation point of view, the MODULEF Library (Begis and Perronnet, 1982) provides FORTRAN modules to apply the f.e.m. to solve bidimensional and tridimensional problems in heat transfer, elasticity, and fluid mechanics. For the effect of numerical integration, the reader is referred to Ciarlet (1978, Chap. 4), where it is said that "most classical finite-difference schemes can be exactly interpreted as finite-element methods with specific finite-element spaces and specific quadrature schemes." As a result, the possible uniform convergence becomes collectively compact (cf. Section 11). Other projection methods, such as least-squares and collocation for partial differential equations are presented briefly in Prenter (1975).

11. Finite-Difference Methods

It is beyond the scope of this book to give a comprehensive treatment of the finite-difference methods. We have chosen to present only a few significant features of these methods, without proofs, but with appropriate references.

11.1. Numerical Differentiation

For a smooth real function x of the real variable t, $x \in C^p$ for example, consider a numerical differentiation formula

$$x^{(k)}(t) \approx (D_h^{(k)}x)(t) := h^{-k} \sum_{i=-r_k}^{s_k} a_{ik}\, x(t + ih)$$

where $a_{ik} \in \mathbb{R}$, $r_k + s_k \geq k$. Using the shift operator S_h^i defined by

$$(S_h^i x)(t) = x(t + ih),$$

we get $D_h^{(k)} = h^{-k} \sum_{i=-r_k}^{s_k} a_{ik} S_h^i$. This numerical differentiation formula is convergent iff $D_h^{(k)}x(t) \to x(t)$ at each t for $x \in C^p$, as $h \to 0$. Simple examples are provided by

$$\partial_h := h^{-1}(S_h^1 - I) \qquad \text{and} \qquad \bar{\partial}_h := h^{-1}(I - S_h^{-1}).$$

It is standard that the function $z \mapsto \chi(z) = \sum_{i=-s_k}^{r_k} a_{ik} z^k$ can be written

$$\chi(z) = z^{-s_k}(z - 1)^k p(z),$$

where $p(z)$ is a polynomial such that $p(1) = 1$. The zeros of $p(z)$ are called the characteristic values of the divided difference $D_h^{(k)}$.

The finite-difference method consists in replacing the derivatives by some convergent numerical differentiation formulas. The differential operator T is then approximated by the matrix $A_h \colon \mathbb{C}^n \to \mathbb{C}^n$, n being the number of grid points.

11.2. Ordinary Differential Equations

We consider the operator associated with the ordinary differential equation (4.33) subjected to (4.34). It is shown in Vainikko (1969b) and Kreiss (1972) that some difference schemes yield a discrete-compact convergence of the resolvents. An interesting feature of the method of finite differences is its connection with the Fredholm method on the integral formulation (Vainikko, 1969c).

Example 4.28 We are concerned with the solutions of the equation

$$u'' = f(t, u), \qquad 0 \leq t \leq 1, \tag{4.55}$$

satisfying the boundary conditions

$$u(0) = u(1) = 0. \tag{4.56}$$

The problem (4.55), (4.56) is equivalent to the integral equation

$$u(t) = \int_0^1 g(t, s)f(s, u(s))\,ds, \tag{4.57}$$

where

$$g(t, s) = \begin{cases} t(s-1) & \text{if } t \le s \\ s(t-1) & \text{if } t \ge s, \end{cases}$$

is the Green function of the differential problem

$$u'', \qquad u(0) = u(1) = 0.$$

$[0, 1]$ is divided into n intervals at the points $t_i = i/n$, $i = 0, \ldots, n$, $h := 1/n$, $u(t_i) := u_i$. The boundary-value problem is replaced by the finite-difference problem

$$\begin{aligned} \frac{u_{i-1,n} - 2u_{in} + u_{i+1,n}}{h^2} &= f(t_i, u_{in}), \qquad i = 1, \ldots, n-1, \\ u_{0n} = u_{nn} &= 0, \end{aligned} \tag{4.58}$$

where u_{in} are the values at $\{t_i\}_0^n$ of the approximate solution u_n.

Exercise

4.82 The system of linear equations

$$\frac{u_{i-1} - 2u_i + u_{i+1}}{h^2} = y_i, \qquad i = 1, \ldots, n-1,$$

$$u_0 = u_n = 0$$

has a unique solution

$$u_i = \frac{1}{n}\sum_{j=1}^{n-1} g(t_i, t_j)y_j, \qquad i = 0, \ldots, n,$$

for any right-hand side $\{y_i\}_1^{n-1}$.

The system (4.58) is therefore equivalent (Exercise 4.82) to

$$u_{in} = \frac{1}{n}\sum_{j=1}^{n-1} g(t_i, t_j)f(t_j, u_{jn}), \qquad i = 0, \ldots, n, \tag{4.59}$$

and (4.59) can be interpreted as the result of the Fredholm discretization applied to (4.57), using the rectangle quadrature rule on $[0, 1]$:

$$I_n(x) = \frac{1}{n}\sum_{j=1}^{n-1} x\left(\frac{j}{n}\right).$$

This explains why the convergence of the finite-difference method may correspond to a discrete-compact convergence on the resolvents. □

We go back to the equation (4.33)–(4.34) approximated by a convergent numerical differentiation process. Vainikko (1975, 1976a,b) showed that, if the characteristic values of $D_h^{(p)}$ are of modulus not equal to 1, the resulting approximation is discretely regularly convergent.

Example 4.29 (Anselone and Ansorge) Let $X := \mathring{C}^2[0, 1] = \{u \in C^2(0, 1), u(0) = u(1) = 0, u'' \text{ is equimonotone (or equitone)}\}$ with

$$\|u\|_X := \|u\|_\infty + \|u'\|_\infty + \|u''\|_\infty,$$
$$Y := C^2(0, 1) \qquad \text{with} \quad \|v\|_Y := \|v\|_\infty$$

Thus, $u_n \to u$ in $X \Leftrightarrow u_n \to u$, $u_n' \to u'$, $u_n'' \to u''$ in Y.

Define $A: X \to Y$ by $(Au)(t) := u''(t) + a(t)u'(t) + b(t)u(t)$ with $a, b \in Y$. To discretize A, $[0, 1]$ is divided into n intervals at the points $t_i = i/n$, $i = 0, \ldots, n$. $h := 1/n$.

$A_n: X \to Y$ is defined by

$$(A_n u)(t) = (\bar{\partial}\partial u)(t) + a(t)\frac{u(t+h) - u(t-h)}{2h} + b(t)u(t)$$
$$\text{for} \quad h \le t \le 1 - h,$$
$$\text{with} \quad (\bar{\partial}\partial u)(t) := \frac{u(t+h) - 2u(t) + u(t-h)}{h^2},$$
$$(A_n u)(t) = (A_n u)(h) \qquad \text{for} \quad 0 \le t \le h,$$
$$(A_n u)(t) = (A_n u)(1-h) \qquad \text{for} \quad 1 - h \le t \le 1.$$

We want to show that $A_n \xrightarrow{r} A$. We first note that $A_n \xrightarrow{p} A$ as $n \to \infty$. Let $\{u_n\}_{\mathbb{N}}$ be a bounded sequence in X such that $A_n u_n \to v$ in Y. Then $\{u_n\}_{\mathbb{N}}$, $\{u_n'\}_{\mathbb{N}}$, and $\{u_n''\}_{\mathbb{N}}$ are bounded in Y. It follows that $\{u_n\}$ and $\{u_n'\}$ are equicontinuous, hence relatively compact in Y by the Ascoli–Arzela theorem. Since $A_n u_n \to v$ in Y,

$$\{\bar{\partial}\partial u_n\} = \{A_n u_n - a u_n' - b u_n + \varepsilon(n)\} \qquad \text{with} \quad \varepsilon(n) \xrightarrow[n \to \infty]{} 0$$

is relatively compact in Y. Hence there exist $N_1 \subset \mathbb{N}$ and $u, x, y \in Y$ such that $u_n \to u$, $u_n' \to x$, $\bar{\partial}\partial u_n \to y$ in Y for $n \in \mathbb{N}_1$. Because differentiation is a closed operator, $x = u'$ and $y = x' = u''$. Hence for $n \in N_1$,

$$u_n \to u \text{ in } Y, \ \{u_n\} \subset X \Rightarrow u(0) = u(1) = 0.$$

Since $\bar{\partial}\partial u_n$ is continuous and u_n'' is monotone, there exists $\delta_n(t) \in [-1/n, 1/n]$ such that $(\bar{\partial}\partial u_n)(t + \delta_n) = u_n''(t)$, $n \in N_1$. Now $\{\bar{\partial}\partial u_n\}$ equicontinuous and

$y = u''$ uniformly continuous imply $u_n'' \to u$ in Y. The equimonotonicity of $\{u_n''\}$ yields the same monotonicity for u''. Therefore

$$\left.\begin{array}{l} u_n \to u,\ u_n' \to u',\ u_n'' \to u'' \quad \text{in } \ Y \\ u'' \text{ monotone}, u(0) = u(1) = 0 \end{array}\right\} \Rightarrow u_n \to u \quad \text{in} \quad X.$$

Together with $Au = v$, this proves that $A_n \xrightarrow{\mathrm{r}} A$.

The analysis of this example given in Vainikko (1976b, pp. 101–104) is the discrete analog of the above presentation.

In this context the notion of (discrete) regular convergence is a very useful tool since it allows work directly on the differential operator. On the contrary, the (discrete) compact convergence requires work on the inverse.

11.3. Elliptic Partial Differential Equations

The convergence of the finite-difference method for p.d.e. and for various schemes has been studied in Grigorieff and Jeggle (1973), Vainikko and Tamme (1976) and in Vainikko (1976a,b, 1978a,b) again by means of the notion of *discrete-regular convergence.*

12. Approximation of a Differential Operator by a Neighboring Operator

The approximation of the differential operator T associated with the ordinary differential equation (4.33)–(4.34) by a finite-rank operator does not preserve the asymptotic distribution of the eigenvalues. We end this chapter by the presentation of a method that allows to approximate this asymptotic distribution: T is approximated, not by a finite-rank operator, but by a neighboring differential operator, according to the following definition.

Definition Let T be the differential operator associated with

$$\mathfrak{L}u := u^{(p)} - \sum_{i=0}^{p-1} a_i(\cdot)u^{(i)}$$

on $[0, 1]$, with p homogeneous boundary conditions. T_n is a *neighboring differential operator* iff it is associated with

$$\mathfrak{L}_n u := u^{(p)} - \sum_{i=0}^{p-1} a_i^{(n)}(\cdot)u^{(i)}$$

on [0, 1] subjected to the same boundary conditions, and such that

$$\|a_i - a_i^{(n)}\|_\infty := \max_{t \in [0,1]} |a_i(t) - a_i^{(n)}(t)| \underset{n \to \infty}{\longrightarrow} 0, \qquad i = 0, \ldots, p-1.$$

In Canosa and Gomes de Oliveira (1970) and in Day (1974), the coefficient functions a_i are approximated by step functions, whereas in Pruess (1973a,b) and Smooke (1978, 1980a,b) they are approximated by piecewise polynomial functions.

Example 4.30 We consider a second-order differential equation on [0, 1]:

$$X = C(0,1), \qquad D = \{u \in X; u'' \in X \text{ and } u(0) = u(1) = 0\},$$
$$T: u \in D \mapsto u'' - a_1 u' - a_0 u,$$
$$T_n: u \in D \mapsto u'' - a_1^{(n)} u' - a_0^{(n)} u,$$

where $a_i, a_i^{(n)} \in X$ for $i = 0$ and 1.

We prove that the condition $\|a_i - a_i^{(n)}\|_\infty \to 0$ for $i = 0, 1$ implies that $\|(T - T_n)R(z)\|_\infty \to 0$ in X for any z in $\rho(T)$. First, it can be shown (Kato, 1976, pp. 192–194) that $T - T_n$ is T-bounded:

$$\|(T - T_n)u\|_\infty \leq \alpha_n \|u\|_\infty + \beta_n \|Tu\|_\infty \qquad \text{for} \quad u \in D.$$

Now from $\|a_i - a_i^{(n)}\|_\infty \to 0$ follows that

$$\alpha_n \to 0, \qquad \beta_n \to 0 \qquad \text{as} \quad n \to \infty.$$

For $z \in \rho(T)$ and for $v \in X$,

$$\begin{aligned}\|(T - T_n)R(z)v\|_\infty &\leq \alpha_n \|R(z)v\|_\infty + \beta_n \|(1 + zR(z))v\|_\infty \\ &\leq [(\alpha_n + |z|\beta_n)\|R(z)\|_\infty + \beta_n]\|v\|_\infty.\end{aligned}$$

Therefore $\|(T - T_n)R(z)\|_\infty \to 0$. The convergence $T_n \to T$ is a particular case of uniform radial convergence such that $\|R_n(z) - R(z)\|_\infty \to 0$ for $z \in \rho(T)$.

CHAPTER 5

Spectral Approximation of a Closed Linear Operator

Introduction

Let X be a Banach space on $\mathbb{C}$, and let T be a closed linear operator in X, with domain $D = \text{Dom } T$. Let λ be an *isolated eigenvalue* of T, of *finite algebraic multiplicity* $m < \infty$. There exists a closed Jordan curve Γ around λ and isolating λ that is such that the domain Δ enclosed by Γ contains no other point of the spectrum $\sigma(T)$ than λ. $P := (-1/2i\pi) \int_\Gamma R(z)\, dz$ is the spectral projection associated with λ; $M := PX$ is the invariant subspace; $m := \dim PX$.

Let T_n, $n = 1, 2, \ldots$, be a sequence of closed operators, with domain $\text{Dom}(T_n) = D$. If Γ lies in $\rho(T_n)$, we may define for T_n the resolvent $R_n(z) := (T_n - z)^{-1}$ for $z \in \Gamma$, and the spectral projection $P_n := (-1/2i\pi) \int_\Gamma R_n(z)\, dz$. $\sigma(T_n) \cap \Delta$ is the spectrum of T_n inside Δ. $M_n := P_n X$ is the invariant subspace of T_n associated with $\sigma(T_n) \cap \Delta$.

Definition The sequence $\{T_n\}_{\mathbb{N}}$ is an *approximation* of T iff

$$\text{for all} \quad x \in D, \qquad T_n x \to Tx, \qquad \text{that is} \quad T_n \xrightarrow{\text{p}} T \quad \text{as} \quad n \to \infty. \tag{5.1}$$

The analysis of convergence of the eigenelements of T_n toward those of T will be made by means of the study of the convergence of $\sigma(T_n) \cap \Delta$ to $\{\lambda\}$ for the eigenvalues, and of the convergence of M_n to M for the invariant subspaces.

The key notion will be the stable convergence for z in $\Delta - \{\lambda\}$ (which implies that $T_n - z \xrightarrow{\text{s}} T - z$, uniformly in z in any compact of $\Delta - \{\lambda\}$)

together with the preservation of the algebraic multiplicities. This leads to the notion of strongly stable convergence in a neighborhood of λ in $Q\sigma(T)$, which can be characterized, in the space $\mathscr{L}(\hat{D}, X)$, by the regular convergence $T_n - z \xrightarrow{r} T - z$ for $z \in \Delta$.

1. Convergence of the Spectrum $\sigma(T_n) \cap \Delta$

1.1. Definitions

We recall the definitions of $\underline{\lim}$, $\overline{\lim}$, and lim for a sequence of sets of points in $\mathbb{C}$, in the sense of Kuratowski (1961) or Hahn (1948). Let $E, E_n \subset \mathbb{C}$, $n \in \mathbb{N}$.

(a) $E := \overline{\lim}\, E_n$ iff, for any $x \in E$, there exists an infinite subset $N_1 \subset \mathbb{N}$ such that for any n in N_1 and x_n in E_n, then $x_n \to x$.

(b) $E := \underline{\lim}\, E_n$ iff, for any $x \in E$, there exists $x_n \in E_n$ such that $x_n \to x$, $n \in \mathbb{N}$.

(c) $E := \lim E_n$ iff $\overline{\lim}\, E_n = \underline{\lim}\, E_n = \lim E_n$.

Let λ be an eigenvalue of T, isolated by Γ; Δ is the domain enclosed by Γ, and $\sigma(T) \cap \Delta = \{\lambda\}$, cf. Fig. 5.1.

Definition The spectrum of T_n in Δ *converges* to λ iff

$$\lim_n [\sigma(T_n) \cap \Delta] = \{\lambda\}. \tag{5.2}$$

The generalization of the definition to the case where Δ contains a finite number of isolated eigenvalues is straightforward. Condition (5.2) expresses the *continuity* of $\sigma(T)$ in the neighborhood Δ of λ when T is subjected to the perturbation $T - T_n$: $T_n = T - (T - T_n)$. This continuity is the union of the lower semicontinuity

$$\{\lambda\} \subset \underline{\lim}[\sigma(T_n) \cap \Delta]$$

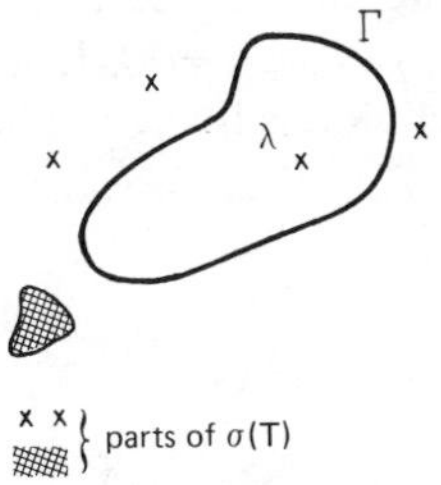

Figure 5.1

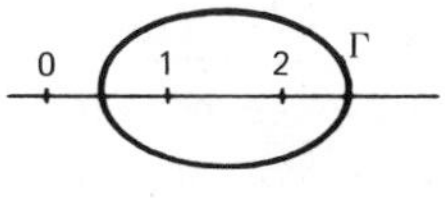

Figure 5.2

and of the upper semicontinuity

$$\{\lambda\} \supset \overline{\lim}[\sigma(T_n) \cap \Delta].$$

The lower semicontinuity means that in any neighborhood of λ, there exists a point of $\sigma(T_n)$ for n large enough; in particular $\sigma(T_n) \cap \Delta$ is not empty. The upper semicontinuity means that λ is the only possible limit of any subsequence of points in $\sigma(T_n) \cap \Delta$.

Condition (5.1) on $T - T_n$ is not sufficient, in general, to imply the continuity (5.2), as we see in the following examples.

Examples of discontinuities of $\sigma(T)$

Example 5.1 (Upper semidiscontinuity) A sequence of eigenvalues of T_n converges to a point in $\rho(T)$. Let $X = l^2$ with the canonical basis $\{e_i\}_{\mathbb{N}}$. $T: x = \sum_{i=1}^{\infty} x_i e_i \mapsto Tx = x_1 e_1$ is a projection on $\{e_1\}$. $\sigma(T) = \{0, 1\}$, where 1 is a simple eigenvalue, isolated by any curve not enclosing 0, cf. Fig. 5.2. We define

$$T_n: x = \sum_{i=1}^{\infty} x_i e_i \mapsto x_1 e_1 + 2x_n e_n, \qquad n = 1, 2, \ldots, \quad \sigma(T_n) = \{0, 1, 2\}.$$

$T_n x \to Tx$ for $x \in X$, and 2 is a point in $\rho(T)$.

Example 5.2 (Lower semidiscontinuity) There exists an eigenvalue of T that is not the limit of any subsequence of eigenvalues of T_n. Let $X = C(0, 1)$, $D = \{x \in X; x' \in X \text{ and } x(1) = x(0)\}$, $T: x \in D \mapsto x' + x$. Then $\sigma(T) = \{1 + 2ik\pi, k \in \mathbb{Z}, i^2 = -1\}$ consists of simple eigenvalues. T is approximated with the following finite difference scheme: $[0, 1]$ is divided into n equal subintervals at $t_i = i/n$, $x(t_i) := x_i$, $i = 0, 1, \ldots, n$. The first derivative is approximated by the *right* divided difference at $t_0, t_2, \ldots, t_{n-1}$ and by the *left* divided difference at t_1. The discrete eigenvalue problem is then

$$x_0 = x_n,$$

$$n(x_1 - x_0) + x_0 = \lambda x_0,$$

$$-n(x_0 - x_1) + x_1 = \lambda x_1 = n(x_1 - x_0) + x_1,$$

$$n(x_{i+1} - x_i) + x_i = \lambda x_i, \qquad i = 2, 3, \ldots, n - 1.$$

This is an $n \times n$ matrix eigenvalue problem: $A_n\xi = \lambda\xi$ with

$$\xi = (x_0, x_1, \ldots, x_{n-1})^{\mathrm{T}}$$

and

$$A_n = \begin{pmatrix} 1-n & n & 0 & & & 0 \\ -n & n+1 & 0 & & & \\ 0 & 0 & 1-n & & n & \\ \vdots & & & \ddots & & \ddots \\ 0 & & 0 & & 1-n & n \\ n & & & & & 1-n \end{pmatrix}$$

If we use a piecewise-linear interpolation, we see that there corresponds to A_n a linear operator T_n with domain $\mathrm{Dom}(T_n) = D$ and such that for $x \in D$, $T_n x \to Tx$. The nonzero eigenvalues of T_n and A_n are $\{1, 1 - n\}$, where 1 has multiplicity 2, and $1 - n$ has multiplicity $n - 2$. The eigenvalues of T, other than 1, are not reached by the approximation T_n.

We now give a sufficient condition for the convergence (5.2).

1.2. *Sufficient Condition for the Convergence of* $\sigma(T_n) \cap \Delta$

Let z be fixed in $\rho(T)$. We suppose that $T_n - z \xrightarrow{s} T - z$. This means that (5.1) is fulfilled, as well as

$$\exists N(z)\colon n > N(z), \qquad z \in \rho(T_n), \qquad \text{and} \qquad \|R_n(z)\| \le M(z). \tag{5.3}$$

We say that the approximation $\{T_n\}_{\mathbb{N}}$ is stable at $z \in \rho(T)$. The pointwise convergence $R_n(z) \xrightarrow{p} R(z)$ follows by Lemma 3.16

Proposition 5.1 *Under condition* (5.3), *the function* $z \mapsto R_n(z)$ *is continuous for* z *in* $\rho(T)$, *uniformly in* n *for* n *large enough.*

Proof Let $z_0 \in \rho(T)$,

$$T_n - z = T_n - z_0 + z_0 - z = (T_n - z_0)[1 - (z - z_0)R_n(z_0)].$$

Formally $R_n(z) = (\sum_{k=0}^{\infty}[(z - z_0)R_n(z_0)]^k)R_n(z_0)$. The series is absolutely convergent if $|z - z_0| < \|R_n(z_0)\|^{-1}$. From (5.3), $\|R_n(z_0)\| < M(z_0)$ for $n > N(z_0)$. Then for any ε, $0 < \varepsilon < 1$, there exists $N(z_0)$ such that for $n > N(z_0)$, $|z - z_0| < \varepsilon/M(z_0)$ implies

$$\|R_n(z) - R_n(z_0)\| \le \|R_n(z_0)\| \sum_{k=1}^{\infty} \varepsilon^k \le M(z_0)\frac{\varepsilon}{1-\varepsilon}. \quad \square$$

Corollary 5.2 *Let K be a compact subset of $\rho(T)$. If (5.3) is fulfilled for all $z \in K$, then*

$$\sup_{z \in K,\, n > N(K)} \|R_n(z)\| \leq M(K).$$

Proof For any $\varepsilon > 0$, the compact set K has a finite cover with sets of diameter less than ε. Hence for each z in K, there exists z_ε such that $|z - z_\varepsilon| < \varepsilon$. There is a finite number of such z_ε and

$$\|R_n(z)\| \leq \|R_n(z) - R_n(z_\varepsilon)\| + \|R_n(z_\varepsilon)\|.$$

Apply the proof of Proposition 5.1 to conclude. □

Note that, if Γ is a Jordan curve around λ, and if (5.3) is fulfilled on Γ, then for $n > N(\Gamma)$, Γ lies in $\rho(T_n)$. The spectral projection P_n associated with

$$\sigma(T_n) \cap \Delta$$

is well defined for all $n > N(\Gamma)$.

1.2.1. Pointwise convergence of the spectral projection P_n

Γ, the closed Jordan curve around λ, is a compact subset of $\rho(T)$, and we suppose that, for all z in Γ, $T_n - z \xrightarrow{s} T - z$.

Proposition 5.3 *If for all z on Γ, $T_n - z \xrightarrow{s} T - z$, then $P_n \xrightarrow{p} P$.*

Proof It is clear from Corollary 5.2 by integration on Γ of $R_n(z) - R(z) = R_n(z)(T - T_n)R(z)$. $(T - T_n)R(z) \in \mathscr{L}(X)$ and for x fixed, the function $z \mapsto (T - T_n)R(z)x$ is continuous for $z \in \rho(T)$. Hence for all $x \in X$ and for $n > N(\Gamma)$,

$$\|(P - P_n)x\| \leq \frac{\text{meas}\,\Gamma}{2\pi} M(\Gamma) \max_{z \in \Gamma}\|(T - T_n)R(z)x\|. \quad \square$$

Corollary 5.4 *If, for all z on Γ, $T_n - z \xrightarrow{s} T - z$, then* $\dim P_n X \geq \dim PX$ *for n large enough.*

Proof Clear from Proposition 3.10 since $\dim PX = m < \infty$. □

This corollary means that if $\{T_n\}$ is stable for all z on Γ, $\sigma(T_n) \cap \Delta$ is not empty for $n > N(\Gamma)$.

1.2.2. Convergence of $\sigma(T_n) \cap \Delta$

We now prove the convergence of $\sigma(T_n) \cap \Delta$ under the stronger hypothesis that $\{T_n\}_{\mathbb{N}}$ is stable in $\Delta - \{\lambda\}$.

Theorem 5.5 *If for all z in $\Delta - \{\lambda\}$, $T_n - z \xrightarrow{s} T - z$, then*

$$\lim_n [\sigma(T_n) \cap \Delta] = \{\lambda\}.$$

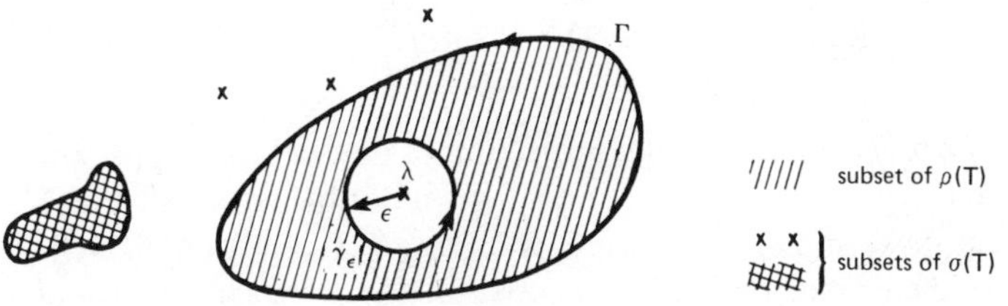

Figure 5.3

Proof Let $\varepsilon > 0$ be given. $\Delta_\varepsilon := \Delta - \{z; |z - \lambda| < \varepsilon\}$ and the circle $\gamma_\varepsilon := \{z; |z - \lambda| = \varepsilon\}$, which isolates λ (cf. Fig. 5.3.), are compact subsets of $\rho(T)$. By hypothesis, for any $\varepsilon > 0$, there exists $N(\varepsilon)$ such that

$$\sup_{z \in \Delta_\varepsilon, n > N(\varepsilon)} \|R_n(z)\| \leq M(\varepsilon).$$

We define

$$P_\varepsilon := \frac{-1}{2i\pi} \int_{\gamma_\varepsilon} R(z)\, dz$$

and $P_{n\varepsilon}$ accordingly. Note that $P_\varepsilon = P$. By Corollary 5.4., $\dim P_{n\varepsilon}X \geq \dim P_\varepsilon X = m$ for all $n > N(\varepsilon)$. This proves the lower semicontinuity of $\sigma(T)$. Now the stability of T_n in Δ_ε shows that $\Delta_\varepsilon \subset \rho(T_n)$ for $n > N(\varepsilon)$. This is the upper semicontinuity of $\sigma(T)$. □

Remark The assumption that λ is of *finite* multiplicity has been made for simplicity, since we are interested mainly in the numerical approximation of points in $Q\sigma(T)$. But it should be noted that $T_n - z \xrightarrow{s} T - z$ for z in $V - D$, V being a neighborhood of D, an arbitrary nonempty compact subset of $\sigma(T)$, implies that $\underline{\lim}_n \operatorname{dist}(\sigma(T_n), D) = 0$ [see the condition called resolvent stability around D in Mills (1979a)].

The stability of the approximation $\{T_n\}$ is only a *sufficient* condition for the convergence of $\sigma(T_n) \cap \Delta$, as shown by the following example, where the approximation $\{T_n\}$ is not stable at the point -1 in $\rho(T)$, but nevertheless the spectrum of T_n converges.

Example 5.3 Let $X = l^2$, with the canonical basis $\{e_i\}_{\mathbb{N}}$. T is the orthogonal projection on $e_1: x = \sum_{i=1}^{\infty} x_i e_i \mapsto Tx = x_1 e_1$.

T is approximated by T_n, defined by $x \mapsto T_n x = x_1 e_1 + \sum_{i=n}^{2n-1} x_{i+1} e_i$, $\sigma(T_n) = \sigma(T) = \{0, 1\}$, and 1 is a simple eigenvalue for both T and T_n.

$$\|(T - T_n)x\| \leq \left\| \sum_{i \geq n} x_{i+1} e_i \right\| \to 0$$

when $n \to \infty$. In the basis $\{e_i\}_1^{2n}$, the matrix A_n associated with T_n is defined by $a_{ij} = 0$ for $i, j = 1, \ldots, 2n$, except for $a_{11} = 1$, $a_{i-1,i} = 1$, $i = n + 2, \ldots, 2n$.

It is easy to check that $\|(T_n + 1)^{-1}\| \geq (n/2)^{1/2}$. So $\{T_n\}$ is not stable at the point -1 in $\rho(T)$.

Example 5.4 Let T be the projection of Example 5.3, approximated now by $T_n: x = \sum_{i=1}^{\infty} x_i e_i \mapsto x_1 e_1 + x_n e_n$. 1 is a double eigenvalue of T_n with the eigenvectors e_1 and e_n. $2 = \dim P_n X > \dim PX = 1$ for all n. When $n \to \infty$, the second eigenvector e_n weakly converges to zero. The spectra of T and T_n are identical, but the multiplicities of 1 differ, causing the nonconvergence of the eigenvectors.

Example 5.4 shows that the convergence of the eigenvalues when considered merely as complex numbers, is not enough. To get the convergence of the eigenvectors, we need the convergence of the eigenvalues with *preservation of the algebraic multiplicities.*

2. Convergence of the Eigenvalues with Preservation of the Multiplicities

For an eigenvalue λ of algebraic multiplicity $m < \infty$, we define the *strongly stable* convergence $T_n - z \overset{\text{ss}}{\to} T - z$ in Δ by the conditions

(i) $T_n - z \overset{\text{s}}{\to} T - z$ for z in $\Delta - \{\lambda\}$,
(ii) $\dim P_n X = m$ for n large enough,

where P_n is the spectral projection of T_n associated with $\sigma(T_n) \cap \Delta$. Similarly $T_n - z \overset{\text{ss}}{\to} T - z$ on Γ if (i) and (ii) hold, but (i) holds for z on Γ only. If $T_n - z \overset{\text{ss}}{\to} T - z$ in Δ (resp. on Γ), it is said that $\{T_n\}_{\mathbb{N}}$ is a strongly stable approximation of T in Δ(resp. on Γ). The strong stability implies the convergence of the eigenvalues with preservation of the multiplicities, as we see now.

Proposition 5.6 *If $T_n - z \overset{\text{ss}}{\to} T - z$ for z in Δ, then $\sigma(T_n) \cap \Delta$ consists, for n large enough, of exactly m eigenvalues, counting their multiplicities.*

Proof The stability of T_n in $\Delta - \{\lambda\}$ implies that $\sigma(T_n) \cap \Delta$ is isolated by Γ for $n > N(\Gamma)$. Hence $\sigma(T_n)$ is separated according to the definition of Chapter 2, Section 7.6; and by Theorem 2.27, $\sigma(T_n) \cap \Delta$ is the spectrum of $P_n T_n P_{n \restriction M_n}$, the part of T_n in $M_n = P_n X$, which has rank $m = \dim P_n X$ for n large enough. Its spectrum consists of m eigenvalues, counting their multiplicities. □

The strong stability of $\{T_n\}_{\mathbb{N}}$ in Δ is then a sufficient condition for the convergence of the eigenvalues in Δ with preservation of the algebraic multiplicities (cf. Example 5.3).

3. Convergence of the Eigenvectors and Invariant Subspaces

Let $\lambda \in Q\sigma(T)$ be of algebraic multiplicity m. We suppose in this section that $T_n - z \overset{\text{ss}}{\to} T - z$ on Γ, a Jordan curve isolating λ: for n large enough, there are exactly m repeated eigenvalues of T_n inside Γ. K_n is the number of *distinct* eigenvalues λ_{in} of T_n inside Γ.

3.1. Convergence of a Sequence of Subspaces of X

Let $G_n, n = 1, 2, \ldots$, be a sequence of subspaces of X converging in some sense to the subspace G. We suppose that $\dim G < \infty$.

We define the unit sphere of G_n (resp. G) by

$$B_{G_n} := \{x \in G_n;\ \|x\| \leq 1\} \qquad (\text{resp.}\quad B_G := \{x \in G;\ \|x\| \leq 1\}.$$

We shall be concerned with two conditions:

(1) $\forall x \in B_G, \exists x_n \in B_{G_n}$ such that $x_n \to x$. (5.4)

(2) Any sequence $\{x_n\}_{\mathbb{N}}$, $x_n \in B_{G_n}$, has a converging subsequence $\exists x \in B_G$ such that $x_n \to x$ for $n \in N_1 \subset \mathbb{N}$. (5.5)

Lemma 5.7 *Under the hypothesis that* $\dim G < \infty$, *condition* (5.4) *implies* $\dim G_n \geq \dim G$, *and condition* (5.5) *implies* $\dim G_n \leq \dim G$.

Proof Because G is finite dimensional, it has a supplementary subspace G': $X = G \oplus G'$. We set $G^* := G'^{\perp} \subset X^*$, and $\alpha := \dim G$. Let $\{x_i\}_1^\alpha$ be a normalized basis of G, and let $\{x_i^*\}_1^\alpha$ be the adjoint basis of G^*: $\langle x_i^*, x_j\rangle = \delta_{ij}$ for $i, j = 1, \ldots, \alpha$. By (5.4), there exists $x_{in} \in B_{G_n}$ such that $x_{in} \to x_i$ for $i = 1, \ldots, \alpha$. Therefore $\langle x_j^*, x_{in}\rangle \to \delta_{ij}$. For n large enough, the α vectors $\{x_{in}\}_1^\alpha$ are independent in G_n. The second implication is clear by Lemma 3.12 applied to G and G_n. □

3.2. Gap Convergence of the Invariant Subspaces

We consider the invariant subspace $M := PX$ (resp. $M_n := P_n X = \bigoplus_{i=1}^{K_n} M_{in}$) associated with the eigenvalue λ of T (resp. all the distinct eigenvalues of T_n inside Γ). ψ (resp. ψ_n) denotes any vector in M (resp. M_n). ψ_n is in general a linear combination of invariant vectors associated with different eigenvalues of T_n inside Γ.

Theorem 5.8 *If* $\{T_n\}$ *is an approximation of* T, *strongly stable on* Γ, *then* $P_n \overset{\text{cc}}{\to} P$ *and* $\Theta(M_n, M) \to 0$.

Proof This is an easy consequence of Propositions 5.3, 3.13, and 3.14. □

We remark that the gap between M and M_n tends to zero as $n \to \infty$, whereas all we know so far about the eigenvalues is that there exist m repeated eigenvalues of T_n inside Γ.

Exercises

5.1 If $T_n - z \xrightarrow{ss} T - z$ on Γ, show that the invariant subspaces M and M_n fulfill conditions (5.4) and (5.5).

5.2 Deduce from $\|(P - P_n)P_n\| \to 0$ that M and M_n fulfill (5.5).

5.3 Under the hypothesis of Exercise 5.1, show that for $\psi \in M$ and for $\psi_n \in M_n$ such that $\|\psi\| = 1$, $\text{dist}(\psi, M_n)$ and $\text{dist}(\psi_n, M) \to 0$ as $n \to \infty$, where $\text{dist}(x, M)$ is defined by $\inf_{y \in M}\|x - y\|$.

We have the following reciprocal of Exercise 5.1.

Theorem 5.9 *Let $\{T_n\}$ be an approximation of T, stable on Γ. We suppose that the invariant subspaces M and M_n fulfill condition* (5.5). *Then $\{T_n\}$ is strongly stable on Γ.*

The proof is left to the reader.

Exercise

5.4 We suppose that $\bar{D} = X$. If $\{T_n\}$ is a strongly stable approximation of T on Γ such that $T_n^* \xrightarrow{p} T^*$, then $\{T_n^*\}$ is also strongly stable on $\bar{\Gamma}$ around $\bar{\lambda}$, and $\|P_n - P\| \to 0$.

3.3. Convergence of the Eigenvectors

We suppose again that $T_n - z \xrightarrow{ss} T - z$ in Δ. Let $\{\lambda_n\}_{\mathbb{N}}$ be a sequence of eigenvalues of T_n converging to λ. To each λ_n is associated an eigenvector φ_n, $\|\varphi_n\| = 1$, $\varphi_n \in E_n := \text{Ker}(T_n - \lambda_n)$. Any vector φ in $E := \text{Ker}(T - \lambda)$ is an eigenvector associated with λ. We set $g := \dim E$ and $g_n := \dim E_n$. Note the difference between M_n and E_n: M_n is the invariant subspace associated with *all* the distinct eigenvalues in Δ, whereas E_n is the eigenspace associated with *one* individual eigenvalue λ_n.

Theorem 5.10 *Let $\{T_n\}_{\mathbb{N}}$ be an approximation of T, strongly stable in Δ. Then for any sequence of eigenvalues λ_n converging to λ and for any sequence of associated normalized eigenvectors $\{\varphi_n\}_{\mathbb{N}}$, there exists a subsequence converging to an eigenvector φ associated with λ. Moreover,*

$$\dim \text{Ker}(T_n - \lambda_n) \leq \dim \text{Ker}(T - \lambda).$$

Proof Let $\varphi_n \in E_n$, $\|\varphi_n\| = 1$. Since $P_n \xrightarrow{cc} P$, the bounded sequence $\{\varphi_n\}_{\mathbb{N}}$ in $E_n \subset M_n$ has a subsequence $\{\varphi_n\}_{N_1 \subset \mathbb{N}}$ converging to φ in M, $\|\varphi\| = 1$.

From the identity for $z \in \Gamma$ and $n \in N_1$,

$$R(z)\varphi - \frac{1}{\lambda - z}\varphi = (R(z) - R_n(z))\varphi + R_n(z)(\varphi - \varphi_n)$$
$$\lambda + \left(\frac{1}{\lambda_n - z} - \frac{1}{\lambda - z}\right)\varphi_n + \frac{1}{\lambda - z}(\varphi_n - \varphi),$$

follows that $R(z)\varphi = (\lambda - z)^{-1}\varphi$ and φ is an eigenvector in E. $g_n \leq g$ follows from Lemma 5.7. □

Example 5.5 Let $X = \mathbb{R}^2$, $\varepsilon \in \mathbb{R}$.

$$A_\varepsilon = \begin{pmatrix} 1 & 0 \\ \varepsilon & 1 \end{pmatrix} \underset{\varepsilon \to 0}{\longrightarrow} A = \begin{pmatrix} 1 & 0 \\ 0 & 1 \end{pmatrix}.$$

The eigenvector of A_ε is

$$e_2 = \begin{pmatrix} 0 \\ 1 \end{pmatrix},$$

which is an eigenvector for A; but any other vector in $\mathbb{R}^2$ is an eigenvector of A and is *not* the limit of any sequence of eigenvectors of A_ε.

Example 5.6 When $m > 1$, we may get only a converging *subsequence* of eigenvectors. Let $X = \mathbb{R}^2$, $\varepsilon \in \mathbb{R}$,

$$A = \begin{pmatrix} 1 + \varepsilon \cos 2/\varepsilon & -\varepsilon \sin 2/\varepsilon \\ -\varepsilon \sin 2/\varepsilon & 1 - \varepsilon \cos 2/\varepsilon \end{pmatrix} \underset{\varepsilon \to 0}{\longrightarrow} A = \begin{pmatrix} 1 & 0 \\ 0 & 1 \end{pmatrix}.$$

A_ε has the eigenvalues $1 \pm \varepsilon$ and eigenvectors $(\sin 1/\varepsilon, \cos 1/\varepsilon)^T$ and $(\cos 1/\varepsilon, -\sin 1/\varepsilon)^T$. These vectors have no limit when $\varepsilon \to 0$. The subsequence defined by $\varepsilon_k := 1/2k\pi$ converges to e_1 and e_2 when $k \to \infty$.

Exercise

5.5 We suppose that λ is simple and $T_n - z \overset{ss}{\to} T - z$ on Γ. Show that

(i) for any eigenvector φ, there exists an eigenvector φ_n such that $\varphi_n \to \varphi$;

(ii) for any eigenvector φ_n, there exists an eigendirection $\{\varphi\}$ (not depending on n) such that $\text{dist}(\varphi_n, \{\varphi\}) \to 0$ as $n \to \infty$.

Under the hypothesis of preservation of the *algebraic* multiplicities, we get, for the *geometric* multiplicities, that $g_n \leq g$ only, in general. As a consequence, (5.4) may not be satisfied for all eigenvectors in E (cf. Examples 5.5 and 5.6). If (5.4) is satisfied for all eigenvectors in E, then $g_n = g$ for n large enough. There is preservation of the geometric multiplicities.

In practice, a multiple eigenvalue is often enough approximated by a set of m simple eigenvalues, and so $g_n = 1$ is smaller than g when the exact eigenvalue has more than one eigenvector.

Example 5.7 We consider the sequence of $N \times N$ matrices $A(\varepsilon)$ defined in Example 1.6. $A(\varepsilon)$ has N simple eigenvalues $\lambda_k(\varepsilon) = 1 + \varepsilon^{1/N} e^{2ik\pi/N}$, $k = 0, 1, \ldots, N-1$, converging to the multiple eigenvalue $\lambda = 1$ of algebraic multiplicity N. λ and each $\lambda_k(\varepsilon)$ have geometric multiplicity equal to 1. Therefore, each eigenvector $\varphi_k(\varepsilon)$ associated with $\lambda_k(\varepsilon)$ converges to e_1, the only eigenvector associated with λ. Indeed, if we set $a_k(\varepsilon) := 1 - \lambda_k(\varepsilon)$, then

$$\varphi_k^{\mathrm{T}}(\varepsilon) = (1, a_k(\varepsilon), a_k^2(\varepsilon), \ldots, a_k^{N-1}(\varepsilon)) \underset{\varepsilon \to 0}{\longrightarrow} e_1^{\mathrm{T}}.$$

To summarize, the strong stability in Δ of the approximation $\{T_n\}$ guarantees

(a) the convergence of the eigenvalues in Δ with preservation of the algebraic multiplicities;

(b) the gap convergence of the invariant subspaces;

(c) for any sequence of eigenvalues such that $\lambda_n \to \lambda$, the convergence of a subsequence of the associated normalized eigenvectors, together with $\dim \operatorname{Ker}(T_n - \lambda_n) \le \dim \operatorname{Ker}(T - \lambda)$.

The hypothesis on the preservation of the algebraic multiplicities cannot be weakened without introducing a numerical instability. For example, for the eigenvectors, the convergence defined in Theorem 5.9 could not be guaranteed any longer (cf. Example 5.4).

3.4. T_n Is of Class $\mathfrak{D}$

We consider the case where the equation in X

$$T\varphi = \lambda\varphi, \qquad \varphi \neq 0, \tag{5.6}$$

is approximated in X_n by the discrete equation

$$\mathscr{T}_n \varphi_n = \lambda_n \varphi_n, \qquad \varphi_n \neq 0. \tag{5.7}$$

Associated with $\mathscr{T}_n$, we defined in Chapter 4, Section 5, the operators from X_n into itself: $\mathscr{R}_n(z) := (\mathscr{T}_n - z)^{-1}$ and

$$\mathscr{P}_n := \frac{-1}{2i\pi} \int_\Gamma \mathscr{R}_n(z)\, dz \qquad \text{for} \quad \Gamma \subset \rho(\mathscr{T}_n).$$

$\mathscr{M}_n := \mathscr{P}_n X_n$ is the invariant subspace of $\mathscr{T}_n$ associated with the set of the eigenvalues of $\mathscr{T}_n$ inside Γ.

We now consider $T_n = \mathscr{T}_n \pi_n$ and the corresponding equation in X:

$$T_n \varphi_n = \lambda_n \varphi_n, \qquad \varphi_n \neq 0. \tag{5.8}$$

(5.7) and (5.8) are equivalent if $\lambda_n \neq 0$. If $\lambda_n = 0$, any vector in $(1 - \pi_n)X$ is an eigenvector for T_n, but not for $\mathscr{T}_n$. We have already seen that, for $z \neq 0$,

$$R_n(z) = (T_n - z)^{-1} = \mathscr{R}_n(z)\pi_n - (1/z)(1 - \pi_n).$$

Therefore, if Γ does not enclose 0,

$$P_n := \frac{-1}{2i\pi}\int_\Gamma R_n(z)\,dz = \mathscr{P}_n\pi_n,$$

$$M_n := P_n X = \mathscr{M}_n, \qquad \text{and} \qquad \dim M_n = \dim \mathscr{M}_n.$$

These considerations justify the definition of the discrete strongly stable stability in Δ by the following conditions:

(i) $T_n - z\pi_n \xrightarrow{\text{d-s}} T - z$ for z in $\Delta - \{\lambda\}$;
(ii) $\dim \mathscr{P}_n X_n = m$ for n large enough.

We write $T_n - z\pi_n \xrightarrow{\text{d-ss}} T - z$ in Δ.

The following generalization of Lemma 3.11 holds. Let $0 \neq \lambda \in Q\sigma(T)$. λ can be isolated by a Jordan curve Γ that does not enclose 0.

Lemma 5.11 *If T_n is of class $\mathfrak{D}$, then, for $\lambda \neq 0$ and for z in Δ, the following are equivalent*:

(*i*) $T_n - z\pi_n \xrightarrow{\text{d-ss}} T - z$;
(*ii*) $T_n - z \xrightarrow{\text{ss}} T - z$.

The proof is left to the reader.

Then if λ is a nonzero eigenvalue of $Q\sigma(T)$, the preceding theory applies to show the spectral convergence.

Exercise

5.6 We suppose that $0 \in Q\sigma(T)$. 0 is isolated by the curve Γ, which defines the domain Δ. Let T_n of class $\mathfrak{D}$ be such that $T_n - z\pi_n \xrightarrow{\text{d-ss}} T - z$ in Δ. Show that $[\sigma(\mathscr{T}_n) \cap \Delta] \to 0$ with preservation of the multiplicities, and $\Theta(\mathscr{M}_n, M) \to 0$, where $\mathscr{M}_n$ (resp. M) is the invariant subspace of $\mathscr{T}_n$ (resp. T) associated with $\sigma(\mathscr{T}_n) \cap \Delta$ (resp. 0).

4. *T* and T_n Are Self-Adjoint in a Hilbert Space *H*

Let H be a Hilbert space on $\mathbb{C}$. Let T and T_n be densely defined *self-adjoint* operators. The preceding theory becomes simplified, because of the properties of self-adjoint operators, listed in Chapter 2, Section 8.

4.1. Specific Results

When T is self-adjoint, $\sigma(T)$ is a part of the real axis.

Theorem 5.12 *Let T and T_n be self-adjoint. If T_n is an approximation of T, $\sigma(T)$ is lower semicontinuous in a neighborhood of any of its points.*

Proof Let λ be given in $\sigma(T)$. For a given $\varepsilon > 0$, we set $z := \lambda + i\varepsilon$. z is in $\rho(T)$.

$$\|R(z)\| = [\text{dist}(z, \sigma(T))]^{-1} = 1/\varepsilon,$$
$$\|R_n(z)\| = [\text{dist}(z, \sigma(T_n))]^{-1} \leq 1/\varepsilon.$$

T_n is stable at z; hence

$$R_n(z) \xrightarrow{\text{p}} R(z) \qquad \text{and} \qquad \|R(z)x\| = \lim_n \|R_n(z)x\|.$$

$$\sup_{x \neq 0} \frac{\|R(z)x\|}{\|x\|} = \frac{1}{\varepsilon}.$$

Let x be fixed such that $\|x\| = 1$ and $\|R(z)x\| \geq 1/2\varepsilon$. For n large enough,

$$\|R_n(z)\| \geq \|R_n(z)x\| \geq 1/3\varepsilon.$$

There exists λ_n in $\sigma(T_n)$ such that $|\lambda_n - z| \leq 3\varepsilon$, that is $|\lambda_n - \lambda| \leq 4\varepsilon$. □

Example 5.8 Let $H = L^2(0, 1)$ with the orthonormal basis $e_i(t) = \sqrt{2} \sin i\pi t$, $i \in \mathbb{N}$. T is defined by $x(t) \mapsto (\cos \pi t)x(t)$, $\sigma(T) = [-1, 1]$. Let E_n be the span of $e_1, \ldots, e_n$. T_n is the Galerkin approximation of T in E_n. The associated matrix $\tilde{A}_n$ has the elements

$$a_{ij} = ((\cos \pi t)e_j(t), e_i(t)) = 0$$

except for

$$a_{i,i+1} = a_{i+1,i} = \tfrac{1}{2} \qquad \text{for} \quad i = 1, \ldots, n-1,$$

$$\sigma(T_n) = \sigma(\tilde{A}_n) \cup \{0\} = \left\{\cos \pi \frac{i}{n+1}, i = 1, \ldots, n\right\} \cup \{0\}.$$

When $n \to \infty$, each point of $[-1, 1]$ is the limit of a sequence of eigenvalues of $\tilde{A}_n$.

Exercise

5.7 Let T_n be self-adjoint. If $T_n \xrightarrow{\text{p}} T$, then $T_n - z \xrightarrow{\text{s}} T - z$ for any z in $\mathbb{C}$ such that $\mathcal{I}m\, z \neq 0$.

Let λ in $Q\sigma(T)$ of multiplicity m be isolated by Γ. We keep the notations of the preceding sections.

Theorem 5.13 *Let T, T_n be self-adjoint such that $T_n \xrightarrow{\text{p}} T$. There exists $N(\Gamma)$ such that, for $n > N(\Gamma)$, $\Gamma \subset \rho(T_n)$ and $P_n \xrightarrow{\text{p}} P$.*

Proof It is based on the spectral representation of T and T_n, as defined in Chapter 2, Section 8. It is proved in Kato (1976, p. 432) that for any real μ that is not an eigenvalue of T, then $E_n(\mu) \xrightarrow{\text{p}} E(\mu)$. Let $\mu_1 < \mu_2$ be the intersections of Γ with the real axis; $P = E(\mu_2) - E(\mu_1)$, cf. Fig. 5.4. For $n > \max(N(\mu_1),$

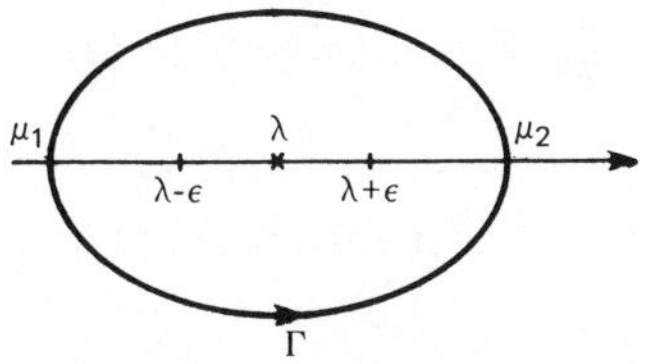

Figure 5.4

$N(\mu_2)$), P_n is defined and $P_n \xrightarrow{p} P$. There are at least m repeated eigenvalues of T_n in $[\mu_1, \mu_2]$. □

We have seen that the stability of $\{T_n\}$ in $\Delta - \{\lambda\}$ implies the convergence of $\sigma(T_n) \cap \Delta$. We show that the converse is true if T_n is supposed to be *normal*.

Lemma 5.14 *If T_n is normal,* $\lim_n[\sigma(T_n) \cap \Delta] = \{\lambda\}$ *implies that condition* (5.3) *holds for any z in $\Delta - \{\lambda\}$.*

Proof Clear from $\|R_n(z)\| = 1/\text{dist}(z, \sigma(T_n))$. □

Lemma 5.15 *Let T, T_n be self-adjoint such that $T_n \xrightarrow{p} T$. If* $\dim P_n X = m$ *for n large enough, then $\sigma(T)$ is upper semicontinuous in the neighborhood of λ and* $\|P_n - P\| \to 0$.

Proof By hypothesis there are m repeated eigenvalues of T_n in $[\mu_1, \mu_2]$ for $n > N$. Now for any $\varepsilon > 0$, there exists $N(\varepsilon)$ such that, for $n > N(\varepsilon)$, there are at least m repeated eigenvalues in $]\lambda - \varepsilon, \lambda + \varepsilon[$, cf. Fig. 5.4. Therefore there is no point of $\sigma(T_n)$ in $]\mu_1, \lambda - \varepsilon]$ or in $[\lambda + \varepsilon, \mu_2[$ for $n > \max(N, N(\varepsilon))$. This proves the upper semicontinuity in $[\mu_1, \mu_2]$. $P_n \xrightarrow{cc} P$ implies

$$\|P - P_n\| \to 0$$

since P and P_n are self-adjoint. □

Proposition 5.16 *Let T and T_n be self-adjoint. The following are equivalent*:

(*i*) $T_n - z \xrightarrow{ss} T - z$ *for z in* Δ,
(*ii*) $T_n \xrightarrow{p} T$ *and* $\dim P_n X = m$ *for n large enough.*

Proof The direct implication is clear. We prove the converse. The continuity of $\sigma(T)$ in $\Delta - \{\lambda\}$ follows from Theorem 5.12 and Lemma 5.15. It is equivalent to the stability by Lemma 5.14. □

The preservation of the multiplicities is then the only point to be proved in case of self-adjoint operators.

Exercise

5.8 Let λ_n be an eigenvalue of multiplicity g_n converging to λ of multiplicity g, T and T_n being self-adjoint. Show that if $T_n - z \xrightarrow{ss} T - z$ in Δ and if $g_n = g$, λ_n is the only eigenvalue approximating λ. This is seldom satisfied in practice unless $g = 1$ (cf. Example 5.6).

4.2. A Sufficient Condition for the Strong Stability

A self-adjoint operator T is said to be *bounded from below* if its numerical range $\{(Tx, x);\ x \in D\}$, which is a subset of the real axis, is bounded from below:

$$(Tx, x) \geq \gamma \|x\|^2 \qquad \text{for all} \quad x \in D \quad \text{and some real } \gamma.$$

Then the spectrum is bounded from below; that is, $\sigma(T) \subset [\gamma, +\infty[$, cf. Kato (1976, p. 278). We suppose that T (resp. T_n) is self-adjoint and bounded from below and that the lower part of its spectrum denoted $\sigma_{\text{inf}}(T)$(resp. $\sigma_{\text{inf}}(T_n)$) contains a finite number of isolated eigenvalues with finite multiplicity. They are numbered μ_i (resp. μ_{in}), $i = 1, \ldots, k$, by increasing value according to their multiplicity.

Proposition 5.17 *Let T and T_n be self-adjoint. If the approximation $\{T_n\}$ is such that its eigenvalues fulfill $\mu_{in} \geq \mu_i$, $i = 1, \ldots, k$, then $\{T_n\}$ is strongly stable in a neighborhood of $\sigma_{\text{inf}}(T)$.*

Proof There is only to prove the preservation of the multiplicities. We suppose that μ_1 is simple. There is only *one* eigenvalue μ_{1n} in $[\mu_1, \mu_2[$ since $\mu_{2n} < \mu_2$ is impossible. □

Exercise

5.9 Prove a similar property when T and T_n are bounded from above and when the upper parts of the spectra consist of isolated eigenvalues with finite multiplicities.

Example 5.9 Let $a(u, v)$ be a coercive hermitian sesquilinear form on a Hilbert space V. Let $\{V_n\}$ be a sequence of finite-dimensional subspaces of H. We suppose that the lower part σ_{inf} of the spectrum consists of isolated eigenvalues with finite multiplicity. We approximate these eigenvalues by the Rayleigh–Ritz method associated with $\{V_n\}$. Then $\mu_{in} \geq \mu_i$ (cf. Chapter 4). The Rayleigh–Ritz method is strongly stable in the neighborhood of σ_{inf}.

Example 5.10 Let T be a bounded self-adjoint operator. It can be considered as bounded from below and from above. We suppose that the lower part σ_{inf} (resp. upper part σ_{sup}) consists of isolated eigenvalues with finite multiplicity. Let $\{E_n\}$ be a sequence of finite dimensional subspaces of H. Let π_n be the orthogonal projection on E_n. The eigenvalues are approximated by those of $\pi_n T \pi_n$. Using the min–max characterization of the eigenvalues, we get $\nu_{in} \leq \nu_i$ (resp. $\mu_{in} \geq \mu_i$) for the largest eigenvalues ν_i in σ_{sup} ordered by decreasing value (resp. lowest eigenvalues μ_i in σ_{inf} ordered by increasing value). This shows the strong stability of the Galerkin method defined by an orthogonal projection in the neighborhood of σ_{inf} and σ_{sup}.

Exercise

5.10 If T is compact self-adjoint and if π_n is an orthogonal projection, show the strong stability of the Galerkin method in $\mathbb{C} - \{0\}$.

5. Strong Stability of an Approximation of a Closed Operator

So far we have given a sufficient condition for the convergence of the eigenvalues with preservation of the multiplicities in terms of the strong stability of T_n in a neighborhood of λ. Its main drawback is that it involves some knowledge about the unknown eigenvalue. This will be partially overcome in this section.

We first characterize in $\mathscr{L}(\hat{D}, X)$ the strongly stable convergence by means of the regular convergence in a neighborhood of λ. Then we show that the various convergences defined in Chapter 3 provide sufficient conditions that, although expressed in terms of T and T_n only, will guarantee the strong stability of $\{T_n\}$ in the neighborhood of *any* isolated eigenvalue of T with finite multiplicity.

For that purpose, we introduce the notion of *strongly stable convergence* in $\rho(T) \cup Q\sigma(T)$. Let $\{T_n\}_{\mathbb{N}}$ be a sequence of operators in $\mathscr{C}(X)$ converging to $T \in \mathscr{C}(X)$.

Definition $T_n - z \xrightarrow{\text{ss}} T - z$ in $\rho(T) \cup Q\sigma(T)$ iff

(i) $T_n - z \xrightarrow{\text{s}} T - z$ for all z in $\rho(T)$;

(ii) for any eigenvalue of T with finite multiplicity m, isolated by the curve Γ, $\dim P_n X = m$ for $n > N(\Gamma)$.

If $T_n - z \xrightarrow{\text{ss}} T - z$ *in* $\rho(T) \cup Q\sigma(T)$, $\{T_n\}$ is said to be a strongly stable approximation of T in $\rho(T) \cup Q\sigma(T)$.

5.1. *Regular Convergence*

We have already proved in Chapter 3, Proposition 3.17, the equivalence $T_n - z \xrightarrow{\text{s}} T - z \Leftrightarrow T_n - z \xrightarrow{\text{r}} T - z$ for all z in $\rho(T)$. We show a stronger characterization in $\mathscr{L}(\hat{D}, X)$ after two preparatory lemmas. We recall that $\hat{D}$ is the domain D considered as a Banach space when equipped with the graph norm (cf. Section 6.2 of Chapter 2).

Lemma 5.18 *If T and T_n in $\mathscr{C}(X)$ are such that* $\dim(\operatorname{Ker} T) < \infty$ *and* $T_n \xrightarrow{\text{r}} T$, *then* $\dim(\operatorname{Ker} T_n) \leq \dim(\operatorname{Ker} T)$.

Proof We set $K := \text{Ker } T$ and $K_n := \text{Ker } T_n$. Let $\{x_n\}_{\mathbb{N}}$ be a bounded sequence in K_n. $T_n x_n = 0$ implies that $x_n \to x$ for $n \in N_1 \subset \mathbb{N}$ by regularity, with $Tx = 0$. Therefore $x \in K$ and Lemma 5.7 applies to K and K_n. $\dim K_n \leq \dim K$ for n large enough. □

Lemma 5.19 *If T and T_n in $\mathscr{C}(X)$ are such that $T_n - z \xrightarrow{s} T - z$ on Γ, then $(T_n - \lambda)(1 - P_n)$ is invertible in $(1 - P_n)X$, and this inverse is uniformly bounded.*

Proof By hypothesis, $P_n \xrightarrow{p} P$. For z in Δ, set

$$S_n(z) := R_n(z)(1 - P_n).$$

$S_n(z)$ is the reduced resolvent of T_n, with respect to the eigenvalues inside Γ. It is holomorphic in z inside Γ. Using $T_n P_n x = P_n T_n x$ for $x \in D$, we get

$$S_n(\lambda)(T_n - \lambda)(1 - P_n) = 1 - P_n = (T_n - \lambda)(1 - P_n)S_n(\lambda).$$

This proves that $(T_n - \lambda)(1 - P_n)$ is invertible in $(1 - P_n)X$, its inverse being $S_n(\lambda)$. By the maximum principle for the norm of $S_n(z)$, holomorphic in z inside Γ, we get

$$\|S_n(\lambda)\| < \max_{z \in \Gamma} \|R_n(z)(1 - P_n)\| \leq \left(\max_{z \in \Gamma} \|R_n(z)\|\right) \|1 - P_n\|. \quad \square$$

Exercise

5.11 Let T, $T_n \in \mathscr{C}(X)$. Show that $T_n - z \xrightarrow{s} T - z$ on Γ implies $S_n(\lambda) \xrightarrow{p} S(\lambda) = S$, the reduced resolvent of T, with respect to λ.

Theorem 5.20 (Vainikko) *The following are equivalent in $\mathscr{L}(\hat{D}, X)$ for all z in $\rho(T) \cup Q\sigma(T)$:*

(*i*) $T_n - z \xrightarrow{r} T - z$,
(*ii*) $T_n - z \xrightarrow{ss} T - z$.

Proof Let $\lambda \in Q\sigma(T)$ of multiplicity m be isolated by Γ. We have to prove the equivalence in $\mathscr{L}(\hat{D}, X)$

$$T_n - z \xrightarrow{r} T - z \text{ in } \Delta \Leftrightarrow T_n - z \xrightarrow{ss} T - z \text{ in } \Delta.$$

The only point remaining to be proved after Proposition 3.17 is that

$$T_n - \lambda \xrightarrow{r} T - \lambda \Leftrightarrow \dim P_n X = m.$$

(1) We first suppose that T and $T_n \in \mathscr{L}(X)$. If (i) holds, $\dim P_n X \geq m$ for n large enough. Let us suppose that $\dim P_n X \geq m + 1$ for $n \in N_1 \subset \mathbb{N}$. There exist at least $m + 1$ eigenvalues of T_n (counting their multiplicities) converging to λ: $\mu_{in} \to \lambda$ for $i = 1, \ldots, m + 1$. Clearly $T_n - \mu_{in} \xrightarrow{r} T - \lambda$ and $\prod_{i=1}^{m+1}(T_n - \mu_{in}) \xrightarrow{r} (T - \lambda)^{m+1}$ (cf. Exercises 3.9 and 3.10 in Chapter 3). Then $\dim(\text{Ker} \prod_{i=1}^{m+1}(T_n - \mu_{in})) \leq \dim(\text{Ker} \prod_{i=1}^{m+1}(T - \lambda))$, and we get the

contradiction $m + 1 \leq m$. Conversely, if (ii) holds, suppose that the bounded sequence $\{x_n\}_{\mathbb{N}}$ is such that $(T_n - \lambda)x_n \to y$ for $n \in N_1 \subset \mathbb{N}$. We write $x_n = P_n x_n + (1 - P_n)x_n$. $P_n \overset{\text{cc}}{\to} P$ and $P_n x_n \to u = Pu$ for $n \in N_2 \subset N_1$. For any $x \in X$, $(T_n P_n - TP)x = P_n(T_n - T)x + (P_n - P)Tx \to 0$, $T_n P_n \overset{\text{p}}{\to} TP$ and $T_n P_n$ is uniformly bounded. Then

$$(T_n - \lambda)P_n x_n - (T - \lambda)u = (T_n - \lambda)P_n(P_n x_n - u) + (T_n P_n - TP)u - \lambda(P_n - P)u \to 0$$

and $(T_n - \lambda)P_n x \to Py = (T - \lambda)u$. If we set $z_n := (T_n - \lambda)(1 - P_n)x_n$, $z_n \to y - (T - \lambda)u$. By Lemma 5.19 and Exercise 5.11, $(1 - P_n)x_n = S_n(\lambda)z_n$ is convergent to $v = S(\lambda)y$. Then for $n \in N_2$, $x_n \to u + v$ such that

$$(T - \lambda)(u + v) = Py + (1 - P)y = y.$$

(2) When T and T_n are in $\mathscr{C}(X)$, we set the problem in $\mathscr{L}(\hat{D}, X)$ where the result holds by (1).

(3) In $\mathscr{C}(X)$ we can only prove that (ii) ⇒ (i) by Exercise 3.11. □

Proposition 5.21 *If the injection* $i: \hat{D} \to X$ *is compact, then, in* $\mathscr{L}(\hat{D}, X)$, $T_n \overset{\text{r}}{\to} T$ *implies* $T_n - z \overset{\text{ss}}{\to} T - z$ *in* $\rho(T) \cup Q\sigma(T)$.

Proof Clear from Proposition 3.19, where it is shown that $T_n - z \overset{\text{r}}{\to} T - z$ for all z in $\mathbb{C}$. Note that $\rho(T) \cup Q\sigma(T) = \mathbb{C}$. □

5.2. *Radial Convergence*

We now deal with a special case of strongly stable convergence provided by the uniform radial convergence.

Let $0 \in \rho(T)$. We first define the *radial convergence* $T_n \overset{\sigma}{\to} T$ iff

(i) $T_n \overset{\text{s}}{\to} T$;

(ii) the following condition is satisfied:

$$\forall \varepsilon > 0, \exists N: \text{for } n > N, \quad r_\sigma[(T - T_n)T^{-1}] < \varepsilon.$$

Note that $(T - T_n)T^{-1} = 1 - T_n T^{-1}$ is bounded with domain X. Therefore its spectral radius is well defined.

We now define the *uniform radial convergence* $T_n - z \overset{\text{u}\sigma}{\to} T - z$ for z in $\rho(T)$ by two conditions:

(i) for any z in $\rho(T)$, $T_n - z \overset{\sigma}{\to} T - z$;

(ii) $\forall \varepsilon > 0$, $\forall K$ compact of $\rho(T)$, $\exists N(K)$: for $n > N(K)$

$$\sup_{z \in K} r_\sigma[(T - T_n)R(z)] < \varepsilon.$$

Exercise

5.12 Show that $\|T - T_n\| \to 0$ or $T_n \overset{cc}{\to} T$ imply $T_n - z \overset{u\sigma}{\to} T - z$ in $\rho(T)$. (*Hint*: Cf. Examples 3.8 and 3.9.)

Theorem 5.22 *The uniform radial convergence $T_n - z \overset{u\sigma}{\to} T - z$ for z in $\rho(T)$ implies the strongly stable convergence $T_n - z \overset{ss}{\to} T - z$ for z in $\rho(T) \cup Q\sigma(T)$.*

Proof Let $\lambda \in Q\sigma(T)$ of multiplicity m, be isolated by Γ. By hypothesis, there exists $N(\Gamma)$ such that

$$\sup_{n > N(\Gamma),\, z \in \Gamma} r_\sigma[(T - T_n)R(z)] < 1.$$

By Proposition 2.20, the spectral radius of the bounded operator $(T - T_n)R(z)$ is upper semicontinuous in $z \in \rho(T)$. Then for $n > N(\Gamma)$,

$$\max_{z \in \Gamma} r_\sigma[(T - T_n)R(z)] < 1.$$

There is only to show the preservation of the algebraic multiplicities: $\dim P_n X = m$. For that purpose, we consider, for $n > N(\Gamma)$, the family of operators $T(t) := T - t(T - T_n)$ depending on the complex parameter t. $T(0) = T$ and $T(1) = T_n$. The spectral projection

$$P(t) = \frac{-1}{2i\pi} \int_\Gamma (T(t) - z)^{-1}\, dz$$

is analytic in t in the disk $\{t \in \mathbb{C}; |t| < 1/r\}$ with

$$r = \sup_{n > N(\Gamma)} \max_{z \in \Gamma} r_\sigma[(T - T_n)R(z)].$$

The condition $r < 1$ ensures that $t = 1$ belongs to this disk. Therefore $\dim P(1)X = m$, that is, $\dim P_n X = m$ for $n > N(\Gamma)$. □

Proposition 5.23 *If $T_n \overset{p}{\to} T$ and if there exists an integer p such that, for all z in $\rho(T)$, $\|[(T - T_n)R(z)]^p\| \to 0$, then $T_n - z \overset{\sigma}{\to} T - z$ in $\rho(T)$ uniformly in z on any compact of $\rho(T)$.*

Proof Let K be a compact subset of $\rho(T)$. We have to prove that $T_n - z \overset{\sigma}{\to} T - z$ uniformly in $z \in K$. This is clear from

$$r_\sigma[(T - T_n)R(z)] \le \|[(T - T_n)R(z)]^p\|^{1/p}. \quad \square$$

Example 5.11 Let $T \in \mathscr{C}(X)$. A neighboring approximation T_n such that $\|(T - T_n)R(z)\| \to 0$ is uniformly radial in $\rho(T)$. See Example 4.30 in Chapter 4.

Example 5.12 The convergences $\|T - T_n\| \to 0$ and $T_n \overset{cc}{\to} T$ are also uniformly radial in $\rho(T)$ (cf. Exercise 5.12). When T is compact, such examples are provided by the projection method and its Sloan variant, if $\pi_n \xrightarrow{\text{p}} 1$. Another example is, in the space of continuous functions, the Nyström method applied to an integral operator with a smooth kernel.

The uniform radial convergence is, in general, a sufficient condition for the strong stability. Its theoretical importance in analytic perturbation theory will appear in full light in Section 6. There is an important case where the uniform radial convergence is a necessary condition for the strong stability: this is when T is *compact* in $\mathscr{L}(X)$.

We now suppose that T is compact. Any nonzero point of $\sigma(T)$ belongs to $Q\sigma(T)$; then $\rho(T) \cup Q\sigma(T) = \mathbb{C} - \{0\}$.

Lemma 5.24 *If T is compact, $T_n - z \xrightarrow{\text{ss}} T - z$ in $\mathbb{C} - \{0\}$ implies that* $r_\sigma(T - T_n) \to 0$.

Proof We suppose that $r_\sigma(T - T_n) \not\to 0$ as $n \to \infty$. There exists a sequence $\{v_n\}_{\mathbb{N}}$, $v_n \in \sigma(T - T_n)$, such that $r_\sigma(T - T_n) = |v_n|$, and $|v_n| \geq c > 0$. By Proposition 2.24, there exists $\{x_p\}_{\mathbb{N}}$ such that for all $p \in \mathbb{N}$, $\|x_p\| = 1$ and $\lim_{p\to\infty} \|(T - T_n)x_p - v_n x_p\| = 0$. $T_n \xrightarrow{\text{p}} T$, $T_n - T$ is uniformly bounded; hence $|v_n| < M$. We may suppose that $v_n \to v \neq 0$. Therefore

$$(T - T_n)x_n - vx_n \to 0, \qquad \|x_n\| = 1.$$

T being compact, $Tx_n \to y$ for $n \in N_1 \subset \mathbb{N}$. Then

$$T_n x_n + vx_n \to y, \qquad n \in N_1.$$

(1) We suppose that $-v \neq 0$ belongs to $Q\sigma(T)$, $-v$ is isolated by the curve Γ. With the notations of Lemma 5.19, $S_n(-v) \xrightarrow{\text{p}} S(-v) =: S$. Then for $n \in N_1$,

$$S_n(-v)(T_n x_n + vx_n) \to Sy,$$

that is,

$$(1 - P_n)x_n \to Sy.$$

Since $P_n \xrightarrow{\text{cc}} P$, $P_n x_n \to t$ for $n \in N_2 \subset N_1$, and $x_n \to t + Sy$. Using $T_n \xrightarrow{\text{p}} T$, we finally get $vx_n \to 0$ for $n \in N_2$, with $v \neq 0$ and $\|x_n\| = 1$, which is a contradiction.

(2) We now suppose that $-v \in \rho(T)$. Similarly

$$R_n(-v) \xrightarrow{\text{p}} R(-v)$$

and

$$R_n(-v)(T_n x_n + vx_n) \to R(-v)y \qquad \text{for} \quad n \in N_1,$$

that is

$$x_n \to R(-v)y \qquad \text{and} \qquad vx_n \to 0. \quad \square$$

Corollary 5.25 *If T and T_n are self-adjoint in a Hilbert space, T being compact, then $T_n - z \xrightarrow{\text{ss}} T - z$ in $\mathbb{C} - \{0\}$ implies that $\|T - T_n\| \to 0$.*

Proof Clear from $r_\sigma(T - T_n) = \|T - T_n\|$. □

Theorem 5.26 (Lemordant) *If T is compact, the following are equivalent:*

(i) $T_n - z \overset{\text{ss}}{\to} T - z$ *in* $\mathbb{C} - \{0\}$;
(ii) $T_n - z \overset{\text{u}\sigma}{\to} T - z$ *in* $\rho(T)$.

Proof The reverse implication is clear by Theorem 5.22. We suppose that (i) holds. Let K be a compact subset of $\rho(T)$; we have to show that

$$\forall \varepsilon > 0, \exists N(K); n > N(K) \Rightarrow r_\sigma[(T - T_n)R(z)] < \varepsilon, \forall z \in K.$$

We suppose that it does not hold; hence

$$\exists \varepsilon > 0, \forall N, \exists n > N(K); r_\sigma[(T - T_n)R(z_n)] \geq \varepsilon \qquad \text{for} \quad z_n \in K.$$

Let $v_n \in \sigma[(T - T_n)R(z_n)]$, such that $r_\sigma[(T - T_n)R(z_n)] = |v_n|$, and $|v_n| \geq \varepsilon$. There exists a sequence $\{x_n\}_{\mathbb{N}}$, $\|x_n\| = 1$, such that

$$(T - T_n)R(z_n)x_n - v_n x_n \to 0 \qquad \text{for} \quad n \in N_1 \subset \mathbb{N}.$$

v_n is bounded and $v_n \to v \neq 0$ for $n \in N_2 \subset N_1$. K is compact; then $z_n \to z \neq 0$ for $n \in N_3 \subset N_2$, and $\|R(z_n) - R(z)\| \to 0$. Therefore $(T - T_n)R(z)x_n - vx_n \to 0$. $TR(z) = 1 + zR(z)$ for all z in $\rho(T)$; hence $z \neq 0$ and

$$R(z) = -1/z + (1/z)TR(z),$$

where $TR(z)$ is compact. Then

$$(T - T_n)R(z)x_n = TR(z)x_n - T_n((1/z)TR(z) - 1/z)x_n.$$

$\{TR(z)x_n\}$ contains a converging subsequence; therefore

$$T_n x_n + zvx_n \to y \qquad \text{for} \quad n \in N_4 \subset N_3.$$

The proof of Lemma 5.24 applies to lead to a contradiction. □

Exercises

5.13 Let $T \in \mathscr{C}(X)$ be such that $R(z_0)$ is compact for some z_0 in $\rho(T)$. Show that $T_n - z \overset{\text{ss}}{\to} T - z$ in $\mathbb{C}$ implies that, for all z in $\mathbb{C}$, $r_\sigma[R_n(z) - R(z)] \to 0$.

5.14 Deduce, from Theorem 5.26 and Exercise 5.13, a necessary and sufficient condition for the strong stability of an approximation T_n of a closed operator T with compact resolvent.

5.3. Connection between T and T^{-1}

An important way of proving the strong stability of an approximation of T in $\mathscr{C}(X)$ is to prove the strong stability of the corresponding approximation of T^{-1} (if it exists) in $\mathscr{L}(X)$. Let $\lambda \in Q\sigma(T)$, $\lambda \neq 0$.

Proposition 5.27 *Let $T, T_n \in \mathscr{C}(X)$ be invertible with $A := T^{-1}$, $A_n := T_n^{-1}$*

in $\mathscr{L}(X)$. We suppose that $T_n \xrightarrow{\text{p}} T$ and $A_n - z \xrightarrow{\text{ss}} A - z$ in $\mathscr{L}(X)$ in a neighborhood of $1/\lambda$. Then $T_n - z \xrightarrow{\text{ss}} T - z$ in $\mathscr{C}(X)$ in a neighborhood of λ.

Proof This is the special case for $z_0 = 0 \in \rho(T)$ of the following property. For $z_0 \in \rho(T)$, set $t := 1/(z - z_0)$, $v := 1/(\lambda - z_0)$. $T_n \xrightarrow{\text{p}} T$ and

$$R_n(z_0) - t \xrightarrow{\text{ss}} R(z_0) - t$$

in the neighborhood of v imply that $T_n - z \xrightarrow{\text{ss}} T - z$ in the neighborhood of λ (see Exercise 3.11 and Proposition 2.28). □

Example 5.13 The conforming finite-element method for T is strongly stable when the projection method for T^{-1} is so, the projection being the elliptic projection defined by the set of basis functions.

We are led to the study of the strong stability in $\mathscr{L}(X)$.

5.4. *Bounded Operators*

Proposition 3.18 can be strengthened as follows.

Proposition 5.28 *$T_n - z \xrightarrow{\text{r}} T - z$ for z in $\rho(T) \cup Q\sigma(T)$ if one of the three properties below holds:*

(*i*) $\|T_n - T\| \to 0$;
(*ii*) $T_n \xrightarrow{\text{cc}} T$;
(*iii*) $T_n \xrightarrow{\text{c}} T$.

Proof Since (i) or (ii) implies (iii), it remains to prove that $T_n \xrightarrow{\text{c}} T$ implies that $T_n - \lambda \xrightarrow{\text{r}} T - \lambda$ for any λ in $Q\sigma(T)$. Let $x_n \in B$, the unit sphere, be such that $(T_n - \lambda)x_n \to y$. $(T - T_n)x_n \to u$ for $n \in N_1 \subset \mathbb{N}$. Thus $(T - \lambda)x_n \to y + u$, $n \in N_1$. We write $x_n = Px_n + (1 - P)x_n$, $Px_n \to v$ for $n \in N_2 \subset N_1$, and $(T - \lambda)(1 - P)x_n \to y + u - (T - \lambda)v$; $(T - \lambda)(1 - P)$ is invertible in $(1 - P)X$; then $\{(1 - P)x_n\}_{N_3}$ is convergent to w, and $\{x_n\}_{N_3}$ converges to $v + w$ such that $(T - \lambda)(v + w) = Py + (1 - P)y + u = y$ since $u = 0$ (Exercise 3.4). □

Exercise

5.15 If $T_n \xrightarrow{\text{c}} T$, using the notations of Section 3, show that $g_n = g$ implies that E and E_n fulfill (5.4) (Goldberg).

Example 5.14 Let T be compact. Let $\{\pi_n\}_{\mathbb{N}}$ be a sequence of projections such that $\pi_n \xrightarrow{\text{p}} 1$. The projection method (resp. the Sloan variant) corresponding to the approximation $\pi_n T$ (resp. $T\pi_n$) is strongly stable in $\mathbb{C} - \{0\}$: $\|(1 - \pi_n)T\| \to 0$ (resp. $T(1 - \pi_n) \xrightarrow{\text{cc}} 0$).

Example 5.15 Let T be an integral operator with a smooth kernel. In the space of continuous functions the Nyström method yields an approximation of T that is strongly stable in $\mathbb{C} - \{0\}$.

Exercise

5.16 Prove the strong stability of the Petrov method defined in Chapter 4, Section 3, under the condition that T is compact and $(\pi_{n\restriction X_n})^{-1}$ is uniformly bounded.

5.5. T_n Is of Class $\mathfrak{D}$

The following generalization of Proposition 3.20 holds.

Theorem 5.29 *If T_n is of class $\mathfrak{D}$, the following are equivalent in $\mathscr{L}(\hat{D}, X)$ for $z \in \rho(T) \cup Q\sigma(T)$:*

(i) $T_n - z\pi_n \xrightarrow{\text{d-ss}} T - z$;
(ii) $T_n - z\pi_n \xrightarrow{\text{d-r}} T - z$

The proof is left to the reader.

We leave as exercises the discrete analogs to Propositions 5.27 and 5.28.

Exercises

5.17 We suppose that $\mathscr{T}_n$ is invertible and that $\mathscr{A}_n := \mathscr{T}_n^{-1} \in \mathscr{L}(X_n)$. Set $A_n := \mathscr{A}_n \pi_n$. Show that $T_n \xrightarrow{\text{p}} T$ and $A_n - z\pi_n \xrightarrow{\text{d-ss}} A - z$ in a neighborhood of $1/\lambda$, $\lambda \neq 0$, imply that

$$T_n - z\pi_n \xrightarrow{\text{d-ss}} T - z$$

in a neighborhood of λ.

5.18 Show that $T_n \xrightarrow{\text{d-c}} T$ implies that $T_n - z\pi_n \xrightarrow{\text{d-r}} T - z$ for z in $\rho(T) \cup Q\sigma(T)$.

5.19 Let T_n be of class $\mathfrak{D}$. We consider the following particular case of d-c convergence: $\|(T - T_n)_{\restriction X_n}\| \to 0$. Prove that $\|(P - P_n)_{\restriction X_n}\| \to 0$, and $\Theta(M, M_n) \to 0$, for a nonzero isolated eigenvalue.

5.20 T is closed with domain D. Let T_n be of class $\mathfrak{D}$. If $0 \in \rho(T)$, prove that $\Theta(T_n, T) \to 0$ iff $\|(T^{-1} - \mathscr{T}_n^{-1})_{\restriction X_n}\| \to 0$. Deduce that T_n is a strongly stable approximation of T in $\rho(T) \cup Q\sigma(T)$.

Example 5.16 The finite-difference method on T is strongly stable when the Fredholm method on T^{-1} is so.

Example 5.17 Let $a(u,v)$ be a coercive continuous form on a Hilbert space such that the associated operator A is bounded but not compact. An example of conforming finite-element methods such that $\pi_n A \pi_n \xrightarrow{\text{d-c}} A$ is given in Descloux *et al.* (1978a). Another is given in Mills (1979b).

Example 5.18 The discrete gap convergence is used to analyze some finite-element methods applied to differential operators in Descloux (1979) and Descloux *et al.* (1981).

5.6. Summary

The relationships between the convergences are summarized in Tables 5.1 and 5.2 (cf. Tables 3.2 and 3.3); z does not appear in the conditions lying in the box.

5.7. Bibliographical Comments

The role of the preservation of the algebraic multiplicities in the numerical approximation of eigenvalues was first studied in Chatelin (1970a). When $T_n - z \overset{\text{ss}}{\rightarrow} T - z$ in Δ, λ is stable under the perturbation $T - T_n$, in the sense of Kato (1976, p. 437).

Table 5.1

Strong Stability in $\rho(T) \cup Q\sigma(T)^a$

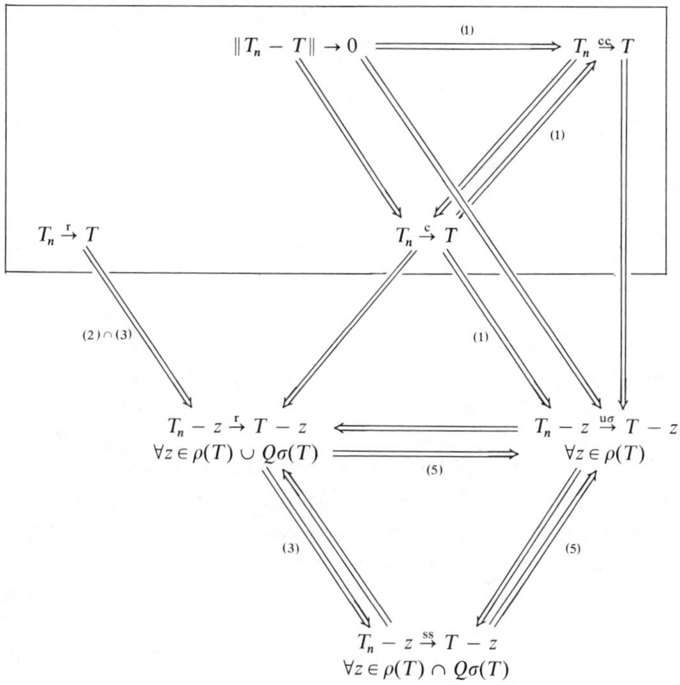

[a] (1), valid if T and T_n are compact; (2), valid if T is anticompact; (3), valid in $\mathscr{L}(\hat{D}, X)$; (5), valid if T is compact.

Table 5.2

Discrete Strong Stability in $\rho(T) \cup Q\sigma(T)$[a]

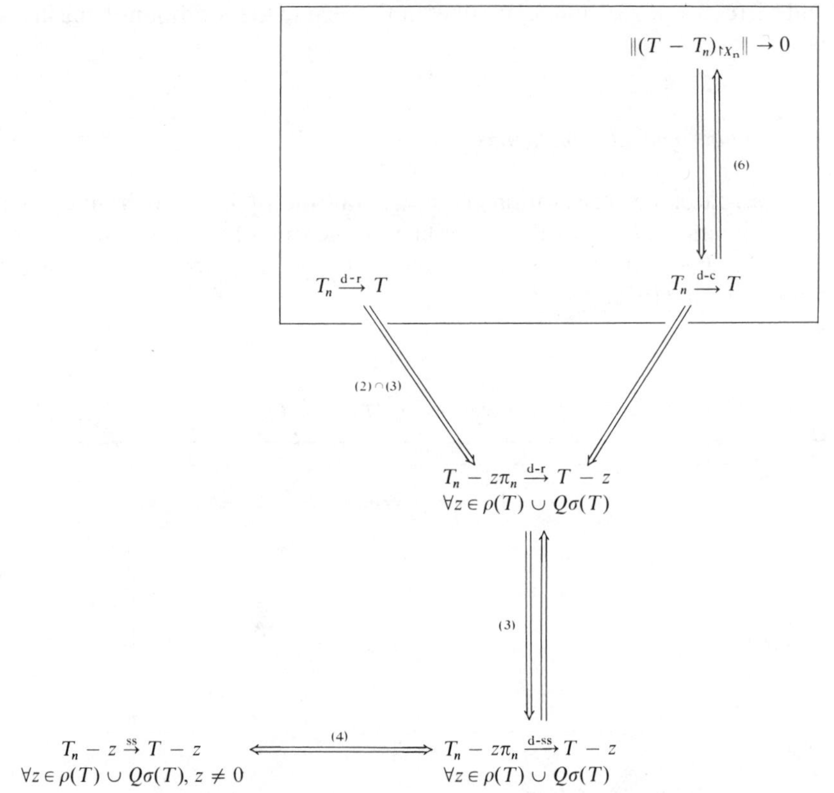

[a] (2), valid if T and T_n are compact; (3), valid in $\mathscr{L}(\hat{D}, X)$; (4), valid if z is nonzero; (6), valid if $T_n = \pi_n^a T$ where π_n^a is an elliptic projection.

The characterization in $\mathscr{L}(\hat{D}, X)$ of the discrete-strong stability in Δ by the discrete regularity in Δ is due to Vainikko (1976a). His proof uses the notion of measure of noncompactness for an operator; the proof given here, which is simpler, is due to Redont (1979b).

The use of the discrete-regular convergence for discretization methods can be found in Grigorieff (1970a,b, 1972, 1973, 1975a,b), Grigorieff and Jeggle (1973), Vainikko (1975, 1976a,b, 1977a,b, 1978a,b) and Vainikko and Karma (1974a,b).

The radial convergence was introduced in a somewhat restricted form by Lemordant in Chatelin and Lemordant (1978) (under the name of strong convergence). When T is compact, the characterization of the strongly stable convergence by the uniform radial convergence is also due to Lemordant (1980).

In Vainikko (1969a) the discrete-compact convergence is introduced and its strong stability is proved, again by means of the measure of noncompactness. In Descloux *et al.* (1978b), and Mills (1979b), the discrete-compact convergence is used for some finite-element method applied to an elliptic differential operator T with a noncompact resolvent (see also Descloux, 1981).

6. Iterative Refinement Method When $T_n - z \overset{u\sigma}{\to} T - z$ in $\rho(T)$

We now deal with special properties of the uniform radial convergence related to the analytic perturbation theory. There are two ways to relate T to T_n for a fixed n: either consider

$$T_n = T - (T - T_n) \qquad \text{or} \qquad T = T_n - (T_n - T).$$

The first way leads to the family of operators $T(t) = T - t(T - T_n)$, depending on the complex parameter t. From the hypothesis

$$r_\sigma((T - T_n)R(z)) \underset{n\to\infty}{\longrightarrow} 0$$

uniformly on any compact subset of $\rho(T)$ follows easily that, for n large enough, $t = 1$ belongs to the analyticity domains of the resolvent, spectral projection, and trace (cf. Exercises 5.21 and 5.22).

The second way leads to the family $T_n(t) = T_n - t(T_n - T)$, which is more interesting from a computational point of view. We have therefore chosen to develop it. The same assumption of uniform radial convergence again implies that $t = 1$ belongs to the analyticity domain of the corresponding functions for n large enough. It enables us to use, for a large class of *numerical* approximation methods, the powerful tool of analytic perturbation theory. $x = R(z)f$, λ, and φ are then the limits of a numerical process that requires us to work with the numerical approximation T_n only.

Exercises

5.21 $T(t) = T - t(T - T_n)$. λ is an eigenvalue of T of multiplicity m, isolated by Γ. We define for z on Γ,

$$R(t, z) = (T(t) - z)^{-1}, \qquad P(t) = \frac{-1}{2i\pi}\int_\Gamma R(t, z)\,dz, \qquad \text{and} \qquad \hat{\lambda}(t) = \frac{1}{m}\operatorname{tr} T(t)P(t).$$

For a fixed n, give the analyticity domain of $R(t, z)$, $P(t)$, and $\hat{\lambda}(t)$.

5.22 We suppose that $T_n - z \overset{u\sigma}{\to} T - z$ in $\rho(T)$. Show that there exists $N(\Gamma)$ such that, for $n > N(\Gamma)$, $t = 1$ belongs to the analyticity domain defined in Exercise 5.21. Give the series expansions of $R_n(z) = (T_n - z)^{-1}$ for z on Γ,

$$P_n = \frac{-1}{2i\pi}\int_\Gamma R_n(z)\,dz \quad \text{and} \quad \hat{\lambda}_n = \frac{1}{m}\operatorname{tr} T_n P_n.$$

6.1. Analytic Perturbation Theory

Let λ_n be an eigenvalue of T_n isolated by Γ. We suppose that λ_n is *simple*, and we define

$$R_n(z) = (T_n - z)^{-1} \quad \text{for} \quad z \text{ on } \Gamma,$$

$$P_n = \frac{-1}{2i\pi}\int_\Gamma R_n(z)\,dz.$$

Lemma 5.30 *If $T_n - z \overset{u\sigma}{\to} T - z$ in $\rho(T)$, then for n large enough, $r_\sigma[(T_n - T)R_n(z)]$ is defined for z in $\rho(T)$. It tends to zero as n tends to infinity, uniformly in z in any compact subset of $\rho(T)$.*

Proof For all $\varepsilon > 0$, all compact K in $\rho(T)$, there exists $N(K)$ such that

$$\sup_{n > N(K)} r_\sigma[(T - T_n)R(z)] < \varepsilon.$$

For $n > N(K)$, $R_n(z) = R(z)\sum_{k=0}^{\infty}[(T - T_n)R(z)]^k$ and $z \in \rho(T_n)$. Therefore

$$\begin{aligned}(T - T_n)R_n(z) &= \sum_{k=1}^{\infty}[(T - T_n)R(z)]^k \\ &= [(T - T_n)R(z)][1 - (T - T_n)R(z)]^{-1},\end{aligned}$$

where the two factors commute. Then for $z \in K$,

$$r_\sigma[(T - T_n)R_n(z)] \le r_\sigma[(T - T_n)R(z)]r_\sigma([1 - (T - T_n)R(z)]^{-1}) \le \frac{\varepsilon}{1 - \varepsilon}$$

(Exercise 2.7).

The desired result follows easily. □

For $T_n(t) = T_n - t(T_n - T)$, let us define

$$R_n(t, z) = (T_n(t) - z)^{-1},$$

$$P_n(t) = \frac{-1}{2i\pi}\int_\Gamma R_n(t, z)\,dz,$$

$$\lambda_n(t) = \operatorname{tr} T_n(t)P_n(t).$$

Lemma 5.30 shows that $t = 1$ belongs to the domain of analyticity of $R_n(t, z)$, $P_n(t)$, and $\lambda_n(t)$ for all n larger than

(a) $N(z)$ defined by $\sup_{n>N(z)} r_\sigma[(T_n - T)R_n(z)] < 1$ for $R_n(t, z)$;
(b) $N(\Gamma)$ defined by $\sup_{n>N(\Gamma)} \max_{z\in\Gamma} r_\sigma[(T_n - T)R_n(z)] < 1$ for $P_n(t)$ and $\lambda_n(t)$.

This is applied now to get the series expansions for $x = R(z)f$ and λ, φ, eigenelements of T, when the approximation T_n is uniformly radially convergent to T.

6.2. *Iterative Refinement of the Solution*

For z fixed in $\rho(T)$, we consider the equation

$$(T - z)x = f. \tag{5.9}$$

We suppose that $T_n - z \overset{\sigma}{\to} T - z$, and approximate (5.9) in X by

$$(T_n - z)x_n = f. \tag{5.10}$$

The solution $x = R(z)f$ is approximated by $x_n = R_n(z)f$. Let n be *fixed* larger than $N(z)$.

Proposition 5.31 *If* $T_n - z \overset{\sigma}{\to} T - z$, *then, for any fixed n larger than* $N(z)$, *the following sequence converges to x as* $k \to \infty$:

$$x^0 := x_n, \qquad x^{k+1} := x^k + R_n(z)(f - (T - z)x^k), \qquad \text{for} \quad k \geq 0. \tag{5.11}$$

Proof Proposition 3.29 with T' taken as T_n yields

$$\begin{aligned} x^{k+1} &= x_n + R_n(z)(T_n - T)x^k \\ &= R_n(z)f + R_n(z)(T_n - z - (T - z))x^k, \end{aligned}$$

which is (5.11), □

Exercises

5.23 Show that $x^k \to x$ when $k \to \infty$ at least as fast as a geometric progression with a quotient arbitrarily close to $r_\sigma[(T_n - T)R_n(z)]$.

5.24 If the approximation T_M is used for the evaluation of Tx^{k-1} ($M \gg n$), show that $x^k \to x_M := (T_M - z)^{-1}f$ as $k \to \infty$, if either $\|T_n - T\| \to 0$ or $T_n \overset{cc}{\to} T$.

Formula (5.11) can be put into the alternative form (5.12).

Proposition 5.32 *The recursion formula* (5.11) *can be written, if* $z \neq 0$,

$x^0 := x_n$,

$$x^{k+1} := \frac{1}{z}(Tx^k - f) + \frac{1}{z}R_n(z)T_n(f - (T - z)x^k) \qquad \textit{for} \quad k \geq 0. \tag{5.12}$$

Proof Use (5.11) and the identity $R_n(z) = (1/z)(R_n(z)T_n - 1)$. □

x^{k+1} is the result of the correction of the iterated solution $\tilde{x}^k = (1/z)(Tx^k - f)$ by r^k, which is the solution of the equation

$$(T_n - z)r^k = T_n(x^k - \tilde{x}^k),$$

since the residual $f - (T - z)x^k$ equals $z(x^k - \tilde{x}^k)$. When $T_n = \pi_n T$, r^k is the solution in $X_n = \pi_n X$ of the equation

$$(\pi_n T - z)r^k = \pi_n T(x^k - \tilde{x}^k).$$

Exercises

5.25 Perform one step of the iteration (5.11) with $T_n = \pi_n T \pi_n$, starting from $x_n = (\pi_n T \pi_n - z)^{-1} f$. We define

$$x_n^{\mathrm{G}} = (\pi_n T - z)^{-1} \pi_n f, \qquad \tilde{x}_n^{\mathrm{G}} = (1/z)(Tx_n^{\mathrm{G}} - f) = x_n^{\mathrm{S}}.$$

Show that the first iterate x^1 satisfies

$$x^1 = \tilde{x}_n^{\mathrm{G}} + (1/z)R_n^{\mathrm{G}}(z)T(1 - \pi_n)f.$$

5.26 Repeat Exercise 5.25, starting from x_n^{G}. Show that the first iterate is

$$x^{1\mathrm{G}} = \tilde{x}_n^{\mathrm{G}} + (1/z)(1 - \pi_n)f.$$

5.27 If T is compact and $\pi_n \xrightarrow{\mathrm{p}} 1$, show that the rate of convergence of (5.12) for $T_n = \pi_n T$ is given by

$$\|x - x^k\| \le c(\|(1 - \pi_n)T\| \; \|(\pi_n T - z)^{-1}\|)^{k+1} \qquad \text{for} \quad k \ge 0.$$

Example 5.19 In Lin Qun (1982a) the iteration (5.12) is considered with $T_n = \pi_n T$, and $x^0 := x_n^{\mathrm{G}} = (\pi_n T - z)^{-1}\pi_n f$ as starting point. Note that $\tilde{x}^0$ is the Sloan solution $\tilde{x}_n$ in that case.

Example 5.20 In Chapter 7, Section 8, we develop the practical application of this method to an integral operator approximated by projection or by numerical quadrature.

Example 5.21 The iterative refinement method is a special case of the iterative defect correction applied with the discretized operator T_n. In our case, the defect, or residual, $f - (T - z)x^k$ is updated by evaluating Tx^k. Various approximations T_M of increasing accuracy can be used to estimate Tx^k. The method is then similar to the variant called *iterative updating defect correction* in Stetter (1978).

Example 5.22 Iterative refinement of type (5.11) is considered in Atkinson (1973, 1976a, Chap. 4) for compact integral operators approximated by projection methods or by the Nyström method. For this latter method, Brakhage (1960) proposes the use of the iterated vector $\tilde{x}^k = (1/z)(Tx^k - f)$

when $z \neq 0$ as an answer to the oscillatory behavior of the residual $f - (T - z)x^k = (T_n - T)y_n^k$. Starting from the identities

$$x = \frac{1}{z}(Tx - f),$$

$$(T_n - z)x = f + (T_n - T)x = f + \frac{1}{z}(T_n - T)(Tx - f),$$

and from the Nyström solution x_n^N he proposes the iteration

$$x'_0 = x_n^N, \qquad \tilde{x}'^k = \frac{1}{z}(Tx'^k - f),$$

$$x'^{k+1} = R_n(z)[f + (T_n - T)\tilde{x}'^k] \qquad \text{for} \quad k \geq 0, \tag{5.13}$$

which is often superior to (5.11). Numerical experiments are given in Atkinson (1973, 1976a, Chap. 4). The evaluation of Tx, where x is a known function, is done by means of the approximation $T_M x$, $M \gg n$.

Example 5.23 The principle of smoothing the residual to improve the correction is the essence of the various multigrid techniques (Brandt, 1977; Hackbusch, 1979; McCormick, 1980; Hemker and Schippers, 1981).

The numerical importance of smoothing the residual will be more fully discussed in Section 6.6 in connection with the iterative defect correction method.

Exercises

5.28 For the Nyström method T_n applied to an intergral operator T with continuous kernel, show that $\|R_n(z)(T_n - T)\|$ may not tend to 0, but

$$c_n(z) := \|[R_n(z)(T_n - T)]^2\| \to 0 \qquad \text{as} \quad n \to \infty.$$

Deduce that

$$\|x - x^{k+2}\| \leq c_n(z)\|x - x^k\|, \qquad k \geq 0,$$

where x^k satisfies (5.11).

5.29 Show that the sequence x'^k defined by (5.13) converges geometrically to x with the rate $(1/|z|)\|R_n(z)(T_n - T)T\|$, which tends to zero as $n \to \infty$ if T is compact and $T_n - z \xrightarrow{s} T - z$.

5.30 Prove that (5.13) is equivalent to

$$x'^{k+1} = x'^k + (1/z)[-1 + R_n(z)T][f - (T - z)x'^k].$$

Check that $(1/z)[-1 + R_n(z)T]$ is an approximation of $R(z)$.

5.31 Show that (5.13) amounts to solve an approximation of the Kantorovitch regularized form of the equation

$$(T - z)(x - x'^k) = d^k := f - (T - z)x'^k.$$

6.3. Iterative Refinement for the Eigenelements

We consider the eigenvalue problem

$$T\varphi = \lambda\varphi, \qquad \varphi \neq 0, \tag{5.6}$$

where we suppose that λ is *simple* and isolated by the curve Γ. We approximate (5.6) in X by

$$T_n\varphi_n = \lambda_n\varphi_n, \qquad \|\varphi_n\| = 1. \tag{5.14}$$

Let n be *fixed*, larger than $N(\Gamma)$, set $r_\Gamma(n) := \max_{z\in\Gamma} r_\sigma[(T - T_n)R_n(z)]$. The disk $\delta_\Gamma(n) = \{t \in \mathbb{C}; |t| < r_\Gamma^{-1}(n)\}$ is the analyticity domain of $\lambda_n(t)$ and $P_n(t)$. Since $t = 1$ belongs to $\delta_\Gamma(n)$, we have $1 = \dim PX = \dim P_nX$. There is only one simple eigenvalue λ_n of T_n inside Γ.

We suppose that D is *dense* in X, so that the adjoint operator T_n^* is uniquely defined. Let φ_n^* be the eigenvector of T_n^* associated with $\bar{\lambda}_n$ and normalized by $\langle\varphi_n^*, \varphi_n\rangle = 1$. Then $P_n = \langle\cdot, \varphi_n^*\rangle\varphi_n$. We also define

$$S_n := \lim_{z\to\lambda_n} R_n(z)(1 - P_n),$$

the reduced resolvent of T_n at λ_n.

We assume that the following condition analogous to (3.17) is fulfilled

$$\langle P_n(t)\varphi_n, \varphi_n^*\rangle \quad \text{is nonzero for} \quad |t| \leq 1. \tag{5.15}$$

Then $P_n(t)\varphi_n \neq 0$ is an eigenvector for $T_n(t)$ such that $P_nP_n(t)\varphi_n \neq 0$. Let $\phi_n(t)$ be the eigenvector normalized by $P_n\phi_n(t) = \varphi_n$. We set $\phi := \phi_n(1)$, the eigenvector for T normalized by $P_n\phi = \varphi_n$. Note that $\phi = \tilde{P}_n^{-1}\varphi_n$ (see Lemma 6.1, Chapter 6).

We give a sufficient condition for (5.15).

Lemma 5.33 *We suppose that $T_n \xrightarrow{p} T$ and that there exists an integer p such that, for all z on Γ, $\|[(T - T_n)R(z)]^p\| \to 0$. Given $r > 1$, then for n large enough, $\lambda_n(t)$ and $\phi_n(t)$ are analytic for $|t| < r$.*

Proof Under the above assumption $T_n - z \xrightarrow{u\sigma} T - z$ on Γ. We set $K_n(z) := (T - T_n)R_n(z)$; then for $n > N$ depending on Γ

$$K_n^p(z) = [(T - T_n)R(z)]^p[1 - (T - T_n)R(z)]^{-p}$$

for z on Γ and $\|K_n^p(z)\| \to 0$ as $n \to \infty$, whereas $K_n^j(z) \xrightarrow{p} 0$, $j = 1, \ldots, p - 1$. For $\xi \in X$, $P_n(P_n(t) - P_n)P_n\xi = \langle\xi, \varphi_n^*\rangle\langle(P_n(t) - P_n)\varphi_n, \varphi_n^*\rangle\varphi_n$; therefore $r_\sigma[(P_n(t) - P_n)P_n] = r_\sigma(P_n(P_n(t) - P_n)P_n) = |\langle P_n(t)\varphi_n, \varphi_n^*\rangle - 1|$ is less than 1 implies $0 < \langle P_n(t)\varphi_n, \varphi_n^*\rangle < 2$. Formally

$$P_n(P_n(t) - P_n)P_n\xi = \frac{-1}{2i\pi}\int_\Gamma P_nR_n(z)\sum_{k=1}^{\infty} t^kK_n^k(z)P_n\xi$$

$$= \frac{-1}{2i\pi}\langle\xi, \varphi_n^*\rangle\int_\Gamma \frac{1}{\lambda_n - z}\left\langle\sum_{k=1}^{\infty} t^kK_n^k(z)\varphi_n, \varphi_n^*\right\rangle\varphi_n\,dz.$$

We get

$$r_\sigma[(P_n(t) - P_n)P_n] \le \frac{\text{meas}\,\Gamma}{2\pi} \max_{z\in\Gamma} \frac{1}{|\lambda_n - z|} \left| \left\langle \sum_{k=1}^{\infty} t^k K_n^k(z)\varphi_n, \varphi_n^* \right\rangle \right|.$$

$$\sum_{k=1}^{\infty} t^k K_n^k(z)\varphi_n = \left(\sum_{k=1}^{\infty} [t^p K_n^p(z)]^k\right)\left(\sum_{j=0}^{p-1} t^j K_n^j(z)\varphi_n\right) + \sum_{j=1}^{p-1} t^j K_n^j(z)\varphi_n.$$

For $j = 1, \dots, p-1$, $K_n^j(z) \overset{p}{\to} 0$ and $P_n \overset{cc}{\to} P$; then

$$\|K_n^j(z)\varphi_n\| \le \|K_n^j(z)P_n\| \to 0.$$

Let $r > 1$ and ε, $0 < \varepsilon < 1$, be given. There exists N such that, for $n > N$,

(i) $\max(r^p\|K_n^p(z)\|, r^j\|K_n^j(z)\varphi_n\|; z \in \Gamma, j = 1, \dots, p-1) < \varepsilon$,
(ii) $|\lambda - \lambda_n| < \varepsilon$.

Then $r_\sigma[(P_n(t) - P_n)P_n] \le \alpha(\varepsilon/(1-\varepsilon)) + \beta\varepsilon$ for $|t| < r$. And ε can be chosen in $]0, 1[$ such that this last constant is less than 1. Therefore (5.15) is fulfilled for $|t| < r$. Note that (i) implies that

$$r_\Gamma(n) \le \|K_n^p(z)\|^{1/p} < \frac{\varepsilon^{1/p}}{r} < \frac{1}{r} < 1. \quad \square$$

Lemma 5.34 *Under the hypotheses of Lemma 5.33, for any fixed n large enough, the following expansions are convergent for* $|t| \le 1$:

$$\lambda_n(t) = \sum_{k=0}^{\infty} t^k v^k \quad \text{and} \quad \phi_n(t) = \sum_{k=0}^{\infty} t^k \eta^k,$$

where v^k *and* η^k *are solutions of the iterative formulas*

$$v^0 = \lambda_n, \qquad v^k = \langle (T - T_n)\eta^{k-1}, \varphi_n^* \rangle \qquad \text{for} \quad k \ge 1,$$

$$\eta^0 = \varphi_n, \qquad \eta^k = S_n\left[(T_n - T)\eta^{k-1} + \sum_{i=1}^{k} v^i \eta^{k-i}\right] \qquad \text{for} \quad k \ge 1.$$

Proof Clear from Lemmas 3.30 and 5.33. $\square$

We define the sequences $\lambda^k := \sum_{i=0}^{k} v^i$ and $\varphi^k := \sum_{i=0}^{k} \eta^i$.

Theorem 5.35 *Under the hypotheses of Lemma 5.33, for any fixed n large enough,* $\lambda = \lim_k \lambda^k$ *and* $\phi = \lim_k \varphi^k$, *where* λ^k *and* φ^k *are solutions of*

$$\lambda^0 = \lambda_n, \quad \lambda^k = \langle T\varphi^{k-1}, \varphi_n^* \rangle, \qquad \varphi^0 = \eta^0 = \varphi_n,$$

$$(T_n - \lambda_n)(\varphi^k - \varphi_n) = (1 - P_n)\left[(T_n - T)\varphi^{k-1} + \sum_{i=1}^{k}\sum_{j=1}^{i} v^j \eta^{i-j}\right], \tag{5.16}$$

where $\lambda^k - \lambda^{k-1} = v^k$ *and* $\varphi^k - \varphi^{k-1} = \eta^k$ *for* $k \ge 1$.

Proof Clear from Proposition 3.31 and Lemma 5.34. $\square$

Corollary 5.36 *Under the hypotheses of Lemma 5.33, the rates of convergence of the sequence λ^k and φ^k are dominated, for n large enough, by*

$$|\lambda - \lambda^k| \le C\frac{q^{k+1}}{1-q}, \qquad \|\phi - \varphi^k\| \le C'\frac{q'^{k+1}}{1-q'},$$

where q and q′ are such that

$$\max_{z\in\Gamma} r_\sigma[(T_n - T)R_n(z)] < q < q' < 1.$$

Proof For $n > N(\Gamma)$, $\lambda_n(t)$ is analytic in $\delta_\Gamma(n)$, and we apply Proposition 3.32 to $T_n(t)$. This yields a rate of convergence q such that $r_\Gamma(n) < q < 1$. We proved in Lemma 5.33 that $\phi_n(t)$ is analytic for $|t| < r$ for n large enough and that $r_\Gamma(n) < 1/r$. For any $q' > 1/r$ we apply the Cauchy inequality. And clearly q can be chosen such that $q < q'$. □

Exercise

5.32 Let T_n be the Galerkin approximation $\pi_n T\pi_n$. Prove that $\lambda^1 = \lambda_n^G$ and $\varphi^1 = \tilde{\varphi}_n^G = \varphi_n^S$.

So far we have considered the case where λ is simple. The case of a single eigenvalue of algebraic multiplicity m or of a group of m close eigenvalues can be treated by adapting the material presented in Section 8.3, Chapter 3. Let $\{\varphi_{in}\}_1^m$ be a basis of M_n, then given $r > 1$, for n large enough, $P_n(t)\varphi_{in} \neq 0$, $i = 1, \ldots, m$, for $|t| < r$ under the assumption of Lemma 5.33. Therefore an analog to Theorem 3.41 can be stated.

6.4. *Practical Aspects of the Computation*

6.4.1. Choice of a starting projection

To establish formulas (5.16) we have used the *spectral* projection P_n of T_n, that is, the knowledge of φ_n and φ_n^*, since $P_n = \langle\cdot, \varphi_n^*\rangle\varphi_n$. This is not necessary however, and moreover, it may not be wise when P_n is ill-conditioned, that is, when $\|\varphi_n^*\|$ is large.

Let $y \in X^*$ be such that $\langle y, \varphi_n\rangle \neq 0$. We set for simplicity $\langle y, \varphi_n\rangle = 1$, and we consider the projection $Q := \langle\cdot, y\rangle\varphi_n$ on the eigendirection $\{\varphi_n\}$ along $K := \operatorname{Ker} Q = \{y\}^\perp \cap X$ (see Exercise 2.3). Clearly $\|Q\| = \|y\| = 1/\operatorname{dist}(\varphi_n, K)$ (see Theorem 2.4 and Exercise 6.32). The operator $\mathring{T}_{n,\lambda_n} = [(1-Q)(T_n - \lambda_n)]_{\restriction K}$ is invertible on $K = (1-Q)X$ (see Chapter 3, Section 8.3), and we call Σ_n the bounded operator $(1-Q)\mathring{T}_{n,\lambda_n}^{-1}(1-Q)$. By considering $Q(P_n(t) - P_n)Q$, it can be shown (Exercise 5.35) that the condition $\langle P_n(t)\varphi_n, y\rangle$ is nonzero for $|t| \le 1$ is satisfied for n large enough under the assumption of Lemma 5.33. Therefore $QP\varphi_n \neq 0$, and we denote by $\hat{\phi}$ the eigenvector of T normalized by $Q\hat{\phi} = \varphi_n$. We now show that $\hat{\phi}$ is the limit of a sequence $\hat{\varphi}^k$ such that $Q\hat{\varphi}^k = \varphi_n$.

Theorem 5.37 *Under the hypotheses of Lemma 5.33, for n fixed large enough,* λ *and* $\hat{\phi}$ *are limits of the sequences* λ^k, $\hat{\varphi}^k$ *defined by*

$$\lambda^0 = \lambda^n, \qquad \lambda^k = \langle T\hat{\varphi}^{k-1} + T_n\hat{\eta}^k, y\rangle,$$

$$\hat{\varphi}^0 = \hat{\eta}^0 = \varphi_n, \qquad \hat{\varphi}^k = \varphi_n + \Sigma_n\left[(T_n - T)\hat{\varphi}^{k-1} + \sum_{i=1}^{k}\sum_{j=1}^{i} v^j\hat{\eta}^{i-j}\right], \tag{5.17}$$

where $\lambda^k - \lambda^{k-1} = v^k$ *and* $\hat{\varphi}^k - \hat{\varphi}^{k-1} = \hat{\eta}^k$ *for* $k \geq 1$.

Proof Let $\lambda_n(t)$, $\hat{\varphi}_n(t)$ be the eigenelements of $T_n(t)$, where $\hat{\phi}_n(t)$ is normalized by $Q\hat{\phi}_n(t) = \varphi_n$. $\hat{\phi}_n(t) = \hat{\gamma}(t)P_n(t)\varphi_n$ with $\hat{\gamma}(t) = 1/\langle P_n(t)\varphi_n, y\rangle$. It follows easily from Exercise 5.35 that $\lambda_n(t)$ and $\hat{\phi}_n(t)$ are analytic functions of t for $|t| \leq 1$. Formulas (5.17) are obtained by identification like formulas (5.16). Let us set $\lambda_n(t) := \sum_{k=0}^{\infty} t^k v^k$ and $\hat{\phi}_n(t) := \sum_{k=0}^{\infty} t^k\hat{\eta}^k$. Because $\lambda_n(t)$ is analytic for $|t| \leq 1$, its series expansion is unique. For $k = 0$, we may choose $v^0 = \lambda^0 = \lambda_n$, $\hat{\eta}^0 = \hat{\varphi}^0 = \varphi_n$. For $k = 1$, we get

$$(T_n - \lambda_n)\hat{\eta}^1 - (T_n - T)\varphi_n = v^1\varphi_n.$$

In $(1 - Q)X$, this yields the equation

$$(1 - Q)(T_n - \lambda_n)\hat{\eta}^1 = (1 - Q)(T_n - T)\varphi_n,$$

which has the unique solution $\hat{\eta}^1 = \Sigma_n(T_n - T)\varphi_n$. Therefore in QX

$$v^1 = \langle (T_n - \lambda_n)\hat{\eta}^1 - (T_n - T)\phi_n, y\rangle = \langle T\varphi_n - T_n\varphi_n + T_n\hat{\eta}^1, y\rangle$$

and

$$\lambda^1 = \lambda_n + v^1 = \langle T\varphi_n + T_n\hat{\eta}^1, y\rangle. \quad \square$$

The corrections $\sum_{k=1}^{\infty} \eta^k$ and $\sum_{k=1}^{\infty} \hat{\eta}^k$ to the approximate eigenvector φ_n belong to two different subspaces, $(1 - P_n)X$ and $(1 - Q)X$, respectively (cf. Fig. 5.5).

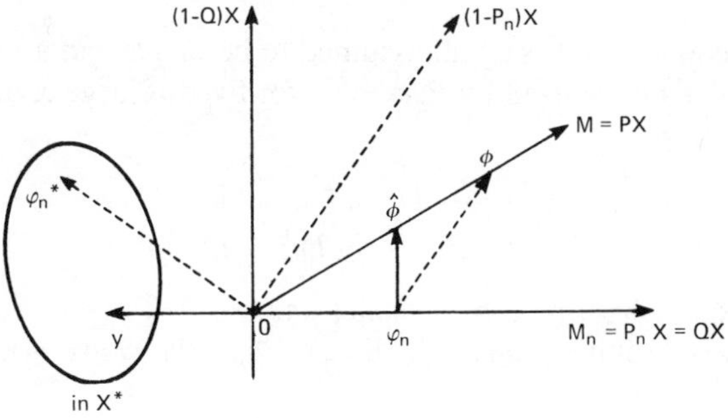

Figure 5.5

Exercise

5.33 Compare the complexity of formulas (5.16) and (5.17).

6.4.2. Condition of the systems to be solved

The practical computation of the sequences x^k, λ^k and φ^k or $\hat{\varphi}^k$ is based on the resolution of the sequences of equations in X:

$$(T_n - z)\zeta^k = f^k, \tag{5.18}$$

$$\begin{aligned} (T_n - \lambda_n)\varepsilon^k &= (1 - P_n)f^k, \qquad P_n\varepsilon^k = 0 \\ (1 - Q)(T_n - \lambda_n)\hat{\varepsilon}^k &= (1 - Q)f^k, \qquad Q\hat{\varepsilon}^k = 0 \end{aligned} \tag{5.19}$$

where f^k is an element of X and $\zeta^k = x^k - x_n$, $\varepsilon^k = \varphi^k - \varphi_n$ or $\hat{\varepsilon}^k = \hat{\varphi}^k - \varphi_n$ for $k \geq 1$.

The exact solution of (5.9) (resp. (5.6)) is the limit of sequences of elements obtained by solving (5.18) (resp. (5.19)), where only the right-hand side f^k (resp. $(1 - P_n)f^k$ or $(1 - Q)f^k$) varies over the iterations. The computation of f^k requires the evaluation of Tg^k, where g^k is a known element of X. If this evaluation cannot be done in closed form, one may use an approximation T_M of higher order than T_n.

Exercises

5.34 Let T_M be an approximation of T. λ_M and ϕ_M are the eigenelements of T_M, where ϕ_M is normalized by $P_n\phi_M = \varphi_n$. If the approximation T_M is used for the evaluation of $T\varphi^{k-1}$ on the right-hand side of (5.16), show that $\lambda^k \to \lambda_M$, $\varphi^k \to \phi_M$ as $k \to \infty$ if $\|T_n - T\| \to 0$ or $T_n \overset{\text{cc}}{\to} T$.

5.35 Prove that Lemma 5.33 is valid for $\phi_n(t)$.

6.5. *Other Iterations When T Is Compact*

The eigenvalue λ of T is again assumed to be *simple* and nonzero. ϕ is the eigenvector of T normalized by $P_n\phi = \varphi_n$ for fixed n, large enough. We consider the iteration

$$\begin{aligned} \mu^0 &:= \lambda_n, \qquad \mu^{k+1} = \langle Tu^k, \varphi_n^* \rangle, \qquad k \geq 0, \\ u^0 &:= \varphi_n, \qquad u^{k+1} = u^k - S_n(Tu^k - \mu^{k+1}u^k), \qquad k \geq 0. \end{aligned} \tag{5.20}$$

Proposition 5.38 *Let T be compact and $\|T_n - T\| \to 0$. For any fixed n large enough, $\lambda = \lim_k \mu^k$ and $\phi = \lim_k u^k$, and the convergence is at least linear:*

$$|\mu^k - \lambda| + \|u^k - \phi\| \leq c\|T - T_n\|^{k+1}, \qquad k \geq 0.$$

Proof We prove the convergence by induction. For $k = 0$, $|\lambda_n - \lambda| + \|\varphi_n - \phi\| \le c\|T - T_n\|$ is proved as follows. $|\lambda_n - \lambda| \le c\|(T - T_n)P\|$ by Theorem 6.6 of Chapter 6. Set $\phi := y_n P\varphi_n$. Then $\gamma_n^{-1} = \langle P\varphi_n, \varphi_n^*\rangle$ and

$$\begin{aligned}|\gamma_n^{-1} - 1| &\le |\langle P\varphi_n - \varphi_n, \varphi_n^*\rangle|\\ &\le \|P - 1\| \operatorname{dist}(\varphi_n, M)\|\varphi_n^*\| \le c\|(T - T_n)P\|,\end{aligned}$$

again by Theorem 6.6. Now

$$\phi - \varphi^n = \gamma_n P\varphi_n - \varphi_n = (\gamma_n - 1)P\varphi_n + P\varphi_n - \varphi_n$$

and

$$\|\phi - \varphi_n\| \le c\|(T - T_n)P\|.$$

We assume that $|\mu^k - \lambda| + \|u^k - \phi\| \le c\|T - T_n\|^{k+1}$ holds. We write the identity

$$\begin{aligned}&(T_n - \lambda_n)(u^{k+1} - \phi)\\ &\quad = (T_n - \lambda_n)u^k - (1 - P_n)(Tu^k - \mu^{k+1}u^k) - (T_n - \lambda_n)\phi\\ &\quad = (T_n - T)(u^k - \varphi) + (\mu^{k+1} - \lambda)u^k + (\lambda - \lambda_n)(u^k - \phi).\end{aligned}$$

Then

$$\mu^{k+1} - \lambda = \langle (T - T_n)(u^k - \phi), \varphi_n^*\rangle \tag{5.21}$$

and

$$u^{k+1} - \phi = (\mu^{k+1} - \lambda)S_n u^k + (\lambda - \lambda_n)S_n(u^k - \phi) + S_n(T_n - T)(u^k - \phi) \tag{5.22}$$

since $P_n(u^{k+1} - \phi) = 0$. We deduce

$$\begin{aligned}|\mu^{k+1} - \lambda| &\le c(|\lambda - \lambda_n| + \|T - T_n\|)\|u^k - \phi\|\\ \|u^{k+1} - \phi\| &\le c[|\mu^{k+1} - \lambda| + (|\lambda - \lambda_n| + \|T - T_n\|)\|u^k - \phi\|]\\ &\le c\|T - T_n\|^{k+2}. \quad \square\end{aligned}$$

The convergence of (5.20) still holds if $T_n \overset{cc}{\to} T$.

Proposition 5.39 *If* $T_n \overset{cc}{\to} T$, *then, for n fixed large enough,*

$$\begin{aligned}&\max(|\mu^{2k+1} - \lambda| + \|u^{2k+1} - \phi\|, |\mu^{2k} - \lambda| + \|u^{2k} - \phi\|)\\ &\qquad \le c\varepsilon_n^k\|(T - T_n)P\|, \qquad k \ge 0,\end{aligned}$$

with $\varepsilon_n = \|(T - T_n)P\| + \|(T - T_n)S_n(T - T_n)\|$.

Proof We first note that $\mu^{2k+1} - \lambda = \langle T(u^{2k} - \phi), \varphi_n^*\rangle$ by definition and that by (5.22)

$$\|u^{2k+1} - \phi\| \le c[|\mu^{2k+1} - \lambda| + |\lambda - \lambda_n|\|u^{2k} - \phi\| + \|u^{2k} - \phi\|].$$

Therefore $|\mu^{2k+1} - \lambda| \leq c\|u^{2k} - \phi\|$ and $\|u^{2k+1} - \phi\| \leq c\|u^{2k} - \phi\|$. It remains to prove that $|\mu^{2k} - \lambda| + \|u^{2k} - \phi\| \leq c\varepsilon_n^k\|(T - T_n)P\|$. It is clear for $k = 0$, we suppose it holds for an arbitrary k. From (5.21) we get $\mu^{2k+2} - \lambda = \langle (T - T_n)(u^{2k+1} - \phi), \varphi_n^* \rangle$ and from (5.22)

$$\begin{aligned}(T - T_n)(u^{2k+1} - \phi) = {} & (\mu^{2k+1} - \lambda)(T - T_n)S_n(u^{2k} - \phi + \phi - \varphi_n) \\ & + (\lambda - \lambda_n)(T - T_n)S_n(u^{2k} - \phi) \\ & + (T - T_n)S_n(T_n - T)(u^{2k} - \phi).\end{aligned}$$

Then

$$\begin{aligned}|\mu^{2k+2} - \lambda| & \leq c(\|u^{2k} - \phi\| + \|\phi - \varphi_n\| + |\lambda - \lambda_n| + \varepsilon_n)\|u^{2k} - \phi\| \\ & \leq c\varepsilon_n\|u^{2k} - \phi\|.\end{aligned}$$

Using again (5.22) we bound similarly $\|u^{2k+2} - \phi\| \leq c\varepsilon_n\|u^{2k} - \phi\|$. We remark that $\varepsilon_n \to 0$ as $n \to \infty$ since $T_n \overset{cc}{\to} T$, □

We may interpret (5.20) in the following way. ϕ is a solution of $(T - \lambda)\phi = 0$, and (5.20) is a natural analog to (5.11) for $(T - z)x - f = 0$. The singularity of $(T - \lambda)^{-1}$ is dealt with by working with the reduced resolvent S_n, an approximation of $S = (T - \lambda)^{-1}(1 - P)$.

An iteration very similar to (5.20) has been proposed by Lin Qun for a self-adjoint compact operator and a projection method $T_n = \pi_n T$ (Lin Qun, 1981a, 1982a). It can be written

$$\begin{aligned}&\mu^0 := \lambda_n, \qquad \mu^{k+1} = \langle Tu^k, \varphi_n^* \rangle, \qquad k \geq 0, \\ &u^0 := \varphi_n, \qquad u^{k+1} = u^k + \lambda_n S_n(u^k - (1/\mu^{k+1})Tu^k), \qquad k \geq 0.\end{aligned} \tag{5.23}$$

Using the identity $\lambda_n S_n = P_n - 1 + S_n T_n$, we remark, as in Proposition 5.32, that (5.23) can be put into the equivalent form

$$u^{k+1} = \frac{1}{\mu^{k+1}} Tu^k + S_n T_n\left(u^k - \frac{1}{\mu^{k+1}} Tu^k\right). \tag{5.24}$$

If we set $\hat{u}^k := (1/\mu^{k+1})Tu^k$, then $u^{k+1} = \hat{u}^k + r^k$, where r^k in the unique solution of $(T_n - \lambda_n)r^k = T_n(u^k - \hat{u}^k)$, such that $P_n r^k = 0$. If $T_n = \pi_n T$, r^k is the Galerkin solution in $X_n = \pi_n X$. This will be used in Chapter 7 in connection with the f.e.m. Note that now $\lambda_n S_n$ is an approximation of

$$\lambda S = \left(\frac{1}{\lambda} T - 1\right)^{-1}(1 - P)$$

associated with the equation $(1 - (1/\lambda)T)\phi = 0$.

Another iteration has also been proposed by Lin Qun (1982):

$$\begin{aligned}&\mu^0 = \lambda_n, \qquad \mu^{k+1} = \langle Tu^k, \varphi_n^* \rangle, \qquad k \geq 0 \\ &u^0 = \varphi_n, \qquad u^{k+1} = u^k - \Sigma_n^{k+1}(Tu^k - \mu^{k+1}u^k), \qquad k \geq 0,\end{aligned} \tag{5.25}$$

where $\Sigma_n^{k+1} = (T_n - \mu^{k+1})^{-1}(1 - P_n)$. The computation is more expensive since the system to solve varies with k.

Exercises

5.36 Consider the eigenvalue problem $\lambda K\varphi = \varphi$, let ψ in X^* be such that $\langle \varphi, \psi \rangle = 1$. Check that (φ, λ) is a zero of the nonlinear operator F defined on $X \times \mathbb{C}$ by $(x, z) \mapsto ((1 - zK)x, \langle x, \psi \rangle - 1)$. Write the Newton's iteration for the equation $F(u) = 0$ with $u = (x, z)$ from an approximate zero $u_0 = (x_0, z_0)$. Write the modified algorithm obtained by letting $z = z_0$ fixed in the derivative F'. Show that the algorithm (5.23) results from a suitable approximation by projection of the latter, together with the choice $z_0 = \lambda_n$, $x_0 = \varphi_n$, $\psi = \varphi_n^*$.

5.37 Establish for the iteration (5.25) the identities

$$\mu^{k+1} - \lambda = \langle (T - T_n)(u^k - \phi), \varphi_n^* \rangle,$$

$$u^{k+1} - \phi = (\mu^{k+1} - \lambda)\Sigma_n^{k+1}\phi + \Sigma_n^{k+1}(T_n - T)(u^k - \phi).$$

Deduce the convergence to λ, ϕ if either $\|T - T_n\| \to 0$ or $T_n \overset{cc}{\to} T$.

5.38 Establish the convergence of the iteration (5.23) when $\|T - T_n\| \to 0$ or $T_n \overset{cc}{\to} T$.

5.39 Study the application of (5.20) on a quasi-decomposable matrix

$$A = \begin{pmatrix} a & v^H \\ u & C \end{pmatrix}$$

with $\|u\|_2$ $\|v\|_2$ small.

We have seen that the iteration (5.20) for an integral compact operator T is convergent when either a projection method or a Nyström method is used to discretize T. Nevertheless when $T_n \overset{cc}{\to} T$, the rate of convergence of (5.20) may be slow, and the iteration (5.24) where the residual at u^k is smoothed by T_n is more advisable (see Ahués *et al.* (1983a) for numerical experiments). Also a variant similar to Brakhage's iteration (5.13) can be devised; $\lambda_n S_n = P_n - 1 + S_n T_n$ is replaced by $P_n - 1 + S_n T$, another approximation of λS, so that the residual at u^k is smoothed by T. This leads to Ahués's iteration (Ahués *et al.*, 1982):

$$\mu^0 = \lambda_n, \qquad \mu^{k+1} = \langle Tu^k, \varphi_n^* \rangle, \qquad k \geq 0,$$

$$u^0 = \varphi_n, \qquad u^{k+1} = \frac{1}{\mu^{k+1}} Tu^k + S_n T\left(u^k - \frac{1}{\mu^{k+1}} Tu^k\right), \qquad k \geq 0. \tag{5.26}$$

Proposition 5.40 *If T is compact and $T_n - z \overset{ss}{\to} T - z$ on Γ, then* (5.26) *converges to λ, ϕ, and*

$$|\mu^k - \lambda| + \|u^k - \phi\| \leq c\|(T - T_n)T\|^{k+1}, \qquad k \geq 0.$$

Proof We first remark that

$$\|(T - T_n)P\| \leq c\|(T - T_n)_{\restriction M}\| \leq c\|(T - T_n)T\|.$$

Then for $k = 0$,

$$\mu^1 - \lambda = \langle T(\varphi_n - \phi), \varphi_n^* \rangle$$

and

$$|\mu^1 - \lambda| + \|u^0 - \phi\| \leq c\|(T - T_n)T\|.$$

Now we suppose that $|\mu^{k+1} - \lambda| + \|u^k - \phi\| \leq c\|(T - T_n)T\|^{k+1}$. It is clear that $|\mu^{k+1} - \lambda| = |\langle T(u^k - \phi), \varphi_n^* \rangle| \leq c\|u^k - \phi\|$. Then we write the identity

$$\begin{aligned}
(T_n - \lambda_n)(u^{k+1} - \phi) &= (T_n - \lambda_n)\frac{Tu^k}{\mu^{k+1}} + Tu^k - \mu^{k+1}\varphi_n \\
&\quad + (P_n - 1)\frac{T^2u^k}{\mu^{k+1}} - (T_n - \lambda_n)\phi \\
&= \frac{1}{\mu^{k+1}}(T_n - T)T(u^k - \phi) \\
&\quad + \frac{\lambda - \mu^{k+1}}{\mu^{k+1}}(T_n - T)\phi - \lambda\phi + \lambda_n\phi \\
&\quad + \frac{1}{\mu^{k+1}}P_nT^2u^k + Tu^k - \frac{\lambda_n}{\mu^{k+1}}Tu^k - \mu^{k+1}\varphi_n \\
&\quad - (\lambda - \lambda_n)\phi + \frac{\mu^{k+1} - \lambda_n}{\mu^{k+1}}Tu^k =: A.
\end{aligned}$$

$$(\mu^{k+1} - \lambda_n)Tu^k = (\mu^{k+1} - \lambda)Tu^k + (\lambda - \lambda_n)T(u^k - \phi) + (\lambda - \lambda_n)\lambda\phi,$$

$$\begin{aligned}
(1 - P_n)A &= \frac{1}{\mu^{k+1}}(T_n - T)T(u^k - \phi) + \frac{\lambda - \mu^{k+1}}{\mu^{k+1}}[(T_n - T)\phi - Tu^k] \\
&\quad + \frac{\lambda - \lambda_n}{\mu^{k+1}}[T(u^k - \phi) + (\lambda - \mu^{k+1})\phi] \\
&= \frac{1}{\mu^{k+1}}(T_n - T)T(u^k - \phi) + \frac{\mu^{k+1} - \lambda_n}{\mu^{k+1}}T(u^k - \phi) \\
&\quad + \frac{\mu^{k+1} - \lambda}{\mu^{k+1}}\lambda\phi.
\end{aligned}$$

By multiplication by S_n we get

$$\begin{aligned}
u^{k+1} - \phi &= \frac{1}{\mu^{k+1}}S_n(T_n - T)T(u^k - \phi) + \frac{\mu^{k+1} - \lambda_n}{\mu^{k+1}}S_nT(u^k - \phi) \\
&\quad + \frac{\mu^{k+1} - \lambda}{\mu^{k+1}}\lambda S_n(\phi - \varphi_n).
\end{aligned}$$

We conclude easily that $\|u^{k+1} - \phi\| \leq c\|(T - T_n)T\|^{k+2}$. □

The iterations (5.20), (5.24), and (5.26) have been defined by means of the projection $P_n = \langle \cdot, \varphi_n^* \rangle \varphi_n$ on $\{\varphi_n\}$. A projection $Q = \langle \cdot, y \rangle \varphi_n$ can be used as well if $\langle \varphi_n, y \rangle = 1$ (cf. Section 6.4).

In Chapter 7, we apply some of the iterative refinement techniques we have described to an integral operator approximated by a finite-rank operator T_n such that either $T_n \overset{cc}{\to} T$ or $\|T_n - T\| \to 0$. The condition of convergence given in Lemma 5.33 will then be satisfied with $p = 2$ or $p = 1$. Because of the special structure of T_n for both projection and approximate quadrature methods, $R_n(z)$, P_n, and S_n can easily be computed from the corresponding matrix operators (see Theorem 4.3 and Corollary 4.4.).

As we have also seen, the discretization of an integral operator often yields a *full* matrix. So a technique that reduces significantly the size of the matirx may prove useful, especially if the memory allocation in the computer is very limited. This technique gives, in particular, a computational scheme to approximate to a given accuracy the eigenelements of certain large matrices (of order M) by calculations with matrices of order n, with $n \ll M$.

We now turn to the presentation of a very powerful method in numerical analysis known as the iterative defect correction method. In order to solve an equation $F(x) = f$, we compute a solution for a nearby equation $\tilde{F}(\tilde{x}) = f$, where F and $\tilde{F}$ are close enough to ensure the existence of a contraction mapping which allows to get x as the limit of a sequence of vectors iteratively computed from $\tilde{x}$.

6.6. *The General Framework of the Iterative Defect Correction Method*

We have already pointed out some connection between the Rayleigh–Schrödinger series and the iterative defect correction (IDC) method described in Stetter (1978). We address this question in more detail now. The presentation is adapted from Ahués *et al.* (1982).

6.6.1. The IDC method

Let F (resp. G): $X \to X$ be a (possibly) nonlinear operator with domain D (resp. Dom G). We are concerned with the equation

$$F(x^*) = 0, \qquad x^* \in D. \tag{5.27}$$

Let $\|\cdot\|_*$ be a norm on X equivalent to $\|\cdot\|$. For $\rho > 0$, we define $B_* := \{x \in X; \|x - x^*\|_* < \rho\}$. We suppose that F is defined on B_*. We assume that we are able to evaluate the direct mapping F and that we know an approximate inverse G according to the following definition.

Definition G is a *local approximate inverse* of F if the following conditions are fulfilled:

(i) $F(B_*) \subseteq \text{Dom}\, G$;
(ii) $G(0) \in B_*$;
(iii) $U := 1 - G \circ F$ is a contraction on B_* with respect to $\|\cdot\|_*$.

The IDC method consists in generating the sequence

$$\begin{aligned} x^0 &:= G(0), \\ x^{k+1} &:= G(0) + x^k - G(F(x^k)), \qquad k \geq 0 \\ &= G(0) + U(x^k). \end{aligned} \tag{5.28}$$

Proposition 5.41 *If G is a local approximate inverse of F, then*

$$\|x^k - x^*\|_* \to 0 \qquad \textit{as} \quad k \to \infty.$$

Proof Set $V(x) := G(0) + U(x)$, x^* is a fixed point of V since $V(x^*) = G(0) + x^* - G(F(x^*)) = x^*$. For any $x \in B_*$,

$$\|V(x) - x^*\|_* = \|V(x) - V(x^*)\|_* \leq l(U)\|x - x^*\|_* < \rho$$

if $l(U) < 1$ is the Lipschitz constant defined by (iii). Hence $V(x) \in B_*$ and $V(B_*) \subset B_*$. Then $\|x^k - x^*\|_* < l^k(U)\|x^0 - x^*\|_*$. F is injective on B_*, so that x^* is the unique zero of F in B_*. □

$x^0 = G(0)$ is an approximation of x^* with defect (or residual) $d^0 := F(x^0)$, the error being $e^* := x^* - G(0) = U(x^*)$. e^* is approximated in turn by $e^1 := x^1 - G(0) = U(x^0) = x^0 - G(d^0)$. This defines the new approximate solution $x^1 = G(0) + x^0 - G(F(x^0))$. Iterating this process yields the IDC method, where $F(x^k)$ is the defect at x^k.

6.6.2. The linear equation

A case of interest is the *linear* equation $(T - z)x = f$, for which F is the affine operator $F(x) = (T - z)x - f$. Because of the linearity of T, a local approximate inverse for F can be derived from an approximate inverse $\tilde{G}$ for $T - z$. Indeed, let $\tilde{G}$ be such that $\tilde{U} = 1 - \tilde{G}(T - z)$ satisfies $r_\sigma(\tilde{U}) < 1$; that is, $r_\sigma[\tilde{G}(T - z)] < 1$. Then the choice $G(x) = \tilde{G}x + u$ for a given u in B_* yields $U(x) = x - \tilde{G}(T - z)x + \tilde{G}f - u$, which is such that $U(x) - U(y) = \tilde{U}(x - y)$; hence U is a contraction with respect to some norm $\|\cdot\|_*$ which is equivalent to $\|\cdot\|$, and G is an approximate inverse for F. The corresponding IDC method is

$$x^0 = u, \qquad x^{k+1} = x^k - \tilde{G}[(T - z)x^k - f], \qquad k \geq 0.$$

Example 5.24 Consider in $X = \mathbb{C}^N$ the linear system of equations $Ax = b$ with unique solution $x^* = A^{-1}b$. We suppose that there exists K such that

$$x^* = Kx^* + Hb \qquad \text{with} \quad H = (I - K)A^{-1}.$$

Upon setting $F(x) = Ax - b$, $G(x) = Ku + Hb + Hx$ for a given u, we find $U(x) = Kx - Ku$, and the IDC method is

$$x^0, \qquad x^{k+1} = Kx^k + Hb, \qquad k \geq 0,$$

which converges to x^* iff $r_\sigma(K) < 1$. U is contracting with respect to the norm $\|\cdot\|_*$, (depending on K) defined in Chapter 2, Section 4.1, where ε is chosen such that $r_\sigma(K) + \varepsilon < 1$. Examples of such an iteration are provided by the Jacobi, Gauss–Seidel, and successive-overrelaxation methods: they correspond to the splitting $A = D(L + I + U)$, where D is the diagonal of A, and L and U are the lower and upper triangles of $D^{-1}A$, respectively. Then $K_J := -(L + U)$ with $Hb = D^{-1}b$, $K_{GS} := -(I + L)^{-1}U$ with $Hb = (I + L)^{-1}D^{-1}b$ and $K_\omega = (I + \omega L)^{-1}((1 - \omega)I - \omega U)$ with $Hb = \omega(I + \omega L)^{-1}D^{-1}b$ and $0 < \omega < 2$.

Example 5.25 We consider again the equation $Ax = b$ and its numerical solution x', exact solution of $A'x' = b$ (cf. Chapter 1, Section 3.5). With the choice $G(x) = A'^{-1}x + A'^{-1}b$, $Ux = x - A'^{-1}Ax = A'^{-1}Hx$ with $H = A' - A$. The IDC method yields

$$x^0 = x' = A'^{-1}b, \qquad x^{k+1} = x' + A'^{-1}Hx^k, \qquad k \geq 0,$$

which converges to x^* iff $r_\sigma(A'^{-1}H) < 1$. This is the iteration defined in Proposition 1.13.

Example 5.26 We now consider in a Banach space X the linear equation (5.9) $(T - z)x = f$ for $z \in \rho(T)$, and its numerical approximation (5.10) $(T_n - z)x_n = f$, and we suppose that $T_n - z \overset{\sigma}{\to} T - z$ (cf. Section 6.2). Let n be fixed larger than $N(z)$. With $F(x) = (T - z)x - f$ and the choice $G(x) = R_n(z)x + x_n$, we find $U = R_n(z)(T_n - T)$ defined in D. The IDC method yields the sequence

$$\begin{aligned} x^0 &= x_n = R_n(z)f, \\ x^{k+1} &= x_n + R_n(z)(T_n - T)x^k, \quad k \geq 0, \\ &= x^k - R_n(z)[(T - z)x^k - f] \\ &= \frac{1}{z}(Tx^k - f) - \frac{1}{z}R_n(z)T_n[(T - z)x^k - f], \end{aligned}$$

which is identical to that defined by (5.11) or (5.12). For n large enough, U can be shown to be a contraction in $\mathscr{L}(\hat{D})$ (cf. Exercise 5.40).

Exercise

5.40 Under the assumption $T_n - z \xrightarrow{\sigma} T - z$, prove that for n fixed large enough U is a contraction in $\mathscr{L}(\hat{D})$. Deduce that $\|x^k - x\| \to 0$.

Example 5.27 We suppose now that T is compact and $T_n - z \xrightarrow{s} T - z$. $z \in \rho(T)$ is nonzero and the choice $\tilde{G} = (1/z)(R_n(z)T - 1)$ implies for $G(x) = \tilde{G}x + x_n$ that

$$U(x) - U(y) = (1/z)R_n(z)(T_n - T)T(x - y).$$

By assumption, $\|(T_n - T)T\| \to 0$ and, for n large enough,

$$\frac{1}{|z|}\|R_n(z)(T_n - T)T\| < 1.$$

The IDC method yields the sequence

$$x^0 = x^n = R_n(z)f,$$

$$x^{k+1} = x^k - \frac{1}{z}(R_n(z)T - 1)[(T - z)x^k - f], \qquad k \geq 0,$$

which is identical to that defined by (5.13) (Exercise 5.30), where T_n is chosen to be the Nyström approximation of T. Here $(1/z)(R_n(z)T - 1)$ is an approximate inverse for $T - z$ different from $R_n(z) = (1/z)(R_n(z)T_n - 1)$ considered in Example 5.26, in particular, it smoothes the residual with T rather than with T_n. Other approximate inverses for $T - z$, T integral are considered in Hemker and Schippers (1981) in connection with multigrid methods.

6.6.3. The eigenvalue problem

We consider the eigenvalue problem (5.6) $T\varphi = \lambda\varphi$, where T is supposed to be *compact* and λ is a nonzero *simple* eigenvalue. Let $\psi \in X^*$ be given such that $\langle \varphi, \psi \rangle \neq 0$, and consider the nonlinear operator

$$F: x \mapsto x - \frac{1}{\langle Tx, \psi \rangle} Tx$$

(resp. $H: x \mapsto Tx - \langle Tx, \psi \rangle x$). Clearly, if φ is normalized by $\langle \varphi, \psi \rangle = 1$, $F(\varphi) = 0$ (resp. $H(\varphi) = 0$) and $\lambda = \langle T\varphi, \psi \rangle$ is nonzero by assumption. Then F is defined around φ. An easy calculation gives for the Fréchet derivative of F at φ, $(D_\varphi F)x = x - (1/\lambda)Tx + (1/\lambda)\langle Tx, \psi \rangle \varphi$ with $\lambda = \langle T\varphi, \psi \rangle$. Set

$$A: x \mapsto \frac{1}{\lambda} Tx - \frac{1}{\lambda}\langle Tx, \psi \rangle \varphi.$$

A is the sum of T and of a rank-1 operator; hence A is compact, and $D_\varphi F = 1 - A$. Then if $1 - A$ is injective, $(D_\varphi F)^{-1} = (1 - A)^{-1} \in \mathscr{L}(X)$. Suppose that $(1 - A)x = 0$; then $(T - \lambda)x = \langle Tx, \psi \rangle \phi$. Because λ is simple, this implies $\langle Tx, \psi \rangle = 0$ and $x = \alpha\varphi$, $\alpha \in \mathbb{C}$, which in turn implies $\alpha\lambda = 0$; hence $\alpha = 0$ and $x = 0$. Therefore F is locally bijective around φ, and φ is an isolated zero. To find a local approximate inverse of F around φ, we consider the singular inhomogenous problem

$$\left(1 - \frac{1}{\lambda} T\right)x = f \tag{5.29}$$

assuming that λ and φ are known. If $f \in (1 - P)X$, (5.29) has for solution the affine manifold $\{\alpha\varphi - \lambda Sf ; \alpha \in \mathbb{C}\}$.

We note that $\varphi - \lambda Sf = \varphi + (P - \lambda S)f$ since $Pf = 0$. We are led to define a local approximate inverse for F under the form

$$G(x) := (P - \lambda S)x + \varphi \qquad \text{for} \quad x \in X.$$

Proposition 5.42 *G is a local approximate inverse of F around φ if ψ is such that $T^*\psi = \bar{\lambda}\psi$.*

Proof

$$V(x) = \varphi + U(x) = x - Px + \lambda Sx + \frac{1}{\langle Tx, \psi \rangle} PTx - \frac{\lambda}{\langle Tx, \psi \rangle} STx.$$

We compute the Fréchet derivative $D_\varphi V$: we get

$$\begin{aligned}(D_\varphi V)x &= x - Px + \lambda Sx + \frac{1}{\lambda} PTx - \frac{1}{\lambda^2} \langle Tx, \psi \rangle PT\varphi - \frac{1}{\lambda} \langle Tx, \psi \rangle ST\varphi \\ &= 0.\end{aligned}$$

V is contracting around φ. □

Of course G is not numerically useful since it assumes the knowledge of λ, φ, and ψ. But it is meaningful in the sense that it provides guidelines on how to define a computable approximate inverse

Example 5.28 We choose

$$G(x) := (P_n - \lambda_n S_n)x + \varphi_n = (1 - S_n T_n)x + \varphi_n.$$

Then

$$U(x) = x + \lambda_n S_n\left(x - \frac{1}{\langle Tx, \varphi_n^* \rangle} Tx\right) - \varphi_n,$$

and the IDC method

$$x^0 = \varphi_n, \qquad x^{k+1} = x^k + \lambda_n S_n\left(x^k - \frac{1}{\langle Tx^k, \varphi_n^* \rangle} Tx^k\right), \qquad k \geq 0,$$

is equivalent to Lin Qun's iteration (5.23) or (5.24).

Example 5.29 $G'(x) := (S_n - (1/\lambda_n)P_n)x + \varphi_n$ is a local approximate inverse of H which defines (5.20). Unlike what happens for the linear equation $(T - z)x = f$, in the case of the eigenvalue problem the sequence defined by (5.20) differs from the sequence φ^k defined by the Rayleigh–Schrödinger expansion $\phi = \sum_{k=0}^{\infty} \eta^k$ in (5.16). Note that φ^k satisfies

$$\varphi^{k+1} = \varphi^k - S_n\left[T\varphi^k - \lambda^1\varphi^k - \sum_{i=1}^{k} v^{i+1}\varphi^{k-i}\right], \qquad k \geq 0.$$

Example 5.30 $\tilde{G}(x) := (1 - S_n T)x + \varphi_n$ is a local approximate for F which defines Ahués's iteration (5.26). It is proved in Ahués *et al.* (1983b) that G(resp. $\tilde{G}$) is a local approximate inverse of F for n large enough under the assumption that $\|T_n - T\| \to 0$ (resp. $T_n - z \overset{\text{ss}}{\to} T - z$ around λ).

6.6.4. Quasi-Newton methods

We consider now the nonlinear equation associated with (5.6) under the form $H(x) := Tx - \langle Tx, \psi \rangle x = 0$, where ψ is again a given vector in X^* such that $\langle \varphi, \psi \rangle = 1$. One may think of modifying Newton's method on H in order to get numerically feasible iterations converging to φ. The original Newton's iteration is

$$x^0, x^{k+1} = x^k - (D_{x^k}H)^{-1}H(x^k), \qquad k \geq 0,$$

with

$$(D_x H)u = Tu - \langle Tx, \psi \rangle u - \langle Tu, \psi \rangle x.$$

We choose $\psi := \varphi_n^*$. The following two examples are taken from Ahués (1982).

Example 5.31 $D_x H$ is approximated by taking x, ψ and T to be respectively φ_n, φ_n^* and T_n. Then Newton's method becomes a method with fixed slope J_n (chord method): $J_n u = (T_n - \lambda_n)u - \langle T_n u, \varphi_n^* \rangle \varphi_n$. Because we seek the difference $x^{k+1} - x^k$ in $(1 - P_n)X$, we may restrict J_n to $(1 - P_n)X$; then $(J_{n\restriction(1-P_n)X})^{-1}(1 - P_n) = S_n$. This defines the iteration

$$x^0 = \varphi_n, \qquad x^{k+1} = x^k - S_n H(x^k)$$

which is (5.20).

Example 5.32 The Jacobian is now approximated by

$$J(x)u = T_n u - \langle Tx, \varphi_n^* \rangle u - \langle T_n u, \varphi_n^* \rangle \varphi_n.$$

With $\mu^{k+1} = \langle Tx^k, \varphi_n^* \rangle$ this yields

$$J(x^k)u = (T_n - \mu^{k+1})u - \langle T_n u, \varphi_n^* \rangle \varphi_n.$$

As previously, $J(x^k)$ may be restricted to $(1 - P_n)X$, then

$$(J(x^k)_{\restriction(1-P_n)X})^{-1}(1 - P_n) = \underline{\Sigma}_n^{k+1},$$

and the iteration

$$x^0 = \varphi_n, \qquad x^{k+1} = x^k - \Sigma_n^{k+1} H(x^k)$$

is (5.25).

6.6.5. The two-level multigrid method

The multigrid method is another example of IDC method, as we show now briefly for the two-level method. Suppose that we want to solve the equation $Tx = f$ in a Banach space, or rather its discretized version in X_h, $\dim X_h < \infty$,

$$T_h x_h = f_h. \tag{5.30}$$

The parameter $h > 0$ represents the diameter of the associated grid. For simplicity of notation we identify X_h with its canonical representation in a given basis of X_h. Similarly T_h is identified with its matrix representation in that basis, etc. Next a coarser grid of diameter $h' > h$ is introduced along with a corresponding discretization $T_{h'}$ and two linear operators p and r. $p\colon X_{h'} \to X_h$ is an injective prolongation and $r\colon X_h \to X_{h'}$ is a surjective restriction. We suppose that the solution of the equation $T_{h'} x_{h'} = f_{h'}$ at the coarse level is computable. We assume also that $\tilde{G}_h = pT_{h'}^{-1}r$ is an approximate inverse for T_h, that is, $r_\sigma(I - pT_{h'}^{-1}rT_h) < 1$, where I denotes the identity on X_h. Then $G_h\colon x \mapsto \tilde{G}_h x + \tilde{G}_h f_h$ is an approximate inverse for F_h, and the IDC method

$$x^0 := pT_{h'}^{-1} r f_h, \qquad x^{k+1} = x^k - pT_{h'}^{-1} r(T_h x^k - f_h), \qquad k \geq 0,$$

is convergent to x_h as $k \to \infty$.

One step of this defect correction iteration is composed with v steps of another iteration on $Tx_h = f_h$ of the relaxation type u^0, $u^{k+1} := Z(u^k) = Ku^k + Hf_h$, where $r_\sigma(K) < 1$. This yields

$$x^0, \qquad x'^k = Z^v(x^k), \qquad x^{k+1} = x'^k - pT_{h'}^{-1} r(T_h x'^k - f_h), \qquad k \geq 0,$$

which converges if $r_\sigma[(I - pT_{h'}^{-1} rT_h)K^v] < 1$.

By an appropriate choice of Z, a rate of convergence much smaller than $r_\sigma(K^v)$ can be achieved (see Hackbusch, 1980b, 1981a). The v steps of the iteration defined by Z represent the smoothing step of the multigrid method.

We now turn to the eigenvalue problem $T\varphi = \lambda\varphi$, where λ is simple. A discretized version is introduced:

$$T_h\varphi_h = \lambda_h\varphi_h, \qquad \varphi_h \in X_h, \tag{5.31}$$

along with the adjoint problem

$$T_h^H\psi_h = \bar{\lambda}_h\psi_h. \tag{5.32}$$

Associated with the coarse grid are two eigenvalue problems analogous to (5.31) and (5.32)

$$T_{h'}\varphi_{h'} = \lambda_{h'}\varphi_{h'}, \qquad T_{h'}^H\psi_{h'} = \bar{\lambda}_{h'}\psi_{h'}$$

with

$$\|\varphi_{h'}\| = \varphi_{h'}^H\psi_{h'} = 1.$$

These coarse grid problems are presumed solved. Moreover, $\lambda_{h'}$ is assumed to be simple. $P_{h'} = \varphi_{h'}\psi_{h'}^H$ is the associated eigenprojection while

$$S_{h'} = (T_{h'} - \alpha I)^{-1}(I - P_{h'})$$

for α a suitable approximation of λ_h. Note that $S_{h'}$ is well defined for h' sufficiently small. (5.31) and (5.32) are solved by employing a smoothing step and a correction step. The smoothing step consists of v Jacobi iterations performed on the relations

$$\varphi_h = \varphi_h - \omega_h(T_h - \alpha I)\varphi_h \tag{5.33}$$

and

$$\psi_h = \psi_h - \omega_h(T_h^H - \bar{\alpha}I)\psi_h. \tag{5.34}$$

Possible starting vectors for the Jacobi iterations are $x^0 = p\varphi_{h'}$ and $y^0 = p\psi_{h'}$. α may be chosen as $\lambda_{h'}$ or as the generalized Rayleigh quotient $\rho(x^k, y^k) := y^{kH}T_hx^k/y^{kH}x^k$, where x^k and y^k are the starting vectors for the kth smoothing step. This smoothing step produces x'^k and y'^k.

The correction step consists of adding $-p\varepsilon$ to the vector x'^k where ε is the solution of a coarse grid system, *viz.*,

$$(T_{h'} - \alpha I)\varepsilon = (I - P_{h'})r(T_hx'^k - \alpha x'^k),$$

$$\psi_{h'}^H\varepsilon = 0.$$

That is, $\varepsilon = S_{h'}r(T_hx'^k - \alpha x'^k)$. The vector y'^k corresponding to (5.34) is likewise corrected by adding $-pS_{h'}^H r(T_h^H y'^k - \bar{\alpha}y'^k)$ (see Ahués and Chatelin, 1983). A complete step (smoothing and correction) is then

$$\begin{aligned} x'^k &= [I - \omega_h(T_h - \alpha I)]^v x^k, \qquad & x^{k+1} &= x'^k - pS_{h'}r(T_hx'^k - \alpha x'^k),\\ y'^k &= [I - \omega_h(T_h^H - \bar{\alpha}I)]^v y^k, \qquad & y^{k+1} &= y'^k - pS_{h'}^H r(T_h^H y'^k - \bar{\alpha}y'^k), \end{aligned} \tag{5.35}$$

and the updated value λ^{k+1} (for the eigenvalue λ_h) is taken to be the Rayleigh quotient $\rho(x^{k+1}, y^{k+1})$ for T_h. Hackbusch (1979) gives proofs of the convergence of (5.35) to φ_h and ψ_h for suitable choices of v, h and h'.

If one is only interested in solving (5.31), one may proceed as follows. Set $F_h(x) := T_h x - (y^H T_h x)x = 0$, where y is a given vector such that $y^H\varphi_h \neq 0$. To define the correction step, consider the IDC method

$$\mu^0 = \lambda_{h'}, \qquad \mu^{k+1} = y^H T_h x^k, \qquad k \geq 0,$$
$$x^0 = P\varphi_{h'}, \qquad x^{k+1} = x^k - pS_{h'} rF_h(x^k), \qquad k \geq 0.$$

Often in practice r and p are such that $rp = I$, the identity on $X_{h'}$. Then a sensible choice for y is $y = r^H\psi_{h'}$, since $y^H\varphi_h = \psi_{h'}^H r\varphi_h$ and $\psi_{h'}^H rp\varphi_{h'} = \psi_{h'}^H\varphi_{h'} = 1$. If $p\varphi_{h'}$ is a reasonable approximation of φ_h, then $y^H\varphi_h$ is likely to be nonzero.

If $rp = I$, then $prpr = pr$ and pr is a projection onto the subspace $pX_{h'}$ of X_h of dimension dim $X_{h'}$. The correction added, in the two-level multigrid method on (5.30) to the vector x'^k is the Galerkin solution obtained by solving in $pX_{h'}$ the residual equation at x'^k

$$T_h x = T_h x'^k - f_h.$$

A similar interpretation obviously holds for (5.31) and (5.32) and the associated residual equations

$$(T_h - \alpha I)x = (I - P_h)(T_h x'^k - \alpha x'^k),$$
$$(T_h^H - \bar{\alpha} I)y = (I - P_h^H)(T_h^H y'^k - \bar{\alpha} y'^k).$$

Note that the possible singularity at α (resp. $\bar{\alpha}$) is removed by means of the eigenprojection P_h (resp. P_h^H).

Example 5.33 The aggregation/disaggregation methods described in Chatelin and Miranker (1982a,b) are a generalization of two-level multigrid methods to matrices which are not necessarily the discretization of continuous operators. In particular to compute the dominant eigenvalue of a large matrix A, the convergence of the *single* sequence of vectors produced by the power method is speeded up by performing from time to time a correction on the current vector by aggregation/disaggregation (Chatelin and Miranker, 1982b).

We conclude this section by giving in Fig. 5.6 a plot of the 12th eigenvector of the biharmonic equation $\Delta^2\varphi = \lambda\varphi$ in the square $\Omega =]0, 1[^2$ and satisfying the homogeneous boundary conditions $\varphi = 0$, $\partial\varphi/\partial n = 0$ on $\partial\Omega$, the boundary of Ω. The computation has been done in Hackbush and Hofmann (1980) by the multigrid method by means of finite-element discretizations.

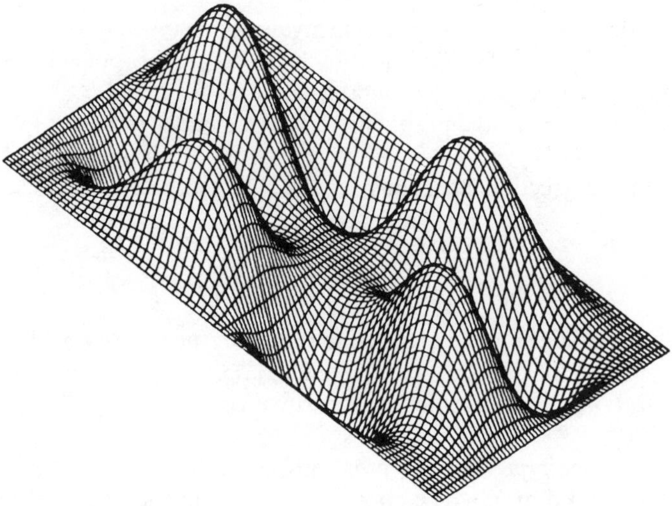

Figure 5.6 Reproduced by courtesy of Professor W. Hackbusch and Dr. G. Hofmann and with permission of ZAMP [*Z. Angew. Math. Phys.* **31**, 730–739 (1980)].

CHAPTER 6

Error Bounds and Localization Results for the Eigenelements

Introduction

We give error bounds on the eigenelements under the very general hypothesis that the approximation T_n of T is strongly stable in a neighborhood Δ of a point λ in $Q\sigma(T)$. Γ is a closed Jordan curve isolating λ and drawn in Δ. λ has algebraic (resp. geometric) multiplicity m (resp. g) and ascent l, $1 \leq l \leq m$. The associated invariant subspace is $M = PX = \text{Ker}(T - \lambda)^l$, the eigenspace is $E = \text{Ker}(T - \lambda)$.

For n large enough, there are exactly m eigenvalues $\{\mu_{jn}\}_1^m$ of T_n (counting their multiplicity) inside Γ. There are K_n distinct values among them, denoted λ_{in}, and λ_n represents any one of them. The invariant subspace for T_n associated with all the eigenvalues inside Γ is

$$M_n = \bigoplus_{i=1}^{K_n} M_{in} = \bigoplus_{i=1}^{K_n} \text{Ker}(T_n - \lambda_{in})^{l_{in}},$$

if each λ_{in} has ascent l_{in}. When T is not self-adjoint, λ is best approximated by $\hat{\lambda}_n$, the arithmetic mean of the eigenvalues of T_n inside Γ converging to λ:

$$\hat{\lambda}_n = \frac{1}{m} \sum_{j=1}^{m} \mu_{jn}.$$

The special case of projection methods is treated in detail; the role of the projection π_n to obtain possible higher orders of convergence for $\lambda - \hat{\lambda}_n$ and $\varphi - \tilde{\varphi}_n$ is shown. In the application to the f.e.m., use of the elliptic projection π_n^a to get the orders of convergence is made when possible. This exemplifies the interest of the abstract formulation of the f.e.m. given in Chapter 4, where the role of π_n^a is stressed.

The second part of this chapter is devoted to error bounds that are computable a posteriori, when the evaluation of Tx for a given x is feasible. The question of the asymptotic behavior of the constants entering these bounds is answered by localization results that extend the Kato–Temple inequalities. These localization results for the eigenelements of T (corresponding to a single eigenvalue or to a group of close eigenvalues) are obtained from the knowledge of the spectrum of a "smaller" part QTQ of T, where Q is a known projection on a finite-dimensional subspace. Proofs are based on the analytic perturbation theory developed in Chapter 3. As an application, a group of close eigenvalues of a quasi-triangular matrix is localized.

1. Theoretical Error Bounds

$(T - T_n)P$ is bounded by the closed graph theorem, and $\varepsilon_n := \|(T - T_n)P\| \to 0$. The quantity ε_n is equivalent to $\max(\|(T - T_n)\psi\|;\ \psi \in M,\ \|\psi\| = 1)$ by Exercise 6.7. Note that if φ is an eigenvector of T, $(T_n - T)\varphi = T_n\varphi - \lambda\varphi$ is the residual vector for T_n computed at λ, φ; ε_n plays the role of the maximum norm of the residual vectors for all vectors in M. We prove five introductory lemmas.

Lemma 6.1 *$P_n \xrightarrow{cc} P$ implies that, for n large enough, $\tilde{P}_n := P_{n\restriction M}$ (resp. $P_{(n)} := P_{\restriction M_n}$) is bijective from M onto M_n (resp. from M_n onto M).*

Proof Let $x \in M$, $\|x\| = 1$, $x = Px$. For n large enough,

$$\big|1 - \|P_n x\|\big| = \big|\|Px\| - \|P_n x\|\big| \leq \|(P - P_n)Px\| \leq \|(P - P_n)P\| \leq \tfrac{1}{2}.$$

Then $\|P_n x\| \leq \frac{1}{2}\|x\|$ for $x \in M$, that is, $\|\tilde{P}_n^{-1}\| \leq 2$. Similarly, let $x_n \in M_n$, $\|x_n\| = 1$, $x_n = P_n x_n$. Then

$$\big|1 - \|Px_n\|\big| = \big|\|P_n x_n\| - \|Px_n\|\big| \leq \|(P_n - P)P_n x_n\| \leq \tfrac{1}{2}$$

for n large enough, since $P_n \xrightarrow{cc} P$ and $\|(P_n - P)P_n\| \to 0$.

$$\|Px_n\| \geq \tfrac{1}{2}\|x_n\| \qquad \text{and} \qquad \|P_{(n)}^{-1}\| \leq 2. \quad \square$$

Consider in $\mathscr{L}(M)$ the operators $F := TP_{\restriction M}$ and $G_n = \tilde{P}_n^{-1}T_n\tilde{P}_n$. F has λ for eigenvalue of multiplicity m, and G_n has the eigenvalues $\{\mu_{jn}\}_1^m$.

Lemma 6.2 $\lambda - \hat{\lambda}_n = (1/m)\operatorname{tr}(F - G_n)$ *and* $\|F - G_n\| = O(\varepsilon_n)$.

Proof The first equality is clear. As for the second, we write $\|F - G_n\| = \max(\|T - \tilde{P}_n^{-1}T_nP_n)\psi\|;\ \ \psi \in M,\ \ \|\psi\| = 1)$ and $(T - \tilde{P}_n^{-1}T_nP_n)\psi = \tilde{P}_n^{-1}P_n(T - T_n)\psi$. This proves the second equality. $\square$

Lemma 6.3 *Let f be a holomorphic function of z in a neighborhood of λ. Then* $(1/m)\,|\mathrm{tr}\, f(F) - f(G_n)| \le \|f(F) - f(G_n)\| \le c\|F - G_n\|$.

Proof $f(F) - f(G_n)$ is the Dunford–Taylor integral

$$\frac{-1}{2i\pi}\int_\Gamma f(z)[(F - z)^{-1} - (G_n - z)^{-1}]\,dz$$

(see Dunford and Schwartz, 1958, Part 1, p. 600). From

$$(F - z)^{-1} - (G_n - z)^{-1} = (F - z)^{-1}(F - G_n)(G_n - z)^{-1}$$

follows that

$$\|f(F) - f(G_n)\| \le c\left(\sup_{z\in\Gamma}|f(z)|\;\|(F - z)^{-1}\|\;\|(G_n - z)^{-1}\|\right)\|F - G_n\|. \quad \square$$

Lemma 6.4 *We have the identities*

$$\varphi - P_n\varphi = S_n(\lambda)(T_n - T)\varphi$$

for any φ in $\mathrm{Ker}(T - \lambda)$ *and*

$$\varphi_n - P\varphi_n = S(\lambda_n)(T - T_n)\varphi_n$$

for any φ_n in $\mathrm{Ker}(T_n - \lambda_n)$.

Proof We start from the identity

$$R(z) - R_n(z) = R_n(z)(T_n - T)R(z) = R(z)(T_n - T)R_n(z).$$

For the eigenvector φ,

$$R(z)\varphi = \frac{1}{\lambda - z}\,\varphi$$

and

$$\varphi - P_n\varphi = \left(\frac{-1}{2i\pi}\int_\Gamma \frac{R_n(z)}{\lambda - z}\,dz\right)(T_n - T)\varphi = S_n(\lambda)(T_n - T)\varphi,$$

where $S_n(\lambda)$ is the value at λ of $S_n(z)$, the reduced resolvent of T_n with respect to its eigenvalues inside Γ: $S_n(z) := R_n(z)(1 - P_n)$ (cf. Chapter 2, Section 7.7). Note that $(T_n - T)\varphi = T_n\varphi - \lambda\varphi$. Similarly,

$$R_n(z)\varphi_n = \frac{1}{\lambda_n - z}\,\varphi_n$$

and

$$P\varphi_n - \varphi_n = \left(\frac{-1}{2i\pi}\int_\Gamma \frac{R(z)}{\lambda_n - z}\,dz\right)(T_n - T)\varphi_n = S(\lambda_n)(T_n - T)\varphi_n$$

with $S(z) := R(z)(1 - P)$; and $(T - T_n)\varphi_n = T\varphi_n - \lambda_n\varphi_n$. $\square$

Exercises

6.1 Give a direct proof of the identities of Lemma 6.4 by using

$$1 - P_n = (T_n - \lambda)^{-1}(1 - P_n)(T_n - \lambda), \qquad 1 - P = (T - \lambda_n)^{-1}(1 - P)(T - \lambda_n).$$

6.2 Prove that

$$\|(1 - P)\psi_n\| \le \|1 - P\| \operatorname{dist}(\psi_n, M) \le c\Theta(M, M_n) \qquad \text{for any} \quad \psi_n \text{ in } M_n, \ \|\psi_n\| = 1,$$

$$\|(1 - P_n)\psi\| \le \|1 - P_n\| \operatorname{dist}(\psi, M_n) \le c\Theta(M, M_n) \qquad \text{for any} \quad \psi \text{ in } M.$$

Lemma 6.5 *We have the identity*

$$(1 - P_n)\psi = \sum_{j=1}^{l} S_n^j(\lambda)(T_n - T)(T - \lambda)^{j-1}\psi \qquad \text{for any} \quad \psi \text{ in } M.$$

Proof $(T - \lambda)^l\psi = 0$ for any ψ in M; and we write

$$(T_n - \lambda)\psi = (T - \lambda)\psi + (T_n - T)\psi,$$

$$\begin{aligned}(1 - P_n)\psi &= (T_n - \lambda)^{-1}(1 - P_n)(T_n - \lambda)\psi \\ &= S_n(\lambda)[(T - \lambda)\psi + (T_n - T)\psi] \\ &= (T_n - \lambda)^{-2}(1 - P_n)(T_n - \lambda)(T - \lambda)\psi + S_n(\lambda)(T_n - T)\psi \\ &= S_n^2(\lambda)(T - \lambda)^2\psi + S_n^2(\lambda)(T_n - T)(T - \lambda)\psi + S_n(\lambda)(T_n - T)\psi \\ &\ \ \vdots\end{aligned}$$

The recursion goes on until $(T - \lambda)^l\psi = 0$ is reached. □

Exercise

6.3 Let λ_{in} of ascent l_{in} be one of the distinct eigenvalues of T_n inside Γ. M_{in} is the associated invariant subspace. Prove that, for any ψ_{in} in M_{in},

$$(1 - P)\psi_{in} = \sum_{j=1}^{l_{in}} S^j(\lambda_{in})(T - T_n)(T_n - \lambda_{in})^{j-1}\psi_{in}.$$

1.1. Bounds on $\lambda - \hat{\lambda}_n$ and $\Theta(M, M_n)$

We estimate the quantities of order ε_n.

Theorem 6.6 *If T_n is a strongly stable approximation of T in Δ, then, for n large enough, the quantities*

$$\|(1 - P)\psi_n\| \qquad \textit{for} \quad \psi_n \in M_n, \ \|\psi_n\| = 1,$$

$$\|(1 - P_n)\psi\| \qquad \textit{for} \quad \psi \in M,$$

$\Theta(M, M_n)$, $\lambda - \hat{\lambda}_n$, *and (if* $\lambda \neq 0$),

$$\frac{1}{\lambda} - \frac{1}{m}\left(\sum_{j=1}^{m} \frac{1}{\mu_{jn}}\right),$$

are at least of the order of ε_n.

Proof P_n is uniformly bounded; therefore

$$\max(\|(1-P)\psi_n\|;\, \psi_n \in M_n,\, \|\psi_n\| = 1,\, \|(1-P_n)\psi\|;\, \psi \in M) \le c\Theta(M, M_n)$$

(cf. Exercise 6.2.). Now $\Theta(M, M_n) = \max(\delta(M, M_n), \delta(M_n, M))$. We first bound $\delta(M, M_n) \le \|(P - P_n)P\|$ from the identity

$$(R(z) - R_n(z))P = R_n(z)(T_n - T)R(z)P = R_n(z)(T_n - T)PR(z)$$

by integration on Γ. Since $\dim M_n = \dim M$ and $\delta(M, M_n) < 1$ for n large enough,

$$\delta(M_n, M) \le \delta(M, M_n)/(1 - \delta(M, M_n))$$

by Proposition 2.12. $\lambda - \hat{\lambda}_n = O(\varepsilon_n)$ is clear from Lemma 6.2, and the last bound is obtained by letting $f(z) = 1/z$ in Lemma 6.3. □

1.2. Bounds on the Individual Eigenvalues and Their Eigenvectors

Let λ_n be one of the distinct eigenvalues of T_n inside Γ, $\lambda_n = \mu_{jn}$ for some j. φ_n is an associated eigenvector, $\|\varphi_n\| = 1$.

Theorem 6.7 *If T_n is a strongly stable approximation of T on Γ, then, for n large enough,*

$$\max_j |\lambda - \mu_{jn}| = O(\varepsilon_n^{1/l}), \qquad \min_j |\lambda - \mu_{jn}| = O(\varepsilon_n^{g/m}),$$

$$\operatorname{dist}(\varphi_n, \operatorname{Ker}(T - \lambda)) = O(\varepsilon_n^{1/l}).$$

Proof The result on the eigenvalues is standard for matrices (see Wilkinson, 1965, p. 81). It can be applied to the $m \times m$ matrices that represent F and G_n in some basis of M.

For the eigenvectors, the proof is in two steps.

(1) We first suppose that T and T_n are in $\mathscr{L}(X)$. $E = \operatorname{Ker}(T - \lambda)$, of finite dimension, has a supplementary subspace W: $X = E \oplus W$. Let Q be the projection on E along W. $(1 - Q)(T - \lambda)_{\restriction M}$ is invertible on W. Let

$$\varphi_n \in \operatorname{Ker}(T_n - \lambda_n),\ \|\varphi_n\| = 1.$$

$$\operatorname{dist}(\varphi_n, \operatorname{Ker}(T - \lambda)) \le \|(1 - Q)\varphi_n\|,$$

with

$$(1 - Q)\varphi_n = [(1 - Q)(T - \lambda)_{\restriction W}]^{-1}(1 - Q)(T - \lambda)(1 - Q)\varphi_n$$

and

$$(T - \lambda)(1 - Q)\varphi_n = (T - T_n)\varphi_n + (\lambda_n - \lambda)\varphi_n.$$

Now

$$(T - T_n)\varphi_n = (T - T_n)(\varphi_n - y) + (T - T_n)y$$

for any y in X. We may choose y to be a vector $\hat{\psi}_{(n)}$ in M such that

$$\|\hat{\psi}_{(n)}\| = 1, \qquad \|\varphi_n - \hat{\psi}_{(n)}\| \leq \Theta(M, M_n) \leq c\varepsilon_n,$$

where c is a generic constant. Therefore

$$\|(T - T_n)\varphi_n\| \leq c\varepsilon_n,$$

thanks to the uniform bound of $T - T_n$. This gives

$$\operatorname{dist}(\varphi_n, \operatorname{Ker}(T - \lambda)) \leq \|(1 - Q)\varphi_n\| \leq c(\varepsilon_n + |\lambda_n - \lambda|).$$

(2) We now suppose that $T, T_n \in \mathscr{C}(X)$. For a given z_0 on Γ, $R(z_0)$ and $R_n(z_0)$ are in $\mathscr{L}(X)$. Their eigenvalues are

$$\nu = \frac{1}{\lambda - z_0} \quad \text{and} \quad \nu_n = \frac{1}{\lambda_n - z_0};$$

the corresponding eigenspaces are identical. On the other hand, the stability of $\{T_n\}$ at z_0 implies that $R_n(z_0)$ is uniformly bounded and

$$(R_n(z_0) - R(z_0))P = R_n(z_0)(T - T_n)R(z_0)P = R_n(z_0)(T - T_n)PR(z_0).$$

The first part of the proof may be applied to $R_n(z_0)$ and $R(z_0)$, and

$$|\nu - \nu_n| = \left|\frac{1}{\lambda - z_0} - \frac{1}{\lambda_n - z_0}\right| = \frac{|\lambda_n - \lambda|}{|\lambda - z_0||\lambda_n - z_0|}. \quad \square$$

Exercises

6.4 Show the alternative formula $\lambda - \hat{\lambda}_n = (1/m)\operatorname{tr} P(T - T_n)P_{(n)}^{-1}$ (de Boor–Swartz).

6.5 With $\varepsilon_n^* := \|(T^* - T_n^*)P^*\|$, prove that

$$\lambda - \hat{\lambda}_n = \frac{1}{m}\operatorname{tr}(T - T_n)P + 0(\varepsilon_n \varepsilon_n^*)$$

(Osborn). Check that for any compact projection P such that $\operatorname{Im} P \subset \operatorname{Dom} T$ then $\operatorname{tr} TP = \operatorname{tr} PTP$.

6.6 Show that $\lambda - \hat{\lambda}_n = (1/m)\operatorname{tr} P_n(T - T_n)\tilde{P}_n^{-1}$ (Vainikko).

6.7 Show that the quantities $\|(T - T_n)_{\restriction M}\|$ and $\|(T - T_n)P\|$ are equivalent when $n \to \infty$.

6.8 For $1 \leq j \leq k \leq l \leq m$, we define $E_k := \operatorname{Ker}(T - \lambda)^k$ and $E_{nj} := \operatorname{Ker}(T_n - \lambda_n)^j$. For $\varphi_{nj} \in E_{nj}$, $\|\varphi_{nj}\| = 1$, show that $\operatorname{dist}(\varphi_{nj}, E_k) = O(\varepsilon_n^{(k-j+1)/l})$.

6.9 For the solution $x = R(z)f$ approximated by $x_n = R_n(z)f$, show that

$$x - x_n = R_n(z)(T_n - T)x \quad \text{with} \quad (T_n - T)x = (T_n - z)x - f,$$

or

$$x - x_n = R(z)(T_n - T)x_n \quad \text{with} \quad (T_n - T)x_n = f - (T - z)x_n.$$

Derive a priori and a posteriori error bounds. Comment on the condition number.

6.10 Prove the identities

$$(T_n - T)x = (T_n - z)(x - x_n); \qquad (T_n - T)x_n = (T - z)(x - x_n);$$

$$(1 - P_n)(T_n - T)\varphi = (T_n - \lambda)(1 - P_n)\varphi; \qquad (1 - P)(T - T_n)\varphi_n = (T - \lambda_n)(1 - P)\varphi_n.$$

Proposition 6.8 *Under the assumption of Theorem* 6.7 *and when* T, $T_n \in \mathscr{L}(X)$, *the rate of convergence of* $\Theta(M, M_n)$ *is bounded from below by the norm of the projected residual* $\|(1 - P_n)(T - T_n)\varphi\|$ *for any* φ *in* $\mathrm{Ker}(T - \lambda)$.

Proof From Theorem 6.6 and Exercise 6.10, we get for any φ in $\mathrm{Ker}(T - \lambda)$

$$\|(1 - P_n)(T_n - T)\varphi\| \le c\|(1 - P_n)\varphi\| \le c\Theta(M, M_n),$$

where c is a generic constant. □

The projected residual is of the order of ε_n unless $(T_n - T)\varphi$ lies nearly in $M_n = P_n X$. This result should be related to the lower bound $\|x - x_n\| \ge c\|(T_n - T)x\|$, which is valid for the solution x under the same assumptions.

1.3. Bibliographical Comments

Error bounds on the eigenelements of a compact operator are given in Vainikko (1967a,b) for Galerkin and perturbed Galerkin methods (see also Krasnoselskii *et al.*, 1972). Error bounds for the Nyström method are given in Anselone (1971) and Atkinson (1975). For a closed operator, under the general hypothesis of strong stability, error bounds are given in Chatelin and Lemordant (1978) and Chatelin (1979).

The arithmetic mean $\hat{\lambda}_n$ was introduced in Bramble and Osborn (1973) and Osborn (1975) for a compact operator under a uniform or collectively compact convergence assumption. For the more general framework of a discrete-regular convergence assumption for a closed operator, the reader is referred to Grigorieff (1975b).

From a practical point of view, a nice numerical study of the approximation of a multiple eigenvalue (with ascent $l > 1$) for an o.d.e. by collocation is given in de Boor and Swartz (1981a).

2. The Projection Method

We have proved that in general $\lambda - \hat{\lambda}_n$ and $\Theta(M, M_n)$ are at least of the order ε_n. There are cases where $\lambda - \hat{\lambda}_n$ is of a *higher* order than ε_n. A fundamental example is provided by the projection method that we study now.

Let $\{\pi_n\}$ be a sequence of projections from X onto X_n such that $\pi_n \xrightarrow{\text{p}} 1$. We approximate the bounded operator T by $T_n = \pi_n T$. The equation

$$T\varphi = \lambda\varphi, \qquad 0 \neq \varphi \in X, \tag{6.1}$$

is therefore approximated in X_n by

$$\pi_n T\varphi_n = \lambda_n \varphi_n, \qquad 0 \neq \varphi_n \in X_n. \tag{6.2}$$

2.1. The Residual

For the eigenvector φ, the residual vector $(T - T_n)\varphi$ is equal to

$$(1 - \pi_n)T\varphi = \lambda(1 - \pi_n)\varphi.$$

More generally, $(T - T_n)P = (1 - \pi_n)TP = (1 - \pi_n)PTP$.

We recall that for any x in X, we have introduced, in Chapter 4, the quantity $\delta_n(x) = \operatorname{dist}(x, X_n) = \|x\|\delta(\{x\}, X_n)$, which satisfies

$$\delta_n(x) \leq \|(1 - \pi_n)x\| \leq (1 + \|\pi_n\|)\, \delta_n(x).$$

$\delta_n(x)$ represents the least possible error when x is approximated by an element of the subspace X_n.

We recall the definition $\delta(M, X_n) = \sup(\operatorname{dist}(\psi, X_n); \psi \in M, \|\psi\| = 1)$.

Lemma 6.9 *For a projection method $T_n = \pi_n T$,*

$$\varepsilon_n = \|(1 - \pi_n)TP\| \leq c\, \delta(M, X_n).$$

Proof Clear from $\|(1 - \pi_n)TP\| \leq \|TP\|\ \|(1 - \pi_n)P\|$. □

Lemma 6.10 *The error bounds on $\lambda - \hat{\lambda}_n$, $\|(1 - P)\psi_n\|$ for $\psi_n \in M_n$ and $\Theta(M, M_n)$ resulting from the approximation* (6.2) *may be established by considering $\pi_n T$, as well as $\pi_n T\pi_n$, for $\lambda \neq 0$.*

Proof Let Γ be a Jordan curve around λ which does not enclose 0. $T_n^{\text{P}} = \pi_n T$ and $T_n^{\text{G}} = \pi_n T\pi_n$ have the same eigenvalues inside Γ and the same corresponding eigenvectors. Moreover $\mathscr{M}_n = M_n^{\text{G}} = M_n^{\text{P}}$ (see Exercises 4.25 and 4.29). The same result holds clearly also for $|\lambda - \mu_{jn}|, j = 1, \ldots, m$, and $\operatorname{dist}(\varphi_n, \operatorname{Ker}(T - \lambda))$. □

The error bounds for the eigenelements can therefore be expressed in terms of $\delta(M, X_n)$ (cf. Theorems 6.6 and 6.7). *The problem of estimating the errors is reduced to a problem in approximation theory*: Evaluate the distance $\delta_n(\psi)$ between a function ψ in M and a subspace X_n in X. This distance depends on the smoothness of the invariant vectors of M and on the approximability properties of the subspaces X_n.

Exercise

6.11 Consider the Petrov method defined by X_n, Y_n (cf. Chapter 4):

$$\pi_n(T - \lambda_n)\varphi_n = 0, \qquad \varphi_n \in X_n,$$

where π_n is a projection on Y_n. Show that, under the assumptions of Proposition 4.8, the errors on the eigenelements can be expressed in terms of $\delta((K - \alpha)^{-1}M, Y_n)$ or $\delta(M, X_n)$.

Since $(T - T_n)\varphi$ is proportional to $(1 - \pi_n)\varphi$ for the projection method, $(T - T_n)\varphi$ lies outside X_n, hence outside M_n. The order ε_n for $\Theta(M, M_n)$ is then optimal. We study now the order of $\lambda - \hat{\lambda}_n$.

2.2. Superconvergence of $\hat{\lambda}_n$

We suppose that, in addition to $\pi_n \xrightarrow{p} 1$ in X, we have $\pi_n^* \xrightarrow{p} 1$ on M^*, the invariant subspace for T^* associated with $\bar{\lambda}$. We define $\alpha_n := \|(1 - \pi_n)P\|$, $\alpha_n^* := \|(1 - \pi_n^*)P^*\|$.

Theorem 6.11 *If $\pi_n \xrightarrow{p} 1$, $\pi_n^* \xrightarrow{p} 1$ on M^*, and $\{\pi_n T\}$ is a strongly stable approximation of T in Δ, then the following convergence rates hold for n large enough:*

$$|\lambda - \hat{\lambda}_n| = O(\alpha_n \alpha_n^*), \qquad \left|\frac{1}{\lambda} - \frac{1}{m}\left(\sum_{j=1}^{m} \frac{1}{\mu_{jn}}\right)\right| = O(\alpha_n \alpha_n^*) \qquad \text{if} \quad \lambda \neq 0.$$

Proof λ (resp. μ_{jn}) are eigenvalues of $T_{\restriction M}$ (resp. $T_{n \restriction M_n}$). In the bases $\{x_i\}_1^m$ and $\{x_i^*\}_1^m$, $T_{\restriction M}$ is represented by the matrix A with coefficients $a_{ij} = \langle Tx_j, x_i^* \rangle$. For n large enough $P_{(n)} = P_{\restriction M_n}$ is bijective, set $x_{in} := P_{(n)}^{-1} x_i$, then $Px_{in} = x_i$. Clearly $\langle x_{in}, x_j^* \rangle = \langle Px_{in}, x_j^* \rangle = \delta_{ij}$; therefore in the bases $\{x_{in}\}_1^m$ and $\{x_i^*\}_1^m$, $T_{n \restriction M_n}$ is represented by the matrix B_n with coefficients $b_{ij}^n = \langle T_n x_{jn}, x_i^* \rangle$. Note that $a_{ij} = \langle TPx_{jn}, x_i^* \rangle = \langle Tx_{jn}, x_i^* \rangle$; hence

$$\begin{aligned} a_{ij} - b_{ij}^n &= \langle (T - T_n)x_{jn}, x_i^* \rangle \qquad \text{for} \quad i, j = 1, \ldots, m, \\ &= \langle (T - T_n)x_j, x_i^* \rangle + \langle (T - T_n)(P_{(n)}^{-1}P - P)x_j, x_i^* \rangle. \end{aligned}$$

Then $\|A - B_n\| \leq c \max_{i,j} |\langle (T - T_n)P_{(n)}^{-1} x_j, x_i^* \rangle|$.

We now specialize T_n to be $\pi_n T$. $P_{(n)}^{-1}P$ (resp. P) is a projection on M_n (resp. M) along $(1 - P)X$. By Lemma 2.16 and Theorem 6.6

$$\|P_{(n)}^{-1}P - P\| \leq c\Theta(M, M_n) \leq c\|(1 - \pi_n)TP\| \leq c\alpha_n.$$

And $\|A - B_n\| \leq c\alpha_n \alpha_n^*$. The desired results follow by Lemmas 6.2 and 6.3 since $z \mapsto z^{-1}$ is holomorphic around $\lambda \neq 0$. □

Exercises

6.12 Prove that $\max_j |\lambda - \mu_{jn}|^l = O(\alpha_n \alpha_n^*)$, $\min_j |\lambda - \mu_{jn}|^m = O[(\alpha_n \alpha_n^*)^g]$, and

$$\text{dist}(\varphi_n, \text{Ker}(T - \lambda)) \leq c \max(\alpha_n, (\alpha_n \alpha_n^*)^{1/l}).$$

6.13 Use Exercise 6.5 to establish directly that $|\lambda - \hat{\lambda}_n| = O(\alpha_n \alpha_n^*)$.

The possible compactness of T (or T^{-1}) has no effect on the convergence rates obtained in Theorem 6.11. The convergence rates are then valid under an assumption of discrete-compact convergence for the Galerkin approximation (see Descloux *et al.*, 1978b; Mills (1979b) or under an assumption of strong stability or regularity (Chatelin, 1979).

We may note that the condition $\pi_n^* \xrightarrow{\text{p}} 1$ on M^* is not required to establish the error bounds, but if π_n^* does tend to 1 on M^*, $\alpha_n^* \to 0$ and $\lambda - \hat{\lambda}_n$ is of higher order than $\Theta(M, M_n)$.

Often enough α_n^* and α_n are of the same order, as we shall see. When $\lambda - \hat{\lambda}_n$ is of order α_n^2, compared to order α_n for $\Theta(M, M_n)$, it is often said that $\hat{\lambda}_n$ is *superconvergent*. As we know, the order α_n for $\text{dist}(\varphi_n, M)$ cannot be improved, but, if we consider for $\lambda_n \neq 0$ the *iterated* eigenvector $\tilde{\varphi}_n = (1/\lambda_n)T\varphi_n \in X$, then it is possible that $\text{dist}(\tilde{\varphi}_n, M) = o(\alpha_n)$, as we see now. The cost of the computation of $\tilde{\varphi}_n$ was discussed in Chapter 4.

2.3. *Superconvergence of the Iterated Eigenvector* $\tilde{\varphi}_n$

When T is compact, the assumption $\lambda_n \neq 0$ is not a restriction: the assumption $\lambda \in Q\sigma(T)$ implies that $\lambda \neq 0$. Therefore $\lambda_n \neq 0$ is satisfied for n large enough.

Proposition 6.12 *If $\pi_n \xrightarrow{\text{p}} 1$ and if T is compact, then, for n large enough,*

$$\text{dist}(\tilde{\varphi}_n, M) \leq c\|T(1 - \pi_n)\|\alpha_n.$$

Proof We recall that $\tilde{\varphi}_n$ is an eigenvector of $T_n^{\text{S}} = T\pi_n$ associated with λ_n. Note that if $\|\varphi_n\| = 1$, $\|\tilde{\varphi}_n\| \leq \|T\|/|\lambda_n|$ is uniformly bounded in n. Then, by Theorem 6.6,

$$\text{dist}(\tilde{\varphi}_n, M) \leq c\|T(1 - \pi_n)P\| \leq c\|T(1 - \pi_n)\| \, \|(1 - \pi_n)P\|. \quad \square$$

T^* is compact; therefore, when $\pi_n^* \xrightarrow{\text{p}} 1$, $\|(1 - \pi_n^*)T^*\| = \|T(1 - \pi_n)\|$ converges to zero. This convergence can be slow, however, unless some restrictions are placed on T and π_n. The case when $\pi_n^* \not\to 1$ is studied in Kulkarni and Limaye (1983a).

Example 6.1 We consider in $X = L^2\,(a, b)$ the integral operator T defined by

$$Tx = \int_b^a k(\cdot, s)x(s)\,ds.$$

Let $\Delta = \{t_i\}_0^n$ be a strict partition of $[a, b]$, $a = t_0 < t_1 < \cdots < t_n = b$. Let $\mathbb{P}_{r,\Delta}$ be the set of functions that reduce to a polynomial of degree less than

$r+1$ on each $\Delta_i := [t_{i-1}, t_i[, i = 1, \ldots, n$. Let π_n be the *orthogonal* projection on $\mathbb{P}_{r,\Delta}$. If $h := \max_i(t_i - t_{i-1}) \to 0$, $\pi_n \xrightarrow{p} 1$, and for x in $C^{r+1}(a, b)$, $\delta_n(x) = \|(1-\pi_n)x\|_2 = O(h^{r+1})$. We suppose that $k \in C^{r+1}([a, b] \times [a, b])$. Then any invariant vector in M is in $C^{r+1}(a, b)$, and $\text{dist}_2(\varphi_n, M) = O(h^{r+1})$. The order h^{2r+2} can be achieved with $\tilde{\varphi}_n$, under our smoothness assumption on k. We define for t in $[a, b]$, $k_t := k(t, \cdot)$. Then for any ψ in M

$$[T(1-\pi_n)\psi](t) = \int_a^b k(t, s)[(1-\pi_n)\psi](s)\, ds$$

$$= ((1-\pi_n)\psi, \overline{k_t}) = ((1-\pi_n)\psi, (1-\pi_n)\overline{k_t}),$$

where $(\cdot, \cdot)$ is the L^2-inner product and $\overline{k}_t$ denotes the function $s \mapsto \overline{k_t(s)}$ for $a \le s \le b$. Therefore

$$\|T(1-\pi_n)\psi\|_2 \le \|T(1-\pi_n)\psi\|_\infty \le \left(\sup_{t \in [a,b]} \|(1-\pi_n)k_t\|_2\right)\|(1-\pi_n)\psi\|_2$$

$$\le ch^{2r+2}.$$

It can even be proved that $\text{dist}_\infty(\tilde{\varphi}_n, M) = O(h^{2r+2})$, where dist_∞ is the distance relative to the max norm $\|\cdot\|_\infty$ (Chapter 7). It is said that $\tilde{\varphi}_n$ is *globally superconvergent* since it is superconvergent at all points t in $[a, b]$. If the kernel is not smooth enough, $\tilde{\varphi}_n$ is no longer globally superconvergent. It may be superconvergent at some distinguished set of points in $[a, b]$ (cf. Chapter 7). □

Exercises

6.14 Consider the equation $(T - z)x = f$ in Example 6.1. Show that if k and f are in C^{r+1} and if $\tilde{x}_n = (1/z)(Tx_n - f)$, then

$$\|x - \tilde{x}_n\|_2 \le ch^{2r+2}$$

(Chandler).

6.15 Give an estimate of $\|\phi - u^1\|$, where $u^1 = \tilde{\varphi}_n + r^0$ has been defined in Chapter 5, by the relation (5.24). When does u^1 improve on $\tilde{\varphi}_n$?

2.4. Bibliographical Comments

The various trace formulas for $\lambda - \hat{\lambda}_n$ explain the possible higher accuracy of $\hat{\lambda}_n$: $\lambda - \hat{\lambda}_n$ is expressed as the scalar product of two vectors in X and X^*, respectively, one of which is of the order of the residual norm. In general, $\lambda - \hat{\lambda}_n$ is of the order of this residual; it is of higher order when the two vectors are nearly orthogonal.

The trace formula given in Exercise 6.5 is used in Osborn (1975) for the finite-element method; that in Exercise 6.6 has been used in Vainikko (1978b) for a finite-difference method; and that in Exercise 6.4 (which is the one used in the proof of Theorem 6.11) is used in de Boor and Swartz (1980) for the collocation method at Gaussian points.

The possible superconvergence for the iterated eigenvector was first noted by Sloan (1976b) for the projection methods. Superconvergence can also be attained with projections such that $\pi_n^* \not\xrightarrow{p} 1$, such as piecewise polynomial interpolation at Gaussian points [more general projections are considered in de Boor and Swartz (1981b)]. We deal with these questions in more detail in Chapter 7.

For approximate quadrature methods, there exists between φ_n^{F} and $\varphi_n^{\mathrm{N}} = (1/\lambda_n)T_n^{\mathrm{N}}\varphi_n^{\mathrm{F}}$ a similar relationship, but no theoretical results have been proved. There is numerical evidence that φ_n^{N} improves on φ_n^{F} even when the kernel is not smooth (cf. the numerical experiments in Section 4.4).

3. One Example: The Finite-Element Method

3.1. The Conforming Finite-Element Method

We consider the weakly posed eigenvalue problem defined in Chapter 4, Section 8.2, by

$$a(u, v) = \lambda(u, v)_H \qquad \text{for all} \quad v \text{ in } V, \tag{6.3}$$

with $V \subset H$, the embedding being continuous but not necessarily compact. a is continuous and coercive on $V \times V$. The operators $A: H \to V$ and $B := A_{\restriction V}$: $V \to V$ are defined in Chapter 4. We set the problem (6.3) in V:

$$\lambda Bu = u, \qquad 0 \neq u \in V.$$

We suppose that $1/\lambda \in Q\sigma(B)$. Let P be the spectral projection of B associated with $1/\lambda$. Let $\{V_n\}$ be a sequence of finite-dimensional subspaces of V, and let π_n^a be the a projection on V_n.

(6.3) is approximated in V_n by

$$a(u_n, v_n) = \lambda(u_n, v_n)_H \qquad \text{for all} \quad v_n \text{ in } V_n, \tag{6.4}$$

which is equivalent to

$$\lambda_n B_n u_n = u_n, \qquad 0 \neq u_n \in V_n \qquad \text{with} \quad B_n = \pi_n^a B.$$

The subscript V below indicates that the norm under consideration is $\|\cdot\|_V$. The assumptions of coercivity and continuity on a ensure that, for any v in V,

$$\max(\|(1 - \pi_n^a)v\|_V, \|(1 - \pi_n^{a\times})v\|_V) \leq c \operatorname{dist}_V(v, V_n),$$

by Lemma 4.25.

We define $\delta_n := \delta_V(M, V_n) = \sup(\text{dist}_V(\psi, V_n);\ \psi \in M,\ \|\psi\| = 1)$ (resp. $\delta_n^\times := \delta_V(M^\times, V_n)$), where M is the invariant subspace for B, associated with $1/\lambda$ (resp. $M^\times$ is the invariant subspace for the a adjoint $B^\times$ associated with $1/\bar{\lambda}$).

Theorem 6.13 *If B_n is a strongly stable approximation of B, then, for n large enough, the following rates of convergence hold:*

$$|\lambda - \hat{\lambda}_n| = O(\delta_n \delta_n^\times), \qquad \Theta_V(M, M_n) = O(\delta_n).$$

Proof It is an easy application of Theorem 6.11 in V, where the notions of adjointness and projection in V are relative to the coercive sesquilinear form a. □

The problem of estimating the error $\Theta_V(M, M_n)$ is the same as the problem of estimating $\|u - u_n\|_V$, according to Exercise 6.16.

Exercises

6.16 Show that an error bound for the approximate solution u_n in V_n of the equation $a(u, v) = (f, v)_H$ for all v in V is $\|u - u_n\|_V \le c\, \delta_n(u)$. Compare $\delta_n(u)$ and $\delta_V(M, V_n)$.

6.17 We suppose that a is hermitian. $1/\lambda \in P\sigma(B)$ is an eigenvalue of B, which is not supposed isolated or of finite multiplicity. $M = \text{Ker}(B - 1/\lambda)$. Prove that

$$\text{dist}(1/\lambda, \sigma(\tilde{B}_n)) \le c\, \delta_V(M, V_n), \qquad \text{with} \quad \tilde{B}_n = \pi_n^a B \pi_n^a.$$

6.18 Give error bounds for the generalized eigenvalue problem defined in Chapter 4, Example 4.25.

Example 6.2 Let Ω be a bounded domain of $\mathbb{R}^N$ with smooth boundary. We set $H = L^2(\Omega)$, $V = H_0^1(\Omega)$, and we consider the problem (6.3) with

$$a(u, v) = \int_\Omega \left[\sum_{i,j=1}^{N} a_{ij}\, \partial_i u\, \partial_j \bar{v} + a_0 u \bar{v} \right] dt$$

of Chapter 4, Example 4.13. Let h be a positive parameter converging to 0. We choose a family $\{V_h\}$ of class $\mathring{S}_{1,k}$, $k \ge 2$ (see Osborn, 1975), which satisfies

(i) $V_h \subset H_0^1(\Omega)$,

(ii) $\inf_{x \in V_h}(\|v - x\|_H + h\|v - x\|_V) \le ch^t \|v\|_t$ holds for all h and v in $H^t(\Omega) \cap H_0^1(\Omega)$, $1 \le t \le k$, where $\|\cdot\|_t$ is the H^t norm.

From condition (ii) follows that

$$\text{dist}_V(v, V_h) \le ch^{k-1} \qquad \text{for} \quad v \in H^k(\Omega). \tag{6.5}$$

The assumptions on a ensure that the elements of $M = PV$ and $M^\times = P^\times V$ are in $H^t(\Omega)$, $t \ge 0$. We conclude readily that

$$|\lambda - \hat{\lambda}_h| = O(h^{2k-2}), \qquad \Theta_V(M, M_h) = O(h^{k-1}).$$

For the gap relative to the L^2 norm, we get from (ii)

$$\Theta_H(M, M_h) = O(h^k).$$

These estimates are given in Bramble and Osborn (1973) and in Osborn (1975). In these papers the problem is stated in H rather than in V (cf. Example 6.3). The reader may look in Ciarlet (1978) at conditions on V_h to get (6.5).

For the Dirichlet problem, a family $\{V_h\}$ satisfying (i) is sometimes difficult to construct because the functions are required to vanish on the boundary $\partial\Omega$, but several methods have been developed for bypassing this difficulty, such as the least-squares method of Bramble and Schatz (1970), the methods of Nitsche (1970–1971), and the Lagrange multipliers method of Babuška (1973).

All these methods are amenable to the approach of the next example.

Example 6.3 We consider again (6.3), but set the problem in H. $\|\cdot\|$ denotes $\|\cdot\|_H$.

$$\lambda A u = u, \qquad 0 \neq u \in H.$$

If we suppose that the embedding of V into H is compact, A is compact and (6.4) defines an approximation $A_h\colon H \to H$ such that $\|A_h - A\| \to 0$. Let A', A'_h be the adjoint operators (with respect to the inner product of H) of A and A_h. M and M' are the invariant subspaces of A and A', respectively. Then, using Exercise 6.5, it easy to prove that

$$|\lambda - \hat{\lambda}_h| \leq c\left[\max_{\substack{\psi \in M,\, \psi^* \in M' \\ \|\psi\| = \|\psi^*\| = 1}} \langle (A - A_h)\psi, \psi^*\rangle + \|(A - A_h)_{\restriction M}\|\, \|(A' - A'_h)_{\restriction M'}\|\right],$$

$$\Theta(M, M_h) \leq c\|(A - A_h)_{\restriction M}\|.$$

Note that in this approach, only the condition $\|A_h - A\| = \|A'_h - A'\| \to 0$ has been used. That the finite-element method is indeed a *projection* method on A in V is *implicitly* used to bound the first term $\langle (A - A_h)\psi, \psi^*\rangle$ (see Osborn, 1975).

We consider now the problem defined in Chapter 4, Example 4.26, and which is more general than (6.3). H_1 and H_2 are two complex Hilbert spaces with respective norms $\|\cdot\|_1$ and $\|\cdot\|_2$. Let be given two continuous sesquilinear forms a and b from $H_1 \times H_2$ into $\mathbb{C}$ such that

$$\inf_{\substack{\|u\|_1 = 1 \\ u \in H_1}} \sup_{\substack{\|v\|_2 = 1 \\ v \in H_2}} |a(u, v)| \geq \alpha > 0, \tag{6.6}$$

$$\sup_{\substack{\|u\|_1 = 1 \\ u \in H_1}} |a(u, v)| > 0 \qquad \text{for all} \quad 0 \neq v \text{ in } H_2.$$

The problem

$$\text{find} \quad \lambda \in \mathbb{C}, 0 \neq u \in H_1 \qquad \text{such that} \quad a(u, v) = \lambda b(u, v) \qquad \text{for all} \quad v \in H_2 \tag{6.7}$$

is equivalent to

$$\lambda T u = u, \qquad 0 \neq u \in H_1,$$

where $T \in \mathscr{L}(H_1)$ has been defined in Example 4.26, as well as the a adjoint $T^\times \in \mathscr{L}(H_2)$. We suppose that $1/\lambda \in Q\sigma(T)$.

Let be given two families of finite-dimensional subspaces, $\{S_{1h}\}$ in H_1 and $\{S_{2h}\}$ in H_2, respectively. (6.7) is approximated by the problem

$$\begin{aligned} &\text{find} \quad \lambda_h \in \mathbb{C}, 0 \neq u_h \in S_{1h} && \text{such that} \\ &a(u_h, v_h) = \lambda_h b(u_h, v_h) && \text{for all} \quad v_h \text{ in } S_{2h}. \end{aligned} \tag{6.8}$$

Concerning (6.8) we assume that

$$\begin{aligned} &\inf_{\substack{u_h \in S_{1h} \\ \|u_h\|_1 = 1}} \sup_{\substack{v_h \in S_{2h} \\ \|v_h\|_2 = 1}} |a(u_h, v_h)| \geq \alpha > 0, \\ &\sup_{\substack{u_h \in S_{1h} \\ \|u_h\|_1 = 1}} |a(u_h, v_h)| > 0 \qquad \text{for all} \quad 0 \neq v_h \text{ in } S_{2h} \quad \text{and} \quad \text{all } h. \end{aligned} \tag{6.9}$$

Under assumptions (6.6) and (6.9), there exist uniquely defined projections $\pi_h^a\colon H_1 \to S_{1h}$ and $\pi_h^{a\times}: H_2 \to S_{2h}$ such that

$$\|(1 - \pi_h^a)u\|_1 \leq c \operatorname{dist}_1(u, S_{1h}) \qquad \text{for all} \quad u \text{ in } H_1 \text{ (see Exercise 4.81).}$$

We define $\delta_h := \delta_1(M, S_{1h})$, $\delta_h^\times := \max_{v \in M^\times} \|(1 - \pi_h^{a\times})v\|_2$.

Proposition 6.14 *In addition to assumptions* (6.6) *and* (6.9), *we suppose that, for any* u *in* H_1, $\operatorname{dist}_1(u, S_{1h}) \to 0$, *and that* $\pi_h^a T$ *is a strongly stable approximation of* T *in a neighborhood of* $1/\lambda$. *Then*

$$\lambda - \hat{\lambda}_h = 0(\delta_h \delta_h^\times), \qquad \Theta(M, M_h) = O(\delta_h).$$

Proof This is again an application of Theorem 6.11, which is left to the reader. Note that $\delta_h^\times$ may or may not tend to 0. $\square$

Exercises

6.19 Prove that (6.6) implies

$$\inf_{\substack{\|v\|_2 = 1 \\ v \in H_2}} \sup_{\substack{\|u\|_1 = 1 \\ u \in H_1}} |a(u, v)| > 0.$$

6.20 Let $N = \dim S_{1h} = \dim S_2 h$, and let $\{e_i^h\}_1^N$ (resp. $\{f_i^h\}_1^N$) be a basis of S_{1h} (resp. S_{2h}). Prove that (6.8) is equivalent to the $N \times N$ generalized matrix eigenvalue problem

$$\tilde{A}^h q_h = \lambda_h \tilde{B}^h q_h$$

where

$$q_h = (q_i^h)_1^N, \qquad u_h = \sum_{i=1}^{N} q_i^h e_i^h, \qquad (\tilde{A}^h)_{ij} = a(e_j^h, f_i^h)$$

and

$$(\tilde{B}^h)_{ij} = b(e_j^h, f_i^h).$$

Show that $\tilde{A}^h$ is invertible under assumptions (6.9).

Example 6.4 The above abstract results are applied in Mills (1979b) to study numerical approximations to the eigenfunctions of singular Weyl–Stone eigenvalue problems. The operator T is noncompact, and the approximation T_h is such that $\|(T - T_h)_{\restriction S_{1h}}\|_1 \to 0$. This is a discrete-compact convergence.

Example 6.5 In Mills (1979c, 1980) the results are extended to reflexive Banach spaces, with application, in the self-adjoint case, to eigenvalues that may have an infinite multiplicity.

Example 6.6 The above abstract setting is also used in Kolata (1979) to analyze the method of Lagrange multipliers and in Mercier *et al.* (1981) to study mixed and hybrid methods. In both cases, the assumption of compactness for T is made.

3.2. A Nonconforming Finite-Element Method

We consider the generalized eigenvalue problem defined in Example 4.25 approximated by the method defined in Example 4.27. We have already explained in Chapter 4 the difficulty of working with the projection π_n^a. If we suppose that $\|A - A_n\|_H \to 0$ (hence A is compact), then a bound similar to that given in Example 6.3 can be established easily in H. Let $A', A_n' : H \to V$ be defined by the relations, for $f \in H$,

$$a(v, A'f) = b(v, f) \qquad \forall v \in V,$$

$$a(v_n, A_n' f) = b(v_n, f) \qquad \forall v_n \in V_n.$$

If $1/\lambda$ is an eigenvalue of A, $1/\bar{\lambda}$ is an eigenvalue of A' with spectral projection $P' = (-1/2i\pi) \int_\Gamma (A' - z)^{-1}\, dz$ and invariant subspace $M' = P'H$. Then, for n large enough.

$$|\lambda - \hat{\lambda}_n| \le c\Bigg[\max_{\substack{u \in M,\, v \in M' \\ \|u\|_H = \|v\|_H = 1}} |b((A - A_n)u, v)| + \|(A - A_n)_{\restriction M}\|_H \|(A' - A_n')_{\restriction M'}\|_H \Bigg].$$

The proof is left to the reader (see Mercier *et al.*, 1981).

3.3. The Iterated Eigenvector

We consider again the eigenvalue problem (6.3) and its approximation (6.4). We suppose that λ is simple. u_n is the eigenvector of $B_n = \pi_n^a B$ associated with $1/\lambda_n$. We consider the iterated eigenvector defined by

$$\tilde{u}_n = \lambda_n B u_n.$$

$\tilde{u}_n$ is the solution of the equation

$$a(\tilde{u}_n, v) = \lambda_n (u_n, v) \qquad \text{for any} \quad v \text{ in } V,$$

which can be solved by using finite elements of higher order. By Proposition 6.12 applied to B, we get

$$\|\tilde{u}_n - u\|_V \leq c \|B(1 - \pi_n^a)\|_V \, \delta_V(M, V_n).$$

$\tilde{u}_n$ improves on u_n when $\|B(1 - \pi_n^a)\|_V \to 0$.

Example 6.7 Consider the Dirichlet problem in Ω, a bounded domain of $\mathbb{R}^2$:

$$\Delta u = \lambda u \quad \text{in } \Omega,$$
$$u = 0 \quad \text{in } \partial\Omega,$$

with $H = L^2(\Omega)$, $V = H_0^1(\Omega)$. We choose piecewise-linear finite elements. Then

$$\|(1 - \pi_n^a)u\|_V \leq h \|u\|_V,$$
$$\|B(1 - \pi_n^a)u\|_V \leq c \|(1 - \pi_n^a)u\|_H \leq ch \|u\|_V.$$

It follows that $\|u - u_n\|_V = 0(h)$ and $\|u - \tilde{u}_n\|_V = 0(h^2)$ if piecewise-quadratic finite elements are used to compute $\tilde{u}_n$. See Lin Qun (1980) and Example 7.5 in Chapter 7.

4. A Posteriori Error Bounds for Bounded Operators

We now give error bounds in terms of the quantity $\eta_n = \|(T - T_n)P_n\|$. $\eta_n \to 0$ since $P_n \overset{\text{cc}}{\to} P$, and η_n is equivalent to

$$\max_{\substack{\psi \in M_n \\ \|\psi\| = 1}} \|(T - T_n)\psi\|,$$

which is computable a posteriori if we can evaluate reasonably easily the term $T\psi$ for any vector ψ in M_n. We assume throughout this section that T is bounded.

4.1. Bounds on $\lambda - \hat{\lambda}_n$ and $\Theta(M, M_n)$

Theorem 6.15 *If $\{T_n\}$ is a strongly stable approximation of T on Γ, then, for n large enough, $\lambda - \hat{\lambda}_n$ and $\Theta(M, M_n)$ are of the order of η_n.*

Proof We recall that

$$\Theta(M, M_n) \le \max(\|(P - P_n)P\|, \|(P - P_n)P_n\|),$$

$$(R(z) - R_n(z))P_n = R(z)(T_n - T)R_n(z)P_n = R(z)(T_n - T)P_n R_n(z).$$

We then get $\delta(M_n, M) \le c\eta_n$ and $\Theta(M, M_n) \le c\eta_n$, by Proposition 2.13. Now we consider in $\mathscr{L}(M_n)$ the operators $F'_n := P_{(n)}^{-1} T P_{(n)}$ and $G'_n := T_{n\restriction M_n}$. Clearly $m(\lambda - \hat{\lambda}_n) = \mathrm{tr}(F'_n - G'_n)$.

Let $\{x_i\}_1^m$ be a basis for M. We set $x_{in} := P_n x_i$, $i = 1, \ldots, m$. Thanks to the assumption $\dim M_n = \dim M$ for n large enough, $\{x_{in}\}_1^m$ is a basis for M_n and the adjoint basis $\{x_{in}^*\}_1^m$ is uniformly bounded in n, by Lemma 3.11. We get

$$\mathrm{tr}(F'_n - G'_n) = \sum_{i=1}^{m} \langle (P_{(n)}^{-1} T P_{(n)} - T_n) x_{in}, x_{in}^* \rangle$$

and

$$|\mathrm{tr}(F'_n - G'_n)| \le c \max_i \|P_{(n)}^{-1} P_{(n)} (T - T_n) x_{in}\| = c\eta_n. \quad \square$$

Exercises

6.21 We define $G''_n := P_{(n)} T_n P_{(n)}^{-1}$ in $\mathscr{L}(M)$. Prove that

$$F - G''_n = P_{(n)}(F'_n - G'_n)P_{(n)}^{-1} \qquad \text{and} \qquad \|F - G''_n\| = O(\eta_n).$$

6.22 Deduce from Exercise 6.21 that

$$\max_j |\lambda - \mu_{jn}|^l = O(\eta_n), \qquad \min_j |\lambda - \mu_{jn}| = O(\eta_n^{g/m}),$$

and

$$\frac{1}{\lambda} - \frac{1}{m}\left(\sum_j^m \frac{1}{\mu_{jn}}\right) = O(\eta_n).$$

6.23 Bound $\|(P - P_n)P\|$ in terms of η_n.

A posteriori error bounds will result from such statements as Theorem 6.15 if we can estimate the constants that enter the bounds. We deal with this question in Sections 5 and 6.

4.2. The Generalized Rayleigh Quotient

If T and T_n are self-adjoint densely defined operators in a Hilbert space, let x_n be an eigenvector of T_n such that $(x_n, x_n) = \|x_n\|^2 = 1$. The Rayleigh quotient $\rho_n := (Tx_n, x_n)$ is known to be an approximation of λ of the order of

$$\frac{1}{\delta(\rho_n)}\|(T-\rho_n)x_n\|^2 \qquad \text{with} \quad \delta(\rho_n) := \operatorname{dist}(\rho_n, \sigma(T) - \{\lambda\})$$

(cf. Section 5).

In the non-self-adjoint case, let λ be an exact isolated eigenvalue with multiplicity m. A possible extension of the notion of Rayleigh quotient is to consider a basis $\{x_{in}\}_1^m$ of M_n and the adjoint basis $\{x_{in}^*\}_1^m$ of M_n^* and set

$$\zeta_n := \frac{1}{m}\sum_{i=1}^{m}\langle Tx_{in}, x_{in}^*\rangle = \frac{1}{m}\operatorname{tr} TP_n.$$

ζ_n is the *generalized Rayleigh quotient* of T based on a given basis in M_n and the corresponding adjoint basis in M_n^*. ζ_n is often computable a posteriori with little additional work (cf. Chapter 7).

Is ζ_n an improvement on $\hat{\lambda}_n$? We have already looked at this question in the case of matrices (Chapter 1, Section 4) with $m = 1$: $\zeta = y^{\mathrm{H}}Ax$, $y^{\mathrm{H}}x = 1$. We know that the answer depends on whether y is an approximate eigenvector for A^{H} or not. The situation is essentially the same in the present context.

We still suppose that T and T_n are densely defined, and set

$$\eta_n^* := \|(T^* - T_n^*)P_n^*\|.$$

η_n^* may or may not tend to zero.

Lemma 6.16 *If T_n is a strongly stable approximation of T on Γ, then, for n large enough,*

$$\lambda - \zeta_n = \lambda - (1/m)\operatorname{tr} TP_n = O(\eta_n\eta_n^*).$$

Proof

$$\begin{aligned} m(\lambda - \hat{\lambda}_n) &= \sum_{i=1}^{m}\langle (P_nT\tilde{P}_n^{-1} - P_nT_nP_n)x_{in}, x_{in}^*\rangle \\ &= \sum_i \langle P_n(T - T_nP_n)\tilde{P}_n^{-1}x_{in}, x_{in}^*\rangle \\ &= \sum_i \langle (T - T_n)\tilde{P}_n^{-1}x_{in}, x_{in}^*\rangle \\ &= \operatorname{tr}(T - T_n)P_n + \sum_i \langle (\tilde{P}_n^{-1}P_n - P_n)x_{in}, (T^* - T_n^*)x_{in}^*\rangle. \end{aligned}$$

Now use

$$m\hat{\lambda}_n = \operatorname{tr} T_nP_n, \qquad \|\tilde{P}_n^{-1}P_n - P_n\| \le c\Theta(M, M_n) \le c\eta_n$$

to get

$$|m\lambda - \operatorname{tr} TP_n| \le c\eta_n\eta_n^*. \quad \Box$$

η_n is of order ε_n. If $\lambda - \hat{\lambda}_n$ is of order ε_n, then ζ_n improves on $\hat{\lambda}_n$ when $\eta_n^* \to 0$.

Example 6.8 When T is an integral operator numerically approximated by T_n of finite rank, the computation of ζ_n requires only matrix computations plus the evaluation of Tx_{in}. Details are given in Chapter 7, Section 8.2. In particular, the knowledge of $\{x_{in}^*\}_1^m$ is *not* required.

Example 6.9 We consider the Nyström approximation of an integral operator with a *smooth* kernel. Then, in an appropriate Banach space, $|||T_n^{\mathrm{N}} - T||| = |||T_n^{\mathrm{N}*} - T^*||| \to 0$ (Chapter 4, Section 7). ζ_n improves on $\hat{\lambda}_n$ (cf. numerical experiments in Section 4.4).

Example 6.10 Consider the method of iterative refinement of the eigenelements defined in Chapter 5, Section 5, when $m = 1$. The first iterate λ^1 is $\lambda^1 = \langle T\varphi_n, \varphi_n^* \rangle = \zeta_n$; this is the generalized Rayleigh quotient based on the eigenvectors φ_n and φ_n^*.

Example 6.11 Let T be self-adjoint in a Hilbert space. Given an approximate eigenvector φ_n, the Rayleigh quotient $\rho_n = (T\varphi_n, \varphi_n)$ is an approximate value of λ that is at least as good as $\zeta_n = (T\varphi_n, \varphi_n^*)$ (and easier to compute). This is because φ_n^* may not be an approximate eigenvector of T when T_n is not self-adjoint (assuming that λ_n is simple).

Exercises

6.24 Let T be self-adjoint, $\rho = (Tx, x)$ with $\|x\| = 1$. For any $\alpha \in \mathbb{C}$ prove that

$$\|(T - \alpha)x\|^2 = \|(T - \rho)x\|^2 + |\alpha - \rho|^2.$$

Deduce that $\min_{\alpha \in \mathbb{C}} \|(T - \alpha)x\| = \|(T - \rho)x\|$.

6.25 Let T be self-adjoint in a Hilbert space. Suppose that the eigenvalue λ_n of T_n associated with φ_n is simple. Show that $\zeta_n = \rho_n$ if T_n is self-adjoint. When T_n is not self-adjoint, but nevertheless $T_n^* \xrightarrow{\mathrm{p}} T^* = T$, compare ζ_n and ρ_n as approximations of λ.

6.26 Let T_n be the Galerkin approximation $\pi_n T \pi_n$ of a bounded operator T. Prove that $\hat{\lambda}_n = \zeta_n$.

That $\hat{\lambda}_n = \zeta_n$ for a Galerkin approximation shows again that $\hat{\lambda}_n$ is an approximation of higher accuracy if $\pi_n^* \xrightarrow{\mathrm{p}} 1$. We look at more specific properties of the projection methods in Section 4.3.

4.3. The Projection Methods

We suppose that λ_n is a *simple* eigenvalue of $T_n = \pi_n T$ with associated eigenvector φ_n, the spectral projection is denoted P_n^{P}. We have considered in Section 2.3 the iterated eigenvector $\tilde{\varphi}_n = (1/\lambda_n)T\varphi_n$. $\tilde{\varphi}_n$ is an eigenvector of

$T_n^S = T\pi_n$, the corresponding spectral projection is denoted P_n^S. We introduce the generalized Rayleigh quotient

$$\tilde{\lambda}_n := \text{tr } TP_n^S.$$

Lemma 6.17 *The following identity holds:*

$$\tilde{\lambda}_n = \text{tr } TP_n^S = \text{tr } TP_n^P.$$

Proof For T_n^G, T_n^P, and T_n^S, the eigenvectors associated with λ_n are respectively φ_n, φ_n, and φ_n^S. We define the adjoint vectors by

$$(\pi_n T\pi_n)^*\varphi_n^* = \bar{\lambda}_n\varphi_n^*, \qquad \langle\varphi_n, \varphi_n^*\rangle = 1,$$

$$(\pi_n T)^*\psi_n = \bar{\lambda}_n\psi_n, \qquad \langle\varphi_n, \psi_n\rangle = 1,$$

and

$$(T\pi_n)^*\psi_n' = \bar{\lambda}_n\psi_n', \qquad \langle\varphi_n^S, \psi_n'\rangle = 1.$$

ψ_n' and φ_n^* are colinear: we set $\psi_n' := \alpha\varphi_n^*$. Then

$$1 = \langle\varphi_n^S, \psi_n'\rangle = \left\langle\frac{1}{\lambda_n} T\varphi_n, \alpha\varphi_n^*\right\rangle = \left\langle\pi_n\varphi_n, \frac{\alpha}{\bar{\lambda}_n} T^*\varphi_n^*\right\rangle$$

$$= \langle\varphi_n, \alpha\varphi_n^*\rangle = \bar{\alpha},$$

that is, $\psi_n' = \varphi_n^*$. Similarly set $\psi_n := (\beta/\bar{\lambda}_n)T^*\varphi_n^*$.

$$1 = \langle\varphi_n, \psi_n\rangle = \left\langle\varphi_n, \frac{\beta}{\bar{\lambda}_n} T^*\varphi_n^*\right\rangle = \left\langle\frac{1}{\lambda_n} T\varphi_n, \beta\varphi_n^*\right\rangle = \bar{\beta};$$

therefore $\psi_n = (1/\bar{\lambda}_n)T^*\varphi_n^*$. Now

$$\tilde{\lambda}_n = \text{tr } TP_n^S = \langle T\varphi_n^S, \varphi_n^*\rangle = \left\langle T\varphi_n, \frac{1}{\bar{\lambda}_n} T^*\varphi_n^*\right\rangle$$

$$= \langle T\varphi_n, \psi_n\rangle = \text{tr } TP_n^P. \quad \square$$

We now give error bounds in terms of $\eta_n := \|(1 - \pi_n)T\varphi_n\|$ and $\eta_n^* := \|(1 - \pi_n^*)T^*\varphi_n^*\|$.

Proposition 6.18 *For n large enough, we have*

$$|\lambda - \lambda_n| = 0(\eta_n\eta_n^*) \qquad \textit{and} \qquad |\lambda - \tilde{\lambda}_n| = O(\|(1 - \pi_n)T\|\eta_n\eta_n^*).$$

Proof For the first bound, we apply Lemma 6.16 to $\pi_n T\pi_n$ with $\zeta_n = \lambda_n$ (since $m = 1$). For the second, we apply Lemma 6.16 to $T_n = \pi_n T$ with $\tilde{\lambda}_n = \text{tr } TP_n^P$. This gives $|\lambda - \tilde{\lambda}_n| \leq c\|(1 - \pi_n)T\varphi_n\| \; \|T^*(1 - \pi_n^*)\psi_n\|$, where ψ_n has been defined in the proof of Lemma 6.17. $\psi_n = (1/\bar{\lambda}_n)T^*\varphi_n^*$ and therefore

$$\|T^*(1 - \pi_n^*)\psi_n\| = \frac{1}{|\lambda_n|}\|T^*(1 - \pi_n^*)T^*\varphi_n^*\| \leq c\|(1 - \pi_n)T\|\eta_n^*. \quad \square$$

$\tilde{\lambda}_n$ improves on λ_n if $\|(1 - \pi_n)T\| \to 0$. This is satisfied if T is compact and $\pi_n \xrightarrow{p} 1$, (cf. numerical experiments in Section 4.4).

Exercises

6.27 Consider the iterative refinement on the eigenvalue applied with $T_n^G = \pi_n T \pi_n$. Show that the first two iterates are such that $\lambda^{1G} = \lambda^{0G} = \lambda_n$, $\lambda^{2G} = \tilde{\lambda}_n$.

6.28 Let T be a compact self-adjoint operator in a Hilbert space, and let π_n be an orthogonal projection. λ_n and φ_n are defined by $\pi_n T\varphi_n = \lambda_n \varphi_n$,

$$\tilde{\varphi}_n = \frac{1}{\lambda_n} T\varphi_n \qquad \text{and} \qquad \tilde{\rho}_n = \frac{(T\tilde{\varphi}_n, \tilde{\varphi}_n)}{(\tilde{\varphi}_n, \tilde{\varphi}_n)}.$$

Prove that

$$\lambda = \tilde{\rho}_n + O(\|(1 - \pi_n)T\|^4)$$

and

$$\lambda = \tilde{\lambda}_n + O(\|(1 - \pi_n)T\|^3).$$

4.4. Numerical Experiments

We report some of the numerical experiments given in Redont (1979a) and in Kulkarni and Limaye (1983a). We consider in $X = C(0, 1)$ the integral operators defined by the following kernels:

(1) $k_1(s, t) = e^{st}$ for $0 \le s, t \le 1$,
(2) $k_2(s, t) = |\cos s - \cos t|$ for $0 \le s, t \le 1$.

Both functions are symmetric in s, t. k_1 is smooth, but k_2 has discontinuous first derivatives.

For the first example, the three dominant eigenvalues are

$$\lambda_1 = 1.353030, \qquad \lambda_2 = 0.105983, \qquad \lambda_3 = 3.560749 \times 10^{-3}.$$

For the second example, the dominant eigenvalue is $\lambda_1 = 0.164565$. $[0, 1]$ is divided into n intervals at $t_i = i/n$, $i = 0, \ldots, n$. The integral operators are approximated by the following:

(1) Projection methods where π_n is the piecewise-linear interpolation at $\{t_i\}_0^n$. This defines a collocation method at $\{t_i\}_0^n$.

(2) Approximate quadrature methods, where the quadrature formula is the Gauss formula with two points on each $[t_i, t_{i+1}]$, $i = 0, \ldots, n - 1$.

v is the size of the corresponding discretization matrix $\tilde{A}_n$: $n = v$ or $v/2$, respectively. i is the rank of the eigenvalue λ_i: $i = 1, 2, 3$.

In the computations, the vector Tx for a given x has been estimated by use of the Nyström approximation of T of size 100.

4.4.1. Comparisons between various methods

For projection (resp. approximate quadrature) methods we consider the approximations T_n^{G}, T_n^{P}, and T_n^{S} (resp. T_n^{F} and T_n^{N}). With $\mathcal{M} \in \{\mathrm{G, P, S, F, N}\}$, the efficiency of each method with respect to a common size v is evaluated through $\eta_n^{\mathcal{M}} := \|(T - T_n^{\mathcal{M}})\varphi_n^{\mathcal{M}}\|_\infty$, $\|\varphi_n^{\mathcal{M}}\|_\infty = 1$. Results for the kernels k_1 and k_2 are listed in Table 6.1. $\rho(\mathcal{M}, \mathcal{M}')$ is the ratio $\eta_n^{\mathcal{M}}/\eta_n^{\mathcal{M}'}$. Recall that $\eta_n^{\mathrm{P}} = \eta_n^{\mathrm{G}}$.

Table 6.1

Values of $\eta_n^{\mathcal{M}}$ and Corresponding Ratios

			v				
$k(s,t)$	i	$\mathcal{M}$	4	8	16	30	50
k_1	1	η_n^{F}	8.4×10^{-2}	4.1×10^{-2}	1.9×10^{-2}	8.3×10^{-3}	3.6×10^{-3}
		η_n^{N}	1.5×10^{-4}	9.3×10^{-6}	5.8×10^{-7}	4.7×10^{-8}	5.7×10^{-9}
		$\rho(\mathrm{N, F})$	1.8×10^{-3}	2.3×10^{-4}	3.0×10^{-5}	5.7×10^{-6}	1.6×10^{-6}
		η_n^{G}	4.2×10^{-3}	1.5×10^{-3}	3.4×10^{-4}	9.2×10^{-5}	2.7×10^{-5}
		η_n^{S}	3.1×10^{-3}	1.0×10^{-3}	2.2×10^{-4}	5.9×10^{-4}	2.1×10^{-5}
		$\rho(\mathrm{S, G})$	0.74	0.67	0.65	0.64	0.78
		$\rho(\mathrm{N, G})$	3.6×10^{-2}	6.2×10^{-3}	1.7×10^{-3}	5.1×10^{-4}	2.1×10^{-4}
	2	η_n^{F}	2.0×10^{-2}	1.2×10^{-2}	6.3×10^{-3}	2.9×10^{-3}	1.2×10^{-3}
		η_n^{N}	7.5×10^{-4}	4.8×10^{-5}	3.0×10^{-6}	2.4×10^{-7}	3.0×10^{-8}
		$\rho(\mathrm{N, F})$	3.7×10^{-2}	4.0×10^{-3}	4.8×10^{-5}	8.3×10^{-5}	2.5×10^{-5}
		η_n^{G}	4.0×10^{-3}	7.8×10^{-4}	1.8×10^{-4}	4.8×10^{-5}	1.4×10^{-5}
		η_n^{S}	3.5×10^{-2}	6.4×10^{-3}	1.4×10^{-3}	3.7×10^{-4}	1.3×10^{-4}
		$\rho(\mathrm{S, G})$	8.7	8.2	7.8	7.7	9.3
		$\rho(\mathrm{N, G})$	0.19	6.5×10^{-2}	1.7×10^{-2}	5.0×10^{-3}	2.1×10^{-3}
	3	η_n^{F}	8.2×10^{-3}	2.6×10^{-3}	6.8×10^{-4}	1.9×10^{-4}	6.9×10^{-5}
		η_n^{N}	4.4×10^{-3}	2.9×10^{-4}	1.9×10^{-5}	1.5×10^{-6}	1.8×10^{-7}
		$\rho(\mathrm{N, F})$	0.54	0.11	2.8×10^{-2}	7.9×10^{-3}	2.6×10^{-3}
		η_n^{G}	1.0×10^{-3}	1.9×10^{-4}	4.3×10^{-5}	1.2×10^{-5}	3.2×10^{-6}
		η_n^{S}	0.24	4.1×10^{-2}	8.7×10^{-3}	2.3×10^{-3}	8.1×10^{-4}
		$\rho(\mathrm{S, G})$	240	216	202	192	253
		$\rho(\mathrm{N, G})$	4.4	2.3	0.44	1.2×10^{-1}	5.6×10^{-2}
k_2	1	η_n^{F}	6.6×10^{-2}	2.8×10^{-2}	1.2×10^{-2}	5.0×10^{-3}	2.1×10^{-3}
		η_n^{N}	9.7×10^{-3}	2.6×10^{-3}	6.9×10^{-5}	1.7×10^{-4}	3.7×10^{-5}
		$\rho(\mathrm{N, F})$	0.15	0.93×10^{-1}	5.7×10^{-2}	3.4×10^{-2}	1.8×10^{-2}
		η_n^{G}	1.9×10^{-2}	4.1×10^{-3}	9.9×10^{-4}	2.8×10^{-4}	7.0×10^{-5}
		η_n^{S}	9.0×10^{-3}	1.7×10^{-3}	9.7×10^{-4}	1.0×10^{-4}	3.6×10^{-5}
		$\rho(\mathrm{S, G})$	0.47	0.41	0.37	0.36	0.51
		$\rho(\mathrm{N, G})$	0.51	0.63	0.70	0.61	0.53

Table 6.2

Ratio $|\lambda - \lambda_n^{\mathcal{M}}|/\eta_n^{\mathcal{M}}$

			ν				
$k(s,t)$	i	$\mathcal{M}$	4	8	16	30	50
k_1	1	F	7.6×10^{-4}	1.0×10^{-4}	1.4×10^{-5}	2.5×10^{-6}	7.1×10^{-7}
		N	0.44	0.44	0.44	0.44	0.44
		G	0.67	0.66	0.64	0.63	0.75
		S	1.0	0.99	0.99	0.99	0.99
	2	F	1.4×10^{-2}	1.5×10^{-3}	1.8×10^{-4}	3.2×10^{-5}	9.2×10^{-6}
		N	0.39	0.39	0.39	0.39	0.39
		G	6×10^{-2}	8.10^{-2}	8.2×10^{-2}	8.1×10^{-2}	9.9×10^{-2}
		S	6.9×10^{-3}	9.8×10^{-3}	1.0×10^{-2}	1.0×10^{-2}	1.0×10^{-2}
	3	F	3.6×10^{-2}	7.9×10^{-3}	1.9×10^{-3}	5.4×10^{-4}	1.9×10^{-4}
		N	6.7×10^{-2}	6.9×10^{-2}	6.9×10^{-2}	6.9×10^{-2}	6.9×10^{-2}
		G	6.7×10^{-2}	3.0×10^{-3}	8.4×10^{-3}	1.1×10^{-2}	1.5×10^{-2}
		S	2.8×10^{-4}	1.4×10^{-5}	4.1×10^{-5}	5.3×10^{-5}	5.6×10^{-5}
k_2	1	F	9.9×10^{-2}	5.4×10^{-2}	3.0×10^{-2}	1.8×10^{-2}	1.3×10^{-2}
		N	0.68	0.56	0.51	0.55	0.74
		G	0.44	0.38	0.36	0.36	0.59
		S	0.90	0.92	0.94	1.0	1.2

The effectiveness of the bound $\lambda - \lambda_n^{\mathcal{M}} = O(\eta_n^{\mathcal{M}})$ is checked through the ratio $|\lambda - \lambda_n^{\mathcal{M}}|/\eta_n^{\mathcal{M}}$. Results are listed in Table 6.2. A comparison between $\|\varphi - \varphi_n^{\mathrm{G}}\|_\infty$ and $\|\varphi - \varphi_n^{\mathrm{S}}\|_\infty$ is done in Table 6.3.

4.4.2. Comparison between T_n^{P} and T_n^{N}

The errors $\lambda - \lambda_n$ and $\lambda - \zeta_n$ are computed for T_n^{P} and T_n^{N} with $\zeta_n = \langle T\varphi_n, \varphi_n^* \rangle$. Taking into account that the operators are self-adjoint in $L^2(0,1)$, the Rayleigh quotient $\rho_n = (T\varphi_n, \varphi_n)$ has also been computed. Results are listed in Table 6.4. Note that for T_n^{P}, $\zeta_n = \mathrm{tr}\, TP_n^{\mathrm{P}} = \tilde{\lambda}_n$.

4.4.3. Computational comments

We see in Table 6.1 that for a given size ν for the discrete problem, the Galerkin and Sloan approximations are as effective, but the Nyström method is superior to the Fredholm approximation. Nyström is also superior to Galerkin at least when the kernel is smooth. Results in Table 6.1 also show that η_n^{S} does not improve on η_n^{G} ($\pi_n^* \nrightarrow 1$), but η_n^{N} improves on η_n^{F} for both kernels.

Table 6.3

Comparison between φ_n^G and φ_n^S

	k_1		k_2	
v	$\|\varphi - \varphi_n^G\|_\infty$	$\|\varphi - \varphi_n^S\|_\infty$	$\|\varphi - \varphi_n^G\|_\infty$	$\|\varphi - \varphi_n^S\|_\infty$
4	4.9×10^{-3}	2.4×10^{-5}	7.3×10^{-2}	1.2×10^{-2}
6	1.8×10^{-3}	2.2×10^{-5}	3.0×10^{-2}	4.8×10^{-3}
8	9.5×10^{-4}	2.0×10^{-5}	1.6×10^{-2}	2.4×10^{-3}
10	5.8×10^{-4}	1.9×10^{-5}	1.0×10^{-2}	1.4×10^{-3}
12	3.9×10^{-4}	1.8×10^{-5}	7.1×10^{-3}	9.7×10^{-4}
16	2.2×10^{-4}	1.7×10^{-5}	4.1×10^{-3}	5.5×10^{-4}
20	1.3×10^{-4}	1.7×10^{-5}	2.4×10^{-3}	4.0×10^{-4}
30	6.4×10^{-5}	1.6×10^{-5}	1.2×10^{-3}	2.2×10^{-4}

Table 6.4

Comparative Improvement Obtained with ζ_n and ρ_n

$k(s, t)$	i	$\mathcal{M}$	v	$\lambda - \lambda_n$	$\lambda - \zeta_n$	$\lambda - \rho_n$	$\lambda - \zeta_n/\lambda - \rho_n$
k_1	1	P	6	-1.9×10^{-3}	8.1×10^{-8}	5.8×10^{-7}	0.14
			8	-1.0×10^{-3}	2.6×10^{-8}	5.7×10^{-8}	0.46
		N	4	6.4×10^{-5}	1.1×10^{-7}	1.4×10^{-9}	78
			6	1.3×10^{-5}	4.6×10^{-9}	*[a]	—
			8	4.1×10^{-6}	*[a]	*[a]	—
	2	P	6	-1.1×10^{-4}	-1.3×10^{-7}	2.1×10^{-6}	-0.062
			8	-6.3×10^{-5}	-3.2×10^{-8}	8.8×10^{-7}	-0.036
		N	4	2.9×10^{-4}	1.9×10^{-6}	-1.6×10^{-7}	-11.9
			6	5.8×10^{-5}	8.1×10^{-8}	-6.7×10^{-9}	-12.1
			8	1.8×10^{-5}	8.4×10^{-9}	*[a]	—
	3	P	6	6.4×10^{-6}	2.2×10^{-8}	7.1×10^{-6}	3.1×10^{-3}
			8	5.7×10^{-7}	2.1×10^{-9}	2.1×10^{-6}	10^{-3}
		N	4	2.9×10^{-4}	1.7×10^{-6}	-5.2×10^{-5}	-0.033
			6	6.2×10^{-4}	1.1×10^{-7}	-2.2×10^{-6}	-0.050
			8	2.0×10^{-5}	1.3×10^{-8}	-2.3×10^{-7}	-0.056
k_2	1	P	4	-8.2×10^{-3}	-5.7×10^{-4}	4.5×10^{-4}	-1.27
			6	-3.0×10^{-3}	-1.9×10^{-4}	5.5×10^{-5}	-3.45
			10	-9.5×10^{-4}	-5.0×10^{-5}	-3.5×10^{-6}	14.3
		N	4	6.6×10^{-3}	-5.5×10^{-3}	6.5×10^{-5}	-86
			6	2.7×10^{-3}	-2.5×10^{-3}	1.5×10^{-5}	-167
			10	9.3×10^{-4}	-9.1×10^{-4}	2.2×10^{-6}	-413

[a] The asterisk indicates that the corresponding error is less than $|\lambda - \lambda_{100}^N|$, where λ_{100}^N is the eigenvalue of the Nyström approximation of size 100 used in the computations of ζ_n and ρ_n.

In Table 6.2, we see that all the ratios $|\lambda - \lambda_n^{\mathcal{M}}|/\eta_n^{\mathcal{M}}$ are constant, except $|\lambda - \lambda_n^{\mathrm{F}}|/\eta_n^{\mathrm{F}}$. This is because (with $\lambda_n^{\mathrm{F}} = \lambda_n^{\mathrm{N}}$ and $\eta_n^{\mathrm{N}} \ll \eta_n^{\mathrm{F}}$) the bound $\lambda - \lambda_n^{\mathrm{F}} = O(\eta_n^{\mathrm{N}})$ is more realistic than the bound $\lambda - \lambda_n^{\mathrm{F}} = O(\eta_n^{\mathrm{F}})$.

In Table 6.3, for T_n^{G}, $\tilde{\varphi} = \varphi_n^{\mathrm{S}} = \varphi^{1\mathrm{G}}$ improves on φ_n, and in Table 6.4, for T_n^{P}, $\zeta_n = \tilde{\lambda}_n$ improves on λ_n.

4.5. *Bibliographical Comments*

Various computable improvements of the eigenvalue λ based on the Rayleigh quotient or on the generalized Rayleigh quotient have been proposed in Linz (1970, 1972), Rakotch (1975, 1978), Spence (1978, 1979), Ghemires (1979) and Kulkarni and Limaye (1983a). Iterative refinement of the eigenvalues and eigenvectors are studied in Chapter 7, Part B.

5. Localization of a Group of Eigenvalues of T

It is of importance, from a numerical point of view, to know the asymptotic behavior of the constants appearing in the error bounds. In this section, we first deal with the localization of one or more eigenvalues of T, that is, the determination of a region of the complex plane where these eigenvalues lie. The localization is based on some given data. For a simple eigenvalue of a closed operator T with domain dense in X, we may formulate the problem in the following way: From the knowledge of the vectors $x \in \operatorname{Dom} T$, $y \in \operatorname{Dom}(T^*)$, and the complex number $\zeta := \langle Tx, y\rangle/\langle x, y\rangle$, is it possible to find the radii of the smallest disks centered at ζ (resp. x) and including λ (resp. φ) such that $T\varphi = \lambda\varphi$? In practice, x (resp. y) may be chosen as an eigenvector φ_n of T_n (resp. φ_n^* of T_n^*), where T_n is an approximation of T.

When T is self-adjoint in a Hilbert space, the answer is given by the Kato–Temple inequalities (Kato, 1949) that we present first.

5.1. *T Is Self-Adjoint*

Let H be a Hilbert space, T a closed self-adjoint operator with domain D dense in H. For $x \in D$ and $\|x\| = 1$, we define the Rayleigh quotient $\rho := (Tx, x)$, and set $\varepsilon := \|(T - \rho)x\|$.

Lemma 6.19 (Kato) *Let a, b be two real numbers such that*

(i) $a < \rho < b$,
(ii) *the open interval $]a, b[$ contains no point of $\sigma(T)$.*

Then $(b - \rho)(\rho - a) \leq \varepsilon^2$.

Proof The quadratic expression $(\lambda - a)(\lambda - b) = \lambda^2 - (a + b)\lambda + ab$ is nonnegative when λ changes over $\sigma(T)$. The same is true when multiplying by $d(E(\lambda)x, x) = d\|E(\lambda)x\|^2 \geq 0$ and integrating over $\mathbb{R}$. Since

$$\int \lambda^2 \, d(E(\lambda)x, x) = (T^2x, x) = \|Tx\|^2,$$

$$\int \lambda \, d(E(\lambda)x, x) = (Tx, x) = \rho,$$

and

$$\int d(E(\lambda)x, x) = \|x\|^2 = 1,$$

we get

$$\|Tx\|^2 - (a + b)\rho + ab \geq 0 \qquad \text{and} \qquad \|Tx\|^2 = \varepsilon^2 + \rho^2,$$

that is, $(b - \rho)(\rho - a) \leq \varepsilon^2$. □

Corollary 6.20 (Krylov–Weinstein) *For any x in D, $\|x\| = 1$, there exists $\lambda \in \sigma(T)$ such that $|\lambda - \rho| \leq \varepsilon$.*

Proof From Lemma 6.19, for every a such that $a < \rho$, the interval $]a, b[$ contains a point of $\sigma(T)$ if $b > \beta = \rho + \varepsilon^2/(\rho - a)$. Since $\sigma(T)$ is a closed set, $]a, \beta]$ contains also a point of $\sigma(T)$. There is only to let $a = \rho - \varepsilon$ to get the desired conclusion. □

When λ is an isolated eigenvalue of T of finite multiplicity m, the Krylov–Weinstein inequality may be improved if we have some information about the distance of ρ to the set $\sigma(T) - \{\lambda\}$. We define $\delta(\rho) := \text{dist}(\rho, \sigma(T) - \{\lambda\})$, and θ the acute angle between x and the eigenspace $M = PX$ associated with λ.

Theorem 6.21 (Kato–Temple) *We suppose that there exists an open interval $]\underline{\lambda}, \bar{\lambda}[$ containing no point of $\sigma(T)$ except λ, an eigenvalue of finite multiplicity, and such that $\underline{\lambda} < \rho < \bar{\lambda}$. Then*

$$\rho - \frac{\varepsilon^2}{\bar{\lambda} - \rho} \leq \lambda \leq \rho + \frac{\varepsilon^2}{\rho - \underline{\lambda}}$$

and

$$\sin \theta \leq \frac{2}{\bar{\lambda} - \underline{\lambda}} \left[\left(\rho - \frac{\underline{\lambda} + \bar{\lambda}}{2} \right)^2 + \varepsilon^2 \right]^{1/2}.$$

Proof For the eigenvalue inequalities we apply Lemma 6.19:

(1) If $\lambda < \rho < \bar{\lambda}$ with $a = \lambda$, $b = \bar{\lambda}$, we get

$$\rho - \frac{\varepsilon^2}{\bar{\lambda} - \rho} \leq \lambda < \rho.$$

(2) If $\underline{\lambda} < \rho < \lambda$ with $a = \underline{\lambda}$, $b = \lambda$, we get

$$\rho < \lambda \leq \rho + \frac{\varepsilon^2}{\rho - \underline{\lambda}}.$$

(3) In the general case, we get the stated inequality.

We now turn to

$$\cos^2 \theta = \|Px\|^2 = (Px, x) = ((E(\lambda) - E(\lambda^-))x, x).$$

The quadratic expression

$$(z - \underline{\lambda})(z - \bar{\lambda}) = z^2 - (\underline{\lambda} + \bar{\lambda}) + \underline{\lambda}\bar{\lambda}$$

is nonnegative, except at λ, when z changes over $\sigma(T)$. By multiplying by $d(E(z)x, x)$ and integrating over $\mathbb{R}$, we get

$$\|Tx\|^2 + (\underline{\lambda} + \bar{\lambda})\rho + \underline{\lambda}\bar{\lambda} - (\lambda - \underline{\lambda})(\lambda - \bar{\lambda}) \cos^2 \theta \geq 0,$$

and

$$\sin^2 \theta = 1 - \cos^2 \theta \leq 1 - \frac{\|Tx\|^2 + (\underline{\lambda} + \bar{\lambda})\rho + \underline{\lambda}\bar{\lambda}}{(\lambda - \underline{\lambda})(\bar{\lambda} - \lambda)}.$$

Since

$$(\lambda - \underline{\lambda})(\bar{\lambda} - \lambda) = \left(\frac{\bar{\lambda} - \underline{\lambda}}{2}\right)^2 - \left(\lambda - \frac{\underline{\lambda} + \bar{\lambda}}{2}\right)^2,$$

this quantity can be bounded in the denominator by $[(\bar{\lambda} - \underline{\lambda})/2]^2$. The desired inequality follows. □

The larger $\bar{\lambda} - \rho$ and $\rho - \underline{\lambda}$ are compared to ε, the better the bounds on the eigenvalues are compared to Krylov–Weinstein inequality.

Corollary 6.22 *If* $\varepsilon < \delta(\rho)$, *then* $|\lambda - \rho| \leq \varepsilon^2/\delta(\rho)$ *and* $\sin \theta \leq \varepsilon/\delta(\rho)$.

Proof We apply Theorem 6.21 with $\underline{\lambda} = \rho - \delta(\rho)$ and $\bar{\lambda} = \rho + \delta(\rho)$. □

Note that if $\varepsilon \geq \delta(\rho)$, we still have $|\lambda - \rho| \leq \varepsilon$ and $\sin \theta \leq 1$.

Exercises

6.29 Show that the bound on the eigenvalue is optimal in terms of the data x and $\rho = (Tx, x)$, by considering in $X = \mathbb{R}^2$ the matrix

$$A = \begin{pmatrix} 0 & \varepsilon \\ \varepsilon & a \end{pmatrix}.$$

6.30 Prove that Theorem 6.21 improves on Corollary 6.20 only if ε is such that $\varepsilon^2 < (\bar{\lambda} - \rho)(\rho - \underline{\lambda})$.

6.31 Use Corollary 6.20 to get an upper bound for $\delta(\rho)$.

When Tx is computable for a given x, the Kato–Temple inequalities give a localization for λ that depends on $\underline{\lambda}$ and $\bar{\lambda}$. These numbers are often computed by means of the Krylov–Weinstein inequality.

When λ is not well separated from the rest of $\sigma(T)$, we may want to localize a *group* of close eigenvalues, rather than each of them individually. This can be done as follows.

Instead of one vector x, we now consider an r-dimensional subspace V of H, $V \subset D$. Let Q be the orthogonal projection on V. As a first step, we suppose that we know the spectral decomposition of QTQ, the part of T in V: $\{\rho_i\}_1^r$ are the repeated eigenvalues, and $\{x_i\}_1^r$ is the associated orthonormalized basis of eigenvectors:

$$(Tx_i, x_j) = \rho_i \delta_{ij}, \qquad i, j = 1, \ldots, r, \quad \rho_1 \leq \rho_2 \leq \cdots \leq \rho_r.$$

We set $\varepsilon_i := \|(T - \rho_i)x_i\|$, $i = 1, \ldots, r$.

We suppose that there exists an open interval $]\underline{\lambda}, \bar{\lambda}[$ such that

(i) it contains no point of $\sigma(T)$ except r repeated eigenvalues ordered by increasing magnitude: $\mu_1 \leq \mu_2 \leq \cdots \leq \mu_r$;

(ii) $\underline{\lambda} < \rho_k < \bar{\lambda}$ for $k = 1, \ldots, r$.

Theorem 6.23 *Under the above hypotheses,*

$$\rho_k - \sum_{i=1}^{r} \frac{\varepsilon_i^2}{\bar{\lambda} - \rho_i} \leq \mu_k \leq \rho_k + \sum_{i=1}^{r} \frac{\varepsilon_i^2}{\rho_i - \underline{\lambda}}, \qquad k = 1, \ldots, r.$$

Proof See Kato (1949). □

Note that the denominators do not contain such quantities as $\rho_i - \rho_j$, which may be small or even zero.

The r eigenvalues of T in $]\underline{\lambda}, \bar{\lambda}[$ have been localized from the knowledge of the eigenelements of QTQ, the part of T in V. We suppose now that the x_i are not necessarily the eigenvectors of QTQ, but an arbitrary orthonormalized basis of V.

We set $\varepsilon_i := \|(1 - Q)Tx_i\|$, $i = 1, \ldots, r$. Then $\varepsilon := (\sum_{i=1}^r \varepsilon_i^2)^{1/2}$ represents the *Schmidt norm*

$$|||(1 - Q)TQ||| := \left(\sum_{i=1}^{r} \|(1 - Q)Tx_i\|^2 \right)^{1/2}$$

[see Kato (1976, p. 262); the Schmidt norm is often called Frobenius or Schur norm in matrix theory]. We also define the arithmetic means

$$\hat{\lambda} := \frac{1}{r} \sum_{i=1}^{r} \mu_i \qquad \text{and} \qquad \hat{\rho} := \frac{1}{r} \sum_{i=1}^{r} \rho_i,$$

where ρ_i is now the Rayleigh quotient (Tx_i, x_i) for $i = 1, \ldots, r$.

Corollary 6.24 *Under the hypotheses of Theorem* 6.23, *the following bounds hold for* $|\mu_k - \rho_k|$ *and for* $|\hat{\lambda} - \hat{\rho}|$, *respectively*:

$$\rho_k - \frac{\varepsilon^2}{\min_i(\bar{\lambda} - \rho_i)} \leq \mu_k \leq \rho_k + \frac{\varepsilon^2}{\min_i(\rho_i - \underline{\lambda})}, \qquad k = 1, \ldots, r,$$

and

$$\hat{\rho} - \frac{\varepsilon^2}{\min_i(\bar{\lambda} - \rho_i)} \leq \hat{\lambda} \leq \hat{\rho} + \frac{\varepsilon^2}{\min_i(\rho_i - \underline{\lambda})}.$$

Proof It is an easy consequence of the invariance of the Schmidt norm under unitary transformations. □

The r eigenvalues of T in $]\underline{\lambda}, \bar{\lambda}[$ have been localized from the knowledge of an arbitrary orthonormalized set of vectors $\{x_i\}_1^r$ in V and the associated Rayleigh quotient $\rho_i = (Tx_i, x_i)$.

5.2. *T Is Not Self-Adjoint and λ Is Simple*

Let T be a closed linear operator with domain D dense in the complex Banach space X. Let $x \in D$, $\|x\| = 1$, $y \in \mathrm{Dom}(T^*)$. We suppose that $\langle y, x\rangle \neq 0$ and set for simplicity $\langle y, x\rangle = 1$. $\zeta := \langle Tx, y\rangle$ is the generalized Rayleigh quotient based on x and y. Let Q be the projection of X on $\{x\}$ along $\{y\}^\perp \cap X$: $Q = \langle \cdot, y\rangle x$. We consider the decomposition of T:

$$\begin{aligned} T &= QTQ + (1 - Q)T(1 - Q) + (1 - Q)TQ + QT(1 - Q) \\ &= \underbrace{\zeta Q + (1 - Q)T(1 - Q)}_{\tilde{T}} \quad + \underbrace{(T - \zeta)Q + Q(T - \zeta)}_{\tilde{H}}. \end{aligned}$$

The analytic perturbation theory applied to the family $T(t) = \tilde{T} + t\tilde{H}$, for t complex, will replace the spectral decomposition of T, which no longer holds.

Exercises

6.32 Prove that $\|Q\| = \|y\| \geq 1$.

6.33 Prove that $\tilde{H} \in \mathscr{L}(X)$ and that $\tilde{T} = T - \tilde{H} \in \mathscr{C}(X)$.

6.34 Show that ζ is an isolated eigenvalue of $\tilde{T}$. Show that the associated spectral projection is Q when ζ is simple.

6.35 If ζ is a simple eigenvalue of $\tilde{T}$, prove that $\mathring{T} := [(1 - Q)(T - \zeta)]_{\restriction \{y\}^\perp \cap X} = (\tilde{T} - \zeta)_{\restriction \{y\}^\perp \cap X}$ is invertible and that $\Sigma := (1 - Q)\mathring{T}_\zeta^{-1}(1 - Q) \in \mathscr{L}(X)$ is the reduced resolvent of $\tilde{T}$ associated with ζ.

6.36 $(1 - Q)Tx = (T - \zeta)x = (1 - Q)(T - \zeta)x$.

6.37 $\tilde{H}Q = (1 - Q)\tilde{H}Q$ and $\tilde{H}(1 - Q) = Q\tilde{H}(1 - Q)$.

6.38 Prove that QX and $(1 - Q)X$ are invariant under $\tilde{H}\tilde{R}(z)\tilde{H}\tilde{R}(z)$, with $\tilde{R}(z) := (\tilde{T} - z)^{-1}$ for z in $\rho(\tilde{T})$.

6.39 Let M and N be two invariant closed subspaces for $T \in \mathscr{L}(X)$, such that $X = M \oplus N$. Prove that $r_\sigma(T) = \max(r_\sigma(T_{\restriction M}), r_\sigma(T_{\restriction N}))$.

We prove two preparatory lemmas.

Lemma 6.25 *For any z in $\rho(\tilde{T})$, $r_\sigma(\tilde{H}\tilde{R}(z)) = (r_\sigma[\tilde{H}\tilde{R}(z)\tilde{H}\tilde{R}(z)Q])^{1/2}$, with $\tilde{R}(z) = (\tilde{T} - z)^{-1}$.*

Proof $r_\sigma^2(\tilde{H}\tilde{R}(z)) = r_\sigma(\tilde{H}\tilde{R}(z)\tilde{H}\tilde{R}(z))$ for any z in $\rho(\tilde{T})$. Since QX and $(1 - Q)X$ are invariant under $\tilde{H}\tilde{R}(z)\tilde{H}\tilde{R}(z)$, we get

$$r_\sigma(\tilde{H}\tilde{R}(z)\tilde{H}\tilde{R}(z)) = \max[r_\sigma(\tilde{H}\tilde{R}(z)\tilde{H}\tilde{R}(z)Q), r_\sigma(\tilde{H}\tilde{R}(z)\tilde{H}\tilde{R}(z)(1 - Q))].$$

Now we write

$$\tilde{H}\tilde{R}(z)\tilde{H}\tilde{R}(z)Q = [\tilde{H}\tilde{R}(z)(1 - Q)][\tilde{H}\tilde{R}(z)Q],$$

$$\tilde{H}\tilde{R}(z)\tilde{H}\tilde{R}(z)(1 - Q) = [\tilde{H}\tilde{R}(z)Q][\tilde{H}\tilde{R}(z)(1 - Q)].$$

This proves that the two operators under consideration have the same spectral radius and ends the proof. □

The residual vector associated with ζ and x (resp. $\bar{\zeta}$ and y) is $u := Tx - \zeta x$ (resp. $v := T^*y - \bar{\zeta}y$). $\hat{y} := y/\|y\|$ is normalized and $\hat{v} := v/\|y\|$ is the corresponding residual vector. The norm of the residual vector is $\varepsilon := \|u\|$ (resp. $\varepsilon^* := \|\hat{v}\|$). Let a be an upper bound for $\|\Sigma\|$: $a \geq \|\Sigma\|$. Let Γ be the circle $\{z \in \mathbb{C}; |z - \zeta| = 1/2a\}$.

Lemma 6.26 *If $\Gamma \subset \rho(\tilde{T})$, one has the bound*

$$\max_{z \in \Gamma} r_\sigma^2[\tilde{H}\tilde{R}(z)] \leq 2a \sum_{k=0}^{\infty} (2a)^{-k} |\langle \Sigma^{k+1}u, v\rangle|.$$

Proof $\tilde{R}(z)x = (\zeta - z)^{-1}x$, and $B_z := Q(\tilde{H}\tilde{R}(z))^2 Q$ is a rank-1 operator such that for $\xi \in X$, $B_z\xi = (\zeta - z)^{-1} \langle \xi, y\rangle \langle \tilde{R}(z)u, v\rangle x$. Then

$$r_\sigma[(\tilde{H}\tilde{R}(z))^2 Q] = r_\sigma(B_z) = 2a|\langle \tilde{R}(z)u, v\rangle|.$$

The Laurent expansion of $\tilde{R}(z)$ about ζ is

$$\tilde{R}(z) = \frac{-Q}{z - \zeta} + \sum_{k=0}^{\infty} (z - \zeta)^k \Sigma^{k+1}.$$

Then

$$\max_{z \in \Gamma} |\langle \tilde{R}(z)u, v\rangle| \leq \sum_{k=0}^{\infty} (2a)^{-k} |\langle \Sigma^{k+1}u, v\rangle|. \quad \square$$

We now turn to the proof of localization results. We introduce the following condition:

$$\text{there exists } \tilde{\varepsilon} \text{ such that, for } k \geq 1, |\langle \Sigma^k u, v\rangle| \leq a^k \|y\| \tilde{\varepsilon}, \tag{6.10}$$

and we set $\tilde{r} := a^2\|y\|\tilde{\varepsilon}$. g is the function $r \mapsto (1 - \sqrt{1 - 4r})/2r$ defined for $0 \leq r \leq \frac{1}{4}$.

Theorem 6.27 (Lemordant) *We suppose that ζ is a simple eigenvalue of $\tilde{T}$ and that (6.10) is fulfilled. Then if $\tilde{r} < \frac{1}{4}$, there exists a simple eigenvalue λ of T such that $|\lambda - \zeta| \leq g(\tilde{r})|\langle \Sigma u, v\rangle|$. λ is the only point of $\sigma(T)$ in the disk $\{z \in \mathbb{C}; |z - \zeta| \leq 1/2a\}$. If $\langle Px, y\rangle \neq 0$, where P is the eigenprojection associated with λ, there exists an eigenvector φ normalized by $\langle \varphi, y\rangle = 1$ such that $\|\varphi - x\| \leq g(\tilde{r})\|\Sigma u\|$.*

Proof We first look at the spectrum of $\tilde{T}$ inside Γ.

$$\sigma(\tilde{T}) = \{\zeta\} \cup \sigma[(1 - Q)T(1 - Q)] \qquad \text{and} \qquad \sigma[(1 - Q)T(1 - Q)]$$

lies in the region $\{z \in \mathbb{C}, |z - \zeta| \geq 1/\|\Sigma\|\}$,which contains Γ in its exterior, because of the condition $a \geq \|\Sigma\| \cdot \tilde{T}$ has the simple eigenvalue ζ inside Γ, and Γ lies in $\rho(\tilde{T})$. For $T(t) = \tilde{T} + t\tilde{H}$, we define formally $R(t, z) = (T(t) - z)^{-1}$ for z on Γ,

$$P(t) = \frac{-1}{2i\pi}\int_\Gamma R(t, z)\, dz.$$

$P(0) = Q$ and

$$P(1) = P = \frac{-1}{2i\pi}\int_\Gamma (T - z)^{-1}\, dz \qquad \text{if} \quad \Gamma \subset \rho(T).$$

By Proposition 3.26, $t = 1$ belongs to the analyticity domain of $P(t)$ if $\max_{z\in\Gamma} r_\sigma[\tilde{H}\tilde{R}(z)] < 1$. And by Lemma 6.26 and condition (6.10) we get

$$\max_{z\in\Gamma} r_\sigma^2[\tilde{H}\tilde{R}(z)] \leq 2a^2\|y\|\tilde{\varepsilon}\sum_{k=0}^{\infty}\left(\frac{1}{2}\right)^k = 4a^2\|y\|\tilde{\varepsilon} < 1.$$

Therefore $P(t)$ is analytic for $|t| \leq 1$; $\dim PX = \dim QX = 1$. T (resp. $T(t)$) has a simple eigenvalue λ (resp. $\lambda(t)$) inside Γ, and the expansion

$$\lambda(t) = \zeta + \sum_{k=1}^{\infty} t^k\lambda_k$$

is convergent for $|t| \leq 1$.

To bound $\lambda - \zeta$, we could use the Cauchy inequalities, as in Chapter 3. We get better results, however, by exploiting the special structure of $\tilde{H}$.

By (3.16) of Chapter 3, Section 7

$$\lambda_k = \frac{(-1)^k}{k}\sum_{*}\underbrace{\operatorname{tr}[\tilde{H}\Sigma^{p_1}\cdots\tilde{H}\Sigma^{p_k}]}_{\tau}, \qquad k \geq 1,$$

with $* = \{p_i \geq 0, i = 1, \ldots, k, \sum_{i=1}^k p_i = k - 1\}$ and $\Sigma^0 = -Q$. The brackets in τ contain k terms of the form $\tilde{H}\Sigma^{p_i}$, one of which at least being $\tilde{H}Q$. Hence

$$-\tau = \operatorname{tr}(\tilde{H}\Sigma^{p_1}\cdots\tilde{H}Q)(Q\tilde{H}\cdots\Sigma^{p_k}) = \operatorname{tr}[(Q\tilde{H}\cdots\Sigma^{p_k})(\tilde{H}\Sigma^{p_1}\cdots\tilde{H}Q)].$$

The special structure of $\tilde{H}$ (Exercise 6.37) implies that the operator between brackets is zero for k odd, that is, $\lambda_{2k+1} = 0$, $k \geq 0$, and $\lambda - \zeta = \sum_{k=1}^{\infty} \lambda_{2k}$. Using again the properties of $\tilde{H}$ and rank $Q = 1$, we get

$$\lambda_{2k} = \frac{1}{k} \sum_{\substack{p_1 + \cdots + p_k = 2k-1 \\ p_i \geq 1}} \underbrace{\prod_{i=1}^{k} \operatorname{tr}(-\tilde{H}\Sigma^{p_i}\tilde{H}Q)}_{A}.$$

The number of terms of type A in λ_{2k} is the coefficient of x^{2k-1} in the expansion of $(x + x^2 + \cdots)^k = [x/(1 - x)]^k$, that is, the coefficient of x^{k-1} in the expansion of $(1 - x)^{-k}$. It is equal to

$$\frac{(2k-2)!}{(k-1)!(k-1)!} = \frac{k}{2} 4^k C_k^{1/2}.$$

Each term of type A contains tr $\tilde{H}\Sigma\tilde{H}Q$; therefore

$$|\lambda_{2k}| \leq \frac{4^k}{2} C_k^{1/2} a^{2k-2} \|y\|^{k-1} \tilde{\varepsilon}^{k-1} |\langle \Sigma u, v \rangle|$$

$$\leq \frac{1}{2a^2 \|y\| \tilde{\varepsilon}} C_k^{1/2} (4a^2 \|y\| \tilde{\varepsilon})^k |\langle \Sigma u, v \rangle|.$$

Using the expansion $1 - \sqrt{1 - 4\tilde{r}} = \sum_{k=1}^{\infty} C_k^{1/2} (4\tilde{r})^k$ for $\tilde{r} < \frac{1}{4}$, we get

$$|\lambda - \zeta| \leq \sum_{k=1}^{\infty} |\lambda_{2k}| \leq g(\tilde{r}) |\langle \Sigma u, v \rangle|.$$

We suppose that $\langle Px, y \rangle \neq 0$, that is, $QPx \neq 0$. Therefore there exists an eigenvector φ normalized by $Q\varphi = x$. We write the identities

$$(\tilde{T} - \zeta)(\varphi - x) + (\tilde{H} - \lambda + \zeta)\varphi + (\tilde{T} - \zeta)x = 0,$$

$$(1 - Q)(\varphi - x) = (1 - Q)\varphi = \varphi - x.$$

Then

$$\varphi - x + \Sigma(\tilde{H} - \lambda + \zeta)(\varphi - x + x) = 0,$$

and

$$\varphi - x + [\Sigma\tilde{H}(1 - Q) - (\lambda - \zeta)\Sigma](\varphi - x) = -\Sigma\tilde{H}x.$$

Therefore

$$\varphi - x = -(1 - (\lambda - \zeta)\Sigma)^{-1}\Sigma\tilde{H}x = -(\tilde{T} - \lambda)^{-1}u$$

and

$$\lambda - \zeta = -\langle (\tilde{T} - \lambda)^{-1}u, v \rangle.$$

This yields

$$\|\varphi - x\| \le \|(1 - (\lambda - \zeta)\Sigma)^{-1}\| \, \|\Sigma u\|$$

with

$$|\lambda - \zeta| \, \|\Sigma\| \le g(\tilde{r})a^2 \|y\| \tilde{\varepsilon} = g(\tilde{r})\tilde{r}$$

and

$$\frac{1}{1 - |\lambda - \zeta| \, \|\Sigma\|} \le \frac{1}{1 - g(\tilde{r})\tilde{r}} = g(\tilde{r}). \quad \square$$

We note again that x (resp. y) is *not* required to be an approximate eigenvector of T (resp. T^*): only condition (6.10) has to be satisfied. This is of an extreme practical importance, as we have already seen with almost-triangular matrices (Chapter 1, Lemma 1.23).

If $\tilde{\varepsilon}$ is small enough, $\langle Px, y\rangle$ is nonzero, according to the following Proposition.

Proposition 6.28 *The condition $\langle Px, y\rangle \neq 0$ is satisfied if $\tilde{r} < \frac{1}{8}$.*

Proof For $\xi \in X$, $Q(P - Q)Q\xi = \langle \xi, y\rangle(\langle Px, y\rangle - 1)x$; hence $0 < \langle Px, y\rangle < 2 \Leftrightarrow r_\sigma(Q(P - Q)Q) < 1$. Let Γ be the circle $\{z; |z - \zeta| = 1/2a\}$ which contains λ and ζ. Then

$$\begin{aligned} Q(P - Q)Q &= \frac{-1}{2i\pi} \int_\Gamma Q[R(z) - \tilde{R}(z)]Q \, dz \\ &= \frac{-1}{2i\pi} \int_\Gamma Q\tilde{R}(z) \sum_1^\infty (-\tilde{H}\tilde{R}(z))^k Q \, dz. \end{aligned}$$

Now $Q\tilde{R}(z) = \tilde{R}(z)Q$ and $Q(\tilde{H}\tilde{R}(z))^{2k-1}Q = 0$, and $Q(\tilde{H}\tilde{R}(z))^{2k}Q = B_z^k$, $k \ge 1$, where $B_z = Q(\tilde{H}\tilde{R}(z))^2 Q$. Hence

$$Q\tilde{R}(z) \sum_{k=1}^\infty B_z^k Q = Q\tilde{R}(z)B_z(1 - B_z)^{-1}Q := Q\tilde{R}(z)C_z Q.$$

Recall that $r_\sigma(B_z) = 2a|\langle \tilde{R}(z)u, v\rangle| \le 4\tilde{r} < 1$ under condition (6.10). For $\xi \in X$, $Q\tilde{R}(z)C_z Q\xi = [1/(\zeta - z)]\langle C_z \xi, y\rangle x$, and

$$Q(P - Q)Q\xi = \frac{-1}{2i\pi}\left[\int_\Gamma \frac{1}{\zeta - z} \langle C_z \xi, y\rangle \, dz\right] x.$$

Since $Q(P - Q)Q$ has rank one with eigendirection $\{x\}$, we get

$$r_\sigma[Q(P - Q)Q] \le \frac{2a}{2a} \max_{z \in \Gamma} |\langle C_z x, y\rangle|.$$

Now $|\langle C_z x, y\rangle| = r_\sigma(C_z) = r_\sigma(B_z)(1 - r_\sigma(B_z))^{-1} \le 4\tilde{r}/(1 - 4\tilde{r})$. The condition $4\tilde{r}/(1 - 4\tilde{r}) < 1$ is satisfied if $\tilde{r} < \frac{1}{8}$. Note that $g(\frac{1}{8}) < 1.2$. $\square$

Exercises

6.40 Prove that $\varphi - x + \Sigma u = (\lambda - \zeta)\Sigma(1 - (\lambda - \zeta)\Sigma)^{-1}\Sigma u$.

6.41 Prove that

$$|\lambda - \zeta + \langle \Sigma u, v \rangle| \le (g(\tilde{r}) - 1)|\langle \Sigma u, v \rangle|,$$

$$\|\varphi - x + \Sigma u\| \le (g(\tilde{r}) - 1)\|\Sigma u\|.$$

Check that $g(\tilde{r}) - 1 \sim \tilde{r}$ as $\tilde{r} \to 0$.

6.42 Show directly that, if $\varepsilon\varepsilon^*$ is small enough, then $\langle Px, y \rangle \neq 0$.

6.43 Using the splitting

$$T = [P_n T P_n + (1 - P_n)T(1 - P_n)] + [(1 - P_n)TP_n + P_n T(1 - P_n)]$$

define a convergent iterative scheme for the eigenelements of $T \in \mathscr{L}(X)$ if $T_n - z \overset{ss}{\to} T - z$ on Γ.

Since $1 \le g(\tilde{r}) \le 2$, the bounds of Theorem 6.27 are computable if we know $w = \Sigma u$, the unique solution of the equation $(\tilde{T} - \zeta)w = u$ in $(1 - Q)X$. The knowledge of w can be used to compute the corrections $-\langle w, v \rangle$ and $-w$ to ζ and x, respectively, given in Exercise 6.41.

Again Σ is ill-conditioned if the eigenvalues of T are close, so we give now localization results for a group of eigenvalues.

5.3. Localization of a Group of Eigenvalues

The data are now the two sets of m vectors $\{x_i\}_1^m$ in Dom(T) and $\{y_i\}_1^m$ in Dom(T^*), such that $\|x_i\| = 1$, $\langle y_i, x_j \rangle = \delta_{ij}$ for $i, j = 1, \dots, m$, and the m complex numbers $\zeta_i := \langle Tx_i, y_i \rangle$ for $i = 1, \dots, m$. $\hat{\zeta} = (1/m)\sum_{i=1}^m \zeta_i$ is the generalized Rayleigh quotient. Let V (resp. W) be the span of $\{x_i\}_1^m$ (resp. $\{y_i\}_1^m$). $Q = \sum_{i=1}^m \langle \cdot, y_i \rangle x_i$ is the projection on V along $W^\perp \cap X$. We suppose that

$$\Sigma := (1 - Q)T_{\hat{\zeta}}^{-1}(1 - Q)$$

is bounded, where $T_{\hat{\zeta}} := [(1 - Q)(T - \hat{\zeta})]_{\restriction W^\perp \cap X}$.

We first suppose for simplicity that QTQ is *diagonal*: $\langle Tx_j, y_i \rangle = \zeta_i \delta_{ij}$. This means that $\{x_i\}_1^m$ (resp. $\{y_i\}_1^m$) is a basis of eigenvectors of QTQ (resp. $Q^*T^*Q^*$). We set $\eta := \max_{i=1,\dots,m} |\zeta_i - \hat{\zeta}|$,

$$u_i := (T - \zeta_i)x_i \qquad \text{with} \quad \|x_i\| = 1,$$

$$v_i := (T^* - \bar{\zeta}_i)y_i \qquad \text{with} \quad \|y_i\| \ge 1.$$

Let a be an upper bound for $\|\Sigma\| : a \ge \|\Sigma\|$. Condition (6.10) takes now the form

$$\text{there exists } \tilde{\varepsilon} \text{ such that, for } k \ge 1, \max_{i,j} |\langle \Sigma^k u_i, v_j \rangle| \le \frac{1}{m} a^k \tilde{\varepsilon}. \qquad (6.11)$$

We set $d := 1/(1 - 2a\eta)$ and we suppose that $\eta < 1/4a$. Then $1 \le d < 2$. Note that η is small when the eigenvalues of QTQ are close, that is, when $\hat{\zeta}$ represents well enough each ζ_i, $i = 1, \dots, m$. We set $\tilde{r} := a^2 d\tilde{\varepsilon}$.

Theorem 6.29 *We suppose that Σ is bounded, $\eta < 1/4a$, and (6.11) is fulfilled. Then, if $\tilde{r} < \frac{1}{8}$, the spectrum of T in the disk $\{z \in \mathbb{C}; |z - \hat{\zeta}| \leq 1/2a\}$ consists of m repeated eigenvalues. Their arithmetic mean $\hat{\lambda}$ is such that $|\hat{\lambda} - \hat{\zeta}| \leq (1/2a)[4\tilde{r}/(1 - 4\tilde{r})]$. Each eigenvalue λ in the disk is such that $|\lambda - \hat{\zeta}| \leq \eta + 2\alpha\tilde{\varepsilon}$.*

For a proof see Chatelin (1983).

We have supposed so far that QTQ is diagonal, but for a non-self-adjoint finite-rank operator such as QTQ, the computation of a basis of eigenvectors, even if it exists, is not always numerically stable. It is then advisable to work directly on a triangular form of QTQ. Following Lemordant (1977), we assume QTQ to be *almost-triangular*. QTQ is of rank m, and the basis $\{x_i\}_1^m$ can be chosen as a set of m orthonormalized vectors in V in which QTQ is almost triangular (by the QR algorithm, for example). $\{y_i\}_1^m$ is the adjoint basis in W such that $\langle y_i, x_j \rangle = \delta_{ij}$, $i, j = 1, \ldots, m$. In these bases QTQ has the numbers $\{\zeta_i\}_1^m$ on the diagonal.

η is now replaced by the norm of the difference between $Q(T - \hat{\zeta})Q$ and its strictly upper-triangular part. η is small when QTQ is almost upper-triangular and when the diagonal elements $\{\zeta_i\}_1^m$ are close enough to their arithmetic mean $\hat{\zeta}$.

Under suitable hypotheses on V and W, an extension of Theorem 6.29 can be established. It takes a more complicated form, because of the generality of the framework. It is given in Chatelin (1983) and Lemordant (1977).

If we consider again the case of an almost-triangular matrix of order N, the numbers $a_{ii} = e_i^{\mathrm{H}} A e_i$, $i = 1, \ldots, N$, are approximate eigenvalues, even though the vectors e_i are not approximate eigenvectors of A or A^{H} for $i \neq 1$ and $i \neq N$. Nevertheless condition (6.10) (or (1.17) in its matrix version) is satisfied for $x = y = e_i$ when A is close enough to a triangular matrix with no two diagonal elements equal to a_{ii} (Lemma 1.23).

More generally, condition (6.11) can be shown to be satisfied for the choice of $\{x_i\}$, $\{y_i\}$ being m vectors of the canonical basis, if A is close enough to a triangular matrix with m *close* diagonal elements.

Example 6.12 Let A be the almost-triangular matrix of Example 3.20, Chapter 3, which has three close diagonal elements.

$$A = \begin{pmatrix} -6 & 2 & 3 & 3 & 1 \\ 10^{-6} & 3.9998 & 1 & 2 & 2 \\ 10^{-6} & 10^{-4} & 4.0002 & 1 & 1 \\ 10^{-6} & 10^{-4} & 10^{-4} & 4 & 1 \\ 10^{-6} & 10^{-6} & 10^{-6} & 10^{-6} & 14 \end{pmatrix}.$$

A has the eigenvalues

$$\lambda_1 = -6 + 7 \times 10^{-7}, \quad \lambda_2 = 4.04930, \quad \lambda_3 = 3.97535 + i \times 0.03771,$$
$$\lambda_4 = 3.97535 - i \times 0.03771, \quad \lambda_5 = 14 + 5 \times 10^{-7}.$$

It is easy to localize λ_1 and λ_7 by means of Gershgorin circles centered at -6 and 14, respectively, with radii equal to 4×10^{-6}.

For the remaining group of three eigenvalues, we consider the orthogonal projection Q on the canonical vectors e_2, e_3, e_4. According to the decomposition $\mathbb{C}^5 = \{e_2, e_3, e_4\} \oplus \{e_1, e_5\}$, A is decomposed into four parts:

$$A = \left(\begin{array}{c|c} QAQ & QA(I-Q) \\ \hline (I-Q)AQ & (I-Q)A(I-Q) \end{array}\right)$$

$$= \left(\begin{array}{ccc|cc} 3.9998 & 1 & 2 & 10^{-6} & 2 \\ 10^{-4} & 4.0002 & 1 & 10^{-6} & 1 \\ 10^{-4} & 10^{-4} & 4 & 10^{-6} & 1 \\ \hline 2 & 3 & 3 & -6 & 1 \\ 10^{-6} & 10^{-6} & 10^{-6} & 10^{-6} & 14 \end{array}\right).$$

We compute

$$\max_{x,\, \|x\|_1 = 1} \|(I-Q)AQx\|_1 = 3 + 10^{-6},$$

$$\max_{x,\, \|x\|_\infty = 1} \|(I-Q)A^{\mathrm{H}}Qx\|_\infty = 2,$$

$$\hat{\zeta} = \tfrac{1}{3} \operatorname{tr} QAQ = 4,$$

$$\mathring{A}_4 = [(I-Q)(A-4I)]_{\restriction\{e_1, e_5\}} = \begin{pmatrix} -10 & 1 \\ 10^{-6} & 10 \end{pmatrix},$$

$$\mathring{A}_4^{-1} = \frac{-1}{100 + 10^{-6}} \begin{pmatrix} 10 & -1 \\ -10^{-6} & -10 \end{pmatrix}.$$

The residual vectors for A corresponding to e_i and $\zeta_i = e_i^{\mathrm{H}} A e_i$, are

$$u_i = (A - \zeta_i I)e_i \quad \text{for} \quad i = 2, 3, 4.$$

Similarly

$$v_i = (A^{\mathrm{H}} - \bar{\zeta}_i I)e_i, \qquad i = 2, 3, 4.$$

QAQ is almost triangular. Let U be its strictly upper-triangular part. Then

$$Q(A - 4I)Q - U = \begin{pmatrix} -2 \times 10^{-4} & 0 & 0 \\ 10^{-4} & 2 \times 10^{-4} & 0 \\ 10^{-4} & 10^{-4} & 0 \end{pmatrix},$$

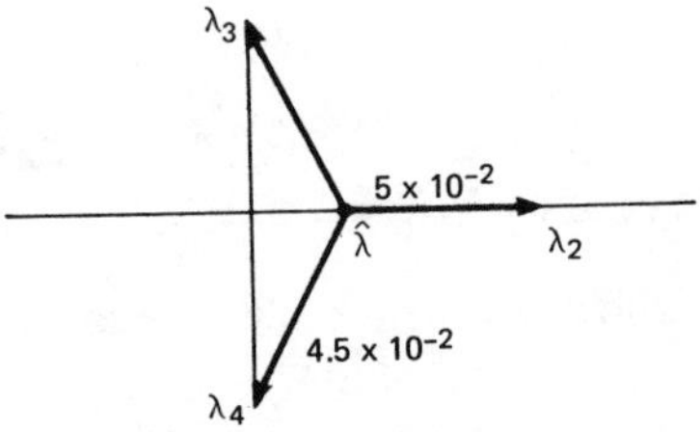

Figure 6.1

and its L^1 norm is $\eta = 4 \times 10^{-4}$. We note that $v_j^{\mathrm{H}}\Sigma u_i = v_j^{\mathrm{H}}(I - Q)\Sigma(I - Q)u_i$. The condition

$$\max_{i,j=2,3,4} |v_j^{\mathrm{H}}\Sigma^k u_i| \leq a^k\tilde{\varepsilon}$$

is satisfied with $a = 0.166$ and $\tilde{\varepsilon} < 5 \times 10^{-6}$. It can be deduced that there are exactly three eigenvalues of A in the disk $\{z; |z - 4| \leq 3\}$ such that

$$\max_i |\lambda_i - 4| \leq 0.126.$$

Their arithmetic mean $\hat{\lambda}$ is such that $|\hat{\lambda} - 4| \leq 4.7 \times 10^{-7}$, cf. Chatelin (1983). The actual errors are, respectively, $\max_{i=2,3,4}|\lambda_i - 4| = 0.05$ and $\hat{\lambda} - 4 = 5.7 \times 10^{-8}$. Note the striking difference in magnitude between $\lambda_i - 4$ and $\hat{\lambda} - 4$, cf. Fig. 6.1.

We insist that approximate eigenvectors for A associated with the cluster of eigenvalues have not been used. We merely work with a basis of the invariant subspace for the 3×3 diagonal block under consideration.

In conclusion of this section, we may say that results such as above yield a localization of the group of m eigenvalues of T from the knowledge of a basis $\{x_i\}_1^m$ of an approximate invariant subspace V of the "part" QTQ of T, in which QTQ is almost triangular. V need not be an approximate invariant subspace for T. It should satisfy, together with W, a weaker condition, of type (6.11).

6. Asymptotic Behavior of the Constants in the Error Bounds

6.1. General Results

The localization results of Section 5 may be used to get some information on the asymptotic behavior of the constants in the error bounds given in Section 4.

If λ is simple, we choose as vector x (resp. y) the eigenvector φ_n of T_n (resp. φ_n^* of T_n^*) normalized by $\langle \varphi_n^*, \varphi_n \rangle = \|\varphi_n\| = 1$. $\zeta_n = \langle T\varphi_n, \varphi_n^* \rangle$. The residual vector is $u_n := (T - \zeta_n)\varphi_n$ (resp. $v_n := (T^* - \bar{\zeta}_n)\varphi_n^*$). Let $\Sigma_n := (1 - P_n)\mathring{T}_{\zeta_n}^{-1}(1 - P_n)$. Application of Theorem 6.27 gives

$$\begin{aligned}|\lambda - \zeta_n| &\leq 2|\langle \Sigma_n u_n, v_n \rangle| \\ &\leq 2\|\Sigma_n u_n\| \, \|v_n\| \leq 2\|\Sigma_n\| \, \|u_n\| \, \|v_n\|.\end{aligned}$$

The constant in the bound of Lemma 6.16 has the asymptotic behavior of $\|\Sigma_n\|$. Similarly, if T is self-adjoint in a Hilbert space, let $\rho_n = (T\varphi_n, \varphi_n)$ be the Rayleigh quotient. The constant in the bound on $|\lambda - \rho_n|$ depends on the asymptotic behavior of $\delta(\rho_n) = \text{dist}(\rho_n, \sigma(T) - \{\lambda\})$.

6.2. *Asymptotic Equalities for the $\perp$-Galerkin and Rayleigh–Ritz Methods*

In the case of the Galerkin method with orthogonal projection applied to a self-adjoint compact operator, we present *asymptotic equalities* for the rates of convergence of the eigenelements. They are also valid for the Rayleigh–Ritz method applied to a self-adjoint elliptic differential operator with compact resolvent.

Let H_n, $n = 1, 2, \ldots,$ be a sequence of finite-dimensional subspaces of a Hilbert space H. We assume that π_n, the orthogonal projection on H_n, is such that for $x \in H$, $\pi_n x \to x$. Let T be a compact self-adjoint operator, and $T_n = \pi_n T \pi_n$ be its Galerkin approximation. By assumption $\|T - T_n\| \to 0$. Let λ be an eigenvalue of T of algebraic multiplicity m. There are m eigenvalues of T_n, numbered according to their multiplicity, say μ_{in}, $i = 1, \ldots, m$, converging to λ. P (resp. P_n) is the orthogonal spectral projection associated with λ (resp. $\{\mu_{in}\}_1^m$). r_n is a real generic constant that tends to 1 when n tends to infinity.

Proposition 6.30 *For n large enough, the following equalities hold:*

$$\|\varphi - P_n\varphi\| = r_n\|(1 - \pi_n)\varphi\|, \qquad \textit{for} \quad \varphi \in \text{Ker}(T - \lambda);$$

$$\|(1 - P)\varphi_{in}\| = r_n\|(1 - \pi_n)P\varphi_{in}\|, \qquad (\lambda - \mu_{in})/\lambda = r_n\|(1 - \pi_n)P\varphi_{in}\|^2,$$

for $\varphi_{in} \in \text{Ker}(T_n - \mu_{in})$, and $i = 1, \ldots, m$, where r_n is a real constant that depends on φ, φ_{in}, λ, μ_{in}, according to the case under consideration. $r_n \to 1$ as $n \to \infty$.

Proof Consider the identity, for $i = 1, \ldots, m$,

$$(T_n - \mu_{in})\pi_n\varphi = \pi_n T(\pi_n - 1)\varphi + (\lambda - \mu_{in})\pi_n\varphi \qquad \text{for} \quad \varphi \in \text{Ker}(T - \lambda).$$

P_n and T_n commute. By left multiplication by $Q_n = 1 - P_n$, we get

$$(T_n - \mu_{in})Q_n\pi_n\varphi = Q_n[\pi_n T(\pi_n - 1)\varphi + (\lambda - \mu_{in})\pi_n\varphi].$$

$(T_n - \mu_{in})^{-1}$ is uniformly bounded on Q_nH. Then

$$\|Q_n\pi_n\varphi\| \leq M[c\|T(\pi_n - 1)\| \, \|(1 - \pi_n)\varphi\| + |\lambda - \mu_{in}| \, \|Q_n\pi_n\varphi\|].$$

$\mu_{in} \to \lambda$ and $(1 - P_n)\pi_n\varphi = (1 - P_n\pi_n)\varphi - (1 - \pi_n)\varphi$. Therefore

$$\frac{1}{\|(1 - \pi_n)\varphi\|} \big| \, \|\varphi - P_n\pi_n\varphi\| - \|(1 - \pi_n)\varphi\| \, \big| \leq \frac{\|Q_n\pi_n\varphi\|}{\|(1 - \pi_n)\varphi\|} \leq c\|T(\pi_n - 1)\|.$$

P_n is the orthogonal projection on P_nH and π_n that on $H_n \supset P_nH$. Hence

$$\|(1 - \pi_n)\varphi\| \leq \|(1 - P_n)\varphi\| \leq \|(1 - P_n\pi_n)\varphi\|,$$

and

$$\|(1 - P_n)\varphi\| = r_n\|(1 - \pi_n)\varphi\|.$$

$r_n \to 1$ as $n \to \infty$, $r_n \geq 1$, and r_n does not depend on φ. We now write

$$\varphi_{in} - P\varphi_{in} = \varphi_{in} - P_n\pi_nP\varphi_{in} + P_n\pi_nP\varphi_{in} - P\varphi_{in},$$

$$\varphi_{in} - P_n\pi_nP\varphi_{in} = (P - P_n\pi_n)(P\varphi_{in} - \varphi_{in}).$$

Since $P - P_n\pi_n = (P - P_n)\pi_n + P(1 - \pi_n)$,

$$\|P - P_n\pi_n\| \to 0.$$

From

$$[1 + (P - P_n\pi_n)](\varphi_{in} - P\varphi_{in}) = (1 - P_n\pi_n)P\varphi_{in}$$

follows

$$\|\varphi_{in} - P\varphi_{in}\| = r_n\|(1 - P_n\pi_n)P\varphi_{in}\| = r_n\|(1 - \pi_n)P\varphi_{in}\|.$$

For $i = 1, \ldots, m$,

$$\begin{aligned}(\lambda - \mu_{in})(\varphi, \varphi_{in}) &= (T\varphi, \varphi_{in}) - (\varphi, T_n\varphi_{in}) = (T(1 - \pi_n)\varphi, \varphi_{in}) \\ &= (T(1 - \pi_n)\varphi, \varphi_{in} - P\varphi_{in}) + (T(1 - \pi_n)\varphi, P\varphi_{in}).\end{aligned}$$

We define

$$A_{in} = (T(1 - \pi_n)\varphi, \varphi_{in} - P\varphi_{in}).$$

Clearly

$$\frac{A_{in}}{\|(1 - \pi_n)\varphi\| \, \|(1 - \pi_n)P\varphi_{in}\|} \to 0$$

and

$$(T(1-\pi_n)\varphi, P\varphi_{in}) = \lambda((1-\pi_n)\varphi, (1-\pi_n)P\varphi_{in}) = \lambda\|(1-\pi_n)P\varphi_{in}\|^2$$

if φ is chosen as $P\varphi_{in}$. We then have the equality

$$\frac{\lambda-\mu_{in}}{\lambda}\,\frac{\|P\varphi_{in}\|^2}{\|(1-\pi_n)P\varphi_{in}\|^2} = 1 + \frac{A_{in}}{\|(1-\pi_n)P\varphi_{in}\|^2} = r_n,$$

where $\|P\varphi_{in}\| \to 1$. □

Exercises

6.44 Prove similar equalities for the Rayleigh–Ritz method applied to a self-adjoint elliptic differential operator with compact resolvent.

6.45 Use Proposition 6.30 to draw conclusions on the optimality of the bounds of Exercise 6.12 on the eigenvalues, when $\pi_n = \pi_n^*$ and $T = T^*$.

6.46 λ is assumed simple. Use the decomposition

$$\pi_n T = T_n = [PT_nP + (1-P)T_n(1-P)] + [(1-P)T_nP + PT_n(1-P)]$$

to prove $(\lambda-\lambda_n)/\lambda = r_n\|(1-\pi_n)\varphi\|^2$, where $r_n \to 1$, $\|\varphi\| = 1$.

For a normal operator T, the equalities for the eigenvectors still hold. For the eigenvalues, we get

$$(\lambda-\mu_{in})/\lambda = z_n\|(1-\pi_n)P\psi_{in}\|^2,$$

where z_n is a complex constant which tends to 1, and $\psi_{in} \in \operatorname{Ker}(T_n^* - \bar{\mu}_{in})$ (see Chatelin, 1975). This means that each eigenvalue μ_{in} lies in the shaded area of the disk

$$\{z;\, |z-\lambda| \le r_n|\lambda|\,\|(1-\pi_n)P\psi_{in}\|^2\},$$

(cf. Fig. 6.2).

The two limit curves are both tangent in λ to the vector $\overrightarrow{0\lambda}$, and when $n \to \infty$, the area narrows on the half-line $[0, \lambda]$. This is the extension to the normal case of the situation in the self-adjoint case where

$$0 \le \lambda - \mu_{in} \le r_n\lambda\|(1-\pi_n)P\varphi_{in}\|^2,$$

for $\lambda > 0$ (cf. Fig. 6.3).

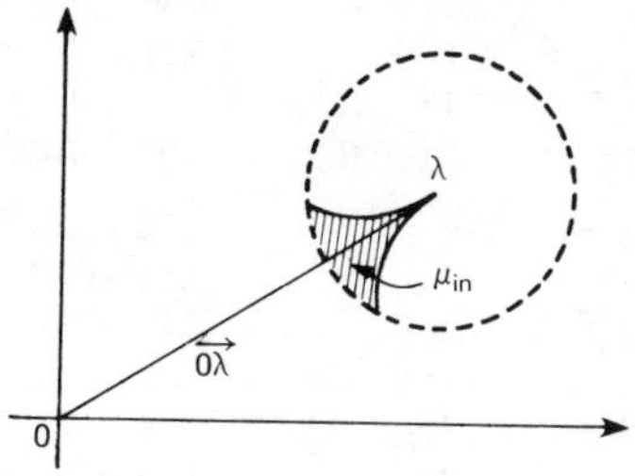

Figure 6.2

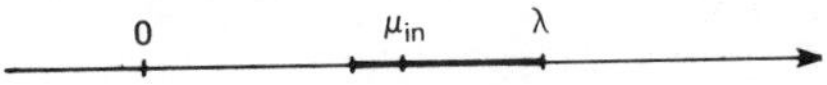

Figure 6.3

Let us see in an example how the constants r_n tend to 1. We consider the Sturm–Liouville problem on [0, 1]:

$$\begin{aligned} x''(t) + \lambda x(t) &= 0, \qquad 0 < t < 1, \\ x(0) = x(1) &= 0. \end{aligned} \tag{6.12}$$

With $H = L^2(0, 1)$, $V = H_0^1(0, 1)$, (6.12) is equivalent to

$$(x, y)_V = \lambda(x, y)_H \qquad \text{for all} \quad y \in V.$$

The eigenelements are $\lambda_i = i^2\pi^2$, $\varphi_i(t) = \sin i\pi t$, $i = 1, 2, \ldots$. We apply the Rayleigh–Ritz method, where V_n is the space of piecewise Hermite polynomials of fifth degree, defined on the points $t_i = i/n$, $i = 0, \ldots, n$. π_n is the V-ortho-

Table 6.5

Asymptotic Behavior of α_i^n and β_i^n

i	n	dim V_n	$\lambda_{in} - i^2\pi^2$	$1 - \alpha_i^n$	$\beta_i^n - 1$
4	4	13	1.32×10^{-2}	9.34×10^{-2}	4.2×10^{-4}
	5	16	5.06×10^{-4}	5.05×10^{-2}	2.3×10^{-4}
	6	19	9.36×10^{-5}	3.81×10^{-2}	7.0×10^{-5}
	7	22	1.95×10^{-5}	2.71×10^{-2}	4.0×10^{-5}
	8	25	4.80×10^{-6}	1.98×10^{-2}	1.0×10^{-5}
	9	28	1.38×10^{-6}	1.50×10^{-2}	2.0×10^{-5}
	10	31	4.57×10^{-7}	1.18×10^{-2}	1.0×10^{-6}
6	4	13	3.69×10^{-1}	1.88×10^{-1}	1.67×10^{-3}
	5	16	5.73×10^{-2}	1.17×10^{-1}	1.30×10^{-3}
	6	19	3.62×10^{-2}	1.01×10^{-1}	2.7×10^{-4}
	7	22	2.99×10^{-3}	6.56×10^{-2}	3.1×10^{-4}
	8	25	8.10×10^{-4}	5.11×10^{-2}	1.3×10^{-4}
	9	28	2.40×10^{-4}	3.93×10^{-2}	7.0×10^{-5}
	10	31	7.00×10^{-5}	3.07×10^{-2}	4.0×10^{-6}
10	4	13	35.9	3.72×10^{-1}	1.34×10^{-3}
	5	16	2.75	2.53×10^{-1}	1.03×10^{-3}
	6	19	2.44	2.12×10^{-1}	3.89×10^{-3}
	7	22	8.75×10^{-1}	1.77×10^{-1}	1.47×10^{-3}
	8	25	3.20×10^{-1}	1.66×10^{-1}	9.1×10^{-4}
	9	28	1.20×10^{-1}	1.19×10^{-1}	7.0×10^{-4}
	10	31	1.10×10^{-1}	1.05×10^{-1}	1.6×10^{-4}

gonal projection on V_n. The simple eigenvalue λ_i is approximated by the simple eigenvalue λ_{in}, P_{in} is the associated eigenprojection. Table 6.5 gives

$$\alpha_i^n := \frac{(\lambda_{in} - \lambda_i)/\lambda_i}{\|(1 - \pi_n)\varphi_i\|_V^2}, \qquad \beta_i^n := \frac{\|(1 - P_{in})\varphi_i\|_V}{\|(1 - \pi_n)\varphi_i\|_V},$$

with $\|\varphi_i\|_V = 1$, for the fourth, sixth, and tenth eigenvalues: $i = 4, 6, 10$. We may note that $0 \le \alpha_i^n \le 1$, whereas $\beta_i^n \ge 1$.

6.3. *Bibliographical Comments*

The localization results of Section 5 are a natural extension to the non-self-adjoint case of the Kato–Temple inequalities (Kato, 1949). Restricted versions have been given in Fiedler and Pták (1964), Pták (1976), and Redont (1979a). They are based on the knowledge of an approximate eigenvector of T.

Theorem 6.27 is an example of the application of the analytic perturbation theory, developed in Chapter 3, to the decomposition

$$T = [QTQ + (1 - Q)T(1 - Q)] + [(1 - Q)TQ + QT(1 - Q)].$$

The interested reader may find in Lemordant (1980) a similar use of the decomposition of T_n induced by the spectral projection P of T to derive convergence rates:

$$T_n = [PT_nP + (1 - P)T_n(1 - P)] + [(1 - P)T_nP + PT_n(1 - P)].$$

CHAPTER 7

Some Examples of Applications

Introduction

In this chapter we present two different kinds of applications of the theory given in the preceding chapters. In Part A we look for optimal convergence rates. As a first step, we consider a Fredholm integral equation of the second kind and the corresponding eigenvalue problem. The projection method uses projections on piecewise polynomials and amounts either to a $\perp$-Galerkin method in L^2 or to a collocation method at Gauss points in C. The aim of the study is twofold:

(1) to treat the eigenvectors (and invariant subspaces) in much the same manner as the solutions;
(2) to do a parallel analysis of the two projection methods cited above, so that many proofs can be factorized.

The crucial role played by the projections in the resulting rates of convergence appears clearly. The similarities and differences between the two methods are discussed in that respect. Superconvergence is obtained for a smooth kernel, as well as for a certain class of possibly discontinuous kernels. This completes the results given in Chapter 6, Section 2. The same analysis is applied in a second step to an ordinary differential equation by means of the interpretations given in Chapter 4, Section 9.4.

In Part B we propose a computational scheme for the iterative refinement of the eigenelements introduced in Chapter 5. Numerical experiments are presented.

A. SUPERCONVERGENCE RESULTS FOR INTEGRAL AND DIFFERENTIAL EQUATIONS

We first consider certain projection methods for the solution of second-kind integral equations of the form

$$(Tx)(t) - zx(t) = f(t), \qquad 0 \le t \le 1, \tag{7.1}$$

where T is the operator defined by

$$x(t) \mapsto \int_0^1 k(t, s)x(s)\, ds, \qquad 0 \le t \le 1. \tag{7.2}$$

Along with (7.1), we consider the eigenvalue problem

$$(T\varphi)(t) = \lambda\varphi(t), \qquad 0 \le t \le 1, \quad \varphi \ne 0, \tag{7.3}$$

where λ is an isolated eigenvalue of T.

(7.1) and (7.3) are regarded as equations in the complex Banach space X (specialized to be L^∞ or L^2 later on). T is supposed to be compact in X with range in C and $z \in \rho(T)$, the resolvent set of T. Let X_n be a finite-dimensional subspace of X and let π_n be a projection onto X_n. Then the projection method consists in approximating (7.1) and (7.3), respectively, by

$$(\pi_n T - z)x_n = \pi_n f, \qquad x_n \in X_n, \tag{7.4}$$

$$\pi_n T\varphi_n = \lambda_n \varphi_n, \qquad \varphi_n \in X_n, \quad \|\varphi_n\| = 1. \tag{7.5}$$

x_n (resp. φ_n) is the projection solution (resp. eigenvector), corresponding to the approximation $T_n^{\mathrm{P}} = \pi_n T$ of T (P for projection).

Given a strict partition $\Delta = \{t_i\}_0^n$ of $[0, 1]$ such that $0 = t_0 < t_1 < \cdots < t_n = 1$, let X_n be the set $\mathbb{P}_{r,\Delta}$ of piecewise polynomials of degree less than $r + 1$ on each subinterval $[t_{i-1}, t_i[$, $i = 1, \ldots, n$, the value at 1 being defined by continuity. We shall consider two types of approximation.

(a) $X = L^2(0, 1)$ and π_n^1 is the orthogonal projection in $L^2(0, 1)$ on $\mathbb{P}_{r,\Delta}$;

(b) $X = L^\infty(0, 1)$ and π_n^2 is the interpolatory map defined for x in $C(0, 1) \subset L^\infty(0, 1)$, so that $\pi_n^2 x$ interpolates x at the $r + 1$ Gauss points $\{\tau_j^i\}_{j=1}^{r+1}$, on each subinterval, $i = 1, \ldots, n$.

$\pi_n^2 x$ is piecewise continuous, so that $\pi_n^2(\pi_n^2 x)$ can be defined and is equal to $\pi_n^2 x$ (Exercise 4.6). π_n^2 defined on $C(0, 1)$ is not a projection. It will be defined as a projection in the space of piecewise-continuous functions C_Δ in Section 2.

Case (a) corresponds to a $\perp$-Galerkin method, and case (b) to a collocation method at the Gauss points $\{\tau_j^i\}$. Let π_n represent any one of the two maps π_n^1 or π_n^2. If $z \neq 0$ (resp. $\lambda_n \neq 0$), we consider the iterated solution $\tilde{x}_n$ (resp. eigenvector $\tilde{\varphi}_n$) given by the formula

$$\tilde{x}_n = \frac{1}{z}(Tx_n - f) \qquad \left(\text{resp.} \quad \tilde{\varphi}_n = \frac{1}{\lambda_n} T\varphi_n\right).$$

Since T is compact, z in $\rho(T)$ is nonzero, and $\lambda_n \neq 0$ is not a restriction either because the isolated eigenvalues of T are nonzero.

$\tilde{x}_n$ and $\tilde{\varphi}_n$ are solutions of the equations

$$(T\pi_n - z)\tilde{x}_n = f, \tag{7.6}$$

$$T\pi_n \tilde{\varphi}_n = \lambda_n \tilde{\varphi}_n, \tag{7.7}$$

corresponding to the approximation $T_n^S = T\pi_n$ of T (S for Sloan). We recall that $\pi_n \tilde{x}_n = x_n$, $\pi_n \tilde{\varphi}_n = \varphi_n$, so that in case (b) the iterated solutions and the solutions themselves agree at the collocation points.

Here, as in the rest of Part A, $\|\cdot\|_v$ is taken to be $\|\cdot\|_2$ in case (a) and $\|\cdot\|_\infty$ in case (b). If k and f are smooth enough, it is known that $\|x_n - x\|_v = O(h^{r+1})$ while $\|\tilde{x}_n - x\|_v = O(h^{2r+2})$ for both cases (a) and (b) (Exercises 4.38 and 4.39). Similar results hold for $\tilde{\varphi}_n$ (Chapter 6, Section 2). Such *superconvergence* is still valid at the partition points $\{t_i\}_0^n$ when k is the Green function of an ordinary differential equation (o.d.e.), as we shall see.

For the $\perp$-Galerkin method, the problem is to find either pointwise or global L^∞-estimate for $x - \tilde{x}_n$ (resp. $\varphi - \tilde{\varphi}_n$). The proof is in four steps:

1. prove the convergence in L^2;
2. bound $x - \tilde{x}_n$ and $\varphi - \tilde{\varphi}_n$ in L^2;
3. bound $x - \tilde{x}_n$ and $\varphi - \tilde{\varphi}_n$ pointwise;
4. deduce global bounds in L^∞.

Properties of the collocation method are deduced in Section 5 from those of the $\perp$-Galerkin method. By doing so, we stress the intimate relationship between the two methods, and we factorize many of the proofs. There is only a small amount of work left to carry superconvergence results from (a) to (b); namely, there is only to bound, uniformly in n, all the derivatives of $\tilde{x}_n$ (resp. $\tilde{\varphi}_n$) up to a certain order. This part uses the technique for the collocation method developed in de Boor and Swartz (1973). Moreover, this part is not even necessary when the kernel is smooth.

1. Definition of the Problem

1.1. Properties of the Integral Operator

$\alpha \geq 0$ and $\gamma \geq -1$ are integers. We are concerned with the following class of kernels. We recall the notation, for $0 \leq t \leq 1$, $k_t(s) := k(t, s)$. Let k_1 (resp. k_2) be the function k defined on the triangle $0 \leq s \leq t \leq 1$ (resp. $0 \leq t \leq s \leq 1$). k_{1t} (resp. k_{2t}) is defined accordingly on $[0, t]$ (resp. $[t, 1]$).

Definition The kernel $k(t, s)$ is of class $\mathfrak{G}(\alpha, \gamma)$ relative to t, with $\alpha \geq 0$ and $\alpha \geq \gamma \geq -1$, iff for $\gamma \geq 0$ and t in $[0, 1]$,

$$k_t \in C^\alpha(0, t) \cap C^\alpha(t, 1) \cap C^\gamma(0, 1) \tag{7.8}$$

uniformly in t on $[0, 1]$. If $\gamma = -1$, k_t has a first-kind discontinuity for $s = t$. γ is called the *continuity* of k and α is called its *order* (of smoothness), both relative to t. If $\gamma = \alpha$, it is said that k is a smooth kernel of continuity α, relative to t.

Theorem 4.1 applies with $p = q = r = 2$ to an integral operator T with kernel k_t of class $\mathfrak{G}(\alpha, \gamma)$ to prove that T is *compact*: $L^2 \to C$. Hence T is compact $L^2 \to L^2$ and $L^\infty \to L^\infty$.

1.2. Abstract Setting and Convergence Results

For z in $\rho(T)$, $R(z) = (T - z)^{-1}$ is bounded and the unique solution of (7.1) is $x = R(z)f$. Let $\lambda \neq 0$ be an isolated eigenvalue of T with algebraic (resp. geometric) multiplicity m (resp. g), and ascent l, $1 \leq l \leq m$, $1 \leq g \leq m$. The associated eigenspace is $E := \text{Ker}(T - \lambda)$ the null space of $T - \lambda$, $\dim E = g$; the invariant subspace is $M := \text{Ker}(T - \lambda)^m$, $\dim M = m$, and $\text{Ker}(T - \lambda)^l = \text{Ker}(T - \lambda)^m$.

Let Γ be a Jordan curve in $\rho(T)$, around λ, which contains neither 0 nor any other eigenvalue of T. $P := (-1/2i\pi) \int_\Gamma R(z)\, dz$ is the *spectral projection* associated with λ, $M = PX$. The operator $T - z$ is invertible on $(1 - P)X$ for any z inside Γ and $S(z) := (T - z)^{-1}(1 - P)$ is the *reduced resolvent* of T with respect to λ at the point z.

Let $\{T_n\}$ be a sequence of operators in $\mathscr{L}(X)$ such that T_n converges to T pointwise. T_n will be either $T_n^{\text{P}} = \pi_n T$ or $T_n^{\text{S}} = T\pi_n$. If $\Gamma \subset \rho(T_n)$, we may define for T_n the resolvent $R_n(z)$ for $z \in \Gamma$, and the spectral projection $P_n := (-1/2i\pi) \int_\Gamma R_n(z)\, dz$. If T_n is strongly stable inside Γ, there are exactly m eigenvalues $\{\mu_{in}\}_{i=1}^m$ of T_n inside Γ (counting their algebraic multiplicities), $\hat{\lambda}_n$ is their arithmetic mean, and λ_n is any of the distinct ones. $M_n := P_n^{\text{P}} X$ is the direct sum of the invariant subspaces of $T_n^{\text{P}} = \pi_n T$ associated with all its distinct eigenvalues inside Γ. Similarly $\tilde{M}_n := P_n^S X$ corresponds to $T_n^S = T\pi_n$. It is easy to check that $\tilde{M}_n = TM_n$ (Exercise 4.25), we say that $\tilde{M}_n$ is the *iterated invariant subspace*.

Let $h := \max_{1 \le i \le n}(t_i - t_{i-1})$. If $h \to 0$, then in case (a) (resp. (b)) for any x in L^2 (resp. in C), $\|\pi_n^1 x - x\|_2 \to 0$ in L^2 (resp. $\|\pi_n^2 x - x\|_\infty \to 0$ in L^∞). Since T is compact in L^2 (resp. L^∞) with range in C, then $\|(\pi_n^1 - 1)T\|_2 \to 0$, $T\pi_n^1 \overset{cc}{\to} T$ in L^2 (resp. $\|(\pi_n^2 - 1)T\|_\infty \to 0$ in L^∞ and $T\pi_n^2 \overset{cc}{\to} T$ in C).

The strong stability of T_n^P and T_n^S inside Γ follows easily (Chapter 5) as well as the convergences

$$
\begin{aligned}
&\|x - x_n\|_\nu \to 0, && \|x - \tilde{x}_n\|_\nu \to 0, && \text{if } f \in C,\\
&|\lambda - \lambda_n| \to 0, && |\lambda - \hat{\lambda}_n| \to 0, &&\\
&\|(1 - P)\varphi_n\|_\nu \to 0, && \|(1 - P)\tilde{\varphi}_n\|_\nu \to 0, &&\\
&\Theta_\nu(M, M_n) \to 0, && \Theta_\nu(M, \tilde{M}_n) \to 0. &&
\end{aligned}
$$

where Θ_ν denotes the gap in the L^ν norm.

We note that this study does not require π_n^2 to be a projection. It merely requires that $\pi_n^2 x \to x$ for x in $C(0, 1)$.

1.3. Error Formulas

A straightforward application of results in Chapter 6 to T_n^P and T_n^S yields the bounds

$$
\begin{aligned}
\|x - x_n\|_\nu &\le c\|(1 - \pi_n)x\|_\nu,\\
\Theta_\nu(M, M_n) &\le c\|(1 - \pi_n)P\|_\nu, \qquad\qquad (7.9)\\
\|(1 - P)\psi_n\|_\nu &\le c\|(1 - \pi_n)P\|_\nu \qquad \text{for} \quad \psi_n \in M_n
\end{aligned}
$$

and

$$
\begin{aligned}
\|x - \tilde{x}_n\|_\nu &\le c\|T(1 - \pi_n)x\|_\nu,\\
\Theta_\nu(M, \tilde{M}_n) &\le c\|T(1 - \pi_n)P\|_\nu, \qquad\qquad (7.10)\\
\|(1 - P)\tilde{\psi}_n\|_\nu &\le c\|T(1 - \pi_n)P\|_\nu \qquad \text{for} \quad \tilde{\psi}_n \in \tilde{M}_n.
\end{aligned}
$$

Let $P_{(n)} := P_{\restriction M_n}: M_n \to M$ be the operator P restricted to M_n. Then

$$
m(\lambda - \hat{\lambda}_n) = \sum_{i=1}^{m} \langle (1 - \pi_n)TP_{(n)}^{-1}x_i, x_i^* \rangle, \qquad (7.11)
$$

where $\{x_i\}_1^m$ (resp. $\{x_i^*\}_1^m$) is a basis of M (resp. the adjoint basis of M^*), as proved in Theorem 6.11.

For any x in M, $P_{(n)}^{-1}x - x = (1 - P)P_{(n)}^{-1}x$. Therefore

$$
\|(1 - P)P_{(n)}^{-1}x\|_\nu \le \|1 - P\|_\nu \operatorname{dist}_\nu(P_{(n)}^{-1}x, M) \le c\Theta_\nu(M, M_n).
$$

We conclude that, for any x in M,

$$\|P_{(n)}^{-1}x - x\|_v \le c\|(1 - \pi_n)P\|_v.$$

The eigenspace $E = \operatorname{Ker}(T - \lambda)$ is finite dimensional, and has therefore a supplementary subspace $W: X = E \oplus W$. Let Q be the projection onto E along W. Then, as proved in Theorem 6.7,

$$\operatorname{dist}_v(\tilde{\varphi}_n, E) \le \|\tilde{\varphi}_n - Q\tilde{\varphi}_n\|_v \le c[\|T(1 - \pi_n)\tilde{\varphi}_n\|_v + |\lambda_n - \lambda|].$$

To establish pointwise estimates we shall need the following lemma.

Lemma 7.1 *The following identities hold:*

$$\tilde{x}_n - x = TR(z)(1 - \pi_n)\tilde{x}_n, \qquad (1 - P)\tilde{\varphi}_n = TS(\lambda_n)(1 - \pi_n)\tilde{\varphi}_n.$$

Proof Straightforward from Lemma 6.4 and Exercise 6.9, with $T_n = T\pi_n$, if we notice that

$$R(z)T = TR(z) \qquad \text{and} \qquad S(\lambda_n)T = TS(\lambda_n). \quad \square$$

We deduce the pointwise equalities for any t in $[0, 1]$:

$$(\tilde{x}_n - x)(t) = \int_0^1 k(t, s)[R(z)(1 - \pi_n)\tilde{x}_n](s)\, ds = (R(z)(1 - \pi_n)\tilde{x}_n, \bar{k}_t),$$

$$[(1 - P)\tilde{\varphi}_n](t) = \int_0^1 k(t, s)[S(\lambda_n)(1 - \pi_n)\tilde{\varphi}_n](s)\, ds$$

$$= (S(\lambda_n)(1 - \pi_n)\tilde{\varphi}_n, \bar{k}_t),$$

where $(\cdot, \cdot)$ is the L^2 inner product, and $\bar{k}_t$ is defined by

$$\bar{k}_t : s \mapsto \bar{k}_t(s) := \overline{k(t, s)} \qquad \text{for} \quad s \text{ in } [0, 1]$$

Let l_t (resp. l_t^1) be the unique solution of

$$(T^* - \bar{z})l_t = \bar{k}_t \qquad (\text{resp. } (T^* - \bar{\lambda}_n)(1 - P^*)_t^1 = (1 - P^*)\bar{k}_t),$$

where the adjoint operators are defined in L^2. Then $l_t = R^*(z)\bar{k}_t$ (resp. $l_t^1 = S^*(\lambda_n)\bar{k}_t$) and the above equalities may be rewritten

$$(\tilde{x}_n - x)(t) = ((1 - \pi_n)\tilde{x}_n, l_t), [(1 - P)\tilde{\varphi}_n](t) = ((1 - \pi_n)\tilde{\varphi}_n, l_t^1). \qquad (7.12)$$

2. Smoothness Properties of the Solutions

2.1. Piecewise Continuous Functions

Given $\Delta = \{t_i\}_0^n$, a strict partition of $[0, 1]$, $0 = t_0 < t_1 < \cdots < t_n = 1$, we set $h_i := t_i - t_{i-1}$, $h := \max_{1 \le i \le n} h_i$, $\bar{\Delta}_i := [t_{i-1}, t_i]$, $i = 1, \ldots, n$. We define $C_\Delta := \prod_{i=1}^n C(\bar{\Delta}_i)$. $f \in C_\Delta$ consists of n components $f_i \in C(\bar{\Delta}_i)$. f is a piecewise-continuous function having (possibly) different left and right values at the partition points t_i. With the norm $\|\cdot\|_\Delta$ defined by $\|f\|_\Delta = \max_i \|f_i\|_\infty$, C_Δ is a Banach space and $C_\Delta \subset L^\infty(0, 1)$ since $\|f\|_\Delta = \|f\|_\infty$ for $f \in C_\Delta$.

We define more generally C_Δ^α for a positive integer α by $C_\Delta^\alpha := \prod_{i=1}^n C^\alpha(\bar{\Delta}_i)$. $f_i \in C^\alpha(\bar{\Delta}_i)$ iff its αth derivative $f_i^{(\alpha)}$ is continuous on $\bar{\Delta}_i$.

For $f \in \mathbb{P}_{r,\Delta}$, if the value at t_i^- is defined by continuity, then $\mathbb{P}_{r,\Delta} \subset C_\Delta$ and the projection π_n is defined from C_Δ onto $\mathbb{P}_{r,\Delta}$ with $f = (f_1, \ldots, f_n) \mapsto \pi_n f = (\pi f_1, \ldots, \pi f_n)$, where πf_i is the projection of $f_i \in C(\bar{\Delta}_i)$ on the polynomials of degree less than $r + 1$ on Δ_i. We remark that π_n^2 defined on C_Δ is now a projection.

We denote by $\|f_i\|_{p, \Delta_i}$ the L^p norm of f_i on $\bar{\Delta}_i$, $1 \le p \le \infty$. We establish bounds where the constants are independent of Δ, that is, of n.

2.2. *Properties of x, φ, $\tilde{x}_n$, and $\tilde{\varphi}_n$*

Theorem 7.2 *Let the kernel of T be of class $\mathfrak{G}(\alpha, \gamma)$ relative to t. Then*

(i) *if $f \in C^\alpha$, the solution x of (7.1) (resp. $\tilde{x}_n$ of (7.6)) is such that $x \in C^\alpha$ (resp. $\tilde{x}_n \in C_\Delta^\alpha$),*

(ii) *the eigenvector φ of (7.3) (resp. $\tilde{\varphi}_n$ of (7.7)) is such that $\varphi \in C^\alpha$ (resp. $\tilde{\varphi}_n \in C_\Delta^\alpha$). Moreover, $M \subset C^\alpha$.*

Proof (i) $x = (1/z)Tx - (1/z)f$, and by induction

$$x = \frac{1}{z^\alpha} T^\alpha x - \left[\frac{1}{z^\alpha} T^{\alpha-1} + \frac{1}{z^{\alpha-1}} T^{\alpha-2} + \cdots + \frac{1}{z}\right] f,$$

where the two terms in x belong to C^α. Now $\tilde{x}_n = (1/z)Tx_n + f$, $x_n \in \mathbb{P}_{r,\Delta} \subset C_\Delta^\infty$. Then, by Theorem 4.1, $Tx_n \in C_\Delta^\alpha$.

(ii) $\lambda\varphi = T\varphi = (1/\lambda)T^2\varphi = (1/\lambda^{\alpha-1})T^\alpha\varphi$ implies $\varphi \in C^\alpha$. Similarly $\varphi_n \in \mathbb{P}_{r,\Delta}$ implies $\tilde{\varphi}_n = (1/\lambda_n)T\varphi_n \in C_\Delta^\alpha$. We now consider $M = \{\varphi \in L^\infty(0, 1); (T - \lambda)^m\varphi = 0\}$. By induction it can be shown that $M \subset C^\alpha$. Similarly $M^* \subset C^\alpha$ if k is of class $\mathfrak{G}(\alpha, \gamma)$ relative to s also. □

Exercise

7.1 If $f \in C_\Delta^\alpha$, then $x \in C_\Delta^\alpha$ under the assumptions of Theorem 7.2.

For $x \in C^\alpha$, we define $\|x\|_{\alpha,p} := \sum_{i=0}^\alpha \|x^{(i)}\|_p$ for $1 \le p \le \infty$.

Proposition 7.3 *Let the kernel of T be of class $\mathfrak{G}(\alpha, \gamma)$ relative to t and s, and let f belong to C^α. Then the following bounds hold for n large enough:*

(i) $\|x\|_{\alpha,\infty} \le c\|f\|_{\alpha,\infty}$,

(ii) $\|Py\|_{\alpha,\infty} \le c\|y\|_1$ *for any y in $L^1(0, 1)$.*

Proof (i) $x = R(z)f$ proves $\|x\|_\infty \le c\|f\|_\infty$. From the proof of Theorem 7.2, we get $\|x^{(\alpha)}\|_\infty \le c(\|f\|_{\alpha,\infty} + \|x\|_\infty)$. Therefore $\|x\|_{\alpha,\infty} \le c\|f\|_{\alpha,\infty}$.

(ii) Let $\{x_i\}_1^m$ (resp. $\{x_i^*\}_1^m$) be a basis of M (resp. the adjoint basis of M^*). P is a finite-rank integral operator defined by $P = \sum_{i=1}^m (\cdot, x_i^*)x_i$. Therefore

$$(Px)(t) = \sum_{i=1}^{m} \left[\int_0^1 \overline{x_i^*(s)}x(s)\,ds\right]x_i(t),$$

and P is defined by the degenerate kernel $p(t, s) := \sum_{i=1}^m x_i(t)\overline{x_i^*(s)}$, where p is a smooth kernel of continuity α relative to t and s. We deduce that, for any y in $L^1(0, 1)$,

$$\|Py\|_{\alpha,\infty} \le \left(\sum_{i=1}^{m} \|x_i\|_{\alpha,\infty}\|x_i^*\|_{\alpha,\infty}\right)\|y\|_1. \quad \square$$

2.3. *Properties of k_t, l_t, and l_t^1*

Exercise

7.2 Prove that if the kernel of T is of class $\mathfrak{G}(\alpha, \gamma)$ relative to t and s, then the kernel $k_n^*(t, s)$ of $T^*(1 - P^*) + \bar{\lambda}_n P^*$ is of class $\mathfrak{G}(\alpha, \gamma)$ relative to t, uniformly in n.

Lemma 7.4 *Let the kernel of T be of class $\mathfrak{G}(\alpha, \gamma)$ relative to t and s. Then the functions k_t, l_t, and l_t^1 belong to C_Δ^α for any t of the partition Δ and to C^γ for $t \notin \Delta$, if $\gamma \ge 0$.*

Proof By (7.8), $k_{t_i} \in C_\Delta^\alpha$ for $t_i \in \Delta$ and $k_t \in C^\gamma$ for $t \notin \Delta$, $\gamma \ge 0$. The kernels of T^* and $T^*(1 - P^*) + \bar{\lambda}_n P^*$ are of class $\mathfrak{G}(\alpha, \gamma)$ relative to t. We may then apply Exercise 7.1 to the equations $(T^* - \bar{z})l_t = \bar{k}_t$ and $(T^* - \bar{\lambda}_n)(1 - P^*)l_t^1 = (1 - P^*)\bar{k}_t$. The result for l_t^1 is uniform in n since $\lambda_n \to \lambda \ne 0$, as $n \to \infty$. Indeed l_t^1 satisfies $l_t^1 = (1/\bar{\lambda}_n^\alpha)T^{*\alpha}l_t^1 - [(1/\bar{\lambda}_n^\alpha)T^{*\alpha-1} + \cdots + 1/\bar{\lambda}^n](1 - P^*)\bar{k}_t$, and $|\lambda_n| > |\lambda| - \varepsilon > 0$ for a given $\varepsilon > 0$ and for n large enough. $\square$

2.4. *Approximation Properties of $\mathbb{P}_{r,\Delta}$*

Let $[a, b]$ be a given bounded interval of $\mathbb{R}$. For $f \in C^\alpha(a, b)$, we define

$$\beta := \min(\alpha, r + 1).$$

For $\beta \ge 1$, $\mathfrak{T}_\beta(f)$ defined by

$$[\mathfrak{T}_\beta(f)](t) := \sum_{j=0}^{\beta-1} \frac{f^{(j)}(a)}{j!}(t - a)^j$$

is the Taylor-series expansion of order $\beta - 1$ for f at the point a for $t \in [a, b]$.

Theorem 7.5 *If $f \in C^\alpha(a, b)$, then for $1 \le p \le \infty$,*

$$\|f^{(j)} - \mathfrak{T}_\beta^{(j)}(f))\|_p \le c(b - a)^{\beta-j}\|f^{(\beta)}\|_p, \qquad 0 \le j \le \beta.$$

Proof The proof is by induction on j. $\|f^{(\beta)} - \mathfrak{T}_\beta^{(\beta)}(f)\|_p = \|f^{(\beta)}\|_p$. For $\beta - 1$,

$$[f^{(\beta-1)} - \mathfrak{T}_\beta^{(\beta-1)}(f)](t) = \int_a^t [f^{(\beta)} - \mathfrak{T}_\beta^{(\beta)}(f)](s)\, ds.$$

Then, by the Hölder inequality, we get

$$\|f^{(\beta-1)} - \mathfrak{T}_\beta^{(\beta-1)}(f)\|_\infty \le (b-a)^{1/q}\|f^{(\beta)} - \mathfrak{T}_\beta^{(\beta)}(f)\|_p$$

with $1/p + 1/q = 1$, and

$$\begin{aligned}\|f^{(\beta-1)} - \mathfrak{T}_\beta^{(\beta-1)}(f)\|_p &\le (b-a)^{1/p}\|f^{(\beta-1)} - \mathfrak{T}_\beta^{(\beta-1)}(f)\|_\infty \\ &\le (b-a)\|f^{(\beta)}\|_p.\end{aligned}$$

We suppose that the inequality holds at the step $j+1$:

$$\|f^{(j+1)} - \mathfrak{T}_\beta^{(j+1)}(f)\|_p \le (b-a)^{\beta-j-1}\|f^{(\beta)}\|_p.$$

Then

$$(f^{(j)} - \mathfrak{T}_\beta^{(j)}(f))(t) = \int_a^t [f^{(j+1)} - \mathfrak{T}_\beta^{(j+1)}(f)](s)\, ds,$$

$$\|f^{(j)} - \mathfrak{T}_\beta^{(j)}(f)\|_\infty \le (b-a)^{1/q}\|f^{(j+1)} - \mathfrak{T}_\beta^{(j+1)}(f)\|_p,$$

$$\|f^{(j)} - \mathfrak{T}_\beta^{(j)}(f)\|_p \le (b-a)^{1/p}\|f^{(j)} - \mathfrak{T}_\beta^{(j)}(f)\|_\infty.$$

The last two inequalities imply

$$\|f^{(j)} - \mathfrak{T}_\beta^{(j)}(f)\|_p \le (b-a)^{\beta-j}\|f^{(\beta)}\|_p. \quad \square$$

Corollary 7.6 *Let π_n be a projection from C_Δ on to $\mathbb{P}_{r,\Delta}$ such that, uniformly in n,*

$$\|\pi_n\|_p < \infty \qquad \text{for} \quad 1 \le p \le \infty.$$

Then for any f in C_Δ^α

$$\|(1-\pi_n)f\|_p \le ch^\beta\|f^{(\beta)}\|_p.$$

Proof On Δ_i, by Theorem 7.5,

$$\begin{aligned}\|(1-\pi)f_i\|_{p,\Delta_i} = \|(1-\pi)(f_i - \mathfrak{T}_\beta(f_i))\|_{p,\Delta_i} &\le c\|f_i - \mathfrak{T}_\beta(f_i)\|_{p,\Delta_i} \\ &\le ch_i^\beta\|f_i^{(\beta)}\|_{p,\Delta_i}.\end{aligned}$$

We first suppose that $1 \le p < \infty$. Then

$$(\|(1-\pi_n)f\|_p)^p = \sum_{i=1}^n (\|(1-\pi)f_i\|_{p,\Delta_i})^p.$$

Therefore

$$\sum_{i=1}^{n} (\|(1-\pi)f_i\|_{p,\Delta_i})^p \le c \sum_{i=1}^{n} h_i^{p\beta} (\|f_i^{(\beta)}\|_{p,\Delta_i})^p$$

$$\le c\left(\max_i h_i\right)^{p\beta} \sum_{i=1}^{n} (\|f_i^{(\beta)}\|_{p,\Delta_i})^p.$$

For $p = \infty$,

$$\|(1-\pi_n)f\|_\infty = \max_i \|(1-\pi)f_i\|_{\infty,\Delta_i}$$

and

$$\|(1-\pi)f_i\|_{\infty,\Delta_i} \le c h_i^\beta \|f_i^{(\beta)}\|_{\infty,\Delta_i}$$

prove that

$$\|(1-\pi_n)f\|_\infty \le c h^\beta \|f^{(\beta)}\|_\infty. \quad \square$$

We define $\beta_1 := \min(\beta, \gamma + 1) \ge 0$.

Corollary 7.7 *Let the kernel k_t be of class $\mathfrak{C}(\alpha, \gamma)$. Then for $1 \le p \le \infty$,*

$$\|(1-\pi_n)k_t\|_p = O(h^\beta) \qquad \text{for} \quad t \in \Delta,$$

and

$$\|(1-\pi_n)k_t\|_p = O(h^{\beta_1}) \qquad \text{for} \quad t \notin \Delta$$

with $\beta = \min(\alpha, r + 1)$ and $\beta_1 = \min(\beta, \gamma + 1)$.

Proof For $t \in \Delta$, $k_t \in C_\Delta^\alpha$ by Lemma 7.4, and the proof of Corollary 7.6 applies: For $1 \le p < \infty$, we get

$$\|(1-\pi_n)k_t\|_p \le c h^\beta [(\|k_{1t}^{(\beta)}\|_{p,[0,t]})^p + (\|k_{2t}^{(\beta)}\|_{p,[t,1]})^p]^{1/p}$$

and

$$\|(1-\pi_n)k_t\|_\infty \le c h^\beta \max(\|k_{1t}^{(\beta)}\|_{\infty,[0,t]}, \|k_{2t}^{(\beta)}\|_{\infty,[t,1]}).$$

For $t \notin \Delta$, we suppose that $t \in]t_{i-1}, t_i[$. For $s < t$ or $s > t$, k_t has continuous derivatives at s up to order α, while for $s = t$, $(k_t)^{(\gamma)}$ is continuous for $\gamma \ge 0$. We then write, for $1 \le p < \infty$,

$$(\|(1-\pi_n)k_t\|_p)^p = \underbrace{\left[\sum_{j \ne i} (\|(1-\pi)(k_t)_j\|_{p,\Delta_j})^p\right]}_{A} + (\|(1-\pi)(k_t)_i\|_{p,\Delta_i})^p.$$

The sum $A^{1/p}$ can easily be bounded in h^β, as above. On Δ_i, we have

$$(\|(1-\pi)(k_t)_i\|_p)^p \le (\|k_t - \mathfrak{T}_{\gamma+1}(k_{1t})\|_{p,[t_{i-1},t]})^p + (\|k_t - \mathfrak{T}_{\gamma+1}(k_{2t})\|_{p,[t,t_i]})^p.$$

Clearly

$$(k_t - \mathfrak{T}_{\gamma+1}(k_{1t}))(s) = \frac{1}{(\gamma+1)!} k_1^{(\gamma+1)}(t, \theta_s)(s-t)^{\gamma+1}$$

for some θ_s such that $t_{i-1} < \theta_s < t$. Together with the similar equality on $[t, t_i]$, this proves that $\|(1-\pi)(k_t)_i\|_{p,\Delta_i}$ can be bounded in $h_i^{\gamma+1}$. The resulting global order is $\beta_1 = \min(\beta, \gamma+1)$. If $\gamma = -1$, $\beta_1 = 0$ and the result is clear. For $p = \infty$, the proof is similar. □

We remark that Corollary 7.7 also applies to l_t and l_t^1 when k_s is also of class $\mathfrak{G}(\alpha, \gamma)$ (Lemma 7.4).

3. Superconvergence Results for the ⊥-Galerkin Method

In this section $X = L^2(0, 1)$, π_n^1 is the orthogonal projection on $\mathbb{P}_{r,\Delta}$, and $\beta^* := \min(\beta, \gamma + 2) \ge 1$. We begin with the following lemma.

Lemma 7.8 *If k_t is of class $\mathfrak{G}(\alpha, \gamma)$ and if $x \in C^\alpha$, then for $t \notin \Delta$,*

$$|(k_t, (1-\pi_n^1)x)| = O(h^{\beta+\beta^*}),$$

with $\beta = \min(\alpha, r+1)$ and $\beta^* = \min(\beta, \gamma+2)$.

Proof $(k_t, (1-\pi_n^1)x) = ((1-\pi_n^1)k_t, (1-\pi_n^1)x)$ by the orthogonality of π_n^1. We first note that if we use Corollary 7.7, we get only the order $\beta + \beta_1$ by writing

$$|(k_t, (1-\pi_n^1)x)| \le \|(1-\pi_n^1)k_t\|_2 \|(1-\pi_n^1)x\|_2 \le ch^{\beta_1+\beta}.$$

The better order $\beta + \beta^*$ is recovered by a careful use of the following bound on $\bar{\Delta}_i$, $i = 1, \ldots, n$, for $x \in C_\Delta$:

$$\|x_i\|_{2,\Delta_i} \le h_i^{1/2}\|x_i\|_{\infty,\Delta_i} \le h_i^{1/2}\|x\|_\infty;$$

$$|(k_t, (1-\pi_n^1)x)| \le \sum_{i=1}^n |((k_t)_i, (1-\pi^1)x_i)|$$

$$\le \sum_{i=1}^n [\|(1-\pi^1)(k_t)_i\|_{2,\Delta_i} \|(1-\pi^1)x_i\|_{2,\Delta_i}].$$

For $i = 1, \ldots, n$,

$$\|(1-\pi^1)x_i\|_{2,\Delta_i} \le ch_i^\beta \|x_i^{(\beta)}\|_{2,\Delta_i} \le ch_i^{\beta+1/2}\|x^{(\beta)}\|_\infty.$$

For $t \notin \Delta$, let $t \in]t_{i-1}, t_i[$. Then $(k_t)_j \in C^\alpha(\bar{\Delta}_j)$ for $j \neq i$ and $(k_t)_i \in C^\gamma(\bar{\Delta}_i)$ for $\gamma \geq 0$. Then

$$\begin{aligned}\|(1-\pi^1)(k_t)_j\|_{2,\Delta_j} &\leq ch_j^\beta \max(\|(k_{1t})_j^{(\beta)}\|_{2,\Delta_j}, \|(k_{2t})^{(\beta)}\|_{2,\Delta_j})\\ &\leq ch_j^{\beta+1/2}\max(\|k_{1t}^{(\beta)}\|_\infty, \|k_{2t}^{(\beta)}\|_\infty),\end{aligned}$$

and

$$\begin{aligned}\|(1-\pi^1)(k_t)_i\|_{2,\Delta_i} &\leq ch_i^{\beta_1}[(\|k_{1t}^{(\beta_1)}\|_{2,[t_{i-1},t]})^2 + (\|k_{2t}^{(\beta_1)}\|_{2,[t,t_i]})^2]^{1/2}\\ &\leq ch_i^{\beta_1+1/2}[(\|k_{1t}^{(\beta_1)}\|_\infty)^2 + (\|k_{2t}^{(\beta_1)}\|_\infty)^2]^{1/2}.\end{aligned}$$

The remark $\min(2\beta, \beta+\beta_1+1) = \beta+\beta^*$ ends the proof. □

3.1. Superconvergence for $\tilde{x}_n$ and $\tilde{\varphi}_n$

Theorem 7.9 *If the kernel of T is of class $\mathfrak{G}(\alpha, \gamma)$ relative to t and s, and if $f \in C^\alpha$, then for n large enough,*

(i) $\max(\|x - x_n\|_2, \Theta_2(M, M_n), \|(1-P)\varphi_n\|_2) = O(h^\beta)$,
(ii) $\max(\|x - \tilde{x}_n\|_2, \Theta_2(M, \tilde{M}_n), \|(1-P)\tilde{\varphi}_n\|_2) = O(h^{\beta+\beta^*})$,
(iii) $\max(|(x - \tilde{x}_n)(t_i)|, |[(1-P)\tilde{\varphi}_n(t_i)]|) = O(h^{2\beta})$, $i = 0, \ldots, n$,
(iv) $\max(\|x - \tilde{x}_n\|_\infty, \|(1-P)\tilde{\varphi}_n\|_\infty) = O(h^{\beta+\beta^*})$,

where $\beta = \min(\alpha, r+1)$ and $\beta^* = \min(\beta, \gamma+2)$.

Proof (i) Since $x \in C^\alpha$ and $M \subset C^\alpha$, then, by Corollary 7.6,

$$\|(1-\pi_n^1)x\|_2 \leq ch^\beta \|x^{(\beta)}\|_2$$

and

$$\|(1-\pi_n^1)P\|_2 \leq ch^\beta \max_{\psi \in M}(\|\psi^{(\beta)}\|_2, \|\psi\|_2 = 1).$$

The result follows by (7.9).

(ii) $\|T(1-\pi_n^1)x\|_2 \leq \|T(1-\pi_n^1)x\|_\infty = \sup_{t\in[0,1]}|(k_t, (1-\pi_n^1)x)|$.

Application of Lemma 7.8 proves that

$$\|T(1-\pi_n^1)x\|_\infty \leq ch^{\beta+\beta^*}\|x^{(\beta)}\|_\infty.$$

$\|T(1-\pi_n^1)P\|_\infty$ can be bounded accordingly. The desired results follow by (7.10).

To prove (iii) and (iv), we write, for a fixed t in $[0, 1]$,

$$((1-\pi_n^1)\tilde{x}_n, l_t) = ((1-\pi_n^1)[(\tilde{x}_n - x) + x], (1-\pi_n^1)l_t),$$

$$((1-\pi_n^1)\tilde{\varphi}_n, l_t^1) = ((1-\pi_n^1)[(1-P)\tilde{\varphi}_n + P\tilde{\varphi}_n], (1-\pi_n^1)l_t^1).$$

For $t \in \Delta$, $\|(1-\pi_n^1)l_t\|_2$ and $\|(1-\pi_n^1)l_t^1\|_2$ are both of order β, by Corollary 7.7, as are the quantities

$$\|\tilde{x}_n - x\|_2, \qquad \|(1-\pi_n^1)x\|_2, \qquad \|(1-P)\tilde{\varphi}_n\|_2,$$

and

$$\|(1 - \pi_n^1)P\tilde{\varphi}_n\|_2,$$

which are at least of order β, by (i) and (ii). This proves (iii). Now for $t \notin \Delta$, $((1 - \pi_n^1)x, l_t)$ and $((1 - \pi_n^1)P\tilde{\varphi}_n, l_t^1)$ are of the order $\beta + \beta^*$ by making use of Lemma 7.8 and of the uniform bound

$$\|(P\tilde{\varphi}_n)^{(\beta)}\|_\infty \leq c\|\tilde{\varphi}_n\|_\infty \leq c.$$

As for the remaining inner products, we deduce from Corollary 7.7 that

$$\max(\|(1 - \pi_n^1)l_t\|_2, \|(1 - \pi_n^1)l_t^1\|_2) \leq ch^{\beta_1}.$$

Therefore, since $\beta_1 \geq 0$,

$$|(\tilde{x}_n - x, (1 - \pi_n^1)l_t)| \leq ch^{\beta_1}\|\tilde{x}_n - x\|_2 \leq ch^{\beta+\beta^*},$$

$$|((1 - P)\tilde{\varphi}_n, (1 - \pi_n^1)l_t^1)| \leq ch^{\beta_1}\|(1 - P)\tilde{\varphi}_n\|_2 \leq ch^{\beta+\beta^*}. \quad \square$$

Exercise

7.3 Use Exercise 6.3 to prove that for any $\tilde{\psi}_n$ in $\tilde{M}_n$, we have

$$[(1 - P)\tilde{\psi}_n](t_i) = O(h^{2\beta}) \quad \text{and} \quad \text{dist}_\infty(\tilde{\psi}_n, M) = O(h^{\beta+\beta^*}),$$

under the assumptions of Theorem 7.9.

3.2. Superconvergence for $\hat{\lambda}_n$

Theorem 7.10 *If the kernel of T is of class $\mathfrak{G}(\alpha, \gamma)$ relative to t and s, then, for n large enough,*

$$|\lambda - \hat{\lambda}_n| = O(h^{2\beta}).$$

Proof We start from the identity (7.11). Let us set $y_{in} := P_{(n)}^{-1}x_i \in \mathbb{P}_{r,\Delta}$, $i = 1, \ldots, m$. The question is to give the order of the inner product

$$((1 - \pi_n^1)Ty_{in}, x_i^*) = ((1 - \pi_n^1)T[(y_{in} - x_i) + x_i], x_i^*)$$

with

$$y_{in} \in M_n, x_i^* \in C^\alpha.$$

The inequality

$$\|y_{in} - x_i\|_2 \leq c\|(1 - \pi_n^1)P\|_2$$

shows that $\|y_{in} - x_i\|_2$ is of the order of $\beta = \min(\alpha, r + 1)$. The order 2β of $((1 - \pi_n^1)Tx_i, x_i^*)$ is clear by application of Corollary 7.6 with $p = 2$. It remains to deal with $((1 - \pi_n^1)T(y_{in} - x_i), x_i^*)$. The order 2β follows readily from

$$|((1 - \pi_n^1)T(y_{in} - x_i), x_i^*)| \leq c\|y_{in} - x_i\|_2\|(1 - \pi_n^1)x_i^*\|_2. \quad \square$$

3.3. *Convergence Rates for* $\lambda - \lambda_n$ *and* $\mathrm{dist}_\infty(\tilde{\varphi}_n, E)$

When the eigenvalue λ is semisimple ($l = 1$), the convergence rates for $\lambda - \lambda_n$ and $\mathrm{dist}_\infty(\tilde{\varphi}_n, E)$ have already been established: they are of the order of $\lambda - \hat{\lambda}_n$ and $\mathrm{dist}_\infty(\tilde{\varphi}_n, M)$, respectively. When $l > 1$, we have the following proposition.

Proposition 7.11 *If the kernel of T is of class $\mathfrak{G}(\alpha, \gamma)$ relative to t and s, then, for n large enough and $l > 1$, $|\lambda - \lambda_n|$ and* $\mathrm{dist}_\infty(\tilde{\varphi}_n, E)$ *are of the order $2\beta/l$.*

The proof is left to the reader. (See Exercise 6.12).

Exercise

7.4 Prove that $\min_i|\lambda - \mu_{in}| = O(h^{2\beta g/m})$.

3.4. *Comments*

We have established, for the ⊥-Galerkin method, convergence rates of order higher than $\beta = \min(\alpha, r + 1)$ with no restriction imposed on the integers $\alpha \geq 0$, $-1 \leq \gamma \leq \alpha$. In practice, when $\alpha \geq 1$, r is chosen such that $r + 1 \leq \alpha$ to take full advantage of the accuracy of functions in $\mathbb{P}_{r,\Delta}$. Therefore $\beta = r + 1$. For a smooth kernel ($\gamma = \alpha \geq r + 1$), $\tilde{x}_n$ and $\tilde{\varphi}_n$ achieve the *double* accuracy h^{2r+2} at any point of $[0, 1]$. When the kernel is not smooth enough ($-1 \leq \gamma < r - 1$), the double accuracy h^{2r+2} is preserved at the partition points, whereas at an arbitrary point of $[0, 1]$ we get the order $r + \gamma + 3$. We note that $\lambda - \hat{\lambda}_n$ achieves the double accuracy in both cases.

4. Connection between the ⊥-Galerkin and Collocation Methods

We introduce the following definitions and notation. Given the $r + 1$ Gauss points $\{\tau_i\}_1^{r+1}$ of the interval $[a, b]$, for $f \in C(a, b)$, let $\pi^2 f$ be the Lagrange interpolant polynomial of degree less than $r + 1$ at the $r + 1$ points $\{\tau_i\}$. $\pi^1 f$ is the least-squares polynomial of degree less than $r + 1$ on $[a, b]$. We define the Legendre polynomial $v\colon s \in [a, b] \mapsto v(s) = \prod_{i=1}^{r+1} (s - \tau_i)$.

Given a function f vanishing at $\{\tau_i\}$, its $(r+1)$th divided difference at the points $\{\tau_i\}$ is denoted $\delta[\tau_1, \ldots, \tau_{r+1}, \cdot]f$, and is abbreviated $\delta^{(r+1)}f$. We have

$$f(s) = v(s)\, \delta[\tau_1, \ldots, \tau_{r+1}, s]f \qquad \text{for} \quad s \notin \{\tau_i\}_1^{r+1},$$

the value for $s \in \{\tau_i\}$ being defined by continuity.

Lemma 7.12 *For f and g in $C(a, b)$,*

$$((1 - \pi^2)f, g) = (v, (1 - \pi^1)g\delta^{(r+1)}\bar{f}).$$

Proof By definition $(1 - \pi^2)f$ vanishes at the $\{\tau_i\}$, $\pi^2 f \in \mathbb{P}_r$ and $\delta^{(r+1)}\pi^2 f = 0$. Then

$$(1 - \pi^2)f = v\delta^{(r+1)}((1 - \pi^2)f) = v\delta^{(r+1)}f.$$

Therefore

$$((1 - \pi^2)f, g) = (v, g\delta^{(r+1)}\bar{f}).$$

Because the $\{\tau_i\}_1^{r+1}$ are the $r + 1$ *Gauss points* on $[a, b]$,

$$\int_a^b v(s)p(s)\, ds = 0$$

for any polynomial p of degree less than $r + 1$. v is orthogonal to $\mathbb{P}_r$, that is, $\pi^1 v = 0$ or

$$((1 - \pi^2)f, g) = ((1 - \pi^1)v, (1 - \pi^1)g\delta^{(r+1)}\bar{f}). \quad \square$$

Corollary 7.13 *Let η be a nonnegative integer. If $g \in C^\eta(a, b)$ and $f \in C^{\eta+r+1}(a, b)$, then*

$$|((1 - \pi^2)f, g)| \le c(b - a)^{r+1+\beta'+1}\|g\|_{\beta',\infty}\|f\|_{\beta'+r+1,\infty},$$

where $\beta' := \min(\eta, r + 1)$.

Proof By Lemma 7.12,

$$|((1 - \pi^2)f, g)| \le \|(1 - \pi^1)v\|_2\|(1 - \pi^1)g\delta^{(r+1)}\bar{f}\|_2.$$

Since $v \in \mathbb{P}_{r+1}$,

$$\|(1 - \pi^1)v\|_2 \le (b - a)^{r+1}\|v^{(r+1)}\|_2 \le (b - a)^{r+1+1/2}\|v^{(r+1)}\|_\infty.$$

Similarly,

$$\begin{aligned} \|(1 - \pi^1)g\delta^{(r+1)}\bar{f}\|_2 &\le (b - a)^{\beta'}\|(g\delta^{(r+1)}\bar{f})^{(\beta')}\|_2 \\ &\le (b - a)^{\beta'+1/2}\|g\|_{\beta',\infty}\|f\|_{\beta'+r+1,\infty}. \quad \square \end{aligned}$$

Lemma 7.12 expresses the relationship between the $\perp$-Galerkin method and the collocation method at Gauss points. This may be related to the fact that in the engineering literature, collocation at Gaussian points is very adequately called *orthogonal* collocation (see Prenter, 1975).

Lemma 7.12 explains the asymmetry of the roles played by f and g, by the presence of the divided difference $\delta^{(r+1)}\bar{f}$. One consequence is expressed by Corollary 7.13, where more smoothness is required from f. The asymmetry is stressed if we write the corresponding bound for π^1, which is symmetric in f and g: if $f, g \in C^\alpha(a, b)$, then $|((1 - \pi^1)f, g)| \leq c(b - a)^{2\beta+1}\|f^{(\beta)}\|_\infty\|g^{(\beta)}\|_\infty$, where $\beta = \min(\alpha, r + 1)$ (cf. Corollary 7.6 and Lemma 7.8). As another consequence, higher-order convergence results on $\tilde{x}_n$ and $\tilde{\varphi}_n$ for collocation at Gauss points will be obtained only if $\alpha \geq r + 1$, in opposition to what happens in the $\perp$-Galerkin method, where no restriction on α is imposed. The *double* accuracy h^{2r+2} will be obtained in case (b) only when $\eta \geq r + 1$, that is, when $\alpha \geq 2r + 2$, whereas it is obtained already for $\alpha \geq r + 1$ in case (a).

5. Superconvergence Results for the Collocation Method at Gauss Points

We consider now the collocation method (b) defined in C_Δ. The collocation points $\{\tau_j^i\}$ are chosen such that the $\{\tau_j^i\}_{j=1}^{r+1}$ are the $r + 1$ Gauss points on each Δ_i, $i = 1, \ldots, n$. π_n^1 (resp. π_n^2) is the orthogonal (resp. interpolatory) projection on $\mathbb{P}_{r,\Delta}$.

When we deal with the L^2 inner product and adjoint operators, the functions under consideration are taken as elements of L^2. The following result is easy.

Proposition 7.14 *If the kernel of T is of class $\mathfrak{G}(\alpha, \gamma)$ relative to t and s and if $f \in C^\alpha$, then $\|x - x_n\|_\infty$, $\|(1 - P)\varphi_n\|_\infty$, and $\Theta_\infty(M, M_n)$ are of the order $\beta = \min(\alpha, r + 1)$.*

If we suppose that $\alpha \geq r + 1$, then $\beta = r + 1$ and we can prove the superconvergence results for $\tilde{x}_n$ and $\tilde{\varphi}_n$ given in Theorem 7.16 by using Corollary 7.13 on the $\bar{\Delta}_i$, $i = 1, \ldots, n$. Before proving this main result, we need some preparation. We recall the Markov inequality for polynomials of degree less than $r + 1$ written, on Δ_i,

$$\|q\|_{r,\infty,\Delta_i} \leq ch_i^{-r}\|q\|_{\infty,\Delta_i} \qquad \text{for any} \quad q \in \mathbb{P}_r(\Delta_i)$$

(de Boor and Swartz, 1973).

Lemma 7.15 *Under the hypothesis of Proposition* 7.14, *if $\alpha \geq r + 1$, then*

$$\|x - x_n\|_{r,\infty,\Delta_i} \leq c(h/h_i)^r h\|x\|_{r+1,\infty} \qquad \textit{for} \quad i = 1, \ldots, n,$$

$$\|(1 - P)\varphi_n\|_{r,\infty,\Delta_i} \leq c(h/h_i)^r h\|P\varphi_n\|_{r+1,\infty} \qquad \textit{for} \quad i = 1, \ldots, n.$$

Proof Since there is no ambiguity, we have omitted the index i for the functions defined on Δ_i.

Since $\alpha \geq r + 1$, x has at least a continuous rth derivative. We define on $\bar{\Delta}_i$ the Taylor expansion of x of order r, $\mathfrak{T}_{r+1}(x) \in \mathbb{P}_r(\bar{\Delta}_i)$:

$$\|x - x_n\|_{r, \infty, \Delta_i} \leq \|x - \mathfrak{T}_{r+1}(x)\|_{r, \infty, \Delta_i} + \|\mathfrak{T}_{r+1}(x) - x_n\|_{r, \infty, \Delta_i}.$$

By application of the Markov inequality to $\mathfrak{T}_{r+1}(x) - x_n$, we get

$$\|\mathfrak{T}_{r+1}(x) - x_n\|_{r, \infty, \Delta_i} \leq ch_i^{-r}\|\mathfrak{T}_{r+1}(x) - x_n\|_{\infty, \Delta_i},$$

$$\|\mathfrak{T}_{r+1}(x) - x_n\|_{\infty, \Delta_i} \leq \|\mathfrak{T}_{r+1}(x) - x\|_{\infty, \Delta_i} + \|x - x_n\|_{\infty, \Delta_i}.$$

By Proposition 7.14, $\|x - x_n\|_{\infty, \Delta_i} \leq ch^{r+1}\|x\|_{r+1, \infty}$. By Theorem 7.5,

$$\|x - \mathfrak{T}_{r+1}(x)\|_{r, \infty, \Delta_i} \leq \|x - \mathfrak{T}_{r+1}(x)\|_{\infty, \Delta_i} \leq ch^{r+1}\|x\|_{r+1, \infty},$$

and the desired bound follows for $\|x - x_n\|_{r, \infty, \Delta_i}$. The proof for $\varphi_n - P\varphi_n$ is quite similar, by writing

$$\varphi_n - P\varphi_n = \varphi_n - \mathfrak{T}_{r+1}(P\varphi_n) + \mathfrak{T}_{r+1}(P\varphi_n) - P\varphi_n$$

and using $\|P\varphi_n\|_{r+1, \infty} < c$ since $\alpha \geq r + 1$. □

5.1. *Superconvergence for $\tilde{x}_n$ and $\tilde{\varphi}_n$*

Since $\alpha \geq r + 1$, we write $\alpha = \eta + r + 1$, where η is a nonnegative integer. We define

$$\beta' := \min(\eta, r + 1), \qquad \beta'_1 := \min(\beta', \gamma + 1), \beta'^* := \min(\beta', \gamma + 2).$$

Theorem 7.16 *Let the kernel of T be of class $\mathfrak{G}(\alpha, \gamma)$ relative to t and s, with $\alpha \geq r + 1$. If $f \in C^\alpha$, then, for n large enough,*

(*i*) $\max(\|x - \tilde{x}_n\|_\infty, \Theta_\infty(M, \tilde{M}_n), \|(1 - P)\tilde{\varphi}_n\|_\infty) = O(h^{r+1+\beta'^*})$,
(*ii*) $\max(|(x - \tilde{x}_n)(t_i)|, |[(1 - P)\tilde{\varphi}_n](t_i)|) = O(h^{r+1+\beta'})$,

for $i = 0, \ldots, n$.

Proof (i) By (7.10),

$$\|x - \tilde{x}_n\|_\infty \leq c\|T(1 - \pi_n^2)x\|_\infty, \qquad \|(1 - P)\tilde{\varphi}_n\|_\infty \leq c\|T(1 - \pi_n^2)P\|_\infty,$$

with

$$\|T(1 - \pi_n^2)x\|_\infty = \sup_{t \in [0, 1]} |((1 - \pi_n^2)x, \bar{k}_t)|.$$

For $t = t_i \in \Delta, \bar{k}_t \in C^\alpha_\Delta$. Therefore, by Corollary 7.13, on Δ_j, $j = 1, \ldots, n$, we get

$$|((1 - \pi^2)x_j, \bar{k}_t)_{\Delta_j}| \leq ch_j^{r+1+\beta'+1}\|(k_t)_j\|_{\beta', \infty, \Delta_j}\|x_j\|_{\beta'+r+1, \infty, \Delta_j}.$$

By summing over j,

$$|((1 - \pi_n^2)x, \bar{k}_t)| \leq ch^{r+1+\beta'}\|x\|_{\beta'+r+1, \infty}.$$

For $t \notin \Delta$, we suppose that $t \in]t_{i-1}, t_i[$. The above bounds are valid on Δ_j for all $j \neq i$. On Δ_i, $(\bar{k}_t)_i \in C^{\gamma+1}$ for $s \in [t_{i-1}, t]$ and $s \in [t, t_{i+1}]$,

$$|((1 - \pi^2)x, \bar{k}_t)_{\Delta_i}| \leq c h_i^{r+1+\beta'_1+1} \|x_i\|_{\beta'_1+r+1, \infty, \Delta_i}.$$

By summing over j,

$$|((1 - \pi_n^2)x, \bar{k}_t)| \leq c h^{r+1+\min(\beta', \beta'_1+1)}.$$

The order $r + 1 + \beta'^*$ results from $\min(\beta', \beta'_1 + 1) = \beta'^*$. $\|T(1 - \pi_n^2)P\|_\infty$ is treated in the same way.

(ii) We first study $x - \tilde{x}_n$. For $t = t_i \in \Delta$, we write by (7.12)

$$(x - \tilde{x}_n)(t) = ((1 - \pi_n^2)\tilde{x}_n, l_t) = ((1 - \pi_n^2)x, l_t) + ((1 - \pi_n^2)(\tilde{x}_n - x), l_t)$$

$$= ((1 - \pi_n^2)x, l_t) + \sum_{j=1}^{n} ((1 - \pi^2)(\tilde{x}_n - x), l_t)_{\Delta_j}.$$

The first inner product is of the same order as $((1 - \pi_n^2)x, \bar{k}_t)$, that is, $r + 1 + \beta'$. As for the second,

$$|((1 - \pi^2)(\tilde{x}_n - x)_j, (l_t)_j)| \leq c h_j^{r+1+\beta'+1} \|(l_t)_j\|_{\beta', \infty, \Delta_j} \|(\tilde{x}_n - x)_j\|_{\beta'+r+1, \infty, \Delta_j}.$$

It remains to bound $\|\tilde{x}_n - x\|_{\beta'+r+1, \infty}$ uniformly in n, with $\beta' + r + 1 \leq \alpha$. First, we note that $\tilde{x}_n - x = (1/z)T(x_n - x)$. Then

$$\|\tilde{x}_n - x\|_{\beta'+r+1, \infty, \Delta_j} \leq c[\|x - x_n\|_\infty + \|x_n - x\|_{\beta'+r, \infty, \Delta_j}].$$

$\|x - x_n\|_\infty = O(h^{r+1})$ is bounded, and x_n on Δ_j is a polynomial of degree less than $r + 1$. Therefore

$$\|x - x_n\|_{\beta'+r, \infty, \Delta_j} \leq \|x\|_{\beta'+r, \infty, \Delta_j} + \|x_n - x\|_{r, \infty, \Delta_j}.$$

By application of Lemma 7.15, we may write for $j = 1, \ldots, n$,

$$(h/h_j)^r h \leq (h/h_j)^{r+1}$$

and

$$|((1 - \pi^2)(\tilde{x}_n - x), l_t)_{\Delta_j}| \leq c h_j^{r+1+\beta'+1} (h/h_j)^{r+1} \|x\|_{\beta'+r+1, \infty, \Delta_j}.$$

By summing over j,

$$|((1 - \pi_n^2)(\tilde{x}_n - x), l_t)| \leq c h^{r+1+\beta'} \|x\|_{\beta'+r+1, \infty}.$$

The analysis of $(1 - P)\tilde{\varphi}_n$ is similar to that of $x - \tilde{x}_n$:

$$[(1 - P)\tilde{\varphi}_n](t) = ((1 - \pi_n^2)P\tilde{\varphi}_n, l_t^1) + ((1 - \pi_n^2)(1 - P)\tilde{\varphi}_n, l_t^1).$$

For the first inner product, $\|P\tilde{\varphi}_n\|_{\beta'+r+1,\infty}$ is uniformly bounded in n by Proposition 7.3 since $\beta' + r + 1 \le \alpha$. As for the second inner product,

$$(1 - P)\tilde{\varphi}_n = \frac{1}{\lambda_n} T(1 - P)\varphi_n \qquad \text{and} \qquad \|((1 - P)\varphi_n)_j\|_{\beta'+r,\infty,\Delta_j}$$

can be bounded by means of Lemma 7.15, much as it was done for the solution. This ends the proof. □

The only nontrivial part of the proof of Theorem 7.16 is be bound $\|\tilde{x}_n - x\|_{\alpha,\infty}$ (resp. $\|(1 - P)\tilde{\varphi}_n\|_{\alpha,\infty}$) uniformly in n. It uses the technique developed in de Boor and Swartz (1973). It should be noted that in the case of a smooth kernel, when $\gamma = \alpha$, $\|\tilde{x}_n\|_{\alpha,\infty}$ (resp. $\|\tilde{\varphi}_n\|_{\alpha,\infty}$) is bounded uniformly by the Lemma 7.17 given below. Then the proof of Theorem 7.16 becomes straightforward.

Lemma 7.17 *If the kernel of T is smooth of continuity α relative to t and s, then the following bounds hold for n large enough*:

$$\|\tilde{x}_n\|_{\alpha,\infty} \le c\|f\|_{\alpha,\infty}, \qquad \|\tilde{\varphi}_n\|_{\alpha,\infty} \le c\|\varphi_n\|_\infty.$$

Proof $\tilde{x}_n = (1/z)Tx_n + f$. Then

$$\|\tilde{x}_n\|_{\alpha,\infty} \le c(\|x_n\|_\infty + \|f\|_{\alpha,\infty}) \le c\|f\|_{\alpha,\infty}.$$

Similarly

$$\tilde{\varphi}_n = \frac{1}{\lambda_n} T\varphi_n, \quad \|\tilde{\varphi}_n\|_{\alpha,\infty} \le \frac{c}{|\lambda_n|}\|\varphi_n\|_\infty, \quad \text{and} \quad \|\varphi_n\|_\infty = 1, \quad |\lambda_n| > |\lambda| - \varepsilon$$

for n large enough. □

5.2. *Convergence Rates for $\lambda - \hat{\lambda}_n$, $\lambda - \lambda_n$, and* $\mathrm{dist}_\infty(\tilde{\varphi}_n, E)$

For the eigenvalues and eigenvectors, the convergence rates are given by the following proposition.

Proposition 7.18 *If the kernel of T is of class $\mathfrak{G}(\alpha, \gamma)$ relative to t and s, with $\alpha \ge r + 1$, then, for n large enough,*

$$|\lambda - \hat{\lambda}_n| = O(\varepsilon_n) \qquad \textit{and} \qquad \max(|\lambda - \lambda_n|, \mathrm{dist}_\infty(\tilde{\varphi}_n, E)) = O(\varepsilon_n^{1/l})$$

if $l > 1$, where $\varepsilon_n = h^{r+1+\beta'}$.

Proof It is analogous to that of Theorem 7.10 and Proposition 7.11. The additional part is to bound uniformly in n all the derivatives of $T(y_{in} - x_i)$ up to the order $\beta' + r + 1$. This can be done by means of Lemma 7.15, since $\|y_{in} - x_i\|_\infty = O(h^{r+1})$ when $\alpha \geq r + 1$. The proof of Theorem 7.16 applies and the order $r + 1 + \beta'$ follows. □

5.3. *Numerical Examples*

Example 7.1 We consider the Fredholm equation of the second kind:

$$\int_0^1 k(t, s)x(s)\,ds - \tfrac{1}{4}x(t) = -\cosh(1), \qquad 0 \leq t \leq 1,$$

with

$$k(t, s) = \begin{cases} -t(1 - s) & \text{if } s \geq t \\ -s(1 - t) & \text{if } s \leq t. \end{cases}$$

The exact solution is $x(t) = \cosh(2t - 1)$.

We choose the uniform partition $\Delta = \{i/5\}_0^5$, $h = \frac{1}{5}$, and on each interval Δ_i, the $r + 1 = 4$ Gauss points. We display in Table 7.1 the values of $x - x_n$ and $x - \tilde{x}_n$ at the partition points t_i, $i = 1, 2, 3, 4$.

We display, in Fig. 7.1., the graphs of the error functions: the collocation error,

$$t \in \Delta_3 \mapsto (x - x_n)(t)$$

and the iterated collocation error,

$$t \in \Delta_3 \mapsto (x - \tilde{x}_n)(t),$$

where $\Delta_3 = [0.4, 0.6]$. We have only *sketched* the two graphs because of the great difference in magnitude between the various values.

Example 7.2 We study the influence of the smoothness properties of f on the computed orders of convergence (Lebbar, 1981a). The kernel is as in Example 7.1; the right-hand side f is chosen such that the solution is

(1) $(t - \frac{1}{2})^4$,
(2) $|t - \frac{1}{2}|^{7/2}$,
(3) $|t - \frac{1}{2}|^{1/2}$.

Table 7.1

Error Values at the Partition Points

i	$(x - x_n)(t_i^-)$	$(x - x_n)(t_i^+)$	$(x - \tilde{x}_n)(t_i)$
1	8×10^{-5}	7×10^{-5}	-5×10^{-12}
2	6×10^{-5}	6×10^{-5}	-7×10^{-12}
3	6×10^{-5}	6×10^{-5}	-7×10^{-12}
4	7×10^{-5}	8×10^{-5}	-5×10^{-12}

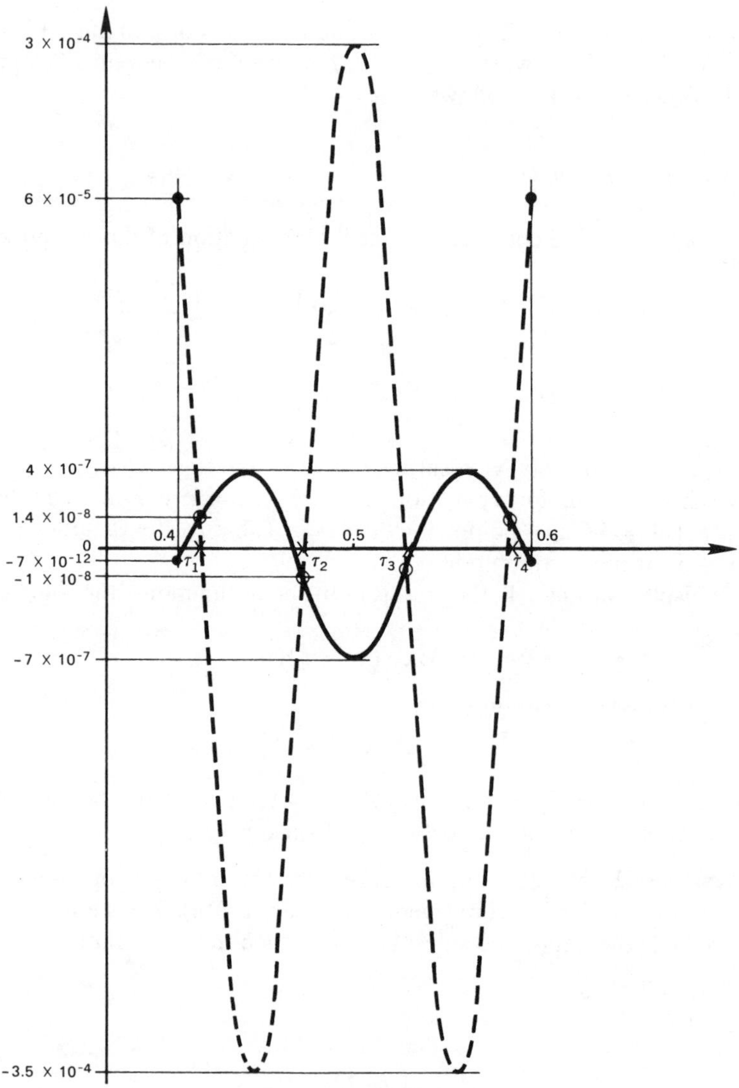

Figure 7.1 Graphs of the errors on Δ_3: [0.4, 0.6]: Dotted line: $x - x_n$. Solid line: $x - \tilde{x}_n$. $r = 3$.

A uniform partition is first chosen on [0, 1], $t_i = i/n$, $i = 0, \ldots, n$, with $n = 2, 4, 8, 16$; n is even so that the partition always contains $\frac{1}{2}$. Piecewise quadratic polynomials are used both for the $\perp$-Galerkin method and for the collocation at the 3 Gauss points. If the error e_n tends to zero at the rate h^ω, then $\omega_n := \log(e_n/e_{2n})/\log 2 \to \omega$ as $n \to \infty$. We compute the three errors

Table 7.2

f Is Smooth on [0, 1]

⊥-Galerkin				Collocation			
n	2	4	8	n	2	4	8
$\omega^a = 3$	2.53	2.81	2.91	$\omega^a = 3$	2.62	2.83	2.92
$\omega^b \geq 5$	4.11	4.35	4.83	$\omega^b \geq 5$	4.59	4.83	4.92
$\omega^c \geq 6$	6.01	6.00	6.00	$\omega^c \geq 6$	7.99	7.99	7.98

$a_n := \|x - x_n\|_\infty$, $b_n := \|x - \tilde{x}_n\|_\infty$, and $c_n := \max_i |(x - \tilde{x}_n)(t_i)|$. The corresponding computed orders of convergence are, respectively, ω_n^a, ω_n^b, and ω_n^c for which we test the convergence to ω^a, ω^b, and ω^c. Results are listed in Tables 7.2, 7.3, and 7.4. When f is smooth, the collocation method gives higher orders for c_n than the projection method; this is not true any longer when f is not smooth enough as seen in Table 7.3.

Table 7.3

f Is Not Smooth "Enough" for Collocation at $t = 1/2$

⊥-Galerkin				Collocation			
n	2	4	8	n	2	4	8
$\omega^a = 3$	2.75	2.89	2.96	$\omega^a = 3$	2.79	2.92	2.96
$\omega^b \geq 5$	4.35	4.66	4.89	$\omega^b \geq 3.5$	4.78	4.56	4.52
$\omega^c \geq 6$	6.05	6.01	6.00	$\omega^c \geq 3.5$	4.61	4.71	4.67

Table 7.4 displays results for the Galerkin method using either a uniform partition or a graded mesh around $\frac{1}{2}$ (Rice, 1969). Such a mesh $\{s_i\}_0^n$, n even, is defined by a real number $q > 1$ in the following way:

$$s_i = \frac{1}{2}\left[1 - \left(1 - \frac{2}{n}i\right)^q\right] \quad \text{for} \quad 0 \leq i \leq \frac{n}{2}$$

and

$$s_i = \frac{1}{2}\left[1 + \left(1 - \frac{2}{n}i\right)^q\right] \quad \text{for} \quad \frac{n}{2} \leq i \leq n.$$

The points are clustered around $\frac{1}{2}$, where x behaves badly and are spread out near the end points 0 and 1, where x is well-behaved. With increasing values of q, the values $\omega^c \geq 6$, $\omega^b \geq 5$ and $\omega^a = 3$ can be recovered, respectively, with $q = 7/4.5 = 1.555$, $q = 2$, and $q = 6$ (Lebbar, 1982). It is worth noting

Table 7.4

f Is "Weakly" Singular at $t = 1/2$

$q = 1$				$q = 1.555$				$q = 2$			
n	2	4	8	n	2	4	8	n	2	4	8
$\omega^a = 0.5$	0.5	0.5	0.5	$\omega^a = 0.77$	0.77	0.77	0.78	$\omega^a = 1$	1.00	1.00	0.99
$\omega^b \geq 2.5$	2.54	2.51	2.50	$\omega^b \geq 3.88$	3.91	3.91	3.89	$\omega^b \geq 5$	4.89	4.89	4.99
$\omega^c \geq 3.5$	4.41	4.46	4.47	$\omega^c \geq 6$	5.67	5.83	5.91	$\omega^c \geq 6$	4.70	5.79	5.95
uniform mesh								graded mesh			

that $\omega^c \geq 6$ is first recovered for the smallest value of q, which means that the corresponding graded mesh has a less severe cluster around $\frac{1}{2}$ than the ones required to get $\omega^b \geq 5$ or $\omega^a = 3$; this fact is of great practical value. The reader is referred to Lebbar (1982) for proofs and more details.

5.4. *Bibliographical Comments*

For a kernel that is the Green function of an o.d.e., the main references on superconvergence results for iterated solutions are Chandler (1979) for the $\perp$-Galerkin method and de Boor and Swartz (1973, 1980, 1981a,b) for the collocation method at Gauss points. The effect of numerical quadrature is also studied in Chandler. For the Galerkin method on a smooth enough kernel, an extra h may be gained at gaussian points (see Richter, 1978). Using the identity of Exercise 6.3, Lebbar (1981b) proves that the orders for $[(1 - P)\tilde{\varphi}_n](t)$ are still valid when $\tilde{\varphi}_n$ is an arbitrary vector in the iterated invariant subspace $\tilde{M}_n = TM_n$.

For weakly singular kernels, somewhat similar superconvergence results hold. The reader is referred to Graham (1980) for the iterated Galerkin method and to Schneider (1979, 1980, 1981) for the iterated collocation and product integration method.

6. Superconvergence at the Partition Points of the Approximate Solution of an Ordinary Differential Equation

We consider the pth-order linear differential equation

$$[\mathfrak{L}(u)](t) = u^{(p)}(t) - \sum_{i=0}^{p-1} a_i(t)u_i(t) = f(t), \qquad 0 \leq t \leq 1.$$

with p homogeneous boundary conditions, which has been presented in Chapter 4, Section 8.1. It can be written

$$Tu = f, \tag{7.13}$$

where T is a closed operator with domain $D \subset X$, the reference Banach space, which will be again L^2 or L^∞ later on. We still assume that $T^{-1} \in \mathscr{L}(X)$. (7.13) is equivalent either to

$$u = Gx, \qquad (1 - K)x = f, \tag{7.14}$$

or, for a given z, $0 \neq -z \in \rho(T)$, to

$$u = A_z y, \qquad (1 - zA_z)y = f, \tag{7.15}$$

by Proposition 4.15 and Example 4.17, respectively.

Lemma 7.19 *If the coefficients a_i belong to C^α, $i = 0, \ldots, p-1$, then*

(i) *the kernel of K is of class $\mathfrak{G}(\alpha, -1)$, relative to t,*
(ii) *the kernel of A_z is of class $\mathfrak{G}(\alpha + 1, p - 2)$, relative to t.*

Proof Let g, k, v be the kernels of G, K, and $A = T^{-1}$, respectively. g_t is of class $\mathfrak{G}(\infty, p-2)$. The kernel k is such that

$$k(t, s) = \sum_{i=0}^{p-1} a_i(t) \frac{\partial^i}{\partial t^i} g(t, s),$$

and (i) follows easily. As for the kernel v, it is related to g in the following way. The equation $A(1 - K)f = Gf$ for any f in X implies

$$\int_0^1 v(t, s)[(1 - K)f](s)\, ds = \int_0^1 g(t, s) f(s)\, ds$$

for any t in $[0, 1]$. Therefore, with obvious definitions, v_t is solution of $(1 - K^*)v_t = g_t$. For a fixed t, $g_t \in C^{p-2}(0, 1)$; hence $v_t \in C^{p-2}(0, 1)$. For $s \neq t$, we look at higher derivatives of v_t at s; the lth derivative $v_t^{(l)}$ is such that

$$v_t^{(l)} = (K^* v_t)^{(l)} + g_t^{(l)},$$

with

$$(K^* v_t)(s) = \int_0^1 \sum_{i=0}^{p-1} \bar{a}^i(\tau) \frac{\partial^i}{\partial \tau^i} \bar{g}(\tau, s) v_t(\tau)\, d\tau, \qquad 0 \leq s \leq 1.$$

For $l = 1$,

$$\frac{\partial}{\partial s}(K^* v_t)(s) = \int_0^1 \sum_{i=0}^{p-1} \bar{a}_i(\tau) \frac{\partial^i}{\partial \tau^i} \frac{\partial}{\partial s} \bar{g}(\tau, s) v_t(\tau)\, d\tau$$
$$+ \bar{a}_{p-1}(s) v_t(s) \left[\frac{\partial^{p-1}}{\partial \tau^{p-1}} \bar{g}(\tau, s) \Big|_{\tau = s^+} - \frac{\partial^{p-1}}{\partial \tau^{p-1}} \bar{g}(\tau, s) \Big|_{\tau = s^-} \right].$$

Similarly, the second derivative at s will depend on $\bar{a}_{p-1}^{(1)}(s)$, $\bar{a}_{p-1}(s)$, $\bar{a}_{p-2}(s)$, and $g_t^{(l)}(s)$. Therefore $v_t^{(l)}$ is continuous at $s \neq t$ for l up to $\alpha + 1$. This proves by induction that the kernel v_t is of class $\mathfrak{G}(\alpha + 1, p - 2)$. To conclude, we note that the kernels of A and A_z have the same smoothness properties. $\square$

The solution u of (7.13) is approximated by the solution u_n of

$$\pi_n[Tu_n - f] = 0, \tag{7.16}$$

when π_n is a projection on $\mathbb{P}_{r,\Delta}$ and u_n belongs to the subset C_n of $C^{p-1}(0, 1)$, which consists of functions that are polynomials of degree less than $p + r + 1$ on each Δ_i, satisfying the boundary conditions.

The choice π_n^1 yields the *method of moments* (case (a)), and the choice π_n^2 yields the *collocation method* at the Gauss points (case (b)). It was first noted by de Boor and Swartz (1973) that the choice of Gauss points as collocation points for an o.d.e. would yield an accuracy higher than anticipated for the solution u_n at the partition points. This fact has become popular under the name of superconvergence, always in the framework of differential operators. It is more recent that a similar property for integral operators has been found for the *iterated* solution $\tilde{x}_n$ (Sloan 1976a; Chandler, 1979). The two results are intimately connected, as we see now.

We gave in Chapter 4 two interpretations of projection methods on (7.13). The first is by means of a Galerkin method on (7.14), and the second is by means of a Petrov method on (7.15). We use the first interpretation to prove the superconvergence by application of the preceding analysis of the Galerkin method.

6.1. *A Galerkin Method on* (7.14)

The key point is the identity (4.47), given in Proposition 4.20:

$$u - u_n = T^{-1}(1 - \pi_n)\tilde{x}_n \qquad \text{with} \quad \tilde{x}_n = Kx_n + f,$$

which is of the same kind as (7.12), where $\tilde{x}_n$ is the iterated solution for K.

Theorem 7.20 *If $a_i \in C^\alpha$, $i = 0, \ldots, p - 1$, if $f \in C^\alpha$, and if $\alpha \geq r + 1$, then, for n large enough, the following rates of convergence hold for the method of moments (resp. collocation method):*

(i) $|(u - u_n)(t_i)| = O(h^{2(r+1)})$ *(resp.* $h^{r+1+\beta'}$*)*, $i = 0, 1, \ldots, n$,
(ii) $\|u - u_n\|_\infty = O(h^{r+1+\beta^*})$ *(resp.* $h^{r+1+\beta'^*}$*)*,

where $\beta^* = \min(r + 1, p)$, $\beta' = \min(\alpha - r - 1, r + 1)$, *and* $\beta'^* = \min(\beta', p)$.

Proof Let v denote the kernel of T^{-1}. v_t is of class $\mathfrak{G}(\alpha + 1, p - 2)$. For any t in $[0, 1]$,

$$(u - u_n)(t) = ((1 - \pi_n)(\tilde{x}_n - x), \bar{v}_t) + ((1 - \pi_n)x, \bar{v}_t).$$

We note that $\beta = \min(\alpha, r + 1) = r + 1$ by assumption.

(1) Case (a): K has a kernel of class $\mathfrak{G}(\alpha, -1)$; hence $\|\tilde{x}_n - x\|_2$ is of order $r + 2$, by Theorem 7.9. Then $|(\tilde{x}_n - x, (1 - \pi_n^1)\bar{v}_t)|$ is of order

$$\begin{aligned} &(r + 2) + (r + 1) &&\text{for} \quad t \in \Delta, \\ r + 2 + {}&\min(r + 1, p - 1) &&\text{for} \quad t \notin \Delta, \end{aligned}$$

by Corollary 7.7; and $|((1 - \pi_n^1)x, (1 - \pi_n^1)\bar{v}_t)|$ is of order

$$\begin{aligned} &2(r + 1) &&\text{for} \quad t \in \Delta, \\ r + 1 + {}&\min(r + 1, p) &&\text{for} \quad t \notin \Delta, \end{aligned}$$

by Lemma 7.8. The desired orders follow.

(2) Case (b): By application of Corollary 7.13 and Theorem 7.16, $|((1 - \pi_n^2)x, \bar{v}_t)|$ is of order

$$\begin{aligned} &\beta' + r + 1 &&\text{for} \quad t \in \Delta, \\ &\beta'^* + r + 1 &&\text{for} \quad t \notin \Delta. \end{aligned}$$

As for the inner product $|((1 - \pi_n^2)(\tilde{x}_n - x), \bar{v}_t)|$, the quantity

$$\|\tilde{x}_n - x\|_{\beta' + r + 1, \infty}$$

for $t \in \Delta$ (resp. $\|\tilde{x}_n - x\|_{\min(\beta', p-1) + r + 1, \infty}$ for $t \notin \Delta$) can be bounded uniformly in n in the way done in the proof of Theorem 7.16, where T has to be replaced by K and z by 1. □

The orders of convergence are $2r + 2$ at the partition points and $r + 1 + p$ globally for the method of moments if $\alpha \geq r + 1 > p$ and for the collocation method if $\alpha \geq 2r + 2$ and $p < r + 1$.

6.2. *A Petrov Method on* (7.15)

For the second interpretation, the identity (4.49)

$$u - u_n = (1/z)(y - \tilde{y}_n) \qquad \text{with} \quad \tilde{y}_n = zA_z y_n + f,$$

was proved in Proposition 4.23. It establishes the superconvergence of the iterated Petrov solution $\tilde{y}_n$ by means of Theorem 7.20.

Exercises

7.5 Prove the identities

$$y - \tilde{y}_n = z(1 - zA_z)^{-1}A_z(\tilde{y}_n - y_n) = zA_z(1 - zA_z)^{-1}(1 - \pi_n)(\tilde{y}_n - y_n).$$

7.6 Prove that

$$\tilde{y}_n - y_n = \tilde{x}_n - x_n. \tag{7.17}$$

7.7 Use (7.17) and Exercise 7.5 to prove *directly* the superconvergence of $\tilde{y}_n$.

Identity (7.17) expresses the connection between the two interpretations of the projection method on (7.13) that we have given by means of Galerkin and Petrov methods.

7. Superconvergence for the Differential Eigenvalue Problem

The eigenvalue problem associated with (7.13) is

$$T\psi = \lambda\psi. \tag{7.18}$$

We assume that $0 \in \rho(T)$. P is the spectral projection associated with λ.

7.1. The Simple Differential Eigenvalue Problem

We have seen in Example 4.16 that (7.18) is equivalent to the generalized eigenvalue problem

$$\psi = G\varphi, \qquad (1 - K)\varphi = \lambda G\varphi. \tag{7.19}$$

The projection method on (7.18) is equivalent to a Galerkin method on (7.19) (Exercise 4.67).

Lemma 7.21 *The following error formula holds*:

$$(1 - P)\psi_n = S(\lambda_n)(\pi_n - 1)\tilde{\varphi}_n \qquad \text{with} \quad \tilde{\varphi}_n = K\varphi_n + \lambda_n G\varphi_n.$$

Proof Let Q be the spectral projection for $(1 - K)\varphi = \lambda G\varphi$:

$$Q = \frac{-1}{2i\pi}\int_\Gamma (1 - K - zG)^{-1}G\,dz \qquad \text{and} \quad P = GQG^{-1}.$$

Then

$$\begin{aligned}(1 - P)\psi_n &= G(1 - Q)\varphi_n \\ &= G(1 - K - \lambda_n G)^{-1}G(1 - Q)G^{-1}(1 - K - \lambda_n G)\varphi_n \\ &= (T - \lambda_n)^{-1}(1 - P)(\varphi_n - \tilde{\varphi}_n) = S(\lambda_n)(\varphi_n - \tilde{\varphi}_n).\end{aligned}$$

The desired result follows from $\pi_n \tilde{\varphi}_n = \varphi_n$. □

This error formula is analogous to (7.11) and is the key to the derivation of the order of convergence of $(1 - P)\psi_n$, as we shall see. The derivation is not as straightforward as for the solution of (7.13), since we have to deal with the *generalized* eigenvalue problem (7.19). Therefore, we need some preparation. If we set $U := (1 - K)^{-1}G$, then

$$(1 - K)\varphi = \lambda G\varphi \Leftrightarrow \varphi = \lambda U\varphi,$$

and Q is the spectral projection for U associated with $1/\lambda$. By assumption $(1 - K)^{-1} \in \mathscr{L}(X)$, and for n large enough, $(1 - \pi_n K)^{-1}$ and $(1 - K\pi_n)^{-1}$ are in $\mathscr{L}(X)$. We define

$$U_n := (1 - \pi_n K)^{-1}\pi_n G, \qquad \tilde{U}_n := (1 - K\pi_n)^{-1}G\pi_n.$$

Exercises

7.8 Prove that $\varphi_n = \lambda_n U_n \varphi_n$, $\pi_n \tilde{\varphi}_n = \varphi_n$ and $\tilde{\varphi}_n = \lambda_n \tilde{U}_n \tilde{\varphi}_n$.

7.9 Prove the identities

$$U - U_n = (1 - \pi_n K)^{-1}(1 - \pi_n)U, \qquad U - \tilde{U}_n = (1 - K\pi_n)^{-1}[K(1 - \pi_n)U + G(1 - \pi_n)].$$

7.10 Study the convergences $U_n \to U$ and $\tilde{U}_n \to U$.

7.11 Show that

$$\|(U - U_n)Q\| \le c\|(1 - \pi_n)Q\|, \qquad \|(U - \tilde{U}_n)Q\| \le c[\|K(1 - \pi_n)Q\| + \|G(1 - \pi_n)Q\|].$$

7.12 Deduce from Exercise 7.11 that

$$\|(1 - Q)\varphi_n\| \le c\|(1 - \pi_n)Q\|, \qquad \|(1 - Q)\tilde{\varphi}_n\| \le c[\|K(1 - \pi_n)Q\| + \|G(1 - \pi_n)Q\|].$$

7.13 Let N_n (resp. $\tilde{N}_n$) be the invariant subspace for U_n (resp. $\tilde{U}_n$) associated with the eigenvalues converging to $1/\lambda$. Check that $N_n \subset \mathbb{P}_{r,\Delta}$. Give bounds for $\Theta(N, N_n)$ and $\Theta(N, \tilde{N}_n)$.

7.14 Prove the alternative formula $U - U_n = (1 - K)^{-1}(1 - \pi_n)(G + KU_n)$.

7.15 Prove that $\lambda - \hat{\lambda}_n$ is of the order of

$$\left| \sum_{i=1}^{m} ((U - U_n)(Q_{\restriction N_n})^{-1}\varphi_i, \varphi_i^*) \right|$$

where φ_i (resp. φ_i^*) belongs to N (resp. $N^* = Q^*X^*$).

Lemma 7.22 *If $a_i \in C^\alpha$, $i = 0, \ldots, p - 1$, then Q has a degenerate kernel of continuity α, relative to both variables.*

Proof We need only to prove that $N \subset C^{\alpha}$ (resp. $N^* \subset C^{\alpha}$). We recall that $(1 - K)^{-1}M = N$, where $M = PX \subset C^{\alpha+1}$ (resp. $M^* \subset C^{\alpha+1}$). For $f \in M$, $h \in N$ is the solution of $(1 - K)h = f$ and $h \in C^{\alpha}$. □

Lemma 7.23 *If $a_i \in C^{\alpha}$, $i = 0, \ldots, p - 1$, then the kernel of*

$$(T - \lambda_n)^{-1}(1 - P)$$

is of class $\mathfrak{G}(\alpha, p - 2)$ relative to t.

Proof Let v^1 be the kernel of $(T - \lambda_n)^{-1}(1 - P)$. For any f in X, the equation $(T - \lambda_n)u = (1 - P)f$ is equivalent to $u = Gx$, and $x = (K + \lambda_n G)x + (1 - P)f$. Therefore

$$\begin{aligned} u &= (T - \lambda_n)^{-1}(1 - P)f = (T - \lambda_n)^{-1}(1 - P)(1 - K - \lambda_n G)x \\ &= (1 - P)Gx = G(1 - Q)x, \end{aligned}$$

for any x in X. We deduce that $v^1(t, s)$ is solution of the equation

$$\int_0^1 v^1(t, s)[(1 - K - \lambda_n G)x](s)\, ds = \int_0^1 g(t, s)[(1 - Q)x](s)\, ds$$

for any x in X and t in $[0, 1]$, which we rewrite

$$(1 - K^* - \bar{\lambda}_n G^*)v_t^1 = (1 - Q^*)g_t.$$

And this equation has a unique solution since Q^* is the spectral projection associated with the eigenvalue problem $(1 - K^*)\varphi^* = \bar{\lambda}G^*\varphi^*$. Noting that the lth derivative satisfies $v_t^{1(l)} = (K^*v_t^1)^{(l)} + \bar{\lambda}_n(G^*v_t^1)^{(l)} + g_t^{(l)} - (Q^*g_t)^{(l)}$, we may apply the proof of Lemma 7.19 to conclude that v_t^1 is of class $\mathfrak{G}(\alpha, p - 2)$. □

Theorem 7.24 *Under the assumptions of Theorem 7.20, for n large enough, the same orders of convergence hold for*

$$|[(1 - P)\psi_n](t_i)|, \qquad i = 0, 1, \ldots, n,$$

and

$$\|(1 - P)\psi_n\|_{\infty},$$

respectively.

Proof The kernel v_t^1 of the reduced resolvent $S(\lambda_n) = (T - \lambda_n)^{-1}(1 - P)$. is of class $\mathfrak{G}(\alpha, p - 2)$ by Lemma 7.23. Then

$$\begin{aligned} [(1 - P)\psi_n](t) &= ((\pi_n - 1)\tilde{\varphi}_n, \bar{v}_t^1) \\ &= ((\pi_n - 1)Q\tilde{\varphi}_n, \bar{v}_t^1) + ((\pi_n - 1)(1 - Q)\tilde{\varphi}_n, \bar{v}_t^1). \end{aligned}$$

(a) *The method of moments.* Q is of continuity α, and $\|K(1 - \pi_n^1)Q\|_2$ is clearly of the order $r + 2$, by Lemma 7.8. We deduce that $\|(1 - Q)\tilde{\varphi}_n\|_2 \le ch^{r+2}$, and the proof of Theorem 7.20 applies.

(b) *The collocation method.* The proof is adapted from the proof of Theorem 7.16. We need that $\|(1 - Q)\varphi_n\|_{\infty} \le ch^{r+1}$, which follows from Exercise 7.12.

In Theorem 7.20, the kernel v_t of T^{-1} is of class $\mathfrak{G}(\alpha + 1, p - 2)$. Nevertheless, the resulting orders are the same for v_t and v_t^1, since with the assumption $\alpha \geq r + 1$ we have $\min(\alpha + 1, r + 1) = \min(\alpha, r + 1) = r + 1$. □

Theorem 7.25 *Under the assumptions of Theorem 7.20,* $|\lambda - \hat{\lambda}_n| = O(h^{2(r+1)})$, *(resp.* $h^{r+1+\beta'}$*) for the method of moments (resp. collocation method).*

Proof We set $Q_{(n)} := Q_{\restriction N_n}$. By Exercise 7.15 we have to find the order of

$$\left|\sum_{i=1}^{m} ((U - U_n)Q_{(n)}^{-1}\varphi_i, \varphi_i^*)\right|$$
$$= \left|\sum_{i=1}^{m} ((1 - \pi_n)(G + KU_n)Q_{(n)}^{-1}\varphi_i, (1 - K^*)\varphi_i^*)\right|.$$

$(1 - K^*)\varphi_i^* \in M^* \subset C^{\alpha+1}$. We set $\varphi_{in} := Q_{(n)}^{-1}\varphi_i \in N_n \subset \mathbb{P}_{r,\Delta}$ and $\eta_{in} := \varphi_{in} - \varphi_i$. $\|\eta_{in}\|_v = \|(Q_{(n)}^{-1}Q - Q)\varphi_i\|_v \leq ch^{r+1}$ for $v = 2$ or ∞.

We first deal with $(1 - \pi_n)G\varphi_{in} = (1 - \pi_n)G(\varphi_i + \eta_{in})$. The result is straightforward by using the proof of Theorem 7.10 and Proposition 7.18. For the collocation method, the derivatives of $G\eta_{in}$ up to the order $\beta' + r + 1$ are bounded in n as done in the proof of Theorem 7.16, using $\|\eta_{in}\|_\infty \leq ch^{r+1}$ and $\eta_{in} = \varphi_{in} - \varphi_i$ with $\varphi_{in} \in \mathbb{P}_{r,\Delta}$.

Now $(1 - \pi_n)KU_n\varphi_{in} = (1 - \pi_n)K(U_n\varphi_{in} - U\varphi_i) + (1 - \pi_n)KU\varphi_i$.

$$\psi_{in} := U_n\varphi_{in} - U\varphi_i = U_n\eta_{in} + (U_n - U)\varphi_i$$

and

$$U_n - U = (1 - \pi_n K)^{-1}(1 - \pi_n)U$$

prove that $\|\psi_{in}\|_v \leq ch^{r+1}$ for $v = 2$ or ∞.

The conclusion follows readily for the projection π_n^1; as for π_n^2 we proceed as earlier by bounding the derivatives of $K\psi_{in}$ up to the order $\beta' + r + 1$, using $\|\psi_{in}\|_\infty \leq ch^{r+1}$ and $\psi_{in} = U_n\varphi_{in} - U\varphi_i$ with $U_n\varphi_{in} \in N_n \subset \mathbb{P}_{r,\Delta}$. □

As for the solution in Section 6, the superconvergence for the eigenelements can be proved by using the integral formulation (7.19). Because (7.19) is a generalized eigenvalue problem, results for the integral eigenvalue problem cannot be applied in a straightforward manner. We have seen, however, that it is not difficult to put the problem into a form to which this theory applies.

If one prefers to work directly with a simple eigenvalue problem, one may choose to work on the alternative integral formulation

$$\psi = A\theta, \qquad \theta = \lambda A\theta. \tag{7.20}$$

The projection method on (7.18) is equivalent to a Petrov method on (7.20) (Exercise 4.70), which in turn is equivalent to a Galerkin method in $X_n' = TX_n$, with the projection $\pi_n' = (\pi_{n \restriction X_n'})^{-1}\pi_n$. Error bounds in terms of $1 - \pi_n'$ can be easily written, and one faces this time the problem of the relation between π_n and π_n'. In turn this problem can be easily solved, because of the intimate connection between the Petrov method on (7.20) and the Galerkin method on (7.19).

Exercises

7.16 Use the identity $(1 - P)\psi_n = (1/\lambda_n)(1 - P)\tilde{\theta}_n$ of Exercise 4.71 to prove the superconvergence of the iterated Petrov eigenvector $\tilde{\theta}_n$.

7.17 Prove that $\theta_n - \tilde{\theta}_n = \varphi_n - \tilde{\varphi}_n$.

7.18 Prove that

$$(1 - P)\tilde{\theta}_n = \lambda_n A(1 - \lambda_n A)^{-1}(1 - P)(\theta_n - \tilde{\theta}_n) = \lambda_n A(1 - \lambda_n A)^{-1}(1 - P)(1 - \pi_n)(\theta_n - \tilde{\theta}_n).$$

Deduce a *direct* proof of the superconvergence of $\tilde{\theta}_n$.

7.19 Prove that $\pi'_n = (1 - K)(1 - \pi_n K)^{-1}\pi_n$. (*Hint*: Use the proof of Proposition 4.8.)

The identity of Exercise 7.17 is analogous to (7.17) for the eigenvectors. As for the eigenvalues, we know that $\lambda - \hat{\lambda}_n$ is of the order of

$$\left| \sum_{i=1}^{m} ((1 - \pi'_n)AP_{(n)}^{-1}\theta_i, \theta_i^*) \right|,$$

where θ_i (resp. θ_i^*) belongs to M (resp. M^*). For that purpose, we study $(1 - \pi'_n)AP_{(n)}^{-1}$. We first write

$$AP_{(n)}^{-1} = G[P_{(n)}(1 - K)]^{-1} = G[(1 - K)Q_{(n)}]^{-1} = GQ_{(n)}^{-1}(1 - K)^{-1}.$$

Then $\pi'_n = (1 - K)(1 - \pi_n K)^{-1}\pi_n$, by Exercise 7.19. Therefore

$$\begin{aligned}(1 - \pi'_n)G &= (1 - K)[(1 - K)^{-1} - (1 - \pi_n K)^{-1}\pi_n]G \\ &= (1 - K)(U - U_n) = (1 - \pi_n)(G + KU_n),\end{aligned}$$

by Exercise 7.14. We conclude that $\lambda - \hat{\lambda}_n$ is of the order of

$$\left| \sum_{i=1}^{m} ((1 - \pi_n)(G + KU_n)Q_{(n)}^{-1}(1 - K)^{-1}\theta_i, \theta_i^*) \right|,$$

where the vectors $(1 - K)^{-1}\theta_i$ belong to N (Exercise 7.20), and Theorem 7.25 applies.

Exercise

7.20 Use $Q = (1 - K)^{-1}P(1 - K)$ to prove that $\theta \in M$ implies $(1 - K)^{-1}\theta \in N$ and $\theta^* \in M^*$ implies $(1 - K^*)\theta^* \in N^*$.

7.2. The Generalized Differential Eigenvalue Problem

We consider now a second differential form $\mathfrak{N}$ of order at most $p - 1$,

$$[\mathfrak{N}(u)](t) = \sum_{i=0}^{p-1} b_i(t)u^{(i)}(t),$$

and the generalized differential eigenvalue problem

$$T\psi = \lambda \mathfrak{N}\psi. \tag{7.21}$$

The main difference between (7.21) and (7.18) is that $\mathfrak{N}$ is not invertible with contrast with the identity operator 1. The error formulas will then be obtained in a slightly different way.

(7.21) is equivalent to

$$\psi = G\varphi, \qquad (1 - K)\varphi = \lambda L\varphi, \qquad \text{with} \quad L := \mathfrak{N}G, \tag{7.22}$$

and to

$$\xi = (1 - K)\varphi, \qquad \xi = \lambda \mathfrak{N}A\xi, \qquad \text{with} \quad \mathfrak{N}A = L(1 - K)^{-1}. \tag{7.23}$$

The spectral projections corresponding to (7.21), (7.22) and (7.23) are denoted P', Q' and R. The following relationships hold:

$$Q' = G^{-1}P'G, \qquad R = (1 - K)Q'(1 - K)^{-1} = TP'T^{-1}.$$

Finally, we define

$$V := (1 - K)^{-1}L \qquad \text{and} \qquad V_n := (1 - \pi_n K)^{-1}\pi_n L,$$

with associated invariant subspaces N and N_n.

Exercises

7.21 Prove that $\varphi = \lambda V\varphi$ and $\varphi_n = \lambda_n V_n \varphi_n$ if $\psi_n = G\varphi_n$ is the solution of

$$\pi_n[T\psi_n - \lambda_n \mathfrak{N}\psi_n] = 0, \; \psi_n \in C_n.$$

7.22 Prove that $V - V_n = (1 - K)^{-1}(1 - \pi_n)(L + KV_n) = (1 - \pi_n K)^{-1}(1 - \pi_n)V$.

Lemma 7.26 *The following error formula holds:*

$$(1 - P')\psi_n = (T - \lambda_n \mathfrak{N})^{-1}(1 - R)(\pi_n - 1)\tilde{\varphi}_n$$

with

$$\tilde{\varphi}_n = K\varphi_n + \lambda_n L\varphi_n.$$

Proof

$$\begin{aligned}
(1 - P')\psi_n &= G(1 - Q')\varphi_n, \\
(1 - Q')\varphi_n &= (\lambda_n(1 - K)^{-1}L - 1)^{-1}(1 - Q')(\lambda_n(1 - K)^{-1}L - 1)\varphi_n \\
&= [\lambda_n L - (1 - K)]^{-1}(1 - K)(1 - Q')(1 - K)^{-1} \\
&\quad \times [\lambda_n L\varphi_n - (1 - K)\varphi_n] \\
&= (1 - K)^{-1}(\lambda_n L(1 - K)^{-1} - 1)^{-1}(1 - R)(\tilde{\varphi}_n - \varphi_n),
\end{aligned}$$

and $(\lambda_n \mathfrak{N}A - 1)^{-1}(1 - R) = \lambda_n S(\mathfrak{N}A, 1/\lambda_n)$, where $S(\cdot, \cdot)$ is the reduced resolvent of $\mathfrak{N}A$ with respect to $1/\lambda$ at the point $1/\lambda_n$. Formally we write

$$\begin{aligned}
G(1 - K)^{-1}(\lambda_n \mathfrak{N}A - 1)^{-1} &= [\lambda_n(\mathfrak{N}G(1 - K)^{-1} - 1)(1 - K)G^{-1}]^{-1} \\
&= (\lambda_n \mathfrak{N} - T)^{-1}
\end{aligned}$$

We conclude that

$$(1 - P')\psi_n = (T - \lambda_n \mathfrak{N})^{-1}(1 - R)(\varphi_n - \tilde{\varphi}_n)$$
$$= (T - \lambda_n \mathfrak{N})^{-1}(1 - R)(\pi_n - 1)\tilde{\varphi}_n,$$

where $(T - \lambda_n \mathfrak{N})^{-1}(1 - R) = \lambda_n AS(\mathfrak{N}A, 1/\lambda_n)$ is well defined as the product of A by the reduced resolvent $S(\cdot, \cdot)$. □

Exercise

7.23 Check that the formula of Lemma 7.26 gives the formula of Lemma 7.21 if $\mathfrak{N} = 1$.

To conclude on the orders of $(1 - P)\psi_n$ and $\lambda - \hat{\lambda}_n$, it remains only to study the properties of the kernels of R, Q', $(T - z\mathfrak{N})^{-1}$ for z such that $(T - z\mathfrak{N})^{-1} \in \mathcal{L}(X)$, and $(T - \lambda_n \mathfrak{N})^{-1}(1 - R)$.

Lemma 7.27 *If $a_i, b_i \in C^\alpha$, $i = 0, \ldots, p - 1$, then*

(*i*) $(T - z\mathfrak{N})^{-1}$ *has a kernel of class $\mathfrak{G}(\alpha + 1, p - 2)$, relative to t,*
(*ii*) *L has a kernel of class $\mathfrak{G}(\alpha, -1)$, relative to t,*
(*iii*) *R and Q' have a degenerate kernel of continuity α, relative to both variables,*
(*iv*) $(T - \lambda_n \mathfrak{N})^{-1}(1 - R)$ *has a kernel of class $\mathfrak{G}(\alpha, p - 2)$, relative to t.*

Proof (i) $(T - z\mathfrak{N})u = \mathfrak{L}u - z\mathfrak{N}u = u^{(p)} - \sum_{i=0}^{p-1} (a_i - zb_i)u^{(i)}$. The kernel of $(T - z\mathfrak{N})^{-1}$ is then of the same class than the kernel of T^{-1} given by Lemma 7.19.

(ii) The kernel of L is $\sum_{i=0}^{p-1} b_i(t)\, \partial^i g(t, s)/\partial t^i$.

(iii) We first deal with R, the spectral projection associated with $\mathfrak{N}A$. For any integer $q \geq 1$, let the space $X^{(q)}$ be defined by

$$X^{(q)} := \{f \in C^{q-1}(0, 1);\ f^{(q-1)} \text{ is absolutely continuous and } f^{(q)} \in X\},$$

X being L^2 or C.

For $q \leq \alpha$, A is an operator $X^{(q)} \to X^{(q+p)}$ and $\mathfrak{N}A\colon X^{(q)} \to X^{(q+1)}$ for $q < \alpha$. Therefore any invariant function ξ for $\mathfrak{N}A$, which is solution of $(\mathfrak{N}A - \lambda)^m\xi = 0$, belongs to C^α. Now an invariant function φ for V is such that $\varphi = (1 - K)^{-1}\xi$. This proves that $\varphi \in C^\alpha$.

(iv) Let v^2 be the kernel of $(T - \lambda_n \mathfrak{N})^{-1}(1 - R)$. For any f in X, the equation $(T - \lambda_n \mathfrak{N})u = (1 - R)f$ is equivalent to $u = Gx$, and

$$x = (K + \lambda_n L)x + (1 - R)f.$$

Hence

$$u = (T - \lambda_n \mathfrak{N})^{-1}(1 - R)(1 - K - \lambda_n L)x = (1 - P')Gx = G(1 - Q')x$$

if we prove that $u = (1 - P')u$. But $A(1 - R) = (1 - P')A$; hence

$$u = A(1 - R)(1 - \lambda_n \mathfrak{R}A)^{-1}(1 - R)f = (1 - P')A(1 - \lambda_n \mathfrak{R}A)^{-1}(1 - R)f.$$

Then for any x in X and any t in $[0, 1]$ we get

$$\int_0^1 v^2(t, s)[(1 - K - \lambda_n L)x](s)\,ds = \int_0^1 g(t, s)[(1 - Q')x](s)\,ds,$$

or

$$(1 - K^* - \bar{\lambda}_n L^*)v_t^2 = (1 - Q'^*)g_t.$$

The conclusion follows as in the proof of Lemma 7.23 by (ii) and (iii). □

It is now straightforward to prove that, when $a_i, b_i \in C^\alpha$ for $\alpha \geq r + 1$, and $i = 0, \ldots, p - 1$,

(i) the orders of convergence for $[(1 - P)\psi_n](t_i)$ and $\|(1 - P)\psi_n\|_\infty$, are those given in Theorem 7.24,

(ii) the orders of convergence of $\lambda - \hat{\lambda}_n$ are those given in Theorem 7.25.

7.3. *Bibliographical Comments*

Collocation at Gauss points has been thoroughly studied in de Boor and Swartz (1973, 1980, 1981a,b) as well as collocation at Lobatto points (1981b). Superconvergence results for other projection methods on a differential equation are given in Locker and Prenter (1983) for the least-squares method, and in Douglas and Dupont (1973, 1974) for the f.e.m. (see also Hemker, 1975; Ahués and Telias, 1982a).

B. ITERATIVE REFINEMENT FOR THE EIGENELEMENTS

We are interested in this part in the computation of the iterative refinement of the eigenelements. We first study the case of an integral operator.

8. *T* Is an Integral Operator

The computation of the eigenelements λ_n, φ_n of the approximation T_n of an integral operator T requires the resolution of a *full* matrix eigenvalue problem, which may be expensive in computer time and storage. Under certain conditions, the accuracy resulting from λ_n and φ_n may be improved with a relatively small amount of extra work, when using extrapolation or iterative refinement techniques. They are especially well suited if the desired accuracy requires the use of a matrix that is too large for the capacity of the available computer.

8.1. Extrapolation

Baker (1971) showed that, when T is approximated by the Nyström method, under the hypotheses of Theorem 4.13, the sequences $\{\lambda_n^{\mathrm{N}}\}$ and $\{\varphi_n^{\mathrm{N}}\}$ (N for Nyström) can be extrapolated. If the kernel is smooth enough, this technique is easy to implement and gives very good results (Redont, 1979b). If the kernel is not smooth, the method is no longer founded theoretically.

The study of extrapolation methods is beyond the scope of this book, and we merely send the reader to a few references and the bibilography therein (Baker, 1971, 1977; Bulirsch and Stoer, 1966; Lin Qun and Liu Jiaquan, 1980; Stetter, 1978).

8.2. Iterative Refinement: A Computational Scheme

We assume throughout the section that the eigenvalue λ is *simple*. We suppose that T is compact, and we approximate T either by projection or by approximate quadrature. This yields a uniform or collectively compact convergence. Therefore the sequences defined in Chapter 5 by formulas (5.16) converge for fixed n, large enough.

The computational feasibility of formulas (5.16) is based on the following:

(1) If, for a given f in X, Tf is not computable in closed form, Tf is approximated by $T_M f$, where T_M is an approximation of higher order than T_n (Chapter 5, Section 6.4.2).

(2) T_n, P_n, and S_n are related to $\mathcal{T}_n, \mathcal{P}_n$, and $\mathcal{S}_n$ by the formulas given in Theorem 4.3 and Corollary 4.4 of Chapter 4, Section 6.1.

The problem is then reduced to the solution of a sequence of equations in the subspace X_n, of dimension n (say). Let us suppose that we are given a basis in X_n (and the adjoint basis in X_n^*). $\mathcal{T}_n$ is represented by the matrix A_n; I is the $n \times n$ identity matrix. We solve in $\mathbb{C}^n$

$$A_n u_n = \lambda_n u_n$$

and

$$(A_n^{\mathrm{H}} - \bar{\lambda}_n I) v_n = 0 \qquad v_n^{\mathrm{H}} u_n = 1.$$

The spectral projection $\mathscr{P}_n$ is then represented by the matrix $\mathbf{P}_n = u_n v_n^{\mathrm{H}}$. The iterative scheme is based on the resolution in $\mathbb{C}^n$ of the sequence of equations

$$(A_n - \lambda_n I)\xi_k = (I - \mathbf{P}_n)b_k, \qquad v_n^{\mathrm{H}}\xi_k = 0, \qquad k \geq 1, \tag{7.24}$$

where b_k is known.

(7.24) is a system of $n + 1$ equations in n unknowns of rank n. It has a unique solution $\xi_k = \mathbf{S}_n(I - \mathbf{P}_n)b_k$; $\mathbf{S}_n$ is the matrix representing $\mathscr{S}_n$ in the chosen basis.

To solve (7.24), one may perform the Gauss decomposition with partial pivoting of the $(n + 1) \times n$ matrix, and the corresponding calculations on the right-hand side; then delete the identically zero equation to get a regular triangular system of n equations in n unknowns (see Kulkarni and Limaye, 1982).

Exercises

7.24 Check that at all stages of the computation of formulas (5.16) the knowledge of φ_n^* is not required.

7.25 The condition of the system (7.24) depends on $\|\mathbf{S}_n\|_2$ and $\|v_n\|_2$, where $\|\cdot\|_2$ is the euclidean norm in $\mathbb{C}^n$.

8.3. The Particular Case of Projection Methods

One may iterate with T_n chosen to be either T_n^{P}, T_n^{G}, or T_n^{S}. There are special relationships between the iterates. We have already seen in Chapter 6 that

$$\begin{aligned} \lambda_n &= \lambda^{0\mathrm{G}} = \lambda^{1\mathrm{G}} && \text{(Exercise 5.32)}, \\ \lambda^{1\mathrm{S}} &= \lambda^{1\mathrm{P}} && \text{(Lemma 6.17)}, \\ &= \lambda^{2\mathrm{G}} && \text{(Exercise 6.27)}, \\ \varphi^{0\mathrm{S}} &= \varphi^{1\mathrm{G}} \end{aligned}$$

It is left to the reader to prove by induction that

$$\lambda^{k\mathrm{P}} = \lambda^{k\mathrm{S}}, \qquad \varphi^{k\mathrm{S}} = (1/\lambda_n)T\varphi^{k\mathrm{P}} \qquad \text{for} \quad k \geq 1.$$

It is remarkable that T_n^{S} and T_n^{P} yield the same iterated eigenvalues.

8.4. Rates of Convergence

As proved in Corollary 5.36, the series λ_n^k (resp. φ_n^k) is dominated by a geometrical series with ratio q (resp. q'). Under the assumption that $\|T - T_n\| \to 0$ (resp. $T_n \xrightarrow{\mathrm{cc}} T$), more specific results can be proved (see

Redont, 1979a; Kulkarni and Limaye, 1983a,b). The case of the collectively compact convergence is interesting: since $\|(T - T_n)U(T - T_n)V\| \to 0$ for any bounded operators U and V, one may prove a "staggered" or "serrated" convergence that is dominated overall by the geometric convergence with ratio q (or q'). The numerical experiments given in Example 7.3 show this phenomenon very clearly for the first few iterates.

Example 7.3 We report some numerical experiments given in Redont (1979a). We consider, in $X = C(0, 1)$, the integral operators defined by the following four kernels:

k_1 and k_2 as defined in Chapter 6, Section 4.4;

$$k_3(s, t) = \begin{cases} s(1 - t) & \text{for} \quad s \le t \\ t(1 - s) & \text{for} \quad t \le s. \end{cases}$$

$$k_4(s, t) = \begin{cases} \frac{1}{2}|s - t| & \text{for} \quad s \le t \\ 2|s - t| & \text{for} \quad t \le s. \end{cases}$$

k_1, k_2, and k_3 are symmetric in s and t; k_1 is smooth and k_2, k_3, k_4 have discontinuous first derivatives.

The eigenvalues associated with k_1 and k_2 have been given in Chapter 6, Section 4.4. The first and third dominant eigenvalues of the integral operators defined by k_3 and k_4 are, for k_3,

$$\lambda_1 = 1/\pi^2 = 0.101321,$$

$$\lambda_3 = 1/9\pi^2 = 0.001257,$$

and for k_4,

$$\lambda_1 = 0.360319,$$

$$\lambda_3 = -0.100387$$

The approximation operators are T_n^{P} and T_n^{N}, where the projection and the quadrature formula have been defined in Chapter 6, Section 4.4. We recall that $\|T_n^{\mathrm{P}} - T\|_\infty \to 0$ and $T_n^{\mathrm{N}} \overset{\mathrm{cc}}{\to} T$. If T is defined by the kernel k_1,

$$|||T_n^{\mathrm{N}} - T||| \to 0$$

in C_*^∞.

Results are listed in Tables 7.5, 7.6, and 7.7, with

$$\alpha = \lambda - \lambda^k, \qquad \beta = |\lambda - \lambda^k|/|\lambda - \lambda^{k-1}|,$$

$$\gamma = \|\phi - \varphi^k\|_2, \qquad \delta = \|\phi - \varphi^k\|_2/\|\phi - \varphi^{k-1}\|_2.$$

The geometric convergence of the series λ^k and φ^k when the convergence $T_n \to T$ is uniform in some norm is clearly indicated in Tables 7.5 and 7.6, as well as in Fig. 7.2.

Table 7.5

Iteration with T_n^{P}

$k(s, t)$	v	i		k = 0	1	2
k_1	6	1	$\|\alpha\|$	2.0×10^{-3}	8.0×10^{-8}	1.2×10^{-10}
			β	—	4.0×10^{-5}	1.5×10^{-3}
		3	$\|\alpha\|$	6.4×10^{-6}	2.2×10^{-8}	9.4×10^{-11}
			β	—	3.4×10^{-3}	4.3×10^{-3}
	10	1	$\|\alpha\|$	6.0×10^{-4}	1.0×10^{-8}	4.6×10^{-12}
			β	—	1.7×10^{-3}	4.6×10^{-4}
		3	$\|\alpha\|$	3.3×10^{-7}	3.1×10^{-10}	7.5×10^{-14}
			β	—	0.9×10^{-3}	2.4×10^{-4}
	16	1	$\|\alpha\|$	2.2×10^{-4}	1.4×10^{-9}	2.2×10^{-13}
			β	—	6.4×10^{-4}	1.6×10^{-4}
		3	$\|\alpha\|$	3.6×10^{-7}	1.0×10^{-11}	1.6×10^{-14}
			β	—	2.8×10^{-3}	1.6×10^{-3}
k_2	6	1	$\|\alpha\|$	3.0×10^{-3}	2.0×10^{-4}	1.5×10^{-4}
			β	—	6.7×10^{-2}	0.75
	10	1	$\|\alpha\|$	9.6×10^{-4}	5.0×10^{-5}	3.5×10^{-5}
			β	—	5.2×10^{-2}	0.7
	16	1	$\|\alpha\|$	3.5×10^{-4}	2.0×10^{-5}	8.2×10^{-6}
			β	—	5.7×10^{-2}	4.1×10^{-1}

Table 7.6

Iteration with T_n^{N}

$k(s, t)$	v	i		k = 0	1	2	3
k_1	6	1	$\|\alpha\|$	1.3×10^{-5}	4.6×10^{-9}	3.7×10^{-13}	4.4×10^{-16}
			β	—	3.5×10^{-4}	8×10^{-5}	1.2×10^{-3}
			γ	8.1×10^{-6}	1.6×10^{-9}	1.2×10^{-13}	8.3×10^{-16}
			δ	—	2.0×10^{-4}	0.75×10^{-4}	6.9×10^{-3}
		3	$\|\alpha\|$	6.2×10^{-3}	1.1×10^{-7}	3.0×10^{-9}	3.8×10^{-11}
			β	—	1.8×10^{-3}	2.7×10^{-2}	1.3×10^{-2}
			γ	2.5×10^{-3}	3.1×10^{-5}	4.2×10^{-7}	3.7×10^{-9}
			δ	—	1.2×10^{-2}	1.4×10^{-2}	0.9×10^{-2}

Table 7.7

Iteration with T_n^{N}

$k(s,t)$	v	i		k = 0	1	2	3	4	5
k_3	16	1	α	9.0×10^{-5}	-1.1×10^{-4}	-2.5×10^{-7}	1.2×10^{-7}	9.3×10^{-10}	-2.6×10^{-10}
			β	—	1.2	2.2×10^{-3}	0.49	7.8×10^{-3}	0.28
			γ	7.2×10^{-4}	2.0×10^{-6}	7.8×10^{-7}	6.9×10^{-9}	1.7×10^{-9}	2.6×10^{-11}
			δ	—	2.8×10^{-3}	0.39	8.9×10^{-3}	0.24	1.5×10^{-2}
		3	α	9.1×10^{-5}	-1.1×10^{-4}	-2.2×10^{-6}	1.0×10^{-6}	7.2×10^{-8}	-1.6×10^{-8}
			β	—	1.2	2.0×10^{-2}	0.46	2.2×10^{-2}	0.23
			γ	6.8×10^{-3}	1.7×10^{-4}	6.0×10^{-6}	5.0×10^{-7}	9.6×10^{-7}	
			δ	—	2.5×10^{-2}	0.36	8.3×10^{-2}	0.19	
k_4	30	1	α	-1.6×10^{-3}	1.6×10^{-3}	-1.6×10^{-5}	7.1×10^{-6}	-2.4×10^{-7}	6.7×10^{-8}
			β	—	1	1.0×10^{-2}	0.45	3.4×10^{-2}	0.28
			γ	3.5×10^{-3}	4.5×10^{-5}	1.6×10^{-5}	5.8×10^{-7}	1.5×10^{-7}	9.1×10^{-9}
			δ	—	1.2×10^{-2}	0.40	3.6×10^{-2}	0.26	6.9×10^{-2}
		3	α	-1.6×10^{-3}	1.5×10^{-3}	5.6×10^{-5}	-2.1×10^{-5}		
			β	—	0.95	3.6×10^{-2}	0.38		
			γ	1.4×10^{-2}	5.9×10^{-4}	2.3×10^{-4}			
			δ	—	4.1×10^{-2}	0.38			

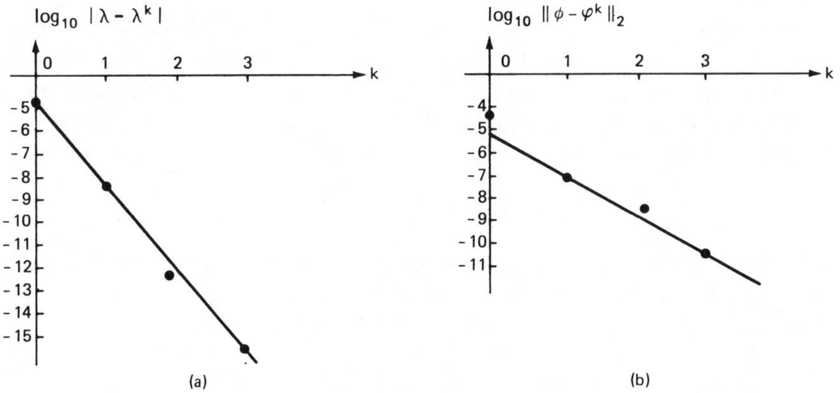

Figure 7.2 Geometric convergence when $\|T_n - T\| \to 0$. Iteration with T_n^N, Nyström approximation of T defined by k_1. (a) Results for the dominant eigenvalue λ_1 and (b) the associated eigenvector ϕ_1

The staggered convergence of the first terms in the series λ^k and φ^k when $T_n \overset{cc}{\to} T$ appears in Table 7.7 and Fig. 7.3. One may note that the slope of the steps decreases. Also, the first iterated eigenvector φ^1 improves on φ_n, whereas the first iterated eigenvalue λ^1 does not improve on λ_n.

In all cases, the rate of improvement decreases when the rank of the eigenvalue increases, because of the condition of the system (7.24). The approximation of λ, ϕ by iterative refinement from λ_n, φ_n with a moderate n, versus the approximation by λ_n and φ_n with an n giving the same accuracy, may result, for the few dominant eigenelements, in a time-saving factor of 10–20.

The convergence of the series λ^k and φ^k has been proved in Chapter 5, Theorem 5.35, under the condition that n is large enough. We study the influence of this condition in Example 7.4.

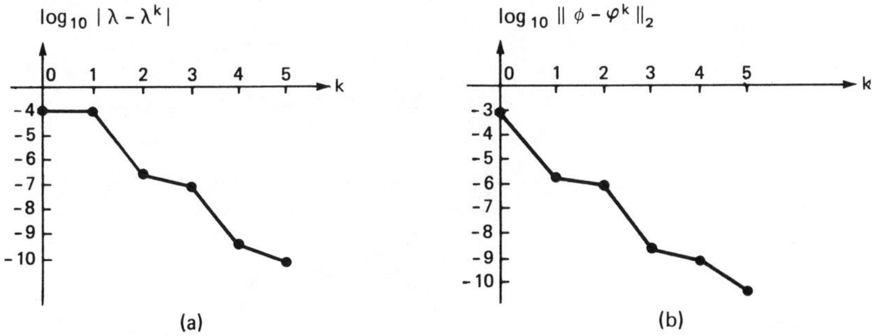

Figure 7.3 "Staggered" covergence when $T_n \overset{cc}{\to} T$. Iteration with T_n^N, Nyström approximation of T defined by k_3. (a) Results for the dominant eigenvalue λ_1 and (b) the associated eigenvector ϕ_1.

Example 7.4 We are concerned with the numerical spectrum of the Schrödinger operator for the potentials $gr^{-(s+2)}$, $-2 \leq s \leq 0$ and $g < 0$. Following Dumont-Lepage *et al.* (1980), and using a method of separation of variables and two integral transformations (Fourier–Fock), one comes up with the eigenvalue problem in l^2

$$(1 - \lambda T)\varphi = 0, \qquad 0 \neq \varphi \in l^2.$$

The operator T depends on the parameters s and l; the nonnegative integer l is the angular momentum. In the canonical basis $\{e_i\}_{\mathbb{N}}$ of l^2, T is represented by the infinite matrix A; the coefficients of A are, up to a constant factor, equal to

$$a_{ij}(s, l) = \frac{(-1)^{i+j}}{[(i+l)(j+l)]^{1/2}} \left(\frac{(i-1)!\,(j-1)!}{(i+2l)!\,(j+2l)!}\right)^{1/2} \times \sum_{\sigma=1}^{i} \left(\frac{(2l+1-s)_{\sigma-1}}{(\sigma-1)!} \frac{(s+1)_{i-\sigma}}{(i-\sigma)!} \frac{(s+1)_{j-\sigma}}{(j-\sigma)!}\right),$$

where $(a)_n := \Gamma(a+n)/\Gamma(a)$ (Pochhammer symbol), $i, j \in \mathbb{N}$.

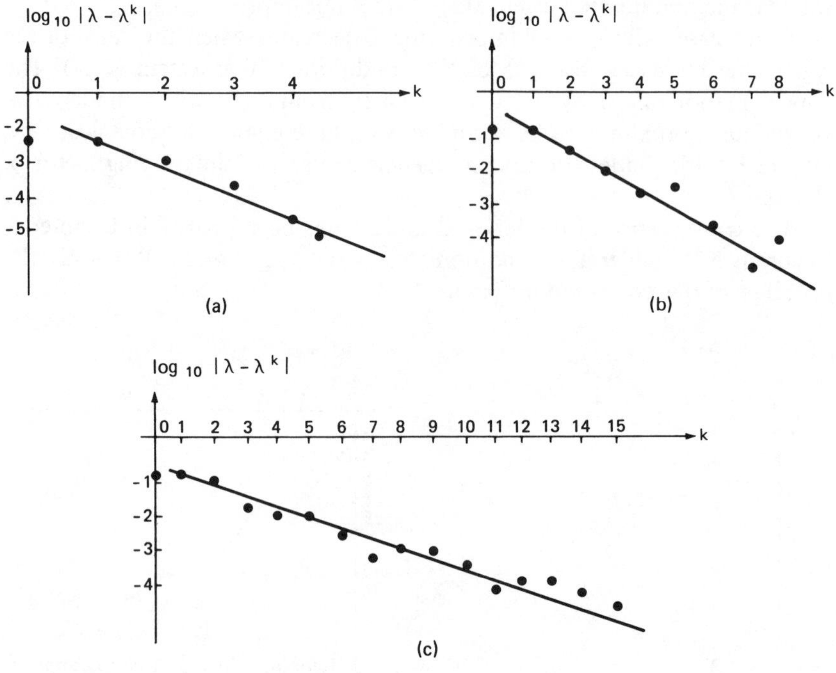

Figure 7.4 Convergence of λ^k. (a) $n = 15$, $s = -0.5$. (b) $n = 30$, $s = -0.05$. (c) $n = 30$, $s = -0.01$.

For $-2 < s < 0$, the operator T is compact and self-adjoint, whereas for $s = -2$ or $s = 0$, T is no longer compact.

We report some of the numerical experiments given in Kulkarni and Limaye (1982). T is approximated by the $\perp$-Galerkin method defined on the subspace spanned by $\{e_i\}_1^n$.

Results concerning the iterative refinement for the dominant eigenvalue are presented in Fig. 7.4. In the calculations, Tx has been evaluated by means of the truncated matrix of size 100, and l is chosen to be zero. For $s = -0.5$, the convergence obtained with $n = 15$ is good, whereas for $s = -0.05$ and $s = -0.01$, the convergence attained with $n = 30$ is unsatisfactory. $n = 30$ is too small when the compact T is too close to a *noncompact* operator.

8.5. *Bibliographical Comments*

Various authors have proposed improved values for the eigenelements of integral operators: Chu and Spence (1981), Linz (1970, 1972), Rakotch (1975, 1978), Spence (1978, 1979a). Iterative refinement of the solution is studied in Atkinson (1973, 1976a).

For the eigenvalue problem, a number of other iterations can be proposed, based on defect correction (Ahués and Chatelin, 1983, Ahués *et al.*, 1983a). For multigrid methods on an integral equation of the second kind, the reader is referred to Hackbusch (1981) and Hemker and Schippers (1981).

9. *T* Is a Differential Operator

For projection methods, we may define an iterative refinement by means of formulas (5.24) applied to the integral formulation. We consider the elliptic partial differential equation (4.41) introduced in Chapter 4, Section 8. Let T be the differential operator associated with the bounded coercive form $a(\cdot, \cdot)$ defined on $V \times V$. The eigenvalue problem

$$T\psi = \lambda\psi, \qquad 0 \neq \psi \in V,$$

is equivalent to

$$\psi = \lambda A\psi, \qquad A = T^{-1}, \qquad \text{if} \quad \lambda \neq 0.$$

We suppose that $\lambda \neq 0$ is *simple*.

Let V_n be a subspace of V. The finite-element approximation of ψ is defined by

$$a(\psi_n, v) = \lambda_n(\psi_n, v)_H \qquad \text{for all} \quad v \in V_n.$$

Let π_n^a be the elliptic projection. We suppose that A is compact (that is, the injection from V into H is compact). Then $\|(\pi_n^a - 1)A\|_H \to 0$. Let $P_n = (\cdot, \psi_n^*)_H \psi_n$ be the spectral projection of $\pi_n^a A$ associated with $1/\lambda_n$.

The series $\{\mu^k\}$ and $\{u^k\}$ are defined by

$$\mu^0 = \lambda_n, \qquad \mu^{k+1} = 1/(Au^k, \psi_n^*)_H, \qquad\qquad k \geq 0,$$

$$u^0 = \psi_n, \qquad P_n u^k = \psi_n, \qquad \hat{u}^k = \mu^{k+1} Au^k, \qquad u^{k+1} = \hat{u}^k + r^k, \qquad k \geq 0,$$

where r^k is the unique solution of

$$(\lambda_n \pi_n^a A - 1)r^k = \lambda_n \pi_n^a A(u^k - \hat{u}^k), \qquad P_n r^k = 0.$$

The convergence to λ, ψ (normalized by $P_n \psi = \varphi_n$) is at least geometrical in $\|(1 - \pi_n^a)A\|_H$, by Exercise 5.38. What computations are required to get μ^k and u^k?

Set $v^k := Au^k$, v^k is the solution of $Tv^k = u^k$, or

$$a(v^k, v) = (u^k, v) \qquad \text{for all} \quad v \in V. \tag{7.25}$$

Then compute $(v_n^k, \psi_n^*)_H$ and $\hat{u}^k = \mu^{k+1} v^k$. r^k is the unique solution in V_n of

$$\begin{aligned} a(r^k, v) - \lambda_n (r^k, v)_H &= \lambda_n (u^k - \hat{u}^k, v)_H \qquad \text{for all} \quad v \in V_n, \\ P_n r^k &= 0. \end{aligned} \tag{7.26}$$

The evaluation Tu^k, which is required in the case of an integral operator T, is replaced here by the resolution of the differential equation (7.25). An approximation of v^k can be obtained by using a f.e.m. with higher-order elements.

The above formulas, like formulas (5.24), require the knowledge of P_n, that is, of ψ_n and ψ_n^*. This is not necessary, and an H-orthogonal projection on $\{\psi_n\}$ may be used (cf. Exercise 7.26 below).

Exercise

7.26 Prove the following analogous version of Exercise 5.38 for the problem $\psi = \lambda A \psi$, where μ^k and u^k are solutions of

$$\mu^0 = \lambda_n, \qquad \mu^{k+1} = 1/(Au^k, \psi_n)_H,$$

$$u^0 = \psi_n, \qquad (u^k, \psi_n)_H = 1, \qquad \hat{u}^k = \mu^{k+1} Au^k, \qquad u^{k+1} = \hat{u}^k + r^k,$$

with

$$(\lambda_n \pi_n^a A - 1)r^k = \lambda_n \pi_n^a A(u^k - \hat{u}^k), \qquad (r^k, \psi_n)_H = 0.$$

μ^k (resp. u^k) converges to λ (resp. ψ normalized by $(\psi, \psi_n)_H = 1$). The rate of convergence is given by

$$|\mu^k - \lambda| + \|u^k - \psi\|_H \leq c\|(1 - \pi_n^a)A\|_H^{k+1}, \qquad k \geq 1.$$

Example 7.5 Consider the Dirichlet eigenvalue problem in Ω, a bounded domain of $\mathbb{R}^N$,

$$\begin{aligned} -\Delta\psi &= \lambda\psi && \text{in } \Omega, \\ \psi &= 0 && \text{on } \partial\Omega. \end{aligned} \tag{7.27}$$

$H = L^2(\Omega)$, $V = H_0^1(\Omega)$, and $a(u, v) = (\nabla u, \nabla v)_H$. The variational formulation of (7.27) is $a(\psi, v) = \lambda(\psi, v)_H$, $\forall v \in V$.

Let λ, ψ be a pair of exact eigenelements. λ_n and ψ_n are the corresponding piecewise-linear finite-element approximations. h is the triangulation diameter. It is well known that

$$|\lambda - \lambda_n| + \|\psi - \psi_n\|_H + h\|\psi - \psi_n\|_V = O(h^2),$$

where ψ, ψ_n are normalized by $\|\psi_n\|_H = (\psi, \psi_n)_H = 1$.

The iterative refinement requires, for $k \geq 0$, the following:

(1) Solve the Dirichlet equation

$$\begin{aligned} -\Delta v^k &= u^k && \text{in } \Omega, \\ v^k &= 0 && \text{on } \partial\Omega. \end{aligned} \tag{7.28}$$

(2) Compute $\mu^{k+1} = 1/(v^k, \psi_n)_H$ and $u^k = \mu^{k+1}v^k$.

(3) Solve the equation in V_n, the set of piecewise-linear finite-elements

$$a(r^k, v) - \lambda_n(r^k, v)_H = \lambda_n(u^k - \hat{u}^k, v)_H \qquad \text{for all} \quad v \in V_n,$$

$$(r^k, \psi_n)_H = 0.$$

(4) Compute $u^{k+1} = u^k + r^k$.

It is left to the reader to prove that

$$|\mu^k - \lambda| + \|\psi - u^k\|_H \leq c(h^2)^{k+1} \qquad \text{for} \quad k \geq 1.$$

The solution v^k of (7.28) can be approximated by using higher-order finite elements, for example, piecewise-quadratic elements. This has been done by Lin Qun and Xie Ganquan in the numerical experiment that we report now (Lin Qun, 1980).

Let Ω be the square $]0, 1[\times]0, 1[$ in $\mathbb{R}^2$, with the triangulation shown in Fig. 7.5. Let λ be the smallest eigenvalue $2\pi^2$ of (7.27):

$$\lambda = 19.739\ldots, \qquad \lambda_n = 22.865, \qquad \tilde{\lambda}_n = \frac{(\nabla\tilde{\psi}_n, \nabla\tilde{\psi}_n)_H}{(\tilde{\psi}_n, \tilde{\psi}_n)_H} = 19.817,$$

where $\tilde{\psi}_n$ is the solution of

$$\begin{aligned} -\Delta\tilde{\psi}_n &= \lambda_n\psi_n && \text{in } \Omega, \\ \tilde{\psi}_n &= 0 && \text{on } \partial\Omega, \end{aligned}$$

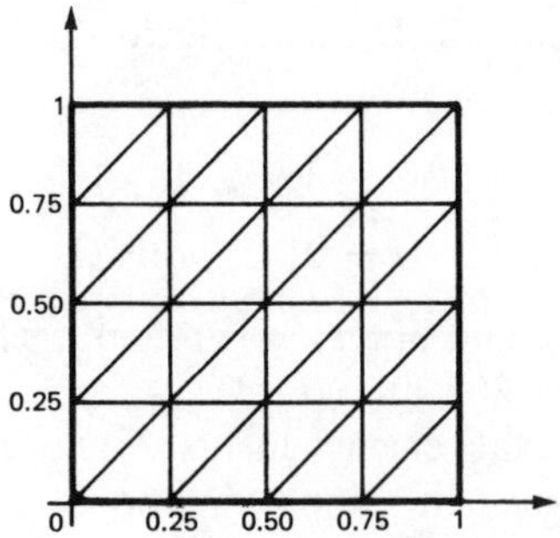

Figure 7.5 Triangulation of $]0, 1[\times]0, 1[$.

and $\tilde{\lambda}_n$ is the Rayleigh quotient based on $\tilde{\psi}_n$. The relative errors are 0.16 and 4×10^{-3}, respectively. This is a variation on the scheme proposed above. One proves that

$$|\lambda - \tilde{\lambda}_n| = O(h^4) \qquad \text{and} \qquad \|\psi - \tilde{\psi}_n\|_V = O(h^2).$$

Bibliographical Comments

An iterative refinement method is proposed in Lin Qun (1981a) for nonlinear eigenvalue problems in relation with the f.e.m. Ghemires (1979) deals with the finite-difference method (see also McCormick, 1981; Nakamura, 1976). For refinement techniques on differential nonlinear equations, the reader is referred to Brandt (1977), Hackbusch (1980b), Lin Qun (1982a), Muroya (1979), and Stetter (1978); see also Oliveira Aleixo (1980).

APPENDIX

Discrete Approximation Theory

In the numerical methods of computing the eigenvalues of an operator T, the approximation is not always defined in X or even in a *subspace* X_n of X. It may be defined in a space *different* from X. The finite-difference method for a differential operator is a good example, where the approximation is defined in some $\mathbb{C}^n$. This point of view of a *discrete* approximation theory was the first to be developed, as it is the most natural to study the finite-difference methods. We give below an overview of the axiomatic presentation of Stummel (1970, 1972) and Vainikko (1974, 1978a).

1. Discrete Approximation of a Banach Space

X and $\mathscr{X}_n, n \in \mathbb{N}$, are Banach spaces, respectively, equipped with the norms $\|\cdot\|$ and $\|\cdot\|_n$.

Notation In what follows, the latin letters $x, y, \ldots$ represent elements of X and the greek letters $\xi, \eta, \ldots$ represent elements of $\mathscr{X}_n$.

The sequences $\{\xi_n\}_{\mathbb{N}}$ and $\{\eta_n\}_{\mathbb{N}}$ are *equivalent* iff

$$\|\xi_n - \eta_n\|_n \to 0 \qquad \text{as} \quad n \to \infty.$$

The equivalence relation is denoted $\{\xi_n\} \sim \{\eta_n\}$.

We set $\mathscr{X} := \prod_{n \in \mathbb{N}} \mathscr{X}_n$. $\sim$ is an equivalence relation on the space $\mathscr{X}$, and we denote by $\mathscr{Y}$ the factor space of equivalent bounded sequences.

Definition The sequence of spaces $\{\mathcal{X}_n\}_{\mathbb{N}}$ *discretely approximates* the space X iff there exists a linear map $\mathcal{P}: X \to \mathcal{Y}$ such that if

$$x \in X \qquad \text{and} \qquad \{\xi_n\}_{\mathbb{N}} \in \mathcal{P}x,$$

then $\|\xi_n\|_n \to \|x\|$ as $n \leq \infty$.

Let $\mathcal{Y}$ be equipped with the factor norm $\|\cdot\|_{\mathcal{Y}}$ defined by

$$\|\zeta\|_{\mathcal{Y}} := \inf_{\{\xi_n\}\in\zeta} \sup_{n\in\mathbb{N}} \|\xi_n\|_n \qquad \text{for} \quad \zeta \in \mathcal{Y}.$$

The map $\mathcal{P}$ of the above definition is isometric from X into $\mathcal{Y}$ since $\|x\| = \|\mathcal{P}x\|_{\mathcal{Y}}$ for all $x \in X$.

By choosing in every equivalence class $\mathcal{P}x$ a sequence $\{r_n(x)\}_{\mathbb{N}}$, we obtain the following equivalent definition: $\{\mathcal{X}_n\}_{\mathbb{N}}$ discretely approximates X iff there exist (possibly) nonlinear maps $r_n: X \to \mathcal{X}_n$, $n \in \mathbb{N}$, such that

(i) $\|r_n(x)\|_n \to \|x\|$ as $n \to \infty$ for any $x \in X$,
(ii) $\|r_n(\alpha x + \beta y) - \alpha r_n(x) - \beta r_n(y)\|_n \to 0$ for any $x, y \in X$, and $\alpha, \beta \in \mathbb{C}$.

r_n is called a *restriction* operator. The operators r_n are not uniquely defined by the map $\mathcal{P}$. If $\{r_n\}$ and $\{r_n'\}$ are two systems of restrictions corresponding to $\mathcal{P}$, then $\|r_n(x) - r_n'(x)\|_n \to 0$ as $n \to \infty$ for any $x \in X$. The r_n need not be linear; they are merely *asymptotically linear*, that is, they satisfy (ii).

If $r_n \in \mathcal{L}(X, \mathcal{X}_n)$, then the r_n are uniformly bounded. Even if there exists a linear r_n, however, it is not necessarily the easiest one to use, as we shall see in Example A.2.

Example A.1 Let Ω be a bounded domain in $\mathbb{R}^k$. $X = C(\Omega)$. Let Ω_n be the set of grid points $\{t_1^{(n)}, t_2^{(n)}, \ldots, t_{N(n)}^{(n)}\}$ in Ω such that the distance of any $t \in \Omega$ to Ω_n tends to 0 as $n \to \infty$. We define $\mathcal{X}_n = \mathbb{C}^{N(n)}$ with the max norm $\|\cdot\|_n$, and

$$r_n: x \in X \mapsto r_n x = (x(t_1^{(n)}), \ldots, x(t_{N(n)}^{(n)}))^{\mathrm{T}}.$$

r_n is linear on X and $\|r_n x\|_n \to \|x\|$ for all continuous x.

Example A.2 It is often the case in practice that there exists a very "natural" restriction, internally connected with the discretization method, which is linear on a subspace X' of X, dense in X, and such that $\|r_n(x)\|_n \to \|x\|$ for any $x \in X'$. The corresponding map $\mathcal{P}$, which is isometric from X' into $\mathcal{Y}$, can be uniquely extended into an isometry from X into $\mathcal{Y}$, but the extended restriction r_n is only asymptotically linear on X in general.

We give an example with $X = L^2(\Omega)$. As we have seen in Example A.1, the natural restriction r_n to consider for discretization methods, such as approximate quadratures or finite differences, is the operator that maps a continuous function into the vector of its values at the mesh points. But r_n is not bounded as an operator in $L^2(\Omega)$.

We consider $\mathscr{X}_n = \mathbb{C}^{N(n)}$ with the norm $\|\cdot\|_n$ defined by

$$\|\xi_n\|_n := \left(\sum_{i=1}^{N(n)} w_{in}|\xi_{in}|^2\right)^{1/2}$$

in association with the quadrature formula $\sum_{i=1}^{N(n)} w_{in}x(t_i^{(n)})$ converging to $\int_\Omega x(t)\,dt$ for all continuous functions x. Hence $\|r_n(x)\|_n \to \|x\|$ for $x \in C(\Omega)$, and therefore the subspace X' is $C(\overline{\Omega})$.

The values of $r_n(x)$ for $x \notin X'$ may be defined through $\mathscr{P}$ as already indicated; but to $\mathscr{P}$ also corresponds a restriction r_n', which is linear on X. Let $\{\sigma_i^{(n)}\}_1^{N(n)}$ be elementary subsets of Ω, $t_i^{(n)} \in \sigma_i^{(n)}$. Then

$$r_n'\colon x \mapsto (r_n'x)_i = \frac{1}{\operatorname{meas}\sigma_i^{(n)}}\int_{\sigma_i^{(n)}} x(t)\,dt, \qquad i = 1, \ldots, N(n).$$

It can be shown that $\|r_n'x\|_n \to \|x\|$ for $x \in L^2(\Omega)$ and $\|r_n(x) - r_n'x\|_n \to 0$ for $x \in X'$. r_n and r_n' are therefore associated with the same $\mathscr{P}$ in X', and

$$r_n' \in \mathscr{L}(X, \mathscr{X}_n).$$

For $x \notin X'$, we may set $r_n(x) := r_n'x$.

Example A.3 In Aubin (1972, p. 5) a slightly different point of view is given with the use not only of *linear* restriction operators $r_n \in \mathscr{L}(X, \mathscr{X}_n)$, but also of linear prolongation operators $p_n \in \mathscr{L}(\mathscr{X}_n, X)$. p_n is an isomorphism from $\mathscr{X}_n$ into a closed subspace X_n of X. p_n and r_n are connected by the property

$$p_n r_n x \to x \qquad \text{for} \quad x \in X.$$

As an example we choose $X = C(a, b)$ (cf. Chapter 3, Example 3.1). $[a, b]$ is divided into $n - 1$ intervals at the points $t_i^{(n)}$, $i = 1, \ldots, n$, $t_1^{(n)} = a$, $t_n^{(n)} = b$. $\mathscr{X}_n = \mathbb{C}^n$ is equipped with the max norm, and we define

$$r_n\colon x \mapsto (x(t_1^{(n)}), x(t_2^{(n)}), \ldots, x(t_n^{(n)}))^{\mathrm{T}},$$

$$p_n\colon (x(t_i^{(n)}) \mapsto \text{its piecewise-linear interpolant at } (t_i, x(t_i^{(n)}))_1^n.$$

If we suppose that $\max_{i=1,\ldots,n-1}|t_{i+1}^{(n)} - t_i^{(n)}| \to 0$, then $p_n r_n x \to x$ for all x in X. p_n is an isomorphism from $\mathbb{C}^n$ into the subspace X_n of piecewise-linear continuous functions.

Example A.4 In Example A.3, one has the property $r_n p_n = 1_n$, the identity operator on $\mathscr{X}_n$. Hence $\pi_n = p_n r_n$ is a projection from X onto $X_n = p_n \mathscr{X}_n$, such that $\pi_n \to 1$ pointwise in X. If the spaces $\mathscr{X}_n$ are finite dimensional, so are the subspaces X_n, and this corresponds to the particular framework for discrete approximation theory presented in Chapter 3.

2. Discrete Approximation of a Closed Operator

The sequence $\{\xi_n\}_{\mathbb{N}}$, $\xi_n \in \mathscr{X}_n$, *discretely converges* (in short, d-converges) to $x \in X$ iff $\|r_n x - \xi_n\|_n \to 0$ as $n \to \infty$; it is denoted $\xi_n \xrightarrow{\text{d}} x$. It is *discretely relatively compact* (in short, d-relatively compact) iff each subsequence $\{\xi_n\}_{N_1 \subset \mathbb{N}}$ has a d-converging subsequence $\{\xi_n\}_{N_2 \subset N_1}$.

Let T be a closed linear operator in X with domain $D \subseteq X$. We suppose for simplicity that the spaces $\mathscr{X}_n$, $n \in \mathbb{N}$, which discretely approximate X, are finite dimensional. This is almost always the case in practice. $\mathscr{T}_n$ is a linear operator from $\mathscr{X}_n$ into itself, with domain $\mathscr{D}_n \subseteq \mathscr{X}_n$. We suppose that $r_n D \subseteq \mathscr{D}_n$ for all n. If T (resp. $\mathscr{T}_n$) is bounded on X (resp. $\mathscr{X}_n$), $D = X$ (resp. $\mathscr{D}_n = \mathscr{X}_n$).

We now define various types of convergence of the sequence $\{\mathscr{T}_n\}_{\mathbb{N}}$ toward T that can be considered:

Discrete-pointwise convergence $\mathscr{T}_n \xrightarrow{\text{d-p}} T$ iff, for all x in D,

$$\|r_n Tx - \mathscr{T}_n r_n x\|_n \to 0 \qquad \text{as} \quad n \to \infty.$$

Discrete-stable convergence $\mathscr{T}_n \xrightarrow{\text{d-s}} T$ iff

(i) $\mathscr{T}_n \xrightarrow{\text{d-p}} T$,
(ii) $\exists N$: for $n > N$, $\mathscr{T}_n^{-1} \in \mathscr{L}(\mathscr{X}_n)$ and $\|\mathscr{T}_n^{-1}\|_n \leq M$.

Discrete-compact convergence $\mathscr{T}_n \xrightarrow{\text{d-c}} T$ for *T compact* iff

(i) $\mathscr{T}_n \xrightarrow{\text{d-p}} T$,
(ii) for any sequence $\{\xi_n\}_{\mathbb{N}}$ such that $\xi_n \in \mathscr{X}_n$, $\|\xi_n\|_n \leq c$, the sequence $\{\mathscr{T}_n \xi_n\}_{\mathbb{N}}$ is d-relatively compact.

Discrete-regular convergence $\mathscr{T}_n \xrightarrow{\text{d-r}} T$ iff

(i) $\mathscr{T}_n \xrightarrow{\text{d-p}} T$,
(ii) any sequence $\{\xi_n\}_{\mathbb{N}}$, where $\xi_n \in \mathscr{D}_n$, $\|\xi_n\|_n \leq c$, and which is such that $\mathscr{T}_n \xi_n \xrightarrow{\text{d}} y$, for $n \in N_1 \subset \mathbb{N}$, is itself such that $\xi_n \xrightarrow{\text{d}} x \in D$, for $n \in N_2 \subset N_1$ and $Tx = y$.

The various types of convergence introduced in Chapter 3 for approximations T_n of class $\mathfrak{D}$ are clearly particular cases of the above definitions, and the two fundamental characterizations given in Proposition 3.20 and Theorem 5.29 can easily be proved in the more general setting of this appendix.

The interested reader can find more references on works in the context of discrete approximation theory in the bibliographical comments given in Chapter 3, Section 6, and Chapter 5, Section 5.7. Another up-to-date reference is Anselone and Ansorge (1981).

References

Adams, R. A. (1975). *Sobolev Spaces.* Academic Press, New York.

Ahlberg, J. H., and Ito, T. (1975). A collocation method for two-point boundary value problems. *Math. Comp.* **29**, 761–776.

Ahués, M. (1982). Raffinement des éléments propres d'un opérateur compact sur un espace de Banach par des méthodes de type Newton à jacobien approché. Unpublished manuscript, Univ. de Grenoble.

Ahués, M., and Chatelin, F. (1983). The use of defect correction to refine the eigenelements of compact integral operators. *SIAM J. Numer. Anal.* (to appear).

Ahués, M., and Telias, M. (1982a). Petrov–Galerkin schemes for the steady state convection–diffusion equation. In *Finite Elements in Water Resources* (K. P. Holz, U. Meissner, W. Zulke, C. A. Brebbia, G. Pinder and W. Gray, eds.), pp. 2–3, 2–12. Springer-Verlag, Berlin and New York.

Ahués, M., and Telias, M. (1982b). Quasi-Newton iterative refinement techniques for the eigenvalue problem of compact linear operators. R.R. IMAG No. 325, Univ. de Grenoble.

Ahués, M., d'Almeida, F., and Telias, M. (1982). On the defect correction method with applications to iterative refinement techniques. R.R. IMAG No. 324, Univ. de Grenoble.

Ahués, M., Chatelin, F., d'Almeida, F., and Telias, M. (1983a). Iterative refinement techniques for the eigenvalue problem of compact integral operators. In *Treatment of Integral Equations by Numerical Methods* (C. T. H. Baker and G. F. Miller, eds.), pp. 373–386. Academic Press, London.

Ahués, M., d'Almeida, F., and Telias, M. (1983b). Two defect correction methods for the eigenvalue problem of compact operators in Banach spaces. *J. Integral Equations* (submitted).

Ahués, M., d'Almeida, F., and Telias, M. (1983c). Iterative refinement for approximate eigenelements of compact operators. *RAIRO Anal. Numér.* (to appear).

Albrecht, J., and Collatz, L., eds. (1980). *Numerical Treatment of Integral Equations.* Birkhaeuser, Basel.

Anderssen, R. S., and Prenter, P. M. (1981). A formal comparison of methods proposed for the numerical solution of first kind integral equations. *J. Austral. Math. Soc. Ser. B* **22**, 491–503.

Anderssen, R. S., de Hoog, F. R., and Lukas, M. A., eds. (1980). *The Application and Numerical Solution of Integral Equations.* Sijthoff & Noordhoff, Alphen aan den Rijn, The Netherlands.

Andrew, A. L. (1973). Eigenvectors of certain matrices. *Linear Algebra Appl.* **7**, 151–162.

Andrew, A. L. (1979). Iterative computation of derivatives of eigenvalues and eigenvectors. *J. Inst. Math. Appl.* **24**, 209–218.

Andrew, A. L., and Elton, G. C. (1971). Computation of eigenvectors corresponding to multiple eigenvalues. *Bull. Austral. Math. Soc.* **4**, 419–422.

Andrushkin, R. I. (1975). On the approximate solution of K-positive eigenvalue problems $Tu - \lambda Su = 0$. *J. Math. Anal. Appl.* **50**, 511–529.

Anselone, P. M. (1971). *Collectively Compact Operator Approximation Theory.* Prentice-Hall, Englewood Cliffs, New Jersey.

Anselone, P. M. (1976). Nonlinear operator approximation. In *Moderne Methoden der numerischen Mathematik*, (J. Albrecht and L. Collatz, eds.), pp. 17–24. Birkhaüser, Basel.

Anselone, P. M., and Ansorge, R. (1979). Compactness principle in non linear operator approximation theory. *Numer. Funct. Anal. Optim.* **1**, 589–618.

Anselone, P. M., and Ansorge, R. (1981). A unified framework for the discretization of nonlinear operator equations. Tech. Rep. 81/5, Angew. Math., Univ. Hamburg.

Anselone, P. M., and Gonzalez-Fernandez, M. J. (1965). Uniformly convergent approximate solutions of Fredholm integral equations. *J. Math. Anal. Appl.* **10**, 519–536.

Anselone, P. M., and Krabs, W. (1979). Approximate solution of weakly singular integral equations. *J. Integral Equations* **1**, 61–75.

Anselone, P. M., and Lee, J. W. (1974). Spectral properties of integral operators with nonnegative kernels. *Linear Algebra Appl.* **9**, 67–87.

Anselone, P. M., and Lee, J. W. (1976). Double approximation methods for the solution of Fredholm integral equations. In *Numerische Methoden der Approximations Theorie*, (L. Collatz, H. Werner, and G. Meinardus, eds.), pp. 9–34. Birkhaeuser, Basel.

Arnold, D. N., and Wendland, W. L. (1982). On the asymptotic convergence of collocation methods. Prepr. 665, Math., Techn. Hochschule Darmstadt.

Arnoldi, W. E. (1951). The principle of minimized iterations in the solution of the matrix eigenvalue problem. *Quart. Appl. Math.* **9**, 17–29.

Astrakhantsev, G. P. (1971). An iterative method for solving elliptic net problems. *Ž. Vyčisl. Mat. i Mat. Fiz.* **11**, 439–448 [*U.S.S.R. Computational Math. and Math. Phys.* **11**, 171–182].

Astrakhantsev, G. P. (1976). The iterative improvement of eigenvalues. *Ž. Vyčisl. Mat. i Mat. Fiz.* **16**, 131–139 [*U.S.S.R. Computational Math. and Math. Phys.* **16**, 123–132].

Atkinson, K. E. (1967a). The numerical solution of Fredholm integral equations of the second kind. *SIAM J. Numer. Anal.* **4**, 337–348.

Atkinson, K. E. (1967b). The numerical solution of the eigenvalue problem for compact integral operators. *Trans. Amer. Math. Soc.* **129**, 458–465.

Atkinson, K. E. (1972). The numerical solution of Fredholm integral equations of the second kind with singular kernels. *Numer. Math.* **19**, 248–259.

Atkinson, K. E. (1973). Iterative variants of the Nyström method for the numerical solution of integral equations. *Numer. Math.* **22**, 17–31.

Atkinson, K. E. (1975). Convergence rates for approximate eigenvalues of compact integral equations. *SIAM J. Numer. Anal.* **12**, 213–222.

Atkinson, K. E. (1976a). *A Survey of Numerical Methods for the Solution of Fredholm Integral Equations of the Second Kind.* SIAM, Philadelphia, Pennsylvania.

Atkinson, K. E. (1976b). An automatic program for linear Fredholm integral equations of the second kind. *ACM Trans. Math. Software* **2**, 154–171.

Atkinson, K. E., Graham, I. G., and Sloan, I. H. (1982). Piecewise continuous collocation for integral equations. Tech. Rep., Math., Univ. of New South Wales, Kensington.

Aubin, J. P. (1972). *Approximation of Elliptic Boundary Value Problems.* Wiley (Interscience), New York.

Aziz, A. K., ed. (1972). *The Mathematical Foundations of the Finite Element Method with Applications to Partial Differential Equations.* Academic Press, New York.

Babuška, I. (1971). Error bounds for finite element method. *Numer. Math.* **16**, 322–333.

Babuška, I. (1973). The finite element method with Lagrangian multipliers. *Numer. Math.* **20**, 179–192.

Babuška, I., and Aziz, A. K. (1972). Survey lectures on the mathematical foundations of the finite element method. In *The Mathematical Foundations of the Finite Element Method with Applications to Partial Differential Equations* (A. K. Aziz, ed.), pp. 5–359. Academic Press, New York.

Babuška, I., and Osborn, J. E. (1978). Numerical treatment of eigenvalue problems for differential equations with discontinuous coefficients. *Math. Comp.* **32**, 991–1023.

Babuška, I., and Rheinboldt, W. (1978). Error estimates for adaptive finite element computations. *SIAM J. Numer. Anal.* **15**, 736–754.

Baker, C. T. H. (1971). The deferred approach to the limit for eigenvalues of integral equations. *SIAM J. Numer. Anal.* **8**, 1–10.

Baker, C. T. H. (1977). *The Numerical Treatment of Integral Equations.* Oxford Univ. Press (Clarendon), London and New York.

Baker, C. T. H., and Hodgson, G. S. (1971). Asymptotic expansions for integration formulae in one and more dimensions. *SIAM J. Numer. Anal.* **8**, 473–480.

Banach, S., and Steinhaus, H. (1927). Sur le principe de la condensation des singularités. *Fund. Math.* **9**, 51–57.

Bank, R. E. (1980). Analysis of a multilevel inverse iteration procedure for eigenvalue problems. Res. Rep. No. 199, Computer Science, Yale Univ., Connecticut.

Bank, R. E., and Rose, D. J. (1981). Analysis of a multilevel iterative method for nonlinear finite element equations. Res. Rep. No. 202, Computer Science, Yale Univ., Connecticut.

Bartels, R. H., and Stewart, G. W. (1972). Algorithm 432, solution of the matrix equation $AX + XB = C$. *Comm. ACM* **15**, 820–826.

Bathé, K.-J., and Wilson, E. L. (1972). Large eigenvalue problems in dynamic analysis. *ASCE J. Engrg. Mech. Div.* **98**, 1471–1485.

Bathé, K.-J., and Wilson, E. L. (1973a). Eigensolution of large structure systems with small band width. *ASCE J. Engrg. Mech. Div.* **99**, 467–480.

Bathé, K.-J., and Wilson, E. L. (1973b). Solution methods for eigenvalue problems in structural mechanics. *Internat. J. Numer. Methods Engrg.* **6**, 213–226.

Bathé, K.-J., and Wilson, E. L. (1976). *Numerical Methods in Finite Element Analysis.* Prentice-Hall, Englewood Cliffs, New Jersey.

Bathé, K.-J. and Ramaswamy, S. (1980). An accelerated subspace iteration method. *Comput. Methods Appl. Mech., Engrg.* **23**, 313–331.

Bauer, F. L. (1957). Das Verfahren der Treppeniteration und verwandte Verfahren zur Lösung algebraisher Eigenwertprobleme. *Z. Angew. Math. Phys.* **8**, 214–235.

Bauer, F. L. (1958). On modern matrix iteration processes of Bernouilli and Graeffe type. *J. Assoc. Comput. Mach.* **5**, 246–257.

Bauer, F. L., and Fike, C. T. (1960). Norms and exclusion theorems. *Numer. Math.* **2**, 137–141.

Bavely, A. C., and Stewart, G. W. (1979). An algorithm for computing reducing subspaces by block diagonalization. *SIAM J. Numer. Anal.* **16**, 359–367.

Begis, D., and Perronnet, A. (1982). The Club MODULEF, a library of computer procedures for finite element analysis. Rep. INRIA-MODULEF 73, INRIA, Le Chesnay.

Berger, D., Gruber, R., and Troyon, F. (1976). A finite element approach to the computation of the magnetohydrodynamic spectrum of straight noncircular plasma equilibria. *Comput. Phys. Commun.* **11**, 313–323.

Berger, W. A., Miller, H. G., Kreuzer, K. G., and Dreizler, R. M. (1977). An iterative method for calculating low lying eigenvalues of an Hermitian operator. *J. Phys. A.* **10**, 1089–1095.

Berger, W. A., Kreuzer, K. G., and Miller, H. G. (1980). An algorithm for obtaining an optimalized projected Hamiltonian and its ground state. *Z. Physik. A* **298**, 11–12.

Birkhoff, G., and Gulati, S. (1974). Optimal few-point discretization. *SIAM J. Numer. Anal.* **11**, 700–728.

Birkhoff, G., de Boor, C., Swartz, B., and Wendroff, B. (1966). Rayleigh–Ritz approximation by piecewise polynomials. *SIAM J. Numer. Anal.* **3**, 188–203.

Björck, Å. (1967a). Solving linear least squares problems by Gram–Schmidt orthogonalization. *BIT* **7**, 1–21.

Björck, Å. (1967b). Iterative refinement of linear least squares solution: I. *BIT* **7**, 251–278.

Björck, Å. (1968). Iterative refinement of linear least squares solution: II. *BIT* **8**, 8–30.

Björck, Å., and Golub, G. H. (1973). Numerical methods for computing angles between linear subspaces. *Math. Comp.* **27**, 579–594.

Björck, Å., and Plemmons, R. J. (1980). *Large Scale Matrix Problems.* American Elsevier, New York.

Bland, S. (1970). The two-dimensional oscillating airfoil in a wind tunnel in subsonic flow. *SIAM J. Appl. Math.* **18**, 830–848.

Blum, E. K., and Geltner, P. B. (1978). Numerical solution of eigentuple-eigenvector problems in Hilbert space by a gradient method. *Numer. Math.* **31**, 231–246.

Bowdler, H., Martin, R. S., Reinsch, C., and Wilkinson, J. H. (1968). The QR and QL algorithms for symmetric matrices. *Numer. Math.* **11**, 293–306.

Brakhage, H. (1960). Über die numerische Behandlung von Integralgleichungen nach der Quadraturformelmethode. *Numer. Math.* **2**, 183–196.

Brakhage, H. (1961). Zur Fehlerabschätzung für die numerische Eigenwertbestimmung bei Integralgleichungen. *Numer. Math.* **3**, 174–179.

Bramble, J. H., and Osborn, J. E. (1973). Rate of convergence estimates for nonselfadjoint eigenvalue approximations. *Math. Comp.* **27**, 525–549.

Bramble, J. H., and Schatz, A. H. (1970). Rayleigh–Ritz–Galerkin methods for Dirichlet's problem using subspaces without boundary conditions. *Comm. Pure Appl. Math.* **23**, 653–675.

Brandt, A. (1977). Multilevel adaptive solutions to boundary value problems. *Math. Comp.* **31**, 333–390.

Brezinski, C. (1975). Computation of the eigenelements of a matrix by the ε-algorithm. *Linear Algebra Appl.* **11**, 7–20.

Brezzi, F. (1974). On the existence, uniqueness and approximation of saddle-point problems arising from Lagrangian multipliers. *RAIRO Anal. Numér.* **2**, 129–151.

Brezzi, F. (1975). Sur la méthode des éléments finis hybrides pour le problème biharmonique. *Numer. Math.* **24**, 103–131.

Browder, F. E. (1967). Approximation-solvability of nonlinear functional equations in normed linear spaces. *Arch. Rational Mech. Anal.* **26**, 33–42.

Browder, F. E., and Petryshyn, W. V. (1968). The topological degree and Galerkin approximations for noncompact operators in Banach spaces. *Bull. Amer. Math. Soc.* **74**, 641–646.

Bruhn, G., and Wendland, W. L. (1967). Über die näherungweise Lösung von linearen Funktionalgleichungen. In *Funktionalanalysis Approximationstheorie Numerische Mathematik* (L. Collatz, G. Meinardus, and H. Unger, eds.), pp. 136–164. Birkhaeuser, Basel.

Brunner, H. (1981). The application of the variation of constants formulas in the numerical analysis of integral and integro-differential equations. *Utilitas Mathematica.* **19**, 255–290.

Buckner, H. (1952). *Die Praktische Behandlung von Integralgleichungen.* Springer-Verlag, Berlin and New York.

Bulirsçh, R., and Stoer, J. (1966). Asymptotic upper and lower bounds for results of extrapolation methods. *Numer. Math.* **8**, 93–101.

Bunch, J. R., and Nielsen, C. P. (1978). Rank-one modification of the symmetric eigenproblem. *Numer. Math.* **31**, 31–48.

Butscher, W., and Kammer, W. E. (1976). Modification of Davidson's method for the calculation of eigenvalues and eigenvectors of large real symmetric matrices: "Root-homing procedure." *J. Comput. Phys.* **20**, 313–325.

Buurema, H. J. (1970). A geometric proof of convergence for the QR method. Ph.D. Thesis, Univ. of Groningen.

Cachard, F. (1981). Etude numérique de réseaux de file d'attente. Thèse Doct.-Ing., Univ. de Grenoble.

Canosa, J., and Gomes de Oliveira, R. (1970). A new method for the solution of the Schrödinger equation. *J. Comput. Phys.* **5**, 188–207.

Canuto, C. (1978). Eigenvalue approximations by mixed-methods. *RAIRO Anal. Numér.* **12**, 27–50.

Chan, S. P., Feldman, H., and Parlett, B. N. (1977). A program for computing the condition numbers of matrix eigenvalues without computing eigenvectors. *ACM Trans. Math. Software* **3**, 186–203.

Chan, T. F., and Keller, H. B. (1982). Arc-length continuation and multigrid techniques for nonlinear elliptic eigenvalue problems. *SIAM J. Sci. Stat. Comp.* **3**, 173–194.

Chandler, G. A. (1979). Superconvergence of numerical solutions of second kind integral equations. Ph.D. Thesis, Australia Natl. Univ., Canberra.

Chang, P. W., and Finlaysson, B. A. (1978). Orthogonal collocation on finite elements for elliptic equations. *Math. Comput. Simulation* **20**, 83–92.

Chatelin,* F. (1970a). Méthodes d'approximation des valeurs propres d'opérateurs linéaires dans un espace de Banach. I. Critère de stabilité. *C. R. Hebd. Séances Acad. Sci. Ser. A* **271**, 949–952.

Chatelin,* F. (1970b). II. Bornes d'erreur. *C. R. Hebd. Séances Acad. Sci. Ser. A* **271**, 1006–1009.

Chatelin,* F. (1971a). Etude de la stabilité de méthodes d'approximation des éléments propres d'opérateurs linéaires. *C. R. Hebd. Séances Acad. Sci. Ser. A* **272**, 673–675.

Chatelin, F. (1971b). Perturbation d'une matrice hermitienne ou normale. *Numer. Math.* **17**, 318–337.

Chatelin,* F. (1972a). Etude de la continuité du spectre d'un opérateur linéaire. *C. R. Hebd. Séances Acad. Sci. Ser. A* **274**, 328–331.

Chatelin,* F. (1972b). Error bounds in QR and Jacobi algorithms applied to hermitian or normal matrices. In *Information Processing 71*, Vol. 2, pp. 1254–1257. North-Holland Publ., Amsterdam.

Chatelin, F. (1973). Convergence of approximate methods to compute eigenelements of linear operators. *SIAM J. Numer. Anal.* **10**, 939–948.

Chatelin,* F. (1975). La méthode de Galerkin. Ordre de convergence des éléments propres. *C. R. Hebd. Séances Acad. Sci. Ser. A* **278**, 1213–1215.

Chatelin, F. (1978). Numerical computation of the eigenelements of linear integral operators by iterations. *SIAM J. Numer. Anal.* **15**, 1112–1124.

Chatelin, F. (1979). Sur les bornes d'erreur a posteriori pour les éléments propres d'opérateurs linéaires. *Numer. Math.* **32**, 233–246.

Chatelin, F. (1981). The spectral approximation of linear operators with applications to the computation of eigenelements of differential and integral operators. *SIAM Rev.* **23**, 495–522.

Chatelin, F. (1983). A posteriori bounds for the eigenvalues of matrices. *Computing* (to appear).

Chatelin, F., and Lebbar, R. (1981). The iterated projection solution for the Fredholm integral equation of second kind. *J. Austral. Math. Soc. Ser. B* **22**, 443–455 (Special issue on integral equations).

* Original publication under Chatelin-Laborde.

Chatelin, F., and Lebbar, R. (1983). Superconvergence results for the iterated projection method applied to a second kind Fredholm integral equation and eigenvalue problem. *J. Integral Equations* (to appear).

Chatelin, F., and Lemordant, J. (1975). La méthode de Rayleigh–Ritz appliquée à des opérateurs différentiels elliptiques—ordres de convergence des éléments propres. *Numer. Math.* **23**, 215–222.

Chatelin, F., and Lemordant, J. (1978). Error bounds in the approximation of eigenvalues of differential and integral operators. *J. Math. Anal. Appl.* **62**, 257–271.

Chatelin, F., and Miranker, W. L. (1982). Acceleration by aggregation of successive approximation methods. *Linear Algebra Appl.* **43**, 17–47.

Chatelin, F., and Miranker, W. L. (1983). Aggregation/disaggregation for eigenvalue problems. *SIAM J. Numer. Anal.* (submitted).

Chen, N.-F. (1975). The Rayleigh quotient iteration for non-normal matrices. Ph.D. Thesis, Univ. of California, Berkeley.

Cheney, W. (1966). *Introduction to Approximation Theory.* McGraw-Hill, New York.

Cheung, L. M., and Bishop, D. M. (1977). The group-coordinate relaxation method for solving the generalized eigenvalue problem for large real symmetric matrices. *Comput. Phys. Commun.* **12**, 247–250.

Christiansen, S., and Hansen, E. B. (1978). Numerical solution of boundary value problems through integral equations. *Z. Angew. Math. Mech.* **58**, T14–T25.

Christiansen, J., and Russel, R. D. (1978). Error analysis for spline collocation methods with application to knot selection. *Math. Comp.* **32**, 415–419.

Chu, K. W., and Spence, A. (1981). Deferred correction for the integral equation eigenvalue problem. *J. Austral. Math. Soc. Ser. B* **22**, 478–490.

Ciarlet, P. G. (1978). *The Finite Element Method for Elliptic Problems.* North-Holland Publ., Amsterdam.

Ciarlet, P. G. (1982). *Introduction à l'Analyse Numérique Matricielle et à l'Optimisation.* Masson, Paris.

Ciarlet, P. G., and Raviart, P. A. (1972). General Lagrange and Hermite interpolation in $\mathbb{R}^n$ with applications to finite element methods. *Arch. Rational. Mech. Anal.* **46**, 177–199.

Ciarlet, P. G., Schultz, M. H., and Varga, R. S. (1968). Numerical methods of high order accuracy for non-linear boundary value problems. III. Eigenvalue problems. *Numer. Math.* **12**, 120–133.

Cline, A. K., Golub, G. H., and Platzman, G. W. (1976). Calculation of normal modes of oceans using a Lanczos method. In *Sparse Matrix Computations* (J. R. Bunch and D. J. Rose, eds.), pp. 409–426. Academic Press, New York.

Cline, A. K., Moler, C. B., Stewart, G. W., and Wilkinson, J. H. (1979). An estimate for the condition number of a matrix. *SIAM J. Numer. Anal.* **16**, 368–375.

Clint, M., and Jennings, A. (1970). The evaluation of eigenvalues and eigenvectors of real symmetric matrices by simultaneous iterations. *Comput. J.* **13**, 76–80.

Clint, M., and Jennings, A. (1971). A simultaneous iteration method for the unsymmetric eigenvalue problem. *J. Inst. Math. Appl.* **8**, 111–121.

Cochran, J. A. (1972). *The Analysis of Linear Integral Equations.* McGraw-Hill, New York.

Coddington, E. A., and Levinson, N. (1955). *Theory of Ordinary Differential Equations.* McGraw-Hill, New York.

Collatz, L. (1937). Konvergenzbeweis und Fehlerabschätzung für das Differenzenverfahren bei Eigenwertproblemen gewöhnlicher Differentialgleichungen zweiter und vierter Ordnung. *Deutsche Math.* **2**, 189–215.

Collatz, L. (1966a). *The Numerical Treatment of Differential Equations*, 3rd ed. Springer-Verlag, Berlin and New York.

Collatz, L. (1966b). *Functional Analysis and Numerical Mathematics.* Academic Press, New York.

Coope, J. A. R., and Sabo, D. W. (1977). A new approach to the determination of several eigenvectors of a large Hermitian matrix. *J. Comput. Phys.* **23**, 404–424.

Corr, R. B., and Jennings, A. (1973). Implementation of simultaneous iteration for vibration analysis. *Comput. & Structures* **3**, 497–507.

Corr, R. B., and Jennings, A. (1976). A simultaneous iteration algorithm for symmetric eigenvalue problems. *Internat. J. Numer. Methods Engrg.* **10**, 647–663.

Courant, R., and Hilbert, D. (1953). *Methods of Mathematical Physics*, Vols. 1 and 2. Wiley (Interscience), New York.

Crandall, S. H. (1951). Iterative procedures related to relaxation methods for eigenvalue problems. *Proc. Roy. Soc. London Ser. A* **207**, 416–423.

Cruickshank, D. M., and Wright, K. (1978). Computable error bounds for polynomial collocation methods. *SIAM J. Numer. Anal.* **15**, 134–151.

Cubillos, P. O. (1980). On the numerical solution of Fredholm integral equations of the second kind. Ph.D. Thesis, Univ. of Iowa.

Cullum, J. (1978). The simultaneous computation of a few of the algebraically largest and smallest eigenvalues of a large, symmetric, sparse matrix. *BIT* **18**, 265–275.

Cullum, J., and Donath, W. E. (1974). A block Lanczos algorithm for computing the q algebraically largest eigenvalues and a corresponding eigenspace for large, sparse symmetric matrices. *Proc. IEEE Conf. Decision Contr., Phoenix, Ariz.*, 505–509.

Cullum, J., and Willoughby, R. (1977). The equivalence of the Lanczos and the conjugate gradient algorithms. Tech. Rep. RC 6903, IBM Research Center, Yorktown Heights.

Cullum, J., and Willoughby, R. (1978). The Lanczos tridiagonalization and the conjugate gradient with local ε-orthogonality of the Lanczos vectors. Tech. Rep. RC 7152, IBM Research Center, Yorktown Heights.

Cullum, J., and Willoughby, R. A. (1979a). Fast modal analysis of large, sparse but unstructured symmetric matrices. *Proc. IEEE Conf. Decision Contr., San Diego, Calif.*, 45–53.

Cullum, J., and Willoughby, R. A. (1979b). Lanczos and the computation in specified intervals of the spectrum of large, sparse real symmetric matrices. In *Sparse Matrix Proceedings 1978* (I. S. Duff and G. W. Stewart, eds.), pp. 220–225. SIAM, Philadelphia, Pennsylvania.

Cullum, J., and Willoughby, R. A. (1980a). The Lanczos phenomenon—an interpretation based upon conjugate gradient optimization. *Linear Algebra Appl.* **29**, 63–90.

Cullum, J., and Willoughby, R. A. (1980b). Computing eigenvectors (and eigenvalues) of large, symmetric matrices using Lanczos tridiagonalization. *Proc. Numerical Analysis Conf.* (G. A. Watson, ed.), Lecture Notes in Mathematics, Vol. 773, pp. 46–63. Springer-Verlag, Berlin and New York.

Dahlquist, G., and Björck, Å. (1974). *Numerical Methods.* Prentice-Hall, Englewood Cliffs, New Jersey.

Dahmen, W. (1980). On multivariate B-splines. *SIAM J. Numer. Anal.* **17**, 179–191.

d'Almeida, F. (1980). Etude numérique de la stabilité dynamique des modèles macroéconomiques—Logiciel pour MODULECO. Thèse 3ème Cycle, Univ. de Grenoble.

Daniel, J. W., Gragg, W. B., Kaufman, L., and Stewart, G. W. (1976). Reorthogonalization and stable algorithms for updating the Gram–Schmidt QR factorization. *Math. Comp.* **30**, 772–795.

Davidson, E. R. (1975). The iterative calculation of a few of the lowest eigenvalues and corresponding eigenvectors of large real symmetric matrices. *J. Comput. Phys.* **17**, 87–94.

Davis, C. (1963). The rotation of eigenvectors by a perturbation. I. *J. Math. Anal. Appl.* **6**, 159–173.

Davis, C. (1965). The rotation of eigenvectors by a perturbation. II. *J. Math. Anal. Appl.* **11**, 20–27.
Davis, C., and Kahan, W. (1968). The rotation of eigenvectors by a perturbation. III. *SIAM J. Numer. Anal.* **7**, 1–46.
Davis, C., Kahan, W., and Weinberger, H. (1982). Norm preserving dilations and their applications to optimal error bounds. *SIAM J. Numer. Anal.* **19**, 445–469.
Davis, G. J., and Moler, C. B. (1978). Sensitivity of matrix eigenvalues. *Internat. J. Numer. Methods Engrg.* **12**, 1367–1373.
Davis, P. J., and Rabinowitz, P. (1974). *Methods of Numerical Integration*. Academic Press, New York.
Day, W. B. (1974). More bounds for eigenvalues. *J. Math. Anal. Appl.* **46**, 523–532.
Dean, P. (1956). The spectral distribution of a Jacobian matrix. *Proc. Cambridge Phil. Soc.* **52**, 752–755.
Dean, P. (1960). Vibrational spectra of diatomic chains. *Proc. Roy. Soc. Ser. A* **254**, 507–521.
Dean, P. (1964). Vibrations of glass-like disordered chains. *Proc. Phys. Soc.* **84**, 727–744.
Dean, P. (1966). The constrained quantum mechanical harmonic oscillator. *Proc. Phys. Soc.* **62**, 277–286.
Dean, P. (1967). Atomic vibrations in solids. *J. Inst. Math. Appl.* **3**, 98–165.
Dean, P. (1972). The vibrational properties of disordered systems: numerical studies. *Rev. Modern Phys.* **44**, 127–168.
de Boor, C. (1968). On uniform approximation by splines. *J. Approx. Theory* **1**, 219–235.
de Boor, C. (1972). On calculating with *B*-splines. *J. Approx. Theory* **6**, 50–62.
de Boor, C. (1976). A bound on the L_∞-norm of L_2-approximation by splines in terms of a global mesh ratio. *Math. Comp.* **30**, 765–771.
de Boor, C., and Rice, J. R. (1979). An adaptive algorithm for multivariate approximation giving optimal convergence rates. *J. Approx. Theory* **25**, 337–359.
de Boor, C., and Swartz, B. (1973). Collocation at Gaussian points. *SIAM J. Numer. Anal.* **10**, 582–606.
de Boor, C., and Swartz, B. (1977). Comments on the comparison of global methods for linear two-point boundary value problems. *Math. Comp.* **31**, 916–921.
de Boor, C., and Swartz, B. (1980). Collocation approximation to eigenvalues of an ordinary differential equation: The principle of the thing. *Math. Comp.* **35**, 679–694.
de Boor, C., and Swartz, B. (1981a). Collocation approximation to eigenvalues of an ordinary differential equation: numerical illustrations. *Math. Comp.* **36**, 1–19.
de Boor, C., and Swartz, B. (1981b). Local piecewise polynomial projection methods for an ode which give high -order convergence at knots. *Math. Comp.* **36**, 21–33.
Dehesa, J. S. (1978). The asymptotic eigenvalue density of rational Jacobi matrices. I. *J. Phys. A* **9**, 223–226.
Dehesa, J. S. (1980). The eigenvalue density of rational Jacobi matrices. II. *Linear Algebra Appl.* **33**, 41–55.
de Hoog, F. R., and Weiss, R. (1973). Asymptotic expansions for product integration. *Math. Comp.* **27**, 295–306.
Delves, L. M., and Abd-Elal, L. F. (1977). The fast Galerkin algorithm for the solution of linear Fredholm equations, algorithm 97. *Comput. J.* **20**, 374–376.
Delves, L. M., and Walsh, J., eds. (1974). *Numerical Solution of Integral Equations*. Oxford Univ. Press (Clarendon), London and New York.
Delves, L. M., Abd-Elal, L. F., and Hendry, J. A. (1979). A fast Galerkin algorithm for singular kernel equations. *J. Inst. Math. Appl.* **23**, 139–166.
de Pree, J. D., and Higgins, J. A. (1970). Collectively compact sets of linear operators. *Math. Zeitschrift* **115**, 366–370.
de Pree, J. D., and Klein, H. S. (1974). Characterization of collectively compact sets of linear operators. *Pacif. J. Math.* **55**, 45–54.

Descloux, J. (1979). Error bounds for an isolated eigenvalue obtained by the Galerkin method. *J. Appl. Math. Phys.* **30**, 167–176.

Descloux, J. (1981). Essential numerical range of an operator with respect to a coercive form and the approximation of its spectrum by the Galerkin method. *SIAM J. Numer. Anal.* **18**, 1128–1133.

Descloux, J., and Geymonat, G. (1979). On the essential spectrum of an operator relative to the stability of a plasma in toroidal geometry. Rep. Math. Dept., Ecole Polytechn. Féd. de Lausanne.

Descloux, J., and Nassif, N. R. (1982). Stability analysis with error estimates for the approximation of the spectrum of self-adjoint operators on unbounded domains by finite element and finite difference methods. Application to Schrödinger's equation. Rep. Math. Dept., Ecole Polytechn. Féd. de Lausanne.

Descloux, J., and Tolley, M. D. (1981). Approximation of the Poisson problem and of the eigenvalue problem for the Laplace operator by the method of the large singular finite elements. Res. Rep. No. 81-01, Angew. Math., Eidg. Techn. Hochschule Zürich.

Descloux, J., Nassif, N., and Rappaz, J. (1977). Various results on spectral approximation. Rep. Math. Dept., Ecole Polytechn. Féd. de Lausanne.

Descloux, J., Nassif, N., and Rappaz, J. (1978a). On spectral approximation. Part 1. The problem of convergence. *RAIRO Anal. Numér.* **12**, 97–112.

Descloux, J., Nassif, N., and Rappaz, J. (1978b). Part 2, Error estimates for the Galerkin method. *RAIRO Anal. Numér.* **12**, 113–119.

Descloux, J., Luskin, M., and Rappaz, J. (1981). Approximation of the spectrum of closed operators—The determination of normal modes of a rotating basin. *Math. Comp.* **36**, 137–154.

Diaz, J. B., and Metcalf, F. T. (1968). A functional equation for the Rayleigh quotient for eigenvalues, and some applications. *J. Math. Mech.* **17**, 623–630.

Dietrich, G. (1978). On the efficient and accurate solution of the skew-symmetric eigenvalue problem. An arrangement of new and already known algorithmic formulations. *Comput. Methods Appl. Mech. Engrg.* **14**, 209–235.

Dieudonné, J. (1960). *Foundations of Modern Analysis*. Academic Press, New York.

Domb, C., Maradudin, A. A., Montroll, E. W., and Weiss, G. H. (1959a). Vibration frequency of spectra of disordered lattices. I. Moments of the spectra for disordered linear chains. *Phys. Rev.* **115**, 18–24; II. Spectra of disordered one-dimensional lattices. *Phys. Rev.* **115**, 24–34.

Domb, C., Maradudin, A. A., Montroll, E. W., and Weiss, G. H. (1959b). The vibration spectra of disordered lattices. *J. Phys. Chem. Solids* **8**, 419–422.

Dongarra, J. J., Moler, C. B., Bunch, J. R., and Stewart, G. W. (1979). *LINPACK User's Guide*. SIAM, Philadelphia, Pennsylvania.

Dongarra, J. J., Moler, C. B., and Wilkinson, J. H. (1981). Improving the accuracy of computed eigenvalues and eigenvectors. Tech. Rep. ANL 81–43, Argonne Nat. Lab., Illinois.

Douglas, J., and Dupont, T. (1973). Superconvergence for Galerkin methods for the two point boundary problem via local projections. *Numer. Math.* **21**, 270–278.

Douglas, J., and Dupont, T. (1974). Galerkin approximations for the two point boundary problem using continuous piecewise polynomial spaces. *Numer. Math.* **22**, 99–109.

Douglas, J., Dupont, T., and Wheeler, M. F. (1974). An L^∞-estimate and a superconvergence result for a Galerkin method for elliptic equations based on tensor products of piecewise polynomials. *RAIRO Anal. Numér.* **2**, 61–66.

Douglas, J., Dupont, T., and Wahlbin, L. (1975). Optimal L_∞-error estimates for Galerkin approximations to solution of two-point boundary value problems. *Math. Comp.* **29**, 475–483.

Dowson, H. R. (1978). *Spectral Theory of Linear Operators*. Academic Press, New York.

Duff, I. S. (1977). A survey of sparse matrix research. *Proc. IEEE* **65**, 500–535.

Duff, I. S., ed. (1979). *Conjugate Gradient Methods and Similar Techniques*. Tech. Rep. R-9636, AERE Harwell.

Duff, I. S. (1980). Recent developments in the solution of large sparse linear equations. In *Computing Methods in Applied Sciences and Engineering* (R. Glowinski and J.-L. Lions, eds.), pp. 407–426. North-Holland Publ., Amsterdam.

Duff, I. S. (1981). A sparse future. In *Sparse Matrices and Their Uses*. (I. S. Duff, ed.). Academic Press, New York.

Duff, I. S. (1982). A survey of sparse matrix software. Report CSS 21, AERE Harwell. To appear in *Sources and Development of Mathematical Software* (W. R. Cowell, ed.). Prentice-Hall, Englewood Cliffs, New Jersey.

Duff, I. S., and Reid, J. K. (1975). On the reduction of sparse matrices to condensed forms by similarity transformations. *J. Inst. Math. Appl.* **15**, 217–224.

Duff, I. S., and Reid, J. K. (1979). Performance evaluation of codes for sparse matrix problems. In *Performance Evaluation of Numerical Software* (L. D. Fosdick, ed.), pp. 121–135. North-Holland Publ., Amsterdam.

Dumont-Lepage, M. C., Gani, N., Gazeau, J. P., and Ronveaux, A. (1980). Spectrum of potentials $gr^{-(s+2)}$ via SL(2, $\mathbb{R}$) acting on quaternions. *J. Phys. A* **13**, 1243–1257.

Dunford, N., and Schwartz, J. T. (1958). *Linear Operators. Part I: General Theory*. Wiley (Interscience), New York.

Dunford, N., and Schwartz, J. T. (1963). *Linear Operators. Part II: Spectral Theory, Selfadjoint Operators in Hilbert Spaces*. Wiley (Interscience), New York.

Dupont, T. (1976). A unified theory of superconvergence for Galerkin methods for two-point boundary problems. *SIAM J. Numer. Anal.* **13**, 362–368.

Edwards, J. T., Licciardello, D. C., and Thouless, D. J. (1979). Use of the Lanczos method for finding complete sets of eigenvalues of large sparse matrices. *J. Inst. Math. Appl.* **23**, 277–283.

Eggermont, P. P. (1982a). Collocation as a projection method and superconvergence for Volterra integral equations of the first kind. Rep. Math. Dept., Univ. of Delaware.

Eggermont, P. P. (1982b). Collocation for Volterra integral equations of the first kind with iterated kernel. Rep. Math. Dept., Univ. of Delaware.

Einarsson, B. (1979). Bibliography on the evaluation of numerical software. *J. Comput. Appl. Math.* **5**, 145–159.

Elman, H. (1982). Iterative methods for large, sparse nonsymmetric systems of linear equations. Res. Rep. No. 229, Computer Science Dept., Yale Univ., Connecticut.

Erdelyi, I. (1965). An iterative least square algorithm suitable for computing partial eigensystems. *SIAM J. Numer. Anal.* **2**, 421–436.

Erdös, P., and Feldheim, E. (1936). Sur le mode de convergence de l'interpolation de Lagrange. *C. R. Hebd. Séances Acad. Sci.* **203**, 913–915.

Evequoz, H. (1980). Approximation spectrale liée à l'étude de la stabilité magnétohydrodynamique d'un plasma par une méthode d'éléments finis non conformes. Thèse Math. Dept., Ecole Polytechn. Féd. de Lausanne.

Evequoz, H., and Jaccard, Y. (1981). A nonconforming finite element method to compute the spectrum of an operator relative to the stability of a plasma in toroidal geometry. *Numer. Math.* **36**, 455–465.

Faddeev, D. K., and Faddeeva, V. N. (1963). *Computational Methods of Linear Algebra*. Freeman, San Francisco, California.

Fairweather, G. (1978). *Finite Element Galerkin Methods for Differential Equations*. Dekker, New York.

Fan, K. (1949). On a theorem of Weyl concerning eigenvalues of linear transformations. *Proc. Nat. Acad. Sci. USA* **35**, 652–655.

Feler, M. G. (1974). Calculation of eigenvectors of large matrices. *J. Comput. Phys.* **14**, 341–349.

Fenner, T. I., and Loizou, G. (1974). Some new bounds on the condition numbers of optimally scaled matrices. *J. Assoc. Comput. Mach.* **21**, 514–524.

Fichera, G. (1978). *Numerical and Quantitative Analysis.* Pitman, London.

Fiedler, M., and Pták, V. (1964). Estimates and iteration procedures for proper values of almost decomposable matrices. *Czechoslovak Math. J.* **89**, 593–608.

Finlaysson, B. A. (1972). *The Method of Weighted Residuals.* Academic Press, New York.

Fix, G. J. (1972). Effects of quadrature errors in finite element approximation of steady state, eigenvalue and parabolic problems. In *The Mathematical Foundations of the Finite Element Method with Applications to Partial Differential Equations* (A. K. Aziz, ed.), pp. 525–556. Academic Press, New York.

Fix, G. J. (1973). Eigenvalue approximation by the finite element method. *Adv. in Math.* **10**, 300–316.

Fix, G. J. (1976). Hybrid finite element methods. *SIAM Rev.* **18**, 460–484.

Fix, G. J., and Heiberger, R. (1972). An algorithm for the ill-conditioned generalized eigenvalue problem. *SIAM J. Numer. Anal.* **9**, 78–88.

Forsythe, G. E., and Henrici, P. (1960). The cyclic Jacobi method for computing the principal values of a complex matrix. *Trans. Amer. Math. Soc.* **94**, 1–23.

Forsythe, G. E., and Wasow, W. (1960). *Finite Difference Methods for Partial Differential Equations.* Wiley (Interscience), New York.

Fox, L., and Goodwin, E. T. (1953). The numerical solution of non singular linear integral equations. *Philos. Trans. Roy. Soc. London* **245**, 501–534.

Francis, J. G. F. (1961–1962). The QR transformation: a unitary analogue to the LR transformation. Parts I and II. *Comput. J.* **4**, 265–271, 332–345.

Frank, R., and Veberhuber, C. W. (1978). Iterated defect correction for differential equations. Part I: Theoretical results. *Computing* **20**, 207–228.

Fredholm, I. (1900). Sur une nouvelle méthode pour la résolution du problème de Dirichlet. *Kung. Vet.-Akad. Förh. Stockholm* pp. 39–46.

Fredholm, I. (1903). Sur une classe d'équations fonctionnelles. *Acta Math.* **27**, 365–390.

Fromme, J., and Golberg, M. (1978). Unsteady two dimensional airloads acting on oscillating thin airfoils in subsonic ventilated wind tunnels. Rep. NASA Contract, Univ. of Nevada, Las Vegas.

Gantmacher, F. R. (1959). *The Theory of Matrices.* Chelsea, New York.

Garabedian, P. R. (1967). *Partial Differential Equations.* Wiley, New York.

Geier, E. (1977). Eigenvalue and eigenvector calculation by simultaneous vector iteration (in German). *Z. Angew. Math. Mech.* **57**, T279–T281.

Gekeler, E. (1974). On the eigenvectors of a finite-difference approximation to the Sturm–Liouville eigenvalue problem. *Math. Comp.* **28**, 973–979.

Georg, K. (1979). On the convergence of an inverse iteration method for nonlinear elliptic eigenvalue problems. *Numer. Math.* **32**, 69–74.

Geradin, M. (1971). Error bounds for eigenvalue analysis by elimination of variables. *J. Sound Vibr.* **19**, 111–132.

Geradin, M. (1979). On the Lanczos method for solving large structural eigenvalue problems. *Z. Angew. Math. Mech.* **59**, T127–T129.

Gerschgorin, S. (1931). On bounding the eigenvalues of a matrix (in German). *Izv. Akad. Nauk SSSR Ser. Mat.* **1**, 749–754.

Ghemires, T. (1979). Utilisation du quotient de Rayleigh dans la méthode aux différences finies. Thèse 3ème Cycle, Univ. de Grenoble.

Glazman, I., and Liubitch, Y. (1972). *Analyse Linéaire dans les Espaces de Dimensions Finies.* Mir, Moscow.

Godunov, S. K., and Propkopov, G. P. (1970). A method of minimal iterations for evaluating the eigenvalues of an elliptic operator (in Russian). *Ž. Vyčisl. Mat. i Mat. Fiz.* **10**, 1180–1190.

Godunov, S. K., and Ryabenki, V. S. (1964). *Theory of Difference Schemes*. North-Holland Publ., Amsterdam.

Gohberg, I. C., and Krein, M. G. (1960). The basic propositions on defect numbers. root numbers and indices of linear operators. *Amer. Math. Soc. Transl.* **13**, 185–264.

Goldberg, S. (1966). *Unbounded Linear Operators: Theory and Applications*. McGraw-Hill, New York.

Goldberg, S. (1974). Perturbations of semi-Fredholm operators by operators converging to zero compactly. *Proc. Amer. Math. Soc.* **45**, 93–98.

Golub, G. H., and Plemmons, R. J. (1980). Large scale geodetic least squares adjustment by dissection and orthogonal decomposition. *Linear Algebra Appl.* **34**, 3–27.

Golub, G. H., and Wilkinson, J. H. (1976). Ill-conditioned eigensystems and the computation of the Jordan canonical form. *SIAM Rev.* **18**, 578–619.

Golub, G. H., Nash, S., and Van Loan, C. (1979). A Hessenberg–Schur method for the problem $AX + XB = C$. *IEEE Trans. Automat. Control* **AC-24,** 909–913.

Goos, G., and Hartmanis, J., eds. (1978). *EISPACK-Matrix Eigensystem Routines-Guide Extension*, Lecture Notes in Computer Science, Vol. 51. Springer-Verlag, Berlin and New York.

Gordon, R. G. (1969). New method for constructing wave-functions for bound states and scattering. *J. Chem. Phys.* **51**, 14–25.

Gordon, R. G. (1971). Quantum scattering using piecewise analytic solutions. In *Methods in Computational Physics* (B. Adler, S. Fernbach, and M. Rottenberg, eds.), Vol. 10, pp. 81–109. Academic Press, New York.

Gose, G. (1979). The Jacobi method for $Ax = \lambda Bx$. *Z. Angew. Math. Mech.* **59**, 93–101.

Gower, J. C. (1980). A modified Leverrier–Faddeev algorithm for matrices with multiple eigenvalues. *Linear Algebra Appl.* **31**, 61–70.

Graham, I. G. (1980). The numerical solution of Fredholm integral equations of the second kind. Ph.D. Thesis, Univ. of New South Wales, Kensington.

Graham, I. G. (1981). Collocation methods for two dimensional weakly singular integral equations. *J. Austral. Math. Soc. Ser. B* **22**, 460–477.

Graham, I. G. (1982). Galerkin methods for second-kind integral equations with singularities. *Math. Comp.* **39**, 519–533.

Graham, I. G., and Sloan, I. H. (1979). On the compactness of certain integral operators. *J. Math. Anal. Appl.* **68**, 580–594.

Grégoire, J. P., Nedelec, J. C., and Planchard, J. (1976). A method of finding the eigenvalues and eigenfunctions of selfadjoint elliptic operators. *J. Comput. Methods Appl. Mech. Engrg.* **8**, 201–214.

Gregory, R. T., and Karney, D. L. (1969). *A Collection of Matrices for Testing Computational Algorithms*. Wiley (Interscience) New York.

Griffiths, D. F., and Lorenz, J. (1978). An analysis of the Petrov–Galerkin finite element method. *J. Comput. Methods Appl. Mech. Engrg.* **14**, 65–92.

Grigorieff, R. D. (1970a). Die Konvergenz des Rand-und Eigenwertproblems linearer gewöhnlicher Differenzengleichungen. *Numer. Math.* **15**, 15–48.

Grigorieff, R. D. (1970b). Über die Koerzitivität gewöhnlicher Differenzenoperatoren und die Konvergenz des Mehrschrittverfahren. *Numer. Math.* **15**, 196–218.

Grigorieff, R. D. (1972). Über die Fredholm-Alternative bei linearen approximationsregulären Operatoren. *Applicable Anal.* **2**, 217–227.

Grigorieff, R. D. (1973). Zur Theorie linearer approximationsregulärer Operatoren I, II. *Math. Nachr.* **55**, 233–249, 250–263.

Grigorieff, R. D. (1975a). Über diskrete Approximation nichtlinearer Gleichungen 1. Art. *Math. Nachr.* **69**, 253–272.

Grigorieff, R. D. (1975b). Diskrete Approximation von Eigenwertproblemen. I. Qualitative Konvergenz. *Numer. Math.* **24**, 355–374.

Grigorieff, R. D. (1975c). Diskrete Approximation von Eigenwertproblemen. II. Konvergenzordnung. *Numer. Math.* **24**, 415–433.

Grigorieff, R. D. (1975d). Diskrete Approximation von Eigenwertproblemen. III. Asymptotische Entwicklung. *Numer. Math.* **25**, 79–97.

Grigorieff, R. D., and Jeggle, H. (1973). Approximation von Eigenwertproblemen bei nichtlinearer Parameterabhängigkeit. *Manuscripta Math.* **10**, 245–271.

Gruber, R. (1975). HYMMIA—band matrix package for solving eigenvalue problems. *Comput. Phys. Comm.* **10**, 30–41.

Gruber, R. (1978). Finite hybrid elements to compute the ideal magnetohydrodynamic spectrum of an axiasymmetric plasma. *J. Comput. Phys.* **26**, 378–388.

Gupta, K. K. (1973). Eigenproblem solution by a combined Sturm sequence and inverse iteration technique. *Internat. J. Numer. Methods Engrg.* **7**, 17–42.

Gupta, K. K. (1974). Eigenproblem solution of damped structural systems. *Internat. J. Numer. Methods Engrg.* **8**, 877–911.

Gupta, K. K. (1976a). On a finite dynamic element method for free vibration analysis of structures. *Comput. Method Appl. Mech. Engrg.* **9**, 105–120.

Gupta, K. K. (1976b). On a numerical solution of the supersonic panel flutter eigenproblem. *Internat. J. Numer. Methods Engrg.* **10**, 637–645.

Gupta, K. K. (1978a). On a numerical solution of the plastic buckling problem of structures. *Internat. J. Numer. Methods Engrg.* **12**, 941–947.

Gupta, K. K. (1978b). Development of a finite dynamic element for free vibration analysis of two-dimensional structures. *Internat. J. Numer. Methods Engrg.* **12**, 1311–1327.

Güssman, B. (1980). L_∞-bounds of L_2-projections on splines. In *Quantitative Approximation* (R. A. de Vore and K. Scherer, eds.), 153–162. Academic Press, New York.

Hackbusch, W. (1979). On the computation of approximate eigenvalue and eigenfunctions of elliptic operators by means of a multigrid method. *SIAM J. Numer. Anal.* **16**, 201–215.

Hackbusch, W. (1980). Multigrid solutions to linear and nonlinear eigenvalue problems for integral and differential equations. Rep. 80–3, Math. Inst., Univ. zu Köln.

Hackbusch, W. (1981a). On the convergence of multigrid iterations. *Beitr. Numer. Math.* **9**, 213–239.

Hackbusch, W. (1981b). Error analysis of the nonlinear multigrid method of the second kind. *Appl. Matem.* **26**, 18–29.

Hacksbusch, W., and Hofmann, G. (1980). Results of the eigenvalue problem for the plate equation. *Z. Angew. Math. Phys.* **31**, 730–739.

Hadamard, J. (1908). Mémoire sur le problème d'analyse relatif à l'équilibre des plaques élastiques encastrées. *Memoires savants étrangers, Acad. Sci. Paris* **33**, 1–128.

Hämmerlin, G. (1976). Zur numerischen Behandlung von homogenen Fredholmschen Integralgelichungen 2. Art mit Splines. In *Spline Functions Karlsruhe 1975* (K. Böhmer, G. Meinardus, and W. Schempp, eds.) Lecture Notes in Mathematics, Vol. 501, pp. 92–98. Springer-Verlag, Berlin and New York.

Hämmerlin, G., and Schumaker, L. L. (1979). Error bounds for the approximation of Green's kernels by splines. *Numer. Math.* **33**, 17–22.

Hämmerlin, G., and Schumaker, L. L. (1980). Procedures for kernel approximation and solution of Fredholm integral equations of the second kind. *Numer. Math.* **34**, 125–141.

Hahn, H. (1948). *Reelle Funktionen: Punktfunktionen.* Chelsea, New York.

Hairer, E. (1978). On the order of iterated defect correction: An algebraic proof. *Numer. Math.* **29**, 409–443.

Hall, G. G. (1977). On the eigenvalues of molecular graphs. *Molecular Phys.* **33**, 551–557.

Halmos, P. H. (1950). Normal dilations and extensions of operators. *Summa Brasil. Math.* **2**, 125–134.

Hanson, R. J. (1972). Integral equations of immunology. *Comm. ACM* **15**, 883–890. *Harwell Subroutine Library Manual.* Harwell, Oxfordshire.

Hashimoto, M. (1970). A method of solving large matrix equations reduced from Fredholm equations of the second kind. *J. Assoc. Comput. Mach.* **17**, 629–636.

Hemker, P. W. (1975). Galerkin's method and Lobatto points. Rep. NW 24/75, Stichting Math. Cent., Amsterdam.

Hemker, P. W. (1982a). The defect correction principle. BAIL II short course lecture notes.

Hemker, P. W. (1982b). Extensions of the defect correction principle. BAIL II short course lecture notes.

Hemker, P. W., and Schippers, H. (1981). Multigrid methods for the solution of Fredholm equations of the second kind. *Math. Comp.* **36**, 215–232.

Hendershott, M. C. (1973). Ocean tides. *Trans. Amer. Geophys. Union* **54**, 76–86.

Henrici, P. (1962). Bounds for iterates, inverses, spectral variation and fields of values of non-normal matrices. *Numer. Math.* **4**, 24–40.

Henrici, P. (1963). Bounds for eigenvalues of certain tridiagonal matrices. *SIAM J. Appl. Math.* **11**, 289–290; **12**, 497.

Herbold, R. J., Schultz, M. H., and Varga, R. S. (1969). The effect of quadrature errors in the numerical solution of boundary value problems by variational techniques. *Aequationes Math.* **3**, 247–270.

Herman, H. (1975). Extension of Lanczos' method of fundamental eigenvalue approximation. *Trans. ASME Ser. E, J. Appl. Mech.* **42**, 484–489.

Hestenes, M. R., and Karush, W. (1951). A method of gradients for the calculation of the characteristic roots and vectors of a real symmetric matrix. *J. Res. Nat. Bur. Standards Sect. B.* **47**, 471–478.

Hestenes, M. R., and Stein, M. L. (1973). The solution of linear equations by minimization. *J. Optim. Theory Appl.* **11**, 335–359.

Hestenes, M. R., and Stiefel, E. (1952). Method of conjugate gradients for solving linear systems. *J. Res. Nat. Bur. Standards Sect. B.* **49**, 409–436.

Hirai, I., Yoshimura, T., and Takamura, K. (1973). On a direct eigenvalue analysis for locally modified structures. *Internat. J. Numer. Methods Engrg.* **6**, 441–456.

Hodges, D. H. (1979a). A theoretical technique for analyzing aeroelastic stability of bearingless rotors. *AIAA J.* **17**, 400–407.

Hodges, D. H. (1979b). Vibration and responses of nonuniform rotating beams with discontinuities. *AIAA–ASME Dyn. Mater. Conf. St. Louis, Missouri*, 29–38.

Hodges, D. H. (1979c). Aeromechanical stability analysis for bearingless rotor helicopters. *J. Amer. Helicopter Soc.* **24**, 2–9.

Hodges, D. H. (1980). Torsion of pretwisted beams due to axial loading. *J. Appl. Mech.* **47**, 393–397.

Householder, A. S. (1964). *The Theory of Matrices in Numerical Analysis.* Ginn (Blaisdell), Boston, Massachusetts.

Householder, A. S., and Bauer, F. L. (1960). On certain iterative methods for solving linear systems. *Numer. Math.* **2**, 55–59.

Houstis, E. N., and Papatheodorou, T. S. (1978). A collocation method for Fredholm integral equation of the 2nd kind. *Math. Comp.* **32**, 159–173.

Houstis, E. N., Lynch, R. E., Rice, J. R., and Papatheodorov, T. S. (1978). Evaluation of numerical methods for elliptic partial differential equations. *J. Comput. Phys.* **27**, 323–350.

Howson, W. P., and Williams, F. W. (1973). Natural frequencies of frames with axially loaded Timoshenko members. *J. Sound Vibration* **26**, 503–515.

Hsiao, G. C., and Wendland, W. L. (1977). A finite element method for some integral equations of the first kind. *J. Math. Anal. Appl.* **58**, 449–481.

Hsiao, G. C., and Wendland, W. L. (1981). The Aubin–Nistche lemma for integral equations. *J. Integral Equations* **3**, 299–315.

Huang, L. (1978). Some perturbation problems for generalized eigenvalues (in Chinese). *Beijing Daxue Xuebao* **4**, 20–25.

Hubbard, B. E. (1962). Bounds for eigenvalues of the Sturm–Liouville problem by finite difference methods. *Arch. Rational Mech. Anal.* **28**, 171–179.

Hughes, T. J. R. (1976). Reduction scheme for some structural eigenvalue problems by a variational theorem. *Internat. J. Numer. Methods Engrg.* **10**, 845–852.

Huseyin, K. (1976). Standard forms of eigenvalue problems associated with gyroscopic systems. *J. Sound Vibration* **45**, 29–37.

Huseyin, K., and Roorda, J. (1971). The loading-frequency relationship in multiple eigenvalue problems. *Trans. ASME Ser. E, J. Appl. Mech.* **38**, 1007–1011.

Ikebe, Y. (1972). The Galerkin method for the numerical solution of Fredholm integral equations of the second kind. *SIAM Rev.* **14**, 465–491.

IMSL Library 3 Reference Manual (1975). Internat. Math. Statist. Libraries, Houston, Texas.

Irons, B. M. (1963). Eigenvalue economizers in vibration problems. *J. Roy. Aero. Soc.* **67**, 526–528.

Isaacson, E., and Keller, H. (1966). *Analysis of Numerical Methods.* Wiley, New York.

Ishihara, K. (1977). Convergence of the finite element method applied to the eigenvalue problem $\Delta u + \lambda u = 0$. *Publ. Res. Inst. Math. Sci.* **13**, 47–60.

Ishihara, K. (1978). A mixed finite element method for the biharmonic eigenvalue problems of plate bending. *Publ. Res. Inst. Math. Sci.* **14**, 399–414.

Ishihara, K. (1979). On the mixed finite element approximation for the buckling of plates. *Numer. Math.* **33**, 195–210.

Ivanov, V. V. (1976). *The Theory of Approximate Methods and Their Application to the Numerical Solution of Singular Integral Equations.* Noordhoff, Groningen, The Netherlands.

Iwai, Z., and Kubo, Y. (1979). Determination of eigenvalues in Marshall's model reduction. *Internat. J. Control* **30**, 823–836.

Ixaru, L. G. (1972). The error analysis of the algebraic method for solving the Schrödinger equation. *J. Comput. Phys.* **9**, 159–163.

Jaccard, Y. (1980). Approximation spectrale par la méthode des éléments finis conformes d'une classe d'opérateurs non compacts et partiellement réguliers. Thèse Math. Dept., Ecole Polytechn. Féd. de Lausanne.

Jacobi, C. G. J. (1846). Über ein leichtes Verfahren die in der Theorie der Säcularstörungen vorkommenden Gleichungen numerisch aufzulösen. *Crelle J. Reine Angew. Math.* **30**, 51–94.

James, R. L. (1971). Uniform convergence of positive operators. *Math. Z.* **120**, 124–142.

Jeggle, H. (1972). Über die Approximation non linearen Gleichungen zweiter Art und Eigenwertprobleme in Banach–Räumen. *Math. Z.* **124**, 319–342.

Jeggle, H., and Wendland, W. L. (1977). On the discrete approximation of eigenvalue problems with holomorphic parameter dependence. *Proc. Roy. Soc. Edinburgh Sect. A* **78**, 1–29.

Jennings, A. (1973). Mass condensation and simultaneous iteration for vibration problems. *Internat. J. Num. Methods Engrg.* **6**, 543–552.

Jennings, A. (1977a). *Matrix Computations for Engineers and Scientists*, Wiley, New York.

Jennings, A. (1977b). Matrices. Ancient and modern. *Bull. Inst. Math. Appl.* **13**, 117–123.

Jennings, A. (1977c). Influence of the eigenvalue spectrum on the convergence rate of the conjugate gradient method. *J. Inst. Math. Appl.* **20**, 61–72.

Jennings, A. (1980). Eigenvalue methods for vibration analysis. *Shock Vibration Digest* **12**, 3–19.

Jennings, A. (1981). Eigenvalue methods and the analysis of structural vibrations. In *Sparse Matrices and Their Uses* (I. S. Duff, ed.), pp. 109–138. Academic Press, New York.

Jennings, A., and Agar, T. J. S. (1979). Progressive simultaneous inverse iteration for symmetric eigenvalue problems. CE Report, Queen's Univ., Belfast.

Jennings, A., and Orr, D. R. L. (1971). Application of the simultaneous iteration method to undamped vibration problems. *Internat. J. Numer. Methods Engrg.* **3**, 13–24.

Jennings, A., and Stewart, W. J. (1975). Simultaneous iteration for partial eigensolution of real matrices. *J. Inst. Math. Appl.* **15**, 351–361.

Jennings, A., Halliday, J., and Cole, M. J. (1978). Solution of linear generalized eigenvalue problems containing singular matrices. *J. Inst. Math. Appl.* **22**, 401–410.

Jensen, P. S. (1972). The solution of large eigenproblems by sectioning. *SIAM J. Numer. Anal.* **9**, 534–545.

Jenssen, O. (1972). Eigenfunctions and spectrum of the hard-sphere collision operator. *Phys. Norveg.* **6**, 180–191.

Jespersen, D. (1978). Ritz–Galerkin methods for singular boundary value problems. *SIAM J. Numer. Anal.* **15**, 813–834.

Johnsen, T. L. (1973). On the computation of natural modes of an unsupported vibrating structure by simultaneous iterations. *Comput. Methods Appl. Mech. Engrg.* **2**, 305–322.

Johnsen, T. L. (1978). A numerical method for eigenreduction of nonsymmetric real matrices. *Comput. & Struct.* **8**, 399–402.

Kagiwada, H. H., Kalaba, R. E., and Vereeke, B. J. (1968). The invariant imbedding numerical method for Fredholm integral equations with degenerate kernels. *J. Approx. Theory* **1**, 355–364.

Kågström, B. (1977a). Methods for the numerical computation of matrix functions and the treatment of ill conditioned eigenvalue problems. Report UMINF-59.77, Univ. of Umeå.

Kågström, B. (1977b). Bounds and perturbation bounds for the matrix exponential. *BIT* **17**, 39–57.

Kågström, B. (1981). How to compute the Jordan normal form—the choice between similarity transformations and methods using chain relations. Report UMINF-91.81, Univ. of Umeå.

Kågström, B., and Ruhe, A. (1980a). An algorithm for numerical computation of the Jordan normal form of a complex matrix. *ACM Trans. Math. Software* **6**, 398–421.

Kågström, B., and Ruhe, A. (1980b). Algorithm 560 JNF, an algorithm for numerical computation of the Jordan normal form of a complex matrix. *ACM Trans. Math. Software* **6**, 437–443.

Kahan, W., and Parlett, B. N. (1976). How far should you go with the Lanczos process? In *Sparse Matrix Computations* (J. R. Bunch and D. J. Rose, eds.), pp. 131–144. Academic Press, New York.

Kahan, W., Parlett, B. N., and Jiang, E. (1981). Residual bounds on approximate eigensystems of nonnormal matrices. *SIAM J. Numer. Anal.* **19**, 470–484.

Kaniel, S. (1966). Estimates for some computational techniques in linear algebra. *Math. Comp.* **20**, 369–378.

Kantorovitch, L. V. (1934). On a new method of approximate solution of partial differential equations (in Russian). *Dokl. Akad. Nauk SSSR* **4**, 532–536.

Kantorovitch, L. V. (1948). Functional analysis and applied mathematics (in Russian). *Usp. Mat. Nauk* **3**, 89–185.

Kantorovitch, L. V., and Akilov, G. P. (1964). *Functional Analysis in Normed Spaces.* Pergamon, Oxford.

Kantorovitch, L. V., and Krylov, V. I. (1955). *Approximate Methods of Higher Analysis.* Wiley (Interscience), New York.

Karma, O. O. (1971). Asymptotic error estimations for characteristic values of holomorphic Fredholm operator functions (in Russian). *Ž. Vyčisl. Mat. i Mat. Fiz.* **11**, 559–568.

Karpel, M., and Newman, M. (1975). Accelerated convergence for vibration modes using the substructure coupling method and fictitious coupling masses. *Israel J. Tech.* **13**, 55–62.

Karpilovskaia, E. B. (1953). On the convergence of an interpolation method for ordinary differential equations (in Russian). *Usp. Mat. Nauk* **8**, 111–118.

Karpilovskaia, E. B. (1963). On the convergence of the collocation method (in Russian). *Dokl. Akad. Nauk SSSR* **151**, 766–769.

Karpilovskaia, E. B. (1965). On the convergence of the subregion method for integro-differential equations (in Russian). *Ž. Vyčisl. Mat. i Mat. Fiz.* **5**, 124–132.

Kato, T. (1949). On the upper and lower bounds of eigenvalues. *J. Phys. Soc. Japan* **4**, 334–339.

Kato, T. (1958). Perturbation theory for nullity, deficiency and other quantities of linear operators. *J. Analyse Math.* **6**, 261–322.

Kato, T. (1976). *Perturbation Theory for Linear Operators*, 2nd ed. Springer-Verlag, Berlin and New York.

Kaucher, E., and Rump, S. M. (1982). E-methods for fixed point equations $f(x) = x$. *Computing* **28**, 31–42.

Keldyš, M. V. (1951). On the characteristic values and characteristic functions of certain classes of nonself-adjoint equations (in Russian). *Dokl. Akad. Nauk SSSR* **77**, 11–14.

Keller, H. B. (1965). On the accuracy of finite difference approximations to the eigenvalues of differential and integral operators. *Numer. Math.* **7**, 412–419.

Keller, H. B. (1975). Approximation methods for nonlinear problems with application to two-point boundary value problems. *Math. Comp.* **29**, 464–474.

Kikuchi, F. (1980a). On a mixed finite element scheme for linear buckling analysis of plates. In *Computational Methods in Nonlinear Mechanics* (J. T. Oden, ed.), pp. 289–302. North-Holland Publ., Amsterdam.

Kikuchi, F. (1980b). Numerical analysis of a mixed finite element method for plate buckling problems. ISAS Rep. No. 584, Univ. of Tokyo, **45**, 165–190.

Kleinert, P. (1979). Cluster approximation for the spectral density of mixed diatomic systems. *Phys. Status Solidi (B)* **91**, 455–465.

Kolata, W. G. (1978). Approximation in variationally posed eigenvalue problems. *Numer. Math.* **29**, 159–171.

Kolata, W. G. (1979). Eigenvalue approximation by the finite element method: the method of Lagrange multipliers. *Math. Comp.* **33**, 63–76.

Kolata, W. G., and Osborn, J. E. (1979). Nonselfadjoint spectral approximation and the finite element method. In *Functional Analysis Methods in Numerical Analysis* (M. Z. Nashed, ed.), Lecture Notes in Mathematics, Vol. 101, pp. 115–133. Springer-Verlag, Berlin and New York.

Kondrashev, V. I. (1945). On some properties of functions of spaces L_p (in Russian). *Dokl. Akad. Nauk SSSR* **48**, 563–566.

Krasnoselskii, M. A., Vainikko, G. M., Zabreiko, P. P., Rutitskii, Ya. B., and Stetsenko, V. Ya. (1972). *Approximate Solutions of Operator Equations.* Wolters-Noordhoff, Groningen, The Netherlands.

Krasnoselskii, M. A., Zabreiko, P. P., Pustilnik, E. I., and Sobolevskiy, P. E. (1976). *Integral Operators in Spaces of Summable Functions.* Noordhoff Int., Leyden, The Netherlands.

Kreiss, H. O. (1972). Difference approximation for boundary and eigenvalue problems for ordinary differential equations. *Math. Comp.* **26**, 605–624.

Kreuzer, K. G., Miller, H. G., Dreizler, R. M., and Berger, W. A. (1980). Extension of an iterative method to obtain low-lying eigenstates of unbounded Hermitian operators. *J. Phys. A.* **13**, 2645–2652.

Kreuzer, K. G., Miller, H. G., and Berger, W. A. (1981). The Lanczos algorithm for self-adjoint operators. *Phys. Lett. A.*, 429–432.

Krylov, A. N. (1931). On the numerical solution of equations which in technical questions are determined by the frequency of small vibrations of material systems (in Russian). *Izv. Akad. Nauk SSSR Otd. Mat. Estest.* **1**, 491–539.

Krylov, V. I. (1962). *Approximate Calculation of Integrals.* Macmillan, New York.

Kublanovskaya, V. N. (1961). On some algorithms for the solution of the complete eigenvalue problem (in Russian). *Ž. Vyčisl. Mat. i Mat. Fiz.* **1**, 555–570.

Kulkarni, R. P. (1982). Convergence and computation of approximate eigenelements. Ph.D. Thesis, Math. Dept., Indian Inst. Technology, Bombay.

Kulkarni, R. P., and Limaye, B. V. (1981). On error bounds in strong approximations for eigenvalue problems. *J. Austral. Math. Soc. Ser. B* **22**, 270–283.

Kulkarni, R. P., and Limaye, B. V. (1982). On Chatelin's algorithm for the computation of the eigenelements by iterations. Rep. Math. Dept., Indian Inst. Technol., Bombay.

Kulkarni, R. P., and Limaye, B. V. (1983a). On the steps of convergence of approximate eigenvectors in the Rayleigh–Schrödinger series. *Numer. Math.* (to appear).

Kulkarni, R. P., and Limaye, B. V. (1983b). Geometric and semi-geometric approximation of spectral projections. *J. Math. Anal. Appl.* (to appear).

Kulisch, U., and Miranker, W. L. (1981). *Computer Arithmetic in Theory and Practice.* Academic Press, New York.

Kuratowski, C. (1961). *Topologie*, 3rd ed. Polska Akad. Nauk, Warsaw.

Kuttler, J. R. (1972). Remarks on a Stekloff eigenvalue problem. *SIAM J. Numer. Anal.* **9**, 1–5.

Kuttler, J. R. (1979). Dirichlet eigenvalues. *SIAM J. Numer. Anal.* **16**, 332–338.

Kuttler, J. R., and Sigillito, V. G. (1968). Inequalities for membrane and Stekloff eigenvalues. *J. Math. Anal. Appl.* **23**, 148–160.

Laasonen, P. (1959). A Ritz method for simultaneous determination of several eigenvalues and eigenvectors of a big matrix. *Ann. Acad. Sci. Fenn. Ser. A* I *Math.* **265**, 3–16.

Lanczos, C. (1950). An iterative method for the solution of the eigenvalue problem of linear differential and integral operators. *J. Res. Nat. Bur. Standards Sect. B* **45**, 255–282.

Lanczos, C. (1952). Solution of systems of linear equations by minimized iterations. *J. Res. Nat. Bur. Standards Sect. B* **49**, 33–53.

Lanczos, C. (1961). *Linear Differential Operators.* Van Nostrand, New York.

Laurent, P. J. (1972). *Approximation et Optimisation.* Hermann, Paris.

Lebbar, R. (1981a). Sur les propriétés de superconvergence des solutions approchées de certaines équations intégrales et différentielles. Thèse 3ème Cycle, Univ. de Grenoble.

Lebbar, R. (1981b). Superconvergence at the knots for the generalized eigenvectors of differential and integral operators. R.R. IMAG No. 272, Univ. de Grenoble.

Lebbar, R. (1982). Superconvergence with adaptive mesh for weakly singular equations. Unpublished manuscript, Univ. de Grenoble.

Lebedev, V. I. (1977). An iterative method with Chebyshev parameters for finding the maximum eigenvalue and corresponding eigenfunction. *U.S.S.R. Computational Math. and Math. Phys.* **17**, 92–101.

Lee, J. W., and Prenter, P. M. (1978). An analysis of the numerical solution of Fredholm integral equations of the first kind. *Numer. Math.* **30**, 1–23.

Lehmann, N. J. (1966). On optimal eigenvalue localization in the solution of symmetric matrix problems. *Numer. Math.* **8**, 42–55.

Lemordant, J. (1977). Localisation d'un groupe de valeurs propres. Note d'étude LA No. 46, Centre de Calcul Scientifique de l'Armement, CELAR, Bruz.

Lemordant, J. (1979). Etude de l'ensemble des schémas d'approximation spectrale stable d'un opérateur compact. Sém. Anal. Numér. No. 319, Univ. de Grenoble.

Lemordant, J. (1980). Localisation de valeurs propres et calcul de sous-espaces invariants. Thèse d'État, Univ. de Grenoble.

Lerou, R. J. L., and Dekker, H. (1981). Exact computation of high-order perturbational eigensolutions and its application to the analysis of a spectral degeneracy in a bistable diffusion process. *Phys. Lett. A* **83**, 371–375.

Lewis, J. G. (1977). Algorithms for sparse matrix eigenvalue problems. Ph.D. Thesis, Report CS-77-595, Computer Sci. Dept., Stanford Univ., Stanford, California.

Lewis, J. G., and Grimes, R. G. (1981). Practical Lanczos algorithms for solving structural engineering problems. In *Sparse Matrices and Their Uses* (I. S. Duff, ed.), pp. 349–355. Academic Press, New York.

Lin Qun (1979). Some problems about the approximate solution for operator equations (in Chinese). *Acta Math. Sinica* **22**, 219–230.

Lin Qun (1980). How to increase the accuracy of lower order elements in nonlinear finite element methods. In *Computing Methods in Applied Sciences and Engineering* (R. Glowinski and J.-L. Lions, eds.), pp. 41–47. North-Holland Publ., Amsterdam.

Lin Qun (1981a). Iterative corrections for nonlinear eigenvalue problem of operator equations. Res. Rep. IMS-1, Institute of Mathematical Sciences, Chengdu Branch of Academia Sinica.

Lin Qun (1981b). Iterative corrections for nonlinear operator equations with applications to difference method. *J. Sys. Sci. Math. Sci.* **1**, 139–146.

Lin Qun (1981c). Deferred corrections for equations of the second kind. *J. Austral. Math. Soc. Ser. B* **22**, 456–459.

Lin Qun (1982a). Iterative refinement of finite element approximations for elliptic problems. *RAIRO Anal. Numér.* **16**, 39–47.

Lin Qun (1982b). Personal communication.

Lin Qun, and Jang Li-Shang (1979). Investigation of the system $\Delta u_i = \sum_{j=1}^{n} u_j (\partial u_i/\partial x_j) + f_i$, $i = 1, \ldots, n$. Tech. Rep., Inst. Math., Acad. Sinica, Beijing.

Lin Qun, and Liu Jiaquan (1980). Extrapolation method for Fredholm integral equations with non-smooth kernels. *Numer. Math.* **35**, 459–464.

Linz, P. (1970). On the numerical computation of eigenvalues and eigenvectors of symmetric integral equations. *Math. Comp.* **24**, 905–909.

Linz, P. (1972). Error estimates for the computation of eigenvalues of selfadjoint operators. *BIT* **12**, 528–533.

Linz, P. (1977). A general theory for the approximate solution of operator equations of the 2nd kind. *SIAM J. Numer. Anal.* **14**, 543–554.

Linz, P. (1979). *Theoretical Numerical Analysis.* Wiley, New York.

Lions, J.-L., and Magenes, E. (1968). *Problèmes aux Limites Non Homogènes et Applications*, Vol. 1. Dunod, Paris.

Lo, W. S. (1973). Spectral approximation theorems for bounded linear operators. *Bull. Austral. Math. Soc.* **8**, 279–287.

Locker, J., and Prenter, P. M. (1978). Optimal L^2- and L^∞-error estimates for continuous and discrete least squares methods for boundary value problems. *SIAM J. Numer. Anal.* **15**, 1151–1160.

Locker, J., and Prenter, P. M. (1979). On least squares methods for linear two-point boundary value problems. In *Functional Analysis Methods in Numerical Analysis* (M. Z. Nashed, ed.), Lecture Notes in Mathematics, Vol. 701, pp. 149–168. Springer-Verlag, Berlin and New York.

Locker, J., and Prenter, P. M. (1980). Regularization with differential operators. I. General theory. *J. Math. Anal. Appl.* **74**, 504–529.

Locker, J., and Prenter, P. M. (1981). Regularization with differential operators. II. Weak least squares finite element solutions to first kind integral equations. *SIAM J. Numer. Anal.* **17**, 247–267.

Locker, J., and Prenter, P. M. (1983). Representors and superconvergence of least squares finite element approximates. *Numer. Funct. Anal. Optim.* (to appear).

Longsine, D. E., and McCormick, S. F. (1980). Simultaneous Rayleigh quotient minimization methods for $Ax = \lambda Bx$. *Linear Algebra Appl.* **34**, 195–234.

Lorentz, G. G. (1966). *Approximation of Functions.* Holt, New York.

Luskin, M. (1979). Convergence of a finite element method for the approximation of normal modes of the oceans. *Math. Comp.* **33**, 493–519.

Luthey, Z. A. (1974). Piecewise analytical solutions method for the radial Schrödinger equation. Ph.D. Thesis, Harvard Univ., Cambridge, Massachusetts.

McCormick, S. F. (1980). Mesh refinement for integral equations. In *Numerical Treatment of Integral Equations* (J. Albrecht and L. Collatz, eds.), pp. 183–190. Birkhaeuser, Basel.

McCormick, S. F. (1981). A mesh refinement method for $Ax = \lambda Bx$. *Math. Comp.* **36**, 485–498.

McCormick, S. F., and Noe, T. (1977). Simultaneous iteration for the matrix eigenvalue problem. *Linear Algebra Appl.* **16**, 43–56.

McLaurin, J. W. (1974). General coupled equation approach for solving the biharmonic boundary value problem. *SIAM J. Numer. Anal.* **11**, 14–33.

Marek, I. (1971). Approximations of the principal eigenelements in K-positive nonselfadjoint eigenvalue problems, *Math. Systems Theory* **5**, 204–215.

Marchuk, G. I., and Agoškov, V. I. (1977). The selection of coordinate functions in the generalized Bubnov–Galerkin method (in Russian). *Dokl. Akad. Nauk SSSR* **232**, 1253–1256.

Masur, E. F. (1973). Bounds and error estimates in structural eigenvalue problems. *J. Struct. Mech.* **1**, 417–438.

Mead, D. J., and Parthan, S. (1979). Free wave propagation in two-dimensional plates. *J. Sound Vibr.* **64**, 325–348.

Meirovitch, L. (1974). A new method of solution of the eigenvalue problem for gyroscopic systems. *AIAA J.* **12**, 1337–1342.

Mercier, B., and Rappaz, J. (1978). Eigenvalue approximation via nonconforming and hybrid finite element methods. Rep. No. 33, Math. Appl., École Polytechnique, Palaiseau.

Mercier, B., Osborn, J., Rappaz, J., and Raviart, P.-A. (1981). Eigenvalue approximation by mixed and hybrid methods. *Math. Comp.* **36**, 427–453.

Mérigot, M. (1974). Régularité des fonctions propres du laplacien dans un cône. *C. R. Hebd. Séances Acad. Sci. Ser. A* **279**, 503–505.

Miesch, A. (1980). Scaling variables and interpretation of eigenvalues in principal component analysis of geologic data. *J. Internat. Assoc. Math. Geol.* **12**, 523–538.

Mika, J. (1971). Fundamental eigenvalues of the linear transport equation. *J. Quant. Spectrosc. Radiat. Transfer* **11**, 879–891.

Mikhlin, S. G. (1964). *Integral Equations and Their Applications to Some Problems of Mechanics, Mathematical Physics and Engineering*, 2nd ed. Pergamon, Oxford.

Mikhlin, S. G. (1971). *The Numerical Performance of Variational Methods.* Wolters-Noordhoff, Groningen, The Netherlands.

Mikhlin, S. G., and Smolitskii, K. L. (1967). *Approximate Methods for Solutions of Differential and Integral Equations.* Amer. Elsevier, New York.

Miller, H. G., and Berger, W. A. (1979). An investigation of pseudoconvergence in an iterative method for calculating the low-lying eigenvectors of a Hermitian matrix. *J. Phys. A.* **12**, 1693–1698.

Mills, W. H. (1979a). The resolvent stability for spectral convergence with application to the finite element approximation of noncompact operators. *SIAM J. Numer. Anal.* **16**, 695–703.

Mills, W. H. (1979b). Optimal error estimates for the finite element spectral approximation of noncompact operators. *SIAM J. Numer. Anal.* **16**, 704–718.

Mills, W. H. (1979c). Finite element error estimates for singular variational eigenvalue problems. Res. Rep., Math. Dept., Pennsylvania State Univ., University Park.

Mills, W. H. (1980). Convergence and errors for projective finite element approximation of variational eigenvalue problems. Res. Rep. No. 8015, Math. Dept., Pennsylvania State Univ., University Park.

Mindlin, R. D. (1956). Simple modes of vibrations of crystals. *J. Appl. Phys.* **27**, 1462–1466.

Miranker, W. L. (1971). Galerkin approximations and the optimization of difference schemes for boundary value problems. *SIAM J. Numer. Anal.* **8**, 486–496.

Miyoshi, T. (1976). A mixed finite element method for the solution of the von Kármán equations. *Numer. Math.* **26**, 255–269.

Mock, M. S. (1976). Projection methods with different trial and test spaces. *Math. Comp.* **30**, 400–416.

Moiseiwitsch, B. L. (1977). *Integral Equations*. Longman, London.

Moler, C. B., and Stewart, G. W. (1973). An algorithm for generalized matrix eigenvalue problems. *SIAM J. Numer. Anal.* **10**, 241–256.

Moore, E. H. (1919–1920). On the reciprocal of the generalized algebraic matrix (abstract). *Bull. Amer. Math. Soc.* **26**, 394.

Moro, G., and Freed, J. H. (1981). Calculation of ESR spectra and related Fokker–Planck forms by the use of the Lanczos algorithm. *J. Chem. Phys.* **74**, 3757–3773.

Moszyński, K. (1980). On the approximation of the spectral density function of a self-adjoint operator. Rep. No. 206, Inst. Math., Polish Academy of Sciences.

Muda, Y. (1973). A new relaxation method for obtaining the lowest eigenvalue and eigenvector of a matrix equation. *Internat. J. Numer. Methods Engrg.* **6**, 511–519.

Munteanu, M. J., and Schumaker, L. L. (1973). Direct and inverse theorems for multidimensional spline approximation. *Indiana Univ. Math. J.* **23**, 461–470.

Muroya, Y. (1979). On a posteriori error estimates for Galerkin approximations to the solutions of two-point boundary value problems. *Mem. School Sci. Engrg. Waseda Univ.* **43**, 163–169.

Mysovskih, I. P. (1957). Computation of the eigenvalues of integral equations by means of iterated kernels (in Russian). *Dokl. Akad. Nauk SSSR* **115**, 45–48.

Mysovskih, I. P. (1964a). On error bounds for approximate methods of estimation of eigenvalues of hermitian kernels. *Amer. Math. Soc. Transl.* **35**, 237–250.

Mysovskih, I. P. (1964b). On error bounds for eigenvalues calculated by replacing the kernel by an approximating kernel. *Amer. Math. Soc. Transl.* **35**, 251–262.

NAG Library Reference Manual. Numerical Algorithms Group, Oxford.

Nakamura, S. (1976). Analysis of the coarse-mesh rebalancing effect on Chebyshev polynomial iterations. *Nuclear Sci. Engrg.* **61**, 98–106.

Nash, J. C. (1974). The Hermitian matrix eigenproblem $Hx = eSx$. *Comput. Phys. Comm.* **8**, 85–94.

Nash, J. C. (1979). *Compact Numerical Methods for Computers: Linear Algebra and Function Minimization*. Wiley, New York.

Nashed, M. Z., and Wahba, G. (1974). Convergence rates of approximate least-squares solutions of linear integral and operator equations of the first kind. *Math. Comp.* **28**, 69–80.

Natanson, I. P. (1964). *Theory of Functions of a Real Variable*, Vols. 1 and 2. Ungar, New York.

Natterer, F. (1977). Uniform convergence of Galerkin's method for splines on highly non uniform meshes. *Math. Comp.* **31**, 457–468.

Nau, R. W. (1976). Computation of upper and lower bounds to the frequencies of clamped cylindrical shells. *Internat. J. Earthquake Engrg. Struct. Dynam.* **4**, 553–559.

Nedelec, J. C. (1977). *Approximation des équations intégrales en mécanique et en physique*. Lecture Notes, Math. Appl., École Polytechnique, Palaiseau.

Nelson, P., and Elder, I. T. (1977). Calculation of eigenfunctions in the context of integration-to-blowup. *SIAM J. Numer. Anal.* **14**, 124–136.

Nemat-Masser, S., and Lang, K.-W. (1979). Eigenvalue problems for heat conduction in composite material. *Iranian J. Sci. Tech.* **7**, 243–260.

Nesbet, R. K. (1981). Large matrix techniques in quantum chemistry and atomic physics. In *Sparse Matrices and Their Uses* (I. S. Duff, ed.), pp. 161–174. Academic Press, New York.

Nikolai, P. J. (1979). Algorithm 538 eigenvectors and eigenvalues of real generalized symmetric matrices by simultaneous iteration. *ACM Trans. Math. Software* **5**, 118–125.

Nikolskii, S. M. (1975). *Approximation of Functions of Several Variables and Imbedding Theorems.* Springer-Verlag, Berlin and New York.

Nisbet, R. M. (1972). Acceleration of the convergence in Nesbet's algorithm for eigenvalues and eigenvectors of large matrices. *J. Comput. Phys.* **10**, 614–619.

Nitsche, J. (1970). Über ein Variationprinzip zur Lösung von Dirichlet Problemen bei Verwendung von Teilräumen, die keinen Randbedingungen unterworfen sind. *Abh. Math. Sem. Univ. Hamburg* **36**, 9–15.

Nitsche, J. (1975). L_∞-convergence of finite element approximation. *Second Conf. Finite Elem., Rennes.*

Nitsche, J. (1977). On projection methods for the plate equation. In *Numerical Analysis* (J. Descloux and J. Marti, eds.), pp. 49–61. Birkhaeuser, Basel.

Noble, B. (1971). A bibliography on "Methods for solving integral equations" Tech. Rep. 1176, 1177, MRC, Univ. of Wisconsin, Madison, Wisconsin.

Noble, B. (1973). Error analysis of collocation methods for solving Fredholm integral equations. In *Topics in Numerical Analysis* (J. J. H. Miller, ed.), pp. 211–232. Academic Press, London.

Nyström, E. J. (1930). Über die praktische Auflösung von Integralgleichungen mit Anwendungen auf Randwertaufgaben. *Acta Math.* **54**, 185–204.

Oden, J. T., and Reddy, J. N. (1976). *An Introduction to the Mathematical Theory of Finite Elements.* Wiley (Interscience), New York.

Ojaiva, I. U., and Newman, M. (1970). Vibration modes of large structure by an automatic matrix-reduction method. *AIAA J.* **8**, 1234–1239.

O'Leary, D. P., Stewart, G. W., and Vandergraft, J. S. (1979). Estimating the largest eigenvalue of a positive definite matrix. *Math. Comp.* **33**, 1289–1292.

Oliveira Aleixo, F. (1980). Collocation and residual correction. *Numer. Math.* **36**, 27–31.

Onega, R. J., and Karcher, K. E. (1976). Nonlinear dynamics of a pressurized water reactor core. *Nucl. Sci. Engrg.* **61**, 276–282.

Orbach, O., and Crowe, C. M. (1971). Convergence promotion in the simulation of chemical processes with recycle: The dominant eigenvalue method. *Canad. J. Chem. Engrg.* **49**, 509–513.

Ortega, J. M. (1972). *Numerical Analysis: A Second Course.* Academic Press, New York.

Ortega, J. M., and Rheinboldt, W. C. (1971). *Iterative Solution of Nonlinear Equations in Several Variables.* Academic Press, New York.

Osborn, J. E. (1967). Approximation of the eigenvalues of a class of unbounded, nonself-adjoint operators. *SIAM J. Numer. Anal.* **4**, 45–54.

Osborn, J. E. (1975). Spectral approximation for compact operators. *Math. Comp.* **29**, 712–725.

Osborn, J. E. (1976). Approximation of the eigenvalues of a non self-adjoint operator arising in the study of the stability of stationary solutions of the Navier–Stokes equations. *SIAM J. Numer. Anal.* **13**, 185–197.

Ostrowski, A. M. (1958–1959). On the convergence of the Rayleigh quotient iteration for the computation of the characteristic roots and vectors. *Arch. Rational Mech. Anal.* **1**, 233–241; **2**, 423–428; **3**, 325–340, 341–347, 472–481; **4**, 153–165.

Paige, C. C. (1970). Practical use of the symmetric Lanczos process with reorthogonalization. *BIT* **10**, 183–195.

Paige, C. C. (1971). The computation of eigenvalues and eigenvectors of very large sparse matrices. Ph.D. Thesis, London Univ.

Paige, C. C. (1972). Computational variants of the Lanczos method for the eigenproblem. *J. Inst. Math. Appl.* **10**, 373–381.

Paige, C. C. (1974). Eigenvalues of perturbed Hermitian matrices. *Linear Algebra Appl.* **8**, 1–10.

Paige, C. C. (1980). Accuracy and effectiveness of the Lanczos algorithm for the symmetric eigenproblem. *Linear Algebra Appl.* **34**, 235–258.

Paige, C. C. (1981). Properties of numerical algorithms related to computing controllability. *IEEE Trans. Automat. Control* **AC-26**, 130–138.

Paine, J. W., and Anderssen, R. S. (1980). Uniformly valid approximation of eigenvalues of Sturm–Liouville problems in geophysics. *Geophys. J. R. Austral. Soc.* **63**, 441–465.

Palmer, T. W. (1969). Totally bounded sets of precompact linear operators. *Proc. Amer. Math. Soc.* **20**, 101–106.

Papathomas, T., and Wing, O. (1976). Sparse Hessenberg reduction and the eigenvalue problem for large sparse matrices. *IEEE Trans. Circuits and Systems* **CAS-23**, 739–744.

Parlett, B. N. (1964). The origin and development of methods of LR type. *SIAM Rev.* **6**, 275–295.

Parlett, B. N. (1965). Convergence of the QR algorithm. *Numer. Math.* **7**, 187–193.

Parlett, B. N. (1968). Global convergence of the basic QR algorithm on Hessenberg matrices. *Math. Comp.* **22**, 803–817.

Parlett, B. N. (1973). Présentation géométrique des méthodes de calcul des valeurs propres. *Numer. Math.* **21**, 223–233.

Parlett, B. N. (1974). The Rayleigh quotient iteration and some generalizations for nonnormal matrices. *Math. Comp.* **28**, 679–693.

Parlett, B. N. (1980a). *The Symmetric Eigenvalue Problem*. Prentice-Hall, Englewood Cliffs, New Jersey.

Parlett, B. N. (1980b). A new look at the Lanczos algorithm for solving symmetric systems of linear equations. *Linear Algebra Appl.* **29**, 323–346.

Parlett, B. N. (1981). Comment résoudre $(K - \lambda M)z = 0$? In *Méthodes Numériques pour les Sciences de l'Ingénieur* (E. Absi, R. Glowinski, P. Lascaux, and H. Veysseyre, eds.), Tome 1, pp. 97–106. Dunod, Paris.

Parlett, B. N., and Kahan, W. (1969). On the convergence of a practical QR algorithm. In *Information Processing* 68, Vol. 1, pp. 114–118. North-Holland Publ., Amsterdam.

Parlett, B. N., and Poole, W. G. (1973). A geometric theory for the QR, LU and power iterations. *SIAM J. Numer. Anal.* **10**, 389–412.

Parlett, B. N., and Reid, J. K. (1981). Tracking the progress of the Lanczos algorithm for large symmetric eigenproblems. *IMA J. Numer. Anal.* **2**, 135–156.

Parlett, B. N., and Reinsch, C. (1969). Balancing a matrix for calculation of eigenvalues and eigenvectors. *Numer. Math.* **13**, 293–304.

Parlett, B. N., and Scott, D. S. (1979). The Lanczos algorithm with selective orthogonalization. *Math. Comp.* **33**, 217–238.

Parlett, B. N., and Taylor, D. (1981). A look ahead Lanczos algorithm for unsymmetric matrices. Techn. Rep. PAM-43, Center for Pure and Appl. Math., Univ. of California, Berkeley.

Penrose, R. (1955). A generalized inverse for matrices. *Proc. Cambridge Philos. Soc.* **51**, 406–413.

Percell, P., and Wheeler, M. F. (1980). A C^1 finite element collocation method for elliptic equations. *SIAM J. Numer. Anal.* **17**, 605–622.

Pereyra, V., and Scherer, G. (1973). Eigenvalues of symmetric tridiagonal matrices: A fast, accurate and reliable algorithm. *J. Inst. Math. Appl.* **12**, 209–222.

Peters, G., and Wilkinson, J. H. (1970a). Eigenvectors of real and complex matrices by LR and QR triangularizations. *Numer. Math.* **16**, 181–204.

Peters, G., and Wilkinson, J. H. (1970b). The least squares problem and pseudoinverses. *Comput. J.* **13**, 309–316.

Peters, G., and Wilkinson, J. H. (1970c). $Ax = \lambda Bx$ and the generalized eigenproblem. *SIAM J. Numer. Anal.* **7**, 479–492.

Petryshyn, W. V. (1963). On a general iterative method for the approximate solution of linear operator equations. *Math. Comp.* **17**, 1–10.

Petryshyn, W. V. (1967a). On the eigenvalue problem $Tu - \lambda Su = 0$ with unbounded and non symmetric operators T and S. *Philos. Trans. Roy. Soc. London Ser. A* **262**, 413–458.

Petryshyn, W. V. (1967b). Projection methods in nonlinear numerical functional analysis. *J. Math. Mech.* **17**, 353–372.

Petryshyn, W. V. (1968). On projectional-solvability and the Fredholm alternative for equations involving linear A-proper operators. *Arch. Rational Mech. Anal.* **30**, 270–284.

Pfeifer, E. (1979). Discrete convergence of multi-step methods in eigenvalue problems for ordinary second order differential equations. *U.S.S.R. Computational Math. and Math. Phys.* **19**, 64–73.

Phillips, J. L. (1972). The use of collocation as a projection method for solving linear operator equations. *SIAM J. Numer. Anal.* **9**, 14–28.

Phillips, D. R. (1978). The existence of determining equations and their application for finding fixed points of nonlinear operators and error bounds for eigenvalues estimates of compact linear operators and finite matrices. Ph.D. Thesis, Univ. of Maryland, College Park.

Pitkäranta, J. (1979). On the differential properties of solutions to Fredholm equations with weakly singular kernels. *J. Inst. Math. Appl.* **24**, 109–119.

Platzman, G. W. (1972a). North Atlantic ocean: Preliminary description of normal modes. *Science* **178**, 156–157.

Platzman, G. W. (1972b). Two dimensional free oscillations in natural basins. *J. Phys. Oceanogr.* **2**, 117–138.

Platzman, G. W. (1975). Normal modes of the Atlantic and Indian oceans. *J. Phys. Oceanogr.* **5**, 201–221.

Platzman, G. W. (1978). Normal modes of the world ocean. Part I. Design of finite-element barotropic model. *J. Phys. Oceanogr.* **8**, 323–343.

Platzman, G. W. (1979). A Kelvin wave in the eastern north Pacific ocean. *J. Geophys. Res.* **84**, 2525–2528.

Pokrzywa, A. (1978). On the asymptotical behaviour of spectra in the method of orthogonal projections. Rep. No. 161, Inst. Math., Polish Academy of Sciences.

Pokrzywa, A. (1979). Method of orthogonal projections and approximation of the spectrum of a bounded operator. *Studia Math.* **65**, 21–29.

Pokrzywa, A. (1980). Spectra of compressions of an operator with compact imaginary part. *J. Operator Theory* **3**, 151–158.

Polansky, O. E., and Gutman, I. (1979). On the calculation of the largest eigenvalue of molecular graph. *MATCH* **5**, 149–159.

Polskii, N. I. (1962). Projection methods in applied mathematics. *Dokl. Akad. Nauk SSSR* **143**, 787–790 [*Soviet Math. Dokl.* **3**, 488–491].

Prenter, P. M. (1973). A collocation method for the numerical solution of integral equations. *SIAM J. Numer. Anal.* **10**, 570–581.

Prenter, P. M. (1975). *Splines and Variational Methods.* Wiley (Interscience), New York.

Prikazchikov, V. G. (1975). Strict estimate of convergence rate in an iterative method for calculation of eigenvalues. *U.S.S.R. Comput. Math. and Math. Phys.* **15**, 1330–1333.

Proskurowski, W. (1978). On the numerical solution of the eigenvalue problem of the Laplace operator by a capacity matrix method. *Computing* **20**, 139–151.

Pruess, S. (1973a). Estimating the eigenvalues of Sturm–Liouville problems by approximating the differential equation. *SIAM J. Numer. Anal.* **10**, 55–68.

Pruess, S. (1973b). Solving linear boundary value problems by approximating the coefficients. *Math. Comp.* **27**, 551–561.

Pruess, S. (1975). High order approximation to Sturm–Liouville eigenvalues. *Numer. Math.* **24**, 241–247.

Pták, V. (1976). Non discrete mathematical induction and iterative existence proofs. *Linear Algebra Appl.* **13**, 223–238.

Raffenetti, R. C. (1979). A simultaneous coordinate relaxation for large, sparse matrix eigenvalue problems. *J. Comput. Phys.* **32**, 403–419.

Raju, I. S., Rao, G. V., and Murthy, T. V. G. K. (1974). Eigenvalues and eigenvectors of large order banded matrices. *Comput. & Structures* **4**, 549–558.

Rakotch, E. (1975). Numerical solution for eigenvalues and eigenfunctions of a hermitian kernel and error estimates. *Math. Comp.* **29**, 794–805.

Rakotch, E. (1976). Numerical solution with large matrices of Fredholm's integral equation. *SIAM J. Numer. Anal.* **13**, 1–7.

Rakotch, E. (1978). Improved error estimates for numerical solutions of symmetric integral equations. *Math. Comp.* **32**, 399–404.

Rall, L. B. (1969). *Computational Solution of Nonlinear Operator Equations.* Wiley, New York.

Ramamurti, V. (1973). Application of simultaneous iteration method to torsional vibration problems. *J. Sound Vibr.* **29**, 331–340.

Ramaswamy, S. (1980). On the effectiveness of the Lanczos method for the solution of large eigenvalue problems. *J. Sound Vibr.* **73**, 405–418.

Ramsden, J. N., and Stoker, J. R. (1969). A semi-automatic method for reducing the size of vibration problems. *Internat. J. Numer. Methods Engrg.* **1**, 339–349.

Rannacher, R. (1979). Nonconforming finite element methods for eigenvalue problems in linear plate theory. *Numer. Math.* **33**, 23–42.

Rao, C. R., and Mitra, S. K. (1971). *Generalized Inverse of Matrices and Its Applications.* Wiley, New York.

Rappaz, J. (1977). Approximation of the spectrum of a noncompact operator given by the magnetohydrodynamic stability of a plasma. *Numer. Math.* **28**, 15–24.

Rappaz, J. (1982). Some properties on the stability related to the approximation of eigenvalue problems. In *Computing Methods in Applied Sciences and Engineering* (R. Glowinski and J.-L. Lions, eds.), pp. 167–174. North-Holland Publ., Amsterdam.

Rayleigh, Lord (Strutt, J. W.) (1894–1896). *The Theory of Sound.* Vol. 1 and 2. Macmillan, London and New York.

Rayleigh, Lord (1899). On the calculation of the frequency of vibration of a system in its gravest mode, with an example from hydrodynamics. *Philos. Mag.* **47**, 566–572.

Redont, P. (1979a). Application de la théorie de la perturbation des opérateurs linéaires à l'obtention de bornes d'erreur sur les éléments propres et à leur calcul. Thèse Doct.-Ing., Univ. de Grenoble.

Redont, P. (1979b). Sur la convergence régulière d'opérateurs. Unpublished manuscript, Univ. de Grenoble.

Reed, M., and Simon, B. (1978). *Analysis of Operators.* Academic Press, New York.

Regińska, T. (1977). Approximate methods for solving differential equations on infinite intervals. *Apl. Mat.* **22**, 92–109.

Regińska, T. (1980). External approximation of eigenvalue problems. Rep. No. 229, Inst. Math., Polish Academy of Sciences.

Regińska, T. (1981). Eigenvalue approximation. In *Computational Mathematics.* Banach Center Publ., Vol. 10, Warzsaw.

Reid, J. K. (1976). A survey of sparse matrix computation. In *Sparse Matrix Techniques* (V. A. Barker, ed.), Lecture Notes in Mathematics, Vol. 572, pp. 41–48. Springer-Verlag, Berlin and New York.

Reid, J. K. (1980). A survey of sparse matrix computation. In *Electric Power Problems: The Mathematical Challenge* (A. M. Erisman, K. W. Neves, and M. H. Dwarakanath, eds.), pp. 41–69. SIAM, Philadelphia, Pennsylvania.

Rellich, F. (1936–1942). Störungstheorie der Spektralzerlegung. *Math. Ann.* **113**, 600–619 (1936); **113**, 667–685 (1936); **116**, 555–570 (1939); **117**, 356–382 (1940); **118**, 462–484 (1942).

Rheinboldt, W. C. (1976). On measures of ill-conditioning for nonlinear equations. *Math. Comp.* **30**, 104–111.

Rheinboldt, W. C. (1980). On a theory of mesh-refinement processes. *SIAM J. Numer. Anal.* **17**, 766–778.

Rice, J. R. (1969). On the degree of convergence of non linear spline approximation. In *Approximation with Special Emphasis on Spline Functions* (I. J. Schoenberg, ed.), pp. 349–365. Academic Press, New York.

Richter, G. R. (1978). Superconvergence of piecewise polynomial Galerkin approximations, for Fredholm integral equations of the second kind. *Numer. Math.* **31**, 63–70.

Richtmyer, R. D., and Morton, K. W. (1967). *Difference Methods for Initial Value Problems*, 2nd ed. Wiley (Interscience), New York.

Riddell, I. J., and Delves, L. M. (1980). The comparison of routines for solving Fredholm integral equations of the second kind. *Comput. J.* **23**, 274–285.

Riehl, J. P., Diestler, D. J., and Wagner, A. F. (1974). Comparison of perturbative and direct numerical integration techniques for the calculation of phase shifts for elastic scattering. *J. Comput. Phys.* **15**, 212–225.

Riesz, F., and Sz.-Nagy, B. (1955). *Leçons d'Analyse Fonctionnelle*, 3rd ed. Akadémiai Kiadó, Budapest.

Ritz, W. (1909). Über eine neue Methode zur Lösung Gewisser Variationsprobleme der Mathematischen Physik. *J. Reine Angew. Math.* **135**, 1–61.

Rivlin, T. J. (1974). *The Chebychev Polynomials.* Wiley, New York.

Roark, A. L. (1971). On the eigenproblem for convolution integral equations. *Numer. Math.* **17**, 54–61.

Roark, A. L., and Shampine, L. F. (1965). On a paper of Roark and Wing. *Numer. Math.* **7**, 394–395.

Roark, A. L., and Wing, G. M. (1965). A method for computing the eigenvalues of certain integral equations. *Numer. Math.* **7**, 159–170.

Rodrigue, G. (1973). A gradient method for the matrix eigenvalue problem $Ax = \lambda Bx$. *Numer. Math.* **22**, 1–16.

Rothblum, U. G. (1975). Algebraic eigenspace of nonnegative matrices. *Linear Algebra Appl.* **12**, 281–292.

Ruge, J. (1981). Multigrid methods for differential eigenvalue and variational problems and multigrid simulation. Ph.D. Thesis, Math. Dept., Colorado State Univ., Fort Collins, Colorado.

Ruhe, A. (1970a). An algorithm for numerical determination of the structure of a general matrix. *BIT* **10**, 196–216.

Ruhe, A. (1970b). Perturbation bounds for means of eigenvalues and invariant subspaces. *BIT* **10**, 343–354.

Ruhe, A. (1970c). Properties of a matrix with a very ill-conditioned eigenproblem. *Numer. Math.* **15**, 57–60.

Ruhe, A. (1974a). SOR-methods for the eigenvalue problem with large sparse matrices. *Math. Comp.* **28**, 695–710.

Ruhe, A. (1974b). Iterative eigenvalue algorithms for large symmetric matrices. In *Eigenwerte Probleme* (L. Collatz, ed.), pp. 97–115. Birkhaeuser, Basel.

Ruhe, A. (1975). Iterative eigenvalue algorithms based on convergent splittings. *J. Comput. Phys.* **19**, 110–120.

Ruhe, A. (1977). Numerical methods for the solution of large sparse eigenvalue problems. In *Sparse Matrix Techniques* (V. A. Barker, ed.), Lecture Notes in Mathematics, Vol. 572, pp. 130–184. Springer-Verlag, Berlin and New York.

Ruhe, A. (1979a). Implementation aspects of band Lanczos algorithms for computation of eigenvalues of large sparse symmetric matrices. *Math. Comp.* **33**, 680–687.

Ruhe, A. (1979b). Eigenvalues in a APL environment, an algorithm based on Rayleigh quotient iteration. *APL. Quote Quad.* **10**, 29–30.

Ruhe, A. (1980). The relation between the Jacobi algorithm and inverse iteration and a Jacobi algorithm based on elementary reflections. *BIT* **20**, 88–96.

Ruhe, A., and Ericsson, T. (1980). The spectral transformation Lanczos method in the numerical solution of large, sparse, generalized symmetric eigenvalue problems. *Math. Comp.* **35**, 1251–1268.

Ruhe, A., and Wiberg, T. (1972). The method of conjugate gradient used in inverse iteration. *BIT* **12**, 543–554.

Rump, S. M., and Böhm, H. (1982). Least significant bit evaluation of arithmetic expressions in single-precision. *Computing* (to appear).

Russel, R. D. (1974). Collocation for systems of boundary value problems. *Numer. Math.* **23**, 119–133.

Russel, R. D. (1977). A comparison of collocation and finite differences for two-point boundary value problems. *SIAM J. Numer. Anal.* **14**, 19–39.

Russel, R. D., and Shampine, L. F. (1972). A collocation method for boundary value problems. *Numer. Math.* **19**, 1–28.

Russel, R. D., and Varah, J. M. (1975). A comparison of global methods for linear two-point boundary value problems. *Math. Comp.* **29**, 1007–1019.

Rutishauser, H. (1958). Solution of eigenvalue problem with the LR-transformation. *Appl. Math. Ser. Nat. Bur. Standards* **49**, 47–81.

Rutishauser, H. (1969). Computational aspects of F. L. Bauer's simultaneous iteration method. *Numer. Math.* **13**, 4–13.

Rutishauser, H. (1970). Simultaneous iteration method for symmetric matrices. *Numer. Math.* **16**, 205–223.

Saad, Y. (1975). Shifts of origin for the QR algorithm. In *Information Processing 74*, pp. 527–531. North-Holland Publ., Amsterdam.

Saad, Y. (1979a). Calcul de vecteurs propres d'une grande matrice creuse par la méthode de Lanczos. In *Méthodes Numériques pour les Sciences de l'Ingénieur* (E. Absi and R. Glowinski, eds.). Dunod, Paris.

Saad, Y. (1979b). Etude de la convergence du procédé d'Arnoldi pour le calcul des éléments propres de grandes matrices creuses non symétriques. Sém. Anal. Numér. No. 321, Univ. de Grenoble.

Saad, Y. (1980a). On the rates of convergence of the Lanczos and the block-Lanczos methods. *SIAM J. Numer. Anal.* **17**, 687–706.

Saad, Y. (1980b). Variations on Arnoldi's method for computing eigenelements of large unsymmetric matrices. *Linear Algebra Appl.* **34**, 269–295.

Saad, Y. (1981). Krylov subspace methods for solving large unsymmetric linear systems. *Math. Comp.* **37**, 105–126.

Saad, Y. (1982a). The Lanczos biorthogonalization algorithm and other oblique projection methods for solving large unsymmetric systems. *SIAM J. Numer. Anal.* **19**, 485–509.

Saad, Y. (1982b). Practical use of Krylov subspace methods for solving indefinite and unsymmetric linear systems. To appear in *SIAM J. Sci. Stat. Comp.*

Saad, Y. (1982c). Projection methods for solving large sparse eigenvalue problems. Techn. Rep. No. 224, Computer Science, Yale Univ., Connecticut.

Sacks-Davis, R. (1975). Real norm-reducing Jacobi-type eigenvalue algorithm. *Austral. Comput. J.* **7**, 65–69.

Sakaguchi, R. L., and Tabarrok, B. (1970). Calculation of plate frequencies from complementary energy functions. *Internat. J. Numer. Methods Engrg.* **2**, 283–293.

Sala, I. (1963). On the numerical solution of certain boundary value problems and eigenvalue problems of the 2nd and 4th order with the aid of integral equations. *Acta Polytech. Scand. Ser. D* **9**, 1–24.

Samet, A., Lermit, J., and Noh, K. (1975). On the intermediate eigenvalues of symmetric sparse matrices. *BIT* **15**, 185–191.

Schäfer, E. (1980). Spectral approximation for compact integral operators by degenerate kernel methods. *Numer. Funct. Anal. Optim.* **2**, 43–63.

Schippers, H. (1981). The automatic solution of Fredholm equation of the second kind. Techn. Rep., Stichting Math. Cent., Amsterdam.

Schlessinger, S. (1957). Approximating eigenvalues and eigenfunctions of symmetric kernels. *SIAM J. Appl. Math.* **5**, 1–14.

Schneider, C. (1979). Regularity of the solution of a class of weakly singular Fredholm integral equations of the second kind. *Integral Equations Operator Theory* **2**, 62–68.

Schneider, C. (1980). Produktintegration mit nichtäquidistanten Stützstellen. *Numer. Math.* **35**, 35–43.

Schneider, C. (1981). Product integration for weakly singular integral equations. *Math. Comp.* **36**, 207–213.

Schrödinger, E. (1926). Quantisierung als eigenwertproblem. IV. Störungtheorie mit Anwendung auf den Starkeffekt der Balmerlinien. *Ann. Physik* **80**, 437–490.

Schwarz, H. R. (1974a). The eigenvalue problem $(A - \lambda B)x = 0$ for symmetric matrices of high order. *Comput. Methods Appl. Mech. Engrg.* **3**, 11–28.

Schwarz, H. R. (1974b). A method of coordinate overrelaxation for $(A - \lambda B)x = 0$. *Numer. Math.* **23**, 135–151.

Schwarz, H. R. (1977). Two algorithms for treating $Ax = \lambda Bx$. *Comput. Methods Appl. Mech. Engrg.* **12**, 181–199.

Scott, D. S. (1979). How to make the Lanczos algorithm converge slowly. *Math. Comp.* **33**, 239–247.

Scott, D. S. (1981a). Solving sparse symmetric generalized eigenvalue problems without factorization. *SIAM J. Numer. Anal.* **18**, 102–110.

Scott, D. S. (1981b). The Lanczos algorithm. In *Sparse Matrices and Their Uses* (I. S. Duff, ed.), pp. 139–159. Academic Press, New York.

Scott, R. (1976). Optimal L^∞-estimates for the finite element method on irregular meshes. *Math. Comp.* **30**, 681–697.

Sebe, T., and Nachamkin, J. (1969). Variational buildup of nuclear shell model bases. *Ann. Physics* **51**, 100–123.

Seneta, E. (1980). Computing the stationary distribution for infinite Markov chains. *Linear Algebra Appl.* **34**, 259–269.

Shavitt, I. (1970). Modification of Nesbet's algorithm for the iterative evaluation of eigenvalues and eigenvectors of large matrices. *J. Comput. Phys.* **6**, 124–130.

Shavitt, I., Bender, C. F., Pipano, A., and Hosteny, R. P. (1973). The iterative calculation of several of the lowest or highest eigenvalues and corresponding eigenvectors of very large symmetric matrices. *J. Comput. Phys.* **11**, 90–108.

Simpson, A. (1973a). A generalization of Kron's eigenvalue procedure. *J. Sound Vibr.* **26**, 129–139.

Simpson, A. (1973b). Kron's method: a consequence of the minimization of the primitive Lagrangian in the presence of displacement constraints. *J. Sound Vibr.* **27**, 377–386.

Simpson, A. (1973c). Eigenvalue and vector sensitivities in Kron's method. *J. Sound Vibr.* **31**, 73–87.

Šindler, A. A. (1967). Certain theorems in the general theory of approximate methods of analysis and their application to the methods of collocation, moments and Galerkin. *Siberian Math. J.* **8**, 302–314.

Šindler, A. A. (1969). The rate of convergence of an enriched collocation method for ordinary differential equations. *Siberian Math. J.* **10**, 160–163.

Singh, S. R. (1976). Some convergence properties of the Bubnov–Galerkin method. *Pacific J. Math.* **65**, 217–221.

Sloan, I. H. (1976a). Convergence of degenerate-kernel methods. *J. Austral. Math. Soc. Ser. B* **19**, 422–431.

Sloan, I. H. (1976b). Error analysis for a class of degenerate kernel methods. *Numer. Math.* **25**, 231–238.

Sloan, I. H. (1976c). Iterated Galerkin method for eigenvalue problems. *SIAM J. Numer. Anal.* **13**, 753–760.

Sloan, I. H. (1976d). Improvement by iteration for compact operator equations. *Math. Comp.* **30**, 758–764.

Sloan, I. H. (1978). On the numerical evaluation of singular integrals. *BIT* **19**, 91–102.

Sloan, I. H. (1980a). On choosing the points in product integration. *J. Math. Phys.* **21**, 1032–1039.

Sloan, I. H. (1980b). A review of numerical methods for Fredholm equations of the second kind. In *The Application and Numerical Solution of Integral Equations* (R. S. Anderssen, F. R. de Hoog, and M. A. Lukas, eds.), pp. 51–74. Sijthoff & Noordhoff, Alphen aan den Rijn. The Netherlands.

Sloan, I. H. (1980c). The numerical solution of Fredholm equations of the second kind by polynomial interpolation. *J. Integral Equations* **2**, 265–279.

Sloan, I. H. (1981). Analysis of general quadrature methods for integral equations of the second kind. *Numer. Math.* **38**, 263–278.

Sloan, I. H. (1983). Superconvergence and the Galerkin method for integral equations of the second kind. In *Treatment of Integral Equations by Numerical Methods* (C. T. H. Baker and G. F. Miller, eds.), pp. 197–208, Academic Press, London.

Sloan, I. H., and Burn, B. J. (1979). Collocation with polynomials for integral equations of the second kind: a new approach to the theory. *J. Integral Equations* **1**, 77–94.

Sloan, I. H., Noussair, E., and Burn, B. J. (1979). Projection method for equations of the second kind. *J. Math. Anal. Appl.* **69**, 84–103.

Smith, B. T., Boyle, J. M., Garbow, B. S., Ikebe, Y., Klema, V. C., and Moler, C. B. (1976). *Matrix Eigensystem Routines—EISPACK Guide*. Lecture Notes in Computer Science, Vol. 6, 2nd ed. Springer-Verlag, Berlin and New York.

Smooke, M. D. (1978). Piecewise analytical perturbation series solutions of the radial Schrödinger equation. Ph.D. Thesis, Harvard Univ., Cambridge, Massachusetts.

Smooke, M. D. (1980a). Error estimates for piecewise analytical perturbation series solutions of the radial Schrödinger equation. I. One-dimensional case. Rep. SAND 80-8611, Sandia Livermore Lab., Livermore, California.

Smooke, M. D. (1980b). Error estimates for piecewise analytical perturbation series solutions of the radial Schrödinger equation. II. Multidimensional case. Rep. SAND 80-8610, Sandia Livermore Lab., Livermore, California.

Sobolev, S. L. (1937). On a boundary problem for semi harmonic equations (in Russian). *Mat. Sb.* **2**, 467–500.

Sobolev, S. L. (1956). Some remarks on numerical solution of integral equations (in Russian). *Izv. Akad. Nauk SSSR Ser. Mat.* **20**, 413–436.

Spence, A. (1975). On the convergence of the Nyström method for the integral equation eigenvalue problem. *Numer. Math.* **25**, 57–66.

Spence, A. (1978–1979). Error bounds and estimates for eigenvalues of integral equations. *Numer. Math.* **29**, 133–147 (1978); **32**, 139–146 (1979).

Spence, A. (1979). Product integration for singular integrals and singular integral equations. In *Numerische Integration* (G. Hämmerlin, ed.), pp. 288–300. Birkhaeuser, Basel.

Spence, A., and Moore, G. (1980). A convergence analysis for turning points of nonlinear compact operator equations. In *Numerical Treatment of Integral Equations* (J. Albrecht and L. Collatz, eds.), pp. 203–212. Birkhaeuser, Basel.

Spence, D. A. (1972). An eigenvalue problem for elastic contact with finite friction. *Proc. Cambridge Phil. Soc.* **73**, 249–268.

Srinivasan, R. S., and Sankaran, S. (1975). Vibration of cantilever cylindrical shells. *J. Sound. Vibr.* **40**, 425–430.

Srivastava, B. P. (1975). Calculation of bounds on the eigenvalue spectrum of anharmonic phonon collision operator. *Phys. Lett. A* **54**, 222–224.

Stetter, H. J. (1973). *Analysis of Discretization Methods for Ordinary Differential Equations.* Springer-Verlag, Berlin and New York.

Stetter, H. J. (1978). The defect correction principle and discretization methods. *Numer. Math.* **29**, 425–443.

Stewart, G. W. (1971). Error bounds for approximate invariant subspaces of closed linear operators. *SIAM J. Numer. Anal.* **8**, 796–808.

Stewart, G. W. (1972). On the sensitivity of the eigenvalue problem $Ax = \lambda Bx$. *SIAM J. Numer. Anal.* **9**, 669–686.

Stewart, G. E. (1973a). *Introduction to Matrix Computations.* Academic Press, New York.

Stewart, G. W. (1973b). Error and perturbation bounds for subspaces associated with certain eigenvalue problems. *SIAM Rev.* **15**, 727–764.

Stewart, G. W. (1975a). The numerical treatment of large eigenvalue problems. In *Information Processing 74*, pp. 666–672. North-Holland Publ., Amsterdam.

Stewart, G. W. (1975b). The convergence of the method of conjugate gradients at isolated extreme points of the spectrum. *Numer. Math.* **24**, 85–93.

Stewart, G. W. (1975c). Gerschgorin theory for the generalized eigenvalue problem $Ax = \lambda Bx$. *Math. Comp.* **29**, 600–606.

Stewart, G. W. (1975d). Methods of simultaneous iteration for calculating eigenvectors of matrices. In *Topics in Numerical Analysis II* (J. J. H. Miller, ed.), pp. 185–196. Academic Press, New York.

Stewart, G. W. (1976a). Simultaneous iteration for computing invariant subspaces of non-Hermitian matrices. *Numer. Math.* **25**, 123–136.

Stewart, G. W. (1976b). A bibliographical tour of the large, sparse generalized eigenvalue problem. In *Sparse Matrix Computations* (J. R. Bunch and D. J. Rose, eds.), pp. 113–130. Academic Press, New York.

Stewart, G. W. (1979). Perturbation bounds for the definite generalized eigenvalue problem. *Linear Algebra Appl.* **23**, 69–85.

Stewart, W. J., and Jennings, A. (1981). A simultaneous iteration algorithm for real matrices. *ACM Trans. Math. Software* **7**, 184–198.

Stoer, J., and Bulirsch, R. (1980). *Introduction to Numerical Analysis.* Springer-Verlag, Berlin and New York.

Strakhovskaya, L. G. (1977). An iterative method for evaluating the first eigenvalue of an elliptic operator. *U.S.S.R. Computational Math. and Math. Phys.* **17**, 88–101.

Strang, G. (1972). Approximation in the finite element method. *Numer. Math.* **19**, 81–98.

Strang, G. (1976). *Linear Algebra and Its Applications.* Academic Press, New York.

Strang, G., and Fix, G. J. (1973). *An Analysis of the Finite Element Method.* Prentice-Hall, Englewood Cliffs, New Jersey.

Stroud, A. H. (1971). *Approximate Calculation of Multiple Integrals.* Prentice-Hall, Englewood Cliffs, New Jersey.

Strygin, V. V. (1973). Application of Bubnov–Galerkin method to find autooscillations (in Russian). *Prikl. Mat. Meh.* **37**, 1015–1019.

Strygin, V. V., and Cygankov, A. I. (1974). Application of collocation and difference method to find autooscillations of differential-difference equations (in Russian). *Ž. Vyčisl. Mat. i Mat. Fiz.* **14**, 691–698.

Stummel, F. (1970). Diskrete Konvergenz linearer Operatoren. Part I. *Math. Ann.* **190**, 45–92.

Stummel, F. (1971). Diskrete Konvergenz linearer Operatoren. Part II. *Math. Z.* **120**, 231–264.

Stummel, F. (1972). Diskrete Konvergenz linearer Operatoren. Part III. In *Linear Operators and Approximation*, pp. 196–216. Birkhaeuser, Basel.

Stummel, F. (1973). Discrete convergence of mappings. In *Topics in Numerical Analysis* (J. J. H. Miller, ed.), pp. 285–310. Academic Press, New York.

Stummel, F. (1977). Approximation methods for eigenvalue problems in elliptic differential equations. In *Numerik und Anwendungen von Eigenwertaufgaben und Verzweigungsproblemen* (E. Bohl, L. Collatz, and K. P. Hadeler, eds.), pp. 133–165. Birkhaüser, Basel.

Stummel, F. (1980). Basic compactness properties of nonconforming and hybrid finite element spaces. *RAIRO Anal. Numér.* **14**, 81–115.

Swartz, B., and Wendroff, B. (1974). The relation between the Galerkin and collocation methods using smooth splines. *SIAM J. Numer. Anal.* **11**, 994–996.

Symm, H. J., and Wilkinson, J. H. (1980). Realistic bounds for a simple eigenvalue and its associated eigenvector. *Numer. Math.* **35**, 113–126.

Symm, H. J., and Wilkinson, J. H. (1981). Error bounds for computed invariant subspaces. Res. Rep. No. 81–02, Angew. Math., Eidg. Techn. Hochschule Zürich.

Szegö, G. (1975). *Orthogonal Polynomials*, Colloquium Publ. No. 23. Amer. Math. Soc., Providence, Rhode Island.

Sz.-Nagy, B. (1946/47). Perturbations des transformations autoadjointes dans l'espace de Hilbert. *Comment. Math. Helv.* **19**, 347–366.

Sz.-Nagy, B. (1951). Perturbations des transformations linéaires fermées. *Acta Sci. Math. (Szeged)* **14**, 125–137.

Sz.-Nagy, B., and Foiaş, C. (1967). Forme triangulaire d'une contraction et factorisation de la fonction caractéristique. *Acta Sci. Math. (Szeged)* **28**, 201–212.

Sz.-Nagy, B., and Foiaş, C. (1970). *Harmonic analysis of operators on Hilbert space.* Akadémiai Kiadó, Budapest and North-Holland Publ., Amsterdam.

Szyld, D. B., and Widlund, O. B. (1978). Applications of conjugate gradient type methods to eigenvalue calculations. Tech. Rep., Courant Inst., New York Univ.

Tamme, E. E. (1977). On regular convergence of difference approximations for Dirichlet problem (in Russian). *Eesti NSV Tead. Akad. Toimetised Füüs.-Mat.* **26**, 3–8.

Taylor, A. E. (1958). *Introduction to Functional Analysis.* Wiley, New York.

Temple, G. (1928). The computation of characteristic numbers and characteristic functions. *Proc. London Math. Soc.* **29**, 257–280.

Temple, G. (1952). The accuracy of Rayleigh's method of calculating the natural frequencies of vibrating systems. *Proc. Roy. Soc. London Ser. A* **211**, 204–224.

Temple, G., and Bickley, W. G. (1933). *Rayleigh's Principle and Its Applications to Engineering.* Constable, London.

Tsunematsu, T., and Takeda, T. (1978). A new iterative method of solution of a large-scale generalized eigenvalue problem. *J. Comput. Phys.* **28**, 287–293.

Uebe, G., and Fisher, J. (1974). Computation of the eigenvalues of large econometric models. *Comput. and Oper. Res.* **1**, 313–339.

Underwood, R. (1975). An iterative block Lanczos method for the solution of large sparse symmetric eigenproblems. Ph.D. Thesis, Rep. STAN-CS-75-496, Comput. Sci. Dept., Stanford Univ., Stanford. California.

Urabe, M. A. (1967). Numerical solution of multi-point boundary value problems in Chebyshev series—Theory and the method. *Numer. Math.* **9**, 341–366.

Urabe, M. A. (1969). Numerical solution of boundary value problems in Chebyshev series—A method of computation and error estimation. In *Conference on the Numerical Solution of Differential Equations* (J. L. Morris, ed.), Lecture Notes in Mathematics, Vol. 109, pp. 40–86. Springer-Verlag, Berlin and New York.

Urabe, M. A. (1975). A posteriori componentwise error estimation of approximate solutions to non linear equations. In *Interval Mathematics* (K. Nickel, ed.), Lecture Notes in Computer Science, Vol. 29, pp. 99–117. Springer-Verlag, Berlin and New York.

Vainikko, G. M. (1964). Asymptotic error bounds for projection methods in the eigenvalue problem. *U.S.S.R. Computational Math. and Math. Phys.* **4**, 9–36.

Vainikko, G. M. (1965a). On the convergence of Galerkin method (in Russian). *Tartu Riikl. Ül. Toimetised* **177**, 148–152.

Vainikko, G. M. (1965b). On the stability and convergence of the collocation method. *Differential Equations* **1**, 186–194.

Vainikko, G. M. (1966). The convergence of the collocation method for nonlinear differential equations. *U.S.S.R. Computational Math. and Math. Phys.* **6**, 35–42.

Vainikko, G. M. (1967a). Galerkin's perturbation method and the general theory of approximate methods for nonlinear equations (in Russian). *Ž. Vyčisl. Mat. i Mat. Fiz.* **7**, 723–751.

Vainikko, G. M. (1967b). The rate of convergence of approximate methods in the problem of eigenvalues (in Russian). *Ž. Vyčisl. Mat. i Mat. Fiz.* **7**, 977–987.

Vainikko, G. M. (1967c). Rapidity of convergence of approximation methods in the eigenvalue problem. *U.S.S.R. Computational Math. and Math. Phys.* **7**, 18–32.

Vainikko, G. M. (1968). On the convergence speed of the method of moments for ordinary differential equations (in Russian). *Sib. Mat. Ž.* **9**, 21–28.

Vainikko, G. M. (1969a). The compact approximation principle in the theory of approximation methods (in Russian). *Ž. Vyčisl. Mat. i Mat. Fiz.* **9**, 739–761.

Vainikko, G. M. (1969b). A difference method for ordinary differential equations (in Russian). *Ž. Vyčisl. Mat. i Mat. Fiz.* **9**, 1057–1074.

Vainikko, G. M. (1969c). The connection between mechanical quadrature and finite difference methods (in Russian). *Ž. Vyčisl. Mat. i. Mat. Fiz.* **9**, 259–270.

Vainikko, G. M. (1970a). On the rate of convergence of certain approximation methods of Galerkin type in an eigenvalue problem. *Amer. Math. Soc. Transl.* **36**, 249–259.

Vainikko, G. M. (1970b). On the convergence of collocation method for multidimensional integral equations (in Russian). *Tartu Riikl. Ül. Toimetised* **253**, 244–257.

Vainikko, G. M. (1971). On the stability of the collocation method (in Russian). *Tartu Riikl. Ül. Toimetised* **281**, 190–196.

Vainikko, G. M. (1974a). On the approximation of fixed points of compact operators (in Russian). *Tartu Riikl. Ül. Toimetised* **342**, 225–236.

Vainikko, G. M. (1974b). Discretely compact sequences (in Russian). *Ž. Vyčisl. Mat. i Mat. Fiz.* **14**, 572–583.

Vainikko, G. M. (1975). Convergence of the difference method when seeking the periodic solutions of ordinary differential equations (in Russian). *Ž. Vyčisl. Mat. i Mat. Fiz.* **15**, 87–100.

Vainikko, G. M. (1976a). *Analysis of Discretization Methods* (in Russian). Univ. of Tartu Publ., Estland.

Vainikko, G. M. (1976b). *Funktionalanalysis der Diskretisierungsmethoden.* Teubner, Leipzig.

Vainikko, G. M. (1977a). Über die Konvergenz und Divergenz von Näherungsmethoden bei Eigenwertproblemen. *Math. Nachr.* **78**, 145–164.

Vainikko, G. M. (1977b). Über die Konvergenzbegriffe für lineare Operatoren in der Numerischen Mathematik. *Math. Nachr.* **78**, 165–183.

Vainikko, G. M. (1978a). Approximative methods for nonlinear equations (two approaches to the convergence problem). *Nonlinear Anal., Theory, Method and Appl.* **2**, 647–687.

Vainikko, G. M. (1978b). Foundations of finite difference method for eigenvalue problems. In *The Use of Finite Element Method and Finite Difference Method Geophys.* (V. Bucha and H. Nedoma, eds.), pp. 173–192. Czech. Acad. Sci., Prague.

Vainikko, G. M., and Karma, O. O. (1974a). The convergence of approximate methods for solving linear and nonlinear operator equations (in Russian). *Ž. Vyčisl. Mat. i Mat. Fiz.* **14**, 828–837.

Vainikko, G. M., and Karma, O. O. (1974b). The convergence rate of approximate methods in the eigenvalue problem when the parameter appears nonlinearly (in Russian). *Ž. Vyčisl. Mat. i Mat. Fiz.* **14**, 1393–1408.

Vainikko, G. M., and Pedas, A. (1971). On solution of integral equations with logarithmical singularities by quadrature formulae methods (in Russian). *Tartu Riikl. Ül. Toimetised* **281**, 201–210.

Vainikko, G. M., and Pedas, A. (1981). The properties of solutions of weakly singular integral equations. *J. Austral. Math. Soc. Ser. B* **22**, 424–434.

Vainikko, G. M., and Tamme, E. E. (1976). Convergence of difference methods in the periodic solution problem for equations of elliptic type (in Russian). *Ž. Vyčisl. Mat. i Mat. Fiz.* **16**, 652–664.

Vainikko, G. M., and Uba, P. (1981). A piecewise polynomial approximation to the solution of an integral equation with weakly singular kernel. *J. Austral. Math. Soc. Ser. B* **22**, 435–442.

Vainikko, G. M., and Umanskii, Y. B. (1968). Regular operators. *Funct. Anal. Appl.* **2**, 175–176.

Vandergraft, J. S. (1971). Generalized Rayleigh method with applications to finding eigenvalues of large matrices. *Linear Algebra Appl.* **4**, 363–368.

Van Dooren, P. M. (1981). The generalized eigenstructure problem in linear system theory. *IEEE Trans. Automat. Control* **AC-26**, 111–129.

Van Kempen, H. P. M. (1966). On the convergence of the classical Jacobi method for real symmetric matrices with non distinct eigenvalues. *Numer. Math.* **9**, 11–18 and 19–22.

Van Loan, C. F. (1975). A general matrix eigenvalue algorithm. *SIAM J. Numer. Anal.* **12**, 819–834.

Van Veldhuizen, M. (1976). A refinement process for collocation approximations. *Numer. Math.* **26**, 397–407.

Varah, J. M. (1968a). The calculation of the eigenvectors of a general complex matrix by inverse iteration. *Math. Comp.* **22**, 785–791.

Varah, J. M. (1968b). Rigorous machine bounds for the eigensystem of a general complex matrix. *Math. Comp.* **22**, 793–801.

Varah, J. M. (1970). Computing invariant subspaces of a general matrix when the eigensystem is poorly conditioned. *Math. Comp.* **24**, 137–149.

Varah, J. M. (1972). Invariant subspace perturbations for a nonnormal matrix. In *Information Processing 71*, Vol. 2, pp. 1251–1253. North-Holland Publ., Amsterdam.

Varah, J. M. (1973). On the numerical solution of ill-conditioned linear systems with applications to ill-posed problems. *SIAM J. Numer. Anal.* **10**, 257–267.

Varga, R. S. (1962). *Matrix Iterative Analysis.* Prentice-Hall, Englewood Cliffs, New Jersey.

Varga, R. S. (1965). Minimal Gershgorin sets. *Pacific J. Math.* **15**, 719–729.

Vorobyev, Y. V. (1965). *Method of Moments in Applied Mathematics.* Gordon & Breach, New York.

Wachpress, E. L. (1966). *Iterative solution of elliptic systems and application to the neutron diffusion equations of reactor physics*. Prentice-Hall, Englewood Cliffs, New Jersey.

Wahba, G. (1973). Convergence rates for certain approximate solutions to Fredholm integral equations of the first kind. *J. Approx. Theory* **7**, 167–185.

Wahba, G. (1976). On the optimal choice of nodes in the collocation projection method for solving linear operator equations. *J. Approx. Theory* **16**, 175–186.

Wang, J. Y. (1976). On the numerical computation of eigenvalues and eigenfunctions of compact integral operators using spline functions. *J. Inst. Math. Appl.* **18**, 177–188.

Ward, R. C. (1976). The QR algorithm and Hyman's method on vector computers. *Math. Comp.* **30**, 132–142.

Ward, R. C., and Gray, L. J. (1978a). Eigensystem computation for skew-symmetric matrices and a class of symmetric matrices. *ACM Trans. Math. Software* **4**, 278–285.

Ward, R. C., and Gray, L. J. (1978b). Algorithm 530, an algorithm for computing the eigensystem of skew-symmetric matrices and a class of symmetric matrices. *ACM Trans. Math. Software* **4**, 286–289.

Watkins, D. S. (1982). Understanding the QR algorithm. *SIAM Rev.* **24**, 427–440.

Weinberger, H. F. (1961). Error bounds in the Rayleigh–Ritz approximation of eigenvectors. In *Part. Diff. Eq. Cont. Mech.* (R. Langer, ed.), pp. 39–53. The Univ. of Wisconsin Press, Madison, Wisconsin.

Weinberger, H. F. (1974). *Variational Methods for Eigenvalue Approximations*. SIAM, Philadelphia, Pennsylvania.

Weinstein, A. (1937). *Etude des spectres des équations aux dérivées partielles de la théorie des plaques élastiques*. Mémoire de Sci. Math. **88**, Gauthier-Villars, Paris.

Weinstein, A. (1963). The intermediate problem and the maximum-minimum theory of eigenvalues. *J. Math. Mech.* **12**, 235–245.

Weinstein, A., and Stenger, W. (1972). *Methods of Intermediate Problems for Eigenvalues: Theory and Ramifications*. Academic Press, New York.

Weiss, R. (1974). The application of implicit Runge–Kutta and collocation methods to boundary value problems. *Math. Comp.* **28**, 449–464.

Wendland, W. L. (1980). On Galerkin collocation methods for integral equations of elliptic boundary value problems. In *Numerical Treatment of Integral Equations* (J. Albrecht and L. Collatz, eds.), pp. 244–275. Birkhaeuser, Basel.

Wendland, W. L., Stephan, E., and Hsiao, G. C. (1979). On the integral equation method for the plane mixed boundary value problem of the laplacian. *Math. Meth. in Appl. Sc.* **1**, 265–321.

Widlund, O. (1977). On best error bounds for approximation by piecewise polynomial functions. *Numer. Math.* **27**, 327–338.

Widlund, O. (1978). A Lanczos method for a class of non symmetric systems of linear equations. *SIAM J. Numer. Anal.* **15**, 801–812.

Wielandt, H. (1956). Error bounds for eigenvalues of symmetric integral equations. *Proc. Sympos. Appl. Math. 6th., Providence*. Amer. Math. Soc., Providence, Rhode Island.

Wielandt, H. (1967). *Topics in the Theory of Matrices*. Univ. of Wisconsin Press, Madison, Wisconsin.

Wilkinson, J. H. (1962). Note on the quadratic convergence of the cyclic Jacobi process. *Numer. Math.* **4**, 296–300.

Wilkinson, J. H. (1965). *The Algebraic Eigenvalue Problem*. Oxford Univ. Press (Clarendon), London and New York.

Wilkinson, J. H. (1968). Global convergence of tridiagonal QR algorithm with origin shifts. *Linear Algebra Appl.* **1**, 409–420.

Wilkinson, J. H. (1969). Global convergence of QR algorithm. In *Information Processing 68*, Vol. 1, pp. 130–133. North-Holland Publ., Amsterdam.

Wilkinson, J. H., and Reinsch, C. H. (1971). *Handbook for Automatic Computation. Linear Algebra*, Vol. 2. Springer-Verlag, Berlin and New York.
Wing, G. M. (1965). On a method for obtaining bounds on the eigenvalues of certain integral equations. *J. Math. Anal. Appl.* **11**, 160–175.
Wittenbrink, K. A. (1973). High order projection methods of moment- and collocation-type for nonlinear boundary value problems. *Computing* **11**, 255–274.
Wolfe, M. A. (1969). The numerical solution of non-singular integral and integro-differential equations by iteration with Chebyshev series. *Comput. J.* **12**, 193–196.
Wright, G. C., and Miles, G. A. (1971). An economical method for determining the smallest eigenvalues of large linear systems. *Internat. J. Numer. Methods Engrg.* **3**, 25–33.
Yamamoto, T. (1979). Componentwise error estimates for approximate solutions of non linear equations. *Inform. Process.* **2**, 121–126.
Yamamoto, T. (1980). Error bounds for computed eigenvalues and eigenvectors. *Numer. Math.* **34**, 189–199.
Yamamoto, Y., and Ohtsubo, H. (1976). Subspace iteration accelerated by using Chebyshev polynomials for eigenvalue problems with symmetric matrices. *Internat. J. Numer. Methods Engrg.* **10**, 935–944.
Yosida, K. (1960). *Lectures on Differential and Integral Equations.* Wiley (Interscience), New York.
Yosida, K. (1965). *Functional Analysis.* Springer-Verlag, Berlin and New York.
Young, A. (1954). The application of approximate product integration to the numerical solution of integral equations. *Proc. Roy. Soc. London Ser. A* **224**, 561–573.
Zaanen, A. C. (1960). *Linear Analysis.* North-Holland Publ., Amsterdam.
Zabreiko, P. P., Koshelev, A. I., Krasnoselskii, M. A., Mikhlin, S. G., Rakovshchik, L. S., and Stetsenko, V. Ya. (1975). *Integral. Equations—A Reference Textbook.* Noordhoff Int., Leyden, The Netherlands.
Zarubin, A. G. (1979). The speed of convergence of projection methods for linear equations. *U.S.S.R. Computational Math. and Math. Phys.* **19**, 265–272.
Zerbi, G., Pieseri, L., and Cabassi, F. (1971). Vibrational spectrum of chain molecules with conformational disorder: polyethylene. *Molecular Phys.* **22**, 241–256.
Zienkiewicz, O. C. (1977). *The Finite Element Method.* McGraw-Hill, New York.
Zlámal, M. (1978). Superconvergence and reduced integration in the finite element method. *Math. Comp.* **32**, 663–685.
Zlatev, Z. (1982). Use of iterative refinement in the solution of sparse linear systems. *SIAM J. Numer. Anal.* **19**, 381–399.
Zwart, P. B. (1973). Multivariate splines with nondegenerate partitions. *SIAM J. Numer. Anal.* **10**, 665–673.

Solutions to Exercises

Mario Ahués
Universidad de Chile
Santiago, Chile

Chapter 1

1.1 $\xi_1 = (x, y_1) = \frac{5}{4}$, $\xi_2 = (x, y_2) = \frac{3}{4}$.

1.2 $1 = \|I\| = \|AA^{-1}\| \leq \|A\| \, \|A^{-1}\| = K(A)$.

1.3 $\|Qx\|_2^2 = (Qx)^{\mathrm{H}}Qx = x^{\mathrm{H}}Q^{\mathrm{H}}Qx = x^{\mathrm{H}}x = \|x\|_2^2$ since $Q^{\mathrm{H}}Q = I$. Since $\|Qx\|_2 = \|x\|_2$, one has $\|Q\|_2 = 1$. But Q^{H} is also unitary, so $\|Q^{\mathrm{H}}\| = 1$.

1.4 $(I - P)^2 = I - 2P + P^2 = I - P$ since $P^2 = P$; $P(I - P)x = Px - P^2x = 0$, so $(I - P)x \in W$ and $\mathrm{Im}(I - P) \subset W$. If $Px = 0$, then $(I - P)x = x$, so $x \in \mathrm{Im}(I - P)$; hence $W = \mathrm{Im}(I - P)$. $(I - P)Px = Px - P^2x = 0$, so $M \subset \mathrm{Ker}(I - P)$. But if $(I - P)x = 0$, then $Px = x$, so $x \in M$; thus $M = \mathrm{Ker}(I - P)$.

1.5 $(P^{\mathrm{H}}P^{\mathrm{H}}x, y) = (P^{\mathrm{H}}x, Py) = (x, P^2y) = (x, Py) = (P^{\mathrm{H}}x, y)$, so $(P^{\mathrm{H}})^2 = P^{\mathrm{H}}$. For any square matrix A, $\mathrm{Ker}\, A^{\mathrm{H}} = (\mathrm{Im}\, A)^{\perp}$ and $\mathrm{Im}\, A^{\mathrm{H}} = (\mathrm{Ker}\, A^{\mathrm{H}})^{\perp}$.

1.6 If $P = P^{\mathrm{H}}$, then $(x - Px, Px) = (P^{\mathrm{H}}x - P^{\mathrm{H}}Px, x) = (Px - P^2x. x) = (Px - Px, x) = 0$, so P is an orthogonal projection. Conversely, if P is orthogonal, $\mathrm{Im}\, P = (\mathrm{Ker}\, P)^{\perp} = \mathrm{Im}\, P^{\mathrm{H}}$ and $\mathrm{Ker}\, P = (\mathrm{Im}\, P)^{\perp} = \mathrm{Ker}\, P^{\mathrm{H}}$, so $P = P^{\mathrm{H}}$.

1.7 For $k = 1$, $A\varphi = \lambda\varphi$ is true. If $A^{k-1}\varphi = \lambda^{k-1}\varphi$, then $A^k\varphi = A(A^{k-1}\varphi) = A(\lambda^{k-1}\varphi) = \lambda^{k-1}A\varphi = \lambda^k\varphi$.

1.8 Let $\lambda \in \sigma(P)$; then $P\varphi = \lambda\varphi$ for some $\varphi \neq 0$, so $P^2\varphi = \lambda^2\varphi = P\varphi = \lambda\varphi$; hence $\lambda^2 - \lambda = \lambda(\lambda - 1) = 0$.

1.9 If $D^k = 0$, then D is not regular, so $0 \in \sigma(D)$. Let $\lambda \in \sigma(D)$; then $D\varphi = \lambda\varphi$ for some $\varphi \neq 0$ and $D^k\varphi = \lambda^k\varphi = 0$, so $\sigma(D) = \{0\}$ and $r_\sigma(D) = 0$.

1.10 If $Q\varphi = \lambda\varphi$ and $\varphi^{\mathrm{H}}\varphi = 1$, then $1 = \|\varphi\|_2^2 = \|Q\varphi\|_2^2 = |\lambda|^2 \, \|\varphi\|_2^2 = |\lambda|^2$. So $r_\sigma(Q) = 1$.

1.11 Let $\lambda \in \sigma(A)$ be such that $|\lambda| = r_\sigma(A)$. Then $\|A\varphi\| = \|\lambda\varphi\| = |\lambda| \, \|\varphi\| = r_\sigma(A)\|\varphi\|$, φ being an eigenvector associated with λ. Then $\varphi \neq 0$, and

$$r_\sigma(A) = \|A\varphi\|/\|\varphi\| \leq \|A\|.$$

1.12 $A\varphi = \lambda\varphi$, $\varphi^{\mathrm{H}}A\varphi = \lambda\varphi^{\mathrm{H}}\varphi$, $\lambda = \varphi^{\mathrm{H}}A\varphi/\varphi^{\mathrm{H}}\varphi$ since $\varphi \neq 0$.

1.13 If $A^{\mathrm{H}} = A$, then $(\varphi^{\mathrm{H}}A\varphi)^{\mathrm{H}} = \varphi^{\mathrm{H}}A\varphi$, so $\varphi^{\mathrm{H}}A\varphi \in \mathbb{R}$.

1.14 If $A\varphi_1 = \lambda_1\varphi_1$, $A^{\mathrm{H}}\psi_2 = \bar{\lambda}_2\psi_2$, and $\lambda_1 \neq \lambda_2$, then $\psi_2^{\mathrm{H}}A\varphi_1 = \lambda_2\psi_2^{\mathrm{H}}\varphi_1 = \lambda_1\psi_2^{\mathrm{H}}\varphi_1$, so $(\lambda_1 - \lambda_2)(\varphi_1, \psi_2) = 0$; hence $\varphi_1 \perp \psi_2$. If $A^{\mathrm{H}} = A$, then $\psi_2 = \varphi_2$ is an eigenvector associated with λ_2.

1.15 N distinct eigenvalues imply N linearly independent eigenvectors. So A is diagonalizable.

1.16 It is true for $N = 1$. Suppose it is true for $N = n - 1$. Let A be an $n \times n$ hermitian matrix, $\lambda \in \sigma(A)$, and φ be a unitary eigenvector. Let U be an $n \times (n-1)$ matrix with orthonormal columns belonging to $\{\varphi\}^\perp$. Then $(xU)^H(xU) = I$; hence $U^H x = 0$ and $x^H U = 0$. We have

$$(xU)^H A(xU) = \begin{pmatrix} \lambda & 0 \\ 0 & U^H AU \end{pmatrix}.$$

But $U^H AU$ is an $(n-1) \times (n-1)$ hermitian matrix.

1.17 There will exist $N = \sum_{i=1}^K m_i$ linearly independent eigenvectors.

1.18 For any orthogonal projection P, $\|x - Px\|_2^2 + \|Px\|_2^2 = \|x\|_2^2$, so $\|Px\|_2 \le \|x\|_2$. For $x \in \operatorname{Im} P$, $Px = x$, so $\|P\| = 1$. If A is hermitian, two eigenvectors corresponding to different eigenvalues will be orthogonal. Hence $i \ne j \Rightarrow M_i \perp M_j$ and $\mathbb{C}^N = \textcircled{\perp}_{i=1}^K M_i$. This proves that $M_i^\perp = \textcircled{\perp}_{j \ne i} M_j$.

1.19 Let $\{\varphi, \varphi_2, \ldots, \varphi_N\}$ be the right eigenvectors and $\{\psi, \psi_2, \ldots, \psi_N\}$ the left ones. Since $\psi \perp \varphi_i$ for $i = 2, \ldots, N$, we get $\operatorname{Ker} P = \{\psi\}^\perp$, so $P\xi = \lambda(\xi)\varphi$, where λ is a linear functional such that $\lambda(\varphi) = 1$ and $\lambda(\xi) = 0$ for $\xi \perp \psi$, so $\lambda(\xi) = \alpha(\xi, \psi)$ and $1 = \lambda(\varphi) = \alpha(\varphi, \psi) = \alpha$. We obtain $P\xi = (\xi, \psi)\varphi = (\psi^H\xi)\varphi$. Hence $\|P\xi\|_2 = |\psi^H\xi|\,\|\varphi\|_2 = |\psi^H\xi| \le \|\psi\|_2\,\|\xi\|_2$, so $\|P\|_2 \le \|\psi\|_2$. But $\|P\psi\|_2 = \|\psi^H\psi\varphi\|_2 = \|\psi\|_2^2$, so $\|P\|_2 = \|\psi\|_2$. Also $1 = |\psi^H\varphi| \le \|\psi\|_2\,\|\varphi\|_2 = \|\psi\|_2$. Finally $P\xi = (\psi^H\xi)\varphi = \varphi(\psi^H\xi) = (\varphi\psi^H)\xi$ shows that $\varphi\psi^H$ is the canonical matrix of P.

1.20 Let $\varphi_1, \ldots, \varphi_N$ be the linearly independent eigenvectors of A, and let $V = (\varphi_1, \ldots, \varphi_N)$. The columns of V^{H-1} are N linearly independent eigenvectors of A^H. P_i^H is a projection on $\operatorname{Ker}(A^H - \bar{\lambda}_i I)$ parallel to $\bigoplus_{j \ne i} \operatorname{Ker}(A^H - \bar{\lambda}_j I)$. $\{\bar{\lambda}_i\}$ are the eigenvalues of A^H.

1.21 Let $\varphi = \sum_{i \notin I} \alpha_i \varphi_i$; then $(\varphi, \psi_j) = 0$ for $j \in I$, so $\varphi \in \{\psi_i; i \in I\}^\perp$. Let now

$$\varphi = \sum_{i=1}^N \alpha_i \varphi_i \in \{\psi_i; i \in I\}^\perp.$$

Then $0 = (\varphi, \psi_j) = \alpha_j$ for $j \in I$. Hence $\varphi = \sum_{i \notin I} \alpha_i \varphi_i$.

$$P\left(\sum_{j \in I} \alpha_j \varphi_j\right) = \sum_{i \in I}\left[\psi_i^H \sum_{j \in I} \alpha_j \varphi_j\right]\varphi_i = \sum_{i,j \in I} \alpha_j \psi_i^H \varphi_j \varphi_i = \sum_{i,j \in I} \alpha_j \delta_{ij} \varphi_i = \sum_{j \in I} \alpha_j \varphi_j,$$

$$P\left(\sum_{j \notin I} \alpha_j \varphi_j\right) = \sum_{i \in I, j \notin I} \alpha_j (\psi_i^H \varphi_j)\varphi_i = 0, \qquad P\xi = \left(\sum_{i \in I} \varphi_i \psi_i^H\right)\xi.$$

1.22 Let A verify $AA^H = A^H A$ and $a_{ij} = 0$ for $i > j$. The element in position i, j is equal in AA^H and in $A^H A$; thus $\sum_{k=1}^N a_{ik}\bar{a}_{jk} = \sum_{k=1}^N \bar{a}_{ki} a_{kj}$. For $i = j$, one gets $\sum_{k=i}^N |a_{ik}|^2 = \sum_{k=1}^i |a_{ki}|^2$. For $i = 1$, $|a_{11}|^2 = \sum_{k=1}^N |a_{ik}|^2$, so $a_{1k} = 0$ for $k > 1$. Suppose $a_{ik} = 0$ for $k > 1$ and $1 \le i \le j - 1$. For $i = j$, $|a_{jj}|^2 = \sum_{k=j}^N |a_{jk}|^2$, so $a_{jk} = 0$ for $k > 1$.

1.23 If A is normal, take Q unitary such that $T = Q^H AQ$ is upper triangular. Then the normal property of A implies $QTT^H Q^H = QT^H TQ^H$, so $TT^H = T^H T$ and T must be diagonal. Conversely, if D is diagonal, Q unitary, and $A = QDQ^H$, then $D^H D = DD^H$ and $A^H A = QD^H DQ^H = QDD^H Q^H = AA^H$.

1.24 Let $D = (\mu_i \delta_{ij})$ and Q be unitary such that $A = QDQ^H$. $\{\mu_i\}$ are the eigenvalues of A. $\|A\|_2 = \|QDQ^H\|_2 = \|D\|_2 = \max_i |\mu_i| = r_\sigma(A)$.

1.25 If $A = 0$, then A is nilpotent. If A is nilpotent, $r_\sigma(A) = 0$; but A being normal, $r_\sigma(A) = \|A\|_2$, so $\|A\|_2 = 0$; hence $A = 0$.

1.26 For any matrix A, $\|A\|_2^2 = \max_{\|x\|_2 = 1} x^H A^H Ax$. $A^H A$ being hermitian is unitarily similar to a diagonal matrix $D = (\eta_i \delta_{ij})$, where $\{\eta_i\}$ are the eigenvalues of $A^H A$, so $\|A\|_2^2 = \max_{\|x\|_2 = 1} x^H QDQ^H x = \max_{\|y\|_2 = 1} y^H Dy = \max_i |\eta_i| = r_\sigma(A^H A)$.

1.27 Characteristic polynomial $p(\lambda) = (\lambda - 1)^3$. Eigenvectors associated with $\lambda = 1$: $\varphi_1 = (0, 1, 1)^T, \varphi_2 = (1, 1, 0)^T$. Principal vector: $\varphi_3 = (0, 0, 1)^T$. Jordan form:

$$J = \begin{bmatrix} 1 & 0 & 0 \\ 0 & 1 & 1 \\ 0 & 0 & 1 \end{bmatrix}.$$

1.28 Let φ be an eigenvector corresponding to λ_i and $\varphi_1, \ldots, \varphi_p$ be the principal vectors following φ: $(A - \lambda_i I)\varphi_1 = \varphi$, $(A - \lambda_i I)\varphi_2 = \varphi_1$, etc. So $(A - \lambda_i I)^j \varphi_j = \varphi$, $1 \le j \le p$. Since $(A - \lambda_i I)^{l_i} \varphi_j = 0, j = 1, \ldots, p$, we get $p \le l_i - 1$. On the other hand, there are g_i linearly independent eigenvectors associated with λ_i. The first column of J_i being $\lambda_i e_1$ (e_1 is the first canonical vector of size m_i), we conclude that there are $g_i - 1$ zeros on the first upper diagonal of J_i.

1.29 Let $\{\varphi_1, \ldots, \varphi_N\}$ be the Jordan basis of A, $V = (\varphi_1, \ldots, \varphi_N)$, $J = V^{-1}AV$. One shows that $A^n = VJ^nV^{-1}$, and since

$$J = \begin{pmatrix} J_1 & & \\ & \ddots & \\ & & J_K \end{pmatrix}, \quad J^n = \begin{pmatrix} J_1^n & & \\ & \ddots & \\ & & J_K^n \end{pmatrix}.$$

Furthermore, for each box

$$J_i^n = \begin{pmatrix} J_{i1}^n & & \\ & \ddots & \\ & & J_{ig_i}^n \end{pmatrix},$$

where

$$J_{ij} = \begin{pmatrix} \lambda_i & 1 & & & \\ & \ddots & \ddots & & \\ & & \lambda_i & 1 & \\ & & & \ddots & \ddots \\ & & & & \lambda_i \end{pmatrix} = \lambda_i I + N$$

with N nilpotent, $N^{l_i} = 0$. $J_{ij}^n = \sum_{k=0}^{n} C_n^k \lambda_i^{n-k} N^k = \sum_{k=0}^{l_i} C_n^k \lambda_i^{n-k} N^k$. The sequence $s_n := C_n^k \lambda_i^{n-k}$, for a fixed k, satisfies $\lim_{n\to\infty}(s_{n+1}/s_n) = \lambda_i$, so $r_\sigma(A) < 1$ implies $\lim_{n\to\infty} A^n = 0$. Conversely, if $\lim_{n\to\infty} A^n = 0$, let λ, φ be eigenelements of A, $\|\varphi\| = 1$, so $|\lambda|^n = \|\lambda^n \varphi\| = \|A^n \varphi\| \le \|A^n\| \to 0$ if $n \to \infty$, so $|\lambda| < 1$.

1.30 $r_\sigma(A^n) = \max_i |\lambda_i^n| = (\max_i |\lambda_i|)^n = (r_\sigma(A))^n \le \|A^n\|$, so $r_\sigma(A) \le \|A^n\|^{1/n}$. Let $\varepsilon > 0$, $A_\varepsilon := (r_\sigma(A) + \varepsilon)^{-1} A$, so $r_\sigma(A_\varepsilon) < 1$ and $\lim_{n\to\infty} A_\varepsilon^n = 0$. For n sufficiently large $\|A_\varepsilon^n\| < \varepsilon$, so $\|A^n\|^{1/n} < \varepsilon^{1/n}(r_\sigma(A) + \varepsilon)$; hence $\lim_{n\to\infty} \|A^n\|^{1/n} \le r_\sigma(A)$.

1.31 Let λ be an eigenvalue of A with algebraic multiplicity m and geometric multiplicity g. The Jordan box J_λ associated with λ has then m columns and g blocks. Let p_i be the number of columns of the block $i (1 \le i \le g)$. $\sum_{i=1}^{g} p_i = m$. Let n_{ij} be the index of column j in block i $(1 \le i \le g,\ 1 \le j \le p_i)$ so that $n_{i,p_i} + 1 = n_{i+1,1}$, $1 \le i \le g - 1$. We know that $p_i \le l$, $1 \le i \le g$, where l is the ascent of λ. $\varphi_{n_{i1}}$, $i = 1, \ldots, g$, are the eigenvectors and $\varphi_{n_{ij}}$, $i = 1, \ldots, g$, $j = 2, \ldots, p_i$, are the principal vectors associated with λ. We have then $A\varphi_{n_{i1}} = \lambda \varphi_{n_{i1}}$, $A\varphi_{n_{ij}} = \lambda \varphi_{n_j} + \varphi_{n.j-1}$, $2 \le j \le p_i$. $P(\sum_1^N \alpha_k \varphi_k) = \sum_{i=1}^{g} \sum_{j=n_{i1}}^{n_{ip_i}} \alpha_j \varphi_j$, so for $x = \sum_1^N \alpha_k \varphi_k \in \mathbb{C}^N$, $APx = \lambda Px + Dx$, where $Dx = \sum_{i=1}^{g} \sum_{j=n_{i2}}^{n_{ip_i}} \alpha_j \varphi_{j-1}$, since $APx \in M$, $PAPx = APx = \lambda Px + Dx$. We note that D verifies $D\varphi_k = 0$ if $k \notin \{n_{ij}; 1 \le i \le g, 2 \le j \le p_i\}$, $D\varphi_{n_{lk}} = \varphi_{n_{l,k-1}}$ for $2 \le k \le p_l$, $1 \le l \le g$. For instance, $D\varphi_{n_{i1}} = 0$ for $1 \le i \le g$ and

$$D^r \varphi_{n_{lk}} = \begin{cases} 0 & \text{if } 1 \le k \le r, \\ \varphi_{n_{l,k-r}} & \text{if } r < k \le p_l. \end{cases}$$

So $D^{p_l}\varphi_{n_{lk}} = 0$, $1 \le k \le p_l$, $1 \le l \le g$, and since $\max_{1\le i\le g} p_i \le l$, then $D^l = 0$. We have shown that if $\lambda_1, \ldots, \lambda_K$ are the distinct eigenvalues of A and if $P_1, \ldots, P_K$ are the corresponding spectral projections, then $AP_i = P_i A P_i = \lambda_i P_i + D_i$, where $D_i^{l_i} = 0$. It is easily seen that $P_i P_j = \delta_{ij} P_i$ and $\sum_1^K P_i = I$. Summing $AP_i = \lambda_i P_i + D_i$ over $i = 1, \ldots, K$ we find $A = \sum_1^K (\lambda_i P_i + D_i)$. Then $P(\sum_1^N \alpha_k \varphi_k) = \sum_{k\in T} \alpha_k \varphi_k$, so we have $PA(\sum_1^N \alpha_k \varphi_k) = \sum_{k\in T} \alpha_k A\varphi_k = A(\sum_{k\in T} \alpha_k \varphi_k) = AP(\sum_1^N \alpha_k \varphi_k)$ where $T := \{n_{ij};\ 1 \le i \le g, 1 \le j \le p_i\}$.

1.32 We proved $D\varphi_k = 0$ if $k \notin T$ and $D\varphi_{n_{lk}} = \varphi_{n_{l,k-1}}$ for $1 \le l \le g$, $2 \le k \le p_l$. If P' is the spectral projection corresponding to an eigenvalue other than λ, we have $P'D = 0$ and $DP' = 0$. But $PD = DP = D$, so in general $D_i P_j = P_j D_i = \delta_{ij} D_i$. Also, if D' is the nilpotent matrix associated with an eigenvalue other than λ, then $\operatorname{Im} D'$ is generated by basis vectors other than φ_k, $k \in T$, and since $D\varphi_k = 0$ for $k \notin T$, we have $DD' = 0$.

1.33 $A = \sum_1^K (\lambda_j P_j + D_j)$, $\sum_1^K P_j = I$, so $(A - \lambda_i I) = \sum_{j\ne i} (\lambda_j - \lambda_i) P_j + \sum_1^K D_j$. $(A - \lambda_i I)^{l_i} = \sum_{j\ne i} (\lambda_j - \lambda_i) P_j + \sum_{j\ne i} \sum_{t=0}^{l_i - 1} C_{l_i}^t (\lambda_j - \lambda_i)^t D_j^{l_i - t}$ because $P_j^{l_i} = P_j$, $D_i^{l_i} = 0$, $P_j D_i = D_i P_j = \delta_{ij} D_j$, $D_i D_j = \delta_{ij} D_j^2$, so $\operatorname{Im}(A - \lambda_i I)^{l_i} \subseteq \bigoplus_{j\ne i} M_j$. But $M_i = \operatorname{Ker}(A - \lambda_i I)^{l_i}$, so $\mathbb{C}^N = M_i \oplus \operatorname{Im}(A - \lambda_i I)^{l_i}$ and also $\mathbb{C}^N = \bigoplus_{j=1}^K M_j$, so $\operatorname{Im}(A - \lambda_i I)^{l_i} = \bigoplus_{j\ne i} M_j$.

1.34 Similar to solution of Exercise 1.20.

1.35 Let λ be a simple eigenvalue of A with right eigenvector φ and left eigenvector ψ. Let $\lambda' \ne \lambda$ be another eigenvalue with $\varphi_1^0, \ldots, \varphi_g^0$ as right eigenvectors; φ_i^j, $i = 1, \ldots, g$, $j = 1, \ldots, t_i$, as principal vectors: $(A - \lambda' I)\varphi_i^j = \varphi_i^{j-1}$. We know that $\psi^{\mathrm{H}}\varphi_i^0 = 0$, $i = 1, \ldots, g$, so $\psi^{\mathrm{H}}(A - \lambda' I)\varphi_i^1 = \psi^{\mathrm{H}}\varphi_i^0 = 0$, $i = 1, \ldots, g$, so $(\lambda - \lambda')\psi^{\mathrm{H}}\varphi_i^1 = 0$ and then $\psi^{\mathrm{H}}\varphi_i^1 = 0$, $i = 1, \ldots, g$. Suppose $\psi^{\mathrm{H}}\varphi_i^j = 0$ for $1 \le j \le t_i - 1$, $i = 1, \ldots, g$. We find easily $\psi^{\mathrm{H}}\varphi_i^{j+1} = 0$. This shows that

$$\psi \perp \bigoplus_{\substack{\delta\in\sigma(A)\\ \delta\ne\lambda}} M_\delta$$

where M_δ is the invariant subspace associated with δ; hence

$$\bigoplus_{\substack{\delta\in\sigma(A)\\ \delta\ne\lambda}} M_\delta \subset \{\psi\}^\perp.$$

But both subspaces have dimension $N - 1$, so they must be equal.

1.36 If $J = V^{-1}AV$, then $J^{\mathrm{H}} = V^{\mathrm{H}}A^{\mathrm{H}}(V^{\mathrm{H}})^{-1}$. The columns of V are the Jordan basis of A, and the columns of $(V^{\mathrm{H}})^{-1}$ are the adjoint basis of the Jordan basis in which A^{H} takes the form J^{H}. It can be considered as a Jordan basis with the reverse ordering, starting from the last vector. For example, $A = \begin{pmatrix} 1 & 1 \\ 0 & 1 \end{pmatrix}$ is already in the Jordan form, so the Jordan basis is the canonical one $\{e_1, e_2\}$. The adjoint basis is again $\{e_1, e_2\}$ and precisely $A^{\mathrm{H}} = \begin{pmatrix} 1 & 0 \\ 1 & 1 \end{pmatrix}$. e_1(resp. e_2) is the eigen- (resp. principal) vector for A; e_2 (resp. e_1) is the eigen- (resp. principal) vector for A^{H}.

1.37 If A is hermitian, P is an orthogonal projection and $\varphi = \psi$. $\{\varphi\}^\perp = (\operatorname{Ker}(A - \lambda I))^\perp = \operatorname{Im}(A - \lambda I)$, so $\Pi = I - P$ is the orthogonal projection on $\operatorname{Im}(A - \lambda I)$.

1.38 Since $\lambda(\varepsilon) = \lambda - \varepsilon\psi^{\mathrm{H}}L\varphi + \sum_{i\ge 2} v_i \varepsilon^i$ converges geometrically,

$$\lim_{\varepsilon\to 0} \frac{\lambda(\varepsilon) - \lambda}{\varepsilon} = -\psi^{\mathrm{H}}L\varphi.$$

1.39 In Theorem 1.7 we found $v_1 = -\psi^{\mathrm{H}}L\varphi/\psi^{\mathrm{H}}\varphi$, so

$$|\lambda - \lambda'| \le \frac{\|\psi\|_2\,\|\varphi\|_2}{|\psi^{\mathrm{H}}\varphi|}\varepsilon + O(\varepsilon^2),$$

which shows that when no normalization condition is imposed, the condition number for λ is $\|\psi\|_2\,\|\varphi\|_2/|\psi^{\mathrm{H}}\varphi|$. Since $\cos(\pi/2 - \theta) = |\psi^{\mathrm{H}}\varphi|/\|\psi\|_2\,\|\varphi\|_2$, the condition number equals $1/\sin\theta$.

1.40 From

$$\nu_k = -\psi^{\mathrm{H}} L \varphi_{k-1}, \qquad k \geq 1,$$

$$\eta_k = S\left(\sum_{i=1}^{k-1} \nu_i \eta_{k-i} + L\eta_{k-1}\right), \qquad k \geq 2, \qquad \eta_0 = \varphi, \qquad \eta_1 = SL\varphi,$$

we find $\lambda' = \lambda + \psi^{\mathrm{H}} H\varphi - \psi^{\mathrm{H}} HSH\varphi - \psi^{\mathrm{H}} LS(-\psi^{\mathrm{H}} L\varphi[SL\varphi] + LSL\varphi)\varepsilon^3 + \cdots$, so $\lambda' = \lambda + \psi^{\mathrm{H}} H\varphi - \psi^{\mathrm{H}} HSH\varphi + O(\varepsilon^3)$. Then if $\psi^{\mathrm{H}} H\varphi = 0$, $s = \|S\|_2$, $\varepsilon = \|H\|_2$, $\varepsilon' = \|H\varphi\|_2$, we get $|\lambda - \lambda'| \leq s\varepsilon\varepsilon' \|\psi\|_2$.

1.41 $V = (\varphi, \varphi_1, \ldots, \varphi_{N-1})$ is regular, and $(V^{-1})^{\mathrm{H}} = (\psi, \psi_1, \ldots, \psi_{N-1})$.

$$V^{-1}AV = \begin{pmatrix} \lambda & & & 0 \\ & \mu_1 & & \\ & & \ddots & \\ 0 & & & \mu_{N-1} \end{pmatrix}, \qquad V^{-1}PV = \left(\begin{array}{c|c} 1 & 0 \\ \hline 0 & 0 \end{array}\right)$$

$$V^{-1}(I - P)V = I - V^{-1}PV = \left(\begin{array}{c|c} 0 & 0 \\ \hline 0 & I_{N-1} \end{array}\right)$$

where I_{N-1} is the identity in $\mathbb{C}^{N-1}$,

$$V^{-1}(A - \lambda I)V = \left(\begin{array}{c|ccc} 0 & & 0 & \\ \hline & \mu_1 - \lambda & & 0 \\ 0 & & \ddots & \\ & 0 & & \mu_{N-1} - \lambda \end{array}\right),$$

$$S = V \left(\begin{array}{c|ccc} 0 & & 0 & \\ \hline & \dfrac{1}{\mu_1 - \lambda} & & 0 \\ 0 & & \ddots & \\ & 0 & & \dfrac{1}{\mu_{N-1} - \lambda} \end{array}\right) V^{-1}.$$

Then

$$SH\varphi = V \left(\begin{array}{c|ccc} 0 & & 0 & \\ \hline & \dfrac{1}{\mu_1 - \lambda} & & 0 \\ 0 & & \ddots & \\ & 0 & & \dfrac{1}{\mu_{N-1} - \lambda} \end{array}\right) \begin{pmatrix} \psi^{\mathrm{H}} H\varphi \\ \psi_1^{\mathrm{H}} H\varphi \\ \vdots \\ \psi_{N-1}^{\mathrm{H}} H\varphi \end{pmatrix} = \sum_{j=1}^{N-1} \frac{\psi_j^{\mathrm{H}} H\varphi}{\mu_j - \lambda} \varphi_j.$$

1.42 $\|\psi_1\| = 1$, $\|\psi_2\| = \sqrt{1 + 10^{20}} \approx 10^{10}$, $\|\psi_3\| = 10^{10}\sqrt{1 + 10^{-20}} \approx 10^{10}$.

1.43 Since Q unitary implies $\|Qx\|_2 = \|x\|_2$, we get $\|Q^{\mathrm{H}} SQ\|_2 = \max_{\|x\|_2 = 1} \|Q^{\mathrm{H}} SQx\|_2 = \max_{\|x\|_2 = 1} \|SQx\|_2 = \max_{\|y\|_2 = 1} \|Sy\|_2 = \|S\|_2$.

1.44 Let $(\mathbf{Y}', \mathbf{Z}')$ be another basis for $\mathbb{C}^N$ such that $\mathbf{Y}'$ is a basis for $\mathbf{Y}$. Then

$$(\mathbf{Y}, \mathbf{Z}) = (\mathbf{Y}', \mathbf{Z}') \begin{pmatrix} A_1 & A_{12} \\ 0 & A_2 \end{pmatrix} = (\mathbf{Y}' A_1, \mathbf{Y}' A_{12} + \mathbf{Z}' A_2)$$

where A_1 is $r \times r$ and both A_1 and A_2 are invertible. $\mathbf{X}_k = \mathbf{Y} C_k + \mathbf{Z} D_k = \mathbf{Y}' C_k' + \mathbf{Z}' D_k'$ with $C_k' = A_1 C_k + A_{12} D_k$, $D_k' = A_2 D_k = (I + A_{12} D_k C_k^{-1} A_1^{-1}) A_1 C_k$. $D_k' C_k' = A_2 F_k (I + A_{12} F_k)^{-1}$ with $F_k = D_k C_k^{-1} A_1^{-1} \to 0$.

1.45 $\mathbf{X}_k = \mathbf{Y}(I + \mathbf{Z}D_k C_k^{-1})C_k$ for k large enough.

1.46 If $\mathbf{X}_k$ spans X_k and $\mathbf{X}'_k$ spans X'_k, then the augmented matrix $(\mathbf{X}_k, \mathbf{X}'_k)$ spans $X_k + X'_k$.

1.47 If $X_k \to Y$, then C_k may be chosen such that $C_k \to I$. Then $\mathbf{X}_k \to \mathbf{Y}$, and the matrix limit of $\mathbf{X}_k$ is unique.

1.48 The minor of the $(1, N)$ element of the unreduced H is nonzero. Therefore rank $H \geq N - 1$. $H - \lambda I$ is also an unreduced Hessenberg matrix; hence for any eigenvalue λ the null space of $H - \lambda I$ has dimension 1. H is nondefective iff its eigenvalues are simple. Now a hermitian Hessenberg matrix is tridiagonal and diagonalizable.

1.49 It can be easily shown inductively that the power method may be written as follows $q_0 = x/\|x\|_2$, $q_k = A^k x/\|A^k x\|_2$. For the QR method we establish $A_{k+1} = Q_k^H A_k Q_k$, $A_{k+1} = \mathbf{Q}_k^H A \mathbf{Q}_k$, $\mathbf{R}_k = \mathbf{Q}_k^H A^k$, $\mathbf{Q}_k = A\mathbf{Q}_{k-1}R_k^{-1}$, so we have $\mathbf{Q}_k^H = \mathbf{Q}_k^{-1} = R_k \mathbf{Q}_{k-1}^H A^{-1}$ and $e_N^H \mathbf{Q}_k^H = e_N^H R_k \mathbf{Q}_{k-1}^H A^{-1} = r_{NN}^{(k)} e_N^H \mathbf{Q}_{k-1}^H A^{-1}$, where $r_{NN}^{(k)}$ is the element of R_k in position (N, N) which is nonzero since R_k is regular. On the other hand, $\mathbf{Q}_k^H = \mathbf{R}_k (A^{-1})^k = \mathbf{R}_k A^{-k}$, so $e_N^H \mathbf{Q}_k^H = e_N^H \mathbf{R}_k A^{-k} = s_{NN}^{(k)} e_N^H A^{-k}$, where $s_{NN}^{(k)}$ is the element of $\mathbf{R}_k$ in position (N, N) which is nonzero since $\mathbf{R}_k$ is regular. This shows that $\mathbf{Q}_k e_N = (A^H)^{-k} e_N / \|(A^H)^{-k} e_N\|_2$; that is, the Nth column of $\mathbf{Q}_k$ is computed by the power method on $(A^H)^{-1}$ with starting vector e_N.

1.50 $\sigma_k = a_{NN}^k$, $A_k - \sigma_k I = Q_k R_k$, $Q_k^H(A_k - \sigma_k I) = R_k$, $Q_k^H = R_k(A_k - \sigma_k I)^{-1}$. $Q_k e_N = (A_k^H - \bar{\sigma}_k I)^{-1} R_k^H e_N = (A_k^H - \bar{\sigma}_k I)^{-1} \bar{r}_{NN}^{(k)} e_N$, where $r_{NN}^{(k)} \neq 0$ is the element of R_k in position (N, N). Hence $Q_k e_N = (A_k^H - \bar{\sigma}_k I)^{-1} e_N / \|(A_k^H - \bar{\sigma}_k I)^{-1} e_N\|_2$. One Rayleigh quotient iteration with $q_{k-1} = e_N$ and $\rho_{k-1} = \bar{a}_{NN}^k$ gives

$$q_k = \frac{(A_k^H - \bar{a}_{NN}^k I)^{-1} e_N}{\|(A_k^H - \bar{a}_{NN}^k I)^{-1} e_N\|} = Q_k e_N$$

and

$$\rho_k = \frac{e_N^H Q_k^H A_k^H Q_k e_N}{e_N^H Q_k^H Q_k e_N} = \frac{e_N^H A_{k+1}^H e_N}{e_N^H e_N} = \bar{a}_{NN}^{k+1}$$

since $Q_k^H Q_k = I$ and $e_N^H e_N = 1$.

1.51 Let $\varepsilon = \|H\|$. $y_i = A'^{-1} H y_{i-1} = (A'^{-1} L)^i x' \varepsilon^i$, where $L = (1/\varepsilon)H$. Then

$$x = x' \sum_{i=0}^{\infty} \varepsilon^i (A'^{-1} L)^i$$

and

$$\|x - x_k\| = \left\| \sum_{i=k+1}^{\infty} (A'^{-1} L)^i \varepsilon^i \right\| \|x'\| \leq \frac{\|x'\| \, \|A'^{-1}\|^{k+1} \varepsilon^{k+1}}{1 - \|A'^{-1}\| \varepsilon}.$$

1.52 $A' = A'^{-1} = I$, $x' = (1, 1, 1, 1)^T$, $x_1 = x' + Hx' = 0.889x'$, $x_2 = x' + Hx_1 = 0.90132x'$, $x_3 = x' + Hx_2 = 0.89995x'$, $\varepsilon = \|H\|_\infty = 0.111$, $\|x_3 - x\|_\infty \leq \varepsilon^4/(1 - \varepsilon) < 1.71 \times 10^{-4}$.

1.53 $\psi'^H x = 0$ is equivalent to $P'x = 0$ since $\operatorname{Ker} P' = \{\psi'\}^\perp$. But $(A' - \lambda' I)_{\restriction\{\psi'\}^\perp}$ is bijective, so $x = (A' - \lambda' I)^{-1}_{\restriction\{\psi'\}^\perp}(I - P')b = S'b = S'(I - P')b$ is the only solution of the system.

1.54 (1.13) is a system of $n + 1$ equations in n unknowns of rank n. Perform the Gauss decomposition of the $(n + 1) \times n$ matrix, and the corresponding calculations on the right-hand side. Then delete the last equation, which is identically zero, to get a regular triangular system of n equations in n unknowns. Use pivoting for numerical stability.

1.55 See solution to Exercise 1.37.

1.56 Take $\lambda = \sum_{j=0}^{\infty} v_j \varepsilon^j$, $\hat{\varphi} = \sum_{j=0}^{\infty} \hat{\eta}_j \varepsilon^j$, $\lambda_k = \sum_{j=0}^{k} v_j$, $\hat{\varphi}_k = \sum_{j=0}^{k} \hat{\eta}_j$. $L = -(1/\varepsilon)H$. Formally, $A\hat{\varphi} = \lambda\hat{\varphi}$ is equivalent to

$$A'\hat{\eta}_0 + \sum_{j=1}^{\infty} \varepsilon^j (A'\hat{\eta}_j + L\hat{\eta}_{j-1}) = v_0 \hat{\eta}_0 + \sum_{i=1}^{\infty} \varepsilon^i \left(\sum_{j=0}^{i} v_j \hat{\eta}_{i-j} \right),$$

which gives $A'\hat{\eta}_0 = \nu_0\hat{\eta}_0$. Choose $\hat{\eta}_0 = \varphi'$, $\nu_0 = \lambda'$, so $\lambda_0 = \lambda'$ and $\hat{\varphi}_0 = \hat{\varphi}'$. Next, compare the coefficients of ε in both series. This gives

$$(A' - \lambda' I)\hat{\eta}_1 = \nu_1\varphi' - L\varphi' \qquad (*)$$

so $\hat{\eta}_1 = \nu_1\Sigma'\varphi' - \Sigma' L\varphi' = -\Sigma' L\varphi'$ because $\Sigma'\varphi' = 0$ since $I - Q$ is a projection on $\{y\}^\perp$ parallel to $\{\varphi'\}$. From $(*)$ we get also $y^H(A' - \lambda' I)\hat{\eta}_1 = \nu_1 - y^H L\varphi'$, so $\nu_1 = y^H A'\hat{\eta}_1 + y^H L\varphi'$ because $y^H\hat{\eta}_1 = -y^H\Sigma' L\varphi' = 0$ since $y^H\Sigma' = 0$ because $I - Q$ projects on $\{y\}^\perp$. Now proceed inductively: $y^H\hat{\eta}_k = 0$ for all k, and $\hat{\eta}_k = \Sigma'(\sum_{j=1}^{k-1}\nu_j\hat{\eta}_{k-j} - L\hat{\eta}_{k-1})$, $\nu_k = y^H A'\hat{\eta}_k + y^H L\hat{\eta}_{k-1}$, for $k \geq 2$. Finally set $\varepsilon = 1$.

1.57 The overall complexity is the same. The main difference is that in (1.11), λ_k depends on φ_{k-1} only. In (1.14), however, λ_k depends on $\hat{\varphi}_{k-1}$ and $\hat{\eta}_k$, so $\hat{\varphi}_k$ has to be computed before λ_k.

1.58 $A = A' - H$ with $A' = \mathrm{diag}(1, 2, 3, 4)$. For $\lambda' = 1$, $\varphi' = \psi' = e_1$, $P' = e_1 e_1^T$. The iteration is

$$\lambda_0 = 1, \qquad \lambda_k = e_1^T A\varphi_{k-1},$$

$$\varphi_0 = \eta^0 = e_1 \qquad \begin{cases} (A - I)(\varphi_k - e_1) = (I - P')\left[H\varphi_{k-1} + \sum_{i=1}^{k}\sum_{j=1}^{i}\nu_j\eta_{i-j}\right], \\ e_1^T\varphi_k = 1. \end{cases}$$

For example,

$$\lambda_1 = \lambda_0, \qquad \varphi_1 = e_1 + \begin{pmatrix} 0 \\ x_2 \\ x_3 \\ x_4 \end{pmatrix} \quad \text{where} \quad \begin{cases} x_2 = -0.1, \\ 2x_3 = -0.1, \\ 3x_4 = -0.1. \end{cases}$$

Hence $x_2 = -0.1$, $x_3 = -0.05$, $x_4 = -0.033$. And $\lambda_2 = e_1^T A\varphi_1 = 0.9684$.

1.59 If A is almost upper triangular, the perturbation H should be taken as the lower triangular part. Then A' is triangular, has known eigenvalues, and (1.13) is easy to solve, to refine on the first and last diagonal element of A'. For the other diagonal elements, see Chapter 3, Section 8.4.

1.60 $P' = e_1 e_1^T, \qquad H = \begin{pmatrix} 0 & -5 \\ -\frac{1}{25} & 0 \end{pmatrix}, \qquad \varphi_k = e_1 + \begin{pmatrix} 0 \\ x_k \end{pmatrix}.$

$\lambda = 0.8282$, $\lambda_0 = 1$, $\varphi_0 = \eta^0 = e_1$, $x_k = e_2^T[H\varphi_{k-1} + \sum_{i=1}^{k}\sum_{j=1}^{i}\nu_j\eta_{i-j}]$ yields

$$\lambda_1 = 1, \qquad x_1 = -\tfrac{1}{25}, \qquad \lambda_2 = 0.8, \qquad x_2 = -\tfrac{1}{25},$$

$$\lambda_3 = 0.8, \qquad x_3 = -\tfrac{4}{5}\tfrac{1}{25} = -0.032, \qquad \lambda_4 = 0.84, \qquad x_4 = x_3,$$

$$\lambda_5 = 0.84, \qquad x_5 = -0.0352, \qquad \lambda_6 = 0.8384, \qquad x_6 = x_5, \qquad \lambda_7 = \lambda_6, \qquad \text{etc.}$$

1.61 Any x in $\{y\}^\perp$ is such that $Qx = 0$; i.e., $(I - Q)x = x$; hence

$$(I - Q)(A - \zeta I)(I - Q) = (\tilde{A} - \zeta I)(I - Q).$$

1.62 $\tilde{A}x = \zeta Qx + (I - Q)A(I - Q)x = \zeta x$ since $Qx = x$ and $(I - Q)x = 0$. Also, we have $y^H\tilde{A} = \zeta y^H Q + y^H(I - Q)A(I - Q) = \zeta y^H$ since $y^H Q = y^H x y^H = y^H$ and $y^H(I - Q) = 0$. So x is a right eigenvector of $\tilde{A}$ and y a left eigenvector of $\tilde{A}$, both corresponding to the eigenvalue ζ. If ζ is simple, $Q = xy^H$ defines the spectral projection.

$$(\tilde{A} - \zeta I)(I - Q)\Sigma = (I - Q)(\tilde{A} - \zeta I)(\tilde{A} - \zeta I)^{-1}(I - Q) = I - Q.$$

1.63 $\tilde{H}\xi = (y^H\xi)(A - \zeta I)x + (y^H(A - \zeta I)\zeta)x = (A - \zeta I)xy^H\xi + xy^H(A - \zeta I)\xi$ since $y^H\xi$ and $y^H(A - \zeta I)\xi$ are complex numbers. But $xy^H = Q$, so $\tilde{H}\xi = [(A - \zeta)Q + Q(A - \zeta I)]\xi$.

1.64 $A = \tilde{A} + \tilde{H}$. Set $H = -\tilde{H}$, $\varepsilon = 1$ in Exercise 1.54 (see the solution) and the interpretation follows. $\tilde{H}$ is such that $Q\tilde{H}Q = (I - Q)\tilde{H}(I - Q) = 0$. Therefore $\eta_{2k} = 0$ for $k \geq 1$ and $v_{2k-1} = 0$ for $k \geq 2$.

1.65 The rate of convergence $\lambda_k \to \lambda$ is linear in (1.16), whereas the convergence $\rho_k \to \lambda$ is quadratic (at least) in the Rayleigh quotient iteration.

1.66 $x = y = e_1$, $Q = e_1e_1^T$, $\zeta = a$, $\sigma = \|(C - aI)^{-1}\|_2$. If a is not an eigenvalue of C and if $r' = \|u\|_2\|v\|_2\sigma^2 < \frac{1}{4}$, then there exists a simple eigenvalue λ of A and an associated eigenvector φ normalized by $e_1^T\varphi = 1$ such that

$$|\lambda - a| \leq g(r')|v^H(C - aI)^{-1}u| \leq 2\sigma\|u\|_2\|v\|_2,$$
$$\|\varphi - e_1\|_2 \leq g(r')\|(C - aI)^{-1}u\|_2 \leq 2\sigma\|u\|_2.$$

1.67 If A is hermitian, $\tilde{A}$ is hermitian if we take $y = x$, so Q and $I - Q$ are orthogonal projections and hence hermitian. $u = v$, $|u^H\Sigma u| \leq \sigma\|u\|_2^2$ with

$$\sigma = \|\Sigma\|_2 = [\text{dist}(\zeta, \sigma[(I - Q)A]_{\restriction\{x\}^\perp})]^{-1}.$$

1.68 Left to the reader.

1.69 Let $B = (x_i\delta_{ij})$, so $B^{-1} = (x_i^{-1}\delta_{ij})$. Then $B^{-1}A = (x_i^{-1}a_{ij})$ and $B^{-1}AB = ((x_j/x_i)a_{ij})$. Hence $B^{-1}AB$ and A share the same diagonal. Since $\sigma(A) = \sigma(B^{-1}AB)$, apply Theorem 1.19 to $B^{-1}AB$.

1.70 Choose first $x_1 = \frac{1}{3}$, $x_2 = x_3 = 10^{-4}$ and find $G_1 = \{z \in \mathbb{C}; |z - 1| \leq 9 \times 10^{-8}\}$, $G_2 = \{z \in \mathbb{C}; |z - 2| \leq 0.3335\}$, $G_3 = \{z \in \mathbb{C}; |z - 3| \leq 0.4001\}$, which are disjoint. So G_1 localizes an eigenvalue close to 1 with precision 9×10^{-8}. Next choose $x_1 = x_3 = 10^{-4}$, $x_2 = \frac{1}{2}$; G_1, G_2, G_3 will be disjoint and $G_2 = \{z \in \mathbb{C}; |z - 2| \leq 4.2 \times 10^{-8}\}$. Finally, choose $x_1 = x_2 = 10^{-4}$ and $x_3 = \frac{1}{4}$ to find three Gershgorin disks which are disjoint and

$$G_3 = \{z \in \mathbb{C}; |z - 3| \leq 8.8 \times 10^{-8}\}.$$

The bounds given by Exercise 1.40 are also of order 10^{-8} since $\|H\|_2 \leq \sqrt{3}\|H\|_\infty = 3\sqrt{3} \times 10^{-4}$ and $\|(A - I)^{-1}_{\restriction\{e_1\}^\perp}\|_2 = \frac{1}{2}$.

1.71 Apply the Perron–Frobenius theorem: there exists an eigenvector of B associated with the eigenvalue $r_\sigma(B)$ of B which has positive components.

1.72 If A is not regular, then $0 \in \sigma(A)$ and we can find i such that $|a_{ii} - 0| \leq \sum_{j\neq i}|a_{ij}|$. That is, if $|a_{ii}| > \sum_{j\neq i}|a_{ij}|$ for all i, then A must be regular.

1.73 Since $a_{ii} > \sum_{j\neq i}|a_{ij}|$ for all i and since $\forall\lambda \in \sigma(A) \subseteq \mathbb{R}$, $\exists i$ such that $-\sum_{j\neq i}|a_{ij}| \leq \lambda - a_{ii} \leq \sum_{j\neq i}|a_{ij}|$; hence $0 < a_{ii} - \sum_{j\neq i}|a_{ij}| \leq \lambda$. So all the eigenvalues of A are positive.

1.74 An almost diagonal matrix can be written under the form

$$\begin{pmatrix} a & v^H \\ u & C \end{pmatrix},$$

where $\|u\|$ $\|v\|$ are small enough. If $\max(\|u\|_1, \|v\|_1)$ is small enough, there exists a simple eigenvalue λ of A such that $|\lambda - a| \leq \max(\|u\|_1, \|v\|_1)$ by Corollary 1.20. See Exercise 1.66 for the application of Corollary 1.18. The latter gives a sharper bound.

1.75 $\Delta = VDV^{-1}$, $D = (\mu_i\delta_{ij})$, $A = \Delta + H$. Let φ be an eigenvector associated with the eigenvalue λ of A and $\|\varphi\| = 1$. Let $r = \Delta\varphi - \lambda\varphi = \Delta\varphi - A\varphi = -H\varphi$. By Proposition 1.15 there is an eigenvalue μ_i of Δ such that $|\lambda - \mu_i| \leq \|V\|\,\|V^{-1}\|\,\|r\|$. But $\|r\| = \|H\varphi\| \leq \|H\|$.

1.76 Since $Q = xy^H$ and $\|x\|_2 = 1$, we have $\|Q\|_2 = \|y\|_2$. Then, if $a = \sigma$ and $\tilde{\varepsilon} = \varepsilon\varepsilon^*$, we get $\tilde{r} = \sigma^2\|y\|_2\varepsilon\varepsilon^* = r'$ and Corollary 1.18 follows. If $a = \sigma$ and $\tilde{\varepsilon} = r/\sigma^2\|y\|_2$, then $\tilde{r} = r$ and Theorem 1.17 follows.

1.77 $\|I - Q\|_1 = 1$; $\Sigma = (I - Q)\hat{A}_{a_{ii}}^{-1}(I - Q)$: $\hat{A}_{a_{ii}}^{-1} = (R - H)^{-1} = R^{-1}\sum_{i \geq 0}(HR^{-1})^i$.

$$\|\Sigma\|_1 \leq \|I - Q\|_1^2 \|R^{-1}\|_1 \frac{1}{1 - \|HR^{-1}\|_1} \leq \frac{s_1}{1 - s_1\varepsilon_1} = \frac{2}{3}a < a.$$

We recall that $A = R - H$. The eigenvector of R associated with a_{ii} can be written under the form

$$x = (\times \quad \cdots \quad \times \quad \underset{\substack{\uparrow \\ i}}{1} \quad 0 \quad \cdots \quad 0)^{\mathrm{T}}$$

where $\times$ denotes a (possibly) nonzero component. Hence $\|\varphi - x\| \leq c\varepsilon$, where φ is an eigenvector of A associated with a_{ii}. Therefore $e_i^{\mathrm{T}}\varphi \neq 0$ for ε small enough, and φ may be normalized such that $e_i^{\mathrm{T}}\varphi = 1$.

1.78 With $A = \begin{pmatrix} 0 & -\alpha \\ \alpha & 1/\beta \end{pmatrix}$, $x = y = e_1$, $\zeta = 0$, $Q = e_1e_1^{\mathrm{T}}$, $\Sigma = \begin{pmatrix} 0 & 0 \\ 0 & \beta \end{pmatrix}$, $v^{\mathrm{H}}\Sigma^k u = -\alpha^2\beta^k$. Therefore $a = |\beta|$, $\tilde{e} = \alpha^2$, $\tilde{r} = (\alpha\beta)^2$, and the bounds of Theorem 1.22 are sharp.

1.79 See Wilkinson (1965). The QR algorithm yields a block triangular matrix $\tilde{R}$ with diagonal blocks of size at most 2, and a unitary matrix $\tilde{Q}$ such that $A\tilde{Q} - \tilde{Q}\tilde{R} = E$. Set $F = \tilde{Q}^{-1}E$, then $\tilde{Q}^{-1}A\tilde{Q} = \tilde{R} + F$. Since the computed $\tilde{Q}$ is almost unitary, F can be computed as $\tilde{Q}^{\mathrm{H}}E$.

1.80
$$\begin{aligned}
\|Ax - \lambda x\|_2^2 &= (x^{\mathrm{H}}A^{\mathrm{H}} - \bar{\lambda}x^{\mathrm{H}})(Ax - \lambda x) \\
&= x^{\mathrm{H}}A^{\mathrm{H}}Ax - \lambda x^{\mathrm{H}}A^{\mathrm{H}}x - \bar{\lambda}x^{\mathrm{H}}Ax + \bar{\lambda}\lambda \\
&= x^{\mathrm{H}}A^{\mathrm{H}}Ax + (x^{\mathrm{H}}Axx^{\mathrm{H}}A^{\mathrm{H}}x - x^{\mathrm{H}}Axx^{\mathrm{H}}A^{\mathrm{H}}x) \\
&\quad - \lambda x^{\mathrm{H}}A^{\mathrm{H}}x - \bar{\lambda}x^{\mathrm{H}}Ax + \bar{\lambda}\lambda \\
&= x^{\mathrm{H}}A^{\mathrm{H}}Ax - x^{\mathrm{H}}Axx^{\mathrm{H}}A^{\mathrm{H}}x + x^{\mathrm{H}}Ax[x^{\mathrm{H}}A^{\mathrm{H}}x - \bar{\lambda}] - \lambda[x^{\mathrm{H}}A^{\mathrm{H}}x - \bar{\lambda}] \\
&= x^{\mathrm{H}}A^{\mathrm{H}}(Ax - xx^{\mathrm{H}}Ax) + |x^{\mathrm{H}}Ax - \lambda|^2,
\end{aligned}$$

which is minimal when $\lambda = x^{\mathrm{H}}Ax$.

1.81
$$\mu_k = \min\{\max\{x^{\mathrm{H}}Ax; x \in V_k, \|x\|_2 = 1\}; V_k \subseteq \mathbb{C}^N, \dim V_k = k\}$$
$$\mu_k' = \min\{\max\{u^{\mathrm{H}}A_{N-1}u; u \in V_k', \|u\|_2 = 1\}; V_k' \subseteq \mathbb{C}^{N-1}, \dim V_k' = k\}.$$

If $x \in V_k' \times \{0\}$, then $x = \begin{pmatrix} u \\ 0 \end{pmatrix}$ with $u \in V_k'$. $x^{\mathrm{H}}Ax = u^{\mathrm{H}}A_{N-1}u$ and $W_k = V_k' \times \{0\}$ is a subspace of $\mathbb{C}^N$ of dimension k. Furthermore $\|x\|_2 = \|u\|_2$. Hence

$$\max_{\substack{x \in W_k \\ \|x\|_2 = 1}} x^{\mathrm{H}}Ax = \max_{\substack{u \in V_k' \\ \|u\|_2 = 1}} u^{\mathrm{H}}A_{N-1}u,$$

so we get

$$\mu_k \leq \max_{\substack{u \in V_k' \\ \|u\|_2 = 1}} u^{\mathrm{H}}A_{N-1}u$$

for all $V_k' \subset \mathbb{C}^{N-1}$ with $\dim V_k' = k$; hence $\mu_k \leq \mu_k'$. On the other hand,

$$\mu_{k+1} = \min\{\max\{x^{\mathrm{H}}Ax; x \in V_{k+1}, \|x\|_2 = 1\}; V_{k+1} \subseteq \mathbb{C}^N, \dim V_{k+1} = k + 1\}.$$

Let $V_k' \subseteq \mathbb{C}^{N-1}$ be of dimension k and $W_{k+1} \subseteq \mathbb{C}^N$ be of dimension $k + 1$ such that $V_k' \times \{0\} \subseteq W_{k+1}$. Then

$$\max_{\substack{x \in V_k' \times \{0\} \\ \|x\|_2 = 1}} x^{\mathrm{H}}Ax \leq \max_{\substack{x \in W_{k+1} \\ \|x\|_2 = 1}} x^{\mathrm{H}}Ax,$$

so $\mu_k' \leq \max\{x^{\mathrm{H}}Ax; x \in W_{k+1}, \|x\|_2 = 1\}$; hence $\mu_k' \leq \mu_{k+1}$.

1.82 $\varphi_n = V_n\xi_n$, so $\pi_n(AV_n\xi_n - \lambda_n V_n\xi_n) = 0$. That is, $AV_n\xi_n - \lambda_n V_n\xi_n \in X_n^\perp$, so

$$V_n^H(AV_n\xi_n - \lambda_n V_n\xi_n) = 0,$$

and we get $(V_n^H A V_n)\xi_n = \lambda_n(V_n^H V_n)\xi_n$. This is a generalized eigenvalue problem.

1.83 If $x = Q_n\xi$, $y = Q_n\eta$, and $\pi_n A_{\restriction X_n} x = y$, then $\pi_n(AQ_n\xi - Q_n\eta) = 0$, so $AQ_n\xi - Q_n\eta \in X_n^\perp$ and $Q_n^H A Q_n \xi = \eta$. Thus $Q_n^H A Q_n$ represents $\pi_n A_{\restriction X_n}$ in the basis Q_n.

1.84 Let λ_n and $\varphi_n \in X_n$ be eigenelements of $A_n = \pi_n A$. Then $A_n\varphi_n = \lambda_n\varphi_n$; but $\varphi_n = \pi_n\varphi_n$, so also $\pi_n A \pi_n \varphi_n = \lambda_n\varphi_n$ and (λ_n, φ_n) are eigenelements of $A_n\pi_n = \pi_n A_n \pi_n$, and conversely. Let now λ_n, $\varphi_n \in X_n$ be eigenelements of $\mathscr{A}_n = \pi_n A_{\restriction X_n}$, so $\mathscr{A}_n\varphi_n = \lambda_n\varphi_n$; that is, $\pi_n A_{\restriction X_n}\varphi_n = \lambda_n\varphi_n$ or $\pi_n A\varphi_n = \lambda_n\varphi_n$. Then (λ_n, φ_n) are eigenelements of $A_n = \pi_n A$, and conversely.

1.85 $B_n^H = (Q_n^H A Q_n)^H = Q_n^H A^H Q_n = Q_n^H A Q_n = B_n$. B_n represents $\mathscr{A}_n$. Since $\pi_n^H = \pi_n$, $\lambda_n = \varphi_n^H A_n \varphi_n = \varphi_n^H \pi_n A \varphi_n = (\pi_n^H\varphi_n)^H A\varphi_n = (\pi_n\varphi_n)^H A\varphi_n = \varphi_n^H A\varphi_n$. λ_n is the Rayleigh quotient for A based on φ_n.

1.86 Take the projected system $\pi_n(Ax_n - b) = 0$, and define $x_n = Q_n\xi_n$, $\eta_n = Q_n^H b$, that is, $Q_n Q_n^H b = Q_n\eta_n$, where $Q_n Q_n^H$ is the matrix of the orthogonal projection π_n on X_n. Since $AQ_n\xi_n - b \in X_n^\perp$, we obtain $Q_n^H A Q_n \xi_n = Q_n^H b$; that is, $B_n\xi_n = \eta_n$.

1.87 Let $\mu_1 \geq \mu_2 \geq \cdots \geq \mu_N$ be the eigenvalues of the hermitian matrix A with associated eigenvectors $\varphi_1, \ldots, \varphi_N$ such that $\|\varphi_i\|_2 = 1$, M is generated by $\varphi_1, \ldots, \varphi_r$, W is generated by $\varphi_{r+1}, \ldots, \varphi_N$, and P is the orthogonal projection on M. We know that $M \perp W$, $\mathbb{C}^N = M \oplus W$. U is generated by $x_1, \ldots, x_r$ such that the $\{Px_i\}_1^r$ are independent. Let $X_n = A^n U$ and π_n be the orthogonal projection on X_n. Let $\mu_1^{(n)}$ be the dominant eigenvalue of $A_n = \pi_n A$:

$$\mu_1^{(n)} = \max_{\substack{x\in X_n\\ \|x\|_2=1}} x^H\pi_n Ax = \max_{\substack{x\in X_n\\ \|x\|_2=1}} x^H Ax \leq \mu_1 = \max_{\substack{x\in \mathbb{C}^N\\ \|x\|_2=1}} x^H Ax.$$

We then have

$$0 \leq \mu_1 - \mu_1^{(n)} = \min_{\substack{x\in X_n\\ x\neq 0}} \frac{x^H(\mu_1 I - A)x}{\|x\|_2^2}.$$

Both $\{Px_i\}_1^r$ and $\{\varphi_i\}_1^r$ are bases of M, so $\varphi_1 = \sum_{k=1}^r t_k Px_k = P(\sum_{k=1}^r t_k x_k)$, $u_1 := \sum_1^r t_k x_k \in U$, and $Pu_1 = \varphi_1$. Then $u_1 = Pu_1 + (I - P)u_1 = \varphi_1 + (I - P)u_1$ and $v_1 := (I - P)u_1 = \sum_{j=r+1}^N \alpha_j\varphi_j \in W$. Then $u_1 = \varphi_1 + v_1$ and $\varphi_1 \perp v_1$. Since $u_1 \in U$, $A^n u_1 \in X_n$ and $\hat{x} := (1/\mu_1^n)A^n u_1 = \varphi_1 + (1/\mu_1^n)A^n v_1 \in X_n$, so

$$\hat{x} = \varphi_1 + \frac{1}{\mu_1^n}\sum_{j=r+1}^N \alpha_j\mu_j^n\varphi_j \qquad \text{and} \qquad \|\hat{x}\|_2^2 \geq \|\varphi_1\|_2^2 = 1.$$

We then get $\mu_1 - \mu_1^{(n)} \leq \hat{x}^H(\mu_1 I - A)\hat{x}/\|\hat{x}\|_2^2 \leq \hat{x}^H(\mu_1 I - A)\hat{x}$. We find

$$\hat{x}^H(\mu_1 I - A)\hat{x} = \frac{\mu_1}{\mu_1^{2n}}\sum_{j=r+1}^N |\alpha_j|^2\mu_j^{2n} - \frac{1}{\mu_1^{2n}}\sum_{j=r+1}^N |\alpha_j|\mu_j^{2n+1} := C.$$

Since $-\mu_{r+1} \leq -\mu_{r+2} \leq \cdots \leq -\mu_N$,

$$C \leq \frac{\mu_1 - \mu_N}{\mu_1^{2n}}\sum_{j=r+1}^N |\alpha_j|^2\mu_j^{2n} \leq (\mu_1 - \mu_N)\left(\frac{\mu_{r+1}}{\mu_1}\right)^{2n}\sum_{j=r+1}^N |\alpha_j|^2.$$

But $\sum_{j=r+1}^N |\alpha_j|^2 = \|v_1\|_2^2 = \|u_1 - \varphi_1\|_2^2 = \|u_1 - Pu_1\|_2^2 = \tan^2\theta_1$ since $\|Pu_1\|_2 = \|\varphi_1\| = 1$, where θ_1 is the acute angle between u_1 and φ_1.

1.88 $p(z) = (z/\lambda_1)^{n-1} \in \mathbb{P}_{n-1}$ and $p(\lambda_1) = 1$. It is a holomorphic function of z in $D = \{z; |z| \leq |\lambda_2|\}$. Hence $\max_{z\in D}|p(z)| \leq |\lambda_2/\lambda_1|^{n-1}$ by the maximum principle.

1.89 See Exercise 1.86. We solve $\pi_n A x_n = \pi_n b$ for $x_n \in X_n$. Thus if $Ax^* = b$,

$$x_n = (\pi_n A_{\restriction X_n})^{-1} \pi_n A x^*.$$

Put $x^* = \pi_n x^* + (I - \pi_n)x^*$, and obtain $x_n - x^* = ((\pi_n A_{\restriction X_n})^{-1}\pi_n A - I)(I - \pi_n)x^*$. If

$$\|(\pi_n A_{\restriction X_n})^{-1}\|_2 \le M,$$

then, since $\|\pi_n\|_2 = 1$, $\|x_n - x^*\|_2 \le (1 + M\|A\|_2)\,\mathrm{dist}(x^*, X_n)$.

1.90 $\mathrm{dist}(z^*, K_n) = \min_{q \in \mathbb{P}_{n-1}} \|z^* - q(A)r_0\|_2$. Since $r_0 = b - Ax_0 = Ax^* - A(x^* - z^*) = Az^*$,

$$\mathrm{dist}(z^*, K_n) = \min_{q \in \mathbb{P}_{n-1}} \|z^* - q(A)Az^*\|_2 = \min_{\substack{p \in \mathbb{P}_n \\ p(0)=1}} \|p(A)z^*\|_2.$$

1.91 $\pi_n(Az^{(n)} - r_0) = 0$, $\pi_n(b - r_0 + Az^{(n)} - b) = 0$, $\pi_n(A(x_0 + z^{(n)}) - b) = 0$, so $x^{(n)} = x_0 + z^{(n)}$. Then, since $z^* = x^* - x_0$, we get $z^* + x^{(n)} - z^{(n)} = x^*$, so $x^* - x^{(n)} = z^* - z^{(n)}$.

1.92 $A = VDV^{-1}$ with $D = (\lambda_i \delta_{ij})$, $p(A) = Vp(D)V^{-1}$ for each $p \in \mathbb{P}_n$, so $\|p(A)z^*\|_2 \le C \max_{i=1,\dots,N} |p(\lambda_i)|$.

1.93 Cf. the proof of Theorem 1.34. $\gamma = 1 + 2\lambda_{\min}/(\lambda_{\max} - \lambda_{\min}) = (\lambda_{\max} + \lambda_{\min})/(\lambda_{\max} - \lambda_{\min})$.

1.94 $\pi_n'(A\varphi_n - \lambda_n\varphi_n) = 0$, $\varphi_n \in X_n$, is equivalent to $\varpi_n(A\varphi_n - \lambda_n\varphi_n) = 0$, $\varphi_n \in X_n$.

1.95 Set $\varphi_n = P_n u_n$. $(Q_n^{\mathrm{H}} A P_n)u_n = \lambda_n (Q_n^{\mathrm{H}} P_n)u_n$ is equivalent to $(Q_n^{\mathrm{H}} P_n)^{-1}(Q_n^{\mathrm{H}} A P_n)u_n = \lambda_n u_n$ (cf. Exercise 1.94).

Chapter 2

2.1 Let $x \in M$ and $\varphi_x(f) = \langle f, x\rangle$ for all $f \in X^*$. Clearly $|\varphi_x(f)| \le \|x\|\,\|f\|$, so φ_x is continuous on X^* and $\mathrm{Ker}\,\varphi_x = \varphi_x^{-1}\{0\}$ is closed. Finally $M^\perp = \bigcap_{x \in M} \mathrm{Ker}\,\varphi_x$ is closed.

2.2 For $x = x_M + x_N \in M \oplus N = X$ and $f \in M^\perp \cap N^\perp$ we have $\langle f, x\rangle = \langle f, x_M\rangle + \langle f, x_N\rangle = 0$, so $f = 0$. Let $\pi_M : x \in X \mapsto x_M \in M \subseteq X$ and $\pi_N = 1 - \pi_M$: $x \in X \mapsto x_N = x - x_M \in N \subseteq X$. For $f \in X^*$ let $f_M = f \circ \pi_M$, $f_N = f \circ \pi_N$. Clearly π_M, π_N, f_N, f_M are continuous and $f_M \in M^\perp$, $f_n \in N^\perp$, and $f = f_M + f_N$, so $X^* = M^\perp \oplus N^\perp$.

2.3 Let $\dim M = 1$. Then $\exists x_0 \in M$, $x_0 \ne 0$, and $\exists f \in X^*$ such that $\langle f, x_0\rangle = 1$. If $N = \mathrm{Ker}\, f$, then N is closed, and for $x \in M \cap N$ we have $x = \lambda x_0$ and $0 = \langle f, x\rangle = \bar\lambda\langle f, x_0\rangle = \bar\lambda$, so $x = 0$. Also $u := x - \langle x, f\rangle x_0 \in N$, $v := \langle x, f\rangle x_0 \in M$, and $u + v = x$. The map $\pi_M : x \in X \mapsto \langle x, f\rangle x_0 \in M \subset X$ is continuous since $\|\pi_M x\| \le \|f\|\,\|x_0\|\,\|x\|$. Suppose now that every subspace M_n of X of dimension n has a supplementary subspace. Let M be generated by $n+1$ independent vectors $y_1, \dots, y_n, y_{n+1}$. Then $M_n = \{y_1, \dots, y_n\}$ has a supplementary subspace N_n, and $y_{n+1} = y_{n+1}^M + y_{n+1}^N$, where $y_{n+1}^M \in M_n$, $y_{n+1}^N \in N_n$, and $y_{n+1}^N \ne 0$. $\{y_1, \dots, y_n, y_{n+1}^N\}$ is still a basis of M. This means we may suppose $M = M_n \oplus \{x_0\}$ with $x_0 \in N_n$. There exists $f \in X^*$ such that $\langle f, x_0\rangle = 1$ and $\langle f, x_n\rangle = 0\ \forall x_n \in M_n$. Let $N = N_n \cap \mathrm{Ker}\, f$. If $x \in M \cap N$, then $x = \lambda x_0 + x_n$ with $\lambda \in \mathbb{C}$ and $x_n \in M_n$. So $0 = \langle f, x\rangle = \bar\lambda\langle f, x_0\rangle = \bar\lambda$ and $x = x_n$. But $N \subseteq N_n$, so $x \in M_n \cap N_n$ and then $x = 0$. We showed $M \cap N = \{0\}$. For all $x \in X$ the following identity holds: $x = \langle x, f\rangle x_0 + x_n + x - \langle x, f\rangle x_0 - x_n$, where $x_n \in M_n$. Hence $u := \langle x, f\rangle x_0 + x_n \in M_n$, $v := x - \langle x, f\rangle x_0 - x_n$ verifies $\langle f, v\rangle = \langle f, x\rangle - \langle f, x\rangle\langle f, x_0\rangle - \langle f, x_n\rangle = \langle f, x\rangle - \langle f, x\rangle = 0$ since $\langle f, x_0\rangle = 1$, $\langle f, x_n\rangle = 0$. $v \in N$; hence $X = M \oplus N$ because the functions $x \mapsto x_n$, $x \mapsto \langle x, f\rangle x_0$ are continuous.

2.4 $\forall x \in M$ and $\forall f \in M^\perp$, $\langle f, x\rangle = 0$, so $x \in (M^\perp)^\perp$. $\forall x \in (M^\perp)^\perp$ and $\forall f \in M^\perp$, $\langle x, f\rangle = 0$; if $x \notin M$, then $\exists f \in X^*$ such that $\langle f, x\rangle \ne 0$, $\langle f, y\rangle = 0\ \forall y \in M$; hence $f \in M^\perp$, so we get a contradiction. Then $x \in M$.

2.5 $\forall x \in \operatorname{Dom} T$, $x \neq 0$, $\|Tx\|/\|x\| = \|T(x/\|x\|)\|$. The rest follows from the definition of sup and inf.

2.6 $\int_a^s k(s, t)x(t)\,dt = \int_a^b \hat{k}(s, t)x(t)\,dt$ where

$$\hat{k}(s, t) = \begin{cases} k(s, t), & a \le t \le s, \\ 0, & s < t \le b. \end{cases}$$

Note that $\hat{k}$ may be discontinuous along $a \le t = s \le b$.

2.7 $\|(TU)^k\|^{1/k} = \|T^k U^k\|^{1/k} \le \|T^k\|^{1/k}\|U^k\|^{1/k}$. If $U \in \mathscr{L}(X, Y)$ and $T \in \mathscr{L}(Y, X)$, then $TU \in \mathscr{L}(X)$ and $UT \in \mathscr{L}(Y)$. Write

$$(UT)^k = U(TU)^{k-1}T, \qquad \lim_k \|(UT)^k\|^{1/k} \le \lim_k (\|U\|\,\|T\|)^{1/k}(\|TU\|^{k-1})^{1/k};$$

therefore $r_\sigma(UT) \le r_\sigma(TU)$. Similarly $r_\sigma(TU) \le r_\sigma(UT)$.

2.8 If $\forall \varepsilon > 0\ \exists \delta > 0$ such that $\|x\| < \delta \Rightarrow \|Tx\| < \varepsilon$, then for $x = u - v$, $\|u - v\| < \delta \Rightarrow \|Tu - Tv\| < \varepsilon$, so T is uniformly continuous. Let $\varepsilon > 0$ and $\delta > 0$ be such that $\|x\| < \delta \Rightarrow \|Tx\| < \varepsilon$. Take $r = \delta/2$ and for $y \in X$, $y \neq 0$, set $x = (r/\|y\|)y$. Since $\|x\| = \delta/2 < \delta$, we have $\|Tx\| < \varepsilon$; that is, $\|Ty\| \le (2\varepsilon/\delta)\|y\|$ (which is obviously true even for $y = 0$). So T is bounded. Conversely, if $\|Tx\| \le M\|x\|$ for all $x \in X$, then for $\varepsilon > 0$ take $0 < \delta < \varepsilon/M$ and the continuity of T follows.

2.9 Since $\{0\}$ is closed in X and T is continuous, then $T^{-1}\{0\} = \operatorname{Ker} T$ is closed in X.

2.10 All norms in X are equivalent. Take $\{x_1, \ldots, x_N\}$ a basis of X. For $x = \sum_{j=1}^N \alpha_j x_j$ we have $\|Tx\| \le M\|x\|$ with $M = \max_{1 \le i \le N}\{\|Tx_i\|\}$ and $\|x\| = \sum_j |\alpha_j|$.

2.11 If $y_1 = Tx_1$, $y_2 = Tx_2$, then $y_1'' = x_1$, $y_1(a) = y_1(b) = 0$, and $y_2'' = x_2$, $y_2(a) = y_2(b) = 0$. So we have $(y_1 + \lambda y_2)'' = x_1 + \lambda x_2$, $(y_1 + \lambda y_2)(a) = (y_1 + \lambda y_2)(b) = 0$; hence $y_1 + \lambda y_2 = T(x_1 + \lambda x_2)$ and T is linear. Also

$$y'(u) = \int_a^u x(t)\,dt + K_1, \qquad y(s) = \int_a^s \left[\int_a^u x(t)\,dt\right] du + (s - a)K_1 + K_2.$$

$$y(a) = 0 \Rightarrow K_2 = 0, \qquad y(b) = 0 \Rightarrow K_1 = -\frac{1}{b - a}\int_a^b \left[\int_a^u x(t)\,dt\right] du,$$

$$y(s) = \int_a^s du \int_a^u x(t)\,dt - \frac{s - a}{b - a}\int_a^b du \int_a^u x(t)\,dt.$$

Let

$$k(s, t) = \begin{cases} (s - b)(t - a)/(b - a), & a \le t \le s, \\ (s - a)(t - b)/(b - a), & s \le t \le b. \end{cases}$$

Then

$$\int_a^b k(s, t)x(t)\,dt = \int_a^s (s - b)(t - a)x(t)\frac{dt}{b - a} + \int_s^b (s - a)(t - b)x(t)\frac{dt}{b - a}$$

$$= \frac{s - b}{b - a}\int_a^s (t - a)x(t)\,dt + \frac{s - a}{b - a}\int_s^b (t - b)x(t)\,dt.$$

Integrating by parts we find $\int_a^s du \int_a^u x(t)\,dt = s\int_a^s x(t)\,dt - \int_a^s tx(t)\,dt$ and also $\int_a^b du \int_a^u x(t)\,dt = b\int_a^b x(t)\,dt - \int_a^b tx(t)\,dt$.

So $y(s)$ can be written under the form

$$y(s) = \int_a^s (s-t)x(t)\,dt + \frac{s-a}{b-a}\left[\int_a^s (t-b)x(t)\,dt + \int_s^b (t-b)x(t)\,dt\right]$$

$$= \int_a^s \left[(s-t) + \frac{s-a}{b-a}(t-b)\right]x(t)\,dt + \frac{s-a}{b-a}\int_s^b (t-b)x(t)\,dt$$

$$= \frac{s-b}{b-a}\int_a^s (t-a)x(t)\,dt + \frac{s-a}{b-a}\int_s^b (t-b)x(t)\,dt.$$

That is $y(s) = \int_a^b k(s,t)x(t)\,dt$. T is bounded in $C(a,b)$ since k is continuous on $[a,b]^2$.

2.12 Let $X = Y$ be the space of real polynomials and $Tx(t) = x'(t)$. Clearly, $\operatorname{Im} T = Y$, but $\operatorname{Ker} T = \{\text{constants}\} \neq \{0\}$; T^{-1} does not exist.

2.13 Let $Tx(t) = 0.1\int_0^1 e^{st}x(s)\,ds$. Then $\|T\|_\infty = 0.1e < 0.3 < 1$, $(1-T)^{-1} \in \mathscr{L}(X)$, and there is a unique solution. By the method of successive approximations $\|x - x_k\|_\infty \leq (0.3)^k/(1-0.3) < 0.1$ is satisfied for $k \geq 2$.

2.14 $$\left\|\sum_{j=0}^k K^j\right\| \leq \sum_{j=0}^k \|K^j\| \leq 1 + \sum_{j=1}^k M^j \frac{(b-a)^j}{(j-1)!} \leq 1 + M(b-a)e^{M(b-a)},$$

$\sum_{j=0}^\infty K^j$ converges in $\mathscr{L}(X)$ and $K^j \to 0$ in $\mathscr{L}(X)$ for $j \to \infty$. We then get

$$\left(\sum_{j=0}^k K^j\right)(1-K) = (1-K)\sum_{j=0}^k K^j = 1 - K^{k+1} \to 1$$

in $\mathscr{L}(X)$. $\sum_{j=0}^\infty K^j = (1-K)^{-1}$ and the unicity of the solution of Volterra's equation $(1-K)x = y$ follows.

2.15 $x \in \operatorname{Ker} T \Leftrightarrow Tx = 0 \Leftrightarrow \langle y, Tx\rangle = 0\ \forall y \in Y^* \Leftrightarrow \langle T^*y, x\rangle = 0\ \forall y \in Y^* \Leftrightarrow \langle w, x\rangle = 0$ $\forall w \in \operatorname{Im} T^* \Leftrightarrow x \in (\operatorname{Im} T^*)^\perp$. Changing T by T^* and using $T^{**} = T$ we find $(\operatorname{Ker} T^*)^\perp = \operatorname{Im} T$ (we are dealing with closed subspaces since they are finite dimensional). Then $\operatorname{Im} T = Y \Leftrightarrow$ $\operatorname{Ker} T^* = \{0\} \Leftrightarrow T^*$ regular (since $\dim X = \dim Y$).

2.16 $\|T\|^2 = \sup_{\|x\|=1}(Tx, Tx) \leq \sup_{\|x\|=1}(x, T^*Tx) \leq \sup_{\|x\|=1}\|T^*T\|\ \|x\|^2 = \|T^*T\|$. But $\|T^*T\| \leq \|T^*\|\ \|T\| = \|T\|^2$, so the equality follows.

2.17 The linearity of T is easy to check. Let $\{x_n\} \subseteq l^2$ be a bounded sequence: $\|x_n\|^2 = \sum_{i\geq 1}|x_n^i|^2 \leq M$. Let $y_n = Tx_n$: $y_n^i = \sum_{j\geq 1} a_{ij}x_n^j$, $i, n = 1, 2, \ldots$. The vertical sequence $\{y_k^1\}_{k\in\mathbb{N}}$ is bounded in $\mathbb{C}$, so it contains a convergent subsequence $\{y_{n_1(k)}^1\}_{k\in\mathbb{N}}$. The vertical sequence $\{y_{n_1(k)}^2\}_{k\in\mathbb{N}}$ is also bounded in $\mathbb{C}$, so it contains a convergent subsequence $\{y_{n_2(k)}^2\}_{k\in\mathbb{N}}$, and so on. Finally, the diagonal sequence $\{y_{n_k(k)}\}_{k\in\mathbb{N}}$ is a convergent subsequence of $\{y_n\}$ in l^2. Let $\alpha = \sum_{i,j\geq 1}|a_{ij}|^2$, we have

$$\|Tx\|^2 = \left(\sum_i\left(\sum_j a_{ij}x_je_i\right), \sum_k\left(\sum_l a_{kl}x_le_k\right)\right) = \sum_i\left(\sum_j a_{ij}x_j\right)\left(\sum_l \bar{a}_{il}\bar{x}_l\right)$$

$$= \sum_i |((a_{ij})_{j=1}^\infty, (x_j)_{j=1}^\infty)|^2 \leq \sum_i \|(a_{ij})_{j=1}^\infty\|^2\|(x_j)_{j=1}^\infty\|^2 = \alpha\|x\|^2.$$

Hence $\|T\| \leq \alpha^{1/2}$.

2.18 If $\dim \operatorname{Im} T < \infty$, $T \in \mathscr{L}(X,Y)$, and $B \subseteq X$ is bounded, then TB is bounded in $\operatorname{Im} T$; $\overline{TB}$ the closure of TB is compact in $\operatorname{Im} T$ and hence in Y.

2.19 If $\|T_n - T\| \to 0$, then $\|T_n\| \leq M$. Let $x \in X$, $\varepsilon > 0$, and write $Tx = Tx - T_nx + T_nx$. For sufficiently large n, $\|Tx\| \leq \|Tx - T_nx\| + \|T_nx\| \leq (\varepsilon + M)\|x\|$, so T is continuous. Let $\{x_n\}$ be a bounded sequence in X: $\|x_n\| \leq C$. Then $\{T_1x_n\}$ has a convergent subsequence $\{T_1x_{n_1(k)}\}$ and $\{T_2x_{n_1(k)}\}$ has a convergent subsequence $\{T_2x_{n_2(k)}\}$ and so on. We deduce that $\{T_kx_{n_{k-1}(p)}\}_{p\in\mathbb{N}}$ has a convergent subsequence $\{T_kx_{n_k(p)}\}_{p\in\mathbb{N}}$. Let $\varepsilon > 0$; then $\|Tx_{n_k(k)} - Tx_{n_q(q)}\| \leq$

$\alpha + \beta + \gamma$, where $\alpha = \|Tx_{n_k(k)} - T_p x_{n_k(k)}\|$, $\beta = \|T_p x_{n_k(k)} - T_p x_{n_q(q)}\|$, $\gamma = \|T_p x_{n_q(q)} - Tx_{n_q(q)}\|$. Fixing k and q we choose p such that $\alpha < \varepsilon/3$ and $\gamma < \varepsilon/3$ independently of k and q. Then we choose k_0 and q_0 such that $k > k_0$ and $q > q_0$ imply $\beta < \varepsilon/3$.

2.20 $\dim(\operatorname{Im} TA) = \dim T(\operatorname{Im} A) \le \dim TX = \dim(\operatorname{Im} T) < +\infty$. $\dim(\operatorname{Im} AT) = \dim A(\operatorname{Im} T) \le \dim(\operatorname{Im} T) < +\infty$. Set $M = \operatorname{Im} T$ and $N = \operatorname{Im} A$. Then $(AT)_{\restriction N} = (A_{\restriction M})(T_{\restriction N})$ and $(TA)_{\restriction M} = (T_{\restriction N})(A_{\restriction M})$. In given bases in M and N, $A_{\restriction M}$ and $T_{\restriction N}$ are represented by (possibly) rectangular matrices for which the result is clear.

2.21 Let $\{x_i\}$, $\{y_i\}$ be two bases of M and $\{x_i^*\}$, $\{y_i^*\}$ their adjoints. There exists an invertible $m \times m$ matrix (v_{ij}) such that $y_i = \sum_i v_{ij} x_i$ and an invertible $m \times m$ matrix (u_{ij}) such that $y_j^* = \sum_i u_{ij} x_i^*$. Since $\sum_k \bar{u}_{ki} v_{kj} = \delta_{ij}$, $\sum_i \langle y_i^*, Ty_i\rangle = \sum_i \langle x_i^*, Tx_i\rangle$.

2.22 We already know that finite-rank operators are compact. Conversely, if P is a compact projection and $B_M = \{x \in M; \|x\| \le 1\}$, where $M = PX$ is itself a Banach space since it is closed, then $PB_M = B_M$ is relatively compact. So M has a relatively compact unit sphere, which means $\dim M < +\infty$.

2.23 $\forall x^* \in X^*$ and $\forall y \in X$, $\langle x^*, Py\rangle = \langle x^*, P^2 y\rangle = \langle P^* x^*, Py\rangle = \langle P^* P^* x^*, y\rangle$, $(P^*)^2 = P^*$. $M^* := P^* X^* = \operatorname{Ker}(1 - P^*) = (\operatorname{Im}(1 - P))^\perp = (\operatorname{Ker} P^\perp) = N^\perp$. $N^* := (1 - P)^* X^* = \operatorname{Ker} P^* = (\operatorname{Im} P)^\perp = M^\perp$.

2.24 If $z(z - 1) \ne 0$,

$$\left(\frac{1}{z} 1 + \frac{1}{z(z-1)} P\right)(z1 - P) = 1 - \frac{1}{z} P + \frac{1}{z-1} P - \frac{1}{z(z-1)} P = 1.$$

2.25 Let $x \in H$ and $y \in M$. $\|x - y\|^2 = \|x - P_M x + P_M x - y\|^2 = \|x - P_M x\|^2 + \|P_M x - y\|^2$ since $x - P_M x \in M^\perp$ and $P_M x - y \in M$. So $\|x - P_M x\| \le \|x - y\|$ $\forall y \in M$.

2.26 $\|P\| = \sup_{\|x\|=1} \|Px\|$. Since $Px = x$ for $x \in PX$, we have certainly $\|P\| \ge 1$. If P is an orthogonal projection, then

$$\|x\|^2 = \|x - Px + Px\|^2 = \|x - Px\|^2 + \|Px\|^2, \qquad \|Px\| \le \|x\|, \qquad \text{and} \qquad \|P\| = 1.$$

If P is not an orthogonal projection, we can find x such that $\|Px\| > \|x\|$ and $\|P\| > 1$.

2.27 $a_{01} = 1, a_{11} = 0, b_{01} = b_{11} = 0, a_{02} = a_{12} = 0, b_{02} = b_{12} = 0, x_1(t) = 1, x_2(t) = t, \gamma_1(s) = -s, \gamma_2(s) = 1, \beta_1(s) = -s, \alpha_1(s) = 0, \beta_2(s) = s, \alpha_2(s) = s - 1$.

2.28 $a_{01} = 1$, $a_{11} = 0$, $b_{01} = 0$, $b_{11} = -1$, $x_1(t) = 1$, $x_2(t) = t$, $\gamma_1(s) = -s$, $\gamma_2(s) = 1$, $\beta_1(s) = \beta_2(s) - s$, $\alpha_1(s) = \beta_2(s)$, $\alpha_2(s) = \beta_2(s) - 1$.

2.29 $a_{01} = 1, a_{11} = -1, b_{01} = b_{11} = 1, a_{02} = a_{12} = 0, b_{02} = b_{12} = 1, x_1(t) = 1, x_2(t) = t, \beta_1(s) = -\frac{2}{3}(s + 1), \beta_2(s) = (s + 1)/3, \alpha_1(s) = (s - 2)/3, \alpha_2(s) = (s - 2)/3$.

2.30 If $T \in \mathscr{C}(X, Y)$ and T^{-1} exists, then $T^{-1} \in \mathscr{C}(Y, X)$. If in addition $\operatorname{Im} T = Y$, then $\operatorname{Dom}(T^{-1}) = Y$, so $T^{-1} \in \mathscr{L}(Y, X)$.

2.31 $\|x\|_D = \sup_t[|x(t)| + |x''(t) + tx'(t) + t^2 x(t)|]$ for $x \in C^2(0, 1)$. In general, $\|Tx\|_Y \le \|x\|_X + \|Tx\|_Y = \|x\|_D$, so $\|T\|_{\hat{D}\to Y} \le 1$.

2.32 $\int_0^1 y(t)x'(t)\,dt = \int_0^1 z(t)x(t)\,dt$ $\forall x \in D$ becomes

$$x(1)y(1) + \int_0^1 (-y'(t))x(t)\,dt = \int_0^1 z(t)x(t)\,dt.$$

So $T^* x = -x'$ with $\operatorname{Dom} T^* = \{x \in X; x \text{ absolutely continuous}, x' \in X, x(1) = 0\}$.

2.33
$$\int_0^1 y(-x'' + 2x' - x) = -x'y\Big|_0^1 - \int_0^1 x'y' + 2xy\Big|_0^1 - 2\int_0^1 xy' - \int_0^1 xy$$

$$= x'(0)y(0) - x'(1)y(1) + \int_0^1 x(-y'' - 2y' - y).$$

So $\operatorname{Dom} T^* = \{x \in L^2(0, 1); x(0) = x(1) = 0,\ x' \text{ absolutely continuous},\ x'' \in L^2(0, 1)\}$ and $T^* x = -x'' - 2x' - x$.

2.34 $X = \left\{u \in L^2(\Omega); \frac{\partial u}{\partial x}, \frac{\partial u}{\partial y}, \frac{\partial^2 u}{\partial x^2}, \frac{\partial^2 u}{\partial y^2} \in L^2(\Omega)\right\}$,

Dom $T = \{u \in X;\ u(1, y) = u(0, y),\ u(x, 1) = u(x, 0) = 0,\ (\partial u/\partial x)(0, y) = 0\ \forall (x, y) \in \Omega\}$, and $T = -\Delta$.

$$-\int_0^1\int_0^1 v\frac{\partial^2 u}{\partial x^2} - \int_0^1\int_0^1 v\frac{\partial^2 u}{\partial y^2} = -\int_0^1\left[v(x, y)\frac{\partial u}{\partial x}\bigg|_{x=0}^{x=1} - \int_0^1 \frac{\partial v}{\partial x}\frac{\partial u}{\partial x}\,dx\right]dy$$

$$- \int_0^1\left[v(x, y)\frac{\partial u}{\partial y}\bigg|_{y=0}^{y=1} - \int_0^1 \frac{\partial v}{\partial y}\frac{\partial u}{\partial y}\,dy\right]dx$$

$$= -\int_0^1\left[v(1, y)\frac{\partial u(1, y)}{\partial x} - \frac{\partial v}{\partial x}u(x, y)\bigg|_{x=0}^{x=1} + \int_0^1 \frac{\partial^2 v}{\partial x^2}u\,dx\right]dy$$

$$- \int_0^1\left[v(x, 1)\frac{\partial u(x, 1)}{\partial y} - v(x, 0)\frac{\partial u(x, 0)}{\partial y} + \int_0^1 \frac{\partial^2 v}{\partial y^2}u\,dy\right]dx$$

$$= -\int_0^1 v(1, y)\frac{\partial u(1, y)}{dx}\,dy + \int_0^1\left(\frac{\partial v(1, y)}{\partial x} - \frac{\partial v(0, y)}{\partial x}\right)u(1, y)\,dy$$

$$- \int_\Omega u\,\Delta v - \int_0^1 v(x, 1)\frac{\partial u(x, 1)}{\partial y}\,dx + \int_0^1 v(x, 0)\frac{\partial u(x, 0)}{\partial y}\,dx.$$

Dom $T^* = \{v \in X;\ v(1, y) = 0,\ \partial v(1, y)/\partial x = \partial v(0, y)/\partial x,\ v(x, 1) = v(x, 0) = 0\ \forall (x, y) \in \Omega\}$, and $T^* = -\Delta$.

2.35 See A. Taylor (1958)

2.36 Following Exercise 2.24, if $P \neq 0$ and $P \neq 1$,

$$\|(P - z)^{-1}\| \leq \frac{1}{|z|\,|z - 1|}\|P\| + \frac{1}{|z|} \qquad \forall z \in \mathbb{C} - \{0, 1\},$$

so $\rho(P) = \mathbb{C} - \{0, 1\}$ and $\sigma(P) = \{0, 1\} = P\sigma(P)$. Let $M = PX$. Any $x \neq 0$ is such that $Px = x$ if $x \in M$ and $Px = 0$ if $x \in (1 - P)X$.

2.37 T is quasi-nilpotent iff $r_\sigma(T) = 0$; that is, $\max_{\lambda \in \sigma(T)} |\lambda| = 0 \Leftrightarrow \sigma(T) = \{0\}$.

2.38 $\|x_k - x\| = \|\sum_{p \geq k} T^p y\| \leq \|y\| \sum_{p \geq k} \|T^p\|$. Let $\varepsilon > 0$ such that $r_\sigma(T) + \varepsilon < 1$. There exists k_0 such that for $k > k_0$, $\|T^k\| < (r_\sigma(T) + \varepsilon)^k$, so

$$\|x_k - x\| \leq \|y\| \sum_{p \geq k} (r_\sigma(T) + \varepsilon)^k = \frac{\|y\|}{1 - (r_\sigma(T) + \varepsilon)}(r_\sigma(T) + \varepsilon)^{k+1}.$$

2.39 $(1 - L)^{-1}$ exists if $r_\sigma(L) < 1$. Then $(1 - L)^{-1} = \sum_{k \geq 0} L^k$. Let $L = tT$, so $r_\sigma(tT) = |t| r_\sigma(T) < 1 \Rightarrow |t| < 1/r_\sigma(T) \Rightarrow (1 - tT)^{-1} = \sum_{k \geq 0} t^k T^k$.

2.40 If $r_\sigma(T) = 0$, then $\sigma(T) = \{0\}$, so $T_M\colon M \to M$ is similar to a triangular matrix with zero diagonal since $\sigma(T_M) = P\sigma(T_M) \subseteq \sigma(T)$. So $T_M^k = 0$ for some integer $k \leq \dim M$.

2.41 Let $L = T - z_0$. L is invertible, and the extended spectrum of $L^{-1} = R(z_0)$ is mapped on the spectrum of L by the one-to-one mapping $z \mapsto 1/z$. That is, $\hat{z} = z - z_0 \mapsto 1/\hat{z} = 1/(z - z_0)$ for $z \in \sigma(T)$.

$$R(z_0) - \frac{1}{z - z_0} = \frac{(z - z_0)R(z_0) - 1}{z - z_0} = \frac{(z - z_0)(T - z_0)^{-1} - (T - z_0)(T - z_0)^{-1}}{z - z_0}$$

$$= \frac{1}{z - z_0}(z - T)R(z_0).$$

$$\left(R(z_0) - \frac{1}{z - z_0}\right)^{-1} = -(z - z_0)(T - z)^{-1}(T - z + z - z_0) = z_0 - z - (z - z_0)^2 R(z).$$

$$r_\sigma(R(z_0)) = \max_{\lambda \in \sigma(R(z_0))} |\lambda| = \max_{\lambda \in \sigma(T)} \frac{1}{|\lambda - z_0|} = \frac{1}{\min_{\lambda \in \sigma(T)} |\lambda - z_0|} = \frac{1}{\operatorname{dist}(z_0, \sigma(T))}.$$

2.42 $\sigma(T_2^{-1}) = \varnothing, \sigma(T_3^{-1}) = \left\{\dfrac{b-a}{\log k + 2in\pi}; n \in \mathbb{Z}\right\}$.

2.43 For $x \in D = \operatorname{Dom} T$, $TR(z)x = (T - z + z)R(z)x = x + zR(z)x = R(z)Tx$. By integration on Γ, it follows that $TPx = PTx$ for $x \in D$, and obviously $R(z)P = PR(z)$. The remaining identities follow in the same way.

2.44 $(A - zB)x = Bf$, $y = Bx$, $g = Bf$. Then $x = (A - zB)^{-1}Bf$, so

$$y = Bx = B(A - zB)^{-1}Bf = B(A - zB)^{-1}g,$$

and the result follows.

2.45 Let $X = C^1(0, 1)$ with $\|x\| = \|x'\|_\infty + \|x\|_\infty$. Then $Bx(t) := x'(t) + tx(t)$ is in $\mathscr{L}(X, Y)$, where $Y = C(0, 1)$ with the norm $\| \ \|_\infty$. And $Ax(t) := x''(t) + tx'(t) + x(t)$ is in $\mathscr{C}(X, Y)$ with domain $D = \{x \in X; x'' \in Y, x(0) = x(1) = 0\}$. So the problem becomes $Ax = \lambda Bx$, $x \in D$, $x \neq 0$, $\lambda \in \mathbb{C}$.

2.46 $ix'(t) = \lambda x(t), x'/x = -\lambda i, x(t) = e^{-\lambda it}, x(0) = 1 = e^{-2\pi\lambda i} \Rightarrow \lambda \in \mathbb{Z}. x_k(t) = r^{-kit}, k \in \mathbb{Z}$, $x = R(z)f \Leftrightarrow (T - z)x = f$. So $ix' - zx = f$; hence $x(t) = \alpha(t)e^{-zit}$, where

$$\alpha(t) = -i\int_0^t e^{-zi\xi}f(\xi)\, d\xi + C$$

and

$$C = -\frac{ie^{-2\pi iz}}{1 - e^{2\pi iz}}\int_0^{2\pi} e^{-iz\xi}f(\xi)\, d\xi.$$

The spectral projections are left to the reader. Note that their image spaces are one dimensional.

2.47 Set $U: x \mapsto x' - x$ for $x \in \Delta := \{x \in X;\ x$ absolutely continuous, $x' \in X,\ x(0) = x(1)\} \supset D$. From Example 2.23 with $k = 1$, $\sigma(U) = \{2ij\pi - 1; j \in \mathbb{Z}\}$, the eigenvalues are simple, and the eigenprojections may be deduced from Exercise 2.46. Since $T = -U^2$, T has the same eigenvectors with eigenvalues $\sigma(T) = \{4k^2\pi^2 - 1 + 4ik\pi; k \in \mathbb{Z}\}$. Clearly, $\sigma(T^*) = \overline{\sigma(T)}$, or directly from $T^* = -V^2$, where $V: x \mapsto x' + x$ for $x \in \Delta$, and $\sigma(V) = \{2ik\pi + 1; k \in \mathbb{Z}\}$.

2.48 See Yosida (1965, p. 229).

2.49 $Tx = 0$ has infinitely many independent solutions, for instance, all continuous functions x such that $x(t) = f(t)/a(t)$, where f is a continuous odd function on $[-1, 1]$. For example, $x_n(t) = t^{2n+1}/a(t)$, $x \in \mathbb{N}$. Let λ be any nonzero eigenvalue of T and set $\alpha(\varphi) := \int_{-1}^1 a(t)\varphi(t)\, dt$. Then $T\varphi = \lambda\varphi$ reads $(T\varphi)(s) = \lambda\varphi(s) = \alpha(\varphi)b(s)$; therefore $\varphi(s) = \alpha(\varphi)b(s)/\lambda$, and $\alpha(\varphi) \neq 0$. Hence λ is simple, and b defines the eigendirection. Then

$$\alpha(\varphi) = \int_{-1}^1 a(t)\alpha(\varphi)\frac{1}{\lambda}b(t)\, dt$$

implies $\lambda = \int_{-1}^1 a(t)b(t)\, dt$ since $\alpha(\varphi) \neq 0$.

2.50 Let $x \in \mathbb{R}$ and $z < 0$; $\|(T - z)x\|^2 = ((T - z)x, (T - z)x) = \|Tx\|^2 - 2z(Tx, x) + z^2\|x\|^2 \geq \|Tx\|^2 + z^2\|x\|^2 \geq z^2\|x\|^2$ since $-2z > 0$ and $(Tx, x) > 0$. So $\|(T - z)^{-1}\| \leq 1/|z|$ for $z < 0$. Hence $\sigma(T)$ is nonnegative.

2.51 $T = \sum_k \lambda_k P_k$. Let $\mu_j = \lambda$ be simple, $(T - \mu_j) = \sum_k \lambda_k P_k - \mu_j$. Let $x = x_Q + x_{1-Q}$, where $x_Q \in QH$ and $x_{1-Q} \in (1 - Q)H$. Let $x_Q = \sum_j x_j$, $x_j \in P_jH$,

$$(T - \mu_j)x = Tx_Q - \mu_j(x_Q + x_{1-Q}) = \sum_k \mu_k x_k - \mu_j \sum_k x_k - \mu_j x_{1-Q}$$

$$= \sum_{k \neq j}(\mu_k - \mu_j)x_k - \mu_j x_{1-Q} \in \{x_j\}^\perp.$$

Let

$$x = -\frac{1}{\mu_j}y + \frac{1}{\mu_j}\sum_{k \neq j}\mu_k\frac{(y, x_k)x_k}{\mu_k - \mu_j} + \alpha x_j.$$

Then

$$(T-\mu_j)x = -\frac{1}{\mu_j}Ty + \frac{1}{\mu_j}\sum_{k\neq j}\mu_k^2\frac{(y,x_k)x_k}{\mu_k-\mu_j} + \alpha\mu_j x_j + y - \sum_{k\neq j}\mu_k\frac{(y,x_k)x_k}{\mu_k-\mu_j} - \alpha\mu_j x_j$$

$$= -\frac{1}{\mu_j}\left[\sum_{k\neq j}\mu_k(y,x_k)x_k - \sum_{k\neq j}\mu_k^2\frac{(y,x_k)x_k}{\mu_k-\mu_j}\right] + y - \sum_{k\neq j}\mu_k\frac{(y,x_k)x_k}{\mu_k-\mu_j} = y.$$

2.52 $Tx = -x''$, $(T-z)x = -x'' - zx = f$, $x'' + zx = -f$, $x(0) = x(1) = 0$, $x = R(z)f$. For $z = 0$, $x'' = -f$. From Exercise 2.27 we get $x(t) = -\int_0^1 g(t,s)f(s)\,ds$, where

$$g(t,s) = \begin{cases}(s-1)t, & 0 \le t \le s \le 1,\\ (t-1)s, & 0 \le s < t \le 1.\end{cases}$$

We use the Ascoli–Arzela theorem to prove that $(R(z)f)(t) = -\int_0^1 g(t,s)f(s)\,ds$ is a compact operator. See, for instance, Example 2.16.

Chapter 3

3.1 For $x \in X$, we have

$$\begin{aligned}\|(T-T_n)x\| &= \|(1-\pi_n)Tx + \pi_n Tx - \mathscr{T}_n\pi_n x\|\\ &\le \|(1-\pi_n)Tx\| + \|\mathscr{T}_n\pi_n x - \pi_n Tx\|\\ &= \|(1-\pi_n)Tx\| + \|\mathscr{T}_n\pi_n x - \pi_n T\pi_n x + \pi_n T\pi_n x - \pi_n Tx\|\\ &\le \|(1-\pi_n)Tx\| + \|\mathscr{T}_n(\pi_n x) - \pi_n T(\pi_n x)\| + \|\pi_n T(\pi_n - 1)x\|\\ &\le \|(1-\pi_n)T\|\,\|x\| + \|(T_n - \pi_n T)_{\restriction X_n}\|\,\|\pi_n\|\,\|x\| + \|\pi_n\|\,\|T\|\,\|(\pi_n - 1)x\|,\end{aligned}$$

which tends to 0 as $n \to \infty$ since π_n is uniformly bounded. So $T_n \xrightarrow{p} T$. Take now $x_n \in X_n$ with $\|x_n\| \le c$ for all n.

$$\begin{aligned}\|(T-T_n)x_n\| &= \|Tx_n - \mathscr{T}_n\pi_n x_n\|\\ &\le \|Tx_n - \pi_n Tx_n\| + \|\pi_n Tx_n - \mathscr{T}_n x_n\|\\ &= \|(1-\pi_n)Tx_n\| + \|(\mathscr{T}_n - \pi_n T)x_n\|\\ &\le c(\|(1-\pi_n)T\| + \|(\mathscr{T}_n - \pi_n T)_{\restriction X_n}\|),\end{aligned}$$

which tends to 0 as $n \to \infty$. So $\{(T-T_n)x_n\}$ is a convergent sequence (with limit 0).

3.2 $(T-T_n)x = (T-T_n)\pi_n x + T(1-\pi_n)x$, where $\pi_n x \in X_n$ and $\pi_n \xrightarrow{p} 1$ proves that $T_n \xrightarrow{p} T$. For $x_n \in X_n$, $\|x_n\| \le c$, $\|(T-T_n)x_n\| \le c\|(T-T_n)_{\restriction X_n}\| \to 0$ as $n \to \infty$. $(T-T_n)_{\restriction X_n} = [(1-\pi_n)T]_{\restriction X_n} + [\pi_n T - T_n]_{\restriction X_n}$. Hence I and II imply that $\|(T-T_n)_{\restriction X_n}\| \to 0$.

3.3 Since $\bigcup_{n=1}^{\infty}(T-T_n)B$ is relatively compact, so is $(T-T_n)B$ for each n. Since $T_n \xrightarrow{p} T$ in $\mathscr{L}(X)$ we get $\sup_{y\in(T-T_n)B}\|(T-T_n)y\| \to 0$. But $\sup_{y\in(T-T_n)B}\|(T-T_n)y\| = \sup_{x\in B}\|(T-T_n)^2x\| = \|(T-T_n)^2\|$.

3.4 $x_n \to x$ for $n \in N_1 \subset \mathbb{N}$. T_n is uniformly bounded since $T_n \xrightarrow{p} T$ in $\mathscr{L}(X)$. Then $(T-T_n)x_n = (T-T_n)(x_n - x) + (T-T_n)x$. And $\lim_{n\in N_1}(T-T_n)x_n = 0$. That is, $y = 0$.

3.5 We first note that $\Theta(S,T) = \Theta(S^{-1},T^{-1})$ for any invertible S, T in $\mathscr{C}(X,Y)$; next we state that if $\Theta(S,T) < (1 + \|T^{-1}\|^2)^{-1/2}$, then

$$\|S^{-1} - T^{-1}\| \le \frac{(1+\|T^{-1}\|^2)\delta(S^{-1},T^{-1})}{1-(1+\|T^{-1}\|^2)^{1/2}\delta(S^{-1},T^{-1})}$$

for S^{-1}, T^{-1} in $\mathscr{L}(Y, X)$. These results prove that $\Theta(T_n, T) \to 0 \Rightarrow \|T_n^{-1} - T^{-1}\| \to 0$. Conversely

$$A(y) := \inf_{v \in T(D)} [\|y - v\|^2 + \|T^{-1}y - T_n^{-1}v\|^2]^{1/2} \le \|T^{-1}y - T_n^{-1}y\| \le \|T^{-1} - T_n^{-1}\| \, \|y\|,$$

$$A := \sup_{\substack{y \in T(D) \\ \|y\|^2 + \|T^{-1}y\|^2 = 1}} A(y) \le \|T^{-1} - T_n^{-1}\|.$$

But $A = \delta(T, T_n)$, so $\delta(T, T_n) \to 0$ if $\|T^{-1} - T_n^{-1}\| \to 0$. Similarly $\delta(T_n, T) \to 0$ under the same condition. Hence $\|T^{-1} - T_n^{-1}\| \to 0$ implies $\Theta(T, T_n) \to 0$.

3.6 Use Example 3.8 and identity (2.4) of Section 7.1, Chapter 2.

3.7 We have $\|T_n^{-1}\| \le M$ for n large enough, for a given nonzero vector x in D and $\varepsilon = (1/2M)\|x\|$, there exists N such that $\|T_n^{-1}\| \le M$ and $\|T_n x - Tx\| < \varepsilon$ if $n > N$. So $\|T_n x\| - \varepsilon < \|Tx\|$. But $\|T_n x\| \ge (1/M)\|x\|$; hence $(1/2M)\|x\| < \|Tx\|$. Also we have $T_n^{-1}x \to T^{-1}x$ $\forall x \in X$. Then $T^{-1} \in \mathscr{L}(X)$.

3.8 We use the identity $T_n - z = T_n(1 - zT_n^{-1})$. We know that $(1 - zT_n^{-1}) \in \mathscr{L}(X)$ if $\|zT_n^{-1}\| = |z| \, \|T_n^{-1}\| \le |z|M < 1$, that is, $|z| < 1/M \Rightarrow (T_n - z)^{-1} \in \mathscr{L}(X)$. Also we know that

$$\|(T_n - z)^{-1}\| \le \|T_n^{-1}\| \, \|(1 - zT_n^{-1})^{-1}\| \le M \frac{1}{1 - |z|M},$$

so $T_n - z \xrightarrow{\text{s}} T - z$ since, clearly, $T_n - z \xrightarrow{\text{p}} T - z$.

3.9 For any x in D, $\|(T_n - z_n - (T - z))x\| \le \|(T_n - T)x\| + |z - z_n| \, \|x\| \to 0$; then $T_n - z_n \xrightarrow{\text{p}} T - z$. Now let the bounded sequence $\{x_n\}$ in D be such that

$$(T_n - z_n)x_n \to y, \qquad n \in N_1 \subset \mathbb{N}.$$

Then $(T_n - z)x_n \to y$ for $n \in N_1$. But $T_n - z \xrightarrow{\text{r}} T - z$; therefore $x_n \to x \in D$, $n \in N_2 \subset N_1$, and $(T - z)x = y$. That is, $T_n - z_n \xrightarrow{\text{r}} T - z$. For any x in D, $\|z_n T_n x - zTx\| \le |z_n| \, \|T_n x - Tx\| + |z_n - z| \, \|Tx\| \to 0$. Suppose that $y_n = z_n T_n x_n \to y$ for a bounded sequence $\{x_n\}$ in D. Since $z_n \to z \ne 0$, $T_n x_n = y_n/z_n$ is bounded and

$$\left\| T_n x_n - \frac{y}{z} \right\| \le \left| 1 - \frac{z_n}{z} \right| \|T_n x_n\| + \frac{1}{|z|} \|z_n T_n x_n - y\| \to 0.$$

Since $T_n \xrightarrow{\text{r}} T$, $x_n \to x$ for $n \in N_1 \subset \mathbb{N}$ and $zTx = y$, which proves that $z_n T_n \xrightarrow{\text{r}} zT$.

3.10 First we note that $T_n \xrightarrow{\text{p}} T \Rightarrow \|T_n\| \le M$ and $\|T_n U_n x - TUx\| \le M\|U_n x - Ux\| + \|T_n Ux - TUx\| \to 0$, so $T_n U_n \xrightarrow{\text{p}} TU$. Next, take a bounded sequence $\{x_n\}$ in X. Since $U_n \xrightarrow{\text{p}} U$, $\|U_n\| \le c$, $U_n x_n = u_n$ is a bounded sequence in X. If $T_n U_n x_n \to y$ for $n \in N_1$, this means that $T_n u_n \to y$, $n \in N_1$. But $T_n \xrightarrow{\text{r}} T$, so $\exists N_2 \subset N_1$ and $\exists u \in X$ such that $u_n \to u$ for $n \in N_2$ and $Tu = y$. This means $U_n x_n \to u$, $n \in N_2$, and since $U_n \xrightarrow{\text{r}} U$, $\exists N_3 \subset N_2$ and $\exists x \in X$ such that $x_n \to x$ for $n \in N_3$ and $Ux = u$. $TUx = Tu = y$. Hence $T_n U_n \xrightarrow{\text{r}} TU$.

3.11 For $z \in \rho(T)$, $R(z_0) - t = t(z - T)R(z_0)$ and $(R(z_0) - t)^{-1} = -(1/t) - (1/t^2)R(z)$.

(i) $T_n - z \xrightarrow{\text{s}} T - z$ and $R_n(z)$, therefore $(R_n(z_0) - t)^{-1}$, are uniformly bounded.
(ii) $(R_n(z_0) - t)^{-1}$, therefore $R_n(z)$, are uniformly bounded.
(iii) is proved for $z \ne \lambda$, $t \ne v$ by equivalence with (i), using Proposition 3.17.

Note that (iii) holds for $z = \lambda$ and $t = v$ if $R_n(z_0) \xrightarrow{\text{r}} R(z_0)$. Indeed, if $(R_n(z_0) - v)x_n = v(\lambda - T_n)R_n(z_0)x_n \to y$, then $u_n = R_n(z_0)x_n \to u = R(z_0)y$ and $x_n \to x$ such that $(R(z_0) - v)x = y$.

(iv) Let $\{x_n\}$ be such that $\|x_n\| \le M$, $x_n \in D$, $(T_n - z)x_n = (1/t)(T_n - z_0)(R_n(z_0) - t)x_n \to y$; hence $(R_n(z_0) - t)x_n \to R(z_0)y$ since $R_n(z_0) \xrightarrow{\text{p}} R(z_0)$, $x_n \to x$ by the regularity of $R_n(z_0) - t$, and $(T - z)x = y$. Note that $z = \lambda$ and $t = \nu$ are possible.

3.12 $T_n - z\pi_n \overset{\text{d-r}}{\to} T - z \Leftrightarrow T_n - z\pi_n \overset{\text{d-s}}{\to} T - z \underset{z \neq 0}{\Leftrightarrow} T_n - z \overset{\text{s}}{\to} T - z \Leftrightarrow T_n - z \overset{\text{r}}{\to} T - z.$

3.13 Let the bounded sequence $\{x_n\}$, $x_n \in X_n$, be such that $(T_n - z\pi_n)x_n \to y$, $n \in N_1$. Then $(T_n - T)x_n + (T - z\pi_n)x_n \to y$, $n \in N_1$, $-R(z)(T - T_n)x_n + x_n \to R(z)y$. Since $T_n \overset{\text{d-c}}{\to} T$, $(T - T_n)x_n \to u$ for $n \in N_2 \subset N_1$. So $x_n \to R(z)(y + u) = x$ for $n \in N_2$ and $u = 0$ (Exercise 3.4). Therefore $(T - z)x = y$.

3.14 Let the bounded sequence $\{x_n\}$, $x_n \in D_n$, $\|x_n\|_D \le c$ be such that $(T_n - z)x_n \to y$ in X. Then $x_n \to x$ in X for $n \in N_1 \subset \mathbb{N}$, and $T_n x_n \to y + zx$ in X for $n \in N_1$. By the regularity assumption $T_n \overset{\text{d-r}}{\to} T$ in $\mathscr{L}(\hat{D}, X)$, $x_n \to x$ in $\hat{D}$, which proves that $T_n - z\pi_n \overset{\text{d-c}}{\to} T - z$ in $\mathscr{L}(\hat{D}, X)$.

3.15 We have the following equivalent equations

$$(T_n + \Delta T_n)(x_n + \Delta x_n) = y_n + \Delta y_n$$
$$T_n(1 + T_n^{-1}\Delta T_n)(x_n + \Delta x_n) = y_n + \Delta y_n$$
$$[1 - T_n^{-1}(-\Delta T_n)](x_n + \Delta x_n) = T_n^{-1}(y_n + \Delta y_n).$$

If $\|T_n^{-1}\| \le 1/M$ and $\varepsilon < M$, then $\|T_n^{-1}\|\,\|\Delta T_n\| \le \varepsilon/M < 1$ and

$$\|(1 + T_n^{-1}\Delta T_n)^{-1}\| = \left\| \sum_{k=0}^{\infty} (-T_n^{-1}\Delta T_n)^k \right\| \le \frac{1}{1 - \|T_n^{-1}\Delta T_n\|}.$$

$$\Delta x_n = (1 + T_n^{-1}\Delta T_n)^{-1} T_n^{-1}(y_n + \Delta y_n) - T_n^{-1} y_n$$
$$= T_n^{-1}\Delta y_n + \left[\sum_{1}^{\infty} (-T_n^{-1}\Delta T_n)^k\right] T_n^{-1}(y_n + \Delta y_n).$$

$$\|\Delta x_n\| \le \frac{\alpha}{M} + \frac{1}{M}\frac{\varepsilon}{M - \varepsilon}(\alpha + C) = \beta.$$

3.16 By Lemma 3.22, for all t in δ'_Γ, $T(t)$ is closed with domain D and for $z \in \Gamma$, $T(t) - z$ has a bounded inverse, therefore $\Gamma \subset \rho(T(t))$. Let Δ be the disk $\{t; |t| < 1/\rho\}$ with $\rho < r'_\Gamma$. Clearly, $\Delta \supset \delta'_\Gamma$. We can find $\tau \in \Delta$ and $z_0 \in \Gamma$ with $|\tau| r'_\Gamma \ge |\tau| r_\sigma[HR'(z_0)] \ge 1$. Set $r_0 := r_\sigma[HR'(z_0)]$. Let $\mu_0 \in \sigma[\tau HR'(z_0)]$ such that $|\mu_0| = |\tau| r_0$. $\tau' = \tau/\mu_0$ is such that $|\tau'| = 1/r_0 \le |\tau|$; hence $\tau' \in \Delta$. And

$$\mu_0\left(T' - \frac{\tau}{\mu_0}H - z_0\right)R'(z_0) = \mu_0 - \tau HR'(z_0),$$

where the right-hand-side operator has no bounded inverse. Therefore $z_0 \notin \rho(T' - \tau' H)$ for $\tau' \in \Delta$, and δ'_Γ is maximal.

3.17 A closed operator T with domain D belongs to $\mathscr{L}(\hat{D}, X)$. For $H \in \mathscr{L}(\hat{D}, X)$, $HR'(z) \in \mathscr{L}(X)$ for z on Γ. Let $W = \{H \in \mathscr{L}(\hat{D}, X); \max_{z \in \Gamma} r_\sigma[HR'(z)] < 1\}$. W is a balanced set iff, $\forall \mu \in \mathbb{C}$ and $\forall H \in W$, $|\mu| \le 1 \Rightarrow \mu H \in W$. For $|\mu| \le 1$, $r_\sigma(\mu HR'(z)) \le r_\sigma(HR'(z))$; hence W is balanced. For any z on Γ, $1 - HR'(z)$ is an homeomorphism in X. $(T' - H - z)^{-1} = R'(z)(1 - HR'(z))^{-1}$ belongs to $\mathscr{L}(X, \hat{D})$. This shows that $T' - H \in \mathscr{L}(X)$ and $\Gamma \subset \rho(T' - H)$. Let now W' be a balanced set such that $W \subset W'$, $W' \neq W$. Let $H \in W'$, $H \notin W$. We can find $z_0 \in \Gamma$ and $\mu_0 \in \mathbb{C}$ such that $\rho_0 := r_\sigma(HR'(z_0)) \ge 1$, $\mu_0 \in \sigma[(1/\rho_0)HR'(z_0)]$, and $|\mu_0| = 1$ since $r_\sigma[(1/\rho_0)HR'(z_0)] = 1$. $1/\rho_0|\mu_0| = 1/\rho_0 \le 1$; hence $(1/\rho_0\mu_0)H \in W'$. And

$$\mu_0\left(T' - \frac{1}{\rho_0\mu_0}H - z_0\right)R'(z_0) = \mu_0 - \frac{1}{\rho_0}HR'(z_0),$$

where the right-hand-side operator has no bounded inverse. Therefore $z_0 \notin \rho(T' - (1/\rho_0 \mu_0)H)$. This proves that W is maximal.

This exercise is adapted from Lemordant (1980). It is very similar to Exercise 3.16. It shows how to study the dependence of a family of operators on a perturbation without introducing a complex parameter t.

3.18 Let $t \in \delta'_\Gamma$ and $|t| \leq 1$.

$$P(t) - P' = \frac{-1}{2i\pi} \sum_{k=1}^{\infty} t^k \int_\Gamma R'(z)[HR'(z)]^k \, dz,$$

$$\|P(t) - P'\| \leq \frac{\text{meas } \Gamma}{2\pi} M \frac{\alpha}{1 - \alpha},$$

with $\alpha := \max_{z \in \Gamma} \|HR'(z)\|$ and $M := \max_{z \in \Gamma} \|R'(z)\|$. By assumption

$$\alpha < \left(1 + \frac{\text{meas } \Gamma}{2\pi} M \|\varphi'^*\|\right)^{-1}.$$

Therefore $\|P(t) - P'\| < 1/\|\varphi'^*\|$ and $|\langle P(t)\varphi', \varphi'^*\rangle - 1| < 1$ prove that $\langle P(t)\varphi', \varphi'^*\rangle \neq 0$ for $|t| \leq 1$. We remark that a weaker condition is obtained if $M\|\varphi'^*\|$ is replaced by

$$\max_{z \in \Gamma} \|R'^*(z)\varphi'^*\|.$$

3.19 $B = \begin{pmatrix} 1 & 1 \\ 1 & 2 \end{pmatrix}$, $A = \begin{pmatrix} 1 & 0 \\ 0 & 2 \end{pmatrix}$, $x' = \begin{pmatrix} 1 \\ 0 \end{pmatrix}$, and $H = A - B = \begin{pmatrix} 0 & -1 \\ -1 & 0 \end{pmatrix}$. We clearly see that $0 \in \rho(A) \cap \rho(B)$ and that $A^{-1} = \begin{pmatrix} 1 & 0 \\ 0 & \frac{1}{2} \end{pmatrix}$, $HA^{-1} = \begin{pmatrix} 0 & -\frac{1}{2} \\ -1 & 0 \end{pmatrix}$, and $r' = r_\sigma(HA^{-1}) = \frac{1}{\sqrt{2}} < 1$. We compute $A^{-1}H = \begin{pmatrix} 0 & -1 \\ -\frac{1}{2} & 0 \end{pmatrix}$, $(A^{-1}H)^2 = \begin{pmatrix} \frac{1}{2} & 0 \\ 0 & \frac{1}{2} \end{pmatrix}$. If we assume

$$(A^{-1}H)^{2k} = \begin{pmatrix} \dfrac{1}{2^k} & 0 \\ 0 & \dfrac{1}{2^k} \end{pmatrix}.$$

we conclude by multiplication by $(A^{-1}H)^2$ that

$$(A^{-1}H)^{2(k+1)} = \begin{pmatrix} \dfrac{1}{2^{k+1}} & 0 \\ 0 & \dfrac{1}{2^{k+1}} \end{pmatrix};$$

hence the relationship is valid for all $k \in \mathbb{N}$. In the same way we state for all integer k.

$$(A^{-1}H)^{2k+1} = \begin{pmatrix} 0 & -\dfrac{1}{2^k} \\ -\dfrac{1}{2^{k+1}} & 0 \end{pmatrix}$$

and thus

$$u_{2k} = \begin{pmatrix} \dfrac{1}{2^k} \\ 0 \end{pmatrix} \quad \text{and} \quad u_{2k+1} = \begin{pmatrix} 0 \\ -\dfrac{1}{2^{k+1}} \end{pmatrix} \quad \text{from} \quad u_0 = x' = \begin{pmatrix} 1 \\ 0 \end{pmatrix}.$$

Hence

$$x = \sum_{k=0}^{\infty} u_k = \begin{pmatrix} \sum_0^{\infty} \left(\frac{1}{2}\right)^k \\ -\sum_1^{\infty} \left(\frac{1}{2}\right)^k \end{pmatrix} = \begin{pmatrix} 2 \\ -1 \end{pmatrix},$$

which is the exact solution.

We see that the series converge geometrically with quotient

$$1/\sqrt{2} = r_\sigma(HA^{-1}) = \|(HA^{-1})^2\|^{1/2}.$$

3.20 We note that since $x = R(z)f$, $R(z) = R'(z)\sum_{i=0}^{\infty}(HR'(z))^i$, $x_k = R'(z)Hx_{k-1} + x' = \sum_{i=0}^{k}(R'(z)H)^i x'$, and $x' = R'(z)f$, then

$$x - x_k = R'(z)\sum_{i=0}^{\infty}(HR'(z))^i f - \sum_{i=0}^{k}(R'(z)H)^i R'(z)f.$$

But $(R'(z)H)^i R'(z) = R'(z)(HR'(z))^i$; hence

$$\begin{aligned} x - x_k &= R'(z)\sum_{i=k+1}^{\infty}(HR'(z))^i f = R'(z)(HR'(z))^{k+1}\sum_{i=0}^{\infty}(HR'(z))^i f \\ &= R'(z)(HR'(z))^k HR'(z)\sum_{i=0}^{\infty}(HR'(z))^i f \\ &= R'(z)(HR'(z))^k HR(z)f \\ &= R'(z)(HR'(z))^k Hx. \end{aligned}$$

Given $\varepsilon > 0$ we construct $\|\cdot\|_*$ equivalent to $\|\cdot\|$ such that $\|HR'(z)\|_* \le r_z' + \varepsilon$. Then

$$\|x - x_k\| \le c(r_z' + \varepsilon)^k$$

since $R'(z)$ is uniformly bounded.

3.21 See the solution of Exercise 1.53.

3.22 From (3.23) we get the equation $(1 - P')T'\eta_k - \lambda'\eta_k = B_k\varphi'$ with $Q_k = \langle\cdot, \varphi'^*\rangle\eta_k$, $Q' = P' = \langle\cdot, \varphi'^*\rangle\varphi'$, and $\eta_k \in (1 - P')X$. But we can also compute

$$B_k\varphi' = (1 - P')H\eta_{k-1} - \sum_{i=1}^{k-1} Q_i H\eta_{k-i-1} + \sum_{i=1}^{k-1} Q_i T'\eta_{k-i} \qquad \text{for} \quad k \ge 2.$$

Setting $v_k = \langle -H\eta_{k-1}, \varphi_i^*\rangle$ we find

$$B_k\varphi' = (1 - P')H\eta_{k-1} + \sum_{i=1}^{k-1} v_{k-i}\eta_i + \sum_{i=1}^{k-1} Q_i T'\eta_{k-i}.$$

But $Q_i T'\eta_{k-i} = Q_i T'(1 - P')\eta_{k-i} = Q_i(1 - P')T'\eta_{k-i}$ and $Q_i(1 - P') = Q_i(1 - Q') = 0$. Also we note that $\sum_{i=1}^{k-1} v_{k-i}\eta_i = \sum_{i=1}^{k-1} v_i\eta_{k-i}$.

$$(1 - P')T'\eta_k - \lambda'\eta_k = (1 - P')H\eta_{k-1} + (1 - P')\sum_{i=1}^{k-1} v_i\eta_{k-i},$$

$$(1 - P')(T' - \lambda')(1 - P')\eta_k = (1 - P')\left[H\eta_{k-1} + \sum_{i=1}^{k-1} v_i\eta_{k-i}\right],$$

and

$$\eta_k = S'\left[H\eta_{k-1} + \sum_{i=1}^{k-1} v_i \eta_{k-i}\right], \qquad \text{for} \quad k \geq 2,$$

which is (3.20) since $S'\eta_0 = S'\varphi' = 0$.

3.23 $Q'P(t)Q' = \sum_{i=1}^{m} \langle \cdot, y_i\rangle\langle P(t)x_i, y_i\rangle x_i$. Hence $Q'P(t)Q' - Q'$ is an operator of rank m. Its matrix representation in the bases $\{x_i\}$, $\{y_i\}$ is the diagonal $\langle P(t)x_i, y_i\rangle - 1$. Hence

$$r_\sigma(Q'(P(t) - P')Q') < 1 \Leftrightarrow \max_i |\langle P(t)x_i, y_i\rangle - 1| < 1 \Leftrightarrow 0 < \max_i \langle P(t)x_i, y_i\rangle < 2$$

$$\Rightarrow Q'P(t)x_i \neq 0, \; i = 1, \ldots, m.$$

If $m = 1$ and $Q' = P'$, this gives back the condition in (3.17).

3.24 Set $T' = Q'T'Q' + (1 - Q')T'Q' + Q'T'(1 - Q') + (1 - Q')T'(1 - Q')$. Since $Q'X = M' = P'X$, we note that $(1 - Q')T'Q'X = (1 - Q')T'M' \subseteq (1 - Q')M' = (1 - Q')Q'M' = \{0\}$. Hence $(1 - Q')T'Q' = 0$, and we can write

$$T' = \begin{pmatrix} T'_1 & T'_2 \\ 0 & T'_4 \end{pmatrix},$$

where $T'_1 = Q'T'_{\restriction M'}$, $T'_2 = Q'T'_{\restriction W}$, $T'_4 = (1 - Q')T'_{\restriction W}$. This shows that

$$(T' - z)^{-1} = \begin{pmatrix} (T'_1 - z)^{-1} & -(T'_1 - z)^{-1}T'_2(T'_4 - z)^{-1} \\ 0 & (T'_4 - z)^{-1} \end{pmatrix}$$

and $\sigma(T') = \sigma(T'_1) \cup \sigma(T'_4)$, so $\{0\} \cup \sigma(T') = \sigma(Q'T'Q') \cup \sigma((1 - Q')T'(1 - Q'))$.

To show that these two spectra are disjoint except for zero, it suffices to show that $\sigma(T'_1) \cap \sigma(T'_4) = \emptyset$. Since $P'X = Q'X$, we remark that $\sigma(T'_1)$ is the part of $\sigma(T')$ which lies inside Γ. For any $\mu' \in \sigma(T'_1)$, there exists $z \in \mathbb{C}$ such that $\mu' \notin \sigma(T' + zP') = \sigma(Q'(T' + zP')_{\restriction M'}) \cup \sigma(T'_4)$; hence $\mu' \notin \sigma(T'_4)$. Therefore $\sigma(T'_1) \cap \sigma(T'_4) = \emptyset$. This proves that $\sigma(T'_4)$ is the part of $\sigma(T')$ which lies outside Γ.

3.25 By definition $\sigma(T'_1) = \{\mu'_i\}_1^m$. Then $\{\mu'_i - \hat{\lambda}'\}_1^m = \sigma(U'_{M'})$. $\sigma(T'_4) = \sigma(T') - \{\mu'_i\}_1^m$; therefore the spectrum of $\mathring{T}_{\hat{\lambda}'} := (1 - Q')(T' - \hat{\lambda}')_{\restriction W}$ is $\sigma(T'_4)$ translated by $\hat{\lambda}'$. We conclude that

$$\sigma(U'_{\restriction M'}) \subset \{z; |z - \hat{\lambda}'\} \leq a\},$$
$$\sigma(\mathring{T}'_{\hat{\lambda}'}) \subset \{z; |z - \hat{\lambda}'| \geq b\}.$$

$\hat{\lambda}'$ does not lie in the spectrum of $\mathring{T}'_{\hat{\lambda}'}$ if $b > 0$.

3.26 Simple algebraic manipulation.

3.27 Let b_k be the right-hand side of (3.27). $B_k = \begin{pmatrix} 0 \\ b_k \end{pmatrix}$, $Q_k = \begin{pmatrix} 0 \\ \mathring{\eta}_k \end{pmatrix}$, $H'_1 = H'_2 = 0$. Hence (3.23) becomes

$$(T'_4 - a_{ii}I)\mathring{\eta}_k = -H'_4\mathring{\eta}_{k-1} + \sum_{j=1}^{k-1} \mathring{\eta}_j + T'_2\mathring{\eta}_{k-j} \qquad \text{for} \quad k \geq 2.$$

Since $v_k = T'_2\mathring{\eta}_k$, we get $T'_2\mathring{\eta}_{k-j} = v_{k-j}$. (3.27) follows from $\varphi_k = \sum_{j=1}^{k} \mathring{\eta}_j$. Now (3.27) is also (1.14) with $y = e_1$, and noting that $e_i^{\mathrm{T}}H'\eta_k = 0$.

3.28 Clearly,

$$u' = \begin{pmatrix} \times & \times & \cdots & \times & \overset{\overset{i}{\downarrow}}{1} & 0 & \cdots & 0 \end{pmatrix}^{\mathrm{T}} \quad \text{and } v' = \begin{pmatrix} \times & \times & \cdots & \times & \overset{\overset{i}{\downarrow}}{0} & 1 & 0 & \cdots & 0 \end{pmatrix}^{\mathrm{T}}$$

are a basis of the invariant subspace for the triangular matrix T, associated with a_{ii} and $a_{i+1,i+1}$. Hence if $\varepsilon = \|H\|$ is small enough, $\|u' - u\| \le c\varepsilon$, $\|v' - v\| \le c\varepsilon$ implies $e_i^T u \ne 0$, $e_{i+1}^T v \ne 0$.

3.29 Set $a := \max(|\alpha - \lambda|; \lambda \in \sigma(B))$, $b := \min(|\alpha - \lambda|; \lambda \in \sigma(A))$. If $a < b$, $X_i \to X$ with a rate arbitrarily close to a/b. This is a stronger condition than $\sigma(B) \cap \sigma(A) = \emptyset$, which is the necessary and sufficient condition for the existence and unicity of X. It yields a simpler algorithm since no Schur decomposition for A and B is required.

Chapter 4

4.1 Let B be a bounded set in $X = C(a, b)$. $\exists M_B > 0$ such that $\forall x \in B$, $\|x\|_\infty \le M_B$. Let $C_k > 0$ be such that

$$\sup_{a \le t \le b} \int_a^b |k(t, s)|\, ds \le C_k.$$

Let $x \in B$. Then $\|Tx\|_\infty \le C_k(b - a)M_B$, so TB is a bounded set in X. Now let $t_1, t_2 \in [a, b]$ and $x \in B$:

$$|Tx(t_1) - Tx(t_2)| \le M_B \int_a^b |k(t_1, s) - k(t_2, s)|\, ds,$$

which goes to 0 uniformly in x if $t_1 \to t_2$. Hence TB is equicontinuous. Finally T is a continuous operator since $\|Tx_1 - Tx_2\|_\infty \le C_k(b - a)\|x_1 - x_2\|_\infty$. This proves that T is compact. Let $p = 1$ in Theorem 4.1.

4.2 $$\delta(\{x\}, X_n) = \sup_{\substack{x \in \{x\} \\ \|x\| = 1}} \inf_{y \in X_n} \|x - y\|$$

$$= \inf_{y \in X_n} \left\| \frac{x}{\|x\|} - y \right\| = \frac{1}{\|x\|} \inf_{y \in X_n} \|x - y\|$$

$$= \frac{1}{\|x\|} \delta_n(x).$$

4.3 $\|x - \pi_n x\| = \inf_{y \in E_n} \|x - y\| = \text{dist}(x, E_n)$.

4.4 $(\pi_n x)(t) = \sum_{i=1}^n \chi_i(t) p_i(t)$, where $p_{i \restriction \Delta_i} \in \mathbb{P}_r$ and χ_i is 1 on Δ_i and 0 on $[a, b] \backslash \Delta_i$.

$$\|x - p_i\|^2_{L^2(\Delta_i)} = \inf_{p \in \mathbb{P}_r} \|x - p\|^2_{L^2(\Delta_i)},$$

$$\|x - \pi_n x\|_{L^2(a,b)} = \left\| \sum_{i=1}^n \chi_i x - \sum_{i=1}^n \chi_i p_i \right\|_{L^2(a,b)}$$

$$= \sum_{i=1}^n \|x - p_i\|_{L^2(\Delta_i)} = \sum_{i=1}^n \inf_{p \in \mathbb{P}_r} \|x - p\|_{L^2(\Delta_i)}$$

$$= \inf_{p \in \mathbb{P}_{r,\Delta}} \|x - p\|_{L^2(a,b)} = \text{dist}_2(x, \mathbb{P}_{r,\Delta}).$$

The rest is left ot the reader.

4.5 See Theorem 7.5 and Corollary 7.6.

4.6 On each Δ_i, $x_{\restriction \Delta_i}$ is continuous and its Lagrange interpolation $L_n(x_{\restriction \Delta_i})$ is well defined. And $L_n^2(x_{\restriction \Delta_i}) = L_n(x_{\restriction \Delta_i})$. For $x \in C(a, b)$, $x \mapsto L_n x$ is a map into $\mathbb{P}_{r,\Delta} \subset L^\infty(a, b)$, and $L_n x \to x$ in $L^\infty(a, b)$. Apply Theorem 3.1 to conclude that $\|L_n\|_{C \to L^\infty} < \infty$.

4.7 $\dim \mathbb{P}_{r-1} = r$, $\dim \mathbb{P}_{r-1,\Delta} = nr$. At each t_i, $i = 1, \ldots, n - 1$, we impose $r - 1$ continuity conditions, so we have $(n - 1)(r - 1)$ linearly independent constraints to be satisfied by the elements of $S_{r,\Delta}$. In this way we get $\dim S_{r,\Delta} = nr - (n - 1)(r - 1) = n + r - 1$.

4.8 $I_n(r) = \sum_{i=0}^{n} r(t_i) \underbrace{\int_0^1 \frac{e_i(t)}{\sqrt{t}}\, dt}_{w_i}.$

$$w_0 = \int_0^h \frac{1}{h}(h - t)\frac{dt}{\sqrt{t}} = \frac{4}{3}h^{1/2},$$

$$w_n = \int_{1-h}^1 \frac{1}{h}(t - 1 + h)\frac{dt}{\sqrt{t}} = 2 - \frac{4}{3}\frac{2 - h + \sqrt{1-h}}{1 + \sqrt{1-h}},$$

$$w_i = \int_{(i-1)h}^{ih} \frac{1}{h}(t - (i-1)h)\frac{dt}{\sqrt{t}} + \int_{ih}^{(i+1)h} \frac{1}{h}((i+1)h - t)\frac{dt}{\sqrt{t}}$$

$$= \frac{4}{3}h^{1/2}[(i-1)^{3/2} - 2i^{3/2} + (i+1)^{3/2}] \qquad \text{for} \quad 0 < i < n.$$

For $t \in [t_{j-1}, t_j]$ we have $r(t) - \pi_n r(t) = r(t) - r(t_{j-1})e_{j-1}(t) - r(t_j)e_j(t)$, which we study. $\exists \xi_j^1 \in]t_{j-1}, t[$ such that $r(t) = r(t_{j-1}) + (t - t_{j-1})r'(t_{j-1}) + \frac{1}{2}(t - t_{j-1})^2 r''(\xi_j^1)$. $\exists \xi_j^2 \in]t_{j-1}, t_j[$ such that $r(t_j) = r(t_{j-1}) + hr'(t_{j-1}) + \frac{1}{2}h^2 r''(\xi_j^2)$. $e_{j-1}(t) + e_j(t) = 1$ and $e_j(t) = (1/h)(t - t_{j-1})$. This yields $r(t) = \pi_n r(t) = \frac{1}{2}[(t - t_{j-1})^2 r''(\xi_j^1) - h^2 r''(\xi_j^2)e_j(t)]$. So $\sup_{t\in\Delta_j}|r(t) - \pi_n r(t)| \le (1/n^2)\sup_{\xi\in\Delta_j}|r''(\xi)|$. It follows that

$$|(I - I_n)(r)| \le \sup_{0\le t\le 1}|r(t) - \pi_n r(t)| \int_0^1 \frac{dt}{\sqrt{t}} \le \frac{2}{n^2}\sup_{0\le t\le 1}|r''(t)|.$$

4.9 $x = (x_j)_1^\infty \in l^2$, $b = (b_i)_1^\infty \in l^2$, $(Tx)_i = \sum_{j=1}^\infty a_{ij}x_j$. T is bounded:

$$\|T\|^2 \le \sum_{i,j}|a_{ij}|^2 < \infty.$$

$Tx - x = b$ is solved by the projection method on $E_n = \{e_1, \dots, e_n\}$. $\pi_n(T - 1)x_n = \pi_n b = \sum_{j=1}^n b_j e_j$. That is, $\langle (T - 1)x_n - b, e_j\rangle = 0$, $j = 1, \dots, n$. $x_n = \sum_{i=1}^n x_i^{(n)} e_i$, where the $x_i^{(n)}$ are solutions of

$$\sum_{j=1}^n a_{ij}x_j^{(n)} - x_i^{(n)} = b_i, \qquad i = 1, \dots, n.$$

We need the condition $1 \notin \sigma(A)$ for the solution to be unique. It is satisfied if, for example, $\sum_{i,j}|a_{ij}|^2 < 1$, because then $\|T\| < 1$ and $(1 - T)^{-1}$ exists and is bounded.

4.10 Let $\{e_i\}_1^n$ (resp. $\{f_i\}_1^n$) be a basis of X_n (resp. Y_n), the adjoint basis is $\{e_i^*\}_1^n$ (resp. $\{f_i^*\}_1^n$). $\pi_n x = \sum_{i=1}^n \langle x, f_i^*\rangle f_i$. (4.6) becomes

$$\sum_{j=1}^n \langle Te_j, f_i^*\rangle \xi_{jn} - z\sum_{j=1}^n \langle e_j, f_i^*\rangle \xi_{jn} = \langle f, f_i^*\rangle, \qquad i = 1, \dots, n.$$

(4.7) becomes

$$\sum_{j=1}^n \langle Te_j, f_i^*\rangle u_{jn} = \lambda_n \sum_{j=1}^n \langle e_j, f_i^*\rangle u_{jn}, \qquad i = 1, \dots, n.$$

4.11 From condition (4.8) it follows that π_n' is a continuous operator. $\pi_n'^2 x = \tilde{\pi}_n^{-1}\pi_n\tilde{\pi}_n^{-1}\pi_n x = \pi_n' x, \forall x \in X$, since $\pi_n\tilde{\pi}_n^{-1} = \tilde{\pi}_n\tilde{\pi}_n^{-1} = 1_{Y_n}$. π_n is a surjection on Y_n and $\tilde{\pi}_n^{-1}$ is a bijection from Y_n to X_n, so π_n' projects on X_n. Apply $\tilde{\pi}_n^{-1}$ to Eqs. (4.6) and (4.7).

4.12 $\theta_n = \cos\alpha_n$, where α_n is the acute angle between E_n and F_n, $\tau_n = \sin\alpha_n$. $\theta_n < 1 \Leftrightarrow \tau_n > 0 \Leftrightarrow 1/\tau_n = \|\tilde{\pi}_n^{-1}\| < \infty$

4.13 $\langle x, y\rangle_T = \langle x, Ty\rangle$, $\|Tx_n - f\| = \min_{y\in E_n}\|Ty - f\|$. For $v_n = Tx_n$, $\|v_n - f\| = \min_{w\in TE_n=F_n}\|w - f\|$. $\langle v_n - f, w\rangle = 0, \forall w \in F_n, \langle v_n - f, Tv\rangle = 0, \forall v \in E_n, \langle Tx_n - f, v\rangle_T = 0$, $\forall v \in E_n$, $\pi_n^T(Tx_n - f) = 0$, $x_n \in E_n$, where π_n^T is the T-orthogonal projection on E_n.

4.14 $r = f - (T - z)\tilde{x}$, $v = x - \tilde{x}$. $(T - z)v = r$. Set $w = Tv$, $w - zv = r$,

$$(T - z)w = Tr, \qquad v = \frac{1}{z}(w - r).$$

By projection:

$$(\pi_n T - z)w_n = \pi_n Tr, \quad w_n \in X_n, \qquad v_n = \frac{1}{z}(w_n - r).$$

$$x_n'' = \tilde{x} + \frac{1}{z}(w_n - f + (T - z)\tilde{x})$$

$$= \frac{1}{z}w_n - \frac{1}{z}f + \frac{1}{z}T\tilde{x}.$$

4.15 $\langle (T - z)x_n, e_i\rangle_{L^2} = \langle f, e_i\rangle_{L^2}$, $i = 1, \ldots, n$. That is,

$$\int_a^b \left[\int_a^b k(t, s)x_n(s)\,ds - zx_n(t)\right]\overline{e_i(t)}\,dt = \int_a^b f(t)\overline{e_i(t)}\,dt,$$

$$x_n(s) = \sum_{j=1}^n \xi_j e_j(s).$$

We get

$$\sum_{j=1}^n \xi_j \int_a^b \left[\int_a^b k(t, s)e_j(s)\,ds - ze_j(t)\right]\overline{e_i(t)}\,dt = \int_a^b f(t)\overline{e_i(t)}\,dt, \qquad i = 1, \ldots, n.$$

Since $\{e_j\}$ is orthonormal, and after using quadrature rules with weights $\{w_i\}_1^n$, and nodes $\{t_i\}_1^n$ we get

$$\sum_{j=1}^n \xi_j\left[\sum_{l=1}^n \int_a^b w_l k(t_l, s)e_j(s)\,ds\,\overline{e_i(t_l)} - z\right] = \sum_{l=1}^n w_l f(t_l)\overline{e_i(t_l)}, \qquad i = 1, \ldots, n.$$

$\int e_j(t)\overline{e_i(t)}\,dt = \delta_{ij}$ is approximated by $\sum_{l=1}^n w_l e_j(t_l)\overline{e_i(t_l)} \approx \delta_{ij}$. The inverse of $(w_j\overline{e_i(t_j)})$ is taken to be $(e_j(t_i))$. We deduce that $(\tilde{A}_n - z\tilde{B}_n)\xi_n = (f(t_l))$, which is (4.10) with $\tau_l = t_l$ and $l_j = e_j$.

4.16 Consider the projection π_n on Y_n defined by $(\pi_n x)(t) = \sum_{j=1}^n c_{jn} l_j'(t)$, with

$$\sum_{j=1}^n c_{jn} l_j'(\tau_i) = x(\tau_i), \qquad i = 1, \ldots, n.$$

$\pi_n[(T - z)x_n - f] = 0$, $x_n \in X_n$, defines a Petrov method with right (resp. left) subspace X_n (resp. Y_n). Clearly, the functions $\{l_i'\}_1^n$ can be arbitrary, provided that $\det(l_j'(\tau_i)) \neq 0$.

4.17 Let $\pi_n x = \sum_{i=1}^n x(t_i)e_i$. $\pi_n T\pi_n$ defines the matrix $a_{ij} = \int_a^b k(t_i, s)e_j(s)\,ds$. If $k(t_i, s)$ is approximated by $k(t_i, t_j)$ on $\{s; e_j(s) \neq 0\}$, a_{ij} is approximated by $a_{ij}' = k(t_i, t_j)\int_a^b e_j(s)\,ds = w_j k(t_i, t_j)$, which is the coefficient of the matrix of the Fredholm method.

4.18 Set

$$\vartheta(s) = \frac{1}{z}\sum_{j=1}^n w_j k(s, t_j)x_n(t_j) - \frac{1}{z}f(s).$$

For ϑ to interpolate x_n at the points t_i, one must have $\vartheta(t_i) = x_n(t_i)$, that is,

$$\frac{1}{z}\sum_{j=1}^{n} w_j k(t_i, t_j)x_n(t_j) - \frac{1}{z} f(t_i) = x_n(t_i).$$

So $\{x_n(t_i)\}_{i=1}^{n}$ must satisfy (4.14).

4.19 $F: (z, x) \mapsto (z, (1/z)Tx)$. $F(\lambda, \varphi) = (\lambda, (1/\lambda)T\varphi) = (\lambda, \varphi)$ since $T\varphi = \lambda\varphi$. Define one fixed-point iteration from (λ_n, φ_n).

$$(\lambda_{n+1}, \varphi_{n+1}) := F(\lambda_n, \varphi_n) = \left(\lambda_n, \frac{1}{\lambda_n} T\varphi_n\right).$$

This gives

$$\lambda_{n+1} = \lambda_n \quad \text{and} \quad \varphi_{n+1} = \frac{1}{\lambda_n} T\varphi_n = \tilde{\varphi}_n \quad \text{in} \quad (4.19).$$

4.20 For $Tx = f$ substitute $(T - z_0)x = f - z_0 x$, so $x = (1/z_0)[f - (T - z_0)x]$, which suggests the iterate $\tilde{x}_n = (1/z_0)[f - (T - z_0)x_n]$.

4.21 $T_n\varphi_n = \lambda_n\varphi_n$ becomes $(T_n + \alpha)\varphi_n = (\lambda_n + \alpha)\varphi_n$. Since $T\varphi = 0$, $(T + \alpha)\varphi = \alpha\varphi$, $\tilde{\varphi}_n = (\lambda_n + \alpha)^{-1}(T + \alpha)\varphi_n$.

4.22 $\tilde{x}_n = Tx_n - b$. $(\tilde{x}_n)_i = \sum_{j=1}^{n} a_{ij}x_j^{(n)} - b_i$, $i \in \mathbb{N}$, where $(x_i^{(n)})$ is the solution of

$$\sum_{j=1}^{n} a_{ij}x_j^{(n)} - x_i^{(n)} = b_i, \qquad i = 1, \ldots, n.$$

Therefore $(\tilde{x}_n)_i = x_i^{(n)}$ for $i = 1, \ldots, n$.

4.23 $\pi_n Tx_n - zx_n = \pi_n f$, $\tilde{x}_n = (1/z)(Tx_n - f)$. So $(T\pi_n - z)\tilde{x}_n = T\pi_n(1/z)(Tx_n - f) - Tx_n + f = (1/z)T(\pi_n f + zx_n) - (1/z)T\pi_n f - Tx_n + f = f$.

4.24 $\pi_n[(T - z)x_n - f] = 0 \Rightarrow \pi_n(Tx_n - f) = z\pi_n x_n$. Hence $\pi_n\tilde{x}_n = (1/z)\pi_n(Tx_n - f) = \pi_n x_n$.

4.25 $P_\kappa = \sum_i P_{\kappa,i}$, $\kappa \in \{\alpha, \beta, \gamma\}$, and $\mathscr{P} = \sum_i \mathscr{P}_i$. Apply Theorem 4.3 to each λ_i.

4.26 $T_n^P\varphi_n^G = \pi_n T\varphi_n^G = \lambda_n\varphi_n^G$.

4.27 $T_n^G = T_\alpha$ in Theorem 4.3, $R_n^G(z) = R_\alpha(z)$. $T_n^G\varphi_n^G = \pi_n T\pi_n\varphi_n^G = \pi_n T\varphi_n^G = \lambda_n\varphi_n^G$.

4.28 From (4.9) we get $y_n = [(\pi_n T - z)_{\restriction X_n}]^{-1}\pi_n Tf = \mathscr{R}_n(z)\pi_n Tf$. Therefore

$$x_n' = \frac{1}{z}(y_n - f) = \frac{1}{z}\mathscr{R}_n(z)\pi_n Tf - \frac{1}{z} f = R_n^P(z)f = x_n^P.$$

4.29 $M_n^G = P_n^G X = \mathscr{P}_n X_n = \mathscr{M}_n$. $\mathscr{D}_n^k\mathscr{M}_n \subseteq \mathscr{M}_n$ for $k = 0, 1, \ldots, l - 1$ implies $\mathscr{E}_n\mathscr{M}_n = \mathscr{M}_n$, $M_n^P = P_n^P X = \mathscr{E}_n\mathscr{P}_n\pi_n TX$. If $0 \in \rho(T)$, Im $T = \text{Dom}(T^{-1}) = X$; therefore $M_n^P = \mathscr{E}_n\mathscr{P}_n X_n = \mathscr{E}_n\mathscr{M}_n = \mathscr{M}_n$. If $0 \in \sigma(T)$, consider $\alpha \in \rho(T)$, then $0 \in \rho(T - \alpha)$ and the invariant subspaces are unchanged.

4.30 $P_n^P\varphi = \mathscr{E}_n\mathscr{P}_n\pi_n T\varphi = \lambda\mathscr{E}_n\mathscr{P}_n\pi_n\varphi = \lambda\mathscr{E}_n P_n^G\varphi$. If $\lambda_n \neq 0$ is semisimple, $\mathscr{E}_n = \mathscr{P}_n/\lambda_n$; therefore $P_n^P\varphi = (\lambda/\lambda_n)P_n^G\varphi$.

4.31 $$T_n^S\varphi_n^S = T\pi_n\left(\frac{1}{\lambda_n} T\varphi_n^G\right) = \frac{1}{\lambda_n} TT_n^P\varphi_n^G = T\varphi_n^G = \lambda_n\left(\frac{1}{\lambda_n} T\varphi_n^G\right) = \lambda_n\varphi_n^S.$$

$P_n^S\varphi = T\pi_n\mathscr{E}_n\mathscr{P}_n\pi_n\varphi = T\pi_n\mathscr{E}_n P_n^G\varphi$. If $\lambda_n \neq 0$ is semisimple $P_n^S\varphi = (1/\lambda_n)TP_n^G\varphi$.

4.32 $M_n^S = P_n^S X = T\pi_n\mathscr{E}_n\mathscr{P}_n X_n = T\mathscr{M}_n = T\pi_n\mathscr{M}_n = T_n^S\mathscr{M}_n$.

4.33 $\mathscr{T}'_n\colon X_n \to X_n$, $\langle \mathscr{T}'_n e_j, e_i^* \rangle = w_j k(t_i, t_j)$.

$$\langle x, e_j^* \rangle = x(t_j), \qquad \pi_n x = \sum_{i=1}^{n} \langle x, e_i^* \rangle e_i.$$

$$T_n^F x = \sum_{i=1}^{n} \left[\sum_{j=1}^{n} w_j k(t_i, t_j) x(t_j) \right] e_i,$$

$$(\mathscr{T}'_n \pi_n x)_i = \sum_{j=1}^{n} w_j k(t_i, t_j)(\pi_n x)_j,$$

$$\mathscr{T}'_n \pi_n x = \sum_{i=1}^{n} \left[\sum_{j=1}^{n} w_j k(t_i, t_j) x(t_j) \right] e_i = T_n^F x.$$

$$T_n^N x(s) = \sum_{j=1}^{n} w_j k(s, t_j) x(t_j).$$

It follows that $\pi_n T_n^N = T_n^F$.

4.34 Apply (4.20) to

$$T = \left(\begin{array}{c|c} \mathscr{T}'_n \pi_n & 0 \\ \hline 0 & 0 \end{array} \right) = T_n^F.$$

Then use Exercise 4.27.

4.35 l_n is the ascent of λ_n and $\mathscr{D}'_n = (\mathscr{T}'_n - \lambda_n)\mathscr{P}'_n$. Apply (4.22) to

$$T = \left(\begin{array}{c|c} \mathscr{T}'_n \pi_n & 0 \\ \hline (1 - \pi_n) T_n^N & 0 \end{array} \right) = T_n^N.$$

Then use Exercise 4.32 to show that $M_n^N = T_n^N \mathscr{M}'_n$.

4.36
$$\begin{aligned} \|R_n(z)\pi_n x - R(z)x\| &= \|R_n(z)\pi_n x - R_n(z)x + R_n(z)x - R(z)x\| \\ &\le \|R_n(z)\| \, \|(1 - \pi_n)x\| + \|(R_n(z) - R(z))x\|. \end{aligned}$$

Since $R_n(z) \xrightarrow{\text{p}} R(z)$, $\|R_n(z)\| \le M$, and $\pi_n \xrightarrow{\text{p}} 1$, we get $\|R_n(z)\pi_n x - R(z)x\| \to 0$.

4.37 Since π_n is an orthogonal projection, $\|(1 - \pi_n)x\| \le \|x - x_n^G\|$, so $c_1 = 1$. $zR_n^P(z) = R_n^P(z)\pi_n T - 1$. Since $\pi_n^* = \pi_n$ and T^* is compact we get

$$\|x - x_n^G\| \le \|(1 - \pi_n)x\| + \|R_n^P(z)\| \, \|(1 - \pi_n)T^*\pi_n\| \, \|(1 - \pi_n)x\| \le (1 + \varepsilon_n)\|(1 - \pi_n)x\|,$$

where $\varepsilon_n := \|R_n^P(z)\| \, \|(1 - \pi_n)T^*\pi_n\| \to 0$ as $n \to \infty$.

4.38 If $h \xrightarrow{\text{p}} 0$, then $\pi_n \xrightarrow{\text{p}} 1$. But since T is compact, TB is relatively compact (B being the unit ball in $L^2(a, b)$), so from Theorem 3.2 it follows that $\|(1 - \pi_n)T\|_2 \to 0$. From Proposition 4.6 and Exercise 4.5 it follows that $\|x - x_n^G\|_2 = 0(h^{r+1})$.

4.39 Because $TX \subset C(a, b)$, $L_n T$ is well defined. Now for $x \in C(a, b)$, $(L_n - 1)x \to 0$ in $L^\infty(a, b)$. This proves that $\|(1 - L_n)T\|_\infty \to 0$ and $\|(L_n T - z)^{-1}\|_\infty < M$ for $z \in \rho(T)$. From Proposition 4.6 follows $x_n^G \to x$ for $x \in C(a, b)$. Apply Corollary 7.6 to bound $\|(1 - L_n)x\|_\infty$.

4.40 T is compact from $L^\infty(a, b)$ into $C(a, b)$. For $x \in C(a, b)$, $(L_n - 1)x \to 0$ in $L^\infty(a, b)$. Therefore $TL_n \xrightarrow{\text{cc}} T$ in $C(a, b)$ by Theorem 4.5. From Proposition 4.6 follows $x_n^S \to x$.

4.41 If $f \in C^p(a, b)$ and $k \in C^p([a, b]^2)$, then $x \in C^p(a, b)$. A weaker condition is given in Chapter 7, Theorem 7.2.

4.42 $\pi_n K - \alpha \xrightarrow{\text{s}} K - \alpha$ implies $(\pi_n K - \alpha)^{-1}$ uniformly bounded and $u_n \to u$ since $\pi_n \xrightarrow{\text{p}} 1$.

4.43 At the Chebyshev points τ_i, $\tilde{x}_n(\tau_i) = x_n(\tau_i)$. Therefore $\max_i |(x - x_n)(\tau_i)| \le c \operatorname{dist}_\infty(x, \mathbb{P}_{n-1})$. This could not be derived from Proposition 4.9, which proves a convergence in quadratic mean with weight $\rho(t)$.

4.44 $\vartheta(s) = \int_a^b |k(s, t)|\, dt$, $s \in [a, b]$, is continuous on $[a, b]$; hence there exists $s_0 \in [a, b]$ such that $\vartheta(s) \leq \vartheta(s_0)$ for all $s \in [a, b]$. Let $\{x_m\}$ be a sequence in $C(a, b)$ such that $\|x_m\| = 1$, $x_m(t_i) = 0$, $i = 1, \ldots, n$, and $(Tx_m)(s_0) \geq \vartheta(s_0) - 1/m$ (the existence of such a sequence is guaranteed by measure and integration theory). We then get

$$\|T\| - \frac{1}{m} \leq \|Tx_m\| = \|(T - T_n^N)x_m\| \leq \|T - T_n^N\|.$$

When $m \to \infty$, this yields $\|T - T_n^N\| \geq \|T\|$.

4.45 I. For $x \in X_n$, $x(t) = \sum_i x(t_i)e_i(t)$.

$$(\mathcal{T}'_n - \pi_n T)x = \sum_i \left(\sum_j \left[w_j k(t_i, t_j) - \int_a^b k(t_i, s)e_j(s)\, ds \right] x(t_j) \right) e_i(t),$$

$$\|(\mathcal{T}'_n - \pi_n T)_{\restriction X_n}\|_\infty \leq \max_{i,j} \sup_{s \in D_j} |k(t_i, t_j) - k(t_i, s)| \sum_j \int_a^b |e_j(s)|\, ds \to 0.$$

II. $\|(1 - \pi_n)T\| \to 0$.

4.46 Suppose the following conditions hold:

(1) r continuous on $[a, b] \times [a, b]$,
(2) $\max_{a \leq s \leq b} \int_a^b |\sigma(s, t)|\, dt < +\infty$,
(3) $\lim_{s' \to s} \int_a^b |\sigma(s', t) - \sigma(s, t)|\, dt = 0$, $\forall s \in [a, b]$,
(4) $\sup_n \sum_{i=1}^n \max_{a \leq t \leq b} |e_i(t)| < +\infty$.

Then, for $x \in C(a, b)$ with $\|x\| \leq 1$, we get for $s \in [a, b]$

$$|T_n x(s)| \leq \sum_{i=1}^n |r(s, t_i)|\, |x(t_i)| \max_{a \leq t \leq b} |e_i(t)| \int_a^b |\sigma(s, t)|\, dt$$
$$\leq c\|r(s, \cdot)\|_\infty \|x\|_\infty.$$

Then $\|T_n x\|_\infty \leq c\|x\|_\infty \leq c$, and $\{T_n x\}$ is uniformly bounded for $x \in B$. Moreover, if $s, s' \in [a, b]$

$$T_n x(s') - T_n x(s) = \sum_{i=1}^n \left[r(s', t_i)x(t_i) \int_a^b e_i(t)\sigma(s', t)\, dt - r(s, t_i)x(t_i) \int_a^b e_i(t)\sigma(s, t)\, dt \right]$$
$$= \sum_{i=1}^n \left[(r(s', t_i) - r(s, t_i))x(t_i) \int_a^b e_i(t)\sigma(s', t)\, dt + r(s, t_i)x(t_i) \int_a^b e_i(t)(\sigma(s', t) - \sigma(s, t))\, dt \right].$$

So $|T_n x(s') - T_n x(s)| \leq c_1\|r(s', \cdot) - r(s, \cdot)\|_\infty + c_2 \int_a^b |\sigma(s', t) - \sigma(s, t)|\, dt \to 0$ when $s' \to s$, independently of n. The result follows.

4.47 Let $\varepsilon > 0$ be given, there exists $m(\varepsilon)$ such that $\|T_n^m x - T_n x\| \leq \|T_n^m - T_n\| < \varepsilon$ for $m \geq m(\varepsilon)$, $n \geq 1$, and $x \in B$. So $\{T_n^m x; m \geq m(\varepsilon), n \geq 1, x \in B\}$ has an ε-cover with centers in K, a relatively compact set; therefore a finite cover can be extracted. For each m, $1 \leq m < m(\varepsilon)$, $\{T_n^m x; n \geq 1, x \in B\}$ has a finite ε-cover because K^m is relatively compact. It follows that $\{T_n^m x; m \geq 1, n \geq 1, x \in B\}$ has also a finite ε-cover, which implies it is relatively compact.

4.48 $(T^{-1} - z)^{-1} = -\dfrac{1}{z}\left(T - \dfrac{1}{z}\right)^{-1} T.$

$$P\left(\frac{1}{\lambda}, T^{-1}\right) = -\frac{1}{2\pi i}\int_{\Gamma(1/\lambda)} \left(-\frac{1}{z}\right)\left(T - \frac{1}{z}\right)^{-1} T\, dz, \qquad \left(t = \frac{1}{z}\right)$$

$$= -\frac{1}{2\pi i}\int_{\Gamma(\lambda)} \frac{1}{t}\, T(T - t)^{-1}\, dt$$

$$= -\frac{1}{2\pi i}\int_{\Gamma(\lambda)} \frac{1}{t}\,(T - t + t)(T - t)^{-1}\, dt$$

$$= -\underbrace{\frac{1}{2\pi i}\int_{\Gamma(\lambda)} \frac{dt}{t}}_{=0} - \frac{1}{2\pi i}\int_{\Gamma(\lambda)} (T - t)^{-1}\, dt$$

$$= P(\lambda, T).$$

4.49 Exercise 2.29 shows that

$$\left.\begin{array}{l} u'' = x \\ u(0) - u'(0) = 0 \\ u(1) + u'(1) = 0 \end{array}\right\} \Leftrightarrow u(t) = \int_0^1 g(t, s)x(s)\, ds$$

with

$$g(t, s) = \begin{cases} \frac{1}{3}(t + 1)(s - 2), & t \le s, \\ \frac{1}{3}(s + 1)(t - 2), & t > s. \end{cases}$$

$$(P) \Leftrightarrow \begin{cases} u = Gx \\ x = Kx + f \end{cases} \quad \text{with} \quad k(t, s) = \frac{\partial g(t, s)}{\partial t} = \begin{cases} \frac{1}{3}(s - 2) \\ \frac{1}{3}(s + 1). \end{cases}$$

4.50 $\psi = G\varphi$, $(1 - K)\varphi = \lambda\mathfrak{N}G\varphi$, where K is defined in Proposition 4.15. Both kernels have a first-kind discontinuity along $s = t$.

4.51 This is a well-known result in partial differential equations. See, for instance, Adams (1975).

4.52 For any closed densely defined operator $T\colon D \subseteq H \to H$, $\langle u, v\rangle_D := \langle u, v\rangle_H + \langle Tu, Tv\rangle_H$ defines an inner product in D.

$$\langle u, u\rangle_D = \|u\|_H^2 + \|Tu\|_H^2 > 0, \qquad \forall u \neq 0 \text{ in } D,$$

$$\langle v, u\rangle_D = \overline{\langle u, v\rangle_H} + \overline{\langle Tu, Tv\rangle_H} = \overline{\langle u, v\rangle_D},$$

$$\langle \alpha u + v, w\rangle_D = \alpha\langle u, w\rangle_H + \alpha\langle Tu, Tw\rangle_H + \langle v, w\rangle_H + \langle Tv, Tw\rangle_H$$
$$= \alpha\langle u, w\rangle_D + \langle v, w\rangle_D$$

with associated norm $|||u|||_D = (\langle u, u\rangle_D)^{1/2} = (\|u\|_H^2 + \|Tu\|_H^2)^{1/2}$ for which $\|Tu\|_H^2 \le \|u\|_H^2 + \|Tu\|_H^2 = |||u|||_{\hat{\hat{D}}}^2$, so $T \in \mathscr{L}(\hat{\hat{D}}, H)$. It remains to show that $\hat{\hat{D}}$ is complete. Let $\{u_n\}$ be a Cauchy sequence in $\hat{\hat{D}}$: $|||u_n - u_m|||_{\hat{\hat{D}}}^2 = \|u_n - u_m\|_H^2 + \|Tu_n - Tu_m\|_H^2 \to 0$ as $n, m \to \infty$. So $\{u_n\}$ and $\{Tu_n\}$ are Cauchy sequences in H. Hence $u_n \to u$, and since T is closed, $Tu_n \to Tu$. This shows that $|||u_n - u|||_{\hat{\hat{D}}}^2 = \|u_n - u\|_H^2 + \|Tu_n - Tu\|_H^2 \to 0$; hence u_n converges to u in $\hat{\hat{D}}$. $\hat{\hat{D}}$ is then a Hilbert space.

4.53 $\langle T'u, v\rangle_H = a^*(u, v) = \overline{a(v, u)} = \overline{\langle Tv, u\rangle_H} = \langle u, Tv\rangle_H$

4.54 A is a closed operator with domain H; hence it is bounded. Since T has domain D, A has range D.

4.55 By Lax–Milgram on V, $a(u, v) = (S_a^{-1}u, v)_V$, $b(u, v) = (S_b^{-1}u, v)_V$, where S_a^{-1}, S_b^{-1} and S_a are bounded on V. Therefore

$$a(\psi, v) = \lambda b(\psi, v),\ \forall v \in V \Leftrightarrow S_a^{-1}\psi = \lambda S_b^{-1}\psi \Leftrightarrow S_a S_b^{-1}\psi = (1/\lambda)\psi.$$

4.56 Use Exercise 4.53 and note that $a(Bu, v) = a(u, B^\times v)$ by definition of $B^\times$ and $a(Au, v) = (u, v)_H = a(u, B^\times v) = a(u, A'v)$, $\forall u, v \in V$ since $Bu = Au$ for $u \in V$.

4.57 If a is hermitian, $(Au, v)_H = a(Au, Av) = (u, Av)_H$; hence $A = A'$ and $B^\times = B$. If a is coercive, $(Au, u)_H = a(Au, Au) \geq \alpha\|Au\|_V^2 > 0$ for $0 \neq u \in H$.

4.58 Since the forms $(\mathfrak{M}x, y)_2$ and $(\mathfrak{N}x, y)_2$ are hermitian, positive-definite sesquilinear, it follows that $(x, y)_H := (\mathfrak{M}x, y)_2$ satisfies $(x, x)_H = (\mathfrak{M}x, x)_2 > 0$ for $x \neq 0$ in $\mathscr{D}$, $(\alpha x + x', y)_H = \alpha(\mathfrak{M}x, y)_2 + (\mathfrak{M}x', y)_2 = \alpha(x, y)_H + (x', y)_H$. $(x, y)_H = \overline{(y, x)_H}$, and similarly for $\mathfrak{N}$. $\mathfrak{M}u = \lambda\mathfrak{N}u$, $u \in \mathscr{D} \Leftrightarrow (\mathfrak{M}u, v)_2 = \lambda(\mathfrak{N}u, v)_2$, $\forall v \in V$. Hence $(u, v)_H = \lambda(u, v)_V$, $\forall v \in V$.

4.59 $\pi_n(T\psi_n - \lambda_n\psi_n) = 0$, where π_n is the orthogonal projection on $F_n = \{1, t, \ldots, t^{n-1}\}$.

$$\int_a^b (\mathfrak{L}\psi_n - \lambda_n\psi_n)t^{i-1}\,dt = 0, \qquad i = 1, \ldots, n.$$

$\psi_n \in E_n \Rightarrow \psi_n = \sum_{j=1}^n \xi_{jn}e_j$. Then

$$\sum_{j=1}^n \xi_{jn}\int_a^b \mathfrak{L}e_j(t)t^{i-1}\,dt = \lambda_n\sum_{j=1}^n \xi_{jn}\int_a^b e_j(t)t^{i-1}\,dt, \qquad i = 1, \ldots, n.$$

Let $\xi^{(n)} = (\xi_{jn})$, $a_{ij} = \int_a^b \mathfrak{L}e_j(t)t^{i-1}\,dt$, $b_{ij} = \int_a^b e_j(t)t^{i-1}\,dt$; $A = (a_{ij})$ and $B = (b_{ij})$. The system to be solved is then $A\xi^{(n)} = \lambda_n B\xi^{(n)}$.

4.60 Let $u \in E_n = \mathbb{P}_{n+p-1} \cap (4.34)$. Then $D^p u \in \mathbb{P}_{n-1}$, $\mathfrak{L}u \in TE_n = (1 - K)G^{-1}E_n$. This proves that $D^pE_n \subseteq \mathbb{P}_{n-1}$, $\mathfrak{L}E_n \subseteq (1 - K)\mathbb{P}_{n-1}$. The converse is left to the reader.

4.61 $L_n(T\psi_n - \lambda_n\psi_n) = 0$, $\psi_n = \sum_{j=1}^n \xi_{jn}e_j$, $T\psi_n(t_i) = \lambda_n\psi_n(t_i)$, $\sum_{j=1}^n \xi_{jn}\mathfrak{L}e_j(t_i) = \lambda_n\sum_{j=1}^n \xi_{jn}e_j(t_i)$, $i = 1, \ldots, n$. Set $a_{ij} = \mathfrak{L}e_j(t_i) = e_j^{(p)}(t_i) - \sum_{k=0}^{p-1} a_k(t_i)e_j^{(k)}(t_i)$, $b_{ij} = e_j(t_i)$; $A = (a_{ij})$, $B = (b_{ij})$, $\xi^{(n)} = (\xi_{jn})$. Hence $A\xi^{(n)} = \lambda_n B\xi^{(n)}$.

4.62 For the method of moments (resp. collocation) the matrix is given by $\int_a^b \mathfrak{L}e_j t^{i-1}\,dt$ (resp. $\mathfrak{L}e_j(t_i)$) and the right-hand side is given by $\int_a^b ft^{i-1}\,dt$ (resp. $f(t_i)$).

4.63 See the solution of Exercise 4.16.

4.64 Since the functions in C_n are polynomials of degree less than or equal to $p + r$ on each Δ_i, then $D^pC_n \subseteq \mathbb{P}_{r,\Delta}$. For the converse, remark that (4.34) imposes p linearly independent conditions on the coefficients. The projection π_n defined in Exercise 4.4 is the L^2-orthogonal projection on $F_n = \mathfrak{M}C_n = \mathbb{P}_{r,\Delta}$. $\mathfrak{L}u_n - f \perp F_n \Leftrightarrow \pi_n(\mathfrak{L}u_n - f) = 0$, $u_n \in C_n$.

4.65 Let L_n be the Lagrange interpolation (Exercise 4.6) at the points $\{\tau_i^j\}$. Then

$$(\mathfrak{L}u_n - f)(\tau_i^j) = 0 \Leftrightarrow L_n(\mathfrak{L}u_n - f) = 0.$$

L_n is a map from $C(a, b)$ into $\mathbb{P}_{r,\Delta}$. It is a projection if defined on $C_\Delta(a, b)$. Again the choice of $\mathbb{P}_{r,\Delta}$ is arbitrary.

4.66 The least-squares method on $Tu = f$ is $\pi_n(Tu_n - f) = 0$, $u_n \in X_n$, π_n $\perp$-projection on $\mathfrak{L}X_n$. It is equivalent to

$$u_n = Gx_n, \qquad \pi_n[(1 - K)x_n - f] = 0, \quad x_n \in D^pX_n = Y_n.$$

This is an $\perp$-Petrov method on $1 - K$ with right subspace Y_n and left subspace $\mathfrak{L}X_n$.

4.67 $\pi_n[T\psi_n - \lambda_n\psi_n] = 0$, $\psi_n \in X_n$, π_n projection on Y_n, is equivalent to

$$\psi_n = G\varphi_n, \qquad \pi_n[(1 - K)\varphi_n - \lambda_n G\varphi_n] = 0, \quad \varphi_n \in Y_n.$$

4.68 To the generalized eigenvalue problem $(1 - K)\varphi = \lambda G\varphi$ is associated the resolvent $(1 - K - zG)^{-1}G$ (cf. Chapter 2, Example 2.28). Similarly, to $(1 - \pi_n K)\varphi_n = \lambda_n \pi_n G\varphi_n$ is associated $(1 - \pi_n K - z\pi_n G)^{-1}\pi_n G$. Set

$$A_n(z) := (1 - \pi_n K - z\pi_n G)^{-1},\ A(z) := (1 - K - zG)^{-1},$$

then

$$A_n(z) - A(z) = A_n(z)(\pi_n - 1)(K + zG)A(z).$$

And

$$\|A_n(z)\pi_n G - A(z)G\| \leq \|(A_n(z) - A(z))\pi_n G\| + \|A(z)\|\ \|(1 - \pi_n)G\|,$$

which tends to 0 since $\|(1 - \pi_n)(K + zG)\| \to 0$ as $n \to \infty$.

4.69 Since $T - \lambda_n = (1 - K - \lambda_n G)G^{-1}$ and $1 - P$ commute, $(T - \lambda_n)(1 - P)\psi_n = (1 - P)(1 - K - \lambda_n G)\varphi_n$. So $(1 - P)\psi_n = (T - \lambda_n)^{-1}(1 - P)(1 - K - \lambda_n G)\varphi_n$. Since $S(\lambda_n) = (T - \lambda_n)^{-1}(1 - P)$, we get $(1 - P)\psi_n = S(\lambda_n)(1 - K - \lambda_n G)\varphi_n$ and setting $\tilde{\varphi}_n = (K + \lambda_n G)\varphi_n$: $(1 - P)\psi_n = S(\lambda_n)(\varphi_n - \tilde{\varphi}_n)$. But $(\pi_n K + \lambda_n \pi_n G)\varphi_n = \varphi_n$ (see Exercise 4.67), so $\varphi_n - \tilde{\varphi}_n = (\pi_n - 1)\tilde{\varphi}_n$. Therefore $(1 - P)\psi_n = S(\lambda_n)(\pi_n - 1)\tilde{\varphi}_n$.

4.70 $\pi_n(T\psi_n - \lambda_n\psi_n) = 0$, $\psi_n \in X_n$, π_n projection on Y_n, is equivalent to

$$\psi_n = A\theta_n \Leftrightarrow \theta_n = T\psi_n, \qquad \pi_n(\theta_n - \lambda_n A\theta_n) = 0, \quad \theta_n \in X'_n = TX_n.$$

4.71 If $\tilde{\theta}_n = \lambda_n A\theta_n$, then since $\psi_n = A\theta_n$,

$$\frac{1}{\lambda_n}(1 - P)\tilde{\theta}_n = (1 - P)\frac{\lambda_n}{\lambda_n}A\theta_n = (1 - P)\psi_n.$$

4.72 It follows from the fact that a is coercive. Otherwise, if u_n and $\bar{u}_n$ are two different solutions of (4.50) in V_n, then, for $u_n - \bar{u}_n \neq 0$, one should have $a(u_n - \bar{u}_n, u_n - \bar{u}_n) = 0 \geq \alpha\|u_n - \bar{u}_n\|_V^2$.

4.73 Let $e_1, \ldots, e_n$ be a basis of V_n. Then $u_n = \sum_{j=1}^n \xi_{jn}e_j$ and (4.50) is equivalent to

$$\sum_{j=1}^{n} \xi_{jn}a(e_j, e_i) = (f, e_i)_H, \qquad i = 1, \ldots, n.$$

Setting $a_{ij} = a(e_j, e_i)$, $b_i = (f, e_i)_H$, $\xi^{(n)} = (\xi_{jn})$, $A = (a_{ij})$, $b = (b_i)$, we get the system $A\xi^{(n)} = b$ to be solved.

4.74 (4.41) defines A such that $a(A\psi, v) = (\psi, v)_H = (1/\lambda)a(\psi, v)$, $\forall v \in V$; that is, $a(A\psi - (1/\lambda)\psi, v) = 0$, $\forall v \in V$. The approximation in V_n yields $a(\pi_n^a(A\psi_n - (1/\lambda_n)\psi_n), v_n) = 0$, $\forall v_n \in V_n$. ψ_n is the eigenvector of $\pi_n^a A$ associated with the eigenvalue $1/\lambda_n$. $\pi_n^a\psi = (1/\lambda)\pi_n^a A\psi$ and $\psi_n = (1/\lambda_n)\pi_n^a A\psi_n$, so $\pi_n^a\psi \neq \psi_n$ in general.

4.75 $\sigma(A) = \sigma(B)$ follows from

(i) $0 \in \rho(A) \Rightarrow 0 \in \rho(B)$,
(ii) $0 \in \rho(B) \Rightarrow 0 \in \rho(A)$,
(iii) $0 \neq z \in \rho(A) \Rightarrow z \in \rho(B)$,
(iv) $0 \neq z \in \rho(B) \Rightarrow z \in \rho(A)$.

To prove (i), $0 \in \rho(A) \Rightarrow \|A^{-1}\|_H < \infty$; let $u \in V$, $\|u\|_V^2 \leq c|a(u, u)| \leq c|(A^{-1}u, u)_H| \leq c\|A^{-1}\|_H\ \|u\|_H^2$; then the norms of V and H are equivalent, $H = V$, $A = B$, and $0 \in \rho(B)$.

For (ii), $0 \in \rho(B) \Rightarrow \|u\|_V \leq c\|Au\|_V$, $\forall u \in V$; then $\|u\|_V^2 \leq c\|Au\|_V^2 \leq c|a(Au, Au)| \leq c|(u, Au)_H| \leq c\|u\|_H^2\ \|A\|_H$, the conclusion is as before.

For (iii), $A - z: H \to H$ is a bijection, hence $B - z: V \to V$ is injective. It suffices to prove it is surjective. Let $f \in V$; there exists $u \in H$ such that $(A - z)u = f$ and $u = (1/z)(Au - f) \in V$; i.e., $(B - z)u = f$.

For (iv), $B - z: V \to V$ is a bijection. Since V is dense in H and $B - z$ is continuous for $\|\cdot\|_H$, it suffices to show that $(B - z)^{-1}$ is continuous for $\|\cdot\|_H$. Suppose this is false, there exists $u_n \in V$, $\|u_n\|_H = 1, \lim_n \|(B - z)u_n\|_H = 0$. Let $v_n := (1/z)Bu_n, \lim_n \|v_n - u_n\|_H = (1/|z|) \lim_n \|Bu_n - zu_n\|_H = 0$. There exists N such that $\|v_n\|_H \geq \frac{1}{2}$ and $\|v_n\|_V \geq c$, for $n \geq N$. Then

$$\begin{aligned} \|(B - z)v_n\|_V^2 &\leq c\,|a(Bv_n - zv_n, Bv_n - zv_n)| \\ &\leq c\,|(v_n, Bv_n - zv_n)_H - (u_n, Bv_n - zv_n)_H| \\ &\leq c\|v_n - u_n\|_H \, \|(B - z)v_n\|_V. \end{aligned}$$

We deduce $\lim_n \|(B - z)v_n\|_V = 0$, with $\|v_n\|_V \geq c$; this implies that $(B - z)^{-1}$ is not continuous for $\|\cdot\|_V$, which contradicts $z \in \rho(B)$. The coercivity of a implies that 0 is not an eigenvalue of A, and consequently not an eigenvalue of B. Let $0 \neq \lambda \in Q\sigma(A)$ be of algebraic multiplicity $m < \infty$, then $\mathrm{Ker}(A - \lambda)^m = \{x \in H; (A - \lambda)^m x = 0\} \subset V$ since $\mathrm{Im}\, A \subset V$, and λ is an eigenvalue of B of multiplicity m.

4.76 $\forall u \in V_n, \alpha\|B_n u\|_V^2 \leq \mathscr{Re}\, a(B_n u, B_n u) \leq |(u, B_n u)_H| \leq \|u\|_H \, \|B_n u\|_V$; hence $\alpha\|B_n u\|_V \leq \|u\|_H$. If $z = 0$, then B_n is invertible, and $\|B_n^{-1}\|_V \leq c$ or $\|B_n^{-1}\|_H \leq c$ implies $\|u\|_V \leq (c/\alpha)\|u\|_H$, $\forall u \in V_n$. The norms $\|\cdot\|_V$ and $\|\cdot\|_H$ are uniformly equivalent on V_n. If $z \neq 0$ and $\|R_n(z)\|_V \leq c$, then $\|(B_n - z)u\|_V \geq (1/c)\|u\|_V$ for $u \in V_n$. For $x \in V_n$, $\|B_n x\|_V \leq c\|(B_n - z)B_n x\|_V \leq (c/\alpha)\|(B_n - z)x\|_H$,

$$\|x\|_H = \frac{1}{|z|}\|B_n x - B_n x - zx\|_H \leq \frac{1}{|z|}[\|(B_n - z)x\|_H + \|B_n x\|_H] \leq \frac{1}{|z|}\left(1 + \frac{c}{\alpha}\right)\|(B_n - z)x\|_H.$$

Therefore $\|R_n(z)\|_H \leq (1/|z|)(1 + c/\alpha)$. Conversely, if $\|R_n(z)\|_H \leq c$ for $z \neq 0$, we get similarly $\|R_n(z)\|_V \leq (1/|z|)(1 + c/\alpha)$.

4.77 $A: H \to H$ with range in V is compact, $\|x - \pi_n^a x\|_H \to 0$ for $x \in V$. Let B be the unit ball of H; AB is a relatively compact set of H in V; hence $\|(1 - \pi_n^a)A\|_H \to 0$.

4.78 $a(u, v) = (Tu, v)_H = \overline{a(v, u)} = \overline{(Tv, u)_H} = (u, Tv)_H \Rightarrow T = T'$ in H. $(u, v)_a = a(u, v) = (Tu, v)_H$, so $\|u\|_a^2 = (Tu, u) > 0$ for $u \neq 0$, $u \in D$. A is treated similarly.

4.79 $a(Su, v) = (u, v)_V$ for $u, v \in V$, so $(Su, v)_a = (u, v)_V$ and $(u, v)_a = (S^{-1}u, v)_V$. $a(v, Su) = \overline{a(Su, v)} = \overline{(Su, v)_a} = \overline{(u, v)_V} = (v, u)_V = (Sv, u)_a = a(Sv, u) = \overline{(u, Sv)_a}$. So $S^\times = S$. For S^* defined by $(Su, Sv)_a = (u, Sv)_V = (S^*u, v)_V$, we prove similarly $S^* = S$.

4.80 $(\pi_n^a B \pi_n^a)^\times = \pi_n^{a\times} B^\times \pi_n^{a\times} = \pi_n^a B \pi_n^a$. The matrix A of (4.50) is given in Exercise 4.73. It is hermitian since $a_{ij} = a(e_j, e_i) = \overline{a(e_i, e_j)} = \overline{a}_{ji}$. For $\xi = (x_i)$,

$$\begin{aligned} \xi^{\mathrm{H}} A \xi &= \sum_i \sum_j a_{ij} x_j \bar{x}_i = a\left(\sum_j x_j e_j, \sum_i x_i e_i\right) \\ &= a(x, x) \qquad \text{if} \quad x = \sum_{i=1}^n x_i e_i \in V_n. \end{aligned}$$

$a(x, x) \geq \alpha\|x\|_V^2 > 0$ for $0 \neq x \in V_n$; that is, $\xi^{\mathrm{H}} A \xi > 0$ for $0 \neq \xi \in \mathbb{C}^n$.

4.81 Since H_{1n} and H_{2n} are closed subspaces of H_1 and H_2, respectively, they are Hilbert spaces, and the hypothesis on the restriction of $a(\cdot, \cdot)$ to $H_{1n} \times H_{2n}$ ensures the existence (by Riesz's representation theorem) of an isomorphism $\tilde{A}_n: H_{1n} \to H_{2n}$ uniquely defined by

$$(\tilde{A}_n u_n, v_n)_2 = a(u_n, v_n), \qquad \forall u_n \in H_{1n}, v_n \in H_{2n}.$$

Since

$$\sup_{\substack{v_n \in H_{2n} \\ \|v_n\|_2 = 1}} |a(u_n, v_n)| \geq \alpha\|u_n\|_1, \qquad \forall u_n \in H_{1n},$$

we have $\|\tilde{A}_n^{-1}\| \leq 1/\alpha$. Let π_{2n} be the orthogonal projection from H_2 onto H_{2n}. For $u_n \in H_{1n}$ and $v_n \in H_{2n}$ we have

$$(\pi_{2n} A u_n, v_n)_2 = (Au_n, v_n)_2 = a(u_n, v_n).$$

Then $\tilde{A}_n = \pi_{2n} A_{\restriction H_{1n}}$ and $\|\tilde{A}_n\| \leq \|A\|$.

It is easily seen that $\pi^a_{1n} = \tilde{A}_n^{-1}\pi_{2n}A$ and that for all $u \in H_1$, $\hat{u}_n := \pi^a_{1n}u$ is the unique element of H_{1n} which satisfies $a(\hat{u}_n, v_n) = a(u, v_n)$, $\forall v_n \in H_{2n}$. For all $u \in H_1$, $u_n \in H_{1n}$, $(1 - \pi^a_{1n})u = u - u_n - \tilde{A}_n^{-1}\pi_{2n}A(u - u_n)$, so $\|(1 - \pi^a_{1n})u\|_1 \le (1 + \|A\|/\alpha)\|u - u_n\|_1$; hence $\|(1 - \pi^a_{1n})u\|_1 \le c\delta_{1n}(u)$. The reader can show the existence of π^a_{2n}. Since for all $u \in H_1$, $v \in H_2$, $a(\pi^a_{1n}u, v) = a(\pi^a_{1n}u, \pi^a_{2n}v)$ by definition of π^a_{2n} and $a(u, \pi^a_{2n}v) = a(\pi^a_{1n}u, \pi^a_{2n}v)$ by definition of π^a_{1n}, we conclude $a(u, \pi^a_{2n}v) = a(\pi^a_{1n}u, v)$, which shows $(\pi^a_{1n})^\times = \pi^a_{2n}$. We remark that in general π^a_{2n} is not pointwise converging to 1 in H_2.

4.82

$$u_{i-1} = \frac{1}{n}\sum_{j=1}^{n-1} g(t_{i-1}, t_j)y_j,$$

$$-2u_i = -\frac{2}{n}\sum_{j=1}^{n-1} g(t_i, t_j)y_j,$$

$$u_{i+1} = \frac{1}{n}\sum_{j=1}^{n-1} g(t_{i+1}, t_j)y_j,$$

$$\begin{aligned}
\frac{u_{i-1} - 2u_i + u_{i+1}}{h^2} &= \frac{1}{nh^2}\sum_{j=1}^{n-1} y_j(g(t_{i-1}, t_j) - 2g(t_i, t_j) + g(t_{i+1}, t_j)) \\
&= \frac{1}{h}\Bigg[\sum_{j=1}^{i-1} y_j t_j(t_{i-1} - 1) + \sum_{j=1}^{n-1} y_j t_{i-1}(t_j - 1) \\
&\quad - 2\sum_{j=1}^{i} y_j t_j(t_i - 1) - 2\sum_{j=i+1}^{n-1} y_j t_i(t_j - 1) \\
&\quad + \sum_{j=1}^{i+1} y_j t_j(t_{i+1} - 1) + \sum_{j=i+2}^{n-1} y_j t_{i+1}(t_j - 1)\Bigg] \\
&= \frac{1}{nh}\Bigg[\left(\frac{i-1}{n} - 1\right)\sum_{j=1}^{i-1} jy_j + \frac{i-1}{n}\sum_{j=1}^{n-1}(j - n)y_j \\
&\quad - 2\left(\frac{i}{n} - 1\right)\sum_{j=1}^{i} jy_j - 2\frac{i}{n}\sum_{j=i+1}^{n-1}(j - n)y_j \\
&\quad + \left(\frac{i+1}{n} - 1\right)\sum_{j=1}^{i+1} jy_j + \frac{i+1}{n}\sum_{j=i+2}^{n-1}(j - n)y_j\Bigg] \\
&= y_i.
\end{aligned}$$

The matrix

$$\begin{bmatrix}
-2 & 1 & 0 & 0 & \cdots & & 0 \\
1 & -2 & 1 & 0 & & & \\
0 & 1 & -2 & 1 & \ddots & & \vdots \\
 & & 1 & -2 & 1 & & \\
\vdots & & & & \ddots & \ddots & 0 \\
 & & & \ddots & \ddots & \ddots & 1 \\
0 & \cdots & & & 0 & 1 & -2
\end{bmatrix}$$

is regular since it has nonzero eigenvalues (what are they?).

Chapter 5

5.1 $T_n - z \xrightarrow{ss} T - z$ on Γ implies $P_n \xrightarrow{cc} P$. For $x \in B_M$, $x \neq 0$, we have $P_n x \to Px = x$; hence $P_n x \neq 0$ for n large enough. Set $x_n = (\|x\|/\|P_n x\|)P_n x$, $x_n \in B_{M_n}$, and $x_n \to x$: (5.4) is fulfilled. For (5.5), one may proceed as in Exercise 5.2.

5.2 Since $P_n \xrightarrow{cc} P$, we get $\|(P - P_n)P_n\| \to 0$. For any x_n in B_{M_n},

$$\|(P - P_n)P_n x_n\| = \|Px_n - x_n\| \leq \sup_{\substack{y \in X \\ \|y\| \leq 1}} \|(P - P_n)P_n y\| \to 0.$$

Set $y_n = Px_n$. Given $\varepsilon > 0$, $\|y_n\| \leq 1 + \varepsilon$ for n large enough and $y_n \in B_\varepsilon := \{x \in M; \|x\| \leq 1 + \varepsilon\}$, which is compact. Hence $\exists N_1 \subset \mathbb{N}$ such that $y_n \to y \in B_\varepsilon$, $n \in N_1$; hence $x_n \to y$, $n \in N_1$. But this implies $\|y\| < \|x_n\| + \varepsilon$ for n large enough, which in turn implies $y \in B_M$. This proves (5.5).

5.3 From Theorem 5.8, $\Theta(M_n, M) \to 0$, where

$$\Theta(M_n, M) = \max\left(\sup_{x \in B_M} \operatorname{dist}(x, M_n), \sup_{x \in B_{M_n}} \operatorname{dist}(x, M)\right).$$

$\operatorname{dist}(\psi, M_n) = \|\psi\| \operatorname{dist}((1/\|\psi\|)\psi, M_n) \to 0$, $\operatorname{dist}(\psi_n, M) \to 0$ since $\psi_n \in B_{M_n}$.

5.4 We have $\|R_n(z)\| \leq M$ for $n > N(\Gamma)$ and $z \in \Gamma$. So $\|R_n^*(z)\| = \|(T_n^* - \bar{z})^{-1}\| \leq M$ for $n > N(\Gamma)$ and $z \in \bar{\Gamma}$. The strong stability of $\{T_n^*\}$ on $\bar{\Gamma}$ follows. $P_n^* \xrightarrow{p} P^*$, P, P_n, P_n^*, and P^* are compact, so Proposition 3.7 applies to get $\|P - P_n\| = \|P_n^* - P^*\| \to 0$.

5.5 Remark that for n large enough $\sigma(T_n) \cap \Delta$ consists of only one eigenvalue and that $\dim M_n = 1$. Then use Exercise 5.3.

5.6 $\mathscr{R}_n(z)$ is uniformly bounded in Δ-$\{0\}$ and $\dim \mathscr{P}_n X_n = m$, the multiplicity of 0. Hence the eigenvalues of $\mathscr{T}_n$ inside Γ converge to 0, with total multiplicity m. From the identity

$$(\mathscr{T}_n - z)^{-1}\pi_n - \pi_n(T - z)^{-1} = \mathscr{R}_n(z)(\pi_n T - \mathscr{T}_n \pi_n)R(z)$$

follows that $\mathscr{P}_n \pi_n - \pi_n P \xrightarrow{p} 0$. Then $\mathscr{P}_n \pi_n \xrightarrow{p} P$ and $\dim \mathscr{P}_n \pi_n X = \dim PX$ implies $\mathscr{P}_n \pi_n \xrightarrow{cc} P$ and $\Theta(\mathscr{M}_n, M) \to 0$.

5.7 $\|(T_n - z)^{-1}\| = [\operatorname{dist}(z, \sigma(T_n))]^{-1} \leq |\mathscr{I}m\, z|^{-1} < \infty$.

5.8 Because $g_n = g$, the multiple eigenvalue λ cannot split into several approximating values.

5.9 Similar to the proof of Proposition 5.17 with the condition $\mu_{in} \leq \mu_i$, $i = 1, \ldots, k$.

5.10 For a compact T, $\rho(T) \cup Q\sigma(T) = \mathbb{C} - \{0\}$ by Theorem 2.34. Then apply Example 5.10 and Exercise 5.9.

5.11 For z in Δ, $S_n(z)(T_n - T)S(z) = (1 - P_n)S(z) - S_n(z)(1 - P) = S(z) - S_n(z) - (P_n - P)S(z) - S_n(z)(P_n - P) \to 0$. By using $\|S_n(z) \leq \max_{z \in \Gamma} \|R_n(z)(1 - P_n)\| < \infty$ and $P_n \xrightarrow{p} P$, we conclude that $S_n(z) \xrightarrow{p} S(z)$ for z in Δ, in particular at λ.

5.12 From Example 3.8 we get $\|T - T_n\| \to 0 \Rightarrow T_n - z \xrightarrow{s} T - z$, $z \in \rho(T)$. Then

$$r_\sigma[(T - T_n)R(z))] \leq \|R(z)\| \, \|T - T_n\| \to 0.$$

For every compact $K \subseteq \rho(T)$ $\sup_{z \in K} r_\sigma[(T - T_n)R(z)] \leq \|T - T_n\| \sup_{z \in K}\|R(z)\| \to 0$. So $T_n - z \xrightarrow{u\sigma} T - z$, $\forall z \in \rho(T)$. If $T - T_n \xrightarrow{cc} 0$, then $T_n - z \xrightarrow{s} T - z$, and $r_\sigma[(T - T_n)R(z)] \leq \|[(T - T_n)R(z)]^2\|^{1/2} \to 0$. The result follows.

5.13 For $z_0 \in \rho(T)$, $R_n(z_0) \xrightarrow{p} R(z_0)$. By Exercise 3.11 and Proposition 2.28 we see that $R_n(z_0) - t \xrightarrow{ss} R(z_0) - t$ for $t \in \mathbb{C} - \{0\}$. We then apply Theorem 5.24 and Proposition 2.36.

5.14 Apply Theorem 5.26 to $R(z_0)$ to get $R_n(z_0) - t \xrightarrow{u\sigma} R(z_0) - t$ in $\rho(R(z_0))$.

5.15 See S. Goldberg (1974).

5.16 If $\tilde{\pi}_n = \pi_{n \restriction X_n}$ has a uniformly bounded inverse, then $\pi_n' = \tilde{\pi}_n^{-1}\pi_n$ is a projection on X_n which is pointwise convergent to 1. If T is compact, then $\|(1 - \pi_n')T\| \to 0$ and the result follows.

5.17 Apply first Lemma 5.11, then Proposition 5.27, and again Lemma 5.11.

5.18 By Exercise 3.13, there is only to prove that $T_n \xrightarrow{\text{d-c}} T \Rightarrow T_n - \lambda\pi_n \xrightarrow{\text{d-r}} T - \lambda$ for $\lambda \in Q\sigma(T)$. Adapt the proof of Proposition 5.28 with $x_n \in X_n$, $\|x_n\| \leq 1$.

5.19 For $z \neq 0$, $R_n(z)_{\restriction X_n} = \mathscr{R}_n(z)$; therefore $(R_n(z) - R(z))_{\restriction X_n} = R(z)(T - T_n)_{\restriction X_n}\mathscr{R}_n(z)$ and $\|(P_n - P)_{\restriction X_n}\| \to 0$. Then $\dim PX_n = \dim P_n X_n$ for n large enough; $P(\pi_n - 1) \xrightarrow{\text{cc}} 0$ implies $\dim PX_n = \dim PX$. From $P_n \xrightarrow{\text{p}} P$ we deduce that $P_n \xrightarrow{\text{cc}} P$ and $\Theta(M_n, M) \to 0$. It could also be deduced directly from $T_n \xrightarrow{\text{d-c}} T$.

5.20 Adapt solution of Exercise 3.5.

5.21 In view of Section 7 of Chapter 3, $R(t, z)$ is an analytic function of t in $\delta_z := \{t \in \mathbb{C};\ |t| < 1/r_z\}$, where $r_z = r_\sigma[(T - T_n)R(z)]$. $P(t)$ and $\hat{\lambda}(t)$ are analytic functions of t in $\delta_\Gamma = \{t \in \mathbb{C};\ |t| < 1/r_\Gamma\}$, where $r_\Gamma = \max_{z \in \Gamma} r_\sigma[(T - T_n)R(z)]$.

5.22 Since Γ is compact and $T_n - z \xrightarrow{\text{u}\sigma} T - z$, then for n large enough $r_\Gamma < 1$, $1 \in \delta_\Gamma \subset \delta_z$. The series expansion for $R_n(z)$ is $R_n(z) = R(z)\sum_{k=0}^{\infty}[(T - T_n)R(z)]^k$. For P_n and $\hat{\lambda}_n$, the expansions are computed from (3.15) and (3.16) in Proposition 3.26, Section 7.2 of Chapter 3, by letting $t = 1$, $T' = T$, $H = T - T_n$.

5.23 $[T_n - z - (T_n - T)]x = [1 - (T_n - T)R_n(z)]u = f$ with $u = (T_n - z)x$. Set $u_0 := f$, $u_k := (T_n - T)R_n(z)u_{k-1} + f$, then $x^k = R_n(z)u_k$. By Exercise 2.38,

$$\|u_k - u\| \leq c(\varepsilon)[r_\sigma[(T - T_n)R_n(z)] + \varepsilon]^{k+1}$$

and $\|x^k - x\| \leq \|R_n(z)\|\ \|u_k - u\|$.

5.24
$$(T_M - z)x_M = f \Leftrightarrow [T_n - z - (T_n - T_M)]x_M = f.$$

$$(T_n - T_M)R_n(z) = (T_n - T)R_n(z) + (T - T_M)R_n(z)$$

$$\|T_n - T\| \to 0 \Rightarrow \|(T_n - T_M)R_n(z)\| < 1 \qquad \text{for } n \text{ large enough,}$$

$$T_n \xrightarrow{\text{cc}} T \Rightarrow \|[(T_n - T_M)R_n(z)]^2\| < 1 \qquad \text{for } n \text{ large enough.}$$

5.25 $x_n = R_n^G(z)f$, $x^1 = R_n^G(z)(T_n^G - T)R_n^G(z)f + R_n^G(z)f$. Keeping notations of Theorem 4.3,

$$R_n^G(z)T_n^G = \mathscr{R}_n(z)\pi_n T\pi_n, \qquad R_n^G(z)T = \mathscr{R}_n(z)\pi_n T - \frac{1}{z}(1 - \pi_n)T, \qquad R_n^G(z)f = x_n^G - \frac{1}{z}(1 - \pi_n)f.$$

$$\begin{aligned}
x^1 &= \frac{1}{z}(1 - \pi_n)Tx_n^G + \frac{1}{z}\mathscr{R}_n(z)\pi_n T(1 - \pi_n)f - \frac{1}{z^2}(1 - \pi_n)T(1 - \pi_n)f + R_n^G(z)f \\
&= \tilde{x}_n^G + \frac{1}{z}\left[\mathscr{R}_n(z)\pi_n - \frac{1}{z}(1 - \pi_n)\right]T(1 - \pi_n)f \\
&= \tilde{x}_n^G + \frac{1}{z}R_n^G(z)T(1 - \pi_n)f.
\end{aligned}$$

5.26
$$\begin{aligned}
x_n^G &= R_n^G(z)\pi_n f = R_n^P(z)\pi_n f. \\
x^{1G} &= R_n^G(z)(T_n^G - T)x_n^G + x_n^G \\
&= \left[\mathscr{R}_n(z)\pi_n T(\pi_n - 1) + \frac{1}{z}(1 - \pi_n)T\right]x_n^G + x_n^G \\
&= \tilde{x}_n^G + \frac{1}{z}f - \frac{1}{z}\pi_n Tx_n^G + x_n^G \\
&= \tilde{x}_n^G - \frac{1}{z}(\pi_n T - z)R_n^P(z)\pi_n f + \frac{1}{z}f \\
&= \tilde{x}_n^G + \frac{1}{z}(1 - \pi_n)f.
\end{aligned}$$

5.27 $x - x^k = x - x_n - R_n(z)(T_n - T)x^{k-1}$. But

$$x - x_n = (R(z) - R_n(z))f = R_n(z)(T_n - T)R(z)f = R_n(z)(T_n - T)x.$$

So

$$x - x^k = R_n(z)(T_n - T)(x - x^{k-1}) = [R_n(z)(T_n - T)]^k(x - x^0) = [R_n(z)(T_n - T)]^{k+1}x.$$

Hence

$$\|x - x^k\| \le \|x\|[\|(\pi_n T - z)^{-1}\| \; \|(1 - \pi_n)T\|]^{k+1}.$$

Since $\pi_n \xrightarrow{p} 1$ and T is compact, then $\|(1 - \pi_n)T\| \to 0$ (Theorem 3.2).

5.28 See Exercise 4.44. $C_n(z) \to 0$ since $T_n \xrightarrow{cc} T$. By Exercise 5.27:

$$x - x^{k+2} = [R_n(z)(T_n - T)]^2(x - x^k).$$

5.29
$$\begin{aligned} x'^{k+1} - x &= R_n(z)\left(f + (T_n - T)\left[\frac{1}{z}(Tx'^k - f)\right]\right) - R(z)f \\ &= [R_n(z) - R(z)]f + \frac{1}{z}R_n(z)(T_n - T)(Tx'^k - f) \\ &= \frac{1}{z}R_n(z)(T_n - T)Tx'^k - \frac{1}{z}R_n(z)(T_n - T)(T - z)x - R_n(z)(T_n - T)x \\ &= \frac{1}{z}R_n(z)(T_n - T)T(x'^k - x). \end{aligned}$$

$\|x'^{k+1} - x\| \le (1/|z|)\|R_n(z)(T_n - T)T\| \; \|x'^k - x\|$. $R_n(z)$ is uniformly bounded and $T_n \xrightarrow{p} T$; then, T being compact, $\|(T_n - T)T\| \to 0$ and $(1/|z|)\|R_n(z)(T_n - T)T\| \le c\|(T_n - T)T\| \to 0$.

5.30
$$\begin{aligned} x'^{k+1} &= R_n(z)f + \frac{1}{z}(1 - R_n(z)(T - z))(Tx'^k - f) \\ &= \frac{1}{z}(1 - R_n(z)T(Tx'^k - f) + R_n(z)Tx'^k + x'^k - x'^k \\ &= x'^k + \frac{1}{z}(-1 + R_n(z)T)(f - (T - z)x'^k). \end{aligned}$$

$$\begin{aligned} \tilde{R}_n(z) &:= \frac{1}{z}(-1 + R_n(z)T) = \frac{1}{z}R_n(z)T - \frac{1}{z}R(z)(T - z) \\ &= \frac{1}{z}R_n(z)T - \frac{1}{z}R(z)T + R(z). \end{aligned}$$

$$\|[R(z) - \tilde{R}_n(z)]x\| \le \frac{1}{|z|}\|[R(z) - R_n(z)]Tx\| \to 0 \qquad \text{if} \quad T_n - z \xrightarrow{s} T - z.$$

5.31 Set $d^k = f - (T - z)x'^k$. Then (5.13) becomes $x'^{k+1} = x'^k - (1/z)d^k + (1/z)R_n(z)Td^k$. Now consider the equation $(T - z)(x - x'^k) = d^k$. Its regularized form is

$$(T - z)u = Td^k, \qquad x - x'^k = \frac{1}{z}(u - d^k).$$

Its approximate solution is $u^k = R_n(z)Td^k$. Therefore $x''^{k+1} = x'^k + (1/z)(u^k - d^k) = x'^{k+1}$.

5.32 $\lambda^1 = \langle T\varphi_n, \varphi_n^*\rangle = \langle T\pi_n\varphi_n, \pi_n^*\varphi_n^*\rangle$, with $\varphi_n = \varphi_n^G$,

$= \langle \pi_n T\pi_n\varphi_n, \varphi_n^*\rangle = \lambda_n\langle\varphi_n, \varphi_n^*\rangle = \lambda_n = \lambda_n^G.$

$$\varphi^1 = \varphi_n + S_n(T_n - T)\varphi_n = \varphi_n + S_nT_n\varphi_n - S_nT\varphi_n.$$

But $S_nT_n\varphi_n = 0$, so $\varphi^1 = \varphi_n - (\mathscr{S}_n\pi_n - (1/\lambda_n)(1 - \pi_n))T\varphi_n$ in notations of Corollary 4.4. Since $(1/\lambda_n)\pi_nT\pi_n\varphi_n = \varphi_n$ and $\mathscr{S}_n\pi_nT\varphi_n = \mathscr{S}_n\pi_nT\pi_n\varphi_n = \lambda_n\mathscr{S}_n\varphi_n = 0$, then $\varphi^1 = (1/\lambda_n)T\varphi_n = \tilde{\varphi}_n^G = \varphi_n^S$.

5.33 See the solution of Exercise 1.57.

5.34 $T_M\phi_M = \lambda_M\phi_M$, $P_n\phi_M = \varphi_n$. $T_M = T_n - (T_n - T_M)$, see Exercise 5.24.

5.35 $Q(P_n(t) - P_n)Q\xi = \langle\xi, y\rangle\langle(P_n(t) - P_n)\varphi_n, y\rangle\varphi_n$.

$$r_\sigma[Q(P_n(t) - P_n)Q] < 1 \Leftrightarrow 0 < \langle P_n(t)\varphi_n, y\rangle < 2.$$

$$Q(P_n(t) - P_n)Q\xi = -\frac{\langle\xi, y\rangle}{2i\pi}\int_\Gamma\left\langle R_n(z)\sum_{k=1}^{\infty} t^kK_n^k(z)\varphi_n, y\right\rangle dz,$$

$$r_\sigma\left[(P_n(t) - P_n)Q \le \frac{\text{meas }\Gamma}{2\pi}\|y\|\max_{z\le\Gamma}\left(\|R_n(z)\|\sum_{k=1}^{\infty}|t|^t\|K_n^k(z)\varphi_n\|\right).\right.$$

Use the proof of Lemma 5.33 to conclude that $\langle P_n(t)\varphi_n, y\rangle$ is nonzero for $|t| \le 1$. Then $P_n(t)\varphi_n$ is an eigenvector for $T_n(t)$ such that $QP_n(t)\varphi_n \ne 0$. $\hat{\phi}_n(t)$ normalized by $Q\hat{\phi}_n(t) = \varphi_n$ is well defined and $\hat{\phi} = \hat{\phi}_n(1)$.

5.36 $F(\varphi, \lambda) := ((1 - \lambda K)\varphi, \langle\varphi, \psi\rangle - 1) = 0.$

$$F'_u = \begin{pmatrix} 1 - zK & -Kx \\ \langle\cdot, \psi\rangle & 0 \end{pmatrix}.$$

The Newton's iteration is u_0, $u_{i+1} - u_i = -F'_{u_i}F(u_i)$, $i \ge 0$; that is,

$$(1 - z_iK)x_{i+1} = (z_{i+1} - z_i)Kx_i, \qquad z_{i+1} - z_i = (\langle(1 - z_iK)^{-1}x_i, \psi\rangle)^{-1}.$$

Set $\hat{x}_i = z_{i+1}Kx_i$ and $x_{i+1} = \hat{x}_i + y_i$. y_i is solution of $(1 - z_iK)y_i = z_iK(\hat{x}_i - x_i)$. z is kept fixed equal to z_0 in F'_{u_i}. It yields $(1 - z_0K)x'_{i+1} = (z'_{i+1} - z_0)Kx'_i$, or $x'_{i+1} = z'_{i+1}Kx'_i + y'_i$ with $(1 - z_0K)y'_i = z_0K(\hat{x}'_i - x'_i)$. K is approximated by $K_n = \pi_nK$: $K_n\varphi_n = \lambda_n\varphi_n$ and $K_n^*\varphi_n^* = \bar{\lambda}_n\varphi_n^*$. We choose $z_0 = \lambda_n$, $x_0 = \varphi_n$, $\psi = \varphi_n^*$. Then we approximate z'_{i+1} and y'_i by $\tilde{z}_{i+1}$ and $\tilde{y}_i$ solutions of $\tilde{z}_{i+1}\langle Kx'_i, \varphi_n^*\rangle = \langle(1 - \lambda_n\pi_nK)x'_{i+1}, \varphi_n^*\rangle + \langle\lambda_n\pi_nKx'_i, \varphi_n^*\rangle = 1$, $(1 - \lambda_n\pi_nK)\tilde{y}_i = \lambda_n\pi_nK(\hat{x}'_i - x'_i)$. This defines the algorithm (5.23).

5.37 $$(T_n - \mu^{k+1})(u^{k+1} - \phi) = (T_n - \mu^{k+1})\frac{Tu^k}{\mu^{k+1}} + T_nu^k - \frac{1}{\mu^{k+1}}T_nTu^k - (T_n - \mu^{k+1})\phi.$$

$$= (T_n - T)(u^k - \phi) + (\mu^{k+1} - \lambda)\phi.$$

$P_n(u^{k+1} - \phi) = 0$; therefore $\mu^{k+1} - \lambda = \langle(T - T_n)(u^k - \phi), \varphi_n^*\rangle$. $u^{k+1} - \phi = (\mu^{k+1} - \lambda)\Sigma_n^{k+1}\phi + \Sigma_n^{k+1}(T_n - T)(u^k - \phi)$.

Let Γ be a Jordan curve around λ and λ_n; if μ^{k+1} is inside Γ, then

$$\|\Sigma_n^{k+1}\| = \|(T_n - \mu^{k+1})^{-1}(1 - P_n)\| \le \max_{z\in\Gamma}\|(T_n - z)^{-1}(1 - P_n)\|,$$

and Σ_n^{k+1} is uniformly bounded in k.

We prove the convergence by induction.

(i) $\|T_n - T\| \to 0$. It is easily proved that $\|u^k - \phi\| \le c\|T_n - T\|^{k+1}$ and $|\mu^{k+1} - \lambda| \le c\|T_n - T\|^{k+2}$, for $k \ge 0$.

(ii) $T_n \overset{\text{cc}}{\to} T$. If μ^k is inside Γ, then the quantity $\|(T_n - T)\Sigma_n^k(T_n - T)\| + \|(T_n - T)\Sigma_n^kP\|$ is bounded by $\varepsilon_n := \max_{z\in\Gamma}(\|(T_n - T)R_n(z)(1 - P_n)(T_n - T)\| + \|(T_n - T)R_n(z)(1 - P_n)P\|)$, which tend to zero as $n \to \infty$.

We suppose that $\max(\|u^{2k+1} - \phi\| + |\mu^{2k+1} - \lambda|, \|u^{2k} - \phi\| + |\mu^{2k} - \lambda|) \le c\varepsilon_n^k\|(T - T_n)P\|$ holds for an arbitrary k. Then $|\mu^{2k+2} - \lambda| \le c\|(T_n - T)(u^{2k+1} - \phi)\|$ and $(T_n - T)(u^{2k+1} - \phi) = (\mu^{2k+1} - \lambda)(T_n - T)\Sigma_n^{2k+1}P\phi + (T_n - T)\Sigma_n^{2k+1}(T_n - T)(u^{2k} - \phi)$. Therefore $|\mu^{2k+2} - \lambda| \le c\varepsilon_n^{k+1}\|(T - T_n)P\|$. Similarly, $u^{2k+2} - \phi = (\mu^{2k+2} - \lambda)\Sigma_n^{2k+2}\phi + \Sigma_n^{2k+2}(T_n - T)(u^{2k+1} - \phi)$ proves that $\|u^{2k+2} - \phi\| \le c\varepsilon_n^{k+1}\|(T - T_n)P\|$. The rest of the proof is left to the reader.

5.38 $$(T_n - \lambda_n)(u^{k+1} - \phi) = (T_n - \lambda_n)u^k + \lambda_n(1 - P_n)\left(u^k - \frac{1}{\mu^{k+1}} Tu^k\right) - T_n\phi + \lambda_n\phi$$

$$= T_n u^k - \frac{\lambda_n}{\mu^{k+1}} Tu^k - T_n\phi + \lambda_n\phi$$

$$= (\mu^{k+1} - \lambda)\frac{\lambda_n}{\mu^{k+1}\lambda} Tu^k + \frac{1}{\lambda}(\lambda_n - \lambda)T(u^k - \phi)$$

$$- (T - T_n)(u^k - \phi).$$

$$\mu^{k+1} - \lambda = \langle T(u^k - \phi), \varphi_n^* \rangle \to 0.$$

$$\mu^{k+1} - \lambda = \frac{\lambda_n - \lambda}{\lambda_n}\langle T(u^k - \phi), \varphi_n^* \rangle + \frac{\lambda}{\lambda_n}\langle (T - T_n)(u^k - \phi), \varphi_n^* \rangle,$$

$$u^{k+1} - \phi = (\mu^{k+1} - \lambda)\frac{\lambda_n}{\mu^{k+1}\lambda}[S_n T(u^k - \phi) + \lambda S_n\phi]$$

$$+ \frac{\lambda - \lambda_n}{\lambda} S_n T(u^k - \phi) + S_n(T_n - T)(u^k - \phi).$$

Apply the proofs of Propositions 5.38 and 5.39.

5.39 Set $x^k = \binom{0}{\mathring{x}^k}$, $\mathring{x}^k \in \mathbb{C}^{N-1}$. The defect $Ax^k - \mu^{k+1}x^k$ has its first component equal to zero. Set $Ax^k - \mu^{k+1}x^k = \binom{0}{\mathring{r}^k}$, then $\mathring{x}^{k+1} = \mathring{x}^k - (C - aI)^{-1}\mathring{r}^k \to \mathring{x}$ such that $\varphi = \binom{1}{\mathring{x}}$. This is a fixed system to solve with a varying right-hand side, in contradistinction with the algorithm given in Section 8.4 of Chapter 3. There we had a *scalar* fixed point formulation of the eigenvalue problem (on λ_k), whereas here we have a *vectorial* algorithm (on x^k).

5.40 We set the problem in $\mathscr{L}(\hat{D})$, where $\hat{D}$ is equipped with the graph norm $\|x\|_D = \|x\| + \|Tx\|$, for $x \in D$. $T_n - T \in \mathscr{L}(\hat{D}, X)$ and $R_n(z) \in \mathscr{L}(X, \hat{D})$ so that $U \in \mathscr{L}(\hat{D})$. And $r_\sigma(U) = r_\sigma[(T_n - T)R_n(z)] \to 0$ as $n \to \infty$. For $\varepsilon > 0$, let n be fixed such that $r_\sigma(U) + \varepsilon < 1$. Then $\|U\|_* < 1$, where $\|\cdot\|_*$ (which depends on ε and U) is equivalent to $\|\cdot\|_D$. Therefore $\|x^k - x\| \le \|x^k - x\|_D \le c(\varepsilon, U)(r_\sigma(U) + \varepsilon)^{k+1}$.

Chapter 6

6.1 We know that $1 - P_n = (T_n - \lambda)^{-1}(1 - P_n)(T_n - \lambda)$ and $S_n(z) = (T_n - z)^{-1}(1 - P_n)$. Hence $1 - P_n = S_n(\lambda)(T_n - \lambda)$ and $(1 - P_n)\varphi = S_n(\lambda)(T_n - \lambda)\varphi = S_n(\lambda)(T_n - T)\varphi$ for $\varphi \in \operatorname{Ker}(T - \lambda)$. We also have $S(z) = R(z)(1 - P)$, $1 - P = (T - \lambda_n)^{-1}(1 - P)(T - \lambda_n)$. So $1 - P = S(\lambda_n)(T - \lambda_n)$ and $(1 - P)\varphi_n = S(\lambda_n)(T - T_n)\varphi_n$ for $\varphi_n \in \operatorname{Ker}(T_n - \lambda_n)$.

6.2 For $\psi_n \in P_n X = M_n$ we have $(1 - P)\psi_n = (1 - P)(\psi_n - \psi)$, $\quad \forall \psi \in M = PX$. For $\psi_n \in M_n$, $\|\psi_n\| = 1$,

$$\|(1 - P)\psi_n\| = \|(1 - P)(\psi_n - \psi)\| \le \|1 - P\| \inf_{\psi \in M} \|\psi_n - \psi\| = \|1 - P\| \operatorname{dist}(\psi_n, M)$$

$$\le \|1 - P\| \sup_{\substack{x_n \in M_n \\ \|x_n\| = 1}} \operatorname{dist}(x_n, M) = \|1 - P\|\delta(M_n, M)$$

$$\le \|1 - P\| \max(\delta(M_n, M), \delta(M, M_n)) = \|1 - P\|\Theta(M_n, M).$$

Similarly, for $\psi \in M$ and all $\psi_n \in M_n$, $(1 - P_n)\psi = (1 - P_n)(\psi - \psi_n)$, so

$$\begin{aligned}\|(1 - P_n)\psi\| &\leq \|1 - P_n\| \inf_{\psi_n \in M_n} \|\psi - \psi_n\| = \|1 - P_n\| \operatorname{dist}(\psi, M_n)\\ &\leq \|1 - P_n\| \|\psi\| \operatorname{dist}\left(\frac{1}{\|\psi\|}\psi, M_n\right)\\ &\leq \|\psi\| \|1 - P_n\| \sup_{\substack{x \in M\\ \|x\| = 1}} \operatorname{dist}(x, M_n) = \|\psi\| \|1 - P_n\|\delta(M, M_n) \leq c\Theta(M, M_n)\end{aligned}$$

since $\|P_n\|$ is uniformly bounded.

6.3 Following Lemma 6.5:

$$(T_n - \lambda_{n_i})^{l_{n_i}}\psi_{n_i} = 0 \qquad \text{for} \quad \psi_{n_i} \in M_{n_i},$$

$$(T - \lambda_{n_i})\psi_{n_i} = (T_n - \lambda_{n_i})\psi_{n_i} + (T - T_n)\psi_{n_i}.$$

$$\begin{aligned}(1 - P)\psi_{n_i} &= (T - \lambda_{n_i})^{-1}(1 - P)(T - \lambda_{n_i})\psi_{n_i}\\ &= S(\lambda_{n_i})[(T_n - \lambda_{n_i})\psi_{n_i} + (T - T_n)\psi_{n_i}]\\ &= (T - \lambda_{n_i})^{-2}(1 - P)(T - \lambda_{n_i})(T_n - \lambda_{n_i})\psi_{n_i} + S(\lambda_{n_i})(T - T_n)\psi_{n_i}\\ &= S^2(\lambda_{n_i})(T_n - \lambda_{n_i})^2\psi_{n_i} + S^2(\lambda_{n_i})(T - T_n)(T_n - \lambda_{n_i})\psi_{n_i} + S(\lambda_{n_i})(T - T_n)\psi_{n_i}.\end{aligned}$$

When we reach $(T_n - \lambda_{n_i})^{l_{n_i}}\psi_{n_i} = 0$, we get

$$(1 - P)\psi_{n_i} = \sum_{j=1}^{l_{n_i}} S^j(\lambda_{n_i})(T - T_n)(T_n - \lambda_{n_i})^{j-1}\psi_{n_i}.$$

6.4 We have the following identities

$$\operatorname{tr} T_{n \restriction M_n} = \sum_{i=1}^{m} \mu_{i_n} = m\hat{\lambda}_n, \qquad \operatorname{tr} T_{\restriction M} = m\lambda.$$

$$\begin{aligned}\operatorname{tr} P(T - T_n)P_{(n)}^{-1} &= \operatorname{tr} PTP_{(n)}^{-1} - \operatorname{tr} PT_nP_{(n)}^{-1}\\ &= \operatorname{tr} TPP_{(n)}^{-1} - \operatorname{tr} P_{(n)}T_{n \restriction M_n}P_{(n)}^{-1}\\ &= \operatorname{tr} T_{\restriction M} - \operatorname{tr} T_{n \restriction M_n} = m(\lambda - \hat{\lambda}_n).\end{aligned}$$

6.5 Let $\{x_i\}_1^m$, $\{x_i^*\}_1^m$ be two adjoint bases in M and M^*.

$$\begin{aligned}\operatorname{tr} P(T - T_n)P_{(n)}^{-1} &= \sum_{i=1}^{m} \langle P(T - T_n)P_{(n)}^{-1}x_i, x_i^*\rangle\\ &= \operatorname{tr} P(T - T_n)P + \sum_i \langle (P_{(n)}^{-1}P - P)x_i, (T^* - T_n^*)x_i^*\rangle.\end{aligned}$$

$P_{(n)}^{-1}P$ (resp. P) is a projection on M_n (resp. M) along $(1 - P)X$. By Lemma 2.16 and Theorem 6.6, $\|P_{(n)}^{-1}P - P\| \leq c\Theta(M, M_n) \leq c\varepsilon_n$ for n large enough. Therefore $|\lambda - \hat{\lambda}_n - (1/m)\operatorname{tr} P(T - T_n)P| \leq c\varepsilon_n\varepsilon_n^*$. Let P be an arbitrary compact projection with Im $P \subset$ Dom T, then TP is bounded on X with finite rank, and $TP = (TP)P$. By Exercise 2.20, tr TP = tr$(TP)P$ = tr PTP.

6.6
$$\begin{aligned}\operatorname{tr} P_n(T - T_n)\tilde{P}_{(n)}^{-1} &= \operatorname{tr} P_nT\tilde{P}_n^{-1} - \operatorname{tr} P_nT_n\tilde{P}_n^{-1}\\ &= \operatorname{tr} \tilde{P}_nT_{\restriction M}\tilde{P}_n^{-1} - \operatorname{tr} T_nP_n\tilde{P}_n^{-1}\\ &= \operatorname{tr} T_{\restriction M} - \operatorname{tr} T_{n \restriction M_n} = m(\lambda - \hat{\lambda}_n).\end{aligned}$$

6.7
$$\begin{aligned}\|(T - T_n)_{\restriction M}\| &= \sup_{\substack{x \in M\\ x \neq 0}} \frac{\|(T - T_n)x\|}{\|x\|} = \sup_{\substack{y \in X\\ Py \neq 0}} \frac{\|(T - T_n)Py\|}{\|Py\|} \leq \|(T - T_n)P\|\\ &= \sup_{\substack{x \in X\\ x \neq 0}} \frac{\|(T - T_n)Px\|}{\|x\|} = \|(T - T_n)_{\restriction M}P\| \leq \|P\| \|(T - T_n)_{\restriction M}\|.\end{aligned}$$

6.8 Since $\dim \operatorname{Ker}(T-\lambda)^k$ is finite, there is a closed subspace N of X such that $X = \operatorname{Ker}(T-\lambda)^k \oplus N$. Let Q be the projection on $\operatorname{Ker}(T-\lambda)^k$ along N. The equation $(T-\lambda)^k x = y$ is uniquely solvable in N; hence $[(T-\lambda)^k{}_{\restriction N}]^{-1}$ exists and is bounded from $\operatorname{Im}(T-\lambda)^k$ onto N. We define $\varphi_k = Q\varphi_{nj}$. We see that $\varphi_{nj} - \varphi_k = (1-Q)\varphi_{nj} \in N$, so

$$\|\varphi_{nj} - \varphi_k\| \le c\|(T-\lambda)^k(\varphi_{nj} - \varphi_k)\|.$$

Since $\Theta(M, M_n)$ is of order ε_n we can find $\tilde{\varphi} \in M$ such that $\|\varphi_{nj} - \tilde{\varphi}\| \le c\varepsilon_n$.

$$\|[(T-\lambda)^k - (T_n-\lambda)^k]\varphi_{nj}\| = \left\|\sum_{i=0}^{k-1}(T_n-\lambda)(T-T_n)(T-\lambda)^{k-i-1}[\varphi_{nj} - \tilde{\varphi} + \tilde{\varphi}]\right\| \le c\varepsilon_n.$$

For $j \le k$,

$$\|(T_n-\lambda)^k\varphi_{nj}\| = \left\|\sum_{i=k-j+1}^{k} C_k^i(\lambda_n-\lambda)^i(T_n-\lambda_n)^{k-i}\varphi_{nj}\right\| \le c|\lambda-\lambda_n|^{k-j+1}.$$

By Theorem 6.7, $|\lambda-\lambda_n| \le c\varepsilon_n^{1/l}$, then

$$\begin{aligned}\operatorname{dist}(\varphi_{nj}, E_k) &\le \|\varphi_{nj} - \varphi_k\| \le c\|[(T-\lambda)^k - (T_n-\lambda)^k]\varphi_{nj} + (T_n-\lambda)^k\varphi_{nj}\| \\ &\le c\varepsilon_n^{(k-j+1)/l}.\end{aligned}$$

6.9 $x = R(z)f$, $x_n = R_n(z)f$. $x - x_n = (R(z) - R_n(z))f = R_n(z)(T_n - T)R(z)f = R_n(z)(T_n-T)x$, $(T_n-T)x = (T_n-z)x + zx - Tx = (T_n-z)x - f$. Also $x - x_n = R(z)(T_n-T)R_n(z)f = R(z)(T_n-T)x_n$, $(T_n-T)x_n = (T_n-z)x_n + (z-T)x_n = f - (T-z)x_n$. We then get $\|x-x_n\| \le \|R_n(z)\|\,\|(T_n-T)x\|$ (a priori bound), and $\|x-x_n\| \le \|R(z)\|\,\|(T_n-T)x_n\|$ (a posteriori bound). $\|R_n(x)\|$ and $\|R(z)\|$ play the role of condition numbers.

6.10 Use Exercise 6.9 and Lemma 6.4. We get for $T, T_n \in \mathscr{L}(X)$, $c_1\|(T_n-T)x\| \le \|x-x_n\| \le c_2\|(T_n-T)x\|$; that is,

$$\|x-x_n\| = O(\|(T_n-T)x\|) \quad \text{or} \quad O(\|(T_n-T)x_n\|).$$

$$c_1'\|(1-P_n)(T_n-T)\varphi\| \le \|(1-P_n)\varphi\| \le c_2'\|(T_n-T)\varphi\| \qquad \text{for} \quad \varphi \in \operatorname{Ker}(T-\lambda),$$

$$c_1''\|(1-P)(T-T_n)\varphi_n\| \le \|(1-P)\varphi_n\| \le c_2''\|(T-T_n)\varphi_n\| \qquad \text{for} \quad \varphi_n \in \operatorname{Ker}(T_n-\lambda_n).$$

6.11 $\pi_n(T-\lambda_n)\varphi_n = 0 \Leftrightarrow \pi_n'(T-\lambda_n)\varphi_n = 0$. By Lemma 6.9, this yields a bound in terms of $\|(1-\pi_n')P\| \le c\delta(M, X_n)$. But $\|(1-\pi_n')f\| \le c\|(1-\pi_n)(K-\alpha)^{-1}f\|$ for $f \in X$. Therefore $\|(1-\pi_n')P\| \le c\delta((K-\alpha)^{-1}M, Y_n)$.

6.12 Adapt the proof of Theorem 6.7 with $\|A - B_n\| = 0(\alpha_n\alpha_n^*)$. For the eigenvector φ_n we get

$$\operatorname{dist}(\varphi_n, \operatorname{Ker}(T-\lambda)) \le \|(1-Q)\varphi_n\| \le c(\alpha_n + |\lambda_n - \lambda|) \le c\max(\alpha_n, (\alpha_n\alpha_n^*)^{1/l}).$$

6.13 $\varepsilon_n = \|(1-\pi_n)TP\| \le \|T\|\alpha_n$, $\varepsilon_n^* = \|T^*(1-\pi_n^*)P^*\| \le \|T\|\alpha_n^*$. $\operatorname{tr}(T-T_n)P = \sum_i \langle(1-\pi_n)Tx_i, x_i^*\rangle = \sum_i \langle(1-\pi_n)PTx_i, (1-\pi_n^*)x_i^*\rangle$ proves that $|\operatorname{tr}(T-T_n)P| \le c\alpha_n\alpha_n^*$.

6.14 From Proposition 4.6, $\|x - x_n^S\| \le c\|T(1-\pi_n)x\|$.

$$T(1-\pi_n)x(t) = \int_a^b k(t,s)(1-\pi_n)x(s)\,ds = ((1-\pi_n)x, \bar{k}_t) = ((1-\pi_n)x, (1-\pi_n)\bar{k}_t).$$

So $\|x - x_n^S\| \le ch^{2r+2}$ since

$$|((1-\pi_n)x, (1-\pi_n)\bar{k}_t)| \le \left(\sup_{a \le t \le b}\|(1-\pi_n)k_t\|_2\right)\|(1-\pi_n)x\|_2.$$

6.15 From the solution of Exercise 5.38 follows $\operatorname{dist}(u^1, M) \le \|\phi - u^1\| \le c\|(1-\pi_n)T\|\,\|\phi - \varphi_n\|$ with $\|\varphi_n - \phi\| = \|(P_n - 1)\phi\| \le \|(P_n - P)P\|\,\|\phi\| \le \alpha_n\|\phi\|$. $\|\phi\| = |\langle P\varphi_n, \varphi_n^*\rangle|^{-1}\|P\varphi_n\|$. Since $\|(1-\pi_n)T\| \to 0$, for n large enough $|\langle P\varphi_n, \varphi_n^*\rangle| > \frac{1}{2}$ and $\|\phi\| \le 2\|P\|$.

By Theorem 6.12,

$$\operatorname{dist}(\tilde{\varphi}_n, M) \le \|P\tilde{\varphi}_n - \tilde{\varphi}_n\| = \|(P - P_n)P_n\| \, \|\tilde{\varphi}_n\| \le c\|T(1 - \pi_n)\|\alpha_n\|\tilde{\varphi}_n\|.$$

u^1 may improve on $\tilde{\varphi}_n$ when $\pi_n^* \to 1$.

6.16 From $|a(u, v)| \le \beta\|u\|_V \, \|v\|_V$, $\mathcal{Re}\, a(u, u) \ge \alpha\|u\|_V^2$, and $a(u - u_n, u - u_n) = a(u - u_n, u - v_n)$, $\forall v_n \in V_n$, we get $\|u - u_n\|_V \le (\beta/\alpha)\|u - v_n\|_V$, $\forall v_n \in V_n$. So $\|u - u_n\|_V \le (\beta/\alpha) \inf_{v_n \in V_n}\|u - v_n\|_V = c\delta_n(u)$. Now

$$\delta_V(M, V_n) = \sup_{\substack{u \in M \\ \|u\|_V = 1}} \delta_n(u) = \max_{\substack{u \in M \\ \|u\|_V = 1}} \delta_n(u).$$

6.17 Set $\nu := 1/\lambda$. The norm $\|u\|_a = \sqrt{a(u, u)}$ for $u \in V$ is equivalent to $\|\cdot\|_V$. $\tilde{B}_n = \pi_n^a B \pi_n^a$ is selfadjoint with respect to the inner product $(\cdot, \cdot)_a$. Therefore

$$\begin{aligned}\operatorname{dist}(\nu, \sigma(\tilde{B}_n)) &= \inf_{\substack{y \in V \\ \|y\|_a = 1}} \|(\tilde{B}_n - \nu)y\|_a \le \inf_{\substack{y \in V_n \\ \|y\|_a = 1}} \|(\tilde{B}_n - \nu)y\|_a \\ &\le \inf_{\substack{y \in V_n \\ \|y\|_a = 1}} \|(A - \nu)y\|_a = \inf_{\substack{y \in V_n \\ \|y\|_a = 1}} \|(A - \nu)(y - x)\|_a \qquad \forall x \in M \\ &\le c \sup_{\substack{x \in M \\ \|x\|_V = 1}} \inf_{y \in V_n} \|y - x\|_V \le c \sup_{\substack{x \in M \\ \|x\|_V = 1}} \operatorname{dist}_V(x, V_n) = c\delta_V(M, V_n).\end{aligned}$$

6.18 Since the problem is equivalent to $\lambda B u = u$, with $B \in \mathcal{L}(V)$, and the approximation is defined by means of π_n^a, the error bounds given in Theorem 6.13 hold.

6.19 We define the application

$$v \in H_2 \mapsto |v|_2 := \sup_{\substack{u \in H_1 \\ u \ne 0}} \frac{|a(u, v)|}{\|u\|_1},$$

$|\cdot|_2$ is a norm in H_2 such that $|v|_2 \le \beta\|v\|_2$ since a is continuous. We wish to prove that $|v|_2 \ge \alpha\|v\|_2$ for $\alpha > 0$, and it is sufficient to prove that $(H_2, |\cdot|_2)$ is complete. We define $U\colon H_2 \to H_1^*$ by $\langle u, Uv\rangle_1 = a(u, v)$ for all $u \in H_1$. If $\{v_n\}$ is a Cauchy sequence in $(H_2, |\cdot|_2)$, Uv_n is a Cauchy sequence in H_1^*, and $Uv_n \to l \in H_1^*$. If U is onto, there exists $v \in H_2$ such that $l = Uv$. $U^*\colon H_1 \to H_2^*$ is such that $\langle U^*u, v\rangle_2 = \langle u, Uv\rangle_1 = a(u, v)$ for $u \in H_1$, $v \in H_2$. From (6.6), we conclude that $\|U^*u\|_{H_2^*} \ge \alpha\|u\|_{H_1}$ for $u \in \operatorname{Dom} U^*$, which is equivalent to U is onto.

6.20 $u_h = \sum_{i=1}^N q_i^h e_i^h \Rightarrow a(u_h, f_j^h) = \sum_{i=1}^N q_i^h a(e_i^h, f_j^h)$ and $b(u^h, f_j^h) = \sum_{i=1}^N q_i^h b(e_i^h, f_j^h)$ for $j = 1, \ldots, N$. $a(u_h, v_h) = \lambda_h b(u_h, v_h)$, $\forall v_h \in S_{2h}$ becomes then equivalent to $\tilde{A}^h q_h = \lambda_h \tilde{B}^h q_h$. If $\tilde{A}^h$ is singular, there exists $x_h = \sum_{i=1}^N x_i^h e_i^h \ne 0$ in S_{1h} such that $\tilde{A}^h x_h = 0 = (a(x_h, f_j^h))_{j=1}^N$; hence $a(x_h, v_h) = 0$, $\forall v_h \in S_{2h}$ and (6.9) is not satisfied.

6.21 $F = TP_{\restriction M} = T_{\restriction M}\colon M \to M$, $G_n'' = P_{(n)} T_n P_{(n)}^{-1}\colon M \to M$, $F_n' = P_{(n)}^{-1} T P_{(n)}\colon M_n \to M_n$, $G_n' = T_{n\restriction M_n}\colon M_n \to M_n$.

$$P_{(n)}(F_n' - G_n')P_{(n)}^{-1} = P_{(n)}P_{(n)}^{-1}TP_{(n)}P_{(n)}^{-1} - P_{(n)}T_{n\restriction M_n}P_{(n)}^{-1} = T_{\restriction M} - P_{(n)}T_nP_{(n)}^{-1} = F - G_n''.$$

$$\|F - G_n''\| \le c\|F_n' - G_n'\|$$

and

$$\|F' - G_n'\| = \max(\|P_{(n)}^{-1}P_{(n)}(T - T_n)\psi\|, \psi \in M_n, \|\psi\| = 1) \le c\eta_n.$$

6.22 Apply Wilkinson (1965, p. 81) to the $m \times m$ matrices which represent F and G_n'' in some basis of M.

6.23 By Exercise 6.2, $\|(P - P_n)P\| \le c\Theta(M, M_n) \le c\eta_n$.

6.24 $((T-\alpha)x, \quad (T-\alpha)x) = \|Tx\|^2 - (\alpha+\bar{\alpha})\rho + \alpha\bar{\alpha}, \quad ((T-\rho)x, \quad (T-\rho)x) = \|Tx\|^2 - \rho^2$, $\|(T-\alpha)x\|^2 - \|(T-\rho)x\|^2 = |\alpha-\rho|^2 \geq 0$. Cf. Exercise 1.80.

6.25 If T_n is selfadjoint, $\varphi_n = \varphi_n^*$, $(\varphi_n, \varphi_n) = 1$, $\rho_n = (T\varphi_n, \varphi_n) = \zeta_n$. By Exercise 6.24, $|\lambda - \rho_n| \leq c\|(T-\rho_n)\varphi_n\|^2 \leq c\|(T-\lambda_n)\varphi_n\|^2$ and by Lemma 6.16, $|\lambda - \zeta_n| \leq c\|(T-\lambda_n)\varphi_n\|$ $\|(T-\bar{\lambda}_n)\varphi_n^*\|$ when T_n is not selfadjoint.

6.26 $\zeta_n = \dfrac{1}{m}\sum_{i=1}^m \langle Tx_{in}, x_{in}^*\rangle.$

$$\hat{\lambda}_n = \frac{1}{m}\operatorname{tr} T_nP_n = \frac{1}{m}\sum_{i=1}^m \langle T_nx_{in}, x_{in}^*\rangle$$

$$= \frac{1}{m}\sum_{i=1}^m \langle \pi_nT\pi_nx_{in}, x_{in}^*\rangle = \frac{1}{m}\sum_{i=1}^m \langle Tx_{in}, x_{in}^*\rangle$$

since $\pi_nx_{in} = x_{in}$ and $\pi_n^*x_{in}^* = x_{in}^*$.

6.27 In Exercise 5.32 we found $\lambda^{1G} = \lambda^{0G}$ and $\varphi^{1G} = \varphi_n^S$. So $\lambda^{2G} = \langle T\varphi^{1G}, \varphi_n^*\rangle = \langle T\varphi_n^S, \varphi_n^*\rangle = \langle T\tilde{\varphi}_n, \varphi_n^*\rangle = \tilde{\lambda}_n$.

6.28 Since T is selfadjoint and π_n is orthogonal,

$$\eta_n = \eta_n^* \leq \|(1-\pi_n)T\| \qquad \text{and} \qquad |\lambda - \tilde{\lambda}_n| \leq c\|(1-\pi_n)T\|^3.$$

$$\|(T-\lambda_n)\tilde{\varphi}_n\| = \|(T - T\pi_n)\tilde{\varphi}_n\| \leq \|T(1-\pi_n)\|\ \|(1-\pi_n)T(1/\lambda_n)\varphi_n\| \leq c\|(1-\pi_n)T\|^2,$$

and

$$|\lambda - \tilde{\rho}_n| \leq c\|(T-\tilde{\rho}_n)\tilde{\varphi}_n\|^2 \leq c\|(T-\lambda_n)\tilde{\varphi}_n\|^2 \quad \text{(Exercise 6.24)}$$
$$\leq c\|(1-\pi_n)T\|^4.$$

6.29 $A = \begin{pmatrix} 0 & \varepsilon \\ \varepsilon & a\end{pmatrix}$ has two eigenvalues such that $\lambda(a-\lambda) + \varepsilon^2 = 0$. With $x = e_1$, $\rho = 0$ and $|\lambda| = \varepsilon^2/|a-\lambda| = \varepsilon^2/\delta(\rho)$. Let θ be the acute angle between the eigenvectors and e_1 or e_2; $\tan\theta = |\lambda|/|\varepsilon| = |\varepsilon|/|a-\lambda| = |\varepsilon|/\delta(\rho)$.

6.30 From Corollary 6.20, $-\varepsilon \leq \lambda - \rho \leq \varepsilon$. From Theorem 6.21, $-\varepsilon^2/(\bar{\lambda}-\rho) \leq \lambda - \rho \leq \varepsilon^2/(\rho - \underline{\lambda})$. It then improves on the former if $\varepsilon^2/(\bar{\lambda}-\rho) < \varepsilon$ and $\varepsilon^2/(\rho-\underline{\lambda}) < \varepsilon$, that is, if $\varepsilon < \bar{\lambda} - \rho$ and $\varepsilon < \rho - \underline{\lambda}$, hence if $\varepsilon^2 < (\bar{\lambda}-\rho)(\rho-\lambda)$.

6.31 Suppose we know three vectors x_i, $\|x_i\| = 1$, $i = 1, 2, 3$, such that the Rayleigh quotients $\rho_i = (Tx_i, x_i)$ satisfy $\rho_1 < \rho_2 < \rho_3$. We assume that the disks $\{z; |z-\rho_i| \leq \varepsilon_i = \|(T-\rho_i)x_i\|\}$ are disjoint, and there is no point of $\sigma(T)$ in between. Then there exists $\lambda_i \in \sigma(T)$ such that $|\lambda_i - \rho_i| \leq \varepsilon_i$, $i = 1, 2, 3$. If $\delta_2 := \operatorname{dist}(\rho_2, \sigma(T) - \{\lambda_2\})$, we have $\delta_2 \geq \min(\rho_2 - \rho_1 - \varepsilon_1, \rho_3 - \varepsilon_3 - \rho_2)$.

6.32 $\|Q\| = \sup\limits_{t\in X}\dfrac{\|Qt\|}{\|t\|} = \sup\limits_{t\in X}\dfrac{|\langle t, y\rangle|}{\|t\|} = \|y\|$, $1 = \langle x, y\rangle \leq \|y\|$.

6.33 $(T-\zeta)Q$ is of finite rank and domain X, $Q(T-\zeta)t = \langle (T-\zeta)t, y\rangle\ x = \langle t, (T^* - \bar{\zeta})y\rangle x$ is defined for all t in X. $\tilde{H}$ is then of finite rank and domain X, so $\tilde{H} \in \mathscr{L}(X)$. $T - \tilde{H}$ is closed with domain D [see Kato (1976, p. 164)].

6.34 $\tilde{T}x = \zeta x + (1-Q)T(1-Q)x = \zeta x$. Now $\tilde{T}^* = \bar{\zeta}Q^* + (1-Q^*)T^*(1-Q^*)$, where Q^* is a projection on $\{y\}$ along $\{x\}^\perp$. Similarly, $\tilde{T}^*y = \bar{\zeta}y$. It can be shown as in Exercise 3.24 that $\sigma(\tilde{T}) = \{\xi\} \cup \sigma((1-Q)T_{\restriction \{y\}^\perp \cap X})$, where the two sets are disjoint. ζ is then an isolated eigenvalue of $\tilde{T}$, and if it is simple, Q is the associated spectral projection.

6.35 $(I-Q)(T-\zeta)(1-Q) = (\tilde{T}-\zeta)(1-Q)$. If ζ is a simple eigenvalue of $\tilde{T}$, $\mathring{T}_\zeta$ is regular since $\operatorname{Ker}(\tilde{T}-\zeta) = QX$ has a zero intersection with $(1-Q)X$. This proves that $\Sigma \in \mathscr{L}(X)$. Now the reduced resolvent of $\tilde{T}$ with respect to ζ is $\Sigma(z) = (\tilde{T}-z)^{-1}(1-Q)$.

$$\Sigma(\zeta) = \lim_{z\to\zeta}[(\tilde{T}-z)^{-1}(1-Q)] = \mathring{T}_\zeta^{-1}(1-Q) = \Sigma.$$

6.36 $(1 - Q)Tx = (1 - Q)(T - \zeta)x = Tx - QTx = (T - \zeta)x.$

6.37 $$\tilde{H}Q = (T - \zeta)Q + Q(T - \zeta)Q = (T - \zeta)Q.$$

$$(1 - Q)\tilde{H}Q = (1 - Q)(T - \zeta)Q = (T - \zeta)Q - Q(T - \zeta)Q = (T - \zeta)Q.$$

$$\tilde{H}(1 - Q) = Q(T - \zeta)(1 - Q) = Q(T - \zeta) - Q(T - \zeta)Q = Q(T - \zeta).$$

$$Q\tilde{H}(1 - Q) = Q(T - \zeta).$$

6.38 We recall that Q and $1 - Q$ commute with $\tilde{R}(z)$.

$$\begin{aligned}\tilde{H}\tilde{R}(z)\tilde{H}\tilde{R}(z)QX &= \tilde{H}\tilde{R}(z)\tilde{H}Q\tilde{R}(z)X = \tilde{H}\tilde{R}(z)(1 - Q)\tilde{H}Q\tilde{R}(z)X \\ &= \tilde{H}(1 - Q)\tilde{R}(z)\tilde{H}Q\tilde{R}(z)X = Q\tilde{H}(1 - Q)\tilde{R}(z)\tilde{H}\tilde{R}(z)QX \subseteq QX.\end{aligned}$$

Also

$$\begin{aligned}\tilde{H}\tilde{R}(z)\tilde{H}\tilde{R}(z)(1 - Q)X &= \tilde{H}\tilde{R}(z)\tilde{H}(1 - Q)\tilde{R}(z)X = \tilde{H}\tilde{R}(z)Q\tilde{H}(1 - Q)\tilde{R}(z)X \\ &= \tilde{H}Q\tilde{R}(z)\tilde{H}\tilde{R}(z)(1 - Q)X \\ &= (1 - Q)\tilde{H}Q\tilde{R}(z)\tilde{H}\tilde{R}(z)(1 - Q)X \subseteq (1 - Q)X.\end{aligned}$$

6.39 $\sigma(T) = \sigma(T_{\restriction M}) \cup \sigma(T_{\restriction N}).$

$$\begin{aligned}r_\sigma(T) &= \max_{\lambda \in \sigma(T)} |\lambda| = \max\left(\max_{\lambda \in \sigma(T_{\restriction M})} |\lambda|, \max_{\lambda \in \sigma(T_{\restriction N})} |\lambda| \right) \\ &= \max(r_\sigma(T_{\restriction M}), r_\sigma(T_{\restriction N})).\end{aligned}$$

6.40 In Theorem 6.27 we have written the identity $\varphi - x = -(1 - (\lambda - \zeta)\Sigma)^{-1}\Sigma u$. Hence $\varphi - x + \Sigma u = [1 - (1 - (\lambda - \zeta)\Sigma)^{-1}]\Sigma u$. Writing $1 = (1 - (\lambda - \zeta)\Sigma)(1 - (\lambda - \zeta)\Sigma)^{-1}$ we get $\varphi - x + \Sigma u = -(\lambda - \zeta)\Sigma(1 - (\lambda - \zeta)\Sigma)^{-1}\Sigma u$.

6.41 $\lambda_2 = \mathrm{tr}(-\tilde{H}\Sigma\tilde{H}Q)$; hence $|\lambda - \zeta + \langle \Sigma u, v\rangle| \le \sum_{k=2}^{\infty} |\lambda_{2k}| \le (g(\tilde{r}) - 1)|\langle \Sigma u, v\rangle|$. Following Exercise 6.40 and the last two inequalities of Theorem 6.27 we get

$$\begin{aligned}\|\varphi - x + \Sigma u\| &\le |\lambda - \zeta|\, \|\Sigma\|\, \|(1 - (\lambda - \zeta)\Sigma)^{-1}\|\, \|\Sigma u\| \\ &\le |\lambda - \zeta|\, \|\Sigma\| \frac{1}{1 - |\lambda - \zeta|\, \|\Sigma\|} \|\Sigma u\| \le \frac{g(\tilde{r})\tilde{r}}{1 - g(\tilde{r})\tilde{r}} \|\Sigma u\| \\ &= \left(-1 + \frac{1}{1 - g(\tilde{r})\tilde{r}}\right)\|\Sigma u\| = (g(\tilde{r}) - 1)\|\Sigma u\|.\end{aligned}$$

$(g(\tilde{r}) - 1)/\tilde{r} \to 1$; that is, $g(\tilde{r}) - 1 \sim \tilde{r}$. When u is not small, x is not an approximate eigenvector for T, but the corrected vector $x - \Sigma u$ is one.

6.42 $\varepsilon\varepsilon^*$ is small if at least ε or ε^* is small. We first suppose that ε is small, and consider the decomposition

$$T = \underbrace{\zeta Q + QT(1 - Q) + (1 - Q)T(1 - Q)}_{T'} + \underbrace{(T - \zeta)Q}_{H'} = T' + H'.$$

$\|H'\| = \|y\|\varepsilon$. ζ is a simple eigenvalue of T' with spectral projection Q' such that $Q'x = x$. Therefore, for ε small enough $\|Px - x\| \le c\varepsilon < 1$; that is, $|\langle Px, y\rangle - 1| < 1 \Rightarrow \langle Px, y\rangle \neq 0$.

If ε^* is small, we consider $T = T'' + H''$ with $H'' = QT(1 - Q)$ and $\|H''\| = \|y\|\varepsilon^*$. We conclude similarly that $\|P^*y - y\| < 1$ and $\langle x, P^*y\rangle = \langle Px, y\rangle \neq 0$.

6.43 $\tilde{T}^n = P_nTP_n + (1 - P_n)T(1 - P_n)$; $\tilde{H}^n = (1 - P_n)TP_n + P_nT(1 - P_n)$ with $P_n = \langle \cdot, \varphi_n^*\rangle\varphi_n$, and $T = \tilde{T}^n + \tilde{H}^n$. By assumption $P_n \stackrel{cc}{\to} P$. $\tilde{H}^n \stackrel{p}{\to} 0$, $\|(1 - P_n)(T - T_n)P_n\| \to 0$, $P_nT(1 - P_n) = (P_n - P)T(1 - P_n) + TP(P - P_n)$ show that $\tilde{H}^n \stackrel{cc}{\to} 0$.

$$\zeta_n = \mathrm{tr}\, TP_n = \langle T\varphi_n, \varphi_n^*\rangle.$$

Formulas in Lemma 5.34 define an iterative scheme from $v^0 = \zeta_n, \eta^0 = \varphi_n$, where $Q, T - T_n$, and $R_n(z)$ are replaced, respectively, by P_n, $\tilde{H}^n$, and $(\tilde{T}^n - z)^{-1} = \tilde{R}^n(z)$.

From the structure of $\tilde{H}^n$ follows that $v^{2k+1} = 0, \eta^{2k} = 0$ for $k \geq 1$. From Theorem 6.27 and Proposition 6.28 we get for $a \geq \|\tilde{R}^n(\zeta_n)(1 - P_n)\|$

$$\max_{z \in \Gamma} r_\sigma^2(\tilde{H}^n\tilde{R}^n(z)) = 2a \max_{z \in \Gamma} |\langle \tilde{H}^n\tilde{R}^n(z)\tilde{H}^n\varphi_n, \varphi_n^*\rangle| = \rho_n \to 0,$$

$r_\sigma(P_n(P - P_n)P_n) = \rho_n/(1 - \rho_n) \to 0$ as $n \to \infty$.

6.44 For T and its inverse (if $0 \in \rho(T)$) the spectral projections are identical and

$$\frac{1/\lambda - 1/\mu_{in}}{1/\lambda} = \frac{\mu_{in} - \lambda}{\mu_{in}} = \frac{\lambda}{\mu_{in}}\frac{\mu_{in} - \lambda}{\lambda},$$

where $\lambda/\mu_{in} \to 1$ as $n \to \infty$.

6.45 $\alpha_n = \alpha_n^* = \|(1 - \pi_n)P\| \geq \max_i \|(1 - \pi_n)P\varphi_{in}\|, l = 1, g = m$. Proposition 6.30 shows that the bounds in Exercise 6.12 cannot be improved.

6.46 $\tilde{T}_n = PT_nP + (1 - P)T_n(1 - P)$, $\tilde{H}_n = (1 - P)T_nP + PT_n(1 - P)$, with $P = (\cdot, \varphi)\varphi$, and $T_n = \tilde{T}_n + \tilde{H}_n$. Since $\tilde{H}_n = (1 - P)(T_n - T)P + P(T_n - T)(1 - P)$, $\|\tilde{H}_n\| \to 0$. $\zeta_n = \text{tr}\, T_nP = \lambda + \text{tr}(\pi_n - 1)TP = \lambda + \lambda\|(\pi_n - 1)\varphi\|^2$. Let Γ be a Jordan curve isolating λ. For n large enough, ζ_n is the only eigenvalue of $\tilde{T}_n$ inside Γ since $\|PT_nP - PTP\| = \|P(\pi_n - 1)TP\| \to 0$. $\tilde{\Sigma}_n = (1 - P)[(\tilde{T}_n - \zeta_n)_{\restriction\{\varphi\}^\perp}]^{-1}(1 - P)$.

$$\|P(\pi_n - 1)T\tilde{\Sigma}_n(\pi_n - 1)T\varphi\| \leq c\|(\pi_n - 1)T\|\;\|(\pi_n - 1)\varphi\|^2.$$

Then by Theorem 6.27

$$|\lambda_n - \zeta_n| = |\lambda_n - \lambda - \lambda\|(\pi_n - 1)\varphi\|^2| \leq c\|(\pi_n - 1)T\|\;\|(\pi_n - 1)\varphi\|^2$$

and

$$\left|\frac{\lambda_n - \lambda}{\lambda\|(\pi_n - 1)\varphi\|^2} - 1\right| \leq c\|(\pi_n - 1)T\| \to 0.$$

Chapter 7

7.1 It follows from the identity used in Theorem 7.2

$$x = \frac{1}{z^\alpha}T^\alpha x - \left[\frac{1}{z^\alpha}T^{\alpha-1}f + \cdots + \frac{1}{z}f\right]$$

7.2 If T is defined by (7.2), then T^* is an integral operator with kernel $\overline{k(s, t)}$ which is of class $\mathfrak{G}(\alpha, \gamma)$ relative to t and s. Since P is defined by a smooth kernel $p(s, t)$ of continuity α relative to t and s, so is P^*. The kernel $k_n^*(t, s)$ of $T^*(1 - P^*) + \bar{\lambda}_n P^*$ is the sum of the kernel of $T^*(1 - P^*)$ plus $\bar{\lambda}_n$ times the kernel of P^*. It is then of class $\mathfrak{G}(\alpha, \gamma)$ relative to t, uniformly in n.

7.3 $\tilde{\psi}_n \in \tilde{M}_n$ is associated with the eigenvalue λ_n of ascent l_n.

$$\begin{aligned}
[(1 - P)\tilde{\psi}_n](t) &= \left[\sum_{j=1}^{l_n} S^j(\lambda_n)T(1 - \pi_n)(T\pi_n - \lambda_n)^{j-1}\tilde{\psi}_n\right](t) \\
&= \sum_{j=1}^{l_n} (S^j(\lambda_n)(1 - \pi_n)(T\pi_n - \lambda_n)^{j-1}\tilde{\psi}_n, \bar{k}_t) \\
&= \sum_{j=1}^{l_n} ((1 - \pi_n)(T\pi_n - \lambda_n)^{j-1}\tilde{\psi}_n, l_t^{(j)}),
\end{aligned}$$

where $l_t^{(j)}$ is the solution of $(T^* - \bar{\lambda}_n)^j(1 - P^*)l_t^{(j)} = (1 - P^*)\bar{k}_t$. For $j = 1, \ldots, l_n$, $l_t^{(j)}$ can be shown to be of class $\mathfrak{G}(\alpha, \gamma)$ uniformly in n (see Exercise 7.2). We remark that $(T\pi_n - \lambda_n)^{j-1}\tilde{\psi}_n \in M_n$, and then we can proceed as in Theorem 7.9 to bound $|\langle(1 - \pi_n^1)\xi_n, l_t^{(j)}\rangle|$ with $\xi_n \in \tilde{M}_n$.

7.4 See Exercise 6.12.

7.5
$$\begin{aligned} y - \tilde{y}_n &= (1 - zA_z)^{-1}f - zA_z y_n - f \\ &= [(1 - zA_z)^{-1} - 1]f - zA_z y_n \\ &= [(1 - zA_z)^{-1} - (1 - zA_z)^{-1}(1 - zA_z)]f - zA_z y_n \\ &= z(1 - zA_z)^{-1}A_z f - zA_z y_n \\ &= z(1 - zA_z)^{-1}A_z f - z(1 - zA_z)^{-1}(1 - zA_z)A_z y_n \\ &= z(1 - zA_z)^{-1}[A_z f - (1 - zA_z)A_z y_n]. \end{aligned}$$

But $A_z f = A_z\tilde{y}_n - zA_z^2 y_n$, so

$$\begin{aligned} y - \tilde{y}_n &= z(1 - zA_z)^{-1}[A_z\tilde{y}_n - zA_z^2 y_n - A_z y_n + zA_z^2 y_n] \\ &= z(1 - zA_z)^{-1}A_z(\tilde{y}_n - y_n) \\ &= zA_z(1 - zA_z)^{-1}(\tilde{y}_n - y_n) \end{aligned}$$

because A_z and $(1 - zA_z)^{-1}$ commute. Since $\tilde{y}_n = zA_z y_n + f$, then $\pi_n\tilde{y}_n = \pi_n zA_z y_n + \pi_n f = \pi_n y_n$. Hence $\pi_n(\tilde{y}_n - y_n) = 0$, so $(1 - \pi_n)(\tilde{y}_n - y_n) = \tilde{y}_n - y_n$, and we get

$$y - \tilde{y}_n = zA_z(1 - zA_z)^{-1}(1 - \pi_n)(\tilde{y}_n - y_n).$$

7.6 We remark that $T = (1 - K)G^{-1} = (1 - zA_z)A_z^{-1}$; $u_n = Gx_n = A_z y_n$, $\tilde{x}_n = Kx_n + f$, $\tilde{y}_n = zA_z y_n + f$. So $x_n - \tilde{x}_n = (1 - K)x_n - f = (1 - K)G^{-1}u_n + y_n - \tilde{y}_n - (1 - zA_z)A_z^{-1}u_n = y_n - \tilde{y}_n$.

7.7 We find $y - \tilde{y}_n = zA_z(1 - zA_z)^{-1}(1 - \pi_n)\tilde{x}_n$, a formula similar to that of Lemma 7.1, the kernel of A_z being of class $\mathfrak{G}(\alpha + 1, p - 2)$.

7.8 From Exercise 4.67 we get $\pi_n\tilde{\varphi}_n = \pi_n K\varphi_n + \lambda_n\pi_n G\varphi_n = \varphi_n$ since φ_n satisfies

$$(1 - \pi_n K)\varphi_n - \lambda_n\pi_n G\varphi_n = 0;$$

that is, $\varphi_n = \lambda_n U_n\varphi_n$.

Since $\tilde{\varphi}_n = K\varphi_n + \lambda_n G\varphi_n$ and $\varphi_n = \pi_n\tilde{\varphi}_n$ we get

$$\tilde{\varphi}_n = K\pi_n\tilde{\varphi}_n + \lambda_n G\pi_n\tilde{\varphi}_n, \qquad (1 - K\pi_n)\tilde{\varphi}_n = \lambda_n G\pi_n\tilde{\varphi}_n;$$

therefore $\tilde{\varphi}_n = \lambda_n\tilde{U}_n\tilde{\varphi}_n$.

7.9 $U - U_n = U - (1 - \pi_n K)^{-1}\pi_n G$. But $G = (1 - K)U$, so

$$\begin{aligned} U - U_n &= U - (1 - \pi_n K)^{-1}\pi_n(1 - K)U \\ &= (1 - \pi_n K)^{-1}[1 - \pi_n K - \pi_n(1 - K)]U \\ &= (1 - \pi_n K)^{-1}(1 - \pi_n)U. \end{aligned}$$

$$\begin{aligned} U - \tilde{U}_n &= U - (1 - K\pi_n)^{-1}G\pi_n = U - (1 - K\pi_n)^{-1}(1 - K)U\pi_n \\ &= (1 - K\pi_n)^{-1}[U(1 - \pi_n) + K(U\pi_n - \pi_n U)]. \end{aligned}$$

But $U - KU = G$; hence

$$\begin{aligned} U(1 - \pi_n) + K(U\pi_n - \pi_n U) &= U - G\pi_n - K\pi_n U \\ &= U - G + G - G\pi_n - K\pi_n U = K(U - \pi_n U) + G(1 - \pi_n) \\ &= K(1 - \pi_n)U + G(1 - \pi_n). \end{aligned}$$

So $U - \tilde{U}_n = (1 - K\pi_n)^{-1}[K(1 - \pi_n)U + G(1 - \pi_n)]$.

7.10 For example, $(1 - K)^{-1} \in \mathscr{L}(X)$ and G compact implies U compact. $\pi_n \xrightarrow{p} 1$ implies

(i) $\|(1 - \pi_n)K\| \to 0$ and $(1 - \pi_n K)^{-1}$ is uniformly bounded,
(ii) $K\pi_n \xrightarrow{cc} K$ and $(1 - K\pi_n)^{-1}$ is uniformly bounded,
(iii) $\|U - U_n\| \to 0$,
(iv) $\tilde{U}_n \xrightarrow{cc} U$.

7.11 Since Q is the spectral projection for U, $QU = UQ$,

$$(U - U_n)Q = (1 - \pi_n K)^{-1}(1 - \pi_n)QU,$$

$$(U - \tilde{U}_n)Q = (1 - K\pi_n)^{-1}[K(1 - \pi_n)QU + G(1 - \pi_n)Q].$$

$$\|(U - U_n)Q\| \leq c\|(1 - \pi_n)Q\|,$$

$$\|(U - \tilde{U}_n)Q\| \leq c[\|K(1 - \pi_n)Q\| + \|G(1 - \pi_n)Q\|].$$

7.12 It follows from Theorem 6.6 that $\|(1 - Q)\varphi_n\| \leq c\|(U - U_n)Q\|$ and $\|(1 - Q)\tilde{\varphi}_n\| \leq c\|(U - \tilde{U}_n)Q\|$. Then we apply the preceding exercise.

7.13 M_n, the invariant subspace defined by $\pi_n(T - \lambda_n)\psi_n = 0$, $\psi_n \in X_n$, is a subspace of C_n. Therefore $N_n = D^p M_n \subset D^p C_n = \mathbb{P}_{r,\Delta}$.

$$\Theta(N, N_n) \leq c\|(1 - \pi_n)Q\|, \ \Theta(N, \tilde{N}_n) \leq c(\|K(1 - \pi_n)Q\| + \|G(1 - \pi_n)Q\|).$$

7.14
$$\begin{aligned} U - U_n &= (1 - K)^{-1}G - U_n \\ &= (1 - K)^{-1}\pi_n G + (1 - K)^{-1}(1 - \pi_n)G - U_n \\ &= (1 - K)^{-1}(1 - \pi_n K)U_n + (1 - K)^{-1}(1 - \pi_n)G - (1 - K)^{-1}(1 - K)U_n \end{aligned}$$

since $\pi_n G = (1 - \pi_n K)U_n$. Hence

$$\begin{aligned} U - U_n &= (1 - K)^{-1}[U_n - \pi_n K U_n + (1 - \pi_n)G - U_n + KU_n] \\ &= (1 - K)^{-1}[(1 - \pi_n)KU_n + (1 - \pi_n)G] \\ &= (1 - K)^{-1}(1 - \pi_n)[G + KU_n]. \end{aligned}$$

7.15 See the proof of Theorem 6.11.

7.16 Apply Theorem 7.23.

7.17 We recall that $\psi_n = A\theta_n$, so $(1 - K)G^{-1}\psi_n = \theta_n$. $\tilde{\theta}_n = \lambda_n A\theta_n = \lambda_n\psi_n$, $\psi_n = G\varphi_n$, $\tilde{\varphi}_n = K\varphi_n + \lambda_n G\varphi_n = K\varphi_n + \lambda_n\psi_n$. So

$$\begin{aligned} \varphi_n - \tilde{\varphi}_n &= (1 - K)\varphi_n - \lambda_n\psi_n = (1 - K)G^{-1}\psi_n - \tilde{\theta}_n \\ &= \theta_n - \tilde{\theta}_n. \end{aligned}$$

7.18 $(1 - P)\tilde{\theta}_n = \lambda_n A\theta_n - \lambda_n PA\theta_n = \lambda_n A\theta_n - \lambda_n AP\theta_n$, $\theta_n - \tilde{\theta}_n = (1 - \lambda_n A)\theta_n$, so $\theta_n = (1 - \lambda_n A)^{-1}(\theta_n - \tilde{\theta}_n)$; hence

$$\begin{aligned} (1 - P)\tilde{\theta}_n &= \lambda_n A(1 - \lambda_n A)^{-1}(\theta_n - \tilde{\theta}_n) - \lambda_n AP(1 - \lambda_n A)^{-1}(\theta_n - \tilde{\theta}_n) \\ &= \lambda_n A(1 - \lambda_n A)^{-1}(1 - P)(\theta_n - \tilde{\theta}_n). \end{aligned}$$

We also have $\pi_n(\theta_n - \tilde{\theta}_n) = \pi_n(\theta_n - \lambda_n A\theta_n) = 0$ by definition of θ_n. Hence $(1 - \pi_n)(\theta_n - \tilde{\theta}_n) = \theta_n - \tilde{\theta}_n$, so

$$\begin{aligned} (1 - P)\tilde{\theta}_n &= \lambda_n A(1 - \lambda_n A)^{-1}(1 - P)(1 - \pi_n)(\theta_n - \tilde{\theta}_n) \\ &= \lambda_n A(1 - \lambda_n A)^{-1}(1 - P)(\pi_n - 1)\tilde{\varphi}_n, \end{aligned}$$

a formula similar to that of Lemma 7.1, the kernel of A being of class $\mathfrak{G}(\alpha + 1, p - 2)$.

7.19 Let $g \in X$ and u, u_n be defined by $(1 - K)u = g$, $\pi_n(1 - K)u_n = \pi_n g$, so $(1 - K)u_n = \pi'_n g$. Then

$$\begin{aligned}(1 - \pi'_n)g &= (1 - K)(u - u_n) = (1 - K)(1 - \pi_n K)^{-1}(1 - \pi_n)(1 - K)^{-1}g \\ &= (1 - K)(1 - \pi_n K)^{-1}[1 - \pi_n K - \pi_n(1 - K)](1 - K)^{-1}g \\ &= (1 - K)[1 - (1 - \pi_n K)^{-1}\pi_n(1 - K)](1 - K)^{-1}g \\ &= (1 - (1 - K)(1 - \pi_n K)^{-1}\pi_n)g.\end{aligned}$$

So $\pi'_n g = (1 - K)(1 - \pi_n K)^{-1}\pi_n g$.

7.20 Let $\theta \in M$, $Q(1 - K)^{-1}\theta = (1 - K)^{-1}P\theta = (1 - K)^{-1}\theta \in N$. Let

$$\theta^* \in M^*,\ Q^*(1 - K^*)\theta^* = (1 - K^*)P^*\theta^* = (1 - K^*)\theta^* \in N^*.$$

7.21 $T\psi = \lambda\mathfrak{N}\psi$, $\psi = G\varphi$ is equivalent to $(1 - K)\varphi = \lambda L\varphi$; that is, $\varphi = \lambda(1 - K)^{-1}L\varphi = \lambda V\varphi$. $\psi_n = G\varphi_n$ is the solution of

$$\pi_n(T\psi_n - \lambda_n\mathfrak{N}\psi_n) = 0, \qquad \psi_n \in C_n.$$

$$\pi_n((1 - K)\varphi_n - \lambda_n\mathfrak{N}G\varphi_n) = 0,$$

$$\varphi_n - \pi_n K\varphi_n - \lambda_n\pi_n L\varphi_n = 0,$$

$$(1 - \pi_n K)\varphi_n - \lambda_n\pi_n L\varphi_n = 0,$$

$$\varphi_n = \lambda_n(1 - \pi_n K)^{-1}\pi_n L\varphi_n = \lambda_n V_n\varphi_n.$$

7.22
$$\begin{aligned}V - V_n &= V - (1 - \pi_n K)^{-1}\pi_n(1 - K)V \\ &= (1 - \pi_n K)^{-1}[(1 - \pi_n K)V - \pi_n(1 - K)V] \\ &= (1 - \pi_n K)^{-1}(1 - \pi_n)V.\end{aligned}$$

But also

$$\begin{aligned}V - V_n &= (1 - K)^{-1}L - V_n \\ &= (1 - K)^{-1}\pi_n L + (1 - K)^{-1}(1 - \pi_n)L - V_n \\ &= (1 - K)^{-1}[(1 - \pi_n K)V_n + (1 - \pi_n)L - (1 - K)V_n] \\ &= (1 - K)^{-1}[-\pi_n K V_n + (1 - \pi_n)L + K V_n] \\ &= (1 - K)^{-1}(1 - \pi_n)(L + K V_n).\end{aligned}$$

7.23 If $\mathfrak{N} = 1$, then $L = G$ and $R = P = P'$. Lemma 7.26 takes then the form $(1 - P)\psi_n = (T - \lambda_n)^{-1}(1 - P)(\varphi_n - \tilde{\varphi}_n) = S(\lambda_n)(\pi_n - 1)\tilde{\varphi}_n$.

7.24 Since $\lambda^k\varphi_n = P_n T\varphi^{k-1}$, (5.16) requires to know P_n and S_n, that is, $\mathscr{P}_n$ and $\mathscr{S}_n$ only when T_n has the structure of T_α, T_β, or T_γ given in Chapter 4, Section 6.1.

7.25 See Exercise 1.53.

7.26 Adapt the solution of Exercise 5.38 where P_n is replaced by $Q_n = (\cdot, \psi_n)_H\psi_n$.

Notation Index

Subject Index

D

E

Computer Science and Applied Mathematics

A SERIES OF MONOGRAPHS AND TEXTBOOKS

Editor

Werner Rheinboldt

University of Pittsburgh

HANS P. KÜNZI, H. G. TZSCHACH, AND C. A. ZEHNDER. Numerical Methods of Mathematical Optimization: With ALGOL and FORTRAN Programs, Corrected and Augmented Edition

AZRIEL ROSENFELD. Picture Processing by Computer

JAMES ORTEGA AND WERNER RHEINBOLDT. Iterative Solution of Nonlinear Equations in Several Variables

AZARIA PAZ. Introduction to Probabilistic Automata

DAVID YOUNG. Iterative Solution of Large Linear Systems

ANN YASUHARA. Recursive Function Theory and Logic

JAMES M. ORTEGA. Numerical Analysis: A Second Course

G. W. STEWART. Introduction to Matrix Computations

CHIN-LIANG CHANG AND RICHARD CHAR-TUNG LEE. Symbolic Logic and Mechanical Theorem Proving

C. C. GOTLIEB AND A. BORODIN. Social Issues in Computing

ERWIN ENGELER. Introduction to the Theory of Computation

F. W. J. OLVER. Asymptotics and Special Functions

DIONYSIOS C. TSICHRITZIS AND PHILIP A. BERNSTEIN. Operating Systems

ROBERT R. KORFHAGE. Discrete Computational Structures

PHILIP J. DAVIS AND PHILIP RABINOWITZ. Methods of Numerical Integration

A. T. BERZTISS. Data Structures: Theory and Practice, Second Edition

N. CHRISTOPHIDES. Graph Theory: An Algorithmic Approach

ALBERT NIJENHUIS AND HERBERT S. WILF. Combinatorial Algorithms

AZRIEL ROSENFELD AND AVINASH C. KAK. Digital Picture Processing

SAKTI P. GHOSH. Data Base Organization for Data Management

DIONYSIOS C. TSICHRITZIS AND FREDERICK H. LOCHOVSKY. Data Base Management Systems

JAMES L. PETERSON. Computer Organization and Assembly Language Programming

WILLIAM F. AMES. Numerical Methods for Partial Differential Equations, Second Edition

ARNOLD O. ALLEN. Probability, Statistics, and Queueing Theory: With Computer Science Applications

ELLIOTT I. ORGANICK, ALEXANDRA I. FORSYTHE, AND ROBERT P. PLUMMER. Programming Language Structures

ALBERT NIJENHUIS AND HERBERT S. WILF. Combinatorial Algorithms. Second edition.

AZRIEL ROSENFELD. Picture Languages, Formal Models for Picture Recognition

ISAAC FRIED. Numerical Solution of Differential Equations

ABRAHAM BERMAN AND ROBERT J. PLEMMONS. Nonnegative Matrices in the Mathematical Sciences

BERNARD KOLMAN AND ROBERT E. BECK. Elementary Linear Programming with Applications

CLIVE L. DYM AND ELIZABETH S. IVEY. Principles of Mathematical Modeling

ERNEST L. HALL. Computer Image Processing and Recognition

ALLEN B. TUCKER, JR. Text Processing: Algorithms, Languages, and Applications

MARTIN CHARLES GOLUMBIC. Algorithmic Graph Theory and Perfect Graphs

GABOR T. HERMAN. Image Reconstruction from Projections: The Fundamentals of Computerized Tomography

WEBB MILLER AND CELIA WRATHALL. Software for Roundoff Analysis of Matrix Algorithms

ULRICH W. KULISCH AND WILLARD L. MIRANKER. Computer Arithmetic in Theory and Practice

LOUIS A. HAGEMAN AND DAVID M. YOUNG. Applied Iterative Methods

I. GOHBERG, P. LANCASTER AND L. RODMAN. Matrix Polynomials.

AZRIEL ROSENFELD AND AVINASH C. KAK. Digital Picture Processing, Second Edition, Vol. 1, Vol. 2

DIMITRI P. BERTSEKAS. Constrained Optimization and Lagrange Multiplier Methods

JAMES S. VANDERGRAFT. Introduction to Numerical Computations, Second Edition

FRANÇOISE CHATELIN. Spectral Approximation of Linear Operators

In preparation

LEONARD UHR. Algorithm-Structured Computer Arrays and Networks: Architectures and Processes for Images, Percepts, Models, Information

GÖTZ ALEFELD AND JÜRGEN HERZBERGER. Introduction to Interval Computations. Translated by Jon Rokne

ROBERT KORFHAGE. Discrete Computational Structures, Second Edition

V. A. BARKER AND O. AXELSSON. Finite Element Solution of Boundary Value Problems: Theory and Computation